T0189328

Lecture Notes in Computer Science 12966

More information about this subseries at http://www.springer.com/series/7412

Chunfeng Lian · Xiaohuan Cao · Islem Rekik · Xuanang Xu · Pingkun Yan (Eds.)

Machine Learning in Medical Imaging

12th International Workshop, MLMI 2021
Held in Conjunction with MICCAI 2021
Strasbourg, France, September 27, 2021
Proceedings

 Springer

Editors
Chunfeng Lian
Xi'an Jiaotong University
Xi'an, China

Xiaohuan Cao
United Imaging Intelligence
Shanghai, China

Islem Rekik
Istanbul Technical University
Istanbul, Turkey

Xuanang Xu
Rensselaer Polytechnic Institute
Troy, NY, USA

Pingkun Yan
Rensselaer Polytechnic Institute
Troy, NY, USA

ISSN 0302-9743   ISSN 1611-3349 (electronic)
Lecture Notes in Computer Science
ISBN 978-3-030-87588-6   ISBN 978-3-030-87589-3 (eBook)
https://doi.org/10.1007/978-3-030-87589-3

LNCS Sublibrary: SL6 – Image Processing, Computer Vision, Pattern Recognition, and Graphics

This Springer imprint is published by the registered company Springer Nature Switzerland AG
The registered company address is: Gewerbestrasse 11, 6330 Cham, Switzerland

Preface

The 12th International Workshop on Machine Learning in Medical Imaging (MLMI 2021) was virtually held in Strasbourg, France, on September 27, 2021, in conjunction with the 24th International Conference on Medical Image Computing and Computer-Assisted Intervention (MICCAI 2021).

With the goal of advancing the scientific research within the broad field of machine learning in medical imaging (MLMI), the workshop provides a forum for researchers and clinicians to communicate and exchange ideas. The community has witnessed the rapid development of the field in the past decade. Artificial intelligence (AI) and machine learning (ML) have significantly changed the research landscape of academia and industry. AI/ML now plays a crucial role in the medical imaging field, including, but not limited to, computer-aided detection and diagnosis, image segmentation, image registration, image fusion, image-guided intervention, image annotation, image retrieval, image reconstruction, etc. MLMI 2021 focused on major trends and challenges in this area and presented original works aiming to identify new cutting-edge techniques and their uses in medical imaging. The workshop facilitated translating medical imaging research from bench to bedside. Topics of interests included deep learning, generative adversarial learning, ensemble learning, sparse learning, multi-task learning, multi-view learning, manifold learning, and reinforcement learning, along with their applications to medical image analysis, computer-aided detection and diagnosis, multi-modality fusion, image reconstruction, image retrieval, cellular image analysis, molecular imaging, digital pathology, etc.

The MLMI workshop series has attracted original, high-quality submissions on innovative research and developments in medical image analysis using ML techniques. MLMI 2021 again received a large number of papers (92 in total). Following good practice from previous years, all the submissions underwent a rigorous double-blinded peer review process, with each paper being reviewed by at least two members of the Program Committee, which was composed of 62 experts in the field. Based on the reviewing scores and critiques, 71 papers were accepted for presentation at the workshop and chosen to be included in this Springer LNCS volume. It was a tough decision and many high-quality papers had to be rejected due to the page limit of this volume.

We are grateful to all Program Committee members for reviewing the submissions and giving constructive comments. We also thank all the authors for making the workshop very fruitful and successful.

September 2021

Chunfeng Lian
Xiaohuan Cao
Islem Rekik
Xuanang Xu
Pingkun Yan

Organization

Workshop Organizers

Chunfeng Lian   Xi'an Jiaotong University, China
Xiaohuan Cao   United Imaging Intelligence, China
Islem Rekik   Istanbul Teknik Universitesi, Turkey
Xuanang Xu   Rensselaer Polytechnic Institute, USA
Pingkun Yan   Rensselaer Polytechnic Institute, USA

Steering Committee

Dinggang Shen   ShanghaiTech University, China
Pingkun Yan   Rensselaer Polytechnic Institute, USA
Kenji Suzuki   Tokyo Institute of Technology, Japan
Fei Wang   Visa Research, USA

Program Committee

Sahar Ahmad   University of North Carolina at Chapel Hill, USA
Ulas Bagci   Northwestern University, USA
Zehong Cao   Southern Medical University and United Imaging Intelligence, China
Heang-Ping Chan   University of Michigan Medical Center, USA
Liangjun Chen   University of North Carolina at Chapel Hill, USA
Liyun Chen   Shanghai Jiao Tong University, China
Zhiming Cui   University of Hong Kong, China
Yuanyuan Gao   Rensselaer Polytechnic Institute, USA
Hao Guan   University of North Carolina at Chapel Hill, USA
Yu Guo   Tianjin University, China
Tiancheng He   Cornell University, USA
Jiashuang Huang   Nanjing University of Aeronautics and Astronautics, China
Yuankai Huo   Vanderbilt University, USA
Khoi Huynh   University of North Carolina at Chapel Hill, USA
Caiwen Jiang   ShanghaiTech University, China
Xi Jiang   University of Electronic Science and Technology of China, China
Zhicheng Jiao   Brown University, USA

Contents

Contrastive Representations for Continual Learning of Fine-Grained
Histology Images 1
    Tapabrata Chakraborti, Fergus Gleeson, and Jens Rittscher

Learning Transferable 3D-CNN for MRI-Based Brain Disorder
Classification from Scratch: An Empirical Study 10
    Hao Guan, Li Wang, Dongren Yao, Andrea Bozoki, and Mingxia Liu

Knee Cartilages Segmentation Based on Multi-scale Cascaded Neural
Networks 20
    Junrui Liu, Cong Hua, Liang Zhang, Ping Li, and Xiaoyuan Lu

Deep PET/CT Fusion with Dempster-Shafer Theory for Lymphoma
Segmentation 30
    Ling Huang, Thierry Denœux, David Tonnelet, Pierre Decazes,
    and Su Ruan

Interpretable Histopathology Image Diagnosis via Whole Tissue Slide
Level Supervision 40
    Zhuoyue Wu, Hansheng Li, Lei Cui, Yuxin Kang, Jianye Liu, Haider Ali,
    Jun Feng, and Lin Yang

Variational Encoding and Decoding for Hybrid Supervision of Registration
Network 50
    Dongdong Gu, Xiaohuan Cao, Guocai Liu, Zhong Xue,
    and Dinggang Shen

Multiresolution Registration Network (MRN) Hierarchy with Prior
Knowledge Learning 61
    Dongdong Gu, Xiaohuan Cao, Guocai Liu, Dinggang Shen,
    and Zhong Xue

Learning to Synthesize 7 T MRI from 3 T MRI with Few Data
by Deformable Augmentation 70
    Jie Wei, Yongsheng Pan, Yong Xia, and Dinggang Shen

Rethinking Pulmonary Nodule Detection in Multi-view 3D CT Point
Cloud Representation 80
    Jingya Liu, Oguz Akin, and Yingli Tian

End-to-End Lung Nodule Detection Framework with Model-Based
Feature Projection Block 91
    Ivan Drokin and Elena Ericheva

Learning Structure from Visual Semantic Features and Radiology
Ontology for Lymph Node Classification on MRI 101
    Yingying Zhu, Shuai Wang, Qingyu Chen, Sungwon Lee, Thomas Shen,
    Daniel C. Elton, Zhiyong Lu, and Ronald M. Summers

Improving Joint Learning of Chest X-Ray and Radiology Report by Word
Region Alignment 110
    Zhanghexuan Ji, Mohammad Abuzar Shaikh, Dana Moukheiber,
    Sargur N Srihari, Yifan Peng, and Mingchen Gao

Cell Counting by a Location-Aware Network 120
    Zuhui Wang and Zhaozheng Yin

Exploring Gyro-Sulcal Functional Connectivity Differences Across Task
Domains via Anatomy-Guided Spatio-Temporal Graph Convolutional
Networks 130
    Mingxin Jiang, Shimin Yang, Zhongbo Zhao, Jiadong Yan,
    Yuzhong Chen, Tuo Zhang, Shu Zhang, Benjamin Becker,
    Keith M. Kendrick, and Xi Jiang

StairwayGraphNet for Inter- and Intra-modality Multi-resolution Brain
Graph Alignment and Synthesis 140
    Islem Mhiri, Mohamed Ali Mahjoub, and Islem Rekik

Multi-Feature Semi-Supervised Learning for COVID-19 Diagnosis
from Chest X-Ray Images 151
    Xiao Qi, David J. Foran, John L. Nosher, and Ilker Hacihaliloglu

Transfer Learning with a Layer Dependent Regularization for Medical
Image Segmentation 161
    Nimrod Sagie, Hayit Greenspan, and Jacob Goldberger

Multi-scale Self-supervised Learning for Multi-site Pediatric Brain MR
Image Segmentation with Motion/Gibbs Artifacts 171
    Yue Sun, Kun Gao, Weili Lin, Gang Li, Sijie Niu, and Li Wang

Deep Active Learning for Dual-View Mammogram Analysis 180
    Yutong Yan, Pierre-Henri Conze, Mathieu Lamard, Heng Zhang,
    Gwenolé Quellec, Béatrice Cochener, and Gouenou Coatrieux

Statistical Dependency Guided Contrastive Learning for Multiple Labeling
in Prenatal Ultrasound 190
  Shuangchi He, Zehui Lin, Xin Yang, Chaoyu Chen, Jian Wang,
  Xue Shuang, Ziwei Deng, Qin Liu, Yan Cao, Xiduo Lu, Ruobing Huang,
  Nishant Ravikumar, Alejandro Frangi, Yuanji Zhang, Yi Xiong,
  and Dong Ni

Semi-supervised Learning Regularized by Adversarial Perturbation
and Diversity Maximization 199
  Peng Liu and Guoyan Zheng

TransforMesh: A Transformer Network for Longitudinal Modeling
of Anatomical Meshes 209
  Ignacio Sarasua, Sebastian Pölsterl, Christian Wachinger,
  and for the Alzheimer's Disease Neuroimaging

A Recurrent Two-Stage Anatomy-Guided Network for Registration
of Liver DCE-MRI 219
  Wenjun Shen, Liyun Chen, Dongming Wei, Yuanfang Qiao,
  Yiqiang Zhan, Dinggang Shen, and Qian Wang

Learning Infant Brain Developmental Connectivity for Cognitive Score
Prediction 228
  Yu Li, Jiale Cheng, Xin Zhang, Ruiyan Fang, Lufan Liao, Xinyao Ding,
  Hao Ni, Xiangmin Xu, Zhengwang Wu, Dan Hu, Weili Lin, Li Wang,
  John Gilmore, and Gang Li

Hierarchical 3D Feature Learning for Pancreas Segmentation 238
  Federica Proietto Salanitri, Giovanni Bellitto, Ismail Irmakci,
  Simone Palazzo, Ulas Bagci, and Concetto Spampinato

Voxel-Wise Cross-Volume Representation Learning for 3D Neuron
Reconstruction 248
  Heng Wang, Chaoyi Zhang, Jianhui Yu, Yang Song, Siqi Liu,
  Wojciech Chrzanowski, and Weidong Cai

Diagnosis of Hippocampal Sclerosis from Clinical Routine Head MR
Images Using Structure-constrained Super-Resolution Network 258
  Zehong Cao, Feng Shi, Qiang Xu, Gaoping Liu, Tianyang Sun,
  Xiaodan Xing, Yichu He, Guangming Lu, Zhiqiang Zhang,
  and Dinggang Shen

U-Net Transformer: Self and Cross Attention for Medical Image
Segmentation 267
   Olivier Petit, Nicolas Thome, Clement Rambour, Loic Themyr,
   Toby Collins, and Luc Soler

Pre-biopsy Multi-class Classification of Breast Lesion Pathology
in Mammograms 277
   Tal Tlusty, Michal Ozery-Flato, Vesna Barros, Ella Barkan,
   Mika Amit, David Gruen, Michal Guindy, Tal Arazi, Mona Rozin,
   Michal Rosen-Zvi, and Efrat Hexter

Co-segmentation of Multi-modality Spinal Image Using Channel
and Spatial Attention 287
   Yaocong Zou and Yonghong Shi

Hetero-Modal Learning and Expansive Consistency Constraints
for Semi-supervised Detection from Multi-sequence Data 296
   Bolin Lai, Yuhsuan Wu, Xiao-Yun Zhou, Peng Wang, Le Lu,
   Lingyun Huang, Mei Han, Jing Xiao, Heping Hu, and Adam P. Harrison

STRUDEL: Self-training with Uncertainty Dependent Label Refinement
Across Domains 306
   Fabian Gröger, Anne-Marie Rickmann, and Christian Wachinger

Deep Reinforcement Learning for L3 Slice Localization in Sarcopenia
Assessment 317
   Othmane Laousy, Guillaume Chassagnon, Edouard Oyallon,
   Nikos Paragios, Marie-Pierre Revel, and Maria Vakalopoulou

MIST GAN: Modality Imputation Using Style Transfer for MRI 327
   Jaya Chandra Raju, Kompella Subha Gayatri, Keerthi Ram,
   Rajeswaran Rangasami, Rajoo Ramachandran,
   and Mohanasankar Sivaprakasam

Biased Extrapolation in Latent Space forImbalanced Deep Learning 337
   Suhyeon Jeong and Seungkyu Lee

3DMeT: 3D Medical Image Transformer for Knee Cartilage Defect
Assessment 347
   Sheng Wang, Zixu Zhuang, Kai Xuan, Dahong Qian, Zhong Xue, Jia Xu,
   Ying Liu, Yiming Chai, Lichi Zhang, Qian Wang, and Dinggang Shen

A Gaussian Process Model for Unsupervised Analysis of High
Dimensional Shape Data 356
   Wenzheng Tao, Riddhish Bhalodia, and Ross Whitaker

Standardized Analysis of Kidney Ultrasound Images for the Prediction of Pediatric Hydronephrosis Severity 366
Pooneh Roshanitabrizi, Jonathan Zember, Bruce Michael Sprague, Steven Hoefer, Ramon Sanchez-Jacob, James Jago, Dorothy Bulas, Hans G. Pohl, and Marius George Linguraru

Automated Deep Learning-Based Detection of Osteoporotic Fractures in CT Images 376
Eren Bora Yilmaz, Christian Buerger, Tobias Fricke, Md Motiur Rahman Sagar, Jaime Peña, Cristian Lorenz, Claus-Christian Glüer, and Carsten Meyer

GT U-Net: A U-Net Like Group Transformer Network for Tooth Root Segmentation 386
Yunxiang Li, Shuai Wang, Jun Wang, Guodong Zeng, Wenjun Liu, Qianni Zhang, Qun Jin, and Yaqi Wang

Information Bottleneck Attribution for Visual Explanations of Diagnosis and Prognosis 396
Ugur Demir, Ismail Irmakci, Elif Keles, Ahmet Topcu, Ziyue Xu, Concetto Spampinato, Sachin Jambawalikar, Evrim Turkbey, Baris Turkbey, and Ulas Bagci

Stacked Hourglass Network with a Multi-level Attention Mechanism: Where to Look for Intervertebral Disc Labeling 406
Reza Azad, Lucas Rouhier, and Julien Cohen-Adad

TED-Net: Convolution-Free T2T Vision Transformer-Based Encoder-Decoder Dilation Network for Low-Dose CT Denoising 416
Dayang Wang, Zhan Wu, and Hengyong Yu

Self-supervised Mean Teacher for Semi-supervised Chest X-Ray Classification 426
Fengbei Liu, Yu Tian, Filipe R. Cordeiro, Vasileios Belagiannis, Ian Reid, and Gustavo Carneiro

VoxelEmbed: 3D Instance Segmentation and Tracking with Voxel Embedding based Deep Learning 437
Mengyang Zhao, Quan Liu, Aadarsh Jha, Ruining Deng, Tianyuan Yao, Anita Mahadevan-Jansen, Matthew J. Tyska, Bryan A. Millis, and Yuankai Huo

Using Spatio-Temporal Correlation Based Hybrid Plug-and-Play Priors (SEABUS) for Accelerated Dynamic Cardiac Cine MRI 447
Qingyong Zhu and Dong Liang

Window-Level Is a Strong Denoising Surrogate 457
  Ayaan Haque, Adam Wang, and Abdullah-Al-Zubaer Imran

Cardiovascular Disease Risk Improves COVID-19 Patient Outcome
Prediction 467
  Diego Machado Reyes, Hanqing Chao, Fatemeh Homayounieh,
  Juergen Hahn, Mannudeep K. Kalra, and Pingkun Yan

Self-supervision Based Dual-Transformation Learning for Stain
Normalization, Classification andSegmentation 477
  Shiv Gehlot and Anubha Gupta

Deep Representation Learning for Image-Based Cell Profiling 487
  Wenzhao Wei, Sacha Haidinger, John Lock, and Erik Meijering

Detecting Extremely Small Lesions in Mouse Brain MRI with Point
Annotations via Multi-task Learning 498
  Xiaoyang Han, Yuting Zhai, Ziqi Yu, Tingying Peng, and Xiao-Yong Zhang

Morphology-Guided Prostate MRI Segmentation with Multi-slice
Association 507
  Jianping Li, Zhiming Cui, Shuai Wang, Jie Wei, Jun Feng, Shu Liao,
  and Dinggang Shen

Unsupervised Cross-modality Cardiac Image Segmentation
via Disentangled Representation Learning and Consistency Regularization 517
  Runze Wang and Guoyan Zheng

Landmark-Guided Rigid Registration for Temporomandibular Joint
MRI-CBCT Images with Large Field-of-View Difference 527
  Jupeng Li, Yinghui Wang, Shuai Wang, Kai Zhang, and Gang Li

Spine-Rib Segmentation and Labeling via Hierarchical Matching
and Rib-Guided Registration 537
  Caiwen Jiang, Zhiming Cui, Dongming Wei, Yuhang Sun, Jiameng Liu,
  Jie Wei, Qun Chen, Dijia Wu, and Dinggang Shen

Multi-scale Segmentation Network for Rib Fracture Classification
from CT Images 546
  Jiameng Liu, Zhiming Cui, Yuhang Sun, Caiwen Jiang, Zirong Chen,
  Hao Yang, Yuyao Zhang, Dijia Wu, and Dinggang Shen

Knowledge-Guided Multiview Deep Curriculum Learning for Elbow
Fracture Classification 555
Jun Luo, Gene Kitamura, Dooman Arefan, Emine Doganay,
Ashok Panigrahy, and Shandong Wu

Contrastive Learning of Single-Cell Phenotypic Representations
for Treatment Classification 565
Alexis Perakis, Ali Gorji, Samriddhi Jain, Krishna Chaitanya,
Simone Rizza, and Ender Konukoglu

CorLab-Net: Anatomical Dependency-Aware Point-Cloud Learning
for Automatic Labeling of Coronary Arteries 576
Xiao Zhang, Zhiming Cui, Jun Feng, Yanli Song, Dijia Wu,
and Dinggang Shen

A Hybrid Deep Registration of MR Scans to Interventional Ultrasound
for Neurosurgical Guidance 586
Ramy A. Zeineldin, Mohamed E. Karar, Franziska Mathis-Ullrich,
and Oliver Burgert

Segmentation of Peripancreatic Arteries in Multispectral Computed
Tomography Imaging 596
Alina Dima, Johannes C. Paetzold, Friederike Jungmann,
Tristan Lemke, Philipp Raffler, Georgios Kaissis, Daniel Rueckert,
and Rickmer Braren

SkullEngine: A Multi-stage CNN Framework for Collaborative CBCT
Image Segmentation and Landmark Detection 606
Qin Liu, Han Deng, Chunfeng Lian, Xiaoyang Chen, Deqiang Xiao,
Lei Ma, Xu Chen, Tianshu Kuang, Jaime Gateno, Pew-Thian Yap,
and James J. Xia

Skull Segmentation from CBCT Images via Voxel-Based Rendering 615
Qin Liu, Chunfeng Lian, Deqiang Xiao, Lei Ma, Han Deng, Xu Chen,
Dinggang Shen, Pew-Thian Yap, and James J. Xia

Alzheimer's Disease Diagnosis via Deep Factorization Machine Models 624
Raphael Ronge, Kwangsik Nho, Christian Wachinger,
and Sebastian Pölsterl

3D Temporomandibular Joint CBCT Image Segmentation
via Multi-directional Resampling Ensemble Learning Network 634
Kai Zhang, Jupeng Li, Ruohan Ma, and Gang Li

Vox2Surf: Implicit Surface Reconstruction from Volumetric Data 644
    Yoonmi Hong, Sahar Ahmad, Ye Wu, Siyuan Liu, and Pew-Thian Yap

Clinically Correct Report Generation from Chest X-Rays Using Templates 654
    Pablo Pino, Denis Parra, Cecilia Besa, and Claudio Lagos

Extracting Sequential Features from Dynamic Connectivity Network
with rs-fMRI Data for AD Classification 664
    Kai Lin, Biao Jie, Peng Dong, Xintao Ding, Weixin Bian, and Mingxia Liu

Integration of Handcrafted and Embedded Features from Functional
Connectivity Network with rs-fMRI for Brain Disease Classification 674
    Peng Dong, Biao Jie, Lin Kai, Xintao Ding, Weixin Bian, and Mingxia Liu

Detection of Lymph Nodes in T2 MRI Using Neural Network Ensembles 682
    Tejas Sudharshan Mathai, Sungwon Lee, Daniel C. Elton,
    Thomas C. Shen, Yifan Peng, Zhiyong Lu, and Ronald M. Summers

Seeking an Optimal Approach for Computer-Aided Pulmonary Embolism
Detection 692
    Nahid Ul Islam, Shiv Gehlot, Zongwei Zhou, Michael B. Gotway,
    and Jianming Liang

Correction to: Machine Learning in Medical Imaging C1
    Chunfeng Lian, Xiaohuan Cao, Islem Rekik, Xuanang Xu,
    and Pingkun Yan

Correction to: A Gaussian Process Model for Unsupervised Analysis
of High Dimensional Shape Data C2
    Wenzheng Tao, Riddhish Bhalodia, and Ross Whitaker

Author Index 703

Contrastive Representations for Continual Learning of Fine-Grained Histology Images

Tapabrata Chakraborti[1(✉)], Fergus Gleeson[2], and Jens Rittscher[1]

[1] IBME/BDI, Department of Engineering Science, University of Oxford, Oxford, UK
tapabrata.chakraborty@eng.ox.ac.uk
[2] NCIMI/BDI, Department of Oncology, University of Oxford, Oxford, UK

Abstract. We show how a simple autoencoder based deep network with a contrastive loss can effectively learn representations in a continual/incremental manner with limited labelling. This is of particular interest to the biomedical imaging research community, for whom the visual task is often a binary decision (healthy vs. disease) with limited quantity data and costly labelling. For such applications, the proposed method provides a light-weight option of 1) representing patterns with relatively few training samples using a novel collaborative contrastive loss function 2) update the autoencoder based deep network in an unsupervised fashion for continual learning for new incoming data. We overcome the drawbacks of existing methods through planned technical design, and demonstrate the efficacy of the proposed method on three histology image classification tasks (lung, colon, breast cancer) with SOTA results.

Keywords: Continual learning · Fine-grained histology images · Collaborative contrastive representations · Automated cancer detection

1 Introduction

In recent years, deep learning based decision systems have achieved high accuracy in a range of visual computing tasks. However, medical image analysis problems still present several unique challenges [1], which indicate a need of update in the usual deep learning paradigms: 1) in real life, medical images are acquired in small continual batches [2], depending on incoming patient cohorts; the solution is to train deep networks in an incremental learning setting [3]; 2) it is very costly to gather large amount of good quality expert labelling and annotations [4]; the solution is to perform semi-supervised learning, where the incremental learning phase is unsupervised [5,6], that is the incoming samples are directly compared against samples from the pre-trained image dictionary.

We achieve both of the above objectives in the present paper, starting with the simple premise that a large number of medical image classification tasks involve a binary decision, that is classifying between healthy vs. diseased categories. 1) This provides the intuition that a contrastive loss can be used instead

The authors are funded by the UKRI Innovate UK DART Lung Health Programme.

C. Lian et al. (Eds.): MLMI 2021, LNCS 12966, pp. 1–9, 2021.
https://doi.org/10.1007/978-3-030-87589-3_1

of cross-entropy, which helps to compare two images directly in a representation space, without needing labels for new incremental data in a continual learning scenario. 2) Contrastive learning seeks to learn a representation of the negative (healthy) category and compares the query (healthy/diseased) against that template, this actually makes the class imbalance problem (healthy samples more abundant than diseased) an advantage for contrastive learning, which favors the larger number of negative samples to learn the distribution.

Thus the contribution of this paper is two-fold.

1. We present a lightweight semi-supervised continual learning scheme with a contrastive loss, which first trains an autoencoder with a small number of labelled data in the usual supervised manner, and then contrasts new incoming images directly with the dictionary and updates the knowledge in an unsupervised manner, without the need for further labels.
2. The method overcomes drawbacks of several existing methods and establishes a new state-of-the-art in several histology image classification tasks (lung, colon, breast cancer); but at the same time, the proposed collaborative contrastive (CoCo) method is generalised enough to be used for any other similar tasks.

2   Methodology

In this section, we first present a primer on contrastive learning, then we briefly discuss ways of utilising contrastive loss in continual learning and finally, we describe in details of the proposed collaborative contrast (CoCo) algorithm in a continual learning setting.

2.1   Contrastive Learning

Contrastive Learning [7], as the name suggests, is essentially a dictionary lookup task in an encoded space, in order to "contrast" or differentiate between two samples. Though the decision is binary (whether the samples "match", that is are of same type, or not), the setting is different from a usual neural network based 2-class classification scheme. Whereas a traditionally trained network will directly output the class prediction when presented with a test sample, a network with a contrastive loss will predict how similar the test sample (query q) is to a reference sample (key $k \in \{k_i\}$) in some representation space. This is formally presented below using a dot product based formulation of contrastive loss called InfoNCE [8], which is a variant of Noise-contrastive estimation (NCE) [9].

$$\mathcal{L}_{NCE}^i = -\log \frac{\exp(q \cdot k_i / \tau)}{\sum_i \exp(q \cdot k_i / \tau)} \quad (1)$$

Note that τ is a hyper-parameter; q and k are more likely to be encoded data in a representation space, rather than raw data itself. Since the loss function

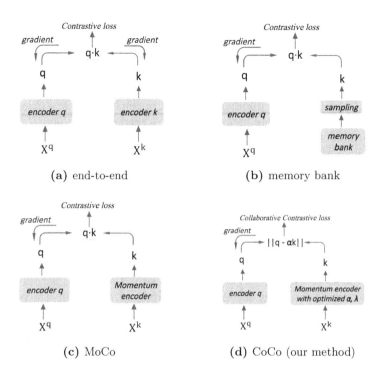

Fig. 1. Comparison of 4 variants of contrastive loss mechanisms in semi-supervised continual learning setting. **a) end-to-end:** both the key (k) encoder and the query (q) enocder are updated by back-propagation. **b) memory bank:** only the query encoder is updated, while keys are sampled from an updating dictionary, keeping memory size constant. **c) MoCo:** key weights are updated as a weighted combination of previous key and query weights, using a momentum constant. **d) Proposed method CoCo:** A new collaborative contrastive loss helps to i) generate a representation of the key dictionary, and ii) gives an optimal value for the momentum coefficient.

compares between two samples, instead of with a label, this method can be used in an unsupervised setting, which is an advantage. The careful reader will note some similarities of the structural setup with that of the discriminator section of an adversarial network.

2.2 Contrastive Loss in Continual Learning

We present several alternative designs for implementing contrastive loss in a continual learning setting in Fig. 1. In each case there are two encoders, which during the "base class" training phase are actually the same. During the incremental learning phase, two copies of the encoder are used, one for the query sample, and one for the keys in the dictionary. The pipelines vary according to the way the encoders and the memory are updated during the continual learning stage. x^q and x^k are the data points before passing through the encoders.

1) *end-to-end.* [10,11] In this design in Fig. 1(a), the error generated by the contrastive loss function is backpropagated through both the key encoder and the query encoder, hence the encoder weights and parameters are updated even in the continual learning stage in a semi-supervised (no labels) fashion. Labels are only used in the initial training phase with base classes. This design is the ideal case, but has a major disadvantage, that of efficient memory management. Though the key encoder gets updated, the keys themselves do not. If the memory is continually augmented by incoming samples, then the size soon becomes impracticable. Also if we try to keep the size same by replacing a fixed number of keys in the dictionary by incoming samples in each batch iteration, the encoder gets susceptible to "catastrophic forgetting", if the incoming samples deviate significantly from the previous dictionary distribution.

2) *memory bank.* [12] In this design in Fig. 1(b), We only update the query encoder, whereas the key encoder is only used for converting the key samples to the representation space; the key encoder itself is not updated. The incoming samples in the continual learning phase are all added to the dictionary space, but a fixed number from the set is sampled as keys, thus managing the memory size. Since weights of the key encoder are not updated, this avoids the possibility of forgetting. On the other hand, the disadvantage of this design is that we lose the benefits of unsupervised learning of the dictionary during continual learning.

3) *momentum contrast.* [13] In this design in Fig. 1(c), the authors try to overcome the disadvantages of the earlier baseline designs, 1) forgetting in the end-to-end design, and 2) and lack of learning in key encoder in the memory bank design. They accomplish this by proposing a compromise: the key encoder still gets updated, but only by a fraction of the new data, in order to limit the effect and avoid forgetting earlier learned representations. This is done by introducing a momentum coefficient $m \in [0,1)$. Let the weights/parameters of the query encoder and the key encoder be θ_q and θ_k respectively. Then in iteration $i+1$, the key encoder is updated based on the previous iteration states as follows.

$$\theta_k^{i+1} \leftarrow m\theta_k^i + (1-m)\theta_q^i. \tag{2}$$

Thus, by controlling the value of the momentum m, the user can specify the speed of update of the key encoder. The authors chose a constant value for m empirically through experimentation, and they advise that this value should be close to 1. This ensures a slow update of the key encoder, thus minimising the chance of sudden forgetting.

2.3 Collaborative Contrast in Continual Learning

Disadvantages of MoCo. Though the momentum contrast scheme succeeds in overcoming the drawbacks of the two baseline designs, it presents a couple of its own limitations.

1. Firstly, the underlying philosophy of contrastive learning is to learn a "reference representation' of the data distribution, and then use that template of the dataset, to decide whether the query sample is a negative or a positive sample depending on the match. The earlier designs do not completely follow the distribution learning ethos of contrastive learning.
2. A second, and a more technical shortcoming, is that the value of the momentum coefficient is set experimentally instead of calculating an optimal value.

We overcome both limitations at one shot in our proposed method as described below.

CoCo: Collaborative Contrast. The starting point of the new formulation is to remain faithful to the basic representation learning premise of contrastive learning. Let all the key outputs of the key encoder be the columns of a key matrix K and corresponding representation vector be α. Then the *collaborative cost function* [14, 15] is given by:

$$p = \|q - K\alpha\|_2^2 + \lambda\|\alpha\|_2^2 \tag{3}$$

Here q is the query sample and λ is the regularization constant. A least squares optimization yields the optimal representation vector as:

$$\hat{\alpha} = \arg\min_{\alpha}\left(\|q - K\alpha\|_2^2 + \lambda\|\alpha\|_2^2\right) = (K^T K + \lambda I)^{-1} K^T q \tag{4}$$

This means the *representation residual* of the i_{th} key when compared with the query becomes:

$$r_i = \frac{\|q - K_i\hat{\alpha}_i\|_2^2}{\|\hat{\alpha}_i\|_2^2} \tag{5}$$

Therefore following Eq. 6, the new *collaborative contrastive loss function* is:

$$\mathcal{L}_{CoCo}^i = -\log \frac{\exp(\frac{\|q - K_i\hat{\alpha}_i\|}{\|\hat{\alpha}_i\|})}{\sum_i \exp(\frac{\|q - K_i\hat{\alpha}_i\|}{\|\hat{\alpha}_i\|})} = -\log \frac{\exp(r_i)}{\sum_i \exp(r_i)} \tag{6}$$

Here the function of the regularization term λ is to ensure the even representation of the weight vector, which results in a lower norm. Thus λ gives a measure of how faithful the new query is to the general distribution of the existing dictionary. Hence in our case, we set the momentum coefficient to be equal to λ, thus automating the calculation of m. The momentum update equation itself is same as Eq. 2, and the class prediction is given by the match having minimum loss.

$$C(q) = C(k_i | \arg\min_i \mathcal{L}_{CoCo}^i) \tag{7}$$

The pseudo-code is provided in Algorithm 1.

Algorithm 1: Collaborative Contrast (CoCo) in Continual Learning

---

1 **for** *each non-benign class D to be contrasted with benign class B* **do**
2 | **Split** into 70% for base class training and 30% for continual learning ;
3 | **Base Train** an autoencoder to classify chosen class P and benign class D ;
4 | **Copy** the encoder into two to form key k encoder and query q encoder ;
5 | **Inititate** the representation vector α and form the dictionary matrix K ;
6 | **Find** initial optimal reconstruction matrix A by Eq. 6. ;
7 | **for** *each query sample q* **do**
8 | | **Find** the optimal α and λ by Eq. 3 and 4 ;
9 | | **Find** the key K_i with minimum loss by Eqs. 5 and 6 ;
10 | | **for** *each key $k \in K$* **do**
11 | | | **Update** key weights with optimal λ using momentum Eq. 2 ;
12 | | | **Update** query encoder with standard back-propagation gradients ;
13 | | **end**
14 | | **Output** the class label of the query sample by Eq. 7;
15 | **end**
16 **end**

---

3 Experiments and Results

In this section, we first present the evaluation datasets and the competing methods. Then we discuss the numerical and qualitative results.

3.1 Experimental Setup

Datasets. We test our method on three classification tasks on histology image data: breast cancer, colon cancer and lung cancer. Medical imaging datasets curated for continual learning are almost non-existent, which is why we have simulated continual learning experiments using standard classification datasets. On the positive side, this gives us the opportunity to compare results against other competing methods more easily.

– *Breast Cancer Histology Dataset (BACH)* [16] consists of 400 high-resolution (2018 × 1356) Hematoxylin and eosin stained microscopy images, with an even distribution over four classes. Each image is labeled as one of four types: 1) normal, 2) benign, 3) in situ carcinoma and 4) invasive carcinoma, according to the predominant tissue type. The main challenge of this dataset is the limited number of images (100) per class. This makes it difficult to train a full deep network, thus the semi-supervised design of the present work is an advantage.
– *Lung + Colon Cancer Histology Dataset (LC25000)* [17] contains 25,000 color images with five classes of 5000 images each. All images are $768 \times 768 pixels$ in size. We run separate experiments on the two colon cancer categories (adenocarcinomas and benign) and the three lung cancer categories (squamous cell carcinomas, adenocarcinomas and benign).

Table 1. Performance Accuracy (%) and standard deviation on Histology Data

Methods	Breast cancer	Colon cancer	Lung cancer
CPC [18]	77.3 ± 2.2	85.6 ± 2.7	86.8 ± 2.6
CMC [19]	79.8 ± 2.0	88.2 ± 2.4	88.4 ± 2.5
end-to-end [10,11]	75.5 ± 2.3	83.7 ± 2.5	83.5 ± 2.1
memory bank [12]	75.9 ± 2.1	84.0 ± 2.1	83.9 ± 2.8
MoCo [13]	82.4 ± 2.5	90.5 ± 2.4	90.2 ± 2.1
CoCo (Our Method)	**86.1 ± 2.3**	**93.3 ± 2.7**	**91.7 ± 2.4**

Competing Methods. Besides the three contrastive continual learning designs described in the methodology section (end-to-end, memory bank, and MoCo), we also compare our CoCo method against two popular baseline contrastive learning algorithms. These are contrastive predictive coding (CPC) and contrastive multiview coding (CMC).

- *Contrastive Predictive Coding (CPC)* [18] learns self-supervised representations by predicting the future in latent space by using powerful autoregressive models. The model uses a probabilistic contrastive loss which induces the latent space to capture information that is maximally useful to predict future samples. CMC is an augmented version of CPC for multiview cases, using CPC as the backbone.
- *Contrastive Multiview Coding (CMC)* [19] learns by contrasting between congruent and incongruent pairs across views. For each set of views, a deep representation is learnt by bringing views of the same scene together in embedding space, while pushing views of different scenes apart. Intuitively, this can be further interpreted as discriminating the joint distribution of views from the product of marginals. In other words, our CMC learns representations by maximizing mutual information between views.

Table 2. Wilcoxon Test of "Benign" vs. other classes

	Breast cancer			Colon cancer	Lung cancer	
	Normal	In-situ	Invasive	Adenocarcinoma	Adeno	Squam
MoCo	81.1	82.7	83.4	90.5	89.8	90.6
Coco	85.6	86.2	86.5	93.3	92.3	89.9
|Difference|	4.5	3.5	3.1	2.8	2.5	0.7
Rank (R)	6	5	4	3	2	1
Sign (S)	1	1	1	1	1	−1

3.2   Results and Analysis

Quantitative Results. Since we operate in a contrastive continual learning setting, for each dataset, the experiments are conducted for two classes at a time (benign vs one of the diseased classes). Also, within the chosen classes, 70% data is randomly selected for the base class training and the remaining 30% for incremental learning. The performance accuracies are then averaged over all the class pair combinations and presented in Table 1, along with the standard deviation. It can be observed that the proposed Coco not only outperforms the benchmark method CPC and CMC, it also yields better results than the recent competitor MoCo (2020), thus achieving state-of-the art performance.

Statistical Analysis. Since MoCo is the closest competitor to our method, we also perform statistical analysis to gauge the level of significance of the improvement in performance. We perform the Wilcoxon signed rank test [20], presented in Table 2. The ranks (R) are allocated according to the magnitude of difference in accuracy between the two methods. The corresponding signs (S) are allocated depending on which method outperforms for that particular experimental setting. The ones for which CoCo is better have sign 1. the rest have sign –1. The Wilcoxon parameter $W = \sum SR$ is calculated and we find $W = 19$. The maximum possible rank value for $n = 6$ binary decision experiments is $n(n+1)/2 = 21$. The Wilcoxon signed rank test states that the null hypothesis (MoCo and CoCo are equally good) may be rejected in one-direction (CoCo better than MoCo) at 5% level of significance if $W \geq 19$, which is true in this case. Hence we can infer that the better performance of CoCo over MoCo is statistically significant.

4   Conclusion

In this paper, we show how a relatively lightweight mechanism can be designed for continual learning in medical image classification tasks, with the background assumption that it is a binary decision (healthy vs diseased), which means a contrastive loss based semi-supervised mechanism can be used. The incorporation of the novel collaborative contrast (CoCo) loss helps to perform unsupervised learning in the incremental learning phase by directly comparing incoming queries to the dictionary keys, thus reducing the need for labelling downstream. CoCo outperforms several competitors including the recent momentum contrast (MoCo) method establishing a new state-of-the-art; it also overcomes drawbacks of existing methods from a design perspective. The SOTA performance has been demonstrated on three histology cancer (lung, breast, colon) datasets in a continual learning scenario, but can be used for other similar tasks.

References

1. Liu, X., et al.: A comparison of deep learning performance against health-care professionals in detecting diseases from medical imaging: a systematic review and meta-analysis. Lancet Dig. Health **1**(6), 271–297 (2019)

2. Gonzalez, C., Sakas, G., Mukhopadhyay, A.: What is Wrong with Continual Learning in Medical Image Segmentation?. arXiv:2010.11008 (2020)
3. Baweja, C., Glocker, B., Kamnitsas, K.: Towards continual learning in medical imaging. arXiv:1811.02496 (2018)
4. Bhalgat, Y., Shah, M., Awate, S.: Annotation-cost Minimization for Medical Image Segmentation using Suggestive Mixed Supervision Fully Convolutional Networks. arXiv:1812.11302 (2018)
5. Ye, M., Zhang, X., Yuen, P.C., Chang, S.-F.: Unsupervised embedding learning via invariant and spreading instance feature. In: Proc, CVPR (2019)
6. Zhuang, C., Zhai, A.L., Yamins, D.: Local aggregation for unsupervised learning of visual embeddings. In Proc, ICCV (2019)
7. Hadsell, R., Chopra, S., LeCun, Y.: Dimensionality reduction by learning an invariant mapping. In Proc, CVPR (2006)
8. Oord, A., Li, Y., Vinyals, O.: Representation Learning with Contrastive Predictive Coding. arXiv:1807.03748v2 (2019)
9. Gutmann, M., Hyvarinen, A.: Noise-contrastive estimation: A new estimation principle for unnormalized statistical models. In Proc, AISTATS (2010)
10. Hjelm, R.D., Fedorov, A., Marchildon, S.L., Grewal, K., Trischler, A., Bengio, Y.: Learning deep representations by mutual information estimation and maximization. In Proc, ICLR (2019)
11. Bachman, P., Hjelm, R.D., Buchwalter, W.: Learning representations by maximizing mutual information across views. arXiv:1906.00910 (2019)
12. Wu, Z., Xiong, Y., Yu, S., Lin, D.: Unsupervised feature learning via non-parametric instance discrimination. In Proc, CVPR (2018)
13. He, K., Fan, H., Wu, Y., Xie, S., Girshick, R.: Momentum Contrast for Unsupervised Visual Representation Learning. In Proc, CVPR (2020)
14. Zhang, L., Yang, M., Feng, X.: Sparse representation or collaborative representation: Which helps face recognition? In Proc, ICCV (2011)
15. Chakraborti, T., McCane, B., Mills, S., Pal, U.: A Generalised Formulation for Collaborative Representation of Image Patches (GP-CRC). In Proc, BMVC (2017)
16. Aresta, G., et al.: BACH: Grand challenge on breast cancer histology images. Medical Image Analysis 56, 122–139 (2019)
17. Borkowski, A.A., Bui, M.M., Thomas, L.B., Wilson, C.P., DeLand, L.A., Mastorides, S.M.: Lung and Colon Cancer Histopathological Image Dataset (LC25000). arXiv:1912.12142 (2019)
18. Henaff, O.J., Razavi, A., Doersch, C., Eslami, S.M., Oord, A.: Data-efficient image recognition with contrastive predictive coding. arXiv:1905.09272 (2019)
19. Tian, Y., Krishnan, D., Isola, P.: Contrastive multiview coding. arXiv:1906.05849 (2019)
20. Rey, D., Neuhauser, M.: Wilcoxon-Signed-Rank Test. Springer, International Encyclopedia of Statistical Science (2011). https://doi.org/10.1007/978-3-642-04898-2

Learning Transferable 3D-CNN for MRI-Based Brain Disorder Classification from Scratch: An Empirical Study

Hao Guan[1], Li Wang[1], Dongren Yao[1], Andrea Bozoki[2], and Mingxia Liu[1(✉)]

[1] Department of Radiology and BRIC, University of North Carolina at Chapel Hill,
Chapel Hill, NC 27599, USA
mxliu@med.unc.edu

[2] Department of Neurology, University of North Carolina at Chapel Hill,
Chapel Hill, NC 27599, USA

Abstract. Reliable and efficient transferability of 3D convolutional neural networks (3D-CNNs) is an important but extremely challenging issue in medical image analysis, due to small-sized samples and the domain shift problem (*e.g.*, caused by the use of different scanners, protocols and/or subject populations in different sites/datasets). Although previous studies proposed to pretrain CNNs on ImageNet, models' transferability is usually limited due to semantic gap between natural and medical images. In this work, we try to answer a key question: *how to learn transferable 3D-CNNs from scratch based on a small (e.g., tens or hundreds) medical image dataset?* We focus on the case of structural MRI-based brain disorder classification using four benchmark datasets (*i.e.*, ADNI-1, ADNI-2, ADNI-3 and AIBL) to address this problem. (1) We explore the *influence of different network architectures* on model transferability, and find that appropriately deepening or widening a network can increase the transferability (*e.g.*, with improved sensitivity). (2) We analyze the *contributions of different parts* of 3D-CNNs to the transferability, and verify that fine-tuning CNNs can significantly enhance the transferability. This is different from the previous finding that fine-tuning CNNs (pretrained on ImageNet) cannot improve the model transferability in 2D medical image analysis. (3) We also study the *between-task transferability* when a model is trained on a source task from scratch and applied to a related target task. Experimental results show that, compared to directly training CNN on related target tasks, CNN pretrained on a source task can yield significantly better performance.

Keywords: Transferability · 3D-CNN · Deep learning · Brain MRI

Electronic supplementary material The online version of this chapter (https://doi.org/10.1007/978-3-030-87589-3_2) contains supplementary material, which is available to authorized users.

1   Introduction

Deep learning (*e.g.*, with convolutional neural networks, CNNs) has been extensively applied in medical image analysis [1–3], but usually suffers from the small-sample-size problem. To address this issue, recent studies proposed to fine-tune pretrained CNNs on ImageNet for medical imaging analysis [4–6]. However, direct transferring pretrained CNNs usually yields sub-optimal performance, due to the fundamental differences between natural and medical images. For instance, medical images (*e.g.*, T1-weighted MRIs) are typically 3-dimensional, whereas CNNs pretrained on ImageNet usually treat all the 2D slices within a subject scan independently. This will lead to loss of information [7,8]. In addition, significant differences in category distributions may also make pretrained CNNs on ImageNet (with $1,000$ classes) unsuitable for analyzing medical images.

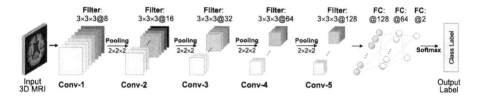

Fig. 1. Architecture of the baseline CNN model for 3D MRI classification, consisting of five convolutional (Conv) layers (with each followed by a pooling layer) and three fully-connected (FC) layers (followed by softmax activation). The term "$3 \times 3 \times 3@8$" denotes a convolutional layer with 8 filters (kernel size: $3 \times 3 \times 3$).

A reasonable alternative is to train 3D-CNNs from scratch on a source medical image dataset/site, and then to fine-tune them on a to-be-analyzed target dataset/site [9,10]. However, training 3D-CNNs from scratch has to face two challenges. *First*, the sample size is often limited (*e.g.*, tens or hundreds). Utilizing inappropriate network architecture will lead to severe over-fitting, which may greatly affect the transferability. *Second*, different sites/datasets may have significant domain shifts in terms of data distribution, caused by different scanners, protocols and populations [11–14]. Thus, it is highly desirable to boost the transferability of 3D-CNNs for medical image analysis.

In this paper, we aim to answer a key question: *how to train transferable 3D-CNNs from scratch with relatively small-sized medical images*? We study the case of structural MRI-based brain disorder classification on 4 benchmark datasets via two tasks: (1) Alzheimer's disease (AD) detection; and (2) mild cognitive impairment (MCI) conversion prediction. We build a baseline 3D-CNN model (see Fig. 1), and evaluate the performance of several variants of the baseline CNN model with different architectures. We then analyze the transferability of the baseline network with transfer learning on different target domains. We also explore the between-task transferability, with the CNN model trained on a source task (*i.e.*, AD detection) and applied to a related task (*i.e.*, MCI conversion prediction). Our empirical findings will provide some insights on how to improve the transferability of CNNs in 3D medical image analysis.

2   Materials and Methodology

2.1   Studied Subjects and MR Image Pre-processing

Four benchmark datasets are used in this work, *i.e.*, Alzheimer's Disease Neuroimaging Initiative (ADNI-1) [15], ADNI-2 [16], ADNI-3 [17] and Australian Imaging Biomarkers and Lifestyle Study of Aging database (AIBL) [18]. Note that the subjects that simultaneously appear in ADNI-1, ADNI-2 and ADNI-3 are removed from ADNI-2 and ADNI-3 to avoid data leakage and guarantee comparison fairness as suggested in [7]. Their domain heterogeneity mainly comes from the use of different scanning parameters (*e.g.*, 1.5T or 3T) and updated scanners. ADNI-1 contains 748 subjects with 1.5T T1-weighted structural MRIs, including 205 AD, 231 cognitively normal (CN), 165 progressive MCI (pMCI) and 147 stable MCI (sMCI) subjects. ADNI-2 has 708 subjects with 3T T1-weighted structural MRIs (*i.e.*, 162 AD, 205 CN, 88 pMCI and 253 sMCI). Note that those subjects with pMCI would convert to MCI within 36 months and sMCI remains stable. ADNI-3 involves 389 subjects with 3T T1-weighted structural MRIs (*i.e.*, 60 AD and 329 CN). AIBL consists of MRIs acquired from 549 subjects (*i.e.*, 71 AD, 447 CN, 11 pMCI and 20 sMCI). The demographic and clinical information of the studied subjects is shown in *Supplementary Materials*.

All brain MRIs are pre-processed through a standard pipeline, including skull stripping, intensity inhomogeneity correction, image re-sampling, and spatial normalization to the Automated Anatomical Labeling (AAL) template.

2.2   Methodology

Architecture of Baseline CNN. Figure 1 illustrates the architecture of the baseline CNN used in this work. This network consists of 5 convolutional (Conv) layers with $3 \times 3 \times 3$ filters, and 3 fully-connected layers with 128, 64 and 2 neurons, respectively. A softmax layer is used for classification. Each Conv layer is composed of a sequence of 3D convolutional filters, followed by batch normalization and ReLU activation function. To reduce the risk of over-fitting, a $2 \times 2 \times 2$ max pooling operation (stride: $2 \times 2 \times 2$) is added after each Conv layer. This is a very basic CNN model that can be flexibly extended in different applications.

Network Training. Considering the relatively larger number of subjects in ADNI-1, we use ADNI-1 as training/source data for model training and validation, whereas the other three datasets are treated as test/target data for transferability evaluation. A 5-fold cross-validation strategy is used. That is, subjects in ADNI-1 are randomly partitioned into 5 folds. For parameter selection, each fold is treated as the validation set in turn, with the rest as the training set.

The trained model is finally applied to target/test data. Such process is repeated five times to avoid bias caused by random partition. For network training, the Adam algorithm is used as the optimizer, with a learning rate of 0.0001. The batch size is set to 2. A drop-out rate of 0.5 is used to avoid over-fitting. We train the network for 30 to 50 epochs, and employ an early stopping strategy when the training or the validation loss continuously changes little for 10 epochs.

Table 1. Results of AD detection achieved by 3D-CNNs with different network depths on three target domains (with models trained on ADNI-1).

Target Domain	Method	AUC (%)	ACC (%)	BAC (%)	SEN (%)	SPE (%)
ADNI-2	CNN-5	93.69 ± 1.34	84.25 ± 2.69	82.65 ± 3.16	69.01 ± 7.23	96.29 ± 1.45
	CNN-7	91.81 ± 0.92	84.69 ± 1.58	83.51 ± 1.65	73.46 ± 2.66	93.56 ± 1.70
	CNN-10	91.38 ± 0.95	83.87 ± 0.70	82.27 ± 0.76	68.64 ± 2.56	95.90 ± 2.03
	CNN-12	91.35 ± 2.48	82.88 ± 2.24	81.42 ± 2.54	68.89 ± 5.53	93.95 ± 1.97
ADNI-3	CNN-5	94.57 ± 0.96	91.00 ± 0.92	78.87 ± 3.68	61.33 ± 8.85	96.41 ± 1.91
	CNN-7	93.51 ± 1.17	90.59 ± 1.91	81.76 ± 3.19	69.00 ± 5.73	94.53 ± 1.78
	CNN-10	91.53 ± 1.05	91.67 ± 1.24	78.59 ± 1.70	59.67 ± 3.61	97.51 ± 1.56
	CNN-12	92.25 ± 2.18	90.54 ± 1.15	80.10 ± 4.56	65.00 ± 9.34	95.20 ± 1.80
AIBL	CNN-5	92.55 ± 0.99	91.00 ± 1.21	82.94 ± 2.99	71.83 ± 7.65	94.05 ± 2.21
	CNN-7	89.25 ± 0.90	89.94 ± 0.76	83.06 ± 1.04	75.77 ± 4.61	90.34 ± 2.73
	CNN-10	90.84 ± 1.29	91.62 ± 1.42	80.33 ± 2.51	64.79 ± 5.63	95.88 ± 1.86
	CNN-12	89.60 ± 1.75	90.50 ± 1.56	81.94 ± 2.33	70.14 ± 5.92	93.74 ± 2.32

Evaluation Metric. Two classification tasks are included: (1) AD detection (*i.e.*, AD vs. CN classification), and (2) MCI conversion prediction (*i.e.*, pMCI vs. sMCI classification). Five metrics are used for performance evaluation, including (1) area under the ROC curve (AUC), (2) classification accuracy (ACC), (3) balanced accuracy (BAC), (4) sensitivity (SEN), and (5) specificity (SPE).

3   Transferability Vs. Different Network Architectures

Due to issues of small-sample-size and high feature dimension of medical images, it is often challenging to design a suitable network architecture to obtain good transferability. We now explore the influence of network capacity (in terms of *network depth* and *network width*) on the transferability of 3D-CNNs.

Influence of Network Depth. To explore the influence of network depth on transferability, we develop three variants of the baseline CNN with different number of convolutional (Conv) layers, including (1) **CNN-7** that contains 7 Conv layers with 8, 16, 32, 64, 64, 128 and 128 filters, respectively; (2) **CNN-10** that consists of 10 Conv layers with 8, 8, 16, 16, 32, 32, 64, 64, 128, and 128, respectively (with detailed architecture shown in *Supplementary Materials*); and (3) **CNN-12** that contains 12 Conv layers with 8, 8, 16, 16, 32, 32, 64, 64, 64, 128, 128 and 128 filters, respectively. The baseline model with 5 Conv layers is called **CNN-5** (see Fig. 1). The FC layers and the training strategy of these variants remain the same as the baseline CNN-5.

The performance of these models on three target/test domains is listed in Table 1. From the results, we can derive the following empirical findings. *First*, adding more layers does not necessarily lead to better AUC and ACC performance. For example, the AUC values of CNN-5 are generally better than those achieved by three deeper CNNs on three test datasets, and there are no significant differences in terms of ACC of four networks (with p-values = 0.5278,

Table 2. Results of AD detection achieved by 3D-CNNs with different network widths on three target domains (with models trained on ADNI-1).

Target Domain	Method	AUC (%)	ACC (%)	BAC (%)	SEN (%)	SPE (%)
ADNI-2	CNN-w1.0	93.69 ± 1.34	84.25 ± 2.69	82.65 ± 3.16	69.01 ± 7.23	96.29 ± 1.45
	CNN-w2.0	94.23 ± 0.55	86.87 ± 2.17	85.59 ± 2.58	74.69 ± 6.13	96.49 ± 1.22
	CNN-w4.0	93.52 ± 0.86	87.63 ± 1.61	86.94 ± 1.80	81.11 ± 4.91	92.78 ± 3.11
	CNN-w5.0	90.71 ± 1.92	83.38 ± 1.49	82.03 ± 1.56	70.49 ± 3.01	93.56 ± 2.14
ADNI-3	CNN-w1.0	94.57 ± 0.96	91.00 ± 0.92	78.87 ± 3.68	61.33 ± 8.85	96.41 ± 1.91
	CNN-w2.0	96.36 ± 0.58	92.70 ± 0.72	83.83 ± 2.10	71.00 ± 5.35	96.66 ± 1.47
	CNN-w4.0	94.71 ± 1.39	90.80 ± 2.56	84.88 ± 1.29	76.33 ± 5.94	93.61 ± 4.10
	CNN-w5.0	91.51 ± 2.10	89.51 ± 1.86	82.76 ± 2.62	73.00 ± 7.11	92.52 ± 3.01
AIBL	CNN-w1.0	92.55 ± 0.99	91.00 ± 1.21	82.94 ± 2.99	71.83 ± 7.65	94.05 ± 2.21
	CNN-w2.0	92.37 ± 0.98	90.42 ± 1.80	85.21 ± 1.47	78.03 ± 4.29	92.39 ± 2.55
	CNN-w4.0	91.82 ± 0.56	88.38 ± 4.55	84.90 ± 1.58	80.00 ± 7.87	89.80 ± 6.50
	CNN-w5.0	88.96 ± 1.84	86.60 ± 2.99	81.41 ± 1.86	72.05 ± 3.75	90.77 ± 2.27

0.5431 and 0.2348 on 3 target domains). *Second*, deeper CNNs tend to produce better sensitivity to precisely detect AD subjects. For example, CNN-7 achieves obvious improvement compared with CNN-5. However, adding too many layers (*e.g.*, > 10) does not necessarily benefit the detection rate of AD. *Besides*, all four networks show fluctuating performance (with larger standard deviations) in terms of sensitivity than other metrics. In contrast, all CNNs achieve relatively stable detection rate of cognitively normal samples. This implies that the cross-domain heterogeneity of AD subjects has more significant influence on the transferability of CNNs. It is interesting to pay more attention to patients when designing cross-domain transfer learning models.

Influence of Network Width. In this work, we refer to the number of filters in each convolutional layer as the network width. To explore the influence of network width on transferability, we develop three variants of the baseline CNN, by increasing the number of filters in each layer using a widening factor w. When w = 2.0, for example, the number of filters in each layer is doubled, and we name this variant as **CNN-w2.0**.

The results of CNNs with different network widths on three test domains are shown in Table 2. From Table 2, we can derive the following empirical findings. *First*, the network width does influence the transferability, and using very small or very large width will not benefit the transferability of CNN models. When the widening factor w> 1.0, the network tends to achieve better overall performance in terms of most metrics. But when w = 5.0, there is a significant performance degradation. This implies that there is a boundary when widening the network to achieve better transferability. *Besides*, widening the network helps increase the sensitivity for AD detection. Compared with the baseline CNN (*i.e.*, CNN-w1.0), wider networks (*e.g.*, CNN-w4.0) tend to have a significant better detection rate (SEN) of AD subjects, and its effect is more pronounced compared to increasing network depth. The possible reason is that using more filters helps capture more local-to-global discriminative features in brain MRIs to identify AD subjects.

4   Transferability Vs. Different Network Components

Influence of Fine-Tuning. A previous study proposed to fine-tune CNNs pretrained on ImageNet for 2D medical image analysis [19]. Although CNNs are pretrained with massive natural images, it is reported that when they are used for medical image analysis, fine-tuning cannot effectively improve their transferability. To analyze the behavior of the proposed 3D-CNN trained from scratch, we use some labeled target data to fine-tune the pretrained baseline CNN and evaluate its performance on the target domain. Specifically, we use ADNI-1 as the source domain to train the baseline CNN, and then fine-tune and test the network on ADNI-2. 20% of the samples in ADNI-2 are randomly selected and used for fine-tuning, while the remaining samples are used as test data for evaluating the transferability. Since the baseline CNN is trained in a 5-fold cross-validation manner, we can obtain five pretrained CNNs. Accordingly, the fine-tuning and evaluation procedures are carried out five times.

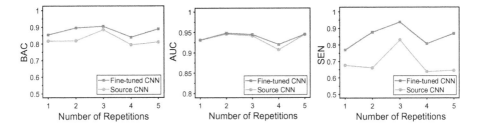

Fig. 2. Performance of fine-tuned and source CNNs in AD detection.

The evaluation results on ADNI-2 in terms of three key metrics are shown in Fig. 2. For comparison, we also report the results achieved by the pretrained CNNs (denoted as source CNN) without fine-tuning. From this figure, we have the following observations. *First*, the overall performance of the fine-tuned CNN is better than the source CNN pretrained on ADNI-1. The underlying reason may be that unlike the pre-trained 2D-CNN on ImageNet, the 3D-CNN trained from scratch using 3D MRIs can learn more disease-related information patterns. This provides a better initialization for the network, which lays the foundation for further network optimization through fine-tuning. *In addition*, fine-tuning makes the network achieve a more balanced accuracy (BAC). As shown in the right of Fig. 2, the most significant improvement comes from the detection of AD subjects (with much higher SEN). These results further support the conclusion in Sect. 3 that AD subjects are more informative in enhancing the transferability of 3D-CNNs, compared with cognitively normal subjects.

Contribution of Different Layers to Mitigating Domain Shift. When we train a 3D-CNN from scratch and apply it to a different domain, an interesting question is to determine the contribution of different layers on mitigating cross-domain differences in data distribution. That is, we'd like to investigate which

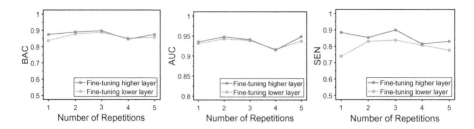

Fig. 3. Performance comparison of fine-tuning higher layer of the pretrained CNN and fine-tuning the first layer of the CNN in AD detection.

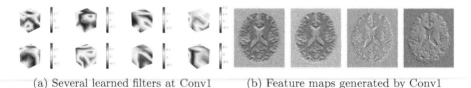

(a) Several learned filters at Conv1   (b) Feature maps generated by Conv1

Fig. 4. Visualization of learned convolution filters and feature maps at the first convolution layer of the fine-tuned CNN in AD detection.

part of a CNN is more easily influenced by the cross-domain data heterogeneity. We address this problem by fine-tuning certain layers while keeping the weights of other layers frozen. If fine-tuning a certain layer can achieve more performance improvement, it is assumed that this layer contributes more to alleviating the domain shift problem. Here, we focus on the low-level part (Conv-1) and the high-level part (Conv-5 and three fully-connected layers) in the baseline CNN. We fine-tune these two parts separately. That is, when fine-tuning the low-level part, the remaining layers are frozen; and vice versa.

Figure 3 reports the results in terms of three key metrics. It can be seen that fine-tuning the high-level part of the CNN produces better SEN and BAC results in most cases, compared with fine-tuning its low-level part. This implies that the high-level part with finer scales contributes more to mitigating the cross-domain data heterogeneity. The low-level part may help extract more domain-independent information. To verify this assumption, we visualize several Conv filters and feature maps in Conv-1 of the fine-tuned CNN in Fig. 4, from which we can see that Conv-1 mainly extracts low-level features (with enhanced edges).

5   Transferability to Related Task

We further investigate the between-task transferability when a model is trained on a source task from scratch and applied to a related target task. Considering that MCI is the prodromal stage of AD [20,21], we develop two CNNs which are trained with AD and CN subjects, and then evaluate their transferability for MCI conversion prediction. 1) CNN-AD1: Baseline CNN and a widened CNN (with w

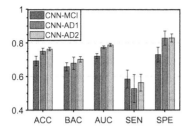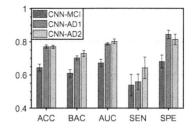

Fig. 5. Results of the baseline network (left) and its widened variant CNN-w4.0 (right), with models tested on ADNI-2 for MCI conversion prediction.

= 4.0) trained with AD and CN samples from ADNI-1. 2) CNN-AD2: Baseline CNN and a widened CNN (with w = 4.0) trained with AD and CN samples from ADNI-1 and ADNI-2. For comparison, we also train a baseline CNN and a widened CNN (with w = 4.0) with MCI samples in ADNI-1. All these networks are tested on ADNI-2 for MCI conversion prediction. The experiment is repeated five times and the results are shown in Fig. 5.

From Fig. 5, one can observe that the networks trained with only AD and CN samples can achieve comparable or even better performance in MCI conversion prediction, compared with CNN-MCI trained on MCI samples. The widened CNNs trained on AD and CN samples achieve greater transferability for MCI conversion prediction, compared with the baseline CNN. Especially, when using more training samples, CNN-AD2 produces much higher SEN values, indicating its superiority in detecting pMCI patients. We also observe that the widened CNN trained on MCI samples suffers some performance decrease which may be attributed to the limited number of training samples. The widened CNN trained on samples from ADNI-1 and ADNI-2 achieves the overall best performance. This may be due to the relatively large amount of training samples and the potential relationship between MCI and AD populations. These results show that it is beneficial to train a CNN model on one task and transfer it to related tasks.

6   Conclusion and Future Work

In this paper, we have conducted an experimental study on how to train a transferable CNN for 3D medical image analysis. Based on the results for brain MRI-based brain disorder classification, we have made some empirical findings. (1) Appropriately adding more layers and widening the network width (*i.e.*, using more filters at each Conv layer) is helpful to improve the transferability, especially for the enhancement of sensitivity. (2) Transfer learning via fine-tuning is beneficial to increase the transferability of the 3D CNN (trained from scratch on a source domain). Fine-tuning high-level layers is helpful to alleviate the domain shift issue. (3) A CNN trained on a specific task (*i.e.*, AD detection) with more training samples can be successfully transferred to a different but related task (*i.e.*, MCI conversion prediction) with small-sized samples. The empirical findings may provide the community with some useful references and techniques on how to leverage 3D-CNNs for various medical image analysis tasks.

In the future, we will study the generalizability of our trained models in identifying other brain disorders such as Parkinson's disease and autism. In addition, we will explore neural architecture search for the analysis of architectures. Since different diseases may affect different brain regions, it is interesting to integrate an attention detection module to CNN, which will also be our future work.

Acknowledgements. This work was supported in part by NIH grants (Nos. AG041721, MH109773 and MH117943).

References

1. Litjens, G., Kooi, T., Bejnordi, B.E., Setio, A.A.A., et al.: A survey on deep learning in medical image analysis. Med. Image Anal. **42**, 60–88 (2017)
2. Mazurowski, M.A., Buda, M., Saha, A., Bashir, M.R.: Deep learning in radiology: an overview of the concepts and a survey of the state of the art with focus on MRI. J. Magn. Reson. Imaging **49**, 939–954 (2019)
3. Guan, H., Liu, Y., Yang, E., Yap, P.T., Shen, D., Liu, M.: Multi-site MRI harmonization via attention-guided deep domain adaptation for brain disorder identification. Med. Image Anal. **71**, 102076 (2021)
4. Morid, M.A., Borjali, A., Del Fiol, G.: A scoping review of transfer learning research on medical image analysis using ImageNet. Computers in Biology and Medicine (2020)
5. Cuingnet, R., Gerardin, E., Tessieras, J., et al.: Automatic classification of patients with Alzheimer's disease from structural MRI: a comparison of ten methods using the ADNI database. NeuroImage **56**(2), 766–781 (2011)
6. Tajbakhsh, N., et al.: Convolutional neural networks for medical image analysis: full training or fine tuning? IEEE Trans. Med. Imaging **35**, 1299–1312 (2016)
7. Wen, J., Thibeau-Sutre, E., Diaz-Melo, M., et al.: Convolutional neural networks for classification of Alzheimer's disease: overview and reproducible evaluation. Med. Image Anal. **63**, 1–19 (2020)
8. Guan, Z., Kumar, R., Fung, Y.R., Wu, Y., Fiterau, M.: A comprehensive study of Alzheimer's disease classification using convolutional neural networks. arXiv:1904.07950 (2019)
9. Liu, M., Zhang, J., Adeli, E., Shen, D.: Landmark-based deep multi-instance learning for brain disease diagnosis. Med. Image Anal. **43**, 157–168 (2018)
10. Guan, H., Liu, M.: Domain adaptation for medical image analysis: a survey. arXiv:2102.09508 (2021)
11. AlBadawy, E.A., Saha, A., Mazurowski, M.A.: Deep learning for segmentation of brain tumors: Impact of cross-institutional training and testing. Med. Phys. **45**(3), 1150–1158 (2018)
12. Pooch, E.H., Ballester, P.L., Barros, R.C.: Can we trust deep learning models diagnosis? The impact of domain shift in chest radiograph classification. arXiv:1909.01940 (2019)
13. Stacke, K., Eilertsen, G., Unger, J., Lundström, C.: A closer look at domain shift for deep learning in histopathology. arXiv:1909.11575 (2019)

14. Wang, M., Zhang, D., Huang, J., Yap, P.T., Shen, D., Liu, M.: Identifying autism spectrum disorder with multi-site fMRI via low-rank domain adaptation. IEEE Trans. Med. Imaging **39**(3), 644–655 (2019)
15. Jack Jr, C.R., Bernstein, M.A., Fox, N.C., et al.: The Alzheimer's disease neuroimaging initiative (ADNI): MRI methods. J. Magn. Reson. Imaging **27**(4), 685–691 (2008)
16. Jack Jr, C.R., et al.: Magnetic resonance imaging in Alzheimer's Disease Neuroimaging Initiative 2. Alzheimer's Dementia **11**(7), 740–756 (2015)
17. Weiner, M.W., et al.: The Alzheimer's Disease Neuroimaging Initiative 3: Continued innovation for clinical trial improvement. Alzheimer's Dementia **13**(5), 561–571 (2017)
18. Ellis, K.A., Bush, A.I., Darby, D., et al.: The Australian Imaging, Biomarkers and lifestyle (AIBL) study of aging: Methodology and baseline characteristics of 1112 individuals recruited for a longitudinal study of Alzheimer's disease. Int. Psychogeriatrics **21**(4), 672–687 (2009)
19. Raghu, M., Zhang, C., Kleinberg, J., Bengio, S.: Transfusion: Understanding transfer learning for medical imaging. In: Advances in Neural Information Processing Systems (NeurIPS), pp. 3347–3357 (2019)
20. Gauthier, S., et al.: Mild cognitive impairment. Lancet **367**(9518), 1262–1270 (2006)
21. Sabbagh, M.N., et al.: Early detection of mild cognitive impairment (MCI) in primary care. J. Prev. Alzheimer's Disease **7**, 165–170 (2020)

Knee Cartilages Segmentation Based on Multi-scale Cascaded Neural Networks

Junrui Liu[1], Cong Hua[1], Liang Zhang[1(✉)], Ping Li[2], and Xiaoyuan Lu[2]

[1] School of Computer Science and Technology, Xidian University, Xian, China
liangzhang@xidian.edu.cn
[2] Shanghai BNC, Shanghai, China
{pli,xylu}@bnc.org.cn

Abstract. Knee arthritis is one of the most common chronic degenerative joint diseases in the world, affecting the quality of life of a considerable part of the Modern population. Therefore, the early detection of knee arthritis is of great significance for diagnosis and treatment. Magnetic resonance imaging (MRI) is one of the most commonly used methods for evaluating joint degeneration in osteoarthritis research. In order to obtain information on knee cartilage degradation from MRI, it is necessary to segment the articular cartilage interface and cartilage surface boundary on the entire joint surface. In this work, we propose a novel cascaded network structure with an effective inception-like multi-scale module for knee joint magnetic resonance images segmentation. Compared with the baseline, a maximum of 1.6% dice score mean promotion is obtained. The code is publicly available at https://github.com/ETVP

Keywords: Knee cartilage segmentation · Cascaded neural network · Medical image · Multi-scale module

1 Introduction

As the most common chronic degenerative joint disease, knee osteoarthritis often leads to chronic disability, mainly affecting the elderly, obese and sedentary people. Knee osteoarthritis (OA) causes severe pain and often leads to a joint arthroplasty in its severe stages. For effective control of clinical treatment progress and alleviation of future disability, early diagnosis is essential [16,21]. Given that OA is a chronic disease, large longitudinal researches have been conducted to study the onset and progression of OA. Currently, one of the major research interests is to assess the composition and morphological changes of articular cartilage tissue [6]. The degeneration and aging process of articular cartilage caused by excessive load of articular cartilage will cause the natural rupture of articular cartilage and lead to joint space narrowing (JSN) and osteophytes [7,12].

Although a variety of imaging methods (such as MRI, OCT, and ultrasound) have been introduced for the enhanced OA diagnosis, Magnetic resonance imaging (MRI) is still the preferred method for diagnosis of knee OA [1,3]. MRI is

Supported by Shanghai BNC.

commonly used in clinical research to evaluate knee joint degeneration, especially the degeneration of the femur (FB), tibia (TB), and respective femur and tibial cartilage (FC, TC). In the concrete analysis, segmentation is needed for these knee joint issues. However, knee joint cartilage is difficult to identify with the MRI scene because of its small volume and similar gray value with the surrounding tissue. Manual segmentation of cartilage tissue is time-consuming, laborious and inefficient, and makes it infeasible for individualized diagnosis and treatment. Moreover, great intraobserver differences in manual segmentation can lead to analysis volatility [17,23]. Therefore, it is essential to develop a method for accurate and automatic segmentation of knee cartilage.

Deep learning (DL) is an advanced machine learning method learning features directly from data, which completely changing the field of medical image analysis methods [25]. To extract feature information on different scales, GoogLeNet [24] uses the inception structure to extract and fuse feature information on different scales. FCN [14] pioneered the use of CNN for image segmentation and significantly improved feature representation. As the most popular models for medical image segmentation, Unet [19] is a typical encoder-decoder structure, in which the encoder gradually reduces the spatial dimension of the feature map, and the decoder gradually repairs the details and spatial dimensions of the object. There is usually a skip connection between the encoder and the decoder, which helps the decoder to accurately reconstruct the target [26]. Although the Unet is widely used in medical image segmentation, only a few methods are used for musculoskeletal research. Prasoon [18] proposed a method to segment tibial cartilage from MRI using three independent and Liu [13] applied 2D Unet combined with SegNet, and 3D deformable convolution to segment 3D MR images. Given the huge GPU memory consumption of 3D CNN, applying 2D methods to segment 3D MRI is still the mainstream choice. Medical image segmentation could be challenging on account of the small size of the targets and the severe imbalance between the foreground and the background.

In order to reduce the loss function in training as much as possible, the network focuses on distinguishing the foreground and the background, which leads to the inability to accurately segment the boundary of the foreground. A commonly used method is cascaded network. *The first* segmentation network only realizes the approximate location of the target to be separated by locate, which narrows the scope and simplifies the search space of the network. It only determines which pixels belong to the foreground and which pixels belong to the background. *The second* network mainly focuses on the image boundary, which makes the segmentation result more accurate. This strategy has been successfully performed in many computer vision problems [20].

In this paper, we proposed a novel two-stage Inception-like Cascaded Network for segmenting knee cartilage. *For the first stage*: 1) segmentation network is used to locate the cartilage to obtain the cartilages rough segmentation result; 2) image processing algorithm is used to expand the cartilages rough segmentation result, so that the expanded rough segmentation result can completely contain the true value of the cartilage; 3) the expanded cartilages rough segmentation result is multiplied by the original MRI to make sure the cartilage MRI eliminates irrelevant information, which greatly solves the problem of the imbalance between the foreground and the background in the segmentation problem. *For the second stage*, the Unet segmentation network combined with the inception-like module is used to finely segment the processed knee joint MRI. The inception-like module combines three scales of convolution kernels and a residual connection, which can extract and fuse feature information on different scales and finally improve the cartilage segmentation result on the boundary. Using our two-stage segmentation strategy, we improve our results step by step. There are two dataset used in this study, one is SKI-10 [9], a public dataset which contains 100 cases, and the other is our private dataset including 33 cases, with cartilages labeled by three experienced radiologists.

2   Method

2.1   The Overall Framework

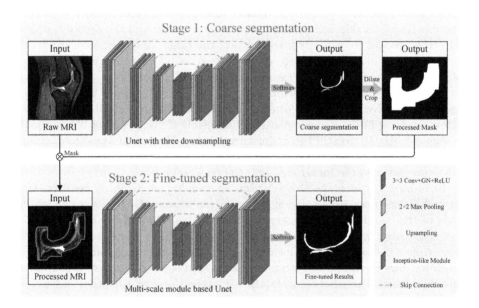

Fig. 1. The overall framework of cartilage segmentation, including a two-stage segmentation network, which can fine-tune the results of coarse segmentation to obtain better results. There is a modified Unet with three downsampling in stage 1; the Multi-scale module using in stage 2 composed of three different sizes of convolution kernel and a residual connection.

Cascaded network is an effective way to improve the performance of an overall network [26]. In cascaded networks, the first network, usually provides valuable information for the second network. Because the knee cartilage occupies a small proportion in the entire MRI, the foreground and background proportions are seriously unbalanced. The shape of the knee cartilage is irregular which makes it difficult to accurately segment the cartilage boundary. In this study, the first network is composed of Unet [19], which is used to pre-segment the knee articular cartilage and generate rough segmentation results to locate the knee articular cartilage; then the carefully constructed fine segmentation network performs fine segmentation on the coarse segmentation results. Our framework is illustrated in Fig. 1. This is a process from coarse to fine, more details are shown below.

2.2 The Two-Stage Cascaded Networks

Traditional cascaded network approach is usually to concatenate the probability map generated by the first network with the original image, then send it to the second network for further fine segmentation. The probability map obtained by the first network can be regarded as the prior probability of the knee cartilage distribution of joint, then the second network can focus on the area of interest to get better results with reference to this prior probability. This method did not leave an impressive result on this current task, and lead to the increased GPU memory consumption and time costs. Unet network has achieved great success in many medical image segmentation tasks, which has a simple structure and an exquisite design, making it easy to have changes on it to experiment with different modules. Unet network structure is employed as the backbone, and we consider filtering the original image based on the final segmentation result of the first network, which can be regarded as using the first network as the attention module, so that the second network reduces the situation where attention is placed on irrelevant information.

The specific method is as follows: first we optimize the segmentation results of the first network, then use traditional image algorithms to refine the boundary: removing the small raised areas and filling the recessed areas and internal voids; after that, we expand the rough segmentation result several times when the expanded segmentation result completely contains the rough segmentation result and also contains the ground truth value as much as possible; then we can multiply the expanded coarse segmentation result with the original MRI image to make the area outside the expanded coarse segmentation result to 0, and the information entering the second fine segmentation network is greatly reduced and useless information is eliminated.

2.3 Inception-Like Convolution Module

In early neural networks, larger convolution kernels often play critical roles in convolution operations and the history of neural network development [11]. However, the large convolution kernel is inefficient, involving a large number of parameters, and makes it impossible to construct a deeper network. Over those years, the VGG Net [22] and GoogLeNet [24] series stacking smaller convolution kernels appear one after another. Although the size of the convolution kernel getting smaller, but due to the existence of the operation of stacking small convolution kernels, the neural network receptive field does not decrease. However, the amount of calculation and parameters of the network is greatly reduced. In addition, there is a consensus that additional stacked small convolution kernel modules can increase the number of layers and handle more nonlinear transformations to achieve better results. In this work, we use 1×1, 3×3, 5×5 multi-scale modules with different sizes of convolution kernels and a residual connection for totally four different branches. Similar to the inception structure, this inception-like structure captures information at different scales by using different sizes of convolution kernels. After the convolution, Group Normalization (GN) is used at the end of each convolution branch, then the feature maps of all branches are concatenated. Our experiment shows that encoder blocks play an important role in capturing image features, so we use this inception-like structure only in the encoder blocks and do not use it in the decoder blocks.

3   Experiments

3.1   Datasets and Implementation Details

We validated our proposed segmentation framework on two datasets. One is our private dataset which contains 33 patients who underwent knee MRI scans in **** Hospital, aged 16–80 years old. All subjects were in the supine position and fixed with an 8-channel knee joint coil. A Philips 3.0T magnetic resonance instrument (Achieva 3.0 TX; Philips Healthcare) was used for image acquisition. The image reconstruction resolution is 512×512, and the layer thickness is 3.0mm. Three experienced radiologists labeled femur, tibia, patella and corresponding cartilages with ITK-SNAP software. We randomly selected 27 cases from the 33 cases of data as the training set, and the remaining 6 cases of images were used as the test set to verify our method. The other dataset is SKI-10, since the the test set and the validation set cannot be obtained, we use the data of 100 cases in the training set to conduct experiments, and randomly divide the data of 100 cases into 5 folds for a k-fold cross-validation test.

Inspired by some previous works [10], we preprocessed MR sequence, including cropping MRI voxel intensity with a window of [0.5%–99.5%], and normalizing voxel intensity with zero-score strategy. Due to the limited amount of data, some simple data augmentation methods were adopted to alleviate the overfitting phenomenon, including random flip (on all three planes independently), small angle rotation ($\pm 15°$ on three planes independently), and slight elastic transformation and intensity jitter.

3.2   Loss Function and Evaluation Metrics

In this work, foreground and background are extremely unbalanced, we adopted the standard dice loss function for training, which can subtly solve the problem of label imbalance by comparing the similarity between the prediction result and the label. Dice loss function can be defined by,

$$\mathcal{L} = \frac{1}{K} \sum_{k=1}^{K} \frac{2 \sum_{i}^{N} P_k(i) L_k(i)}{\sum_{i}^{N} P_k^2(i) + \sum_{i}^{N} L_k^2(i)} \tag{1}$$

Where K is the number of classes, and k represents the k-th class. N is the total number of voxels in the entire image, and i is the i-th voxel. $P_k(i)$ and $L_k(i)$ represent the possibility of prediction and label for class k at voxel i, respectively.

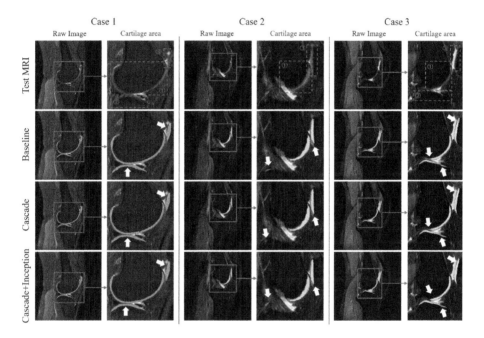

Fig. 2. Difference map of segmentation results in each step on our private dataset. Here we selected three cases from the test set for demonstration, and every two columns come from one case. Red solid boxes in the raw image indicate the areas of articular cartilages should be focusing on, and the three red dotted boxes frame the areas of femoral cartilage, tibial cartilage and patella cartilage, respectively. Green delineation in last three rows illustrates where the voxels were correctly segmented, and red and orange delineation illustrates where the voxels were incorrectly segmented (red means false negative and orange means false positive). Obvious improvement was marked by yellow arrows. (Color figure online)

3.3 Ablation Study

In order to evaluate our proposed Inception-like Cascaded Network framework, we conducted ablation experiments on both two datasets. ***The first*** experiment is to use the Unet network as the baseline segmentation and compare with our cascaded framework; ***the second*** ablation experiment is to add inception-like modules to our cascaded framework for comparison. Results are shown in Table 1 and Table 2, the performance is measured with the dice score. We can see that all the dice score of each cartilage has increased using our Cascaded method both of the two datasets, then we add the inception-like module in the second fine network, dice score has been further improved. If the value of PR is lower than 0.05, we can believe that the difference is statistically significant.

Table 1. Dice similarity coefficients of segmentation results on our private dataset. FC, TC and PC means femoral cartilage, tibial cartilage and patella cartilage respectively.

Cascaded networks	Inception-like module	FC	TC	PC	Mean	PR
×	×	0.794	0.706	0.7569	0.7523	–
✓	×	0.8011	0.7194	0.7707	0.7637	0.007
✓	✓	0.8046	0.7278	0.7723	0.7682	0.040

Table 2. Dice similarity coefficients of segmentation results on public dataset SKI-10. FC and TC means femoral cartilage and tibial cartilage respectively.

Cascaded networks	Inception-like module	FC	TC	Mean
×	×	0.7778	0.7421	0.7560
✓	×	0.7783	0.7464	0.7624
✓	✓	0.7803	0.7486	0.7645

We also take a closer look at the difference maps of each step. As shown in Fig. 2, the last three rows show the difference map of baseline model (Unet), cascaded networks and inception-like cascaded networks, it can be clearly seen that after each step, the predicted results have been significantly improved.

3.4 Comparsion with Other Methods

We have also implemented classical methods in the field of medical image analysis using two datasets, to compare and prove the effectiveness of our proposed framework. Both Unet and Vnet [15] are classic networks. Unet++ [27] is an improved version from Unet. On the basis of Unet, dense connections are added to skip connections. Deeplab v3 [2] is widely used in scene segmentation and has achieved good results. We respectively ,compared with Vnet, Vnet with deformable convolution, Unet++, Deeplab v3 and 2D Unet. The results are as follows: from Table 3 and Table 4, we note that our method achieves competitive results, according to these results, we can also infer that the skip connections between the corresponding encoders and decoders is the key to improve the segmentation performance (Deeplab v3 does not perform well on this task).

Table 3. Comparsion with classic methods using dice similarity coefficients on our private dataset. FC, TC, PC means femoral cartilage, tibial cartilage and patella cartilage respectively.

Methods	FC	TC	PC	Mean
Vnet [15]	0.7592	0.6945	0.7368	0.7301
Deeplab v3 [2]	0.7702	0.6958	0.7551	0.7403
deformable convolution [4]	0.7802	0.7178	0.7387	0.7455
Unet++ [27]	0.8029	**0.7352**	0.7625	0.7668
Ours	**0.8046**	0.7278	**0.7723**	**0.7682**

Table 4. Comparsion with classic methods using dice similarity coefficients on SKI-10 dataset. FC and TC means femoral cartilage and tibial cartilage cartilage respectively.

Method	FC	TC	Mean
Vnet [15]	0.7719	0.7455	0.7587
2.5D Unet [8]	0.7583	0.7231	0.7407
Ours	**0.7803**	**0.7486**	**0.7644**

4   Conclusion

In this work, we explore the performance of cascaded networks and multi-scale modules on medical image segmentation [16]. We propose a novel cascading framework with a significant performance improvement. By adding a novel module we use can achieve further performance improvements. In the future, we can further explore the application of the current popular ViT [5] on this current issue.

References

1. Ambellan, F., Tack, A., Ehlke, M., Zachow, S.: Automated segmentation of knee bone and cartilage combining statistical shape knowledge and convolutional neural networks: data from the osteoarthritis initiative. Med. Image Anal. **52**, 109–118 (2019). https://doi.org/10.1016/j.media.2018.11.009

2. Chen, L.C., Papandreou, G., Schroff, F., Adam, H.: Rethinking atrous convolution for semantic image segmentation. arXiv:1706.05587 (2017)
3. Conaghan, P., Hunter, D., Maillefert, J.F., Reichmann, W., Losina, E.: Summary and recommendations of the oarsi fda osteoarthritis assessment of structural change working group. Osteoarth. Cartilage **19**(5), 606–610 (2011)
4. Dai, J., et al.: Deformable convolutional networks. In: Proceedings of the IEEE International Conference on Computer Vision, pp. 764–773 (2017). https://doi. org/10.1109/ICCV.2017.89
5. Dosovitskiy, A., et al.: An image is worth 16x16 words: Transformers for image recognition at scale. arXiv:2010.11929 (2020)
6. Emery, C.A., et al.: Establishing outcome measures in early knee osteoarthritis. Nat. Rev. Rheumatol. **15**(7), 438–448 (2019). https://doi.org/10.1038/s41584-019-0237-3
7. Górriz, M., Antony, J., McGuinness, K., Giró-i Nieto, X., O'Connor, N.E.: Assessing knee oa severity with cnn attention-based end-to-end architectures. In: International Conference on Medical Imaging with Deep Learning, pp. 197–214. PMLR (2019)
8. Han, X.: Automatic liver lesion segmentation using a deep convolutional neural network method. arXiv:1704.07239 (2017)
9. Heimann, T., Morrison, B.J., Styner, M.A., Niethammer, M., Warfield, S.: Segmentation of knee images: a grand challenge. In: Proc. MICCAI Workshop on Medical Image Analysis for the Clinic, pp. 207–214. Beijing, China (2010)
10. Isensee, F., Petersen, J., Kohl, S.A., Jäger, P.F., Maier-Hein, K.H.: nnu-net: breaking the spell on successful medical image segmentation. arXiv:1904.08128 1, 1–8 (2019)
11. Krizhevsky, A., Sutskever, I., Hinton, G.E.: Imagenet classification with deep convolutional neural networks. Adv. Neural Inf. Process. Syst. **25**, 1097–1105 (2012). https://doi.org/10.1061/(ASCE)GT.1943-5606.0001284
12. Li, Y., Wei, X., Zhou, J., Wei, L.: The age-related changes in cartilage and osteoarthritis. BioMed research international 2013 (2013). https://doi.org/10. 1155/2013/916530
13. Liu, F., Zhou, Z., Jang, H., Samsonov, A., Zhao, G., Kijowski, R.: Deep convolutional neural network and 3d deformable approach for tissue segmentation in musculoskeletal magnetic resonance imaging. Magn. Reson. Med. **79**(4), 2379–2391 (2018). https://doi.org/10.1002/mrm.26841
14. Long, J., Shelhamer, E., Darrell, T.: Fully convolutional networks for semantic segmentation. In: Proceedings of the IEEE Conference on Computer Vision and Pattern Recognition, pp. 3431–3440 (2015). https://doi.org/10.1109/CVPR.2015. 7298965
15. Milletari, F., Navab, N., Ahmadi, S.A.: V-net: Fully convolutional neural networks for volumetric medical image segmentation. In: 2016 Fourth International Conference on 3D vision (3DV), pp. 565–571. IEEE (2016). https://doi.org/10.1109/ 3DV.2016.79
16. Oka, H., et al.: Fully automatic quantification of knee osteoarthritis severity on plain radiographs. Osteoarth. Cartilage **16**(11), 1300–1306 (2008). https://doi.org/ 10.1016/j.joca.2008.03.011
17. Panfilov, E., Tiulpin, A., Klein, S., Nieminen, M.T., Saarakkala, S.: Improving robustness of deep learning based knee MRI segmentation: Mixup and adversarial domain adaptation. In: Proceedings of the IEEE/CVF International Conference on Computer Vision Workshops (2019). https://doi.org/10.1109/ICCVW.2019.00057

18. Prasoon, A., Petersen, K., Igel, C., Lauze, F., Dam, E., Nielsen, M.: Deep feature learning for knee cartilage segmentation using a triplanar convolutional neural network. In: International Conference on Medical Image Computing and Computer-assisted Intervention, pp. 246–253. Springer (2013). https://doi.org/10.1007/978-3-642-40763-5_31

19. Ronneberger, O., Fischer, P., Brox, T.: U-net: Convolutional networks for biomedical image segmentation. In: International Conference on Medical Image Computing and Computer-assisted Intervention, pp. 234–241. Springer (2015). https://doi.org/10.1007/978-3-319-24574-4_28

20. Roth, H.R., et al.: An application of cascaded 3d fully convolutional networks for medical image segmentation. Comput. Med. Imaging Graph. **66**, 90–99 (2018). https://doi.org/10.1016/j.compmedimag.2018.03.001

21. Shamir, L., Ling, S.M., Scott, W., Hochberg, M., Ferrucci, L., Goldberg, I.G.: Early detection of radiographic knee osteoarthritis using computer-aided analysis. Osteoarth. Cartilage **17**(10), 1307–1312 (2009). https://doi.org/10.1016/j.joca.2009.04.010

22. Simonyan, K., Zisserman, A.: Very deep convolutional networks for large-scale image recognition. arXiv:1409.1556 (2014)

23. Stammberger, T., Eckstein, F., Michaelis, M., Englmeier, K.H., Reiser, M.: Inter-observer reproducibility of quantitative cartilage measurements: comparison of b-spline snakes and manual segmentation. Magn. Reson. Imaging **17**(7), 1033–1042 (1999). https://doi.org/10.1016/S0730-725X(99)00040-5

24. Szegedy, C., et al.: Going deeper with convolutions. In: Proceedings of the IEEE Conference on Computer Vision And Pattern Recognition, pp. 1–9 (2015). https://doi.org/10.1109/CVPR.2015.7298594

25. Tiulpin, A., Saarakkala, S.: Automatic grading of individual knee osteoarthritis features in plain radiographs using deep convolutional neural networks. Sci. Rep. **8**(1) (2018). https://doi.org/10.1038/s41598-018-20132-7

26. Zhang, L., et al.: Block level skip connections across cascaded v-net for multi-organ segmentation. IEEE Trans. Med. Imaging **39**(9), 2782–2793 (2020). https://doi.org/10.1109/TMI.2020.2975347

27. Zhou, Z., Siddiquee, M.M.R., Tajbakhsh, N., Liang, J.: Unet++: a nested u-net architecture for medical image segmentation. In: Deep Learning in Medical Image Analysis and Multimodal Learning for Clinical Decision Support, pp. 3–11. Springer (2018). https://doi.org/10.1007/978-3-030-00889-5_1

Deep PET/CT Fusion with Dempster-Shafer Theory for Lymphoma Segmentation

Ling Huang[1,4]([⊠]), Thierry Denœux[1,2], David Tonnelet[3], Pierre Decazes[3], and Su Ruan[4]

[1] Université de technologie de Compiègne, CNRS, Heudiasyc, Compiègne, France
`ling.huang@utc.fr`
[2] Institut universitaire de France, Paris, France
[3] CHB Hospital, Rouen, France
[4] University of Rouen Normandy, Rouen, France

Abstract. Lymphoma detection and segmentation from whole-body Positron Emission Tomography/Computed Tomography (PET/CT) volumes are crucial for surgical indication and radiotherapy. Designing automatic segmentation methods capable of effectively exploiting the information from PET and CT as well as resolving their uncertainty remain a challenge. In this paper, we propose an lymphoma segmentation model using an UNet with an evidential PET/CT fusion layer. Single-modality volumes are trained separately to get initial segmentation maps and an evidential fusion layer is proposed to fuse the two pieces of evidence using Dempster-Shafer theory (DST). Moreover, a multi-task loss function is proposed: in addition to the use of the Dice loss for PET and CT segmentation, a loss function based on the concordance between the two segmentation is added to constrain the final segmentation. We evaluate our proposal on a database of polycentric PET/CT volumes of patients treated for lymphoma, delineated by the experts. Our method get accurate segmentation results with Dice score of 0.726, without any user interaction. Quantitative results show that our method is superior to the state-of-the-art methods.

Keywords: PET/CT · Multi-modality fusion · Lymphoma segmentation · Dempster-shafer theory · Deep learning

1 Introduction

In the clinical diagnosis and radiotherapy planning of lymphoma, PET/CT scanning is an effective imaging tool for tumor segmentation. In PET volumes, the

This work was supported by the China Scholarship Council (grant 201808331005). It was carried out in the framework of the Labex MS2T (Reference ANR-11-IDEX-0004-02).

C. Lian et al. (Eds.): MLMI 2021, LNCS 12966, pp. 30–39, 2021.
https://doi.org/10.1007/978-3-030-87589-3_4

standardized uptake value (SUV) is widely used to locate and segment lymphomas because of its high sensitivity and specificity [11]. Moreover, CT is usually used in combination with PET because of its good representation of anatomical features.

Considering the multiplicity of lymphoma sites (sometimes more than 100) and the wide variation in distribution, shape, type and number of lymphomas, whole-body PET/CT lymphoma segmentation is still challenging although many segmentation methods have been proposed (see a lymphoma patient in Fig. 1). Computer-aided methods for lymphoma segmentation can be classified into three main categories: SUV-threshold-based, region-growing-based and Convolutional-Neural-Network (CNN)-based [7] methods.

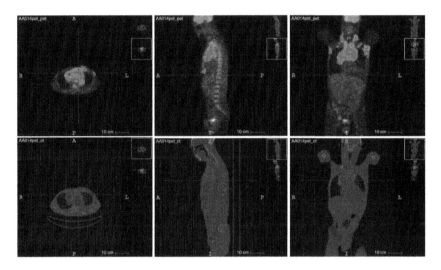

Fig. 1. Examples of PET and CT slices with lymphomas. The first and second row show PET and CT slices of one patient in axial, sagittal and coronal views, respectively.

For PET volumes, it is common to use fixed SUV thresholds to segment lymphoma. This kind of method is fast but it lacks flexibility in boundary delineation and requires clinicians to locate the region of interest. Region-growing-based methods have been proposed to overcome the limitation of boundary delineation of SUV-based methods by taking texture and shape information into account. The common idea is to set the volumes with initializing seeds automatically for segmenting lymphomas [4,8]. However, these methods still require to region of interest manually and are time-consuming.

In recent years, CNNs have achieved great success in computer vision tasks [2,5], as well as in the medical domain. The Fully Connected Network (FCN) model [10] is the first fully convolutional neural network that could be trained end-to-end for pixel-wise classification. UNet [13], a successful modification and extension of FCN, has become the most popular network for medical image segmentation. Based on UNet, many extensions and optimizations have been proposed, such as Deep3D-UNet [18], attention-UNet [12], etc. In [6], Hu et al. propose a Conv3D-based multi-view PET image fusion strategy for lymphoma segmentation. In this approach, 2D and 3D segmentation are performed separately and the results are fused by a Conv3D layer. In [7], Li et al. propose a whole-body PET/CT lymphoma segmentation method that fuses CT and PET slices by concatenating them before training; the method is based on a two-flow architecture (segmentation flow and reconstruction flow). Using a similar approach, Blanc-Durand et al. propose a CNN-based segmentation network for diffuse large B cell lymphoma segmentation by concatenating PET and CT as two channel inputs [1].

It should be noted that the effective fusion of multi-modality information is of great importance in the medical domain. A single-modality image often does not contain enough information and is often tainted with uncertainty. This is why physicians always use PET/CT volumes together for lymphoma segmentation and radiotherapy. Using CNNs, researchers have mainly adopted probabilistic approaches to data fusion, which can be classified into three strategies [17]: image-level, feature-level and decision-level fusion. However, probabilistic fusion is unable to effectively manage conflicts that occur when the same voxel is labeled with two different labels by CT and TEP. Dempster-Shafer theory (DST) [3,14], also known as belief function theory or evidence theory, is a formal framework for information modeling, evidence combination and decision-making with uncertain or imprecise information. Despite the low resolution and contrast of medical images, DST's high ability to describe uncertainty allows us to represent evidence more faithfully than using probabilistic approaches. Researchers from the medical image community have started to actively investigate the use of DST for handling uncertain, imprecision sources of information in different medical tasks, such as medical image retrieval [15], lesion segmentation [9], etc.

In this work, we propose a DST-based PET/CT image fusion model for 3D lymphoma segmentation. To our knowledge, this is the first multi-modality PET/CT volume fusion method using a CNN and DST. The main contributions of this work are (1) a CNN architecture with a DST-based fusion layer that effectively handles uncertainty and conflict when combining PET and CT information; (2) a multi-task loss function making it possible to optimize different segmentation tasks and to increase segmentation accuracy; (3) a 3D segmentation model with end-to-end training for whole-body lymphoma segmentation.

2   Methods

2.1   Network Architecture

Figure 2 shows the workflow of the multi-modality fusion-based lymphoma segmentation framework. It is composed of two modified encoder-decoder (Unet) modules and an evidential fusion layer. To reduce computation cost, we reduce the number of convolution filters of Unet from $(16, 32, 64, 128, 256)$ to $(8, 16, 32, 64, 128)$ to get a "slim UNet". A fusion layer is constructed based on

DST, which will be explained in Sect. 2.3. Two modality images: PET and CT are taken as inputs to our framework. We first feed the prepossessed PET volume into UNet1 and the prepossessed CT volume into UNet2 to independently compute their segmentation probability maps $\mathbf{seg}_{PET}$ and $\mathbf{seg}_{CT}$. These two 3D maps are then transferred to the fusion layer, which computes an evidential segmentation map $\mathbf{seg}_F$. For training, a multi-task loss function is proposed to minimize the Dice loss and mean square loss between three segmentation maps and masks, as explained in Sect. 2.4.

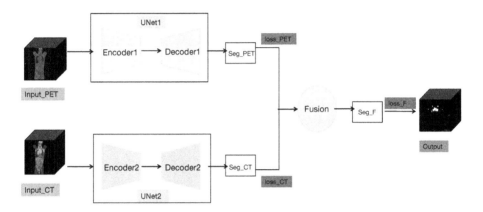

Fig. 2. The global segmentation framework.

2.2 Evidential Fusion Strategy

Let $\Omega = \{\omega_1, \omega_2, ..., \omega_K\}$ be a finite set of possible answers to some question, one and only one of which is true. Evidence about that question can be represented by a mapping m from 2^Ω to $[0,1]$, called a mass function, such that $\sum_{A \subseteq \Omega} m(A) = 1$. For any hypothesis $A \subseteq \Omega$, the quantity $m(A)$ represents a share of a unit mass of belief allocated to the hypothesis that the truth is in A, and which cannot be allocated to any strict subset of A based on the available evidence. Two mass functions m_1 and m_2 representing two independent items of evidence can be combined by Dempster's rule [14] defined as

$$(m_1 \oplus m_2)(A) = \frac{1}{1 - \kappa} \sum_{B \cap C = A} m_1(B) m_2(C), \tag{1}$$

for all $A \subseteq \Omega, A \neq \emptyset$, and $(m_1 \oplus m_2)(\emptyset) = 0$. In Eq. (1), κ represents the degree conflict between m_1 and m_2 defined as

$$\kappa = \sum_{B \cap C = \emptyset} m_1(B) m_2(C). \tag{2}$$

If all focal sets of m are singletons and m is said to be Bayesian, it is equivalent to a probability distribution. In our case, the lymphoma segmentation task can be considered as a two-class classification task; the set of possibilities is $\Omega = \{0,1\}$ where 1 and 0 stand, respectively, for presence and absence of lymphoma in a given voxel. Two Bayesian mass functions m_{PET} and m_{CT} can be obtained from two probability segmentation maps maps seg_{PET} and seg_{CT}. They are aggregated by Dempster's rule (1).

The two segmentation results obtained by PET and CT cannot easily classify boundary voxels due to low signal-to-noise ratio and low contrast, resulting in segmentation uncertainty. Moreover, when seg_{PET} and seg_{CT} assign the same voxel to different classes, there is a high conflict between the two segmentation results. It is difficult to solve these problems by classical fusion methods. In this work, the proposed evidential fusion layer allows us to solve these problems thanks to Dempster's rule. To be specific, for a given voxel, if seg_{PET} gives a probability of 0.26 that it belongs to lymphoma, and seg_{CT} gives a probability of 0.85 that it belongs to lymphoma, the two segmentation results are contradictory and it is unreasonable to simply fuse them by linear combination or majority voting. In our evidential fusion layer, we can reassign the probability that the voxel belongs to lymphoma as 0.67 with Eqs. (1)–(2). This rule takes the conflict into consideration and yields more reliable segmentation results.

2.3   Multi-task Loss Function

Lymphomas segmentation task show sub-optimal performance in CT, which may leads to misleading fusion results in our model. In order to avoid the problem, a multi-task loss function here is proposed to exploit all the available information during training and increase segmentation accuracy. Since lymphomas are visible on both PET and CT, even though they do not have exactly the same shape in these two volumes, they should overlap as much as possible. Here we set PET mask as ground truth for both PET and CT to constrains the overlap rate between PET and CT.

As shown in Fig. 1, we define three loss functions: loss_{CT}, loss_{PET} and loss_F measuring the discrepancy between ground truth and, respectively, CT segmentation maps, PET segmentation maps, and the final segmentation output. For loss_{CT} and loss_{PET}, we use Dice loss,

$$\mathsf{loss}_{PET} = 1 - \frac{2 * \sum_{v=1}^{V} S_1^v G^v}{\sum_{v=1}^{V} S_1^v + \sum_{v=1}^{V} G^v}, \tag{3}$$

$$\mathsf{loss}_{CT} = 1 - \frac{2 * \sum_{v=1}^{V} S_2^v G^v}{\sum_{v=1}^{V} S_2^v + \sum_{v=1}^{V} G^v}, \tag{4}$$

where V is the number of voxels of segmentation outputs, S_1 and S_2 are the PET and CT segmentation outputs and G is the ground truth of lymphoma in PET. For loss_F, we use mean square loss,

$$\mathsf{loss}_F = \sum_{v=1}^{V} (S_f^v - G^v)^2, \tag{5}$$

where S_f is the final segmentation output. The multi-task loss function is defined as

$$\text{loss}_{all} = 0.75 * \text{loss}_{CT} + 0.25 * \text{loss}_{PET} + \text{loss}_F. \tag{6}$$

Here we set the weight of $loss_{CT}$ as 0.75 and the weight of $loss_{PET}$ as 0.25 to enable the model learn more from hard example (CT).

2.4   Implementation Details

All methods were implemented in Python using Tensorflow framework and were trained and tested on a desktop with a 2.20 GHz Intel(R) Xeon(R) CPU E5-2698 v4 and a Tesla V100-SXM2 graphics card with 32 GB GPU memory.

3   Experiments and Analysis

3.1   Dataset and Preprocessing

The experimental dataset consists of 173 labeled cases of real PET/CT volumes, whose labels indicate the ground truth of lymphoma. All PET/CT data were acquired from the Henri Becquerel hospital. The study was approved as a retrospective study by the Henri Becquerel Center Institutional Review Board. All patients' information were de-identified and anonymized prior to analysis. The size of the CT volumes and the corresponding masks vary from $267 \times 512 \times 512$ to $478 \times 512 \times 512$ and their spatial resolution varied from $0.97 \times 0.97 \times 2$ mm^3 to $1.36 \times 1.37 \times 5$ mm^3. The size of the PET volumes and the corresponding masks vary from $276 \times 144 \times 144$ to $407 \times 256 \times 256$ and their spatial resolution varied from $5.3 \times 5.3 \times 2$ mm^3 to $2.73 \times 2.73 \times 3.27$ mm^3. For preprossessing, we first registered the CT volumes and matched them with PET volumes. Then we resized PET, CT and mask volumes into size $128 \times 256 \times 256$. Both PET and CT volumes were normalized into the standard distribution by the linear standardization method. Two kinds of data augmentation were applied here to enrich the training data: deformation and affine transformation. We used 80% of the data for training, 10% for validation and 10% for testing. The Dice score, Precision and Recall were used to evaluate the segmentation results of lymphoma at the voxel level.

3.2   Results and Discussion

The quantitative results of the experiments are represented in Table 1. UNet is the baseline method. We first tested the segmentation performance when using only PET or CT volume. Then we concatenate PET and CT volumes as inputs of the UNet. This is an input-level fusion method. Finally, we used two independent UNets for PET and CT, respectively. The fusion was carried out at decision level by the proposed evidential fusion layer. Compared with mono-modality input, our proposal shows an obvious advantage. Compared with the concatenate-based fusion method, our proposal has 3%, 4%, 6% increase in Dice score, Precision and

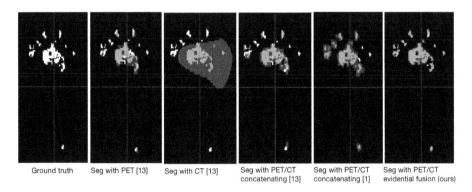

Ground truth   Seg with PET [13]   Seg with CT [13]   Seg with PET/CT   Seg with PET/CT   Seg with PET/CT
                                                          concatenating [13] concatenating [1] evidential fusion (ours)

Fig. 3. Comparison of segmentation results of CT, PET, PET/CT concatenation [13], PET/CT concatenation with [1] and PET/CT evidential fusion. The ground truth is marked in white and the segmentation results in red. (Color figure online)

Recall, respectively. Figure 3 shows the comparison results between the above four input modalities. We can see from the figure that mono-modality input does not allow us to segment small regions, especially for CT. The concatenation of PET and CT solves this problem to some extent [1,13], but the result is still not ideal. The evidential fusion of PET and CT achieves better segmentation result, especially for the small lymphomas located at the top of the images.

Table 1. Performance comparison with the baseline methods on the test set.

Models	Input modality	Dice score	Precision	Recall
UNet (mono-input)	PET	0.67 ± 0.02	0.74 ± 0.05	0.65 ± 0.03
UNet (mono-input)	CT	0.49 ± 0.05	0.71 ± 0.02	0.45 ± 0.02
UNet (concatenate-based fusion)	PET+CT	0.69 ± 0.01	0.67 ± 0.01	0.74 ± 0.05
EUnet (evidential fusion)	PET+CT	$\mathbf{0.72 \pm 0.04}$	$\mathbf{0.71 \pm 0.06}$	$\mathbf{0.80 \pm 0.05}$
Zeng et al. [16]	PET+CT	0.68 ± 0.02	0.68 ± 0.01	0.68 ± 0.01
Hu et al. [6]	PET+CT	0.66 ± 0.03	0.71 ± 0.044	0.67 ± 0.03
Blanc et al.[1]	PET+CT	$\mathbf{0.73 \pm 0.20}$	$\mathbf{0.75 \pm 0.22}$	$\mathbf{0.83 \pm 0.17}$

Since there is no public lymphoma dataset available now, the quantitative comparison with other state-of-the-art methods is difficult, because different datasets and different lymphoma types are segmented. However, we still compare our method with the state-of-the-art in Table 1. In [6], Hu et al. achieve 0.66 ± 0.03 Dice score on a dataset of 109 lymphoma patients with a convolution fusion with 2D slices and 3D volumes. In [1], Blanc-Durand et al. report 0.73 ± 0.20 for Dice score based on the training on 511 lymphoma patients. Though the authors report comparable results with ours, the performance of our model is more stable and needs less training data. We also test the method of [1] with our dataset and achieve 0.64 ± 0.02 in Dice score, where the priory of our proposal is obvious.

A qualitative comparison with our model is shown in Fig. 3. See the fifth image from Fig. 3, their model show sign of slight overfitting but is visually acceptable and our model show visually correct segmentation results. Generally speaking, our model yields better results than state-of-the-art methods with DST-based evidential fusion.

A qualitative comparison is shown in Fig. 4. Here we overlap PET and CT in the same image and paste the segmented mask in this image. Column 1 shows one slice of patient 1 in which a big lymphoma is present. Our model can segment it correctly. Column 2 shows a slice of patient 2 in which multiple lymphomas are present. They are more difficult to segment because some of them are very small. However, our model can still obtain satisfactory results.

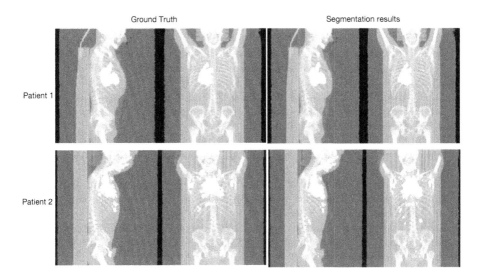

Fig. 4. Illustration of the results of two patients obtained using our EUNet model. From left to right, each column represents the corresponding segmentation results and the ground truth. The lymphomas are showed in white.

4   Conclusion

In this work, a DST-based deep multi-modality medical images fusion strategy has been proposed to deal with the problem of uncertainty and conflict within a deep convolutional neural network. A two-branch segmentation module first processes PET/CT volumes separately. Then the two segmentation maps are fused by the evidential fusion layer. Qualitative and quantitative evaluation results are promising as compared to the baseline and state-of-the-art methods. Future research will aim at improving the network architecture to better exploit the potential of DST to represent and combine uncertain information.

References

1. Blanc-Durand, P., et al.: Fully automatic segmentation of diffuse large b cell lymphoma lesions on 3d fdg-pet/ct for total metabolic tumour volume prediction using a convolutional neural network. Euro. J. Nucl. Med. Mol. Imaging **48**, 1–9 (2020)
2. Bochkovskiy, A., Wang, C.Y., Liao, H.Y.M.: Yolov4: Optimal speed and accuracy of object detection. arXiv:2004.10934 (2020)
3. Dempster, A.P.: Upper and lower probability inferences based on a sample from a finite univariate population. Biometrika **54**(3–4), 515–528 (1967)
4. Desbordes, P., Petitjean, C., Ruan, S.: 3d automated lymphoma segmentation in pet images based on cellular automata. In: 2014 4th International Conference on Image Processing Theory, Tools and Applications (IPTA), pp. 1–6. IEEE (2014)
5. Hershey, S., et al.: Cnn architectures for large-scale audio classification. In: 2017 IEEE International Conference on Acoustics, Speech and Signal Processing (ICASSP), pp. 131–135. IEEE (2017)
6. Hu, H., Shen, L., Zhou, T., Decazes, P., Vera, P., Ruan, S.: Lymphoma segmentation in pet images based on multi-view and conv3d fusion strategy. In: 2020 IEEE 17th International Symposium on Biomedical Imaging (ISBI), pp. 1197–1200. IEEE (2020)
7. Li, H., Jiang, H., Li, S., Wang, M., Wang, Z., Lu, G., Guo, J., Wang, Y.: Densexnet: an end-to-end model for lymphoma segmentation in whole-body pet/ct images. IEEE Access **8**, 8004–8018 (2019)
8. Li, H., Thorstad, W.L., et al.: A novel pet tumor delineation method based on adaptive region-growing and dual-front active contours. Med. Phys. **35**(8), 3711–3721 (2008)
9. Lian, C., Ruan, S., Denœux, T., Li, H., Vera, P.: Joint tumor segmentation in pet-ct images using co-clustering and fusion based on belief functions. IEEE Trans. Image Process. **28**(2), 755–766 (2018)
10. Long, J., Shelhamer, E., Darrell, T.: Fully convolutional networks for semantic segmentation. In: Proceedings of the IEEE Conference on Computer Vision and Pattern Recognition, pp. 3431–3440. Boston, USA (2015)
11. Nestle, U., et al.: Comparison of different methods for delineation of 18f-fdg pet-positive tissue for target volume definition in radiotherapy of patients with non-small cell lung cancer. J. Nuclear Med. **46**(8), 1342–1348 (2005)
12. Oktay, O., et al.: Attention u-net: Learning where to look for the pancreas. arXiv:1804.03999 (2018)
13. Ronneberger, O., Fischer, P., Brox, T.N.: Convolutional networks for biomedical image segmentation. In: Paper presented at: International Conference on Medical Image Computing and Computer-Assisted Intervention. Munich, Germany (2015)
14. Shafer, G.: A Mathematical Theory of Evidence, vol. 42. Princeton University Press (1976)
15. Sundararajan, S.K., Sankaragomathi, B., Priya, D.S.: Deep belief CNN feature representation based content based image retrieval for medical images. J. Med. Syst. **43**(6), 1–9 (2019)
16. Zeng, G., Yang, X., Li, J., Yu, L., Heng, P.A., Zheng, G.: 3d u-net with multi-level deep supervision: fully automatic segmentation of proximal femur in 3d mr images. In: International Workshop on Machine Learning in Medical Imaging. pp. 274–282. Springer (2017). https://doi.org/10.1007/978-3-319-67389-9_32

17. Zhou, T., Ruan, S., Canu, S.: A review: Deep learning for medical image segmentation using multi-modality fusion. Array **3**, 100004 (2019)
18. Zhu, W., et al.: Anatomynet: Deep 3D squeeze-and-excitation u-nets for fast and fully automated whole-volume anatomical segmentation. bioRxiv p. 392969 (2018)

Interpretable Histopathology Image Diagnosis via Whole Tissue Slide Level Supervision

Zhuoyue Wu, Hansheng Li, Lei Cui[(⊠)], Yuxin Kang, Jianye Liu, Haider Ali, Jun Feng[(⊠)], and Lin Yang[(⊠)]

School of Information Science and Technology, Northwest University, Xi'an, China
{leicui,fengjun,linyang}@nwu.edu.cn

Abstract. The deep learning methods supervised by annotating different regions of histopathology images (patch-level labels) have achieved promising outcomes in assisting pathologic diagnosis. However, most clinical data only contains label information for the whole slide image (WSI-level labels), so the methods supervised by WSI-level labels are more necessary than the ones supervised by patch-level labels. Additionally, various methods supervised by WSI-level labels ignore the contextual relations among patches extracted from a WSI, making incorrect predictions for some patches in a WSI and further misclassifying the WSI. In this paper, we propose to utilize an interpretable dual encoder network with a context-capturing RNN module to capture the contextual relations among all patches extracted from a WSI. Besides, we propose to utilize a feature attention module to weigh the importance of each patch automatically. More importantly, visualization of weight for each patch in a WSI demonstrates that our approach matches the concerns of pathologists. Furthermore, extensive experiments demonstrate the superiority of the interpretable dual encoder network.

Keywords: Whole slide image · Contextual relations · Interpretable · Patch-level labels · WSI-level labels.

1 Introduction

Histopathology image diagnosis plays a critical role in treating disease, as it can guide clinic doctors to determine the follow-up treatment plan [1,2]. Recently, deep learning methods for whole slide images (WSIs) diagnosis have continually developed to empower pathologists' efficiency and accuracy. According to which types of labels are available, these deep learning methods can be separated into two classes: methods supervised by patch-level labels [3–6] and WSI-level ones [7–10].

These methods supervised by patch-level labels require manually annotating regions of different tissue types (patch-level labels) for WSIs. Conceivably, patch-level labels' acquisition is time-consuming and labor-intensive, making it

---

J. Feng and L. Yang—Joint corresponding authors.

© Springer Nature Switzerland AG 2021
C. Lian et al. (Eds.): MLMI 2021, LNCS 12966, pp. 40–49, 2021.
https://doi.org/10.1007/978-3-030-87589-3_5

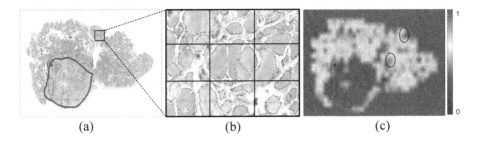

(a)   (b)   (c)

Fig. 1. The Illustration of contextual relations problem. (a) A WSI with malignant lesion outlined by an expert. (b) A benign sub-region in a WSI, which includes nine patches. It is worth noting that the center patch is more likely to have the same properties as its neighbors. (c) All patchs' malignant probabilities in the WSI are obtained by method that ignore contextual relations [10].

challenging to build large training datasets. However, patch-level labels for WSIs can further improve the model performance by modeling context relations among neighboring patches [3]. As an example of context relations, Fig. 1b reveals that the center patch may fall in the benign region with a high probability when its neighboring patches are in the benign region. To model such context relations among neighboring patches, Zanjani et al. [4] utilized Convolutional Neural Network (CNN) to extract features from neighboring patches, and then Conditional Random Field (CRF) is applied to these patches to refine the predicted probability map in the post-processing stage. In parallel to this work, Li et al. [5] proposed a neural conditional random field (NCRF) model that can be trained in an end-to-end manner, avoiding the post-processing stage. It is worth noting that the above methods' tremendous success depends on the patch-level labels for all WSIs.

In contrast, WSI-level labels can easily be obtained from pathological reports. Thus the methods supervised by WSI-level labels are more necessary than the ones supervised by patch-level labels. As a typical WSI-level supervised method, Hou et al. [7] initially took all patches in the WSI as training samples and eliminated the patches with low classification probability iteratively. In [8], the authors extended this approach by clustering patches wisely and eliminating the patches that are less discriminative for the classification task. Chen et al. [10] believe that the eliminated policy is a hard sampling technique which makes a binary decision to select samples and proposed a soft-weighted technique called rectified cross-entropy loss (L_{RCE}). Besides, they introduced an upper transition loss (L_{UT}) to improve the patches' classification accuracy further. However, these methods supervised by WSI-level labels predict each patch independently in the inference stage. From our perspective, the independent predictions ignore the context relations among neighboring patches, leading to inconsistent predictions of patches in the same region. For instance, as shown in the red circle of Fig. 1c, some patches in the benign region are misclassified as malignant ones.

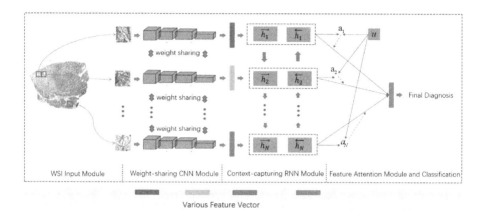

Fig. 2. Overview of the interpretable dual encoder network. It is worth noting that the blue color represents the WSI-level feature vector and other colors represent different patch-level feature vectors.

Overall, we are surprised to find that there has been little discussion about how to model contextual relations among patches when lacking patch-level labels. Thus, this paper regards patches extracted from a WSI as a sequence and proposes to utilize an RNN module to take contextual relations among all patches extracted from the WSI into account when only WSI-level labels are available. Besides, considering that not all patches extracted from a WSI contribute equally to the final diagnosis, we propose to utilize a feature attention module [12] to weigh the importance of each patch automatically. Finally, An interpretable dual encoder network is formed by combining the modules mentioned above. The significant superiority of this network is as follows.

(1) Our network merely requires WSI-level labels that are easy to obtain, which significantly saves a lot of labor cost on annotations.
(2) The context-capturing RNN module of our network is adopted to automatically model the contextual relations among all patches extracted from a WSI, which are different from previous patch-level supervised methods that just model the contextual relations among neighboring patches.
(3) Our network's feature attention module automatically captures the most critical patch for final diagnosis, making we can produce fine heatmaps by visualizing each patch's importance.

2   Methodology

Figure 2 shows an overview of the interpretable dual encoder network. It contains five primary modules: WSI input module, weight-sharing CNN module, context-capturing RNN module, feature attention module, and classification module.

(1) The WSI input module takes a WSI as input and then splits the WSI into non-overlapping patches.
(2) The weight-sharing CNN module acts as a patch's appearance encoder that takes patches extracted from a WSI as input and then codes each patch as a patch-level feature vector.
(3) The context-capturing RNN module jointly processes these patch-level feature vectors and then encodes each patch-level feature vector as context-aware feature vector. By jointly processing these patch-level feature vectors, we make sure that we capture the contextual relations among all patches extracted from a WSI.
(4) The feature attention module merges context-aware feature vectors from each time step into a WSI-level feature vector. It is worth noting that the feature attention module can weigh the importance of each patch automatically.
(5) The classification module uses the WSI-level feature vector for the final diagnosis.

2.1   WSI Input Module

A WSI from the training set is taken as X_i, and meanwhile, its WSI-level label is taken as Y_i. Thus the training set can be expressed as $\{(X_1, Y_1), ..., (X_M, Y_M)\}$, where M is the total number of training samples. The WSI input module takes a WSI X_i as input and then splits the WSI into non-overlapping $512 * 512$ patches. Before passing these non-overlapping patches to the weight-sharing CNN module, we discard non-informative background patches to reduce the computational cost and mark the rest of the foreground patches as $\{x_{i,1}, x_{i,2}, ..., x_{i,N}\}$.

2.2   Weight-Sharing CNN Module

To fully extract the foreground patches' discriminative feature, we employ a modified CNN $f(\cdot)$ as patches' appearance encoder. Specifically, we modify the Resnet [13] by replacing the classification layer with a feature reduction layer that is essentially a three-layer fully connected neural network. The convolution layers of modified Resnet encode a patch into a patch-level feature vector, and further, the feature reduction layer of modified Resnet reduces the patch-level feature vector's dimension to 256. Formally, as shown in Eq. 1, the modified Resnet can map a patch to a 256-dimensional feature vector $v_{i,t}$, where $t = 1, 2, ..., N$. Due to GPU memory limitation, we modify Resnet pre-trained on ImageNet and only update the feature reduction layer's weights during the training process.

$$v_{i,t} = f(x_{i,t}) \tag{1}$$

2.3   Context-Capturing RNN Module

After getting a sequence of patch-level feature vectors $\{v_{i,1}, ..., v_{i,N}\}$, we propose to use a Bidirectional Long Short-Term Memory (BLSTM) [14] to model the contextual relations among them. Note that the BLSTM consists of two independent processing streams, one moving left to right($\overrightarrow{LSTM}(\cdot)$) and the other right to left($\overleftarrow{LSTM}(\cdot)$). As a matter of convenience, we use the index t ($t = 1...N$) to denote the position of the patch-level vector $v_{i,t}$. For the t-th patch-level feature vector $v_{i,t}$, the sub-network $\overrightarrow{LSTM}(\cdot)$ merges $v_{i,t}$ with its previous output $\overrightarrow{h_{i,t-1}}$ to generate the forward feature vector $\overrightarrow{h_{i,t}}$. The above calculation process is shown in Eq. 2. As same as the sub-network $\overrightarrow{LSTM}(\cdot)$, the calculation process of sub-network $\overleftarrow{LSTM}(\cdot)$ is shown in Eq. 3.

$$\overrightarrow{h_{i,t}} = \overrightarrow{LSTM}(v_{i,t}, \overrightarrow{h_{i,t-1}}) \tag{2}$$

$$\overleftarrow{h_{i,t}} = \overrightarrow{LSTM}(v_{i,t}, \overleftarrow{h_{i,t-1}}) \tag{3}$$

Next, Eq. 4 merge the forward vector $\overrightarrow{h_{i,t}}$ with the backward vector $\overleftarrow{h_{i,t}}$ to generate a context-aware feature vector $h_{i,t}$ for the t-th patch-level feature vector.

$$h_{i,t} = [\ \overrightarrow{h_{i,t}},\ \overleftarrow{h_{i,t}}\ ] \tag{4}$$

In this paper, the hyperparameter details of LSTM are as follows. The number of features in the hidden state is 512, and the number of recurrent layers is 3.

2.4   Feature Attention Module and Classification

Given a WSI, not all patches extracted from it contribute equally to the final diagnosis. Therefore, we embed the feature attention module in the interpretable dual encoder network to weigh the importance of each patch automatically. The feature attention module can map the context-aware feature vector set $\{h_{i,1}, ..., h_{i,N}\}$ to a WSI-level feature vector S_i by a weighted sum of these context-aware feature vectors. The details of the weighted sum are as follows. First, fully connected layer projects each patch-level feature vector $h_{i,t}$ to a hidden representation $u_{i,t}$. The fully connected layer is defined in Eq. 5:

$$u_{i,t} = \tanh(W_u h_{i,t} + b_u) \tag{5}$$

As shown in Eq. 6, the importance of context-aware feature vector $h_{i,t}$ is weighed by the similarity between the hidden representation $u_{i,t}$ and a context vector u learned during training.

$$a_{i,t} = \frac{\exp(u_{i,t}^\top u)}{\sum_{t=0}^{N} \exp(u_{i,t}^\top u)} \tag{6}$$

Next, the WSI-level feature vector S_i is formed by a weighted sum of these context-aware feature vectors $\{h_{i,1}, ..., h_{i,N}\}$.

$$S_i = \sum_{t=0}^{N} a_{i,t} h_{i,t} \tag{7}$$

Finally, we use a softmax classifier to predict the WSI-level label $\hat{Y}_i$ for the WSI X_i. The softmax classifier takes the WSI-level feature vector S_i as input and then outputs estimated probabilities $p(y|X_i)$ for all categories.

$$p(y|X_i) = softmax(W_p S_i + b_p) \tag{8}$$

$$\hat{Y}_i = \arg\max p(y|X_i) \tag{9}$$

3   Experiments

3.1   Dataset

To validate the network's effectiveness, we create a digital thyroid frozen section dataset, including 547 WSIs. According to the subsequent surgical plan, these WSIs can be categorized as benign or malignant. As shown in Table 1, the collected dataset includes 97 benign and 154 malignant WSIs for training, 19 benign and 25 malignant WSIs for validation, 85 benign and 167 malignant WSIs for testing. The slides in the training set and testing set only have WSI-level labels. However, the slides in the validation set have both WSI-level labels and patch-level labels.

Table 1. Summary of experimental data

	Benign	Malignant	Total
Train	97	154	251
Val	19	25	44
Test	85	167	252
Total	201	346	547

3.2   Implementation Details

The WSI input module throws the non-informative background patches away as Chen et al. [10]. The interpretable dual encoder network is trained with NVIDIA GeForce GTX 2080 Ti GPU. In the training process, a batch size of 1 WSI is feed into the network. Meanwhile, it is trained in an end-to-end way for 100 epochs, using adam optimizer with a learning rate of 0.0001.

3.3   Performance Comparison

In this sub-section, we conduct ablation experiments to investigate the feature attention module's effectiveness firstly. Then we implement a method supervised by WSI-level labels [10] and compare it with our solution on the different backbone. In the experiments, we used Accuracy, Precision, Recall, and F1-Score as our evaluation criteria.

Table 2. Performance of feature attention module (FA) on different benchmark models

Method				Metrics			
Resnet18 + BLSTM	Resnet50 + BLSTM	Resnet152 + BLSTM	FA	Accuracy	Precision	Recall	F1-score
✓				89.68%	100%	84.43%	91.56%
✓			✓	93.25%	96.30%	93.41%	94.83%
	✓			86.91%	100%	80.24%	89.04%
	✓		✓	90.48%	96.73%	94.27%	95.48%
		✓		90.48%	100%	85.63%	92.26%
		✓	✓	91.67%	97.40%	89.82%	93.46%

Effect of Feature Attention Module. To test whether the additional feature attention module can help improve model performance, we first implement three benchmark models and then embed the feature attention module in each benchmark model. Table 2 illustrates the experimental results with and without the feature attention module. It can be seen that these benchmark models have high precision but low recall for malignant class. Our explanation for this phenomenon is that benchmark models tend to classify some malignant WSIs as benign ones due to sizeable benign lesion regions in these malignant WSIs. In an extreme case, only less than 1% of regions on a misclassified WSI may be malignant while the rest are benign. However, when adding the feature attention module, these models can achieve high accuracy and high recall. Specifically, the feature attention module increased the three benchmark model F1-score by 3.27%, 6.44%, and 1.20%, respectively. Meanwhile, the feature attention module can also increase three benchmark model accuracy by 3.57%, 3.57%, and 1.19%, respectively. It demonstrates that the feature attention module can automatically weigh each patch's importance and further capture the patches most relevant to the final diagnosis.

Comparision with Method Supervised by WSI-level Labels. We compare our method with Chen's method [10] in Table 3. From Table 3, we can observe that our method's performance is better than Chen et al. on the different backbone. The possible reason is that our method models contextual relations among all patches in a WSI, which is different from Chen's method that predicts each patch in a WSI separately. Besides, it is worth noting that Chen's method uses patch-level labels in the validation set to select the best model for testing, whereas we only use WSI-level labels.

To further compare the method proposed by Chen et al. with our method, we visualize the expert annotation in WSI, each patch's malignant probability obtained by Chen's method and each patch's weight obtained by our network's feature attention module, respectively in Fig. 3. We can find that both Chen's model and our model focus on the malignant lesions.

Table 3. Performance of different deep learning algorithms

Methods	Accuracy	Precision	Recall	F1-score
Chen et al.(Resnet18+$L_{RCE} + L_{UT}$) [10]	85.71%	97.12%	80.83%	88.23%
Ours(Resnet18+BLSTM+FA)	93.25%	96.30%	93.41%	94.83%
Chen et al.(Resnet50+$L_{RCE} + L_{UT}$) [10]	88.49%	92.07%	90.42%	91.24%
Ours(Resnet50+BLSTM+FA)	90.48%	96.73%	94.27%	95.48%
Chen et al.(Resnet152+$L_{RCE} + L_{UT}$) [10]	87.30%	96.55%	83.83%	89.74
Ours(Resnet152+BLSTM+FA)	91.67%	97.40%	89.82%	93.46%

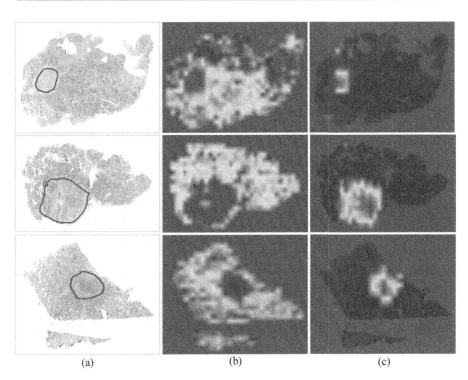

(a)   (b)   (c)

Fig. 3. Prediction visualization of selected examples. (a) WSI with malignant lesion outlined by an expert. (b) Visualizing each patch's malignant probability obtained by Chen's method [10]. (c) Visualizing each patch's weight obtained by our network's feature attention module.

4   Conclusion

In this paper, we find that there has been little discussion about how to model contextual relations among all patches extracted from a WSI when lacking patch-level labels. Thus we propose to utilize an interpretable dual encoder network with a context-capturing RNN module to capture the contextual relations among all patches extracted from a WSI. The comparative experiments on thyroid datasets demonstrate the superiority of our method. Besides, we propose to utilize a feature attention module to weigh the importance of each patch automatically. Extensive ablation experiments demonstrated that the feature attention module could capture the patches most relevant to the final diagnosis.

Acknowledgment. This work was funded by the Natural Science Foundation of Shaanxi Province of China(2021JQ-461).

References

1. Li, Y., Chen, P., Li, Z., Su, H., Yang, L., Zhong, D.: Rule-based automatic diagnosis of thyroid nodules from intraoperative frozen sections using deep learning. Artif. Intell. Med. **108**, 101918 (2020)
2. Srinidhi, C.L., Ciga, O., Martel, A.L.: Deep neural network models for computational histopathology: a survey. Medical Image Analysis, p. 101813 (2020)
3. Kong, B., Wang, X., Li, Z., Song, Q., Zhang, S.: Cancer metastasis detection via spatially structured deep network. In: International Conference on Information Processing in Medical Imaging, pp. 236–248. Springer (2017). https://doi.org/10.1007/978-3-319-59050-9_19
4. Zanjani, F.G., Zinger, S., et al.: Cancer detection in histopathology whole-slide images using conditional random fields on deep embedded spaces. In: Medical imaging 2018: Digital pathology, vol. 10581. International Society for Optics and Photonics, p. 105810I (2018)
5. Li, Y., Ping, W.: Cancer metastasis detection with neural conditional random field. arXiv:1806.07064 (2018)
6. Zhang, Z., et al.: Pathologist-level interpretable whole-slide cancer diagnosis with deep learning. Nat. Mach. Intell. **1**(5), 236–245 (2019)
7. Hou, L., Samaras, D., Kurc, T.M., Gao, Y., Davis, J.E., Saltz, J.H.: Patch-based convolutional neural network for whole slide tissue image classification. In: Proceedings of the IEEE Conference on Computer Vision and Pattern Recognition, pp. 2424–2433 (2016)
8. Zhu, X., Yao, J., Zhu, F., Huang, J.: Wsisa: making survival prediction from whole slide histopathological images. In: Proceedings of the IEEE Conference on Computer Vision and Pattern Recognition, pp. 7234–7242 (2017)
9. Wang, X., et al.: Weakly supervised deep learning for whole slide lung cancer image analysis. IEEE Trans. Cybern. **50**(9), 3950–3962 (2019)
10. Chen, H., et al.: Rectified cross-entropy and upper transition loss for weakly supervised whole slide image classifier. In: International Conference on Medical Image Computing and Computer-Assisted Intervention, pp. 351–359. Springer (2019). https://doi.org/10.1007/978-3-030-32239-7_39
11. Campanella, G., et al.: Clinical-grade computational pathology using weakly supervised deep learning on whole slide images. Nat. Med. **25**(8), 1301–1309 (2019)
12. Zhou, P., et al.: Attention-based bidirectional long short-term memory networks for relation classification. In: Proceedings of the 54th Annual Meeting of the Association for Computational Linguistics (volume 2: Short papers), pp. 207–212 (2016)

13. He, K., Zhang, X., Ren, S., Sun, J.: Deep residual learning for image recognition. In: Proceedings of the IEEE Conference on Computer Vision and Pattern Recognition, pp. 770–778 (2016)
14. Zhang, S., Zheng, D., Hu, X., Yang, M.: Bidirectional long short-term memory networks for relation classification. In: Proceedings of the 29th Pacific Asia Conference on Language, Information and Computation, pp. 73–78 (2015)

Variational Encoding and Decoding for Hybrid Supervision of Registration Network

Dongdong Gu[1,2], Xiaohuan Cao[1], Guocai Liu[2], Zhong Xue[1(✉)],
and Dinggang Shen[3,1(✉)]

[1] Shanghai United Imaging Intelligence Co., Ltd., Shanghai, China
zhong.xue@uii-ai.com
[2] College of Electrical and Information Engineering, Hunan University, Changsha, China
[3] School of Biomedical Engineering, ShanghaiTech University, Shanghai, China
dgshen@shanghaitech.edu.cn

Abstract. Various progresses have been made in improving the accuracy of deep learning-based registration. However, there are still some limitations with current methods because of 1) difficulty of acquiring supervised data, 2) challenge of optimizing image similarity and enforcing deformation regularization, and 3) small number of training samples in such an ill-posed problem. It is believed that prior knowledge about the variability of a population could be incorporated to guide the network training to overcome these limitations. In this paper, we propose a group variational decoding-based training strategy to incorporate statistical priors of deformations for network supervision. Specifically, a variational auto-encoder is employed to learn the manifold for reconstructing deformations from a group of valid samples by projecting deformations into a low dimension latent space. Valid transformations can be simulated to serve as the ground-truth for supervised learning of registration. By working alternatively with the conventional unsupervised training, our registration network can better adapt to shape variability and yield accurate and consistent deformations. Experiments on 3D brain magnetic resonance (MR) images show that our proposed method performs better in terms of registration accuracy, consistency, and topological correctness.

Keywords: Medical image registration · Population-based regularization · Deep learning · Data modeling · Deformation consistency

1 Introduction

Deformable registration aims to determine voxel-wise correspondences between a pair of images while preserving anatomical topology, and has been widely applied for aligning images onto a common atlas for group analysis, fusing multi-modality images of the same subject, or performing longitudinal studies. As deformation fields are high dimensional, and there exist many possible solutions between each image pair, smoothness, consistency, and topology correctness regularizations are necessary to come up with a realistic and practical solution for such an ill-posed problem.

© Springer Nature Switzerland AG 2021
C. Lian et al. (Eds.): MLMI 2021, LNCS 12966, pp. 50–60, 2021.
https://doi.org/10.1007/978-3-030-87589-3_6

Traditional registration approaches typically define a transformation model and similarity measures, and apply iterative optimization to solve the deformation field between two images. Rigorous geometric theorems are used to ensure deformation smoothness and topological correctness, and the optimization is often time-consuming, regardless lacking of flexibility and parameter tuning efforts. In recent years, deep-learning-based registration methods have been investigated to improve computational efficiency while providing comparable accuracy [1, 2]. Supervised learning is based on carefully building the "ground-truth" deformations, which are difficult to obtain (especially through manual checking) [3, 4]. Unsupervised learning is therefore more popular, which does not require ground-truth deformations but learns by maximizing the similarity between images, such as sum of squared difference (SSD) and cross-correlation (CC), and simultaneously by applying deformation regularization losses [5, 6].

However, prior knowledge of anatomical variability, which has been proved to be able to improve robustness and accuracy of registration, is often omitted or has not been studied extensively in the context of deep registration networks. In the literature, several works have been made towards this direction. Shen et al. [7] introduced a fluid-based augmentation method so that deformations can be sampled from a geodesic subspace constructed using LDDMM [8]. For statistical sampling, Bhalodia et al. [9] used the population-level statistics of deformations to regularize the neural networks. Experiments on synthetic data and small organ alignment showed improved robustness and accuracy by applying priors in the learning procedure. More recently, Qin et al. [10] applied variational auto-encoder (VAE) to learn the manifold of biomechanics model of myocardial motion. Similarly, for image segmentation, Myronenko et al. [11] used a VAE to reconstruct the input images, and the VAE decoder is shared with the segmentation network. In registration, these methods have not yet fully considered the distribution and consistency properties of deformations.

In this paper, we propose a deep registration network with group variational decoding to leverage the latent space manifold distribution of deformation fields for supervision of network training. Specifically, a VAE is trained to learn the manifold for reconstructing deformation fields from a group of valid transformations generated by a diffeomorphic registration method SyN [12]. The decoder can reconstruct deformations by statistical sampling in the latent space, and the reconstructed deformations and corresponding warped images are used as the ground-truth for network training. Moreover, the registration network is trained alternatively using supervised and unsupervised learning strategies. Finally, smoothness and pair-wise deformation consistency losses are further incorporated during network training.

Public datasets of 3D brain MR images were used for evaluating the registration performance. Besides Dice similarity coefficient (DSC) and 95% Hausdorff Distance on regions of interest (ROIs), Consistency Distance and Folding Permillage of deformations are also used to evaluate the registration performance. Experimental results show that, compared with state-of-the-art algorithms, the proposed method improves both registration accuracy and deformation consistency.

2   Method

Given a template image T and a subject image S, registration solves a deformation field $\varphi(v)$ so that the warped subject image $S' = S(\varphi(v) + v)$ is similar to the template image, where v is a voxel in T. To solve the registration problem in deep learning, the popular unsupervised learning strategy uses image similarity and field regularization as losses for training and releases the load of simulating ground-truth deformations. But network training is still challenging with relatively small number of samples. In this paper, we employ VAE to generate fields and corresponding images as the supervision information. Together with unsupervised learning and the inverse consistency regularization [13], the performance can be improved in terms of accuracy and consistency.

As shown in Fig. 1, the proposed variational-decoding-based registration network consists of three components. First, group distribution of deformation fields is captured by leveraging a VAE trained using deformations obtained from SyN [12]. Second, deformations are acquired via decoding a random latent space vector conforming to the distribution by VAE. A real image in the training dataset and its deformed versions can be obtained with known synthesized deformations. Then, using alternative supervised and unsupervised learning strategies, images with and without deformations are provided to our deep registration for optimization. Finally, pair-wise inverse consistency is incorporated to enforce deformation consistency properties.

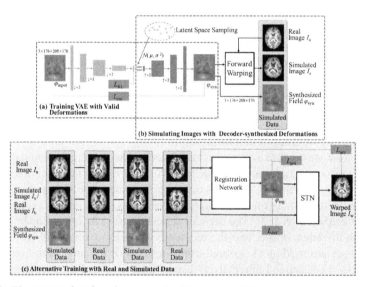

Fig. 1. The proposed registration network with group variational encoding regularization.

Training VAE with Valid Deformations. We leverage VAE to capture the deformation distribution and reconstruct valid deformation fields. For VAE training, the inputs are deformations generated by SyN [12]. VAE projects deformations onto the latent space and recovers them through decoding. Modeling latent space distribution and random sampling of latent vectors allow for generating realistic deformations, which increases deformation variability via supervised learning of the registration network.

Specifically, two convolutional blocks are used at each resolution based on the ResNet structure. Each block includes convolution, ReLU with Group Normalization (GN), and additive skip connection. All convolutions use $3 \times 3 \times 3$ kernels. The initial number of channels is 32 and then doubled after down-sampling. We use 4 resolutions, and the input size is $3 \times 176 \times 208 \times 176$, the same as the deformation field. The encoder endpoint has size $256 \times 22 \times 26 \times 22$, from which the dimension of features is further reduced to 256×1, 128 for mean and 128 for standard deviation (std) of Gaussian distribution. A deformation is then reconstructed by randomly sampling the Gaussian distribution and passing through the decoder. The decoder output has the same size and same number of channels as the input, and sigmoid function is used to replace ReLU.

In VAE training, we down-sample the corresponding brain tissue masks of real images and concatenate with the feature maps of the decoder in each level to better regularize the output deformations. As the images are pre-aligned through affine registration, the tissue masks can help the network capture the brain shape of each training subject, and the decoded deformations can fit the anatomy of input images. We follow the standard framework for training VAE, and the loss function consists of 3 terms:

$$L_{\text{VAE}} = \alpha * L_{\text{KL}} + L_{\text{rec\_MSE}} + L_{\text{rec\_CE}}, \tag{1}$$

where L_{KL} is the KL divergence between the estimated prior normal distribution and $\mathcal{N}(0,1)$ [14], i.e., $L_{\text{KL}} = \frac{1}{N} \sum (\mu^2 + \sigma^2 - \log \sigma^2)$, where $N = 128$ is the size of the latent vectors. α is chosen as 0.0001 empirically. $L_{\text{rec\_MSE}}$ and $L_{\text{rec\_CE}}$ are the mean square error (MSE) and the cross entropy (CE) reconstruction losses between the input field φ_{input} and the predicted field φ_{syn}, respectively,

$$L_{\text{rec\_MSE}} = \frac{1}{|M|} \sum_{v \in M} \varphi_{\text{input}}(v) - \varphi_{\text{syn}}(v)^2, \tag{2}$$

$$L_{\text{rec\_CE}} = -\frac{1}{|M|} \sum_{v \in M} \left[\varphi_{\text{input}}(v) \log(\varphi_{\text{syn}}(v)) + (1 - \varphi_{\text{input}}(v)) \log(1 - \varphi_{\text{syn}}(v)) \right], \tag{3}$$

where v is a voxel, and $|M|$ is the number of voxels.

In VAE, reconstruction can be achieved by randomly sampling from the Gaussian distribution. However, variability and diversity of fields could not be always observed because the sampling tends to be close to the mean values. Therefore, we choose to randomly draw samples with large variability using principal component analysis (PCA) in the latent space [15]. Suppose all the training deformations passed through the encoder yield respective latent vectors and form a latent matrix $L = \{l_1, l_2, \ldots\}$. Using PCA, a latent vector l can be obtained by $l = \bar{l} + M_{128 \times K} b$, where $M_{128 \times K}$ is the eigenvector matrix of covariance matrix of the latent matrix L, and b can be sampled from a uniform distribution defined by the eigenvalues. After sampling l, the deformation fields conforming to the distribution can be synthesized by the well trained decoder. Such a modification increases the variability of VAE decoder for field synthesis.

Simulating Images with Decoder-Synthesized Deformations. As shown in Fig. 1 (b), the input image I_a can be transformed by the synthesized deformation φ_{syn} using the pre-trained VAE, resulting in the forward-warped image $I_s(\varphi_{\text{syn}}(v) + v) = I_a(v)$. Typical cases are shown in Fig. 2. It can be seen that various shapes of images can be simulated with smooth deformations. As the training fields of VAE generated by SyN are diffeomorphic and topologically correct, the transformed images are anatomically plausible. We randomly sampled 20,000 latent vectors and synthesized fields to train the downstream registration network in a supervised learning manner.

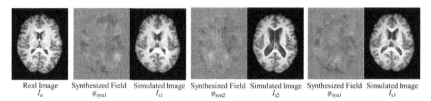

Real Image Synthesized Field Simulated Image Synthesized Field Simulated Image Synthesized Field Simulated Image
I_a φ_{syn1} I_{s1} φ_{syn2} I_{s2} φ_{syn3} I_{s3}

Fig. 2. Typical cases of VAE-simulated deformations and transformed images.

Alternative Training with Real and Simulated Data. The deep registration network in Fig. 1 (c) follows a similar structure of typical CNNs as in [5] and [13]. The network has a U-Net shape with skip-connections. Three resolutions are used, and in each level, we use a series of blocks, each including a GN, a convolution, and a ReLU layer. Downsampling is achieved by using a convolution layer with stride of 2, and the number of channels is doubled. The decoder path uses a reverse fashion. The network training is based on both supervised and unsupervised learning.

For the unsupervised training, the input consists of an image pair (I_a and I_b) with unknown deformation, and the output is the deformation field φ_{reg}. Spatial transformation (STN) is employed to warp image I_a by φ_{reg}. The similarity loss L_{acc} can be calculated by Eq. (4), and MSE between image I_a and the warped image $I_w = I_b(\varphi_{reg}(v) + v)$ is defined over 3D image domain Ω.

$$L_{acc}(I_a, I_b) = \frac{1}{|\Omega|} \sum_{v \in \Omega} \left\| I_a(v) - I_b(\varphi_{reg}(v) + v) \right\|^2. \tag{4}$$

Deformation smoothness loss L_{gra} is introduced to regularize the field: $L_{gra}(\varphi_{reg}) = \frac{1}{|\Omega|} \sum_{v \in \Omega} \left| \nabla \varphi_{reg}(v) \right|$, where $\nabla \varphi(v)$ is gradient of φ at voxel v, and $|\cdot|$ represents L1 norm.

For supervised training with group data distribution, synthesized deformations and corresponding paired images are incorporated to train the registration network. The decoded deformation fields can serve as the ground-truth between images I_a and I_s. In this case, the predicted deformation φ_{reg} should be the same as the decoder-synthesized deformation φ_{syn}, as shown in Fig. 1 (c). So the supervised loss function L_{dvf} is defined as L2 norm of the difference between φ_{syn} and φ_{reg}:

$$L_{dvf} = \frac{1}{|\Omega|} \sum_{v \in \Omega} \left\| \varphi_{reg}(v) - \varphi_{syn}(v) \right\|_2. \tag{5}$$

Meanwhile, we also incorporate image similarity loss to enforce the warped image I_w by φ_{reg} to be similar to the fixed image I_a, similarly as in unsupervised training.

The registration network is trained in an alternative strategy, *i.e.*, using real data and simulated data in different epochs. We use 840 epochs (batch size = 1, about 10,000 iterations) for training and alternatively switch the training strategy after 200, 100, 50, 35, 10 and 5 epochs. Note that the alternation becomes faster as the training performs.

Finally, we incorporated the inverse-consistency training strategy as in [13], which means the bidirectional deformations estimated between an image pair should share the same pathway. Thus, the composition of the forward and backward deformations should be identity or close to identity. We incorporate the consistency constraint to both the supervised and unsupervised learning. For an image pair I_a and I_b/I_s, we first obtain the forward deformation φ_{reg}. Then, the backward deformation φ_{inv} is obtained through the same network by switching the input images. The inverse-consistency loss L_{inv} is incorporated for network training:

$$L_{inv} = \frac{1}{|\Omega|} \sum_{v \in \Omega} \left\| \left(\varphi_{reg} \circ \varphi_{inv} \right)(v) \right\|_2. \tag{6}$$

3 Results

We evaluated the performance of the proposed method using 150 T1 brain MR images from ADNI [16], OASIS [17] and PPMI [18]. We chose 110 images for training and 40 images for testing. Images can be randomly paired for training and testing. To evaluate the performance, we selected 24 labels from 106 brain regions segmented by FreeSurfer [19], including both cortical and sub-cortical structures.

All the images were preprocessed by the following four steps: 1) N3 bias correction; 2) skull stripping; 3) intensity normalization; 4) affine registration to the MNI152 atlas [20]. We selected 3 state-of-the-art registration methods, *i.e.*, FNIRT [21], Demons [22], and SyN [12], as well as different settings of the deep learning algorithm for comparison: DL is a typical unsupervised U-Net architecture; DL-sup is a supervised network trained with the ground-truth deformation generated by SyN [12]; DL-VAE is the proposed group variational decoding registration trained using both real and synthesized data; DL-inv is DL with deformation consistency constraint; and DL-inv5k, DL-inv10k and DL-inv20k are the DL-inv model with different numbers of simulated cases. All the DL models have the same network architecture as introduced previously and are trained for the same number of epochs. The input images are cropped into patches with size 100 × 100 × 100 and stitched back to the image space due to GPU memory limitation. The network was implemented using PyTorch with Adam optimization (learning rate is 0.00001), and NVIDIA Geforce RTX 2080 Ti GPU was used.

Evaluation of Registration Performance. We used "DSC", "95% Hausdorff Distance (HD)", "Consistency Distance" and "Folding Permillage" to evaluate the performance. DSC and HD are calculated using white matter (WM), grey matter (GM) and 22 brain regions (Segs) in the warped and fixed images. Table 1 shows the results, and bold fonts indicate the best accuracy of the proposed method with or without the inverse-consistency constraint, which indicates that, the group variational-decoding strategy is effective to train an accurate registration network using both real and simulated data.

Table 1. Comparison of DSC (%) and 95% HD (mm) of different registration methods.

Method/Metrics	Affine	Demons	SyN	FNIRT	DL	DL-inv	DL-inv5k	DL-inv10k	DL-inv20k	DL-VAE	DL-sup
DSC (GM)	56.1 ± 2.6	67.9 ± 7.9	72.6 ± 2.4	65.6 ± 2.6	71.2 ± 3.9	67.4 ± 3.6	70.2 ± 2.9	71.8 ± 4.5	**74.0 ± 3.5**	**74.4 ± 3.5**	64.2 ± 3.6
HD (GM)	2.8 ± 0.3	1.9 ± 0.8	1.5 ± 0.1	2.2 ± 0.5	1.5 ± 0.2	1.7 ± 0.3	1.6 ± 0.2	1.5 ± 0.3	**1.4 ± 0.2**	**1.3 ± 0.2**	2.0 ± 0.3
DSC (WM)	62.5 ± 3.4	77.0 ± 5.6	80.3 ± 2.2	75.8 ± 1.9	78.5 ± 1.8	75.2 ± 2.3	77.3 ± 1.7	79.2 ± 3.1	**81.2 ± 2.1**	**81.7 ± 2.2**	72.4 ± 2.8
HD (WM)	3.3 ± 0.3	2.2 ± 1.0	1.7 ± 0.3	2.2 ± 0.3	1.8 ± 0.2	2.1 ± 0.2	2.0 ± 0.2	1.7 ± 0.4	**1.5 ± 0.3**	**1.4 ± 0.3**	2.3 ± 0.3
DSC (Seg)	53.5 ± 12.2	59.3 ± 17.7	75.5 ± 6.4	69.8 ± 7.6	73.8 ± 5.1	68.3 ± 7.7	72.6 ± 5.6	70.0 ± 8.5	74.8 ± 5.1△	75.5 ± 4.8	61.1 ± 10.1
HD (Seg)	3.6 ± 1.2	3.4 ± 2.0	1.9 ± 0.7	2.2 ± 0.8	2.0 ± 0.6	2.4 ± 0.8	2.1 ± 0.7	2.3 ± 0.9	2.0 ± 0.6	**1.9 ± 0.6**	3.1 ± 1.3
Time	3.1*	4.5*	21.3*	28.4*	0.03^	0.03^	0.03^	0.03^	0.03^	0.03^	0.03^

The result in "Segs" marked with "Δ" indicates no significant difference between our algorithm and SyN [12] (*p*-value > 0.05).

We also compared the time spent for all the methods in Table 1, including inferring time for deep registration methods using GPU (marked with "^") and optimization time for traditional methods using CPU (marked with "*"), for registering two typical 3D brain MR images with size 176 × 208 × 176 and isotropic 1mm resolution. Despite that deep learning-based methods largely reduce the registration time, the DL methods have similar inferring speed.

Boxplots of DSC in 24 brain ROIs across 40 testing image pairs are shown in Fig. 3. The proposed algorithm achieves the best performance for 9 out of the 24 ROIs in both DSC and HD (marked with "*"), while the performance of the remaining 15 ROIs is comparable (*p*-value > 0.05).

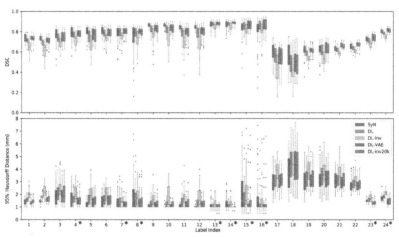

Fig. 3. Boxplots of DSC and 95% HD for 24 brain ROIs across 40 testing cases. The ROIs include: insula (1, 2), cingulum (3, 4), hippocampus (5, 6), amygdala (7, 8), caudate (9, 10), putamen (11, 12), pallidum (13, 14), thalamus (15, 16), ventricle (17, 18), entorhinal (19, 20), and temporal gyrus (21, 22). The structures are in both left and right brain hemispheres. GM (23) and WM (24) are also included for evaluation.

The Consistency Distance results across all the testing cases are shown in Fig. 4 (which also shows the histograms of the Consistency Distance), and it can be seen that DL-inv20k yields better consistent deformation fields. It is worth noting that SyN [12] is based on diffeomorphic framework, and the consistency performance is guaranteed and not reported here. This is also the reason why we choose the deformations obtained from SyN as our prior knowledge when generating the latent space vectors in VAE. We also compare the mean number of folding in Fig. 4. Less folding indicates better topological preservation and better morphological allowable fields. The number of folding decreases greatly in DL-inv20k, indicating that the proposed method yields more reliable deformations for our datasets. By using the group variational-decoding and consistent constraint, the distance is smaller than other compared algorithms.

Notice that the results for Demons, SyN, and FNIRT were obtained by careful parameter tuning. We specially selected some cases that the appearance differences, *e.g.*, the ventricular shapes, are obvious to show the robustness of the proposed registration method. Visualization of typical warped images obtained by using different registration algorithms is presented in Fig. 5. The proposed method did not show obvious faults by comparing the warped image with the fixed image.

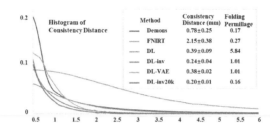

Fig. 4. Distributions of consistency distance and folding permillage.

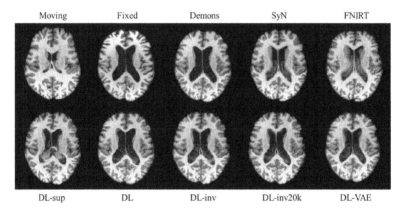

Fig. 5. Typical registration results.

4   Conclusion

We have proposed a hybrid training strategy for deep registration network, where a variational-decoding method with statistical sampling is designed to learn the manifold from a group of training samples and generate deformation fields with variability based on the prior knowledge, which can serve as the ground-truth for unsupervised training. Smoothness and inverse consistency losses are also incorporated. The experimental results show that the proposed method exhibits higher registration accuracy and deformation consistency, compared with state-of-the-art registration methods.

Acknowledgement. This work was partially supported by the National Key Research and Development Program of China (2018YFC0116400) and the National Natural Science Foundation of China (62071176).

References

1. Haskins, G., Kruger, U., Yan, P.: Deep learning in medical image registration: a survey. Mach. Vis. Appl. **31**(1–2), 1–18 (2020). https://doi.org/10.1007/s00138-020-01060-x
2. Boveiri, H.R., Khayami, R., Javidan, R., Mehdizadeh, A.: Medical image registration using deep neural networks: a comprehensive review. Comput. Electr. Eng. **87**, 106767 (2020)
3. Eppenhof, K.A., Pluim, J.P.: Pulmonary CT registration through supervised learning with convolutional neural networks. IEEE Trans. Med. Imaging **38**(5), 1097–1105 (2018)
4. Sokooti, H., de Vos, B., Berendsen, F., Lelieveldt, B.P.F., Išgum, I., Staring, M.: Nonrigid image registration using multi-scale 3D convolutional neural networks. In: Descoteaux, M., Maier-Hein, L., Franz, A., Jannin, P., Collins, D.L., Duchesne, S. (eds.) MICCAI 2017. LNCS, vol. 10433, pp. 232–239. Springer, Cham (2017). https://doi.org/10.1007/978-3-319-66182-7_27
5. Balakrishnan, G., Zhao, A., Sabuncu, M.R., Guttag, J., Dalca, A.V.: VoxelMorph: a learning framework for deformable medical image registration. IEEE Trans. Med. Imaging **38**(8), 1788–1800 (2019)
6. de Vos, B.D., Berendsen, F.F., Viergever, M.A., Staring, M., Išgum, I.: End-to-end unsupervised deformable image registration with a convolutional neural network. In: Cardoso, M.J., et al. (eds.) DLMIA/ML-CDS -2017. LNCS, vol. 10553, pp. 204–212. Springer, Cham (2017). https://doi.org/10.1007/978-3-319-67558-9_24
7. Shen, Z., Xu, Z., Olut, S., Niethammer, M.: Anatomical data augmentation via fluid-based image registration. In: Martel, A.L., et al. (eds.) MICCAI 2020. LNCS, vol. 12263, pp. 318–328. Springer, Cham (2020). https://doi.org/10.1007/978-3-030-59716-0_31
8. Oishi, K., et al.: Atlas-based whole brain white matter analysis using large deformation diffeomorphic metric mapping: application to normal elderly and Alzheimer's disease participants. Neuroimage **46**(2), 486–499 (2009)
9. Bhalodia, R., Elhabian, S.Y., Kavan, L., Whitaker, R.T.: A cooperative autoencoder for population-based regularization of CNN image registration. In: Shen, D., et al. (eds.) MICCAI 2019. LNCS, vol. 11765, pp. 391–400. Springer, Cham (2019). https://doi.org/10.1007/978-3-030-32245-8_44
10. Qin, C., Wang, S., Chen, C., Qiu, H., Bai, W., Rueckert, D.: Biomechanics-informed neural networks for myocardial motion tracking in MRI. In: Martel, A.L., et al. (eds.) MICCAI 2020. LNCS, vol. 12263, pp. 296–306. Springer, Cham (2020). https://doi.org/10.1007/978-3-030-59716-0_29
11. Myronenko, A.: 3D MRI brain tumor segmentation using autoencoder regularization. In: Crimi, A., Bakas, S., Kuijf, H., Keyvan, F., Reyes, M., van Walsum, T. (eds.) BrainLes 2018. LNCS, vol. 11384, pp. 311–320. Springer, Cham (2019). https://doi.org/10.1007/978-3-030-11726-9_28
12. Avants, B.B., Epstein, C.L., Grossman, M., Gee, J.C.: Symmetric diffeomorphic image registration with cross-correlation: evaluating automated labeling of elderly and neurodegenerative brain. Med. Image Anal. **12**(1), 26–41 (2008)
13. Gu, D., et al.: Pair-wise and group-wise deformation consistency in deep registration network. In: Martel, A.L., et al. (eds.) MICCAI 2020. LNCS, vol. 12263, pp. 171–180. Springer, Cham (2020). https://doi.org/10.1007/978-3-030-59716-0_17

14. Kingma, D.P., Welling, M.: Auto-encoding variational bayes. arXiv preprint arXiv:.1312.6114 (2013)
15. Xue, Z., Shen, D., Davatzikos, C.: Statistical representation of high-dimensional deformation fields with application to statistically constrained 3D warping. Med. Image Anal. **10**(5), 740–751 (2006)
16. Mueller, S.G., et al.: Ways toward an early diagnosis in Alzheimer's disease: the Alzheimer's disease neuroimaging initiative (ADNI). Alzheimer's Dement. **1**(1), 55–66 (2005)
17. Marcus, D.S., Wang, T.H., Parker, J., Csernansky, J.G., Morris, J.C., Buckner, R.L.: Open access series of imaging studies (OASIS): cross-sectional MRI data in young, middle aged, nondemented, and demented older adults. J. Cogn. Neurosci. **19**(9), 1498–1507 (2007)
18. Marek, K., et al.: The Parkinson's progression markers initiative (PPMI)–establishing a PD biomarker cohort. Ann. Clin. Transl. Neurol. **5**(12), 1460–1477 (2018)
19. Fischl, B.: FreeSurfer. Neuroimage **62**(2), 774–781 (2012)
20. Mazziotta, J., et al.: A probabilistic atlas and reference system for the human brain: International Consortium for Brain Mapping (ICBM). Philos. Trans. R Soc. B Biol. Sci. **356**(1412), 1293–1322 (2001)
21. Andersson, J.L., Jenkinson, M., Smith, S.: Non-linear registration, aka Spatial normalisation FMRIB technical report TR07JA2. FMRIB Analysis Group of the University of Oxford **2**(1), e21 (2007)
22. Thirion, J.-P.: Image matching as a diffusion process: an analogy with Maxwell's demons. Med. Image Anal. **2**(3), 243–260 (1998)

Multiresolution Registration Network (MRN) Hierarchy with Prior Knowledge Learning

Dongdong Gu[1,2], Xiaohuan Cao[2], Guocai Liu[1(✉)], Dinggang Shen[3,2], and Zhong Xue[2(✉)]

[1] College of Electrical and Information Engineering, Hunan University, Changsha, China
lgc630819@hnu.edu.cn
[2] Shanghai United Imaging Intelligence Co., Ltd., Shanghai, China
zhong.xue@uii-ai.com
[3] School of Biomedical Engineering, ShanghaiTech University, Shanghai, China

Abstract. Deep learning has been extensively used in unsupervised deformable image registration. U-Net structures are often used to infer deformation fields from concatenated input images, and training is achieved by minimizing losses derived from image similarity and field regularization terms. However, the mechanism of multiresolution encoding and decoding with skip connections tends to mix up the spatial relationship between corresponding voxels or features. This paper proposes a multiresolution registration network (MRN) based on simple convolution layers at each resolution level and forms a framework mimicking the ideas of well-accepted traditional image registration algorithms, wherein deformations are solved at the lowest resolution and further refined level-by-level. Multiresolution image features can be directly fed into the network, and wavelet decomposition is employed to maintain rich features at low resolution. In addition, prior knowledge of deformations at the lowest resolution is modeled by kernel-PCA when the template image is fixed, and such a prior loss is employed for training at that level to better tolerate shape variability. The proposed algorithm can be directly used for group analysis or image labeling and potentially applied for registering any image pairs. We compared the performance of MRN with different settings, *i.e.*, w/wo wavelet features, w/wo kernel-PCA losses, using brain magnetic resonance (MR) images, and the results showed better performance for the multiresolution representation and prior knowledge learning.

Keywords: Medical image registration · Multiresolution representation · Prior knowledge · Wavelet decomposition · Convolutional neural network

1 Introduction

In recent years, deep neural networks have been extensively used in deformable image registration, among which VoxelMorph [1] has been widely adopted due to easy training, fast inferring speed, and comparable results to traditional state-of-the-art methods. They have also been applied to align multi-modality images by disentangling the problem to a mono-modal one [2]. In learning-based registration, many deep registration methods

© Springer Nature Switzerland AG 2021
C. Lian et al. (Eds.): MLMI 2021, LNCS 12966, pp. 61–69, 2021.
https://doi.org/10.1007/978-3-030-87589-3_7

employ U-Net structures using multiresolution encoding and decoding paths with skip connections. Basically, a 3D deformation can be inferred by the networks after concatenating the input images, and unsupervised training is achieved by minimizing the losses derived from image similarity and deformation regularization terms.

However, the mechanism of multiresolution encoding and decoding architecture, especially with the skip connection, tends to mix up the spatial relationship between corresponding voxels of the images under registration or the feature maps at different resolutions. This may leave the task of solving deformations entirely to the convolutional layers and lead to more convolutional layers or complex network structures.

Recently, cascaded registration models have been proposed to improve the performance of registration [3–5]. Zhao *et al.* presented a volume tweening network (VTN) [3], which gradually registers a pair of images by using cascaded registration subnetworks, and each time the deformation field between the intermediately warped moving image and the fixed image are refined. Lately, they proposed a recursive cascaded network in [4], so the entire system can be jointly trained. During testing, one cascade network may be iteratively applied multiple times. But clearly, the networks should be different at different levels, especially the first network and the subsequent ones. de Vos *et al.* trained each cascade network one by one by fixing the weights of previous networks [5]. Despite using multiple networks, the effort seeks for more convolutional layers if one looks the cascade networks from end-to-end as a whole. In a typical U-shape network architecture, since the misalignment between feature maps caused by a residual or skip connection is complicated, a simple bilinear up-sampling and concatenating operation may break the symmetry between the down-sampling encoder and up-sampling decoder. Inspired by the Flow-Net [6], an semantic flow estimator layer was adopted in [7], so that the motion between two feature maps in different convolution layers can be compensated.

This paper uses convolutional neural network (CNN) for image registration while adopting the well-accepted idea from traditional image registration algorithms. We incorporate multiresolution representation of images, multilevel or hierarchical registration, and prior knowledge constraints in the training stage. Specifically, we apply elegant CNNs with less layers and parameters to solve the deformation field at the lowest resolution or to refine the fields at higher resolutions. The method is explicitly structured in multiresolution and allows for corresponding image features be directly fed into the subnetwork as inputs, thereby no skip-connection concatenation is needed. Various multiresolution image representations could be used, and herein wavelet decomposition is employed to maintain rich features. CNN at each level can be trained separately, and any image pairs can be well registered using MRN. In the case of group analysis or image labeling, where the fixed image (or template) does not change, a prior knowledge-based loss can be applied to the deformation field at the lowest resolution. Therefore, from a group of valid deformations, the kernel-PCA statistics can be defined to form a new prior knowledge loss.

In experiments, using brain magnetic resonance (MR) images, we compared the performance of MRN with different settings, *i.e.*, w/wo wavelet features, w/wo prior knowledge losses. We used Dice of brain tissues and number of deformation folding to evaluate the effectiveness of the proposed method. The experimental results showed better performance for the multiresolution representation and prior knowledge learning in terms of registration accuracy and topological correctness.

2   Method

2.1   Multiresolution Image Registration Hierarchy

The proposed network addresses several key components for multiresolution registration, including formulating a multi-level structure to first solve the deformations at the lowest resolution and then gradually refine the deformation fields, using multiresolution image features to help better match anatomical structures, and applying prior knowledge or statistics of the deformation fields to regularize the deformations for more robust registration. The CNN-based registration can be summarized in Fig. 1.

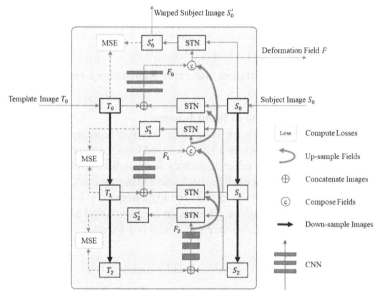

Fig. 1. The structure of MRN. The input template and subject images T_0 and S_0 are fed into the network, which are down-sampled (or transformed using wavelet packet transformation) into different resolution levels. Registration is performed hierarchically at different resolution levels, and the outputs include a deformation field and a warped subject.

As shown in Fig. 1, the input template and subject images (T_0 and S_0) are first down-sampled (half the size) to lower resolution levels (T_1, T_2 and S_1, S_2). F_2 represents the output deformation field at the lowest resolution (level 2), which aligns images T_2 and S_2. The deformed image S_2' can be generated using spatial transformation network (STN) based on F_2. Mean square error (MSE) between the deformed image and template image, and deformation smoothness losses can be applied to train the CNN at this level. The output deformation field F_2 is then up-sampled to level 1 and used to deform the subject image S_1. To train the CNN at level 1, MSE and smoothness losses can be used similarly, and the training can be performed after the lower level network is properly trained. Using the same way, a refinement field F_0 can be solved by training the CNN at level 0. Therefore, the network yields the final deformation field F by solving the deformation F_2 and the refinement fields F_1 and F_0 and then composing them.

It can be seen that the structure of the proposed MRN mimics the traditional registration and gradually refines the deformation fields. The network can be trained level-by-level or even trained independently when available deformations can be generated and properly down-sampled. As the outputs of each CNN are deformation fields, the training and performance of the network can be easily monitored. Moreover, the inputs at different levels are flexible, as long as they reflect multiresolution representations of the images to be registered. Notice that the deformation fields at the lowest resolution reflect the major anatomical variability, and hence different priors can be used at this level. Herein, as an example, we use a kernel-PCA loss to regularize the training at the lowest resolution by assuming the template image is fixed. For the higher two levels, only smoothness losses are used for regularization since the fields only reflect refinements. Finally, to solve the deformations at each level, we use simple CNN (Sect. 2.4) rather than the U-Net structure to eliminate its shortcomings mentioned in Introduction.

2.2 Multiresolution Representation and Prior Knowledge Learning

Multiresolution Representation of Input Images

As mentioned above, MRN allows for using different image representations as inputs. To maintain abundant image information in both low and high frequency, we adopt wavelet packet transform (WPT) [8] to generate images at resolution levels 1 and 2. Figure 2 shows an example of images after WPT. Three selected high-pass bands in the red boxes of the original image are combined to form the high-frequency image (mean absolute values), so the inputs for each image includes an image channel and a high-frequency channel, and the number of input channels of CNNs is 4 for all the levels.

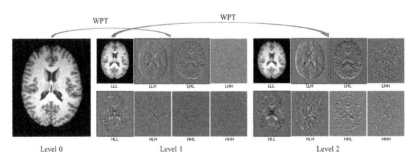

Fig. 2. Multiresolution representation of images with WPT. From left to right: original image as the input at level 0; low pass and combined bands in red boxes as the input of level 1; low pass and combined red bands as the input of level 2. (Color figure online)

Statistical Priors for Learning the CNN at the Lowest Resolution Level

Denoting the deformations at level 2 as $\mathbf{f}_1, \mathbf{f}_2, \ldots, \mathbf{f}_N$, where N is the number of samples available. The prior distribution of the deformations can be modeled by kernel-PCA [9] with Gaussian kernel functions. The samples can be obtained from a traditional method such as SyN [10] after proper down-sampling. According to kernel-PCA, the kernel matrix of N samples can be calculated by:

$$\kappa\left(\mathbf{f}_i, \mathbf{f}_j\right) = \exp\{-\gamma E\left(\mathbf{f}_i, \mathbf{f}_j\right)\}, \; i = 1, \ldots, N; j = 1, \ldots, N, \tag{1}$$

where $E(\mathbf{f}_i, \mathbf{f}_j)$ is the Euclidean distance between two deformation fields $\mathbf{f}_i$ and $\mathbf{f}_j$ normalized by the number of voxels, $E\left(\mathbf{f}_i, \mathbf{f}_j\right) = \frac{1}{|M|}\sum_{v \in M}\|\mathbf{f}_i(v) - \mathbf{f}_j(v)\|_2$, and γ is a constant. M represents the image domain, and $|M|$ is the number of voxels. Then, the centered kernel matrix can be calculated by:

$$\kappa_c = \kappa - 1\kappa - \kappa 1 + 1\kappa 1, \tag{2}$$

where 1 is a square matrix with element of $1/N$. Finally, the eigenvectors and eigenvalues of κ_c form the projection matrix ϕ and the variances λ along principal components. The projection of any new deformation field $\mathbf{f}$ can be calculated by:

$$\boldsymbol{\mu} = \phi^T \mathbf{k}, \tag{3}$$

where $\mathbf{k} = \left[\kappa\left(\mathbf{f}, \mathbf{f}_j\right), j = 1, \ldots, N\right]$. The shape of the input field $\mathbf{f}$ can be regularized by enforcing vector $\boldsymbol{\mu}$ within the range defined by the variance as:

$$L_{\text{kpca}}(\mathbf{f}) = \sum_{i=1,\ldots K} \mu_i^2 / \lambda_i, \tag{4}$$

where K is the number of principal eigenvectors. Traditionally, least squares estimation needs to be performed when reconstructing the constrained deformation field. Herein, with tensor programming of PyTorch, the above loss function can be automatically optimized through the gradient graphs of the software package.

2.3 Network Training Strategies

As seen from Fig. 1, MRN can be trained level-by-level. In addition to the prior loss defined in Eq. (4), we also use the MSE and smoothness loss functions. MSE is defined as the similarity between template images and deformed subject images:

$$L_{\text{MSE},l} = \frac{1}{|M|} \sum_{v \in M} \|T_l(v) - S_l(\mathbf{f}(v) + v)\|^2, \; l = 0, 1, 2. \tag{5}$$

where l represents resolution level. The smoothness loss is defined as:

$$L_{\text{grad}} = \frac{1}{|M|} \sum_{v \in M} \|\nabla \mathbf{f}(v)\|^2. \tag{6}$$

In summary, the total loss for each level l is defined as:

$$L_l = \alpha L_{\text{MSE},l} + \beta L_{\text{Grad},l} + \eta L_{\text{kpca}}, \; l = 0, \; 1, \; 2. \tag{7}$$

η is set to zero for level 0 and 1, so the prior loss does not apply for higher resolutions.

2.4   Algorithm Implementation

The CNNs at each level consist of 6 blocks, and each block includes a $3 \times 3 \times 3$ convolutional layer with no padding followed by ReLU, and the last layer does not have ReLU. The size of output at each level is smaller than that of the input (12 voxel difference) because no-padding operation is used. We cropped input images into patches with size $140 \times 140 \times 140$ and stitched back to the image space to save GPU memory. The output deformation field at the lower level should meet the size of the output of the current level after up-sampling. This gives an exact effective size of $56 \times 56 \times 56$ for each patch. So, we crop overlapping patches by skipping 56 voxels. Another consideration is that the kernel-PCA model is computed from the entire field at level 2, but for patch implementation, we can only use partial deformation field. Thus, the kernel vector k in Eq. (3) is calculated by using the deformation within the corresponding regions of each patch, *i.e.*, $\kappa = \left[\kappa \left(P(\mathbf{f}), P(\mathbf{f}_j) \right), j = 1, \ldots, N \right]$, and $P()$ stands for only picking the deformation within the patch. With bigger patches covering a large area of the brain, distance calculated between two whole deformation fields is approximated by computing via partial fields, avoiding computing the statistics for different patch locations.

The network was implemented using PytTorch with Adam optimization. NVIDIA Geforce RTX 2080 Ti was used for training and testing. We trained the network up to1000 epochs (about 118,000 iterations) with batch size 1. The learning rate was set to 1e-5. The weights of loss terms in Eq. (7) were chosen as $\alpha = 1$, $\beta = 0.01$. η was set to 0.01 for the first half epochs and 0.1 for the rest.

3   Results

3.1   Datasets and Experiment Setting

We evaluated the performance of the proposed method using 150 T1 brain MR images from ADNI [11]. We randomly chose 120 images for training and 30 images for testing. One of the training images was selected as the template image (see Fig. 3 left). The goal was to train MRN by registering 119 sample images onto the template image in an unsupervised fashion. All the images were preprocessed by first applying N3 bias correction and then aligning them globally onto the template image space using affine registration. Skull stripping was performed using BET [12]. The images used for training and testing are with size $180 \times 216 \times 180$ and in isotropic $1 \times 1 \times 1mm^3$ spacing. For CNN inputs, the two images or two sets of WPT data are normalized to the range between 0 and 1 before cropping patches so that the intensities across different patch locations remain consistent for each image.

3.2   Algorithm Comparison

Figure 3 shows an example of the registration results of MRN using WPT inputs and kernel-PCA priors. It can be seen that the warped image at level 2 deforms a lot toward the template image, and the deformation can be further refined in the following levels. The proposed method also yields smooth deformation fields with less folding (will be addressed below). Figure 4 plots more results of different testing subjects for visual assessment. The registration of all the 30 testing samples are successful, and the warped images look similar to the template image.

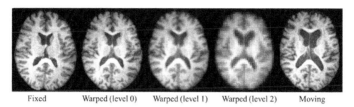

Fixed   Warped (level 0)   Warped (level 1)   Warped (level 2)   Moving

Fig. 3. Illustration of the (intermediate) results of the proposed algorithm. From left to right: the fixed image, the deformed images at level 0, 1, and 2, and the moving image.

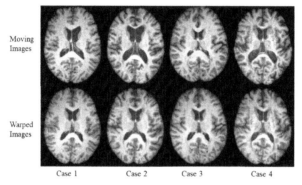

Moving Images

Warped Images

Case 1   Case 2   Case 3   Case 4

Fig. 4. Registration results of different subjects. Top: input moving images; bottom: warped images on the same template space. The fixed image is the same as Fig. 3.

In order to evaluate the performance of the proposed algorithm, we compared different training strategies and methods, including 1) using original images and their down-sampled versions as the network input (MRN-Image); 2) using original images and gradient magnitudes as input at level 0 and WPT data as input at level 1 and 2 (MRN-WPT); 3) with kernel-PCA prior loss based on 2) (MRN-WPT-KPCA); 4) with kernel-PCA prior loss based on 1) (MRN-KPCA); 5) a deep registration network similar to the U-Net architecture in [1] and the baseline used in [13] (Deep Registration); and 6) SyN [10]. The network structures and parameters are the same for all the MRN-based methods. The parameters of deep registration network and SyN are carefully tuned. We tested the methods on 30 testing images respectively.

Dice similarity coefficients (DSC) of white matter (WM), gray matter (GM) and Cerebrospinal Fluid (CSF) with ventricle between the masks on the template image and the ones warped from each subject image are used for quantitative evaluation. We also calculated the smoothness metric, *i.e.* permillage of folding in the deformation fields.

Figure 5 shows results of 30 testing data in terms of DSC for different settings of the compared methods. It can be seen that compared to the Deep Registration method, the DSC of GM, WM and CSF of our proposed MRN were all improved. The methods using

WPT features slightly outperformed others, and in terms of kernel-PCA constraints, they yield similar DSCs, but overall the standard deviations are at least 10% smaller, indicating more consistent and robust results over the testing datasets, and statistical models could generate more variability during network training.

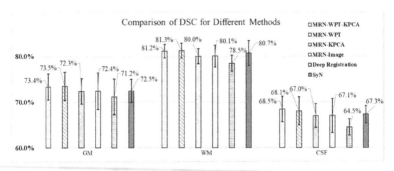

Fig. 5. Comparison of DSC for different methods.

We also counted the number of voxels with incorrect topology (*i.e.*, with negative Jacobian determinants), and the mean and standard deviation values in MRN-Image, MRN-WPT, MRN-KPCA, MRN-WPT-KPCA are 0.49 ± 0.37, 0.53 ± 0.36, 0.46 ± 0.38, and 0.32 ± 0.28, respectively (unit in permillage). The number of folding decreases greatly in MRN-WPT-KPCA, indicting our proposed method can yield reliable deformations, as less folding means better morphologically allowable fields. Moreover, the median of folding in MRN-WPT-KPCA is only 0.2‰, which is significantly smaller than other methods (all > 0.35‰).

One of the drawbacks of the comparison is that as the statistical model losses are only applied in the lowest resolution, the differences might be overwhelmed by the processing at the higher two levels. Additionally, we believe that the statistical losses may help improve the robustness of the registration networks and plan to validate and test on atlas-based applications on bigger datasets with extensive data augmentation.

4   Conclusion

We proposed an MRN using simple convolution layers at each resolution level by mimicking the ideas of well-accepted traditional image registration algorithms. Simple CNNs were used to solve the deformations at each resolution and gradually refine the deformation level-by-level. MRN allows for image features in different resolutions be directly fed into the network, and WPT images are naturally fit into the structure. Additionally, for registering with template image, the prior knowledge of deformations at the lowest resolution is modeled by kernel-PCA, which can be easily embedded into the network training. Experiments using brain MR images showed the advantages of MRN compared with different settings, *i.e.*, w/wo wavelet features, w/wo prior knowledge losses. The results showed better performance for the multiresolution representation and prior knowledge learning.

Acknowledgement. This work was partially supported by the National Key Research and Development Program of China (2018YFC0116400) and the National Natural Science Foundation of China (62071176).

References

1. Balakrishnan, G., Zhao, A., Sabuncu, M.R., Guttag, J., Dalca, A.V.: VoxelMorph: a learning framework for deformable medical image registration. IEEE Trans. Med. Imaging **38**, 1788–1800 (2019)
2. Qin, C., Shi, B., Liao, R., Mansi, T., Rueckert, D., Kamen, A.: Unsupervised deformable registration for multi-modal images via disentangled representations. In: Chung, A.C.S., Gee, J.C., Yushkevich, P.A., Bao, S. (eds.) IPMI 2019. LNCS, vol. 11492, pp. 249–261. Springer, Cham (2019). https://doi.org/10.1007/978-3-030-20351-1_19
3. Zhao, S., et al.: Unsupervised 3D end-to-end medical image registration with volume tweening network. IEEE J. Biomed. Health Inform. **24**, 1394–1404 (2019)
4. Zhao, S., Dong, Y., Chang, E.I., Xu, Y.: Recursive cascaded networks for unsupervised medical image registration. In: International Conference on Computer Vision (ICCV), 2019, pp. 10600–10610. IEEE (2019)
5. de Vos, B.D., Berendsen, F.F., Viergever, M.A., Sokooti, H., Staring, M., Išgum, I.: A deep learning framework for unsupervised affine and deformable image registration. Med. Image Anal. **52**, 128–143 (2019)
6. Ilg, E., Mayer, N., Saikia, T., Keuper, M., Dosovitskiy, A., Brox, T.: Flownet 2.0: evolution of optical flow estimation with deep networks. In: Proceedings of the IEEE Conference on Computer Vision and Pattern Recognition (CVPR), 2017, pp. 2462–2470. IEEE (2017)
7. Li, X., et al.: Semantic flow for fast and accurate scene parsing. In: Vedaldi, A., Bischof, H., Brox, T., Frahm, J.-M. (eds.) ECCV 2020. LNCS, vol. 12346, pp. 775–793. Springer, Cham (2020). https://doi.org/10.1007/978-3-030-58452-8_45
8. Xue, Z., Shen, D., Davatzikos, C.: Statistical representation of high-dimensional deformation fields with application to statistically constrained 3D warping. Med. Image Anal. **10**, 740–751 (2006)
9. Rosipal, R., Girolami, M., Trejo, L.J., Cichocki, A.: Applications: kernel PCA for feature extraction and de-noising in nonlinear regression. Neural Comput. Appl. **10**, 231–243 (2001)
10. Avants, B.B., Epstein, C.L., Grossman, M., Gee, J.C.: Symmetric diffeomorphic image registration with cross-correlation: evaluating automated labeling of elderly and neurodegenerative brain. Med. Image Anal. **12**, 26–41 (2008)
11. Mueller, S.G., et al.: Ways toward an early diagnosis in Alzheimer's disease: the Alzheimer's disease neuroimaging initiative (ADNI). Alzheimer's Dement. **1**, 55–66 (2005)
12. Smith, S.M.: Fast robust automated brain extraction. Hum. Brain Mapp. **17**, 143–155 (2002)
13. Gu, D., et al.: Pair-wise and group-wise deformation consistency in deep registration network. In: International Conference on Medical Image Computing and Computer-Assisted Intervention, pp. 171–180. Springer, Cham (2020) https://doi.org/10.1007/978-3-030-59716-0_17

Learning to Synthesize 7 T MRI from 3 T MRI with Few Data by Deformable Augmentation

Jie Wei[1,2,3], Yongsheng Pan[1,2,3], Yong Xia[1,2(✉)], and Dinggang Shen[3,4(✉)]

[1] National Engineering Laboratory for Integrated Aero-Space-Ground-Ocean Big Data Application Technology, School of Computer Science and Engineering, Northwestern Polytechnical University, Xi'an 710072, Shaanxi, China
yxia@nwpu.edu.cn

[2] Research and Development Institute of Northwestern Polytechnical University in Shenzhen, Shenzhen 518057, China

[3] School of Biomedical Engineering, ShanghaiTech University, Shanghai, China
dgshen@shanghaitech.edu.cn

[4] Shanghai United Imaging Intelligence Co., Ltd., Shanghai, China

Abstract. High-quality magnetic resonance imaging (MRI), which is generally acquired by ultra-high field (7-Tesla, 7 T) MRI scanners, may lead to improved performance for brain disease diagnosis, such as Alzheimer's disease (AD). However, 7 T MRI has not been widely used due to higher cost and longer scanning time. To overcome this, we proposed to utilize the generative adversarial networks (GAN)-based techniques to synthesize the 7 T scans from 3 T scans, for which, the most challenge is that we do not have enough data to learn a reliable mapping from 3 T to 7 T. To address this, we further proposed the Unlimited Data Augmentation (UDA) strategy to increase the learning samples via the deformable registration, which can produce enough paired 3 T and 7 T MR images to learning this mapping. Based on this mapping, we synthesize a 7 T MR scan for each subject in Alzheimer's Disease Neuroimaging Initiative (ADNI), and conduct some experiments to evaluate their effect in two tasks of AD diagnosis, including AD identification and mild cognitive impairment (MCI) conversion prediction. Experimental results demonstrate that our UDA strategy is effective to learn a reliable mapping to high-quality MR images, and the synthetic 7 T scans are possible to increase the performance of AD diagnosis.

1 Introduction

Alzheimer's disease (AD) is a neurodegenerative disorder that could accelerate the patient's cognitive loss and lead to dementia [10]. Previous studies have verified that magnetic resonance imaging (MRI) is a relevant important technique in

This work is mainly completed under the collaboration of J. Wei and Y. Pan. J. Wei and Y. Pan contribute equally.

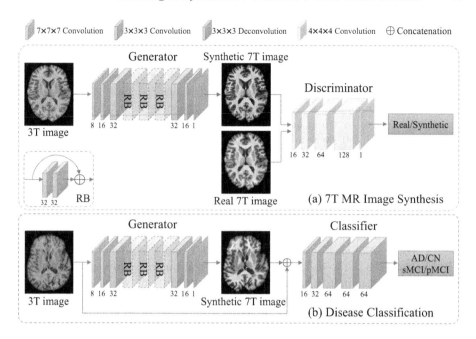

Fig. 1. Framework of our proposed method consisting of two stages, i.e., (a) 7 T MR image synthesis, and (b) disease classification.

AD diagnosis, e.g., identifying AD individuals and predicting the progress of mild cognitive impairment (MCI) [16]. Generally, higher-quality MR scans, which are recognized with a higher signal-to-noise ratio (SNR) and can be acquired with higher magnetic field strength, are possible to result in a more accurate diagnosis of AD. Comparing to 1.5-Tesla (1.5 T) and 3 T MRI, the ultra-high field (7 T) MRI has evident clinical advantages, including higher SNR, better contrast, higher sensitivity to provide detail of anatomy and pathology of the brain, and easier to detect subtle changes [9,11]. However, in addition to producing quality images, 7 T MRI also brings higher costs and longer scanning time.

Currently, the most widely used MRI is 1.5 T and 3 T scanners, of which a lot of scans have been collected, such as Alzheimer's Disease Neuroimaging Initiative (ADNI) [6]. The low quality may lead to an upper bound of analysis. If these scans can be converted to 7 T MR scans, it is possible to boost the diagnosis performance. This produces a specific scenario of image-to-image translation, for which various techniques have been studied, particularly those techniques based on generative adversarial networks (GAN) [3,5,7,19]. Accordingly, it is potential to learn a mapping from 1.5 T or 3 T MR images to 7 T images based on various GANs, e.g., cycle-consistency GAN (cycGAN) [3], pixel2pixel GAN (p2pGAN) [5], L1GAN [2], feature matching GAN (FMGAN) [17], or sense-consistency GAN (SGAN) [14] with various constraints. As 3 T MRI synthesized from 1.5T MRI can help to improve the performance of AD status prediction [18], synthesizing 7 T MRI from 1.5 T/3 T MRI may further promote the performance.

However, learning image-to-image mappings generally requires abundant data collection. It makes the task of synthesizing 7 T MR scans very difficult because 7 T MRI scanners have not been widely applied and there is no public dataset with accessible 7 T MR scans. Even a previous study [15] collected 15 in-house subjects with both 3 T and 7 T MR scans, it is still not enough to learn a reliable mapping. As diagnosis models generally align images based on image appearances by using image registration methods in their pre-processing pipeline, conventional data augmentation such as up-down flip, left-right flip may not be applicable to brain images.

Based on previous studies, we found that the most variability among different images is the regional characteristic, to address which, a deformable augmentation may be effective to enlarge the learning dataset. Accordingly, we propose an effective strategy to do deformable augmentation, so that to increase training samples to learning reliable mappings from 3 T scans to 7 T scans. We register the 3 T MR scan of each subject with paired 3 T and 7 T scans to a target 3 T MRI in the ADNI dataset via deformable registration and apply the deformable field to the corresponding healthy 7 T MR scan. In this way, we can obtain enough subjects with paired 3 T and 7 T MR images.

In this paper, we first design various GAN-based models to learning the mapping from 3 T to 7 T after doing deformable augmentation. Experimental results demonstrate that all of them are promoted by our deformable augmentation strategy, where SGAN achieves the best synthesis performance. Then, we use the mapping learned by SGAN to synthesis a 7 T scan for each 3 T/1.5 T scan in the ADNI dataset, i.e., each subject in ADNI now has a pair of original 3 T/1.5 T scan and a synthetic 7 T scan. Finally, each pair of the original 3 T/1.5 T image and the synthetic 7 T image are combined as a two-channel 3D image and fed into classifiers. Experimental results demonstrate that our synthetic 7 T scans are effective to improve the performance in AD diagnosis.

2   Method

2.1   Dataset

There are two datasets used in this study. The first is an in-house dataset with 15 subjects, each of which has a pair of 3 T and 7 T T1-weighted MR scans. All 3 T and 7 T scans were acquired with Siemens Magnetom Trio 3 T and 7 T MRI scanners, respectively. Specifically, each 3 T scan consists of 224 coronal slices acquired with the 3D magnetization-prepared rapid gradient-echo (MP-RAGE) sequence. The imaging parameters of 3D MP-RAGE sequence were: repetition time (TR) = 1900 ms, echo time (TE) = 2.16 ms, inversion time (TI) = 900 ms, flip angle (FA) = $9°$, and voxel size = $1.0 \times 1.0 \times 1.0$ mm^3. Each 7 T scan consists of 191 sagittal slices acquired with the 3D MP2-RAGE sequence. The imaging parameters of 3D MP2-RAGE sequence were: TR = 6000 ms, TE = 2.95 ms, TI = 800/2700 ms, FA =$4°/4°$, and voxel size = $0.65 \times 0.65 \times 0.65$ mm^3. Each 3T scan was linearly aligned to the MNI standard space with the voxel size of $1.0 \times 1.0 \times 1.0$ mm^3. The corresponding 7 T scan was then aligned to the 3 T scan in the MNI space by using FLIRT. After bias field correction, skull removal, and intensity normalization, the intensity values of the 3 T and 7 T images were rescaled to $[-1, 1]$.

The second dataset is the Alzheimer's Disease Neuroimaging Initiative database (ADNI) [6], where we use the baseline T1 scans of its two subsets, i.e., ADNI-1 and ADNI-2. Subjects in ADNI can be divide into (1) AD patients, (2) CN patients, (3) pMCI patients that would progress to AD within 36 months after baseline, and (4) sMCI patients that would not progress to AD. After removing subjects that exist in both ADNI-1 and ADNI-2 from ADNI-2, there are 200 AD, 231 CN, 171 pMCI, and 150 sMCI subjects in ADNI-1, while there are 165 AD, 209 CN, 89 pMCI, and 256 sMCI subjects in ADNI-2. We apply a similar pre-processing pipeline on MR scans in ADNI. After pre-processing, these scans also have the voxel size of $1.0 \times 1.0 \times 1.0$ mm^3.

2.2   Framework

We propose a two-stage deep learning framework to promote the image quality of T1 scans and boost the performance of AD diagnosis as illustrated in Fig. 1. In the *first* stage, we learn a mapping from 3 T scans to 7 T scans via GAN-based techniques, which is used to promote the quality of scans in ADNI. Especially, to address the small training data problem, we propose the Unlimited Data Augmentation (UDA) with deformable registration. In the *second* stage, based on the synthesized 7 T T1 scans, we develop a deep learning method for AD diagnosis, by learning 3T and 7 T features automatically in a data-driven manner. To the best of our knowledge, this is the first attempt to promote AD diagnosis by mapping common-used 3 T MRI to high-quality 7 T MRI.

2.3   Unlimited Data Augmentation

As we only have a few paired 3 T and 7 T scans, which is not enough to learn a reliable generative model to synthesize 7 T scans, we propose the Unlimited Data Augmentation (UDA) to produce more paired 3 T and 7 T MR scans via the deformable registration [1]. The motivation is that the deformable registration can find a non-linear transformation (i.e., deformation field) to establish anatomical correspondences between each two images, and each group of images still keeps the spatial relationships after being applied with the same deformation field. Based on this, we can produce hundreds of paired training samples with potentially large and complex variety.

As our goal is to find a mapping function from 3 T to 7 T MR scans for the ADNI subjects, we randomly select several scans from ADNI as the fixed templates. We first perform deformable registration to each 3 T MR scan of the in-house dataset and each selected template to obtain a deformation field, and then apply the deformation field to this pair of 3 T and 7 T in-house scans. For example, if we select 30 scans from ADNI as templates, then we can produce 15×30 samples in total, which may be enough to train a reliable mapping.

2.4   Synthesizing 7 T MRI with GAN

In the *first* stage of our framework, we attempt to learn a mapping from 3 T to 7 T MR scans, which is conducted by the GAN-based techniques. We follow the structures in [13], to create our GAN structures, which contains a generator and a discriminator.

The generator is an encoder-decoder network, where the encoder consists of three 3D convolutional layers with strides of $\{1, 2, \text{and } 2\}$ and channels of $\{8, 16, \text{and } 32\}$, respectively, while the decoder consists of two 3D deconvolutional layers (with 32 and 16 channels, respectively) and a 3D convolutional layer with 1 channel. Between the encoder and decoder, we also insert 3 residual blocks (RB), each of which contains two convolutional layers with a shortcut connection. The kernel sizes of the first and last convolutional layers are set to $7 \times 7 \times 7$, while which of the other layers are set to $3 \times 3 \times 3$. All convolutional/deconvolutional layers except the last one are followed by the batch normalization and rectified linear unit (ReLu) activation while the last layer uses the "tanh" activation.

The discriminator is a fully convolutional network with five 3D convolutional layers with the kernel size of $4 \times 4 \times 4$. The channels of these layers are 16, 32, 64, 128, and 1, respectively, where the strides are 2, 2, 2, 2, and 1, respectively. The former four layers are followed by batch normalization and ReLu activation.

During the training phase, the generator attempts to create the mapping to approximate the difference between 3 T and 7 T MR images while the discriminator outputs a binary indicator to distinguish the original and the synthetic 7 T MR images. Besides, with different constraints, the generator and discriminator can form different GANs, such as the basic GAN with only the adversarial loss [3], the cycle-consistency GAN (cycGAN) with the adversarial loss and cycle-consistency loss [19], the pixel-to-pixel GAN (p2pGAN) with the adversarial loss and the mean absolute error (MAE) loss [5], the 3D encoder-decoder with only MAE loss (L1GAN) [2], the feature matching GAN (FMGAN) with the adversarial loss and the feature matching loss [17] and the sense-consistency GAN (SGAN) with the MAE loss and the feature matching loss [14]. Specifically, to adopt the potential location shift, we apply patch-wise training by randomly sampling patches of size $192 \times 192 \times 192$ from the whole volume at the center of each brain with a maximum random shift of $4 \times 4 \times 4$ as the inputs. It should be noted that all these patches can cover the whole brain regions of all subjects. When synthesizing a target 7 T scan during the test phase, we randomly select 50 patches from each 3 T scan, feed them into the generator to obtain 50 synthetic patches, and calculate the average of these synthetic patches as the synthetic 7 T MR scan.

2.5   Disease Classification

After synthesizing 7 T scans for all subjects in ADNI, we further involve these 7 T scans in the classification tasks. Therefore, at the *second* stage, we built a classification model using both original 3 T and synthetic 7 T MR images to identify AD subjects (AD vs. NC classification) and predict MCI progression (pMCI vs. sMCI classification).

Table 1. Results (% except PSNR) of image synthesis achieved by different methods for 3 T scans of subjects in in-house data.

Method	MAE			PSNR			SSIM		
	#1	#10	#20	#1	#10	#20	#1	#10	#20
GAN	49.33	21.14	19.81	16.73	22.37	23.27	39.70	51.11	59.86
cycGAN	9.53	6.81	5.96	27.10	28.89	29.96	74.01	79.84	82.42
p2pGAN	7.14	5.80	5.25	29.03	29.96	30.39	78.74	83.48	84.58
L1GAN	7.18	5.12	4.89	29.35	30.72	31.02	82.32	84.58	85.87
FMGAN	7.00	5.81	5.77	29.18	30.11	30.62	81.47	84.19	84.47
SGAN	**6.55**	**4.81**	**4.53**	**29.79**	**30.91**	**31.40**	**83.90**	**85.02**	**86.29**
	#30	#40	#50	#30	#40	#50	#30	#40	#50
SGAN	4.15	4.08	4.11	31.86	31.96	32.03	87.45	87.51	87.69

The classification model employs the same structure as the disease-image specific neural network (DSNN) [12], which consists of 5 convolutional layers with channels of 16, 32, 64, 64, and 64, respectively. Each layer has the kernel size of $3 \times 3 \times 3$ and stride of 1 and is followed by batch normalization and ReLu activation. Each of the first four convolutional layers is followed by a max-pooling layer to downsample the output with the stride of 2. The last layer is followed by a spatial cosine kernel to utilize the spatial information. While using both original 3 T and synthetic 7 T MR scans as inputs, each pair of 3 T and 7 T images are concatenated along the channel dimension to form a two-channel volume. We also use the same patch-wise training for the DSNN with the patch size of $160 \times 176 \times 160$ and a maximum random shift of $2 \times 2 \times 2$. While testing, we only select the patches at the center of the brain from the 3 T and 7 T scans.

2.6   Implement Details

We trained all networks on a platform with an NVIDIA TITAN Xp GPU (32 GB). For the 7 T MR image synthesis stage, we adopted the Adam optimizer and set the batch size to 1, maximum epoch number to 500, and initial learning rate to 0.001, which was divided by 10 after 100 iterations. For the disease classification stage, we adopted the stochastic gradient descent (SGD) optimizer and set the batch size to 6, maximum epoch number to 300, and initial learning rate to 0.001, which was divided by 10 after 50 iterations. And we apply 3-fold cross-validation on 7 T MR image synthesis and train the model three times for disease classification.

3   Experiments and Results

3.1   Evaluation of Image Synthesis

We evaluate the performance of different 7 T MR image synthesis methods with three commonly used image quality metrics, including (1) the mean absolute

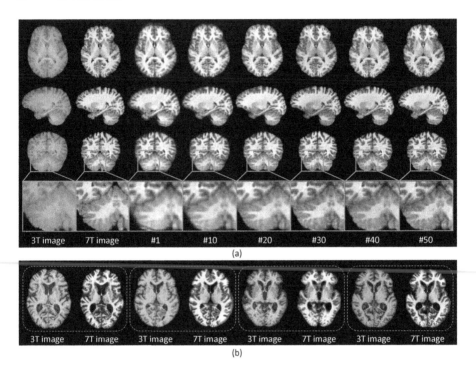

Fig. 2. (a) Comparison of typical synthesis results of SGAN with different expanded factors on one subject from in-house data. (b) The synthesis results of SGAN with the expanded factor of 30 on four subjects from ADNI. The X of $\#X$ is the expanded factor.

error (MAE), (2) the peak signal-to-noise ratio (PSNR), and (3) the structural similarity index measure (SSIM) [4]. Table 1 shows the MAE, PSNR, and SSIM values computed based on the predicted images from different methods, including the basic GAN, cycGAN, p2pGAN, L1GAN, FMGAN, and SGAN, all of which are implemented using the same architecture and details for a fair comparison. The X of $\#X$ is the number of selected fixed templates, which means that the training data was expanded by a factor of X. It can be observed that the performances of all models on all metrics are improved with the increase of the expanded factor. This can be attributed to our effective UDA performed by the deformable registration. Specifically, among these compared models, SGAN achieves the best performance in terms of all three metrics. Moreover, we further evaluate the influence of the expand factor for SGAN by increasing the selected templates. As shown in the last row of Table 1, the benefits from UDA become less and less while the time costs are increasing linearly. Considering the trade-off between performance and efficiency, we think the expanded factor of 30 is the best choice. Therefore, we use SGAN with the expanded factor of 30 to synthesize 7 T MR scans for our second-stage study.

In Fig. 2(a), we further visualize an example of our in-house dataset to display the synthetic 7 T scans of SGAN over expanded factors. From top to bottom are, respectively, the central slice in axial, sagittal, coronal, and enlarged coronal views of the original 3 T and 7 T MR images and the synthetic 7 T MR images. It can be seen that the 7 T MR scan has better contrast than the 3 T scan, and the quality of synthetic images is improving along with the expanded factors. Figure 2(b) gives four examples from ADNI to show the original 3 T and synthetic 7 T scans obtained by SGAN with the expanded factor of 30.

Table 2. Diagnosis results (%) achieved by different methods for ADNI.

Method	AUC	ACC	SEN	SPE	F1S	MCC
	AD vs. CN					
DSNN (3 T)	93.02 ± 0.93	85.62 ± 1.04	78.52 ± 1.95	93.16 ± 1.78	85.90 ± 1.37	70.23 ± 2.12
DSNN (7 T)	91.22 ± 0.91	85.56 ± 0.89	77.42 ± 0.65	94.63 ± 1.46	83.82 ± 1.16	73.37 ± 1.67
DM-DSNN	**94.74 ± 0.29**	**87.40 ± 0.83**	**78.64 ± 3.47**	**95.23 ± 1.08**	**86.07 ± 0.96**	**74.82 ± 1.28**
	pMCI vs. sMCI					
DSNN (3 T)	82.60 ± 0.42	77.58 ± 0.60	65.17 ± 1.83	81.90 ± 0.66	72.21 ± 0.77	44.84 ± 1.58
DSNN (7 T)	80.05 ± 0.67	77.39 ± 0.47	57.12 ± 1.15	84.11 ± 0.66	70.61 ± 0.71	41.23 ± 1.41
DM-DSNN	**83.25 ± 0.30**	**78.61 ± 0.25**	**66.97 ± 1.41**	**84.57 ± 0.39**	**73.55 ± 0.22**	**47.10 ± 0.42**

3.2 Evaluation of Disease Classification

We further evaluated the synthetic 7 T MR scans in both tasks of AD identification (AD vs. CN) and MCI conversion prediction (pMCI vs. sMCI). We use the DSNN in [12] as our classification method, and report both the results that using only 3 T or 7 T modality and using both 3T and 7 T modalities for comparison. Six metrics are computed (as shown in Table 2) to evaluate the classification results, including the area under the receiver operating characteristic (AUC), accuracy (ACC), sensitivity (SEN), specificity (SPE), F1-Score (F1S), and Matthews correlation coefficient (MCC) [8]. Subjects from ADNI-1 and ADNI-2 were used for training and testing the models, respectively.

From Table 2, we can find that the classification results of DM-DSNN that use both original 3 T and synthetic 7 T images outperform each of the single-modality DSNN that use only original 3 T or synthetic 7 T images. It implies that our synthetic 7 T MR images could help to improve the classification performance across external datasets. For instance, the DM-DSNN achieves the best AUC values in both AD vs. CN classification (94.74%), and pMCI vs. sMCI classification (83.25%), while DSNN achieves 93.02%/82.60% while using 3 T scans and 91.22%/80.05% while using synthetic 7 T scans. Besides, while using only a single modality, using our synthetic 7 T images perform comparable but slightly less well than using original 3 T images. It reveals that our synthetic 7 T images can not take place of the real 7 T images even it is useful in disease diagnosis. Nevertheless, we achieve better performance than DSNN, which demonstrates that our two-stage framework is effective to boost diagnosis performance with only T1 MR images.

4 Conclusion

We aim to improve the quality of MRI by learning a mapping from 3T scans to 7 T scans, for which the limited data is the most challenging difficulty. To address this, we propose the Unlimited Data Augmentation (UDA) strategy to increase the training samples by deformable registration. Experiments suggest that this strategy can boost the performance of various GAN-based synthesis techniques, and the performance of the AD diagnosis tasks, i.e., identifying AD subjects and predicting MCI conversion, can be further improved with our synthetic 7 T scans.

Acknowledgment. This work was supported partly by the National Natural Science Foundation of China under Grants 61771397, partly by the CAAI-Huawei MindSpore Open Fund under Grants CAAIXSJLJJ-2020-005B, and partly by the China Postdoctoral Science Foundation under Grants BX2021333.

References

1. Avants, B.B., Epstein, C.L., Grossman, M., Gee, J.C.: Symmetric diffeomorphic image registration with cross-correlation: evaluating automated labeling of elderly and neurodegenerative brain. Med. Image Anal. **12**(1), 26–41 (2008)
2. Cohen, J.P., Luck, M., Honari, S.: Distribution matching losses can hallucinate features in medical image translation. In: Frangi, A.F., Schnabel, J.A., Davatzikos, C., Alberola-López, C., Fichtinger, G. (eds.) MICCAI 2018. LNCS, vol. 11070, pp. 529–536. Springer, Cham (2018). https://doi.org/10.1007/978-3-030-00928-1_60
3. Goodfellow, I.J., Pouget-Abadie, J., Mirza, M., Xu, B., Warde-Farley, D., Ozair, S.: Generative adversarial networks. In: Advances in Neural Information Processing Systems (NIPS), pp. 2672–2680 (2014)
4. Horé, A., Ziou, D.: Image quality metrics: PSNR vs. SSIM. In: 20th International Conference on Pattern Recognition (ICPR) (2010)
5. Isola, P., Zhu, J.Y., Zhou, T., Efros, A.A.: Image-to-image translation with conditional adversarial networks. In: IEEE Conference on Computer Vision and Pattern Recognition (CVPR), pp. 5967–5976 (2017)
6. Jack, C.R., Bernstein, M.A., Fox, N.C., Thompson, P., Weiner, M.W.: The Alzheimer's disease neuroimaging initiative (ADNI): MRI methods. J. Magn. Reson. Imaging **27**(4), 685–691 (2010)
7. Jiang, J., et al.: PSIGAN: joint probabilistic segmentation and image distribution matching for unpaired cross-modality adaptation-based MRI segmentation. IEEE Trans. Med. Imaging **39**(12), 4071–4084 (2020)
8. Koyejo, O.O., Natarajan, N., Ravikumar, P.K., Dhillon, I.S.: Consistent binary classification with generalized performance metrics. In: Advances in Neural Information Processing Systems (NIPS), pp. 2744–2752 (2014)
9. Lian, C., Zhang, J., Liu, M., Zong, X., Hung, S.C., Lin, W., Shen, D.: Multi-channel multi-scale fully convolutional network for 3D perivascular spaces segmentation in 7T MR images. Med. Image Anal. **46**, 106–117 (2018)
10. Long, J.M., Holtzman, D.M.: Alzheimer disease: an update on pathobiology and treatment strategies. Cell **179**(2) (2019)
11. Obusez, E.C., et al.: 7T MR of intracranial pathology: preliminary observations and comparisons to 3T and 1.5T. Neuroimage **168**, 459–476 (2018)

12. Pan, Y., Liu, M., Lian, C., Xia, Y., Shen, D.: Disease-image specific generative adversarial network for brain disease diagnosis with incomplete multi-modal neuroimages. In: Shen, D., et al. (eds.) MICCAI 2019. LNCS, vol. 11766, pp. 137–145. Springer, Cham (2019). https://doi.org/10.1007/978-3-030-32248-9_16
13. Pan, Y., Liu, M., Lian, C., Xia, Y., Shen, D.: Spatially-constrained fisher representation for brain disease identification with incomplete multi-modal neuroimages. IEEE Trans. Med. Imaging **39**(9), 2965–2975 (2020)
14. Pan, Y., Xia, Y.: Ultimate reconstruction: understand your bones from orthogonal views. In: 2021 IEEE 18th International Symposium on Biomedical Imaging (ISBI), pp. 1155–1158 (2021)
15. Qu, L., Zhang, Y., Wang, S., Yap, P.T., Shen, D.: Synthesized 7T MRI from 3T MRI via deep learning in spatial and wavelet domains. Med. Image Anal. **62**, 101663 (2020)
16. Sperling, R., Mormino, E., Johnson, K.: The evolution of preclinical Alzheimer's disease: implications for prevention trials. Neuron **84**(3), 608–622 (2014)
17. Wang, T.C., Liu, M.Y., Zhu, J.Y., Tao, A., Kautz, J., Catanzaro, B.: High-resolution image synthesis and semantic manipulation with conditional GANs. In: 2018 IEEE/CVF Conference on Computer Vision and Pattern Recognition, pp. 8798–8807 (2018)
18. Zhou, X., Qiu, S., Joshi, P.S., Xue, C., Kolachalama, V.B.: Enhancing magnetic resonance imaging-driven Alzheimer's disease classification performance using generative adversarial learning. Alzheimer's Res. Ther. **13**(1) (2021)
19. Zhu, J.Y., Park, T., Isola, P., Efros, A.A.: Unpaired image-to-image translation using cycle-consistent adversarial networks. In: IEEE International Conference on Computer Vision (ICCV), pp. 2242–2251 (2017)

Rethinking Pulmonary Nodule Detection in Multi-view 3D CT Point Cloud Representation

Jingya Liu[1], Oguz Akin[2], and Yingli Tian[1(✉)]

[1] The City College of New York, New York, NY 10031, USA
ytian@ccny.cuny.edu
[2] Memorial Sloan Kettering Cancer Center, New York, NY 10065, USA

Abstract. 3D CT point clouds reconstructed from the original CT images are naturally represented in real-world coordinates. Compared with CT images, 3D CT point clouds contain invariant geometric features with irregular spatial distributions from multiple viewpoints. This paper rethinks pulmonary nodule detection in CT point cloud representations. We first extract the multi-view features from a sparse convolutional (SparseConv) encoder by rotating the point clouds with different angles in the world coordinate. Then, to simultaneously learn the discriminative and robust spatial features from various viewpoints, a nodule proposal optimization schema is proposed to obtain coarse nodule regions by aggregating consistent nodule proposals prediction from multi-view features. Last, the multi-level features and semantic segmentation features extracted from a SparseConv decoder are concatenated with multi-view features for final nodule region regression. Experiments on the benchmark dataset (LUNA16) demonstrate the feasibility of applying CT point clouds in lung nodule detection task. Furthermore, we observe that by combining multi-view predictions, the performance of the proposed framework is greatly improved compared to single-view, while the interior texture features of nodules from images are more suitable for detecting nodules in small sizes.

Keywords: 3D point cloud · Nodule detection · Multi-view feature representation

1 Introduction

Lung cancer is the leading cancer killer and one of the most common cancers around the world [26]. Many efforts have been made on AI-driven computer-aided diagnosis (CAD) systems for lung nodule detection and diagnosis [17,18,28,31]. Most existing 3D nodule detectors [11,31] extract spatial-temporal features from the stacked CT images as 3D volumes via 3D convolutional neural networks (3D CNNs). By extending a two-dimensional feature representation to a three-dimensional space, the significant performance improvement demonstrates the

© Springer Nature Switzerland AG 2021
C. Lian et al. (Eds.): MLMI 2021, LNCS 12966, pp. 80–90, 2021.
https://doi.org/10.1007/978-3-030-87589-3_9

benefits of interpreting the spatial features for nodule detection. Meanwhile, by projecting CT images to multiple planes (i.e., axial, coronal, and sagittal), the multi-view features show the effectiveness to reduce false positives as richer features are obtained from different viewpoints [9,16,22,23]. Although these existing methods can extract spatial-temporal features or multi-view features from the continuous CT images, the data representation with multi-view flexibility that can directly represent 3D spatial features is still remain exploring.

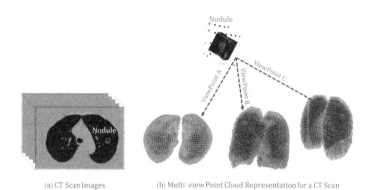

(a) CT Scan Images   (b) Multi-view Point Cloud Representation for a CT Scan

Fig. 1. Illustration of a CT scan in the image (a) and CT point cloud visualized in different views (b). Compared to CT images, as shown in (b), 3D CT point clouds naturally represent pulmonary nodules by multi-view representations in 3D coordinates. View A in (b) has the same viewpoint as the CT image. Note that the background of the lung region is excluded for both CT images and CT point clouds. Best viewed in color.)

Table 1. Comparison between CT image representation and 3D CT point cloud representations.

Visual representation	Geometric invariance	Multi-view perspective	Real-world coordinates	Irregular shape	Grid align	Texture feature	Radiation intensity
Image	No	No	No	No	Yes	Yes	Yes
3D point cloud	Yes	Yes	Yes	Yes	No	No	Yes

Recently, 3D point clouds are considered to be a natural way to capture and represent 3D objects [2,4,12,20,21]. As shown in Fig. 1, unlike CT images (Fig. 1(a)) represented by grids, 3D CT point clouds (Fig. 1(b)) are organized in an irregular manner, thereby provides more flexible feature representations of 3D geometric structure. We summarize the characters of visual representations of 3D CT point clouds and CT images in Table 1. Real-world coordinates are applied to 3D CT point clouds, representing the accurate geometric distribution of radiation intensity. Due to geometric invariance, multi-view representations can be obtained by tilting the 3D CT point cloud heading angles to obtain richer features than the single-view representation. However, the scattered 3D CT point cloud lacks the texture features as CT images that preserve the details of internal nodules. Therefore, this paper attempts to rethink the nodule detection task from multiple perspectives in CT point clouds and investigate the advantages and limitations compared to CT images.

Existing models that apply 3D point clouds to medical image analysis mainly in small areas, such as intra-oral scans [30] and cropped nodule candidate regions [8]. We aim to address the challenging task of predicting nodule regions in the entire lung area. Per our knowledge, this is the first work to explore 3D point cloud data representation for the nodule detection task. The main contributions of this paper are summarized as follows: 1) We explore the possibility of applying 3D CT point cloud and the effectiveness of 3D spatial representations on nodule detection. By rotating the 3D CT point clouds, compared to a single view of stacked CT images, irregular shapes and richer spatial distributions of nodules in lung regions are extracted from multi-views. 2) A nodule proposal optimization schema is proposed to acquire a set of 3D nodule candidate proposals which simultaneously agreed by multi-view feature predictions. The rich multi-level features and point-wise segmentation are further concatenated with multi-view features for final nodule candidate refinement. 3) Experiment results show the advantages of multi-view 3D CT point clouds as spatial representations and the impact of lacking the texture feature representations on small nodule detection. Furthermore, the ablation studies demonstrate that with more multi-views involved, the performance of the proposed framework achieves encouraging improvements, proofing the effectiveness of multi-view 3D features in point clouds.

2   Methods

As shown in Fig. 2, N samples are acquired by combining the original 3D CT point clouds with the rotated multi-view samples, representing irregular geometric distributions from various perspectives. A 3D Sparse Convolution (SparseConv) encoder simultaneously learns the multi-view features by optimizing the nodule proposals in various viewpoints. Last, a final nodule prediction is refined by fusing the multi-level features and semantic segmentation features extracted by the 3D SpaseConv decoder, as well as the multi-view features.

Multi-view Sample Collection: 3D CT point clouds consist of a set of three-dimensional vertices, defined as $\{x, y, z\}$ representing the distribution of points in a world coordinate. Besides, the unique internal texture with the radiation intensities is essential to identify nodule types. We further apply the radiant intensity I of each point to preserve internal intensity-based texture features. The nodule region is represented by (x, y, z, h, w, l), indicating the nodule location $\{x, y, z\}$ and the nodule size in three dimensions $\{h, w, l\}$. During the training, the objective function consists of the BinaryCrossEntropy loss for nodule class probability and $smooth - l1$ loss for nodule region by comparing the predicted nodule candidates and the ground-truths.

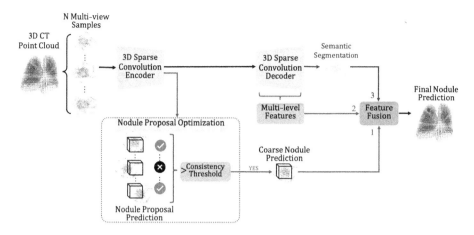

Fig. 2. The pipeline of our proposed 3D point cloud nodule detection framework. Multi-view feature representations are extracted from a 3D sparse encoder-decoder network and further employed to select the coarse nodule predictions by a nodule proposal optimization (details are illustrated in Fig. 3 and explained in Sect. 3 - Nodule Proposal Optimization). Then, the finer candidate prediction is obtained based on the fusion of 1) multi-view features; 2) multi-level features; 3) semantic features of the region of interest (ROI).

By rotating the 3D point clouds along the world coordinate in N angles, the discriminate multi-view features of the same object can be extracted. Therefore, to train a backbone network with robust spatial features from multiple perspectives, we rotate the 3D CT point clouds along the xy-axis, yz-axis, xz-axis, and xyz-axis with a set of rotation angles from $+\theta$ to $-\theta$. As shown in Fig. 3, four sets of $(N-1)$ viewpoints are added, donated as V-xy, V-yz, V-xz, and V-xyz. We implement nine viewpoint samples with the rotation angles of $+45°$ and $-45°$ for V-xy, V-yz, V-xz, and V-xyz. The multi-view features are fed into the backbone network for nodule proposal prediction.

Multi-view and Multi-level Feature Extraction: The irregular shape of the 3D point cloud brings challenges to implement feature extraction with convolutional neural networks (CNN). We follow the baseline PV-RCNN [24] and the extension work, Part-A^2 [25] to extract features by integrating the grids of irregular point clouds as voxels and multi-level point-wise features from the decoder network. A 3D sparse convolution (SparseConv) encoder composed of four layers as es1, es2, es3, and es4. The features are downsampled to $1\times, 2\times, 4\times, 8\times$ as the layers go deeper. A 3D SparseConv decoder comprises four sparse deconvolutional layers as ds1, ds2, ds3, and ds4 for feature upsampling. The lateral connections are conducted through all layers, as the es1 with ds1, es2 with ds2, es3 with ds3, and es4 with ds4, extracting the multi-level point-wise features for final candidate prediction.

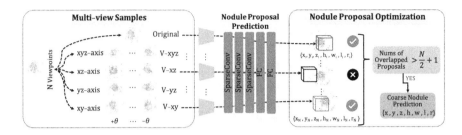

Fig. 3. The detailed illustration of the proposed multi-view samples and nodule prediction optimization schema. Multi-view samples are obtained by rotating the original samples from N viewpoints. By rotating from $+\theta$ to $-\theta$ along the xy-axis, yz-axis, xz-axis, and xyz-axis, $(N-1)$ additional viewpoints are obtained as V-xy, V-yz, V-xz and V-xyz. Features of N views are extracted from the 3D SparseConv encoder and the primary nodule proposals are predicted by a regression head. The coarse nodule predictions are collected if the number of consensus nodule proposals predicted by the multi-view features is greater than a consistency threshold.

Nodule Proposal Optimization: Unlike typical data augmentation methods, we employ the samples from all views to predict and optimize the nodule proposals simultaneously. Reliable preliminary nodule proposals are acquired from multi-views. In the meantime, compared with the ground truth, the 3D SparseConv encoder is updated by the objective function to extract multi-view features for accurate predictions. The distinguish and robust features ensure the prediction of the nodule regions. As shown in Fig. 3, a regression head (RH) is used for nodule proposal prediction, which consists of three SparseConv layers, followed by two fully connected layers to predict the confidence score and proposal region. RH takes the features of N views extracted from the last SparseConv layer of the encoder network and predicts the initial coarse nodule candidate regions. First, we choose a nodule proposal with a confidence score greater than 0.05 and then calculate the number of nodule proposals with a 3D IoU of 0.25 overlap among the multi-view predictions. If the number is higher than a consistency threshold, the nodule proposal is set as the coarse nodule prediction for a finer region regression.

Final Nodule Prediction with Multi-feature Fusion: To guarantee an accurate nodule candidate prediction, rich feature fusion is conducted by combining the multi-view features, multi-level features, and semantic segmentation features, as shown in Fig. 2. First, the point-wise attention benefits the flexible receptive field from the irregular format. We follow PV-RCNN [24], multi-scale features are extracted from ds1, ds2, ds3, and ds4 layers. Second, semantic segmentation maps are predicted by the 3D decoder network and semantic features further extracted by a SparseConv, interpreting the point level classification of nodule candidates. Finally, same as Part-A^2 [25], the multi-level features and the semantic features are concatenated for nodule region prediction along with the multi-view features. A regression head predicts the final nodule candidate region (x, y, z, h, w, l, r) based on the rich multi-feature fusion.

3 Experiments

3.1 Data and Training

We conduct the experiments on the LUNA16 public dataset [1], which contains $1,186$ nodules with sizes ranging from 3 to 30 mm from 888 CT scans. The dataset is officially divided into ten subsets. We follow the same cross-validation protocol by applying nine subsets as training and the remaining subset as testing. One subset from the training subsets is used for validation to monitor the convergence of the training process.

3D CT Point Cloud Reconstruction: A CT scan consists of CT images represented by (x, y) as image coordinates and z as scan depth. To eliminate unrelated regions, we first segment the lung region from the CT images based on the lung region mask and then apply (x, y, z) as the location of each point in the world coordinates, represented by $(x, y, z, radiation\ intensity)$. The air points in the lung regions are discarded based on the standard Hounsfield scale (-1000). Therefore, the CT point clouds precisely represent the nodules, tissues, organs, and lung boundaries in the lung region. Besides, to avoid sampling redundant and non-informative points in the training, we follow Drokin et al. [8] by applying the non-uniform sampling schema, which keeps a high sampling rate at the nodule regions and gradually reduces the sampling rate according to the increasing distance from sample points to the center of nodules.

Model Training: The experiment is modified based on an open source project OpenPCDet [27]. The proposed framework is trained to minimize BinaryCrossEntropy loss and $Smooth$-$L1$ loss. The Adam algorithm is used for optimization with an initial learning rate of $1e^{-4}$ and decreased by 0.1 after 80 and 160 epochs. The batch size is set to 2. The multi-view proposal optimization is conducted after 20 epochs. The model is trained with a total of 200 training epochs.

3.2 Experiments on LUNA16

Compare with State-of-the-Art Nodule Detection Methods: We evaluate the proposed method and compare it with the state-of-the-art image-based nodule detectors in two groups: one-stage methods without false-positive reduction and two-stage frameworks with false-positive reduction. Similar to the LUNA16 challenge, the Free-Response ROC Curve (FROC) [3] is employed to evaluate the sensitivity versus the specificity of the nodule detection at 1/8, 1/4, 1/2, 1, 2, 4, 8 False Positive (FP) levels per scan with the average Competition Performance Metric (CPM) score. As shown in Table 2, the proposed 3D nodule detector reaches comparable results with the one-stage methods trained

Table 2. FROC performance comparison of nodule detection between the proposed 3D point cloud-based method and the state-of-the-art image-based methods on LUNA16 dataset: sensitivity and the corresponding false positives at 1/8, 1/4, 1/2, 1, 2, 4, 8 per scan with and without false positive reduction.

Methods	1/8	1/4	1/2	1	2	4	8	CPM
Without false positive reduction								
Zhu et al. [31]	0.692	0.769	0.824	0.865	0.893	0.917	0.933	0.842
Li et al. [15]	0.739	0.803	0.858	0.888	0.907	0.916	0.920	0.862
Gong et al. [10]	0.713	0.801	0.867	0.917	0.920	0.962	0.971	0.883
Khosravan et al. [13]	0.709	0.836	0.921	0.953	0.953	0.953	0.953	0.897
Liu et al. [17]	0.848	0.876	0.905	0.933	0.943	0.957	0.970	0.919
With false positive reduction								
Dou et al. [7]	0.677	0.737	0.815	0.848	0.879	0.907	0.922	0.827
Dou et al. [6]	0.659	0.745	0.819	0.865	0.906	0.933	0.946	0.839
Wang et al. [29]	0.676	0.776	0.879	0.949	0.958	0.958	0.958	0.878
Ding et al. [5]	0.748	0.853	0.887	0.922	0.938	0.944	0.946	0.891
Kim et al. [14]	0.904	0.931	0.943	0.947	0.952	0.956	0.962	0.942
Ours	0.779	0.848	0.914	0.918	0.939	0.939	0.939	0.895
Ours (w/FPR)	0.794	0.886	0.916	0.934	0.951	0.956	0.959	0.913

on stacked CT images, demonstrating that the spatial features are feasible to employ in the nodule detection task. We observe that performance is improved with a small margin (1.6% CPM) by implementing the false positive reduction to the proposed nodule detector and shows a comparable result with the image-based two-stage frameworks [5–7,29]. We notice that most of the missing nodules are small, with an average size of 5 mm. With few point representations, spatial feature extraction is significantly limited for all the SparseConv encoder layers. For the image-based methods, detailed internal texture features of small nodules are better preserved by the feature pyramid extraction [17,29].

Ablation Study: We investigate the effectiveness of the proposed multi-view feature learning with the nodule proposal optimization schema and the feature fusion on the final nodule candidate predictions. For all ablation studies, the consistency threshold is set to 5 when 9 views involved as the model shows the best performance. Table 3 shows that the CPM is increased by 11.7% when all viewpoints are applied compared with the single-view baseline. As the multi-view features V-xy, V-yz, V-xz, and V-xyz gradually join for training, the sensitivities are raised by 3%, 4.4%, 3.2%, and 1.1% CPM, respectively. The performance is significantly gained by learning the multi-view feature representations through nodule proposal optimization schema. The average sensitivity is further improved 0.7% and 1.4% CPM by multi-level features (MF) and semantic feature (SF) for final nodule prediction, respectively, demonstrating the benefits of irregular

Table 3. Ablation study of the effectiveness of proposed multi-views samples with nodule proposal optimization, multi-level features (MF) and semantic features (SF) by comparing the performance on LUNA16 dataset: sensitivity and the corresponding false positives at 1/8, 1/4, 1/2, 1, 2, 4, 8 per scan.

V-xy	V-yz	V-xz	V-xyz	MF	SF	1/8	1/4	1/2	1	2	4	8	CPM
–	-	–	–	–	–	0.617	0.695	0.765	0.796	0.811	0.811	0.811	0.758
√	–	–	–	–	-	0.632	0.719	0.787	0.819	0.837	0.862	0.862	0.788
√	√	–	–	–	–	0.701	0.749	0.849	0.866	0.879	0.891	0.891	0.832
√	√	√	–	–	–	0.742	0.801	0.863	0.904	0.907	0.917	0.917	0.864
√	√	√	√	–	–	0.756	0.835	0.879	0.902	0.911	0.921	0.921	0.875
√	√	√	√	√	-	0.772	0.843	0.881	0.916	0.919	0.927	0.927	0.883
√	√	√	√	√	√	0.779	0.848	0.914	0.918	0.939	0.939	0.939	0.897

point-wise feature fusion for nodule region regression. We further compare the baseline model to the existing point cloud backbone networks: PointNet [20], PointNet++ [21] and DGCNN [19]. The PV-RCNN substantially outperforms these models by 4.2%, 3.3%, and 3.6% CPM respectively.

4   Discussions and Conclusions

Overall, 3D CT point clouds can be feasible data representations on nodule detection tasks for medical image analysis with a very careful data preprocessing. 3D point clouds naturally present the irregular shape of lung regions in real-world coordinates, which are flexible in employing multi-view features based on the geometric invariance and acquiring the discriminative spatial features from different viewpoints. Multi-view features can be obtained by rotating the 3D point clouds and further aggregated them to acquire the coarse nodule predictions through a nodule proposal optimization schema. By simultaneously learning the multiview features from a 3D SparseConv encoder, combined with the multi-level and semantic segmentation features extracted from a 3D SparseConv decoder, the proposed nodule detector achieves a promising performance on the pulmonary nodule detection task.

The point cloud-based backbone network tends to extract informative spatial relations for nodule prediction. As small nodules contain few points, the lack of spatial feature extraction makes it challenging to detect. In contrast, the image-based methods handle it relatively well due to detailed texture information extracted from image-based CNNs at the low-level layers. There are still opportunities for further improvements on small nodule detection. Future work will extend the 3D point-cloud representations with image-based texture features detecting nodules via a dynamic multi-view nodule proposal optimization schema. It is worth investigating to fully utilize the spatial representations from 3D CT point clouds, which have the potential to benefit the clinical diagnosis, especially in surgical operations.

Acknowledgements. This work was supported in part by the National Science Foundation under award number IIS-2041307 and Memorial Sloan Kettering Cancer Center Support Grant/Core Grant P30 CA008748.

References

1. Aaa, S., et al.: Validation, comparison, and combination of algorithms for automatic detection of pulmonary nodules in computed tomography images: the LUNA16 challenge. Med. Image Anal. **42**, 1–13 (2017). [dataset]
2. Ahmed, S.M., Liang, P., Chew, C.M.: EPN: edge-aware PointNet for object recognition from multi-view 2.5 D point clouds. In: IROS, pp. 3445–3450 (2019)
3. Bandos, A.I., Rockette, H.E., Song, T., Gur, D.: Area under the free-response ROC curve (FROC) and a related summary index. Biometrics **65**(1), 247–256 (2009)
4. Chen, R., Han, S., Xu, J., Su, H.: Point-based multi-view stereo network. In: Proceedings of the IEEE/CVF International Conference on Computer Vision, pp. 1538–1547 (2019)
5. Ding, J., Li, A., Hu, Z., Wang, L.: Accurate pulmonary nodule detection in computed tomography images using deep convolutional neural networks. In: Descoteaux, M., Maier-Hein, L., Franz, A., Jannin, P., Collins, D.L., Duchesne, S. (eds.) MICCAI 2017. LNCS, vol. 10435, pp. 559–567. Springer, Cham (2017). https://doi.org/10.1007/978-3-319-66179-7_64
6. Dou, Q., Chen, H., Jin, Y., Lin, H., Qin, J., Heng, P.A.: Automated pulmonary nodule detection via 3D convnets with online sample filtering and hybrid-loss residual learning. In: International Conference on Medical Image Computing and Computer-Assisted Intervention, pp. 630–638 (2017)
7. Dou, Q., Chen, H., Yu, L., Qin, J., Heng, P.A.: Multilevel contextual 3-D CNNs for false positive reduction in pulmonary nodule detection. IEEE Trans. Biomed. Eng. **64**(7), 1558–1567 (2017)
8. Drokin, I., Ericheva, E.: Deep learning on point clouds for false positive reduction at nodule detection in chest CT scans. arXiv preprint arXiv:2005.03654 (2020)
9. El-Regaily, S.A., Salem, M.A.M., Aziz, M.H.A., Roushdy, M.I.: Multi-view convolutional neural network for lung nodule false positive reduction. Expert Syst. Appl. **162**, 113017 (2020)
10. Gong, Z., Li, D., Lin, J., Zhang, Y., Lam, K.M.: Towards accurate pulmonary nodule detection by representing nodules as points with high-resolution network. IEEE Access **8**, 157391–157402 (2020)
11. Gupta, A., Saar, T., Martens, O., Moullec, Y.L.: Automatic detection of multi-size pulmonary nodules in CT images: large-scale validation of the false-positive reduction step. Med. Phys. **45**(3), 1135–1149 (2018)
12. Han, Z., Wang, X., Liu, Y.S., Zwicker, M.: Multi-angle point cloud-VAE: unsupervised feature learning for 3D point clouds from multiple angles by joint self-reconstruction and half-to-half prediction. In: 2019 IEEE/CVF International Conference on Computer Vision (ICCV), pp. 10441–10450. IEEE (2019)
13. Khosravan, N., Bagci, U.: S4ND: single-shot single-scale lung nodule detection. In: International Conference on Medical Image Computing and Computer-Assisted Intervention, pp. 794–802 (2018)
14. Kim, B.C., Yoon, J.S., Choi, J.S., Suk, H.I.: Multi-scale gradual integration CNN for false positive reduction in pulmonary nodule detection. Neural Netw. (2019)

15. Li, Y., Fan, Y.: DeepSEED: 3D squeeze-and-excitation encoder-decoder convolutional neural networks for pulmonary nodule detection. In: 2020 IEEE 17th International Symposium on Biomedical Imaging (ISBI), pp. 1866–1869. IEEE (2020)

16. Li, Z., Zhang, S., Zhang, J., Huang, K., Wang, Y., Yu, Y.: MVP-Net: multi-view FPN with position-aware attention for deep universal lesion detection. In: Shen, D., et al. (eds.) MICCAI 2019. LNCS, vol. 11769, pp. 13–21. Springer, Cham (2019). https://doi.org/10.1007/978-3-030-32226-7_2

17. Liu, J., Cao, L., Akin, O., Tian, Y.: 3DFPN-HS^2: 3D feature pyramid network based high sensitivity and specificity pulmonary nodule detection. In: Shen, D., et al. (eds.) MICCAI 2019. LNCS, vol. 11769, pp. 513–521. Springer, Cham (2019). https://doi.org/10.1007/978-3-030-32226-7_57

18. Liu, J., Cao, L., Akin, O., Tian, Y.: Accurate and robust pulmonary nodule detection by 3D feature pyramid network with self-supervised feature learning. arXiv preprint arXiv:1907.11704 (2019)

19. Phan, A.V., Le Nguyen, M., Nguyen, Y.L.H., Bui, L.T.: DGCNN: a convolutional neural network over large-scale labeled graphs. Neural Netw. **108**, 533–543 (2018)

20. Qi, C.R., Su, H., Mo, K., Guibas, L.J.: PointNet: deep learning on point sets for 3D classification and segmentation. In: Proceedings of the IEEE Conference on Computer Vision and Pattern Recognition, pp. 652–660 (2017)

21. Qi, C.R., Yi, L., Su, H., Guibas, L.J.: PointNet++: deep hierarchical feature learning on point sets in a metric space. In: Advances in Neural Information Processing Systems, pp. 5099–5108 (2017)

22. Setio, A.A.A., et al.: Pulmonary nodule detection in CT images: false positive reduction using multi-view convolutional networks. IEEE Trans. Med. Imaging **35**(5), 1160–1169 (2016). https://doi.org/10.1109/TMI.2016.2536809

23. Setio, A.A.A., et al.: Pulmonary nodule detection in CT images: false positive reduction using multi-view convolutional networks. IEEE Trans. Med. Imaging **35**(5), 1160–1169 (2016)

24. Shi, S., et al.: PV-RCNN: point-voxel feature set abstraction for 3D object detection. In: Proceedings of the IEEE/CVF Conference on Computer Vision and Pattern Recognition, pp. 10529–10538 (2020)

25. Shi, S., Wang, Z., Shi, J., Wang, X., Li, H.: From points to parts: 3D object detection from point cloud with part-aware and part-aggregation network. IEEE Trans. Pattern Anal. Mach. Intell. (2020)

26. Siegel, R.L., Miller, K.D., Jemal, A.: Cancer statistics, 2019. CA Cancer J. Clin. **69**(1), 7–34 (2019)

27. Team, O.D.: OpenPCDet: an open-source toolbox for 3D object detection from point clouds (2020). https://github.com/open-mmlab/OpenPCDet

28. Usman, M., Lee, B.D., Byon, S.S., Kim, S.H., Lee, B.i., Shin, Y.G.: Volumetric lung nodule segmentation using adaptive ROI with multi-view residual learning. Sci. Rep. **10**(1), 1–15 (2020)

29. Wang, B., Qi, G., Tang, S., Zhang, L., Deng, L., Zhang, Y.: Automated pulmonary nodule detection: high sensitivity with few candidates. In: Frangi, A.F., Schnabel, J.A., Davatzikos, C., Alberola-López, C., Fichtinger, G. (eds.) MICCAI 2018. LNCS, vol. 11071, pp. 759–767. Springer, Cham (2018). https://doi.org/10.1007/978-3-030-00934-2_84

30. Zanjani, F.G., Moin, D.A., Verheij, B., Claessen, F., Cherici, T., Tan, T., et al.: Deep learning approach to semantic segmentation in 3D point cloud intra-oral scans of teeth. In: International Conference on Medical Imaging with Deep Learning, pp. 557–571 (2019)

31. Zhu, W., Liu, C., Fan, W., Xie, X.: DeepLung: deep 3D dual path nets for automated pulmonary nodule detection and classification. In: 2018 IEEE Winter Conference on Applications of Computer Vision (WACV), pp. 673–681 (2018)

End-to-End Lung Nodule Detection Framework with Model-Based Feature Projection Block

Ivan Drokin$^{(\boxtimes)}$ and Elena Ericheva

Intellogic Limited Liability Company (Intellogic LLC), Office 1/334/63, Building 1, 42 Bolshoi blvd., Territory of Skolkovo Innovation Center, 121205 Moscow, Russia
{ivan.drokin,elena.ericheva}@botkin.ai

Abstract. This paper proposes novel end-to-end framework for detecting suspicious pulmonary nodules in chest CT scans. The method's core idea is a new nodule segmentation architecture with a model-based feature projection block on three-dimensional convolutions. This block acts as a preliminary feature extractor for a two-dimensional U-Net-like convolutional network. Using the proposed approach along with an axial, coronal, and sagittal projection analysis makes it possible to abandon the widely used false positives reduction step. The proposed method achieves SOTA on LUNA2016 with 0.959 average sensitivity, and 0.936 sensitivity if the false-positive level per scan is 1/4. The paper describes the proposed approach and represents the experimental results on LUNA2016 as well as ablation studies. The code of the proposed model is available at https://github.com/Botkin-AI/feature-projection-block.

Keywords: Image analysis · Computer vision · Segmentation · Computer detection · Computer-assisted · Lung cancer · Pulmonary nodule · False-positive reduction · Chest CT

1 Introduction and Previous Work

Survival in lung cancer (over five years) is approximately 18.1%[1]. Early-stage lung cancer (stage I) has a five-year survival rate of 60–75%. A recent National Lung Screening Trial (NLST) study revealed that lung cancer mortality can be reduced by at least 20% by using a high-risk annual screening program with low-dose computed tomography (CT) of the chest [1]. Computerized tools, especially image analysis and machine learning, are critical factors for improving diagnostics, facilitating the identification of results that require treatment, and supporting medical experts' workflow [2]. Computer-aided detection (CAD) systems have also shown improvements in radiologists' readability [3–5].

---

[1] 2018 state of lung cancer report: https://www.naaccr.org/2018-state-lung-cancer-report/.

Supported by BOTKIN.AI, Skolkovo.

C. Lian et al. (Eds.): MLMI 2021, LNCS 12966, pp. 91–100, 2021.
https://doi.org/10.1007/978-3-030-87589-3_10

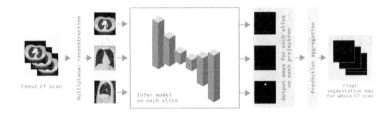

Fig. 1. Scheme of the proposed framework for lung nodule detection

Modern CAD systems are built using a deep learning approach to solve the detection task. CNN models are widely studied. Due to their specificity, CNNs can work efficiently with images and focus on candidate recognition [2]. A multi-contextual 3D residual convolutional neural network (3D Res-CNN) is proposed in [6,7]. The method presented in [8] is based on structural relationship analysis between nodule candidates and vessels. In [9], a rule-based classifier is used to eliminate apparent non-nodules, followed by a multi-view CNN. The CNN from [10] is fed with nodule candidates obtained by combining three candidate detectors specifically designed for solid, subsolid, and large nodules. For each candidate, a set of 2D patches from differently oriented planes is extracted.

Currently, the conventional pipeline in the screening task for CAD consists of several stages—principally detection and cancer classification. A two-stage machine learning algorithm is a popular approach that can assess the risk of cancer associated with a CT scan [11–15]. The first stage uses a nodule detector, which identifies nodules contained in the scan. The second step is used to assess whether nodules are malignant or benign. An evaluation of CAD systems is performed on independent datasets from the LUNA16[2] and ANODE09[3] challenges and the DLCST [16] dataset.

In this work, we have proposed an end-to-end approach for both stages. The detector stage and false-positive reduction stage were developed and inferred in one convolutional neural network. The proposed novel architecture is based on a trainable analog of the Maximum Intensity Projection (MIP) [17] block, which acts as a preliminary feature extractor, followed by a U-Net-like segmentation network.

This end-to-end approach doesn't require any specific data sources or markups but does demand some data preparation. We have described the proposed data processing and augmentation techniques, and an evaluating process that includes current SOTA comparisons.

2   Method

2.1   Overall Pipeline Description

A high-level scheme of our solution is presented in Fig. 1. We propose a framework based on semantic segmentation for lung nodule detection. As a model input, we

---

[2] LUNA16 challenge homepage: https://luna16.grand-challenge.org/.
[3] ANODE09 challenge homepage: https://anode09.grand-challenge.org/.

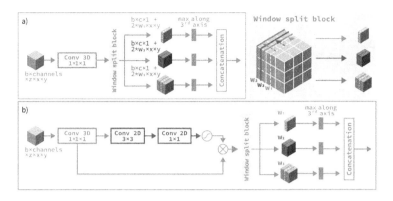

Fig. 2. Scheme of the proposed model-based feature projection. a. Base block design with 3D convolutional bottleneck and maximum feature aggregation. b. Attention-like block design with additional self-attention block

apply not only widely used axial images from the CT study but also sagittal and coronal projections. Radiologists often use multiplanar reconstruction to obtain more accurate results from complex findings analysis. For example, through the use of analysis types such as differentiation between plane and volume objects (e.g., fibrosis and nodules) and the examination of the relation of those findings to adjacent structures such as bronchi and vessels [18]. We use data from all three projections simultaneously and obtain a single model to detect nodules, analyzing any given projection during the training procedure. This approach enables us to train a more robust model without expanding the training dataset. During the inference procedure, we first prepare sagittal and coronal projections of a CT study. After that, we infer the model on each slice independently. Then, we average prediction by applying the reverse data transformation to the axial projection. This approach offers us to obtain a more accurate representation of the three-dimensional shape of the findings using two-dimensional convolutional networks and, as a consequence, reduces false positives. Along with the redesigned segmentation network architecture, this framework's construction allows us to eliminate the false-positive reduction step.

The segmentation network itself consists of two blocks. The first block is a trainable analog of MIP [17], which acts as a preliminary feature extractor, followed by a U-Net-like segmentation network. This approach offers the advantages of both a 3D decoder and a 2D one. Conv3D-based models bring high-quality results. The U-Net-like network enables us to train and infer using a reasonable amount of computing resources.

2.2 Model-Based Feature Projection

MIP is a 2D projection of voxels with the highest attenuation value on every view throughout the volume. This method tends to display bone and contrasting material-filled structures preferentially. Other lower-attenuation structures are

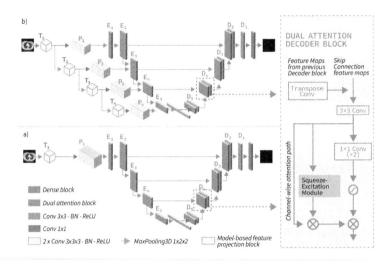

Fig. 3. Scheme of the proposed segmentation model architectures.

not well visualized. The primary clinical application of MIP is to improve the detection of pulmonary nodules and assess their profusion. MIP also helps characterize the distribution of small nodules. In addition, MIP sections of variable thickness are excellent for evaluating vessels' size and location, including the pulmonary arteries and veins [19].

Zheng et al. [20] proposed to use MIPs as the input data for a detection model in a lung nodule screening pipeline. This solution led to more accurate models but still requires a false positive reduction step. We propose to extend the MIP concept to a trainable feature extractor block with 3D-convolutions as the primary step, followed by maximum feature aggregation.

Let's denote a chest CT scan as $D \in R^{d \times n \times n}$. The first axis is aligned with the patient's spine. The two remaining axes are the width and height of the image. Thus, the slab for a given position p and width w is $S(p, w) = \{D(i, j, k) | p = p - w \ldots p + w; j, k = 1 \ldots n\}$. We apply two consecutive blocks of a 3D convolution layer with kernel size $3 \times 3 \times 3$, batch normalization, and ReLU activation to slab $S(p, w)$. After this step, we apply the bottleneck block with kernel size $1 \times 1 \times 1$ to the previous step output, and we get a feature map $M \in R^{c \times d \times n \times n}$. Here we omit the batch axis for simplicity. Let's denote $W = \{w_i\}_{i=1}^{k}$ to be a set of thicknesses for a maximum features aggregation, and c–a central index of the second axis of feature map M. Thus, the output of the maximum feature aggregation block is $F = \left[\max_{j \in [c-w_i; c+w_i]} M_{u,j,l,r} \right]_{i=1}^{k}$, $F \in R^{c \cdot k \times n \times n}$. Feature map F can be used as input in a two-dimensional convolutional network. A self-attention block after the bottleneck layer can extend this design. Figure 2 presents the scheme for the proposed blocks.

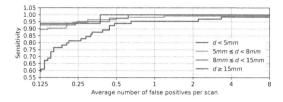

Fig. 4. FROC curves depending on nodule size for AttentionMIP Enc.

2.3 Segmentation Network with Model-Based Feature Projection Blocks

We propose a segmentation network architecture consisting of two blocks: a three-dimensional convolutional preliminary feature extractor and a U-Net-like primary segmentation model. A scheme of the proposed networks is presented in Fig. 3. The feature extractor is a model-based 3D feature projection block on 2D plane. As generally segmentation networks can use any architecture without any restrictions, we decided to redesign UNet [29]. We changed the encoder to ResNest34 [22] and use dual attention decoder blocks [21]. In addition to this base architecture, we propose a three-dimensional encoder model based on blocks of a 3D convolution layer with kernel size $3 \times 3 \times 3$, batch normalization, and ReLU activation. Each block T_i consists of two 3D-blocks, with feature size $32 \cdot i$. P_i is a model-based projection block with a bottleneck size of $4 \cdot i$. For correct subsampling, we use MaxPooling with strides $1 \times 2 \times 2$ after $T_i, i = 1, 2, 3$. Outputs of blocks P_i are concatenated with input features to blocks $E_i, i = 1, 3, 4, 5$.

3 Evaluation

3.1 Experiment on LUNA2016

For evaluation of the proposed framework, we use the commonly accepted baseline LUNA2016. We resample CT scans to 0.8 mm spacing and train networks with all three projections simultaneously. As input, we use a slab with a thickness of 16.8 mm (21 slices). For $W = \{w_i\}_{i=1}^{3}$ we choose 3, 6, 10 respectively. We choose the MinMax scaling function as a preprocessed function, perform a linear map of -1000 HU to 0 and 400 HU to 1, and clip all values outside this range. We use an Adam [23] optimizer with a learning rate of 0.0005 and train each network for five epochs. We also use augmentation techniques: flips, rotations, and zoom of CT scans. We train four networks: 1) with maximum projection blocks, 2) with attention maximum projection blocks, 3) with a three-dimension parallel encoder with maximum projection blocks, and 4) with a three-dimension parallel encoder with attention maximum projection blocks.

Table 1 shows that the proposed framework with various segmentation networks outperforms recently published results.

Table 1. Results of the experiments on LUNA nodule detection track. The sensitivity per FP level per exam

Experiment	0.125 FP	0.25 FP	0.5 FP	1 FP	2 FP	4 FP	8 FP	Mean Sens
Li [24]	0.739	0.803	0.858	0.888	0.907	0.916	0.920	0.862
Wang [25]	0.676	0.776	0.879	0.949	0.958	0.958	0.958	0.878
Drokin [26]	0.725	0.832	0.901	0.933	0.945	0.945	0.945	0.8894
Khosravan [27]	0.709	0.836	0.921	0.953	0.953	0.953	0.953	0.897
Cao [28]	0.848	0.899	0.925	0.936	0.949	0.957	0.960	0.925
MaxMIP	0.8	0.914	0.961	0.974	0.982	0.987	0.987	0.943
AttentionMIP	0.855	0.927	0.961	0.978	0.978	0.982	0.982	0.951
MaxMIP Enc.	0.855	0.936	0.961	0.978	0.978	0.982	0.982	0.953
AttentionMIP Enc.	**0.872**	**0.931**	**0.965**	**0.978**	**0.987**	**0.991**	**0.991**	**0.959**

In Fig. 4, we provide performance analysis for the best model depending on nodule size. The proposed model has a better performance at a nodule range with a diameter of more than 5 mm and has a tendency to display minor sensitivity at a diameter of less than 5 mm at low false-positives per scan rates.

To understand the results better, we decided to add a visualization of the features extracted from feature projection block. Figure 5 a. shows that model-based features tend to ignore normal lung tissue compared to MIPs images, to an extent. Nonetheless they preserve valuable information about lung nodules. Common false-positive detections include pieces of bronchi, blood vessels, fibrosis, etc. Due to this, the property of proposed model-based feature projection blocks allows models to reduce false-positives. Figure 5 b. shows that the model projection block ignores diaphragmatic cupula, and Fig. 5 c. demonstrates that the proposed block is better at noise suppression of CT scanner than MIP projection.

3.2   Ablation Study

As a part of the proposed method analysis, we include the results of an ablation study. We train and evaluate proposed models only on axial data. We also train a model and evaluate a framework with a "naive" block: we use the whole slab as a 21-channel input to proposed a 2D segmentation network without a model-based feature projection block. We train the model with a four-channel input to the segmentation model: the central slice and MIPs with a thickness equal to 5, 10, 15 mm, as was proposed in [20]. Finally, we train the model with a maximum projection block and U-net [29] as a 2D segmentation network. Table 2 shows the results of the ablation study. According to the provided experimental data, all proposed parts of the framework contribute positively to framework performance. As a consequence, the proposed innovations are not redundant.

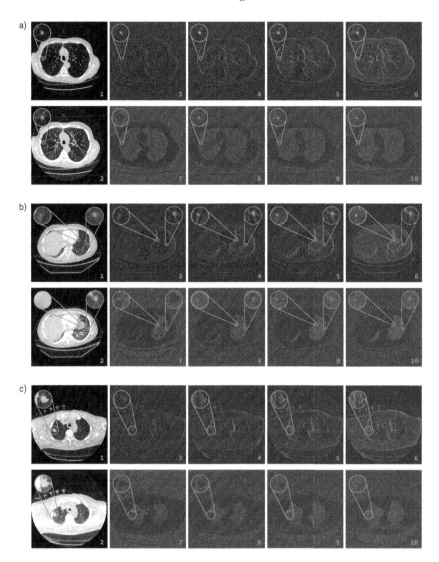

Fig. 5. Feature visualization of proposed network. 1 - central slice; 2 is 21 mm MIP on input data; 3 is a center slice of MaxMIP features, 4–6 is features of MaxMIP with $w_1 = 3, w_2 = 7, w_3 = 10$; 7–10 same as 3–6, but for AttentionMIP. Circles denote true nodules.

Table 2. Results of the ablation study on LUNA2016. The sensitivity per FP level per exam.

Experiment	0.125 FP	0.25 FP	0.5 FP	1 FP	2 FP	4 FP	8 FP	Mean Sens
Only axial data								
MaxMIP	0.631	0.784	0.882	0.937	0.964	0.968	0.972	0.877
AttentionMIP	0.776	0.898	0.941	0.960	0.968	0.972	0.972	0.926
MaxMIP Enc.	0.768	0.858	0.964	0.968	0.972	0.972	0.972	0.925
AttentionMIP Enc.	0.784	0.901	0.933	0.949	0.968	0.976	0.984	0.928
Feature blocks								
MIP	0.686	0.815	0.882	0.913	0.937	0.941	0.941	0.873
Naive	0.729	0.827	0.890	0.941	0.968	0.972	0.972	0.899
U-net								
MaxMIP	0.749	0.862	0.942	0.972	0.980	0.987	0.987	0.925

3.3 Real World Data Evaluation

Our radiologist team analyzed 130 CT scans from Lomonosov Moscow State University Medical Research and Educational Center[4] and Yamalo-Nenets Autonomous Okrug, Russian Federation, diagnosed with the proposed framework. These studies have a wide range of variability and include the analysis of verified malignancies, benign pathologies, solitary and multiple lesions, nodules of different size and localization, associations with other lung pathologies, series with various parameters (series with thin/thick slices, different reconstructions, different protocols: standard Chest CT, low-dose Chest CT, Chest CT + contrast agent), absence/existence of breathing artifacts, etc. During the evaluation step, we followed the LIDC-IDRI markup protocol and we use include and exclude criteria described at LUNA2016 Challenge[5]. This dataset contains 138 nodules. Proposed models achieve a sensitivity of 0.913 at an FP level of 0.1869 per scan.

4   Conclusion and Discussion

We have proposed a novel approach for lung nodule CAD systems based on the U-Net-like segmentation network with a model-based feature projection block as a preliminary feature extractor. Coupled with processing three CT scans projections, this novelty allows us to build a single-model approach and achieve SOTA results on LUNA2016. We have performed an extensive analysis of the proposed solution with the ablation study and the feature analysis. We performed validation on the proposed model using external data. The results shows that the proposed method can be used for lung cancer screening and as a CADe system in real world setup. As the next step of our research, we plan to extend the

---

[4] https://www.msu.ru/en/info/struct/.
[5] https://luna16.grand-challenge.org/Data/.

proposed segmentation framework to other CT scan analysis cases. Moreover, maximum-based feature projection after a three-dimensional convolutional block can be suitable for video analysis, both in the medical and general domains.

References

1. Aberle, D.R., et al.: The national lung screening trial research team: reduced lung-cancer mortality with low-dose computed tomographic screening. New Eng. J. Med. **365**, 395–409 (2011). https://doi.org/10.1056/NEJMoa1102873
2. Greenspan, H., van Ginneken, B., Summers, R.M.: Guest editorial deep learning in medical imaging: overview and future promise of an exciting new technique. IEEE Trans. Med. Imaging **35**(5), 1153–1159 (2016). https://doi.org/10.1109/TMI.2016.2553401
3. Sahiner, B., et al.: Effect of CAD on radiologists' detection of lung nodules on thoracic CT scans: analysis of an observer performance study by nodule size. Acad. Radiol. (2010). https://doi.org/10.1016/j.acra.2009.08.006
4. Wu, N., et al.: Deep neural networks improve radiologists' performance in breast cancer screening. IEEE Trans. Med. Imaging (2019). https://doi.org/10.1109/TMI.2019.2945514
5. Wang, D., Khosla, A., Gargeya, R., Irshad, H., Beck, A.: Deep learning for identifying metastatic breast cancer. arXiv:1606.05718 (2016)
6. Zhang, Z., Li, X., You, Q., Luo, X.: Multicontext 3D residual CNN for false positive reduction of pulmonary nodule detection. Int. J. Imaging Syst. Technol. **29**, 42–49 (2018). https://doi.org/10.1002/ima.22293
7. Jin, H., Li, Z., Tong, R., Lin, L.: A deep 3D residual CNN for false positive reduction in pulmonary nodule detection. Med. Phys. **45**, 2097–2107 (2018). https://doi.org/10.1002/mp.12846
8. Cao, G., Liu, Y., Suzuki, K.: A new method for false-positive reduction in detection of lung nodules in CT images. In: International Conference on Digital Signal Processing (DSP), pp. 474–479, August 2014. https://doi.org/10.1109/ICDSP.2014.6900710
9. El-Regaily, S., Salem, M., Aziz, M., Roushdy, M.: Multi-view convolutional neural network for lung nodule false positive reduction. Expert Syst. Appl. (2019).https://doi.org/10.1016/j.eswa.2019.113017
10. Setio, A., et al.: Pulmonary nodule detection in CT images: false positive reduction using multi-view convolutional networks. IEEE Trans. Med. Imaging. (2016). https://doi.org/10.1109/TMI.2016.2536809
11. Trajanovski, S., et al.: Towards radiologist-level cancer risk assessment in CT lung screening using deep learning. arXiv:1804.01901 (2018)
12. He, K., Zhang, X., Ren, S., Sun, J.: Deep residual learning for image recognition. In: IEEE Conference on Computer Vision and Pattern Recognition (CVPR), pp. 770–778 (2016). https://doi.org/10.1109/CVPR.2016.90
13. Hunar, A., Sozan, M.: A deep learning technique for lung nodule classification based on false positive reduction. J. Zankoy Sulaimani Part A **21**, 107–116 (2019). https://doi.org/10.17656/jzs.10749
14. Tang, H., Zhang, C., Xie, X.: NoduleNet: decoupled false positive reduction for pulmonary nodule detection and segmentation. In: Shen, D., et al. (eds.) MICCAI 2019. LNCS, vol. 11769, pp. 266–274. Springer, Cham (2019). https://doi.org/10.1007/978-3-030-32226-7_30

15. Tang, H., Liu X., Xie X.: An end-to-end framework for integrated pulmonary nodule detection and false positive reduction. arXiv:1903.09880 (2019)
16. Jesper, P., et al.: The danish randomized lung cancer CT screening trialoverall design and results of the prevalence round. J. Thorac. Oncol. **4**, 608–614 (2009). https://doi.org/10.1097/JTO.0b013e3181a0d98f
17. Wallis, J.W., Miller, T.R., Lerner, C.A., Kleerup, E.C.: Three-dimensional display in nuclear medicine. IEEE Trans. Med. Imaging. **8**(4), 297–300 (1989). https://doi.org/10.1109/42.41482
18. Ebner, L., et al.: Maximum-intensity-projection and computer-aided-detection algorithms as stand-alone reader devices in lung cancer screening using different dose levels and reconstruction kernels. AJR Am. J. Roentgenol. (2016). https://doi.org/10.2214/AJR.15.15588
19. Perandini, S., Faccioli, N., Zaccarella, A., Re, T., Mucelli, R.: The diagnostic contribution of CT volumetric rendering techniques in routine practice. Indian J. Radiol. Imaging **20**(2), 92–97 (2010). https://doi.org/10.4103/0971-3026.63043
20. Zheng, S., Guo, J., Cui, X., Veldhuis, R.N.J., Oudkerk, M., van Ooijen, P.M.A.: Automatic pulmonary nodule detection in CT scans using convolutional neural networks based on maximum intensity projection. IEEE Trans. Med. Imaging **39**(3), 797–805 (2020). https://doi.org/10.1109/TMI.2019.2935553
21. Sun, J., Darbeha, F., Zaidi, M., Wang, B.: SAUNet: shape attentive U-Net for interpretable medical image segmentation. arXiv:abs/2001.07645 (2020)
22. Zhang, H., et al.: ResNeSt: split-attention networks. arXiv:abs/2004.08955 (2020)
23. Kingma D., Ba J.: Adam: a method for stochastic optimization. International Conference on Learning Representations (ICLR). arXiv:1412.6980 (2014)
24. Li, Y., Fan, Y.: DeepSEED: 3D squeeze-and-excitation encoder-decoder convolutional neural networks for pulmonary nodule detection. arXiv:1904.03501 (2019)
25. Wang, B., Qi, G., Tang, S., Zhang, L., Deng, L., Zhang, Yongdong: Automated pulmonary nodule detection: high sensitivity with few candidates. In: Frangi, A.F., Schnabel, J.A., Davatzikos, C., Alberola-López, C., Fichtinger, G. (eds.) MICCAI 2018. LNCS, vol. 11071, pp. 759–767. Springer, Cham (2018). https://doi.org/10.1007/978-3-030-00934-2_84
26. Drokin, I., Ericheva, E.: Deep learning on point clouds for false positive reduction at nodule detection in chest CT scans. analysis of images, social networks and texts. In: AIST, Lecture Notes in Computer Science. arXiv:2005.03654. (2020)
27. Khosravan, N., Bagci, U.: S4ND: single-shot single-scale lung nodule detection. In: Frangi, A.F., Schnabel, J.A., Davatzikos, C., Alberola-López, C., Fichtinger, G. (eds.) MICCAI 2018. LNCS, vol. 11071, pp. 794–802. Springer, Cham (2018). https://doi.org/10.1007/978-3-030-00934-2_88
28. Cao, H., et al.: Two-stage convolutional neural network architecture for lung nodule detection. IEEE J. Biomed. Health Inf. **24**(7), 2006–2015 (2020). https://doi.org/10.1109/JBHI.2019.2963720
29. Ronneberger, O., Fischer, P., Brox, T.: U-Net: convolutional networks for biomedical image segmentation. In: Navab, N., Hornegger, J., Wells, W.M., Frangi, A.F. (eds.) MICCAI 2015. LNCS, vol. 9351, pp. 234–241. Springer, Cham (2015). https://doi.org/10.1007/978-3-319-24574-4_28

Learning Structure from Visual Semantic Features and Radiology Ontology for Lymph Node Classification on MRI

Yingying Zhu[3](✉), Shuai Wang[1], Qingyu Chen[2], Sungwon Lee[1],
Thomas Shen[1], Daniel C. Elton[1], Zhiyong Lu[2], and Ronald M. Summers[1](✉)

[1] Imaging Biomarkers and Computer-Aided Diagnosis Laboratory, Clinical Center,
National Institutes of Health, Bethesda, MD 20892, USA
rms@nih.gov
[2] National Center for Biotechnology Information, National Library of Medicine,
National Institutes of Health, Bethesda, MD 20892, USA
[3] Department of Computer Science and Engineering, University of Texas
at Arlington, Arlington, TX 76019, USA
yingying.zhu@uta.edu

Abstract. Medical image classification (for example, lesions on MRI scans) is a very challenging task due to the complicated relationships between different lesion sub-types and expensive cost to collect high quality labelled training datasets. Graph model has been used to model the complicated relationship for medical imaging classification successfully in many previous work. However, most existing graph based models assumed the structure is known or pre-defined, and the classification performance severely depends on the pre-defined structure. To address all the problems of current graph learning models, we proposed to jointly learn the graph structure and use it for classification task in one framework. Besides imaging features, we also use the disease semantic features (learned from clinical reports), and predefined lymph node ontology graph to construct the graph structure. We evaluated our model on a T2 MRI image dataset with 821 samples and 14 types of lymph nodes. Although this dataset is very unbalanced on different types of lymph nodes, our model shows promising classification results on this challenging datasets compared to several state of art methods.

1 Introduction

Accurate classification of lymph nodes is of very important clinical meaning for the diagnosis or prognosis of numerous diseases such as metabolises cancer and can be used as assistance for the radiologist to locate abnormality and diagnosis different diseases. Developing machine learning methods to automatically identify different lymph nodes from MRI scans is a very challenging task due to the similar morphological structures of lymph nodes, very complicated relationship between different types of lymph nodes and the expensive cost to collect large-scale labelled image datasets. Labelling lymph node types from the MRI scans is

© Springer Nature Switzerland AG 2021
C. Lian et al. (Eds.): MLMI 2021, LNCS 12966, pp. 101–109, 2021.
https://doi.org/10.1007/978-3-030-87589-3_11

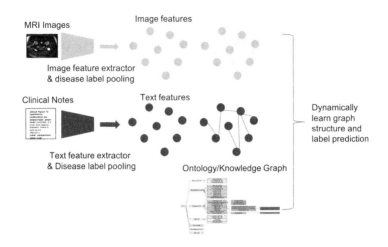

Fig. 1. The framework of the classical graph based classification.

very time consuming and expensive since it requires the well trained radiologist to look into each MRI slices, therefore, it is not realistic to collect a large size labeled dataset to train a deep learning system.

In this work, instead of manually labelling images, we extracted lymph node key words from clinical notes associated with MRI scans by experienced radiologists and used the extracted key words as classification labels for the corresponding MRI images. We extracted 14 different types of lymph nodes from the clinical reports (as shown in Fig. 3) from 821 T2 MRI key slices. It is worth noting that our dataset is highly unbalanced on different types of lymph nodes (some lymph node has more than 80 training images and some lymph node only have less than 10 training images). Considering the unbalanced small size training dataset, it is very challenging to train the classification model with high accuracy and generalizability.

Motivated by recent works on successfully using semantic information for image classification and image captioning problems [10,11], we proposed to leverage the semantic features learned from clinical notes along with a predefined lymph node ontology graph (as shown in Fig. 2) by radiologist for the lymph node classification. We proposed to learn a semantic feature embedding on the clinical reports and used it to learn the semantic relationship between different lymph node types. Based on this semantic embedding space, we also combined it with the ontology graph (shown in Fig. 1 (b)) to construct a knowledge graph to guide our classification task. Besides, most existing graph models assume that the graph structure is known or predefined. When graph structure is unknown, they usually use K-nearest neighbor to construct a graph structure from image features and use it for downstream image classification task [2,8]. Defining graph structure is critical for the downstream graph node classification task, recent work shows that learning a graph structure can significantly boost the performance of graph based classification tasks [12]. In this work, we also proposed a joint graph structure learning and classification framework with prior information from semantic features (learned from clinical reports) and radiologist defined lymph node ontology graph. We evaluated our proposed model on lymph node classification in T2 MRI scans and show consistent improvement compared to several state-of-the-art methods given a small and imbalanced training dataset.

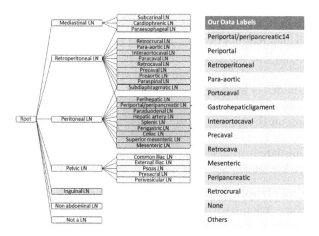

Fig. 2. The predefined ontology graph of a lymph node structure by an experienced radiologist. We use this ontology graph as a prior to construct our knowledge graph. The 14 labels on our dataset is also shown here.

2   Method

Graph Learning for Classification. We define a graph $G = (\mathbf{V}, E)$, where $\mathbf{V}$ is the set of vertices or nodes (we will use nodes throughout the paper), and E is the set of edges. Let $\mathbf{v}_i \in \mathbf{V}$ denote a node and $e_{ij} = (\mathbf{v}_i, \mathbf{v}_j) \in E$ to denote an edge pointing from $\mathbf{v}_j$ to $\mathbf{v}_i$. The adjacency matrix $\mathbf{\Lambda}$ is a $n \times n$ matrix with $\mathbf{\Lambda}_{ij} = 1$ if $e_{ij} \in E$ and $A_{ij} = 0$ if $e_{ij} \notin E$. It is worth noting that we only consider about the un-directed graph in this work, thus $\mathbf{\Lambda}$ is symmetric. It is straightforward to extend our frame work to the directed graph with different graph structure regularization terms. The node of graph has node feature $x_i \in \mathcal{R}^{d \times 1}$, $X = [x_1, \cdots, x_n] \in \mathbb{R}^{n \times d}$ is a node feature matrix for all nodes in the graph. Let $\mathbf{y}_i$ denotes the class labels of node x_i and $\mathbf{Y} = [\mathbf{y}_1, \cdots, \mathbf{y}_n]$ denotes the node class labels of all different nodes on the graph. Conventional graph learning methods usually define a function $f(\mathbf{X}, \mathbf{\Lambda}) = \mathbf{Y}$ with input as node feature vector matrix $\mathbf{X}$ and graph adjacency matrix $\mathbf{\Lambda}$ to estimate the class labels $\mathbf{Y}$ for each node. f can be a simple linear model or a graph neural network model. The classic graph learning solves the following function to learn f,

$$\mathrm{L}(\mathbf{X}, \mathbf{Y}) = \|\mathbf{Y} - f(\mathbf{X}, \mathbf{\Lambda})\|_p + \lambda f(\mathbf{X}, \mathbf{\Lambda})^T \mathbf{f}(\mathbf{X}, \mathbf{\Lambda})$$
$$\text{s. t.} = \mathbf{I} - \mathbf{\Lambda},$$
(1)

where $\|\cdot\|_p$ represents L_p norm, the first term is the label prediction term and the second term is the laplacian constraint term, $\mathbf{I}$ is the identity matrix, is the laplacian matrix of graph. If using a multi-layer deep graph neural network model for f

$$\mathrm{L}(\mathbf{X}, \mathbf{Y}) = \|\mathbf{Y} - f(\mathbf{X}, \mathbf{\Lambda})\|_p + \lambda f(\mathbf{X}, \mathbf{\Lambda})^T \mathbf{f}(\mathbf{X}, \mathbf{\Lambda})$$
$$\mathrm{s.\ t.} = \mathbf{I} - \mathbf{\Lambda} \tag{2}$$

These graph learning methods always assumed that the graph structure is predefined (the adjacency matrix $\mathbf{\Lambda}$ is given), which is not applicable for many cases. For example, the connection or similarity between different patients or different types of lymph nodes is hard to defined. In practice, many previous works just use a K-nearest neighbor to extract the graph adjacency matrix and use it for label propagation on the graph. Recent works have shown that defining an optimal graph structure is crucial for the label classification task on the graph and jointly optimizing label propagation and learning graph structures can significantly improve the performance and generalizability of graph models [2].

Jointly Learning Graph Structure and Classification. We propose to learn the graph structure $\mathbf{\Lambda}$ jointly with graph label propagation, let $\mathbf{\Lambda} = g(\mathbf{X})$ be the function to learn the graph structure, we define the new loss function as,

$$\mathrm{L}(\mathbf{X}, \mathbf{\Lambda}, \mathbf{Y}) = \|\mathbf{Y} - f(\mathbf{X}, \mathbf{\Lambda})\|_p + \lambda_0 f(\mathbf{X}, \mathbf{\Lambda})^T \mathbf{f}(\mathbf{X}, \mathbf{\Lambda}) \tag{3}$$
$$+\lambda_g \|\mathbf{\Lambda} - g(\mathbf{X})\|_p, \mathrm{s.\ t.} = \mathbf{I} - \mathbf{\Lambda}$$

There are many ways to construct the graph structure learning g function, for example, previous work [2] has tried to use Bernoulli sampling to learning the graph structure from discrete nodes, or using sparse and low rank subspace learning to learn the graph structure. We follow the [1,6,13] to use the sparse and low rank subspace learning in our work. However, it is straight forward to extend our work with different graph structure learning methods. We reformulated our objective function as,

$$L(\mathbf{X}, \mathbf{\Lambda}, \mathbf{Y}) = \|\mathbf{Y} - f(\mathbf{X}, \mathbf{\Lambda})\|_p + \lambda_0 f(\mathbf{X}, \mathbf{\Lambda})^T \mathbf{f}(\mathbf{X}, \mathbf{\Lambda}) \tag{4}$$
$$+\lambda_1 \|\mathbf{\Lambda}\|_1 + \lambda_2 \|\mathbf{\Lambda}\|_*, \mathrm{s.\ t.} = \mathbf{I} - \mathbf{\Lambda}, \mathbf{X} = \mathbf{X\Lambda},$$

where $\mathbf{\Lambda}$ is constrained to be low rank using the nuclear norm $\|\|_*$ and sparse using L_1 norm.

Predefined Knowledge Graph. Many previous works have shown that expert knowledge is tremendously helpful for medical data analysis especially when labelled training data-size is small. In this work, we only have 821 labelled MRI slices as training data, which is very small for training a graph neural network. In order to further improve our model, we extracted a knowledge graph from radiologist labelled lymph node ontology graph as shown in Fig. 2. For all labelled training images/nodes, we will construct in-directed edges between them, if they are connected in the predefined ontology graph.

Besides the ontology graph, since we can access the clinical report associated with each MRI slice in our dataset, we also use the MRI reports (shown in Fig. 3) to train a report classification model using pre-trained BERT [4] and apply it to extract a semantic vector by attention pooling for different lymph nodes. Based on the similarity between these semantic features and the ontology graph, we constructed a knowledge graph for node label classification. Denote the adjacency matrix of the knowledge graph as A_{kg}, our problem is further formulated as,

$$
\begin{aligned}
\mathrm{L}(\mathbf{X}, \mathbf{\Lambda}, \mathbf{Y}) ={} & \|\mathbf{Y} - f(\mathbf{X}, \mathbf{\Lambda})\|_p + \lambda_0 f(\mathbf{X}, \mathbf{\Lambda})^T f(\mathbf{X}, \mathbf{\Lambda}) \\
& + \lambda_1 \|\mathbf{\Lambda}\|_1 + \lambda_2 \|\mathbf{\Lambda}\|_*, \text{s. t.} = \mathbf{I} - (\mathbf{\Lambda} + \beta \mathbf{\Lambda}_{kg}), \mathbf{X} = \mathbf{X}\mathbf{\Lambda},
\end{aligned}
\tag{5}
$$

where β is a hyper-parameter to learn for adding the knowledge graph as prior in this problem. In order to solve Eq. 6, we use a Lagrange multiplayer to add the constraints to the objective function,

$$
\begin{aligned}
\mathrm{L}(\mathbf{X}, \mathbf{\Lambda}, \mathbf{Y}) ={} & \|\mathbf{Y} - f(\mathbf{X})\|_p + \lambda_0 f(\mathbf{X})^T (\mathbf{I} - (\mathbf{\Lambda} + \beta \mathbf{\Lambda}_{kg})) \\
& \mathbf{f}(\mathbf{X}) + \lambda_1 \|\mathbf{\Lambda}\|_1 + \lambda_2 \|\mathbf{\Lambda}\|_* + \lambda_3 \|\mathbf{X} - \mathbf{X}\mathbf{\Lambda}\|_2^2
\end{aligned}
\tag{6}
$$

Graph Convolutional Neural Networks. Graph convolutional neural networks have shown great successes on many applications [3,14], our proposed framework can be easily combined with graph convolutional neural networks. Let $\mathbf{H}$ denotes the hidden states of a graph neural network, $\mathbf{W}_l$ as the weights for the hidden layer l, a standard graph convolutional neural network can be formulated $\mathbf{H}_{l+1} = f(\mathbf{W}_l, \mathbf{\Lambda}, \mathbf{H}_l)$, where the input is the image feature matrix $\mathbf{X}$ and output is the predicted labels $\mathbf{Y}$, thus, $\mathbf{H}_0 = \mathbf{X}, \mathbf{H}_{last} = \mathbf{Y}, \mathbf{H} = [\mathbf{H}_0, \cdots, \mathbf{H}_{last}]$. We can rewrite Eq. 7 using graph convolutional neural networks as,

$$
\begin{aligned}
\mathrm{L}(\mathbf{X}, \mathbf{\Lambda}, \mathbf{Y}, \mathbf{W}, \mathbf{H}) ={} & \|\mathbf{Y} - f(\mathbf{X}, \mathbf{W}, \mathbf{\Lambda}, \mathbf{H})\|_p \\
& + \lambda_0 f(\mathbf{X}, \mathbf{W}, \mathbf{\Lambda}, \mathbf{H})^T (\mathbf{I} - (\mathbf{\Lambda} + \beta \mathbf{\Lambda}_{kg})) \mathbf{f}(\mathbf{X}, \mathbf{W}, \mathbf{\Lambda}, \mathbf{H}) \\
& + \lambda_1 \|\mathbf{\Lambda}\|_1 + \lambda_2 \|\mathbf{\Lambda}\|_* + \lambda_3 \|\mathbf{X} - \mathbf{X}\mathbf{\Lambda}\|_2^2.
\end{aligned}
\tag{7}
$$

Optimization Method. It is trivial to optimize both the graph structure and node classification on the same dataset. In order to solve Eq. 7, we proposed a bi-level optimization method to learn the graph structure $\mathbf{\Lambda}$, GNN model weight $\mathbf{W}$ and labels $\mathbf{Y}$ jointly. The graph structure adjacency matrix $\mathbf{\Lambda}$ is learned on the validation dataset, and the GNN model weight $\mathbf{W}$ is learned on the training dataset. Hyperparamters $\lambda_0, \lambda_1, \lambda_2, \lambda_3, \beta$ are also learned on the validation dataset too. The detailed optimization algorithm is shown in Algorithm 1.

Algorithm 1: Optimization algorithm for jointly learn $\mathbf{\Lambda}, \mathbf{W}$

Result: Graph Adjacency Matrix $\mathbf{\Lambda}$, Graph Neural Network weights $\mathbf{W}$

1 Initialize parameter $\mathbf{W}$ randomized, $\mathbf{H}_0 = \mathbf{X}, \mathbf{H}_{last} = \mathbf{Y}, \lambda_0, \lambda_1, \lambda_2, \lambda_3$;

2 $i = 1$, max iteration of outer loop $max_{iter} = 1e5$;

3 max iteration of inner loop $max_k = 1e3$;

4 learning rate $\rho = 1e-3$;

5 **while** $i < max_{iter}$ **do**

6     $\mathbf{W}^{i+1} = \mathbf{W}^i + \mu_w \dfrac{\partial \mathbf{L}^i(\mathbf{X}, \mathbf{\Lambda}^k, \mathbf{W}^i, \mathbf{Y}^i, \mathbf{H}^i)}{\mathbf{Y}}$;

7     $\mu_w = \dfrac{\partial^2 \mathbf{L}^i(\mathbf{X}, \mathbf{\Lambda}^k, \mathbf{W}^i, \mathbf{Y}^i, \mathbf{H}^i)}{\mathbf{W}^2}$;

8     $i = i + 1$;

9     $k = 1$;

10   **while** $k < max_k$ *(Validation Dataset)* **do**

11       $\mathbf{\Lambda}^{k+1,i} = \mathbf{\Lambda}^{k,i} + \mu_A \dfrac{\partial \mathbf{L}^{k,i}(\mathbf{X}, \mathbf{\Lambda}^{k,i}, \mathbf{W}^i, \mathbf{Y}^i, \mathbf{H}^i)}{\mathbf{\Lambda}}$;

12       $\mu_A = \dfrac{\partial^2 \mathbf{L}^i(\mathbf{X}, \mathbf{\Lambda}^k, \mathbf{W}^i, \mathbf{Y}^i, \mathbf{H}^i)}{\mathbf{\Lambda}^2}$;

13       $\mu_A = \rho \mu_A + \mu_A$;

14       $k = k + 1$;

15   **end**

16    $\mu_W = \rho \mu_W + \mu_W$;

17 **end**

---

3   Experiments

Dataset. For model development and validation, we collected large-scale MRI studies from **, performed between Jan 2015 to Sept 2019 along with their associated radiology reports. The majority (63%) of the MRI studies were from the oncology department. This dataset consists of a total of 821 T2-weighted MRI axial slices from 584 unique patients. The lymph node labels were extracted by a radiologist with 8 years of post-graduate experience. The study was a retrospective study and was approved by the Institutional Review Board with a waiver of informed consent. This dataset comprised the reference (gold) standard for our evaluation and comparative analysis.

Benchmark Methods. We implemented several benchmark methods in our experiments. 1) Support Vector Machine (SVM) [7]: applying classical SVM on the extracted multi-scale bounding box features. 2) Structured SVM: constraining the support vector machine to output structural labels constrained by the knowledge graph structure in Fig. 2. 3) Standard Simple Graph Model (SG); 4) SG with Graph Structure Learning (SG+SL); 5) SG with SL and Predefined Knowledge Graph (SG+SL+KG); 6) Deep Neural Graph (GCN); 7) GCN with Graph Structure Learning (GCN+SL); 8) GCN with SL and Predefined Knowledge Graph (GCN+SL+KG); 9) Deep Neural Hyper-Graph (HGCN)[9]; 10) HGCN with Graph Structure Learning (HGCN+SL); 11) HGCN with SL and Predefined Knowledge Graph (HGCN+SL+KG). We use the same lymph

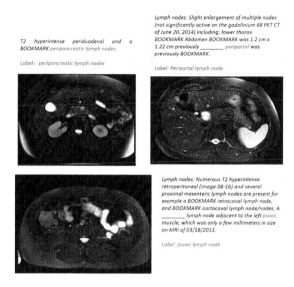

Fig. 3. An example of MRI slice with the labelled bounding boxes for lymph-nodes and graph-cut based detailed annotations of lymph nodes, linked clinical report sentences and radiologist labelled lymph node names.

node image feature embedding framework for all competing methods since we want to show the different classification performance between our methods and all other benchmark methods.

Experiment Setting and Data Processing. We divided the dataset into 10 folds, use one fold as validation dataset and two folds as testing dataset, the left seven folds are used as the training dataset. We run the cross validation for 10 times and report the averaged top-k (k = 1,2,3) classification of accuracy for different types of lymph nodes, F1-score and AUC of binary classification performance. In our dataset, we have the access to the clinical report of each MRI scan. The radiologist describes the lymph node information including the labels, size measurements, and slice numbers in a sentence with hyperlink (called bookmark) referred to the related MRI slices. The radiologist defines a bookmark as a hyperlink connection between the annotation in the image and then writes description in the report. We have one experienced radiologist to extract the lymph node labels from the bookmark linked sentences and use them as the ground truth labels for the lymph nodes in the connected MRI slices.

The size of lymph nodes are measured by four points at maximum dimension of lymph nodes or two points at the maximum dimension of lymph nodes. Based on these key points, we extracted multi-scale bounding boxes around the lymph nodes and extract the features in these bounding boxes using pretrained CNN model on MRI slices. We further use graph-cut to extract the fine contours of the lymph nodes and extract the cut lymph nodes using pretrained CNN model. We concatenated all these multi-scale bounding box features and lymph

Table 1. The top-k ACC & top-k F1 score of multi-class lymph node classification performance by different competing methods

Method	top-1 ACC	top-2 ACC	top-3 ACC	top-1 F1	top-2 F1	top-3 F1
SVM	0.72	0.78	0.85	0.70	0.76	0.83
Structured SVM	0.75	0.81	0.87	0.74	0.80	0.85
SG	0.73	0.80	0.86	0.72	0.79	0.84
SG+SL	0.76	0.82	0.88	0.75	0.81	0.87
SF+SL+KG	0.79	0.85	0.90	0.78	0.84	0.88
GCN	0.75	0.81	0.87	0.74	0.80	0.86
GCN+SL	0.81	0.84	0.89	0.80	0.83	0.88
GCN+SL+KG	**0.83**	**0.86**	**0.91**	**0.82**	**0.86**	**0.90**
Hyper-GCN	0.76	0.83	0.88	0.75	0.82	0.87
Hyper-GCN+SL	0.83	0.87	0.91	0. 82	0.86	0.90
Hyper-GCN+SL+KG	**0.85**	**0.89**	**0.93**	**0.84**	**0.88**	**0.92**

node features and use it as the graph node feature representation. The length of the concatenated multi-scale feature vector is 25088. We use the pre-trained bioBERT to train the clinical notes classification and label attention to extract the semantic label embedding. We used about more than 28000 sentences of de-identified clinical reports from ** hospital and use it to embed the distance between different lymph node names in our dataset. Based on the semantic distance between different lymph nodes, we constructed a semantic embedding graph and combine it with the predefined ontology graph shown in Fig. 2 to refine the final label prediction results.

Quantitative Results. We compared our proposed model to several benchmarks and Table 1 shows the top-k mean accuracy and F1 score of 14 classification results on the 10-fold cross validation. Top-k accuracy has been broadly used for multi-class classification performance measurement in previous work [5]. We show that the simple graph model generally outperforms SVM methods, and the structured SVM improves classical SVM about > 0.03 on both accuracy and F1 score by adding the structured constraints on different classes (extracted from the pre-defined ontology graph). Learning graph structure improves the top-k accuracy of the simple graph model by > 0.03 and F1 score about > 0.02. Using knowledge graph improves the simple graph mode further about > 0.03 on both top-k accuracy and F1 score. The convolutional neural graph also improves the classification top-k accuracy and F1 score consistently compared to the simple graph model. Learning graph structure and using knowledge graph under convolutional neural graph framework, our proposed model achieves the best performance and shows about 0.91 on top-3 accuracy and 0.90 on top-3 F1 score. We also combine the graph learning method with convolutional hyper-graph model and show that it improves the accuracy and F1-score > %2 compared to convolutional graph model. The best top-3 accuracy and F1 score achieved by hyper-graph model is 93% and 0.92%.

Acknowledgment. This research was supported in part by the Intramural Research Program of the National Institutes of Health, Clinical Center and National Library of Medicine.

References

1. Chen, J., Yang, J.: Robust subspace segmentation via low-rank representation. IEEE Trans. Cybern. **44**(8), 1432–1445 (2014)
2. Franceschi, L., Niepert, M., Pontil, M., He, X.: Learning discrete structures for graph neural networks. In: Proceedings of the 36th International Conference on Machine Learning (2019)
3. Hamilton, W.L., Ying, R., Leskovec, J.: Representation learning on graphs: Methods and applications. CoRR abs/1709.05584 (2017)
4. Lee, J., et al.: BioBERT: a pre-trained biomedical language representation model for biomedical text mining. Bioinformatics **36**(4), 1234–1240 (2019)
5. Lu, J., Xu, C., Zhang, W., Duan, L.Y., Mei, T.: Sampling wisely: Deep image embedding by top-k precision optimization. In: Proceedings of the IEEE/CVF International Conference on Computer Vision (ICCV), October 2019)
6. Sui, Y., Wang, G., Zhang, L.: Sparse subspace clustering via low-rank structure propagation. Pattern Recogn. **95**, 261–271 (2019)
7. Tsochantaridis, I., Joachims, T., Hofmann, T., Altun, Y.: Large margin methods for structured and interdependent output variables. J. Mach. Learn. Res. **6**, 1453–1484 (2005)
8. Wu, Z., Pan, S., Chen, F., Long, G., Zhang, C., Yu, P.S.: A comprehensive survey on graph neural networks. CoRR abs/1901.00596 (2019)
9. Yadati, N., Nimishakavi, M., Yadav, P., Louis, A., Talukdar, P.P.: Hypergcn: Hypergraph convolutional networks for semi-supervised classification. CoRR abs/1809.02589 (2018). http://arxiv.org/abs/1809.02589
10. Zareian, A., Karaman, S., Chang, S.: Bridging knowledge graphs to generate scene graphs. In: CVPR (2020)
11. Zhang, D., et al.: Knowledge graph-based image classification refinement. IEEE Access **7**, 57678–57690 (2019)
12. Zhou, Y., Sun, Y., Honavar, V.G.: Improving image captioning by leveraging knowledge graphs. CoRR abs/1901.08942 (2019)
13. Zhu, X., Zhang, S., Li, Y., Zhang, J., Yang, L., Fang, Y.: Low-rank sparse subspace for spectral clustering. IEEE Trans. Knowl. Data Eng. **31**(8), 1532–1543 (2019)
14. Zhu, Y., Zhu, X., Kim, M., Yan, J., Kaufer, D., Wu, G.: Dynamic hyper-graph inference framework for computer-assisted diagnosis of neurodegenerative diseases. IEEE Trans. Med. Imaging **38**(2), 608–616 (2019)

Improving Joint Learning of Chest X-Ray and Radiology Report by Word Region Alignment

Zhanghexuan Ji[1]([✉]), Mohammad Abuzar Shaikh[1]([✉]), Dana Moukheiber[1], Sargur N Srihari[1], Yifan Peng[2], and Mingchen Gao[1]

[1] Department of Computer Science and Engineering, University at Buffalo, The State University of New York, Buffalo, NY, USA
{zhanghex,mshaikh2,danamouk,srihari,mgao8}@buffalo.edu
[2] Population Health Sciences, Weill Cornell Medicine, New York, NY, USA
yip4002@med.cornell.edu

Abstract. Self-supervised learning provides an opportunity to explore unlabeled chest X-rays and their associated free-text reports accumulated in clinical routine without manual supervision. This paper proposes a Joint Image Text Representation Learning Network (JoImTeR-Net) for pre-training on chest X-ray images and their radiology reports. The model was pre-trained on both the global image-sentence level and the local image region-word level for visual-textual matching. Both are bidirectionally constrained on Cross-Entropy based and ranking-based Triplet Matching Losses. The region-word matching is calculated using the attention mechanism without direct supervision about their mapping. The pre-trained multi-modal representation learning paves the way for downstream tasks concerning image and/or text encoding. We demonstrate the representation learning quality by cross-modality retrievals and multi-label classifications on two datasets: OpenI-IU and MIMIC-CXR. Our code is available at https://github.com/mshaikh2/JoImTeR_MLMI_2021.

Keywords: Self-supervised learning · Multi-modality · Attention

1 Introduction

Chest X-ray is the most common medical imaging study globally for conducting clinical routines to assess chest regions. Because of its popularity, large, labeled datasets such as ChestX-ray14 dataset [24], CheXpert [10], OpenI-IU [5], and MIMIC-CXR [11,12], were collected as benchmarks for data-driven deep learning models to archive expert-level performance in analyzing chest regions. Among these biomedical datasets, OpenI-IU and MIMIC-CXR contain radiology reports along with corresponding radiographs. Given the large size of collected images and manual labeling being impractical, the disease labels are usually derived using natural language processing tools applied to the corresponding radiology reports.

Z. Ji and M.A. Shaikh—Equal contributions.

C. Lian et al. (Eds.): MLMI 2021, LNCS 12966, pp. 110–119, 2021.
https://doi.org/10.1007/978-3-030-87589-3_12

Recently, self-supervised representation learning has been explored to extract underlying information from the data by performing proxy tasks that explore the organization of the data itself. This is a promising direction for learning from a large amount of unlabeled biomedical data, where manual labeling is tedious, time-consuming, subjective, and requires domain knowledge. Self-supervised learning provides a great potential to investigate the biomedical data, including both medical images and their associated reports, accumulated during clinical routines. Ideally, both the modalities of the data encode the same medical condition and should be cross-referable.

Self-Attention mechanism was introduced to find the cross references within the same data modality [23]. This concept has contributed tremendously to the recent success of natural language processing models, such as BERT [6]. These models are pre-trained by predicting masked tokens to learn the underlying semantic representations from unlabeled textual data. Once the representation learning models are pre-trained, they can be fine-tuned and used as a backbone for a wide range of downstream natural language processing tasks.

Motivated by the above discussion, we propose to establish the cross references of the chest radiology images and reports to jointly learn the image-text representations. Learning cross-modal visual and textual representation is an essential task that can combine the semantic information contained within images and their descriptive reports [16,17]. These approaches have also been explored in biomedical image analysis [18]. The proposed representation learning mechanism will provide the foundation for a wide range of biomedical vision-and-language tasks, such as clinical inter-modal and intra-modal image-text retrieval, medical visual question answering [1], and automatic clinical report generation [19].

Contributions: We propose JoImTeRNet - a self-supervised pre-training network trained on multimodal inputs. Our network extracts and fuses the representations of the visual and textual modalities using both global image-sentence matching and local attention-based region-phrase matching. Phrases vary from length of one to three words. The proposed local region-phrase alignment enhances the joint representation learning by performing automatic fine-grained matching between image region-of-interests with phrases in reports. The local region-phrase matching is further enhanced using a soft-attention mechanism in the image encoder, without the need for explicit manual bounding box annotation or object detection on images. The quality of the learned representation is tested on the downstream classification and retrieval tasks.

2   Joint Image Text Representation Learning Network

We propose a **Jo**int **Im**age **Te**xt **R**epresentation Learning **Net**work (JoImTeR-Net) shown in Fig. 1. The JoImTeRNet architecture consists of an image and a text encoder. The representations are matched through a list of matching tasks, Text to Image Matching (TIM), Masked Language Modeling (MLM), Phrase to Region Alignment (PRA), and Word to Region Alignment (WRA). The learned

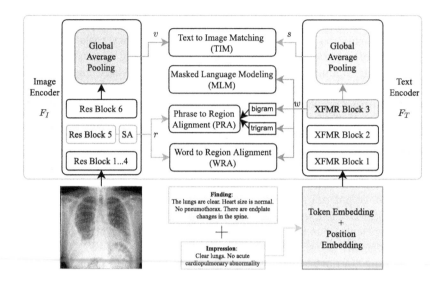

Fig. 1. The architecture of the proposed JoImTeRNet.

image and text representation are mapped to a shared feature space, given the hypothesis that the radiographs and their corresponding report contain consistent semantic meaning.

Given an X-ray image I and its corresponding radiology report T, we first encode them with an image encoder F_I and a text encoder F_T. The image encoder contains one input convolution layer, 6 residual blocks [8] and a global average pooling (GAP) layer. In the meantime, we also get the output from the Soft-Attention (SA) [22] block placed after the ResBlock5 to extract the region features $r \in \mathbb{R}^{D \times M}$, such that $v, r = F_I(\text{I})$ where $v \in \mathbb{R}^D$ is the global image features from GAP. Sentence and word level features s and w are extracted using a Transformer [23] based text encoder F_T, such that $w, s = F_T(\text{T})$, where $w \in \mathbb{R}^{D \times N}$ and $s \in \mathbb{R}^D$. Three transformer layers are deployed in F_T to encode the text report with the self-attention mechanism.

2.1   Matching Images and Sentences

To learn the joint representation of image and text pairs, we use the cross-entropy based matching (CEM) loss [26] and the ranking-based triplet matching (TM) loss [3]. Given a batch of image-text pairs $(\text{I}_i, \text{T}_i)_{i=1}^B$ (B is the batch size) and their corresponding visual features v and sentence features s from F_I and F_T, probability of T_i matching with I_i using softmax as: the image-to-text CEM loss L_{CEM}^{IT} is defined as the negative log posterior probability of the images being matched with their corresponding texts, *i.e.*,

$$L_{CEM}^{\text{IT}} = -\sum_{i=1}^B log(P(\text{T}_i|\text{I}_i)) = -\sum_{i=1}^B log(\frac{e^{\gamma S(\text{I}_i, \text{T}_i)}}{\sum_{j=1}^B e^{\gamma S(\text{I}_i, \text{T}_j)}}) \qquad (1)$$

where γ is the smoothing factor. $P(T_i|I_i)$ is the posterior probability of T_i matching with I_i using softmax. Cosine similarity $S(I_i, T_i) = (v^T s)/(\|v\|\|s\|)$ is used as the similarity score between image-text pairs. During the training, T_i is the correct match to I_i in the batch and all the other $T_j(j \neq i)$ are mismatching texts. Considering that image-text joint representation mapping should be bidirectional, we reverse I and T in Eq. (1) and get the symmetric text-to-image CEM loss as L_{CEM}^{TI}. Thus, the bidirectional CEM loss for globally matching image and text is defined as $L_{CEM}^s = L_{CEM}^{IT} + L_{CEM}^{TI}$.

Although CEM loss is designed to make the similarity between correct image-text pairs relatively higher than other mismatched pairs, it is difficult to set a hard margin between mismatched features. To solve this problem, TM loss [3], a ranking-based criterion, is added to increase the distance of mismatched pairs in the joint embedding space. Given an image I_i as the anchor, T_i is used as the positive paired sample. We then randomly select a mismatching text $T_j(j \neq i)$ within the batch as the negative paired sample. Symmetrically, if T_i is used as the anchor, then I_i and I_j would be positive/negative samples. The bidirectional TM loss for global image-text matching is formed as:

$$L_{TM}^s = L_{TM}^{IT} + L_{TM}^{TI} = \sum_{i,j=1}^{B} \Big[\max(0, S(I_i, T_j) - S(I_i, T_i) + \eta_s) \\ + \max(0, S(I_j, T_i) - S(I_i, T_i) + \eta_s) \Big]$$

(2)

where η_s is the hard margin and S is the cosine similarity the same as in Eq. (1).

2.2   Aligning Image Regions and Report Phrases

Both chest X-rays and their corresponding reports contain lots of fine-grained semantic information. We introduce a region-phrase level matching to align different concepts in the text reports with the regions of the images to further improve the joint representation. We apply region-phrase alignment with both CEM loss and TM loss. The length of a phrase is in the range of 1 to 3 words. Features of words, bigram and trigram phrases are denoted as w, p_2, p_3 respectively.

The cosine similarity between regions and words/phrases is not feasible to calculate directly due to the lack of explicit mapping between them. Instead, an attention-based matching score is deployed to overcome this challenge [7,9,26]. For region-word-level matching, given (I_i, T_i) and their region-word features (r, w), we first calculate the similarity matrix between all possible pairs of region features and word features using dot-product, i.e., $m = w^T r$, where $m \in \mathbb{R}^{N \times M}$, which is further normalized along N words as $\bar{m} = \text{Softmax}_N(m)$. Next, a context feature c is computed as the weighted sum over region features r, weighted by the region-word attention score α as follows:

$$c = \alpha r^T, \text{ where } \alpha_{i,j} = \frac{e^{\gamma_1 \bar{m}_{i,j}}}{\sum_{k=0}^{M-1} e^{\gamma_1 \bar{m}_{i,k}}}$$

(3)

where $c \in \mathbb{R}^{N \times D}$ and $\alpha \in \mathbb{R}^{N \times M}$; γ_1 is a hyper-parameter to tune the required amount of visual attention for a word. Here, the ith vector of c is the attention-weighted representation of all the sub-regions related to the ith word.

The attention-based region-word-level matching score is computed as:

$$S_a(\mathrm{I}, \mathrm{T}) = \log(\sum_{i=1}^{N-1} e^{(\gamma_2 S(c_i, w_i))})^{\frac{1}{\gamma_2}} \tag{4}$$

where $S(c_i, w_i) = (c_i^T w_i)/(\|c_i\| \|w_i\|)$, is the element-wise cosine similarity score between c_i and w_i, γ_2 is the importance magnification hyper-parameter for the most relevant word and context vector pair.

By replacing the cosine similarity score $S(\cdot, \cdot)$ with the region-word matching score $S_a(\cdot, \cdot)$ in Eq. (1) (2), we obtain the bidirectional CEM loss and TM loss for region-word alignment as $L_{CEM}^{p_1} = L_{CEM}^{rw} + L_{CEM}^{wr}$ and $L_{TM}^{p_1} = L_{TM}^{rw} + L_{TM}^{wr}$.

Furthermore, we obtain the phrase features by applying a 1D convolutional layer with kernel size 2 and 3 over w to get bigram $p_2 = \theta_{p_2}^T w$ and trigram $p_3 = \theta_{p_3}^T w$ phrase features respectively [14,27]. Here, $\theta_{p_2}, \theta_{p_3}$ are the convolution kernels of size 2 and 3. Our final cross-entropy with triplet matching (CETM) loss for our image-text joint representation learning is designed as:

$$L_{CETM} = \lambda_{CEM}(L_{CEM}^s + \sum_{i=1}^{3} L_{CEM}^{p_i}) + \lambda_{TM}(L_{TM}^s + \sum_{i=1}^{3} L_{TM}^{p_i}) \tag{5}$$

where λ_{CEM} and λ_{TM} are the loss weight hyper-parameters.

2.3   Downstream Task

In order to demonstrate the performance of joint representation learning, we use the pre-trained image and text encoders as the backbone and test the learned features on multi-label classification. We add projection layers followed by two fully connected layers for multi-label classification. Cross-entropy loss balanced with positive/negative ratio and class-wise weights [25] are used for training.

3   Experiments

3.1   Datasets

MIMIC-CXR v2.0 [11], is a large public dataset consisting of 377,110 chest X-rays associated with 227,835 radiology reports. We limit our study to the frontal-view images and only keep one frontal view image for each report. Following the pre-processing scheme in [3], we extract the *impressions, findings, conclusion* and *recommendation* sections from the raw report, normalized by SciSpaCy [20], and concatenate them. If none of these sections are present, we use the *final report* section. 14 CheXpert labels provided in MIMIC-CXR are used for classification task, where label 1 is considered as positive and all the other labels $(-1, 0)$ and

Table 1. Ablation for selecting the best loss setting. The matching score for OpenI-IU and MIMIC-CXR is computed on 1000 and 1000/3000 test samples respectively. Subscript s, w, p stand for image-text, region-word and region-phrase level matching.

Model setting	MIMIC-CXR												OPENI-IU					
	I2T (1K)			T2I (1K)			I2T (3K)			T2I (3K)			I2T (1K)			T2I (1K)		
	R@1	R@5	R@10	R@1	R@5	R@10	R@1	R@5	R@10	R@1	R@5	R@10	R@1	R@5	R@10	R@1	R@5	R@10
TM_s[3]	5.37	19.43	30.73	5.40	20.23	30.23	2.37	9.70	15.63	2.23	10.20	16.37	1.83	5.70	9.13	1.50	5.67	9.30
$TM_{w,s}$	6.30	21.73	32.23	6.00	20.97	30.90	2.77	10.67	18.07	2.83	10.97	17.83	1.93	6.37	10.17	1.97	6.53	10.30
CEM_s[7]	18.60	43.10	56.10	18.13	43.20	55.97	12.20	31.27	41.80	11.80	30.87	41.10	4.70	12.83	17.73	4.83	12.30	17.87
$CEM_{w,s}$[26]	18.60	44.20	56.27	18.87	43.40	55.67	12.60	31.67	41.80	12.83	31.57	41.27	4.87	13.00	18.00	5.37	13.40	18.33
$CETM_{w,s}$	**19.07**	45.33	57.20	**19.07**	44.70	56.73	**12.77**	31.90	43.00	**12.97**	31.97	42.03	**5.13**	**13.07**	18.73	5.50	13.73	19.20
$CETM_{w,p,s}$	18.93	**46.20**	**58.67**	**19.07**	**45.27**	**58.50**	12.67	**33.20**	**44.07**	12.83	**32.43**	**43.40**	5.07	**13.07**	**18.83**	**5.67**	**13.83**	**19.20**

missing labels are merged as negative. This results in 222,252 image-report pairs with 14 binary labels. We split the dataset into 217,252, 2,000 and 3,000 samples for training, validation and testing respectively.

OpenI-IU. [5] is a public dataset with 3,996 radiology reports and 8,121 associated chest X-ray images, which are manually annotated by human experts using MeSH words. Similar to TieNet [25], only unique frontal images and their corresponding reports which contain either *findings* and/or *impressions* are selected. This yields 3,643 image-report pairs, which are only used as external evaluation sets. For comparison and evaluation purposes, we select the 7 common labels in both OpenI and MIMIC-CXR (Table 3) from the MeSH domain.

3.2 Implementation Details

JoImTeRNet is implemented in Pytorch [21] and all the experiments are carried out on NVIDIA GTX 1080 Ti GPUs. For F_I, we use the basic residual blocks proposed in [8]. We employ 3 layers of Transformer blocks with 8 heads in F_T. The input image is encoded into 256 regions (r) flattened from 16×16 feature map output from Res Block 5 as shown in Fig. 1. The input image is cropped or padded to 2048×2048 then normalized to $[-1, 1]$. Random crop, rotation and color jitter are used for data augmentation. Report input is tokenized by a word-level tokenization scheme, where we collect all the words that appear more than twice in MIMIC-CXR dataset, which results in vocabulary size of $8,410$. The input reports are truncated or padded to the max length of $N = 160$.

Parameter Settings. We pretrain F_I and F_T on MIMIC-CXR training set using the image-text matching task explained in Sect. 2.1 and 2.2 to generate the joint image and text representations. The maximum epoch is set as 30. We employ AdamW [15] optimizer with an initial learning rate of 10^{-4}, which is dropped by 10 times after 20 epochs. L2 weight decay is set as 10^{-4}. For the downstream classification task in Sect. 2.3, we set up two different settings for comparison: randomly initializing the backbone and fine-tuning the pre-trained backbone. The learning rate for the classification head in both settings is set to 10^{-4}. For the randomly initialized setting, we train the backbone using the same learning rate as the classification block for 20 epochs, whereas the pre-trained backbone is fine-tuned with a smaller learning rate of 10^{-5} for only 10

Table 2. Classification AUCs on MIMIC-CXR [11] dataset. "FS" stands for training from scratch (FS). "FT" stands for fine-tuned model. Other comparison experiments are Visualbert [17], Uniter [4], and ClinicalBert [2].

Findings	Image		Image+Report				Report		
	FS	FT	[17]	[4]	FS	FT	[2]	FS	FT
EC	0.738	**0.763**	0.981	0.979	0.989	**0.996**	0.966	0.986	**0.980**
Cardiomegaly	0.794	**0.820**	0.991	0.989	0.989	**0.993**	0.979	0.988	**0.989**
Airspace opacity	0.749	**0.759**	0.991	0.989	0.988	**0.992**	0.978	0.985	**0.986**
Lung lesion	0.696	**0.756**	0.985	0.981	0.996	**0.998**	0.972	**0.989**	0.987
Edema	0.883	**0.895**	0.991	0.990	0.995	**0.996**	0.979	0.993	**0.994**
Consolidation	0.794	**0.810**	0.989	0.988	0.994	**0.998**	0.979	**0.996**	0.995
Pneumonia	0.733	**0.738**	0.977	0.974	0.981	**0.985**	0.962	0.980	**0.984**
Atelectasis	0.808	**0.824**	0.988	0.987	0.988	**0.995**	0.976	0.991	**0.992**
Pneumothorax	0.845	**0.855**	0.992	0.991	0.989	**0.993**	0.979	0.993	**0.994**
Pleural effusion	0.898	**0.904**	0.993	0.992	0.986	**0.997**	0.981	0.989	**0.990**
Pleural others	0.812	**0.839**	0.981	0.973	0.996	**0.999**	0.964	**0.998**	0.993
Fracture	0.641	**0.714**	0.976	0.977	**0.997**	0.990	0.958	**0.997**	0.997
Support devices	0.901	**0.913**	**0.995**	0.994	0.992	**0.995**	0.983	0.988	**0.993**
No findings	0.865	**0.874**	–	–	0.985	**0.989**	–	**0.980**	**0.980**
Avg	0.792	**0.815**	0.987	0.985	0.991	**0.994**	0.974	**0.990**	**0.990**

epochs. Our model pre-trained with the full loss setting CETM$_{wps}$ is used as the backbone for fine-tuning. The batch size is set to 32 for all the experiments. We select the loss hyper-parameters as $\gamma, \gamma_1, \gamma_2 = 2, 1, 1$, $\eta_s, \eta_w = 0.5, 0.5$ and $\lambda_{CEM}, \lambda_{TM} = 2.0, 1.0$.

3.3   Performances

Evaluation Metric. We evaluate the performance of JoImTeRNet by cross-modality retrieval task: given one image (text) as a query, we rank a subset of text (image), including the paired one, based on cosine similarity between the image and text features from JoImTeRNet. Recall@K (R@K) [13] is reported, where $K \in \{1, 5, 10\}$, which measures the fraction of times the correct matching is retrieved among the top K results in the test set. We compute R@K on a subset of 1000 image-text pairs and on the full 3000 samples in our MIMIC-CXR test set. We also report R@K on a subset of 1000 samples in OpenI-IU in order to evaluate JoImTeRNet on the external dataset (Table 1).

Ablation Study for Loss Settings. Ablation studies for different combinations of our losses are listed in Table 1. As we can see, full loss setting CETM$_{wps}$ achieves the highest R@5 and R@10 scores on all the test set, which shows the effectiveness of our multilevel phrase matching loss. In addition, the matching

Table 3. Classification AUCs on OpenI-IU [5] dataset. "FS" stands for training From Scratch (FS). "FT" stands for Fine-tuned model. Other comparison experiments are ChestX-ray14 [24], TieNet [25], Visualbert [17], Uniter [4], and ClinicalBert [2].

Findings	Image			Image+Report					Report				No. of samples
	[24]	FS	FT	[25]	[17]	[4]	FS	FT	[25]	[2]	FS	FT	
Cardiomegaly	0.803	0.924	**0.937**	0.962	0.977	0.978	0.956	**0.985**	0.944	0.969	0.966	**0.987**	315
Edema	0.799	0.937	**0.953**	**0.995**	0.982	0.989	0.922	0.962	**0.984**	0.976	0.947	0.964	40
Consolidation	0.790	0.951	**0.951**	0.989	0.996	**0.998**	0.954	0.975	0.969	**0.982**	0.938	0.975	28
Pneumonia	0.642	0.863	**0.934**	**0.994**	0.990	0.988	0.877	0.949	**0.983**	0.982	0.880	0.943	36
Atelectasis	0.702	0.829	**0.858**	0.972	**0.992**	0.982	0.947	0.978	**0.981**	0.947	0.952	0.971	293
Pneumothorax	0.631	0.926	**0.936**	0.960	0.988	0.983	0.962	**0.989**	0.960	0.973	0.951	**0.989**	22
Pleural effusion	0.890	0.938	**0.957**	0.977	**0.985**	0.983	0.922	0.971	0.968	**0.976**	0.926	0.968	140
No finding	–	0.844	**0.851**	–	–	–	0.883	**0.961**	–	–	0.898	**0.930**	2789
Avg	0.751	0.910	**0.932**	0.978	**0.987**	0.986	0.934	0.973	0.970	**0.972**	0.932	0.971	
W. Avg	0.771	0.893	**0.915**	0.971	**0.985**	0.982	0.943	0.978	0.965	0.964	0.949	**0.975**	

performance degrades when the model is trained on global matching loss only without the region-phrase(word)-level matching, i.e. CEM_s performs worse than CEM_{ws}. Similar results are found when comparing TM_s with TM_{ws}. This result shows that our proposed method for assisting joint representation learning using region-word matching is able to improve the representation ability of the image-text encoder. Moreover, the CETM combination consistently gains performance compared with only CEM loss or TM loss settings, which is just as we expected in Sects. 2.1 and 2.2. Notice that the matching scores are much lower on OpenI, since OpenI contains a large amount of similar reports, e.g. 'No acute disease.', which can have very similar feature representation from our model and thus largely degrade the matching score.

Downstream Image Classification Results. The AUCs from our two settings on both datasets along with other SOTA performances are shown in Table 2 and 3. We can see that the classifier performance finetuned on JoimTerNet backbone (FT) is always higher than training from scratch (FS), which shows the advance of our pre-training method. As shown in Table 2, FT achieves the highest AUCs on most tasks and labels on MIMIC-CXR test set (internal evaluation), even better than some SOTA models [2,4,17] on image-text and text classification. For the external evaluation on OpenI in Table 3, our FT setting extremely improves average AUC on image classification by 18% compared with TieNet [25], and also gains 1% on wAvg AUC than ClinicalBERT [2] on report classification. For the image-text classification, our model is still comparable with other SOTA models, even though our text encoder only contains 3 transformer layers compared with [4,17] which has a 12 layer BERT encoder as the backbone.

4 Conclusion

We propose a joint image-text representation learning network and show its performance on cross-modality retrieval and multi-label classification. We demonstrate the potential of self-supervised learning when it meets the continuously generated biomedical images and reports. We also leverage and show the importance of information contained within the relationship of words, phrases and image regions. Future work includes more complicated downstream tasks regarding both images and text.

Acknowledgment. This research was supported in part by NSF through grant IIS-1910492. It also was supported by the National Library of Medicine under Award No. 4R00LM013001.

References

1. Abacha, A.B., Hasan, S.A., Datla, V.V., Liu, J., Demner-Fushman, D., Müller, H.: VQA-Med: overview of the medical visual question answering task at ImageCLEf 2019. In: CLEF (Working Notes) (2019)
2. Alsentzer, E., et al.: Publicly available clinical BERT embeddings. In: Proceedings of the 2nd Clinical Natural Language Processing Workshop, pp. 72–78. Association for Computational Linguistics (2019)
3. Chauhan, G., et al.: Joint modeling of chest radiographs and radiology reports for pulmonary edema assessment. In: Martel, A.L., et al. (eds.) MICCAI 2020. LNCS, vol. 12262, pp. 529–539. Springer, Cham (2020). https://doi.org/10.1007/978-3-030-59713-9_51
4. Chen, Y.-C., et al.: UNITER: UNiversal image-TExt representation learning. In: Vedaldi, A., Bischof, H., Brox, T., Frahm, J.-M. (eds.) ECCV 2020. LNCS, vol. 12375, pp. 104–120. Springer, Cham (2020). https://doi.org/10.1007/978-3-030-58577-8_7
5. Demner-Fushman, D., et al.: Preparing a collection of radiology examinations for distribution and retrieval. J. Am. Med. Inform. Assoc. **23**(2), 304–310 (2016)
6. Devlin, J., Chang, M.W., Lee, K., Toutanova, K.: BERT: pre-training of deep bidirectional transformers for language understanding. In: Proceedings of the 2019 Conference of the North American Chapter of the Association for Computational Linguistics: Human Language Technologies, Volume 1 (Long and Short Papers), pp. 4171–4186. Association for Computational Linguistics (2019)
7. Fang, H., et al.: From captions to visual concepts and back. In: Proceedings of the IEEE Conference on Computer Vision and Pattern Recognition, pp. 1473–1482 (2015)
8. He, K., Zhang, X., Ren, S., Sun, J.: Deep residual learning for image recognition. In: Proceedings of the IEEE Conference on Computer Vision and Pattern Recognition, pp. 770–778 (2016)
9. Huang, P.S., He, X., Gao, J., Deng, L., Acero, A., Heck, L.: Learning deep structured semantic models for web search using clickthrough data. In: Proceedings of the 22nd ACM International Conference on Information & Knowledge Management, pp. 2333–2338 (2013)
10. Irvin, J., et al.: CheXpert: a large chest radiograph dataset with uncertainty labels and expert comparison. In: Proceedings of the AAAI Conference on Artificial Intelligence, vol. 33, pp. 590–597 (2019)
11. Johnson, A.E., et al.: MIMIC-CXR, a de-identified publicly available database of chest radiographs with free-text reports. Sci. Data **6**(1), 1–8 (2019)
12. Johnson, A.E., et al.: MIMIC-CXR-JPG, a large publicly available database of labeled chest radiographs. arXiv preprint arXiv:1901.07042 (2019)

13. Karpathy, A., Fei-Fei, L.: Deep visual-semantic alignments for generating image descriptions. In: Proceedings of the IEEE Conference on Computer Vision and Pattern Recognition, pp. 3128–3137 (2015)
14. Kim, Y.: Convolutional neural networks for sentence classification. In: Proceedings of the 2014 Conference on Empirical Methods in Natural Language Processing (EMNLP), pp. 1746–1751. Association for Computational Linguistics (2014)
15. Kingma, D.P., Ba, J.: Adam: a method for stochastic optimization. arXiv preprint arXiv:1412.6980 (2014)
16. Li, G., Duan, N., Fang, Y., Gong, M., Jiang, D.: Unicoder-VL: a universal encoder for vision and language by cross-modal pre-training. In: Proceedings of the AAAI Conference on Artificial Intelligence, vol. 34, pp. 11336–11344 (2020)
17. Li, L.H., Yatskar, M., Yin, D., Hsieh, C.J., Chang, K.W.: VisualBERT: a simple and performant baseline for vision and language. arXiv preprint arXiv:1908.03557 (2019)
18. Li, Y., Wang, H., Luo, Y.: A comparison of pre-trained vision-and-language models for multimodal representation learning across medical images and reports. In: 2020 IEEE International Conference on Bioinformatics and Biomedicine (BIBM), pp. 1999–2004. IEEE (2020)
19. Li, Y., Liang, X., Hu, Z., Xing, E.P.: Hybrid retrieval-generation reinforced agent for medical image report generation. In: Advances in Neural Information Processing Systems 31: Annual Conference on Neural Information Processing Systems 2018, NeurIPS 2018, Montréal, Canada, 3–8 December 2018, pp. 1537–1547 (2018)
20. Neumann, M., King, D., Beltagy, I., Ammar, W.: ScispaCy: fast and robust models for biomedical natural language processing. In: Proceedings of the 18th BioNLP Workshop and Shared Task, pp. 319–327. Association for Computational Linguistics (2019)
21. Paszke, A., et al.: Pytorch: an imperative style, high-performance deep learning library. Adv. Neural Inf. Process. Syst. **32**, 8026–8037 (2019)
22. Shaikh, M.A., Duan, T., Chauhan, M., Srihari, S.: Attention based writer independent verification. In: 2020 17th International Conference on Frontiers in Handwriting Recognition (ICFHR), pp. 373–379 (2020)
23. Vaswani, A., et al.: Attention is all you need. In: Advances in Neural Information Processing Systems, pp. 5998–6008 (2017)
24. Wang, X., Peng, Y., Lu, L., Lu, Z., Bagheri, M., Summers, R.M.: ChestX-ray8: hospital-scale chest X-ray database and benchmarks on weakly-supervised classification and localization of common thorax diseases. In: Proceedings of the IEEE Conference on Computer Vision and Pattern Recognition, pp. 2097–2106 (2017)
25. Wang, X., Peng, Y., Lu, L., Lu, Z., Summers, R.M.: TieNet: text-image embedding network for common thorax disease classification and reporting in chest X-rays. In: Proceedings of the IEEE Conference on Computer Vision and Pattern Recognition, pp. 9049–9058 (2018)
26. Xu, T., Zhang, P., Huang, Q., Zhang, H., Gan, Z., Huang, X., He, X.: AttnGAN: fine-grained text to image generation with attentional generative adversarial networks. In: Proceedings of the IEEE Conference on Computer Vision and Pattern Recognition, pp. 1316–1324 (2018)
27. Yang, Z., He, X., Gao, J., Deng, L., Smola, A.: Stacked attention networks for image question answering. In: Proceedings of the IEEE Conference on Computer Vision and Pattern Recognition, pp. 21–29 (2016)

Cell Counting by a Location-Aware Network

Zuhui Wang and Zhaozheng Yin[✉]

Stony Brook University, Stony Brook, NY, USA
{zuwang,zyin}@cs.stonybrook.edu

Abstract. The purpose of cell counting is to estimate the number of cells in microscopy images. Most popular methods obtain the cell numbers by integrating the density maps that are generated by deep cell counting networks. However, these cell counting networks that reply on estimated cell density maps may leave cell locations in a black-box. In this paper, we propose a novel cell counting network leveraging cell location information to obtain accurate cell numbers. Evaluated on four widely used cell counting datasets, our method which uses cell locations to boost the cell density map generation and cell counting, achieves superior performances compared to the state-of-the-art. The source codes will be available in our Github.

Keywords: Cell counting · Set loss · Supervised learning

1 Introduction

Cell counting is about obtaining the exact cell number inside microscopy images. The cell number, as a basic statistic, is valuable for biomedical discovery and patient disease diagnosis. Doctors can infer the disease stage from cell numbers and choose the suitable treatment for patients [1, 2, 4, 16]. For example, tumor cell growth of breast cancer is an important biomarker indicative of patients' prognosis. The number of cancer cells influences the patient's treatment plan [13]. However, cell counting is a tedious and time-consuming task for humans. Meanwhile, manually counting hundreds of cells in microscopy images may be error-prone and contain subjective errors. Figure 1 illustrates some challenges existing in cell counting. Some image regions are crowded with occluded cells, and some image regions have sporadic cells. The cells may exhibit various intensities and irregular shapes. Moreover, the microscopy image itself may have blurring or low image contrast issues. An efficient way to solve this problem is to develop an automatic cell counting system to obtain the cell number rapidly and accurately. Thus, automatically estimating cell numbers from microscopy images, using computer vision and machine learning approaches, has attracted a lot of attention in this community [1–3, 7, 10, 15, 16].

© Springer Nature Switzerland AG 2021
C. Lian et al. (Eds.): MLMI 2021, LNCS 12966, pp. 120–129, 2021.
https://doi.org/10.1007/978-3-030-87589-3_13

1.1 Related Work

To train machine learning-based cell counting algorithms, cell counting datasets have been annotated, in which a single dot is annotated for each cell center in the image. As shown in Fig. 1(b), cell centers are marked as 1, and the rest image regions are set to 0. Based on this annotation format, there are two main categories of methods to count cells with machine learning models:

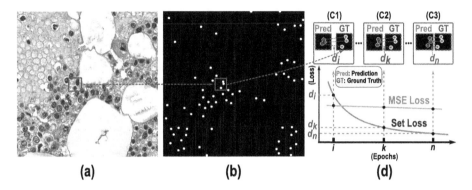

Fig. 1. A microscopy image example from the Modified Bone Marrow dataset [1]. (a) is the microscopy image containing multiple cells. (b) is the corresponding dot annotations. (c1) to (c3) illustrate a local patch of the ground truth of three cells along with their corresponding predictions in different training epochs. (d_i, d_k or d_n stands for the overall distance of all the cells in the patch.) (d) is the comparison between MSE loss and the proposed set loss during the model training. With more epochs, MSE loss varies slightly, but the set loss decreases significantly.

- Detection-based methods: the general idea is to detect all the cells inside images and obtain cell numbers based on the object detection results. Convolutional neural networks (CNNs) were developed to detect cells and then count them [7,10,15]. The detection-based counting methods rely on the accuracy of cell detection algorithms. However, the detection performance is affected severely by cell occlusion, shape and appearance variations of cells, and image background noise. The detection-based methods perform object-wise detection while being blind to the cell densities or global cell count.
- Regression-based method: some deep learning models were trained to regress the cell number from an input image directly [11,17]. An alternative way is to generate density maps, which is a currently popular strategy to do the cell counting task. Most of the state-of-the-art models are based on this density map generation method [1–3,16], from which the final cell number can be calculated by integrating the density map. However, the regression-based methods only focus on regressing the total cell number without considering the cell location information.

1.2   Motivation

The detection-based method detects objects independently, which does not consider the overall cell number or the global cell density map during the object-wise detection. Currently, the performance of detection-based methods [7,10,15] is inferior to that of regression-based methods [1–3,16]. For the regression-based methods, a density map is generated by the model to estimate the cell number. Since cell location is ignored in the regression-based method, there might be a black-box pitfall, i.e., the overall cell number estimation looks correct, but the individual cells might not be correctly detected. This issue is worse when multiple cells overlap together. In addition, overlapped cells are more difficult to count correctly than separate cells due to their high crowdedness. For example, the marked patch in Fig. 1(a) contains one separate cell and two overlapped cells that are difficult to count accurately without domain knowledge. Based on the predicted cells in Fig. 1(c1) to (c3), we can observe that: (1) the predicted cells are easily overlapped together when the ground truth cells are close to each other; (2) the distance between the set of predicted cells and the set of ground truth (d_i, d_k, and d_n) changes accordingly when the predicted cells are separated in different training epochs; (3) the Mean Square Error (MSE) loss on cell density maps is not sensitive to the cell location changes; and (4) the cell estimation loss in terms of set distance (*set loss*) becomes lower when the cell distance is smaller. Our key insight is that considering cell location information will benefit cell counting results. Therefore, it is worth unifying the advantages of regression-based cell counting methods by considering cell locations during density map estimation and cell counting.

1.3   Our Proposal and Contributions

We propose a location-aware cell counting network that unifies cell density map generation, cell counting, and individual cell localization. The main contributions of this paper are summarized below:

- The proposed cell counting network can count the overall cell number using the estimated cell density map that is constrained by individual cell locations. The proposed model surpasses state-of-the-art models on three public benchmarks of real cell images [1,6,11] and achieves competitive performance on a synthetic dataset [8].
- A new loss function consisting of three terms is proposed for the cell counting network: a pixel-wise loss to compare density maps; a set loss to compare the sets of cell locations, and a global scalar loss to compare cell counts.

2   Methodology

In this section, we first introduce the main components of the proposed cell counting network. Then we present the new loss function for training the model.

Cell Counting by a Location-Aware Network   123

2.1   Network Architecture

The network architecture is illustrated in Fig. 2. First, a feature encoder is employed to extract features from input images. Then, a feature enhancement module (attention mechanisms and dilated convolutions) is applied. The enhanced features are input to two branches: (1) one branch uses a feature decoder to generate the cell density map; and (2) the other branch uses a cell localization module to detect cells.

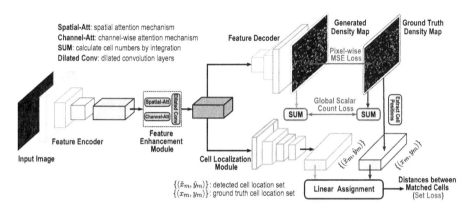

Fig. 2. The architecture of the proposed network for cell counting.

We use the first ten convolutional layers of the pre-trained VGG16 [14] model as the initialized feature encoder. Each convolutional layer uses a 3×3 kernel. The max-pooling operation is followed by every two or three convolutional layers. There are three max-pooling layers inside the encoder. Then, the feature maps are passed through a feature enhancement module that contains a spatial attention mechanism, a channel-wise attention mechanism, and multiple dilated convolutional layers. The spatial attention mechanism is designed to pay attention to the important feature parts spatially. It consists of two convolutional layers, and a sigmoid function is applied after the second convolutional layer. The channel-wise attention mechanism assigns different weights to the feature maps for our tasks. It takes the input feature maps into a global average pooling layer. Then, the outputs are fed into two fully-connected layers, followed by a sigmoid function. Finally, the feature maps after the attention mechanisms are sent to three dilated convolutional layers [9] to extract features at different scales.

The enhanced features are sent to two separate branches. First, they are input to a decoder, which contains three convolutional layers followed by up-sampling layers to generate the final density maps. Predicted cell numbers are obtained by integrating the density maps directly. Second, they are input to a cell localization module that contains five convolutional layers followed by max-pooling layers. The final output of the cell localization module is the set of coordinates of detected cells, $\{(\hat{x}_m, \hat{y}_m)\}$, where m denotes the detected cell indices in the training sample.

2.2 Loss Function

First, we employ the L_2 loss (MSE, Mean Squared Error) to compare the generated density map with the ground truth:

$$L_{\text{MSE}} = \frac{1}{N} \sum_{n=1}^{N} \left\| \hat{\mathbf{Y}}_n - \mathbf{Y}_n \right\|_2^2, \tag{1}$$

where $\hat{\mathbf{Y}}_n$ is the n-th density map generated by the proposed model, $\mathbf{Y}_n$ is the corresponding ground truth density map. N is the total number of training image samples.

Second, to incorporate cell detection into the density-map based cell counting, we propose a set loss between two sets of cell locations. The first set of cell locations $\{(\hat{x}_m, \hat{y}_m)\}$ belongs to the cells detected by the cell localization module, and the second set $\{(x_m, y_m)\}$ is the ground truth locations of cells. A linear assignment (Hungarian) algorithm [5] is employed to find the optimally matched cell pairs between the detection and ground truth. Based on the matched positions, we compute the set loss as:

$$L_{\text{SET}} = \frac{1}{N} \sum_{n=1}^{N} \left(\sum_{m=1}^{M} \left\| (\hat{x}_m, \hat{y}_m)_n - (x_{\delta(m)}, y_{\delta(m)})_n \right\|_1 \right), \tag{2}$$

where $(\hat{x}_m, \hat{y}_m)_n$ is the location of the m-th detected cell in the n-th image. $\delta(m)$ is the matched cell location in the ground truth calculated by the Hungarian algorithm. M is the total number of detected cell locations, and it is set to a fixed number during the model training (e.g., M is the maximum number of cells of an image in the dataset).

Besides, based on the generated density map, we propose the third loss term as a global scalar loss as:

$$L_{\text{COUNT}} = \frac{1}{N} \sum_{n=1}^{N} \left\| \hat{c}_n - c_n \right\|_1, \tag{3}$$

where $\hat{c}_n = \sum \hat{\mathbf{Y}}_n$ is the estimated cell number of the n-th image by integrating the cell density map, and $c_n = \sum \mathbf{Y}_n$ is the ground truth cell number of the n-th image.

The total loss function (L) for our proposed model is:

$$L = L_{\text{MSE}} + \lambda_1 L_{\text{SET}} + \lambda_2 L_{\text{COUNT}}, \tag{4}$$

where λ_1, and λ_2 are weights for the loss components. In the experiment, they are set to 1×10^{-2}, and 1×10^{-3} based on cross validations.

3   Experiment and Results

In this section, we first briefly describe the public datasets and the evaluation metric. Then, we compare the proposed model with state-of-the-art methods. Ablation studies are performed to evaluate the effectiveness of the main model components. The proposed model is implemented by PyTorch [12].

Table 1. Details of four public cell counting datasets. **#Cells/Image** denotes the mean value of cell numbers in each image and the corresponding standard deviation. **#Train** is the number of randomly selected training samples for model training including cross-validation for parameter selection, and the rest samples are used for testing. **#Total** is the total number of images in the dataset.

Datasets	Resolutions	#Cells/Image	#Train : #Total
VGG Cell [8]	256 × 256	174 ± 64	50 : 200
MBM [6]	600 × 600	126 ± 33	15 : 44
ADI [1]	150 × 150	165 ± 44.2	50 : 200
DCC [11]	960 × 960	34 ± 22	100 : 176

3.1   Datasets and Evaluation Metric

Four public benchmark datasets are used to evaluate the proposed model: (1) VGG Cell [8]: a synthetic cell dataset, (2) MBM [6]: a bone marrow dataset, (3) ADI [1]: a dataset of human adipocyte cells, and (4) DCC [11]: the Dublin cell counting dataset. The original images in DCC have various resolutions. We resize all the image samples to their maximum size (i.e., 960×960) for the model training. The statistical details of the four datasets are listed in Table 1.

By following the previous works [1,16], mean absolute error (MAE) is used as the evaluation metric:

$$MAE = \frac{1}{T} \sum_{t=1}^{T} \|\hat{c}_t - c_t\|_1, \tag{5}$$

where T is the number of testing samples, and $\hat{c}_t$ and c_t are the estimated cell counting number and the ground truth cell counting number of the t-th testing image, respectively.

3.2   Performance Comparison and Illustration

We compare the performance of the proposed model with several state-of-the-art methods on the four public datasets. The evaluation results are listed in Table 2. The FCRN-A and FCRN-B are two deep learning models based on Fully Convolutional Networks [16]. Count-ception [1] is a CNN-based model that infers the cell numbers based on redundant counting maps. SAU-Net [3] is a U-Net [2] based model with a self-attention module. In the table, we can observe that the proposed model outperforms others in the three real cell image datasets (MBM, ADI, and DCC). For the VGG Cell dataset, the leading model (i.e., Count-ception [1]) is highly engineered for this synthetic dataset [3], and the gap between our proposed method and the top performed method is small (0.39, i.e., the average difference between [1] and ours is less than one cell). Samples of generated density maps of the proposed method on the four datasets are shown in Fig. 3. Though the four datasets exhibit various cell appearances, shapes, and densities, our model is capable of cell counting with different imaging modalities.

Table 2. Evaluation results of different models on the four datasets, using mean absolute error (MAE) metric.

No.	Datasets	FCRN-A [16]	FCRN-B [16]	U-Net [2]	Count-ception [1]	SAU-Net [3]	Ours
1	VGG Cell [8]	2.90	3.20	4.51	**2.30**	2.60	2.69
2	MBM [6]	21.30	25.39	20.88	8.80	5.70	**4.83**
3	ADI [1]	33.01	30.80	33.44	19.40	14.20	**12.51**
4	DCC [11]	6.92	7.97	8.99	-[a]	3.00	**2.78**

[a]Note: the result is absent in the original publication.

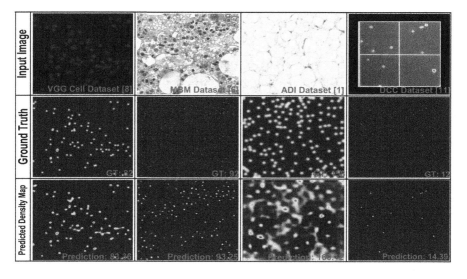

Fig. 3. Examples of input image, ground truth cell density map (cell number in red), and predicted cell density map (cell number in red). (Color figure online)

Table 3. Ablation study on the four datasets, with MAE as the evaluation metric.

No.	Models	VGG	MBM	ADI	DCC
1	Proposed-w/o-CountLoss	3.82	6.87	13.94	5.00
2	Proposed-w/o-SetLoss	4.22	7.52	17.07	6.76
3	Proposed-w/o-MSELoss	14.10	15.65	28.69	15.51
4	Proposed-w/o-AttentionMechanisms	8.08	6.22	21.42	8.11
5	Proposed-w/o-DilatedConvolutions	7.28	6.14	20.61	6.66
6	Proposed	2.69	4.83	12.51	2.78

3.3 Ablation Study

We conduct the ablation study on the four public datasets, and the experimental results are summarized in Table 3. Comparing No. 6 with No. 1 to No. 3, we can observe that each loss term makes its contribution to reduce the MAE. The MSE loss term that compares the cell density maps plays the most important role in the loss function, and the proposed set loss term can also improve the model performance substantially. For details, with the help of set loss, the MAE values can be reduced by 1.53 for VGG, 2.69 for MBM, 4.56 for ADI, and 3.98 for DCC, respectively. Some qualitative examples of the effectiveness of the proposed set loss term are illustrated in Fig. 4, where the predicted density map with set loss shows clear cell locations compared to the predicted density map without the cell location constraint. Besides, the count loss term can further help the proposed model reduce the MAE value.

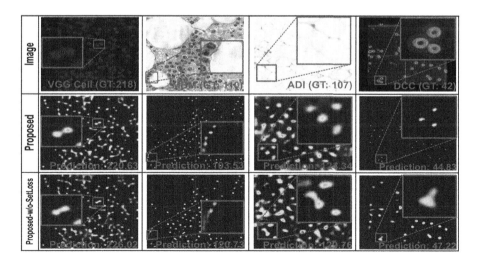

Fig. 4. Examples of the effectiveness of the proposed set loss on the four public datasets.

Accordingly, comparing No. 6 with No. 4 and No. 5, we can observe that the feature enhancement modules (attention mechanisms and dilated convolutions) can reduce the MAE, which verifies that the feature maps enhanced by these two modules are effective for cell counting.

4   Conclusion

In this paper, we proposed a location-aware cell counting network and a set loss function to count cells in microscopy images. The proposed model achieves superior performances on four cell counting benchmark datasets, which demonstrates the effectiveness of using a complementary task (e.g., cell locations) to boost the density-map based cell counting.

Acknowledgment. This project was supported by Stony Brook University - Brookhaven National Laboratory (SBU-BNL) seed grant on annotation-efficient deep learning.

References

1. Cohen, J.P., Boucher, G., Glastonbury, C.A., Lo, H.Z., Bengio, Y.: Count-ception: counting by fully convolutional redundant counting. In: 2017 IEEE International Conference on Computer Vision Workshops, ICCV Workshops 2017, Venice, Italy, 22–29 October 2017, pp. 18–26 (2017)
2. Falk, T., et al.: U-Net: deep learning for cell counting, detection, and morphometry. Nat. Methods **16**(1), 67 (2019)
3. Guo, Y., Stein, J.L., Wu, G., Krishnamurthy, A.K.: SAU-Net: a universal deep network for cell counting. In: Proceedings of the 10th ACM International Conference on Bioinformatics, Computational Biology and Health Informatics, BCB 2019, Niagara Falls, NY, USA, 7–10 September 2019, pp. 299–306. ACM (2019)
4. Hagos, Y.B., Narayanan, P.L., Akarca, A.U., Marafioti, T., Yuan, Y.: ConCORDe-Net: cell count regularized convolutional neural network for cell detection in multiplex immunohistochemistry images. In: Shen, D., et al. (eds.) MICCAI 2019, Part I. LNCS, vol. 11764, pp. 667–675. Springer, Cham (2019). https://doi.org/10.1007/978-3-030-32239-7_74
5. Jonker, R., Volgenant, T.: Improving the Hungarian assignment algorithm. Oper. Res. Lett. **5**(4), 171–175 (1986)
6. Kainz, P., Urschler, M., Schulter, S., Wohlhart, P., Lepetit, V.: You should use regression to detect cells. In: Navab, N., Hornegger, J., Wells, W.M., Frangi, A.F. (eds.) MICCAI 2015, Part III. LNCS, vol. 9351, pp. 276–283. Springer, Cham (2015). https://doi.org/10.1007/978-3-319-24574-4_33
7. Khan, A., Gould, S., Salzmann, M.: Deep convolutional neural networks for human embryonic cell counting. In: Hua, G., Jégou, H. (eds.) ECCV 2016, Part I. LNCS, vol. 9913, pp. 339–348. Springer, Cham (2016). https://doi.org/10.1007/978-3-319-46604-0_25
8. Lempitsky, V.S., Zisserman, A.: Learning to count objects in images. In: Lafferty, J.D., Williams, C.K.I., Shawe-Taylor, J., Zemel, R.S., Culotta, A. (eds.) Advances in Neural Information Processing Systems 23: 24th Annual Conference on Neural Information Processing Systems 2010. Proceedings of a meeting held, 6–9 December 2010, Vancouver, British Columbia, Canada, pp. 1324–1332 (2010)

9. Li, Y., Zhang, X., Chen, D.: CSRNet: dilated convolutional neural networks for understanding the highly congested scenes. In: Proceedings of the IEEE Conference on Computer Vision and Pattern Recognition, pp. 1091–1100 (2018)
10. Lucidi, M., Marsan, M., Visaggio, D., Visca, P., Cincotti, G.: Microscopy direct escherichia coli live/dead cell counting. In: 2018 20th International Conference on Transparent Optical Networks (ICTON), pp. 1–4. IEEE (2018)
11. Marsden, M., McGuinness, K., Little, S., Keogh, C.E., O'Connor, N.E.: People, penguins and petri dishes: adapting object counting models to new visual domains and object types without forgetting. In: 2018 IEEE Conference on Computer Vision and Pattern Recognition, CVPR 2018, Salt Lake City, UT, USA, 18–22 June 2018, pp. 8070–8079. IEEE Computer Society (2018)
12. Paszke, A., et al.: Pytorch: an imperative style, high-performance deep learning library. In: Advances in Neural Information Processing Systems 32: Annual Conference on Neural Information Processing Systems 2019, NeurIPS 2019, Vancouver, BC, Canada, 8–14 December 2019, pp. 8024–8035 (2019)
13. Shah, M.A., Wang, D., Rubadue, C., Suster, D., Beck, A.H.: Deep learning assessment of tumor proliferation in breast cancer histological images. In: 2017 IEEE International Conference on Bioinformatics and Biomedicine, BIBM 2017, Kansas City, MO, USA, 13–16 November 2017, pp. 600–603. IEEE Computer Society (2017)
14. Simonyan, K., Zisserman, A.: Very deep convolutional networks for large-scale image recognition. In: Bengio, Y., LeCun, Y. (eds.) 3rd International Conference on Learning Representations, ICLR 2015, San Diego, CA, USA, 7–9 May 2015, Conference Track Proceedings (2015)
15. Xia, T., Jiang, R., Fu, Y., Jin, N.: Automated blood cell detection and counting via deep learning for microfluidic point-of-care medical devices. CoRR abs/1909.05393 (2019)
16. Xie, W., Noble, J.A., Zisserman, A.: Microscopy cell counting and detection with fully convolutional regression networks. CMBBE Imaging Vis. **6**(3), 283–292 (2018)
17. Xue, Y., Ray, N., Hugh, J., Bigras, G.: Cell counting by regression using convolutional neural network. In: Hua, G., Jégou, H. (eds.) ECCV 2016, Part I. LNCS, vol. 9913, pp. 274–290. Springer, Cham (2016). https://doi.org/10.1007/978-3-319-46604-0_20

Exploring Gyro-Sulcal Functional Connectivity Differences Across Task Domains via Anatomy-Guided Spatio-Temporal Graph Convolutional Networks

Mingxin Jiang[1], Shimin Yang[1], Zhongbo Zhao[1], Jiadong Yan[1], Yuzhong Chen[1],
Tuo Zhang[2], Shu Zhang[3], Benjamin Becker[1], Keith M. Kendrick[1], and Xi Jiang[1(✉)]

[1] School of Life Science and Technology, MOE Key Lab for Neuroinformation, University of
Electronic Science and Technology of China, Chengdu, China
xijiang@uestc.edu.cn
[2] School of Automation, Northwestern Polytechnical University, Xi'an, China
[3] Center for Brain and Brain-Inspired Computing Research, School of Computer Science,
Northwestern Polytechnical University, Xi'an, China

Abstract. One of the most prominent anatomical characteristics of the human brain lies in its highly folded cortical surface into convex gyri and concave sulci. Previous studies have demonstrated that gyri and sulci exhibit fundamental differences in terms of genetic influences, morphology and structural connectivity as well as function. Recent studies have demonstrated time-frequency differences in neural activity between gyri and sulci. However, the functional connectivity between gyri and sulci is currently unclear. Moreover, the regularity/variability of the gyro-sulcal functional connectivity across different task domains remains unknown. To address these two questions, we developed a novel anatomy-guided spatio-temporal graph convolutional network (AG-STGCN) to classify task-based fMRI (t-fMRI) and resting state fMRI (rs-fMRI) data, and to further investigate gyro-sulcal functional connectivity differences across different task domains. By performing seven independent classifications based on seven t-fMRI and one rs-fMRI datasets of 800 subjects from the Human Connectome Project, we found that the constructed gyro-sulcal functional connectivity features could satisfactorily differentiate the t-fMRI and rs-fMRI data. For those functional connectivity features contributing to the classifications, gyri played a more crucial role than sulci in both ipsilateral and contralateral neural communications across task domains. Our study provides novel insights into unveiling the functional differentiation between gyri and sulci as well as for understanding anatomo-functional relationships in the brain.

Keywords: Functional MRI · Functional connectivity · Cortical folding · Graph convolutional network

Electronic supplementary material The online version of this chapter (https://doi.org/10.1007/978-3-030-87589-3_14) contains supplementary material, which is available to authorized users.

1 Introduction

Folding of the cerebral cortex into convex gyri and concave sulci is one of the most prominent anatomical features of the human brain [1, 2]. Gyri and sulci not only serve as the two core building blocks for cortical anatomy, but also provide a foundation to decipher the brain's anatomo-functional relationships. A variety of previous studies have extensively demonstrated that gyri and sulci exhibit fundamental differences in terms of genetics, morphology, structural connections, and function [2–5]. Therefore, there has been increasing interest in the field of brain mapping to investigate functional differentiation between gyri and sulci to achieve a better understanding of anatomo-functional relationships in the human brain together with how it contributes to cognition and behavior in health as well as in mental disorders.

With the development of deep learning methodologies, recent pioneering studies have demonstrated that gyri and sulci have fundamentally different functional MRI (fMRI) derived oscillations in the time-frequency domain in both human [6–8] and macaque [6] brains. However, there are still two major questions to address. First, previous studies focused on the time-frequency characteristics of fMRI signals between gyri and sulci, and merely fed the gyral/sulcal fMRI signals as the input for deep learning models [6–9], while the functional connectivity, i.e., functional dependency of fMRI signals, between gyri and sulci remains to be explored. Second, previous studies merely focused on the gyro-sulcal functional differences in a single task or during resting state, while the regularity/variability of such gyro-sulcal functional characteristics across different task domains is largely unknown.

In order to address these two questions, we proposed a novel anatomy-guided spatio-temporal graph convolutional network (AG-STGCN) model to classify task-based fMRI (t-fMRI) and resting state fMRI (rs-fMRI) data, and to further investigate gyro-sulcal functional connectivity differences across different task domains. The graph convolutional network has shown satisfying ability in classification or prediction [10]. After Yan et al. first proposed the concept of spatio-temporal GCN (STGCN) in action recognition [11], the STGCN model has also been applied to rs-fMRI BOLD signals to model both spatial and temporal characteristics of fMRI [12]. In this study, we extended the STGCN by adding anatomical priors (gyral/sulcal) for graph construction, and adopted it for the first time to t-fMRI/rs-fMRI classification. Firstly, the gyral/sulcal nodes and edges of the graph were constructed under the guidance of cortical anatomy. Next, the constructed graphs across different subjects were fed into the AG-STGCN to classify the t-fMRI and rs-fMRI data. Finally, the gyro-sulcal functional connectivity differences were assessed based on the learned graph edges of AG-STGCN. We hypothesized that t-fMRI and rs-fMRI data, which record task-evoked and intrinsic brain activity respectively, can be well classified based on the gyro-sulcal functional connectivity features [13]. We also hypothesized that gyri and sulci would exhibit different roles in those functional connectivity features [2–5].

2 Methods

2.1 Data Acquisition and Preprocessing

We adopted the 'grayordinate' t-fMRI and rs-fMRI data of 897 healthy subjects (age range 22–35) from the publicly released Human Connectome Project (HCP) S900 dataset [14]. The demographic information of the 897 subjects is detailed in [14]. The t-fMRI data consisted of seven different task paradigms corresponding to seven task domains, including emotion, gambling, language, motor, relational, social and working memory. The rs-fMRI data was acquired with eyes open and relaxed fixation. The major acquisition parameters of the fMRI are: 90×104 matrix, 72 slices, TR $= 0.72$ s, TE $= 33.1$ms, FOV $= 220$mm, flip angle $= 52°$, BW $= 2290$Hz/Px and 2.0mm isotropic voxels. The fMRI data preprocessing was performed using the HCP MR Data preprocessing pipelines version 3 [15]. In this work, we performed seven independent classifications, each of which was based on a pair of one specific t-fMRI and the one rs-fMRI of 800 selected subjects. Note that since the fMRI time points of the seven t-fMRI data (176, 253, 316, 284, 232, 274, 405, respectively) are less than those of the rs-fMRI data (1200), we randomly selected segments of the rs-fMRI time series with the same length of the paired t-fMRI data. The selection of different segments did not affect the classification performance as illustrated in Sect. 3.1.

2.2 Anatomy-Guided Gyro-Sulcal Spatio-Temporal Graph Construction

To construct the anatomy-guided gyro-sulcal spatio-temporal graphs for rs-fMRI/t-fMRI classification, we firstly parcellated the individual cortical surface into 68 regions of interest (ROI) according to the Desikan-Killiany atlas [16] (Fig. 1(a)). Secondly, since each cortical vertex had associated geometric measurement and corresponding fMRI signals, each ROI was further divided into gyral and sulcal parts by assessing the 'sulc' attribute which measures convexity during the cortical development process of each vertex [17]. We sorted the 'sulc' values of all vertices within an ROI (Fig. 1(b)), and defined the 10% of vertices with the highest 'sulc' values as gyri and the 10% with the lowest 'sulc' values as sulci (Fig. 1(c)) [18]. In this way, we obtained 136 regions in the cortical surface including 68 gyri and 68 sulci. Thirdly, the 136 gyral/sulcal regions were defined as the nodes at every spatial graph (at time point τ), and the averaged fMRI signal across all vertices within the gyral/sulcal region was adopted as the representative signal of the node to construct the graph edge. There were two kinds of edges representing the spatial and temporal characteristics of the graph, respectively. The spatial edge was defined as the connection between any pair of nodes in the graph at a time point (Fig. 1(e)), and the temporal edge was the connection between a node at a time point and its corresponding one at the proceeding time point [12]. Considering the balance between resting state and task-based graphs, the value on an edge was defined as the functional connectivity by concatenating rs-fMRI or t-fMRI signals across all subjects and then calculating the Pearson correlation coefficient between the two nodes. Finally, we obtained 800 resting state and 800 task-based graphs as the input of the following AG-STGCN model for classification.

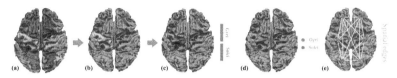

Fig. 1. Illustration of the anatomy-guided gyro-sulcal spatio-temporal graph construction. (a) Cortical parcellation into 68 regions of interest (ROI). Each color represents one ROI. (b) An example brain ROI colored in yellow. (c) Definition of gyri and sulci within the example ROI in (b). (d) Gyri and sulci were used as nodes in a graph. (e) An example spatial graph with six nodes (three gyri and three sulci). The spatial edges are colored in pale yellow. (Color figure online)

2.3 AG-STGCN Modeling for Graph Classification

An AG-STGCN model was adopted to classify the input gyro-sulcal spatio-temporal graphs between rs-fMRI and t-fMRI (Fig. 2). It consisted of a graph convolutional network (GCN), a temporal convolutional network (TCN), a global average pooling (GAP) layer and a fully connected (FC) layer. We first considered the convolution operation of the spatial graph at a single time point. We denoted the input feature map of N nodes $v_{ti}(i = 1, \ldots, N)$ at time point t as $\mathbf{f}_{in}$ and the spatial convolutional kernel as $\mathbf{W}_s$. The GCN output feature map $\mathbf{f}_s$ of the nodes was defined as [11]:

$$\mathbf{f}_s = \mathbf{\Lambda}^{-1/2}(\mathbf{A} + \mathbf{I})\mathbf{\Lambda}^{-1/2}\mathbf{f}_{in}\mathbf{W}_s \tag{1}$$

where $\mathbf{A}$ was the value matrix of spatial edges, and $\mathbf{\Lambda}$ was a diagonal matrix of $\Lambda^{ii} = \sum_j A^{ij} + 1$. When the graph convolution results of each node were obtained, we performed temporal convolution in TCN to add the influence of the time sequence:

$$\mathbf{f}_T = \mathbf{f}_s \otimes \mathbf{W}_T \tag{2}$$

In practice, Eq. (2) was N one-dimensional convolution in regular grid spacing, and $\mathbf{W}_T$ denoted the temporal convolutional kernel. After GCN and TCN layers, the AG-STGCN's output was fed to the global average pooling layer to reduce the dimension of parameters. Finally, we added the fully connected layer with the activation function sigmoid to determine whether the graph belonged to resting state or task.

At the training stage, we adopted an 8-fold cross-validation by randomly splitting the dataset into 87.5% training and 12.5% testing in each fold. We adopted similar hyper parameter settings of the STGCN model as in [12]. We set the length of temporal kernel Γ as 3 and utilized stochastic gradient descent (SGD) as an optimizer. The model was trained with 20 epochs and the batch size was set to 64. The base learning rate was set to 0.01 with a decay rate weight of 0.002. The proposed model was implemented in Pytorch with an NVIDIA GeForce GTX1080Ti GPU.

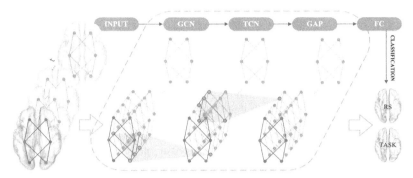

Fig. 2. The architecture of the anatomy-guided spatio-temporal graph convolutional network (AG-STGCN). The input anatomy-guided gyro-sulcal spatio-temporal graphs are classified into resting state (RS) or task-based (TASK) graphs as output by AG-STGCN. Note that each graph is a fully connected graph and is simplified for a better visualization.

2.4 Gyro-Sulcal Functional Connectivity Difference Assessment Based on the Learned Edges of AG-STGCN

We firstly identified the gyro-sulcal functional connectivity features contributing to the classification between task and resting state by assessing the 'edge importance' [12] which measured the importance of spatial graph edges in the classification. The edge importance was defined as a mask matrix $\mathbf{M} \in \mathbb{R}^{136 \times 136}$ all initialized to 1, and was learned during the training process by using $(\mathbf{A} + \mathbf{I}) * \mathbf{M}$ instead of $\mathbf{A} + \mathbf{I}$ in Eq. (1). In this way, $\mathbf{M}$ would change continuously in each iteration and affect the contribution of each edge across all STGC layers to obtain the optimal model, similar to the attention mechanism [19]. Since our graphs were fully connected and undirected, the matrix $\mathbf{M}$ was enforced to be positive and symmetric for better interpretability [12]. In $\mathbf{M}$, the graph edges with larger weights of 'edge importance' tended to have more contributions to the classification. We therefore defined top $k\%$ edges with largest weights out of all 9180 edges (upper/lower triangular matrix of edge importance $\mathbf{M} \in \mathbb{R}^{136 \times 136}$) within each classification as the features. We then assessed the gyro-sulcal functional connectivity difference characteristics by comparing the number of connections and nodes between gyri and sulci in those features in the whole brain as well as within and across hemispheres for investigation of ipsilateral and contralateral neural communications during the task.

3   Results

3.1   Classification Performance Between Task Domain and Resting State

We performed seven independent AG-STGCN classifications, each of which aimed to classify data from the 800 subjects for one specific t-fMRI and one rs-fMRI using 8-fold cross-validation. The average classification accuracy of the seven classifications are reported in Table 1. Our proposed AG-STGCN model has satisfyingly achieved more than 96% classification accuracy across all seven classifications, which not only verifies the effectiveness of this model, but also indicates that there are truly significant gyro-sulcal functional connectivity differences in task compared to resting state. Note that we randomly selected ten different segments of rs-fMRI time series and obtained stable classification accuracy with the mean accuracy of 97.13% and variance of 5.44 in the emotion/resting state classification (Supplemental Fig. 1), indicating that the selection of time series segment in rs-fMRI did not affect the robustness of AG-STGCN. In addition, we compared the classification performances with AG-GCLSTM (Anatomy-Guided Graph Convolutional Long Short Term Memory) [11] which also combined the spatial and temporal information of 4D fMRI data for a fair comparison, and AG-STGCN showed superior classification performance than AG-GCLSTM (Supplemental Table 1).

Table 1. Average classification accuracy of 8-fold cross-validation for the seven classifications.

Classifications	Accuracy (%)	Classifications	Accuracy (%)
Emotion/Resting State	99.13	Relational/Resting State	99.88
Gambling/Resting State	96.75	Social/Resting State	97.94
Language/Resting State	99.75	Working Memory/Resting State	96.63
Motor/Resting State	97.88	**Average Accuracy**	97.88

3.2 Functional Connectivity Differences Between Gyri and Sulci Within a Single Task Domain

We tested different top $k\%$ edges with largest weights from 0.1% to 0.5% with step size 0.1%, and found that the gyro-sulcal functional connectivity difference characteristics (Sect. 2.4) remained the same based on those different numbers of features. Therefore, we randomly selected 0.3% ($k = 0.3$) as a representative to demonstrate the functional connectivity difference between gyri and sulci in detail in Sects. 3.2 and 3.3. As an example, Fig. 3 shows the identified functional connectivity features in emotion/resting state (Fig. 3(a)) and motor/resting state (Fig. 3(b)) classifications. We consistently observed that: 1) the number of gyral nodes is greater than that of sulcal ones; and 2) the number of gyro-gyral connections is greater than that of sulco-sulcal ones, and the number of gyro-sulcal connections is in-between; 3) the gyro-gyral connections are more distributed in the whole-brain while the sulco-sulcal connections are more localized. From existing neuroscience literature, the connectivity features of emotion/resting state classification (Fig. 3(a)) mainly involve the fusiform gyrus (#7), frontal lobe (#3, #12, #14, #27, #28), inferior parietal cortex (#8), postcentral gyrus (#22), precentral gyrus (#24), superior temporal region (#30) and visual cortex (#5, #13, #21), which are all reported to play a functional role in emotion processing. In motor/resting state classification (Fig. 3(b)), the connectivity features mainly involve sensory-motor regions (#17, #22, #24), frontal lobe (#14, #28) and inferior parietal cortex (#8), which have been reported to be involved in motor tasks.

We then statistically compared the number of gyro-gyral (G-G), gyro-sulcal (G-S) and sulco-sulcal (S-S) connections in the functional connectivity features of each of the

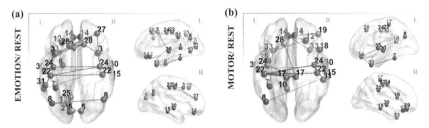

Fig. 3. Functional connectivity features in the emotion/resting state (Fig. 3(a)) and motor/resting state (Fig. 3(b)) classifications. The gyral and sulcal nodes are represented as orange and blue bubbles, respectively. The connections are colored in orange. (Color figure online)

seven task/resting state classifications. In Fig. 4, we observe that there is a significant difference in the number of the three types of connections (three-way ANOVA, $p < 0.05$). Moreover, the number of gyro-gyral connections is significantly larger than that of sulco-sulcal ones, while the number of gyro-sulcal connections is in-between (post-hoc right-tailed paired-samples t-test, $p < 0.05$, Bonferroni correction for multiple comparisons), which are consistent across the seven task domains.

In addition, we divided those functional connectivity features into inter- and intra-hemispheric ones, and found that there was also significant difference in the number of the three types of connections (G-G > G-S > S-S) in both inter-hemispheric and intra-hemispheric functional connections in the seven task domains (Supplemental Table 2).

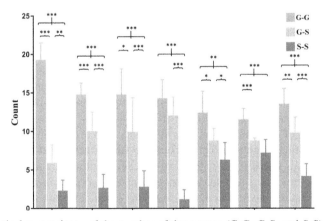

Fig. 4. Statistical comparisons of the number of three types (G-G, G-S, and S-S) of functional connections within each of the seven classifications. * indicates $p < 0.05$, ** indicates $p < 0.01$, *** indicates $p < 0.001$. WM: working memory.

3.3   Regularity and Variability of Functional Connectivity Differences Between Gyri and Sulci Across Task Domains

We investigated the regularity/variability of the functional connectivity features across the seven task domains. Figure 5(a-b) shows all of the functional connectivity features in the seven classifications. For nodes, the left lingual gyrus (#13), left and right medial orbital frontal cortex (#14) are the top three regions most involved across the seven task domains. Note that the top three nodes are all gyral regions. For connections, the top three that are most involved across the seven task domains are: the gyral part of left medial orbital frontal cortex - the gyral part of right medial orbital frontal cortex, left postcentral gyrus - right postcentral gyrus, and left fusiform gyrus - the gyral part of left entorhinal cortex. Note that these connections are all gyro-gyral connections with even two homotopic gyro-gyral connections within medial orbital frontal cortex and postcentral gyrus. Those functional connections as well as nodes in Fig. 5(a-b) are likely to be involved in cognitive functions based on available neuroscience literature. Moreover, our results show that gyral nodes as well as gyro-gyral functional connections are involved more than sulcal ones in the functional connectivity features, indicating more participation of gyri than sulci in neural communications during tasks.

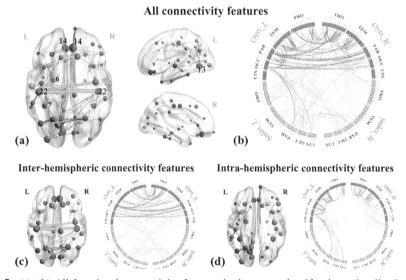

Fig. 5. (a)–(b) All functional connectivity features in the seven classifications visualized on the cortical surface (a) and as a circular plot (b). (c) Inter-hemispheric functional connectivity features across the seven task domains. (d) Intra-hemispheric functional connectivity features across the seven task domains. In each subfigure visualized on the cortical surface, the gyral and sulcal nodes are represented as orange and blue bubbles, respectively. The size of the bubble represents the occurrence probability of the node, and the thickness of the connection is in proportion with the occurrence probability of the functional connection. In each circular plot, the gyral and sulcal regions are colored in orange and blue, and are further organized into five major brain lobes (frontal, temporal, parietal, occipital and cingulate). The functional connections between G-G, G-S, and S-S are represented as orange, green, and blue lines, respectively. (Color figure online)

We further divided the functional connectivity features across all seven tasks in Fig. 5(a-b) into inter-hemispheric (Fig. 5(c)) and intra-hemispheric (Fig. 5(d)) ones. There are 54 G-G and 16 S-S inter-hemispheric connections, accounting for 57.4% and 17% of all inter-hemispheric ones respectively. There are 53 G-G and 8 S-S intra-hemispheric connections, accounting for 55.8% and 8.4% of all intra-hemispheric ones respectively. Again, the gyral nodes as well as gyro-gyral connections are dominant in both inter- and intra- hemispheric connectivity features. Moreover, for those inter-hemispheric connections (Fig. 5(c)), there are 36.17% homotopic ones and the other non-homotopic ones might be related to hemispheric lateralization [20]. For intra-hemispheric connections (Fig. 5(d)), the majority are short-range ones.

4   Conclusion

In this paper, we investigated the functional connectivity difference between gyri and sulci across different task domains by proposing a novel anatomy-guided spatio-temporal graph convolutional network (AG-STGCN). To the best of our knowledge, it was one of the earliest studies to explore the gyro-sulcal functional connectivity difference between task and resting state using advanced deep learning methodologies. Our results demonstrated that there were truly significant gyro-sulcal functional connectivity differences under task compared to resting state. Moreover, there was a greater involvement of gyri than sulci in both ipsilateral and contralateral neural communications during tasks. This study provided novel insights for unveiling the functional segregation between gyri and sulci, as well as for understanding how such brain anatomo-functional relationship contributed to brain cognitive functions. In the future, we plan to apply AG-STGCN on clinical datasets, such as from individuals with autism, in order to identify the gyro-sulcal functional connectivity alterations in mental disorders as potential biomarkers.

Acknowledgements. This work was partly supported by the National Natural Science Foundation of China (61976045), Sichuan Science and Technology Program (2021YJ0247), Key Scientific and Technological Projects of Guangdong Province Government (2018B030335001), National Natural Science Foundation of China (31971288 and 31671005), National Natural Science Foundation of China (62006194), the Fundamental Research Funds for the Central Universities (3102019QD005), High-level researcher start-up projects (06100-20GH020161).

References

1. Rakic, P.: Specification of cerebral cortical areas. Science **241**(4862), 170–176 (1988)
2. Van Essen, D.C.: A tension-based theory of morphogenesis and compact wiring in the central nervous system. Nature **385**(6614), 313–318 (1997)
3. Hilgetag, C.C., Barbas, H.: Developmental mechanics of the primate cerebral cortex. Anat. Embryol. **210**(5–6), 411–417 (2005)
4. Stahl, R., et al.: Trnp1 regulates expansion and folding of the mammalian cerebral cortex by control of radial glial fate. Cell **153**(3), 535–549 (2013)
5. Nie, J.X., et al.: Axonal fiber terminations concentrate on gyri. Cereb. Cortex **22**(12), 2831–2839 (2012)

6. Zhang, S., et al.: Deep learning models unveiled functional difference between cortical gyri and sulci. IEEE Trans. Biomed. Eng. **66**(5), 1297–1308 (2019)
7. Liu, H., et al.: The cerebral cortex is bisectionally segregated into two fundamentally different functional units of gyri and sulci. Cereb. Cortex **29**(10), 4238–4252 (2018)
8. Jiang, M., et al.: Exploring functional difference between gyri and sulci via region-specific 1d convolutional neural networks. In: Liu, M., Yan, P., Lian, C., Cao, X. (eds.) MLMI 2020. LNCS, vol. 12436, pp. 250–259. Springer, Cham (2020). https://doi.org/10.1007/978-3-030-59861-7_26
9. Dvornek, N.C., Ventola, P., Pelphrey, K.A., Duncan, J.S.: Identifying autism from resting-state fMRI using long short-term memory networks. In: Wang, Q., Shi, Y., Suk, H.-I., Suzuki, K. (eds.) Machine Learning in Medical Imaging, pp. 362–370. Springer International Publishing, Cham (2017). https://doi.org/10.1007/978-3-319-67389-9_42
10. Defferrard, M., Bresson, X., Vandergheynst, P.: Convolutional neural networks on graphs with fast localized spectral filtering. In: 30th Conference on Neural Information Processing Systems (2016)
11. Yan, S., Xiong, Y., Lin, D.: Spatial temporal graph convolutional networks for skeleton-based action recognition. In: 32nd AAAI Conference on Artificial Intelligence 2018, pp. 7444–7452 (2018)
12. Gadgil, S., Zhao, Q.Y., Pfefferbaum, A., Sullivan, E.V., Adeli, E., Pohl, K.M.: Spatio-temporal graph convolution for functional mri analysis. In: Medical Image Computing and Computer Assisted Intervention 2020, pp. 528–538 (2020)
13. Zhang, S., Li, X., Lv, J., Jiang, X., Guo, L., Liu, T.: Characterizing and differentiating task-based and resting state fMRI signals via two-stage sparse representations. Brain Imaging Behav. **10**(1), 21–32 (2015). https://doi.org/10.1007/s11682-015-9359-7
14. Van Essen, D.C., Smith, S.M., Barch, D.M., Behrens, T.E.J., Yacoub, E., Ugurbil, K.: The WU-Minn human connectome project: an overview. Neuroimage **80**, 62–79 (2013)
15. Smith, S.M., et al.: Resting-state fMRI in the human connectome project. Neuroimage **80**, 144–168 (2013)
16. Desikan, R.S., et al.: An automated labeling system for subdividing the human cerebral cortex on MRI scans into gyral based regions of interest. Neuroimage **31**(3), 968–980 (2006)
17. Lewis, L.D., Setsompop, K., Rosen, B.R., Polimeni, J.R.: Fast fMRI can detect oscillatory neural activity in humans. Proc. Natl. Acad. Sci. **113**(43), E6679–E6685 (2016)
18. Yang, S.M., et al.: Temporal variability of cortical gyral-sulcal resting-state functional activity correlates with fluid intelligence. Front. Neural Circuits **13**(36), 1–12 (2019)
19. Vaswani, A., et al.: Attention is all you need. In: Advances in Neural Information Processing Systems, vol. 30, pp. 6000–6010 (2017)
20. Tian, L.X., Wang, J.H., Yan, C.G., He, Y.: Hemisphere- and gender-related differences in small-world brain networks: a resting-state functional MRI study. Neuroimage **54**(1), 191–202 (2010)

StairwayGraphNet for Inter- and Intra-modality Multi-resolution Brain Graph Alignment and Synthesis

Islem Mhiri[1,2], Mohamed Ali Mahjoub[1], and Islem Rekik[2(✉)]

[1] BASIRA Lab, Faculty of Computer and Informatics, Istanbul Technical University, Istanbul, Turkey
[2] Université de Sousse, Ecole Nationale d'Ingénieurs de Sousse, LATIS- Laboratory of Advanced Technology and Intelligent Systems, 4023 Sousse, Tunisie
irekik@itu.edu.tr
http://basira-lab.com/

Abstract. Synthesizing multimodality medical data provides complementary knowledge and helps doctors make precise clinical decisions. Although promising, existing multimodal brain graph synthesis frameworks have several limitations. First, they mainly tackle only one problem (intra- or inter-modality), limiting their generalizability to synthesizing inter- and intra-modality simultaneously. Second, while few techniques work on super-resolving low-resolution brain graphs within a single modality (i.e., intra), inter-modality graph super-resolution remains unexplored though this would avoid the need for costly data collection and processing. More importantly, both target and source domains might have different distributions, which causes a domain fracture between them. To fill these gaps, we propose a multi-resolution StairwayGraph-Net (SG-Net) framework to jointly infer a target graph modality based on a given modality and super-resolve brain graphs in both inter and intra domains. Our SG-Net is grounded in three main contributions: (i) predicting a target graph from a source one based on a novel graph generative adversarial network in both inter (e.g., morphological-functional) and intra (e.g., functional-functional) domains, (ii) generating high-resolution brain graphs without resorting to the time consuming and expensive MRI processing steps, and (iii) enforcing the source distribution to match that of the ground truth graphs using an inter-modality aligner to relax the loss function to optimize. Moreover, we design a new Ground Truth-Preserving loss function to guide both generators in learning the topological structure of ground truth brain graphs more accurately. Our comprehensive experiments on predicting target brain graphs from source graphs using a multi-resolution stairway showed the

This project has been funded by the TUBITAK 2232 Fellowship (Project No: 118C288).

Electronic supplementary material The online version of this chapter (https://doi.org/10.1007/978-3-030-87589-3_15) contains supplementary material, which is available to authorized users.

outperformance of our method in comparison with its variants and state-of-the-art method. SG-Net presents the first work for graph alignment and synthesis across varying modalities and resolutions, which handles graph size, distribution, and structure variations. Our Python TIS-Net code is available on BASIRA GitHub at https://github.com/basiralab/SG-Net.

1 Introduction

Multimodal brain imaging has shown tremendous potential for neurological disorder diagnosis, where each imaging modality offers specific information for learning more holistic and informative data representations. Although doctors require many imaging modalities to aid in precise clinical decision, they suffer from restricted medical conditions such as high acquisition cost and processing time [1]. To address these limitations, several deep learning studies have investigated multimodal MRI synthesis [2]. Such methods either synthesize one modality from another (i.e., cross-modality) or map both modalities to a commonly shared domain. Notably, generative adversarial networks (GANs) [3,4] have achieved great success in predicting medical images of different brain image modalities from a given modality. For instance, [5] proposed a joint neuroimage synthesis and representation learning framework with transfer learning for subjective cognitive decline conversion prediction where they imputed missing PET images using MRI scans. Also, [6] aimed to synthesize a missing MRI modality from multiple modalities using conditional GAN. Although significant clinical representations were obtained from the latter studies, more substantial challenges still exist. The brain connectome has a complex non-linear structure that is difficult to capture using linear models [7]. Moreover, many approaches do not make effective use of or even fail to handle non-Euclidean structured data (i.e., geometric data), such as graphs and manifolds [8]. Therefore, a deep learning model that retains graph-based data representation topology provides a relevant research direction to be explored.

Recently, geometric deep learning techniques have shown great potential in learning the deep graph-structure. Particularly, deep graph convolutional networks (GCNs) have imbued the field of network neuroscience research through various tasks such as studying the mapping between human connectome and disease outcome [9]. Recent landmark studies have relied on using GCN to predict a target brain graph from a source brain graph. For instance, [10] introduced MultiGraphGAN architecture, which predicts multiple brain graphs from a single brain graph while preserving the topological structure of each target predicted graph. Moreover, [11] defined a multi-GCN-based generative adversarial network (MGCN-GAN) to synthesize individual structural connectome from a functional connectome. However, all these works can only transfer brain graphs from *one modality* to another while preserving the same resolution, limiting their generezability to *cross-modality* brain graph synthesis at *different resolutions* (i.e., node size). Therefore, using multi-resolution graphs (e.g., super-resolution) remains a significant challenge in designing generalizable and scalable brain graph synthesis models. [12] circumvented this issue by designing a graph neural network for super-resolving low-resolution (LR) functional brain connectomes from a single modality. However, super-resolving brain graphs across modalities (i.e., inter) is strikingly lacking. Furthermore, most previous works still face domain fracture problems resulting in the difference in distribution between the source and target domains. Remarkably, domain alignment remains mostly scarce in brain graph synthesis tasks [9]. While yielding outstanding performance, all the aforementioned methods have tackled only one problem (inter *or* intra-modality), which hinders their generalizability to synthesizing inter and intra-modalities jointly.

To address the challenges above and motivated by the recent development of graph neural network-based solutions, we propose a multi-resolution Stairway-GraphNet (SG-Net) method to jointly predict and super-resolve a target graph modality based on a given modality in both inter- and intra-domains. To do so, prior to the prediction blocks, we propose an inter-modality aligner network based on adversarially regularized variational graph autoencoder (ARVGA) [13] to align the training graphs of the source modality to that of the target one. Second, given the aligned source graphs, we design an inter-modality super-resolution graph GAN (gGAN) to map the aligned source graph from one modality (e.g., morphological) to the target modality (e.g., functional). Note that the alignment step facilitates the training of our super-resolution gGAN since both source and target domains have been aligned by the inter-modality aligner network (i.e., shared mode). To capture the complex relationship in both direct and indirect brain connections, we design the super-resolution generator and discriminator of our inter-modality GAN using edge-based GCNs [14]. Then, we synthesize high-resolution (HR) functional brain graphs from LR functional graphs using GCN-based intra-modality GAN. Besides, to resolve the inherent instability of GANs, we propose a novel ground-truth-preserving (GT-P) loss function to enforce our multi-resolution generators to effectively learn the ground-truth brain graph.

The main contributions of our work are four-fold. **On a methodological level,** StairwayGraphNet presents the first work for graph alignment and synthesis across varying modalities and resolutions, which can also be leveraged for boosting neurological disorder diagnosis. **On a clinical level,** learning multi-resolution brain connectivity synthesis can provide comprehensive brain maps that capture multimodal relationships (functional, structural, etc.) between brain regions, thereby charting brain dysconnectivity patterns in disordered populations [7]. **On a computational level,** our method generates HR graphs without resorting to any computational MRI processing step such as registration and parcellation. **On a generic level,** our framework is a generic method as it can be applied to predict brain graphs derived from any neuroimaging modalities with different resolutions and complex nonlinear distributions.

2   Proposed Method

This section presents the key steps of our stairway graph alignment, prediction, and super-resolution (SG-Net) framework. Fig. 1 presents an overview for the proposed framework in three major steps, which are detailed below.

Problem Definition. A brain graph can be represented as $\mathbf{G}(\mathbf{V}, \mathbf{E})$, where each node in $\mathbf{V}$ denotes a brain region of interest and each edge in $\mathbf{E}$ connecting two ROIs k and l denotes the strength of their connectivity. Each training subject i is represented by three brain graphs $\{\mathbf{G}_i^s(\mathbf{V}_i^s, \mathbf{E}_i^s), \mathbf{G}_i^{t_1}(\mathbf{V}_i^{t_1}, \mathbf{E}_i^{t_1}), \mathbf{G}_i^{t_2}(\mathbf{V}_i^{t_2}, \mathbf{E}_i^{t_2})\}$, where $\mathbf{G}^s$ is the source brain graph (morphological brain network) with n_r nodes, $\mathbf{G}^{t_1}$ is the first target brain graph (LR functional brain network) with $n_{r'}$ nodes

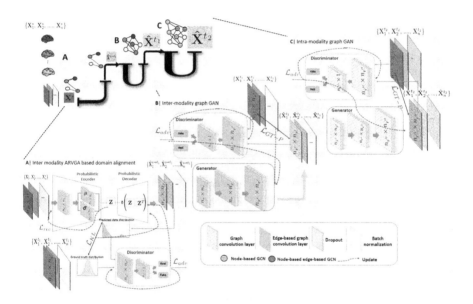

Fig. 1. *Illustration of the Proposed Inter and Intra-modality Multi-resolution Brain Graph Alignment and Synthesis using SG-Net.* **A| Graph-based inter-modality aligner.** We aim to align the training graphs of the source modality $\mathbf{X}^s$ to that of the target one $\mathbf{X}^{t_1}$. Thus, we design an ARVGA model with KL-divergence to bridge the gap between the distributions of the source and target graphs. **B| Adversarial inter-modality graph GAN.** Next, we propose an inter-modality multi-resolution graph GAN to transform the aligned source brain graph $\hat{\mathbf{X}}^{s \rightarrow t_1}$ (e.g., morphological) into the target graph (e.g., functional) with different structural and topological properties. **C| Adversarial intra-modality super-resolution graph GAN.** Then, we super-resolve the predicted functional graphs $\hat{\mathbf{X}}^{t_1}$ into a HR brain graphs $\hat{\mathbf{X}}^{t_2}$ using an intra-modality super-resolution GAN architecture. The aligner and both generators are trained in an end-to-end manner by optimizing a novel Ground Truth-Preserving (GT-P) loss function which guides them in learning the topology of the ground truth brain graphs more effectively.

and $\mathbf{G}^{t_2}$ is the second target brain graph (HR functional brain network) with $n_{r''}$ nodes where $n_r \neq n_{r'} \neq n_{r''}$.

A- Graph-based Inter-modality Aligner Block. Existing cross-domain frameworks are mainly tailored for images. However, since graphs are in the non-Euclidean space, the cross-domain prediction for graphs is exceedingly challenging. Hence, we propose a graph-based inter-modality aligner framework that enforces the predictions of the source graphs model to have the same distribution as the ground truth graphs. Specifically, given a set of n training source brain networks (e.g., morphological connectomes) $\mathbf{X}_{tr}^s$ and a set of n training ground-truth brain networks (e.g., functional connectomes) $\mathbf{X}_{tr}^{t_1}$, for each subject i, our aligner takes $\mathbf{X}_i^s$ as input and outputs $\hat{\mathbf{X}}_i^{s \to t_1}$ which shares the same distribution of $\mathbf{X}_i^{t_1}$ (Fig. 1-A). Drawing inspiration from ARVGA [13], our framework is composed of a variational autoencoder A_{align} and a discriminator D_{align}. The A_{align} comprises a probabilistic encoder that encodes an input as a distribution over the latent space instead of encoding an input as a single point. Indeed, the ARVGA produces a vector of mean μ and standard deviation σ, which provides more continuity in the latent space than the conventional autoencoder. This continuity enables the probabilistic decoder not only to reproduce an input vector but also to generate new data from the latent space [15]. To provide a more effective embedding, we propose an adversarial learning scheme using a discriminator D_{align} that **enforces** the latent representation to match that of the prior distribution of the ground-truth. Specifically, our encoder comprises three edge-based GCN layers adjusted by adding batch normalization and dropout to each layer's output (Fig. 1-A). Indeed, batch normalization efficiently accelerates the network training through a fast convergence of the loss function, and dropout reduces the possibility of overfitting. Therefore, these two operations help optimize and simplify the network training. To refine our inter-modality aligner, we propose an alignment loss function: $\mathcal{L}_{align} = \lambda_{adv}\mathcal{L}_{adv} + \lambda_{rec}\mathcal{L}_{rec} + \lambda_{KL}\mathcal{L}_{KL}$, where $\mathcal{L}_{adv}$ [16] is the adversarial loss quantifying the difference between the generated and ground truth target graphs since both A_{align} and D_{align} are iteratively optimized. $\mathcal{L}_{rec} = \left\| \mathbf{X}^s - \hat{\mathbf{X}}^{s \to t_1} \right\|^2$ denotes the reconstruction loss that tends to enhance the encoding-decoding scheme, where $\mathbf{X}^s$ is the source brain graph and $\hat{\mathbf{X}}^{s \to t_1}$ is the aligned source graphs. $\mathcal{L}_{KL}$ is Kulback-Leibler (KL) divergence loss which acts as a regularization term. $\mathcal{L}_{KL}$, also known as the relative entropy, is an asymmetric measure that quantifies the difference between two probability distributions. Thereby, we use $\mathcal{L}_{KL}$ to minimize the discrepancy between ground-truth and aligned source brain graph distributions. $\mathcal{L}_{KL} = \sum_{i=1}^{n} KL(q_i \| p_i)$, where the KL divergence for subject i is defined as: $KL(q_i \| p_i) = \int_{-\infty}^{+\infty} q_i(x) \log \frac{q_i(x)}{p_i(x)} dx$ where q is the true distribution (ground truth) and p is the aligned.

B- Adversarial Inter-modality Graph Generator Block. Following the inter-modality alignment step, we design an inter-modality generator that handles shifts in graph resolution (i.e., node size variation) coupled with an adversarial discriminator.

i- Inter-modality Multi-resolution Brain Graph Generator. Inspired by the dynamic edge convolution proposed in and [14] the U-net architecture [17] with skip connections, our inter-modality graph generator G_{inter} is composed of three edge-based GCN layers regularized by batch normalization and dropout for each layer's output (Fig. 1-B). G_{inter} takes as input the aligned source graphs to the target distribution $\hat{\mathbf{X}}^{s \to t_1}$ of size $n_r \times n_r$ and outputs the predicted target brain graphs $\hat{\mathbf{X}}_i^{t_1}$ of size $n_{r'} \times n_{r'}$, where $n_r \neq n_{r'}$. Particularly, owing to [14], each edge-based GCN layer includes a unique dynamic filter that outputs edge-specific weight matrix dictating the information flow between nodes k and l to learn a comprehensive vector representation for each node ($\mathbf{z}_k^h \in \mathbb{R}^{1 \times d_h}$ is the embedding of node k in layer h where d_h denotes the output dimension of the corresponding layer). Next, to learn our inter-modality multi-resolution mapping, we define a mapping function $\mathcal{T}_r : \mathbb{R}^{n_r \times d_h} \mapsto \mathbb{R}^{d_h \times d_h}$ that takes as input the embedded matrix of the whole graph $\mathbf{Z}^h$ in layer h of size $n_r \times d_h$ and outputs the generated target graph of size $d_h \times d_h$. We formulate $\mathcal{T}_r$ as follows: $\mathcal{T}_r = (\mathbf{Z}^h)^T \mathbf{Z}^h$. As such, shifting resolution is only defined by fixing the desired target graph resolution d_h. In our case, we set d_h of the latest layer in the generator to $n_{r'}$ to output the predicted target brain graph $\hat{\mathbf{X}}^{t_1}$ of size $n_{r'} \times n_{r'}$ (Fig. 1-B).

ii- Inter-modality Graph Discriminator Based on Adversarial Training. Our inter-modality generator G_{inter} is trained adversarially against a discriminator network D_{inter} (Fig. 1-B). To discriminate between the predicted and ground truth target graph data, we design a two-layer graph neural network [14]. Our discriminator D_{inter} takes as input the real brain graph $\mathbf{X}_i^{t_1}$ and the generator's output $\hat{\mathbf{X}}_i^{t_1}$, and outputs a value between 0 and 1, measuring the generator's output's realness. To boost our discriminator's performances, we adopt the adversarial loss function to maximize the discriminator's output value for the $\mathbf{X}_i^{t_1}$ and minimize it for $\hat{\mathbf{X}}_i^{t_1}$.

C- Adversarial Intra-modality Graph Generator Block. To avoid time-consuming image processing pipelines, we aim to predict intra-modality brain graphs at higher resolutions. Specifically, we design an intra-modality GAN similar to inter-modality GAN, yet without alignment and using node-based GCN layer [18] instead of edge-based GCN (Fig. 1-C). In fact, we used edge-based GCN for the inter-domain generation since simultaneously transferring one modality to another and super-resolving brain graphs are difficult tasks requiring more robust GCN layers. But, as we super-resolve up the network *stairs* (Fig. 1), edge-based GCN becomes highly time consuming and RAM-devouring. Thus, we use node-based GCN for super-resolving brain graphs from the same modality, which is less computationally expensive.

D- Ground Truth-Preserving Loss Function. Conventionally, GAN generators are optimized based on the response of their corresponding discriminators. However, within a few training epochs, the discriminator can easily distinguish real graphs from predicted graphs, and the adversarial loss would be close to 0. In such a case, the generator cannot provide satisfactory results. To overcome this dilemma, we need to enforce the generator-discriminator synchronous learning during the training process. Thus, we propose a new ground truth-preserving (GT-P) loss for both intra and inter-modality prediction blocks (Fig. 1-B and C). Our loss is composed of four sub-losses: adversarial loss [16], $L1$ loss [19], Pearson correlation coefficient (PCC) loss [11], and topological loss. We define our GT-P loss function as follows: $\mathcal{L}_{\text{GT-P}} = \lambda_1 \mathcal{L}_{adv} + \lambda_2 \mathcal{L}_{L1} + \lambda_3 \mathcal{L}_{PCC} + \lambda_4 \mathcal{L}_{top}$. To improve the predicted target brain graph quality, we propose to add an $L1$ loss term minimizing the distance between each predicted subject $\hat{\mathbf{X}}^t$ and its related ground truth $\mathbf{X}^t$. Even robust to outliers, the $L1$ loss only focuses on the element-wise similarity in edge weights between the predicted and real brain graphs and ignores the overall correlation between both graphs. Hence, we include the Pearson correlation coefficient (PCC) in our loss which measures the overall correlation between the predicted and real brain graphs. Since higher PCC indicates a higher correlation between the ground-truth and the predicted graphs, we propose to minimize the PCC loss function as follows: $\mathcal{L}_{PCC} = 1 - PCC$.

Furthermore, each brain graph has its unique topology, we introduce a topological loss function that guides the generator to maintain the nodes' topological profiles while learning the global graph structure. To do so, we define the $L1$ loss between the real and predicted eigenvector centralities (ECs) capturing the centralities of a node's neighbors. Hence, we define our topology loss as $\mathcal{L}_{top} = \|\mathbf{c}^t - \hat{\mathbf{c}}^t\|_1$, where $\hat{\mathbf{c}}^t$ denotes the EC vector of the predicted brain graph and $\mathbf{c}^t$ is the EC vector of the real one.

3   Results and Discussion

Evaluation dataset and parameters. We used three-fold cross-validation to evaluate our SG-Net framework on 150 subjects from the Southwest University Longitudinal Imaging Multimodal (SLIM) public dataset[1] where each subject has T1-w, T2-w MRI and resting-state fMRI (rsfMRI) scans. Our model is implemented using the PyTorch-Geometric library [20]. Following several preprocessing steps [21], a 35×35 cortical morphological network was generated from structural T1-w MR for each subject denoted as $\mathbf{X}^s$. For the resting-state functional MRI images, two separate brain networks with 160×160 (LR) denoted as $\mathbf{X}^{t_1}$ and 268×268 (HR) denoted as $\mathbf{X}^{t_2}$ were produced for each subject using two group-wise whole-brain parcellation approaches proposed in [22] and [23], respectively. For the aligner hyperparameters, we set $\lambda_{adv} = 1$, $\lambda_{rec} = 0.1$ and $\lambda_{KL} = 0.001$. Also, we set both generators hyperparameters as follows: $\lambda_1 = 1$, $\lambda_2 = 1$, $\lambda_3 = 0.1$, and $\lambda_4 = 2$. Moreover, we chose AdamW [24] as our default optimizer and set the learning rate at 0.025 for the A_{align} and the generators networks, and 0.01 for all the discriminators. Finally, we trained our model for 400 epochs using a single Tesla V100 GPU (NVIDIA GeForce GTX TITAN with 32 GB memory).

---

[1] https://github.com/basiralab/SG-Net.

Table 1. *Prediction Results Using Different Evaluation Metrics.* Evaluation of alignment and prediction brain graph synthesis by our SG-Net against seven comparison methods. CC: closeness centrality, BC: betweenness centrality and EC: eigenvector centrality. w/o: without. $\star$: SG-Net significantly outperformed all benchmark methods using two-tailed paired t-test ($p < 0.05$).

Methods	MAE	MAE (BC)	MAE (CC)	MAE (EC)
SG-Net w/o alignment	0.382	0.035	0.76	0.0171
Statistical alignment based-SG-Net	0.345	0.031	0.42	0.0166
SG-Net using VGAE [13] based-alignment	0.191	0.012	0.21	0.0148
SG-Net using ARGA [13] based-alignment	0.136	0.010	0.182	0.0142
GSR-Net [12]	0.253	<u>0.0083</u>	<u>0.1447</u>	**0.0114**
SG-Net w/o PCC	0.115	0.0087	0.1448	0.0115
SG-Net w/o topology	<u>0.102</u>	0.0094	0.1473	0.0120
SG-Net	**0.097***	**0.0073***	**0.1439***	**0.0114***

Evaluation and Comparison Methods. *Brain graph alignment.* To evaluate the performance of our aligner, we carried out four major comparisons. As shown in Table 1 and **supplementary Table 1** in both inter and intra-domains, SG-Net w/o alignment method scored the highest MAE between the real and predicted brain graphs. This demonstrates that the domain alignment improves the quality of the generated graphs. Moreover, according to **supplementary Table 1** and **supplementary Fig. 1**, when using a complex non-linear distribution (our source morphological distribution in Fig. 1), the statistical aligner may not align to the target distribution properly. Undeniably, an *adversarial learning-based* aligner can better adapt to any distribution, thereby achieving better results.

Insights into Topological Measures. To prove the fidelity of the predicted brain graphs to the real ones in topology and structure, we tested SG-Net using various topological measures (eigenvector, closeness, and betweenness). We note that the methods using the topological loss (GSR-Net [12], SG-Net w/o PCC and SG-Net) achieved better results. Specifically, our method produced the lowest MAE between the ground truth and predicted brain graphs in both inter- and intra-domains across all topological measurements ((Table 1 and **supplementary Table 1**), showing that our model can better preserve the topology of the predicted functional connectomes.

Insights into the Proposed GT-P Loss Function. To investigate the efficiency of the proposed GT-P loss, we trained SG-Net with different loss functions. As demonstrated in Table 1) and **supplementary Table 1**, our proposed GT-P loss outperforms its ablated versions and state-of-the-art method. In fact, the $L1$ loss focuses only on minimizing the distance between two brain graphs at the local level. Besides, PCC aims to maximize global connectivity patterns between the predicted and real brain graphs. However, both losses overlook the topological properties of graphs. Therefore, the EC is introduced in our topological loss to quantify a node's influence on a network's information flow. The combination of these complementary losses scored the best results while relaxing both intra-and inter-modality graph hypothesis between source and target domains.

Reproducibility. Besides generating realistic HR functional brain graphs, our framework could also capture the delicate variations in connectivity patterns across subjects. Specifically, we display in (Fig. 2-a) and **supplementary Fig. 2**, the real, predicted, and residual brain graphs for a representative testing subject using five different methods in both inter and intra-domains. The residual graph is calculated by computing the absolute difference between the real and predicted brain graphs. An average difference value of the residual is represented on top of each residual graph achieving a noticeable reduction by SG-Net.

Neuro-biomarkers. Fig. 2-b displays the top 10 strongest connectivities of real and predicted HR functional brain graphs of 3 randomly selected testing subjects. Since brain connectivity patterns differ from an individual to another [25], we note that the top 10 connectivities are not identical. Yet, our model can accurately predict such variations as well as individual trends in HR functional connectivity based on morphological brain graphs derived from the conventional T1-w MRI. This result further indicates that our method is trustworthy for synthesizing *multimodal* brain dysconnectivity patterns in disordered populations [7] from limited neuroimaging resources (only T1-w MRI)

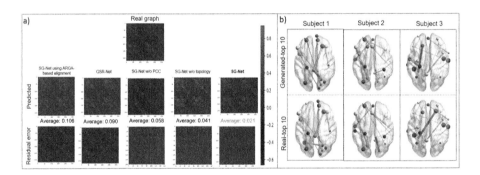

Fig. 2. *Visual Comparison Between the Real and the Predicted Target Brain Graphs.* a) Comparing the ground truth to the predicted brain graphs by SG-Net and five comparison methods using a representative testing subject. We display the residual matrices computed using the absolute difference between ground truth and predicted brain graph connectivity matrices. b) The top 10 strongest connectivities of real and predicted HR functional brain networks of 3 randomly selected testing subjects.

4   Conclusion

In this paper, we proposed the first work for inter- and intra-modality multi-resolution brain graph alignment and synthesis, namely StairwayGraphNet, which nicely handles variations in graph distribution, size, and structure. Our method can also be leveraged for developing precision medicine as well as multimodal neurological disorder diagnosis frameworks. The proposed SG-Net outperforms the baseline methods in terms of alignment, prediction, and super-resolution results. SG-Net not only predicts reliable HR functional graphs from morphological ones but also preserves the topology of the target domain. In future work, we will extend our architecture to predict multiple HR modality graphs from a single source one.

Acknowledgments. This work was funded by generous grants from the European H2020 Marie Sklodowska-Curie action (grant no. 101003403, http://basira-lab.com/normnets/) to I.R. and the Scientific and Technological Research Council of Turkey to I.R. under the TUBITAK 2232 Fellowship for Outstanding Researchers (no. 118C288, http://basira-lab.com/reprime/). However, all scientific contributions made in this project are owned and approved solely by the authors.

References

1. Cao, B., Zhang, H., Wang, N., Gao, X., Shen, D.: Auto-gan: self-supervised collaborative learning for medical image synthesis. In: Proceedings of the AAAI Conference on Artificial Intelligence, pp. 10486–10493 (2020)
2. Yu, B., Wang, Y., Wang, L., Shen, D., Zhou, L.: Medical image synthesis via deep learning. In: Lee, G., Fujita, H. (eds.) Deep Learning in Medical Image Analysis. AEMB, vol. 1213, pp. 23–44. Springer, Cham (2020). https://doi.org/10.1007/978-3-030-33128-3_2
3. Singh, N.K., Raza, K.: Medical image generation using generative adversarial networks: A review, pp. 77–96. A Computational Perspective in Healthcare, Health Informatics (2021)
4. Wang, C.: Dicyc: gan-based deformation invariant cross-domain information fusion for medical image synthesis. Inf. Fus. **67**, 147–160 (2021)
5. Liu, Y., Pan, Y., Yang, W., Ning, Z., Yue, L., Liu, M., Shen, D.: Joint neuroimage synthesis and representation learning for conversion prediction of subjective cognitive decline. In: International Conference on Medical Image Computing and Computer-Assisted Intervention, pp. 583–592 (2020)
6. Zhan, B., et al.: LR-CGAN: latent representation based conditional generative adversarial network for multi-modality MRI synthesis. Biomedical Signal Processing and Control **66**, 102457 (2021)
7. van den Heuvel, M.P., Sporns, O.: A cross-disorder connectome landscape of brain dysconnectivity. Nature Rev. Neurosci. **20**, 435–446 (2019)
8. Bronstein, M.M., Bruna, J., LeCun, Y., Szlam, A., Vandergheynst, P.: Geometric deep learning: going beyond euclidean data. IEEE Sig. Process. Mag. **34**, 18–42 (2017)
9. Bessadok, A., Mahjoub, M.A., Rekik, I.: Graph neural networks in network neuroscience. arXiv preprint arXiv:2106.03535 (2021)
10. Bessadok, A., Mahjoub, M.A., Rekik, I.: Topology-aware generative adversarial network for joint prediction of multiple brain graphs from a single brain graph. In: International Conference on Medical Image Computing and Computer-Assisted Intervention, pp. 551–561 (2020)

11. Zhang, L., Wang, L., Zhu, D.: Recovering brain structural connectivity from functional connectivity via multi-gcn based generative adversarial network. In: International Conference on Medical Image Computing and Computer-Assisted Intervention, pp. 53–61 (2020)
12. Isallari, M., Rekik, I.: GSR-Net: Graph super-resolution network for predicting high-resolution from low-resolution functional brain connectomes. In: International Workshop on Machine Learning in Medical Imaging, pp. 139–149 (2020)
13. Pan, S., Hu, R., Long, G., Jiang, J., Yao, L., Zhang, C.: Adversarially regularized graph autoencoder for graph embedding. arXiv preprint arXiv:1802.04407 (2018)
14. Simonovsky, M., Komodakis, N.: Dynamic edge-conditioned filters in convolutional neural networks on graphs. In: Proceedings of the IEEE conference on computer vision and pattern recognition, pp. 3693–3702 (2017)
15. Zemouri, R.: Semi-supervised adversarial variational autoencoder. Mach. Learn. Know. Extr. **2**, 361–378 (2020)
16. Goodfellow, I.J., et al.: Generative adversarial networks. arXiv preprint arXiv:1406.2661 (2014)
17. Ronneberger, O., Fischer, P., Brox, T.: U-net: Convolutional networks for biomedical image segmentation. In: International Conference on Medical Image Computing and Computer-Assisted Intervention, pp. 234–241 (2015)
18. Kipf, T.N., Welling, M.: Semi-supervised classification with graph convolutional networks. arXiv preprint arXiv:1609.02907 (2016)
19. Gürler, Z., Nebli, A., Rekik, I.: Foreseeing brain graph evolution over time using deep adversarial network normalizer. In: International Workshop on Predictive Intelligence In Medicine, pp. 111–122 (2020)
20. Fey, M., Lenssen, J.E.: Fast graph representation learning with pytorch geometric. arXiv preprint arXiv:1903.02428 (2019)
21. Fischl, B.: Neuroimage. Freesurfer **62**, 774–781 (2012)
22. Dosenbach, N.U., et al.: Prediction of individual brain maturity using FMRI. Science **329**, 1358–1361 (2010)
23. Shen, X., Tokoglu, F., Papademetris, X., Constable, R.T.: Groupwise whole-brain parcellation from resting-state FMRI data for network node identification. Neuroimage **82**, 403–415 (2013)
24. Loshchilov, I., Hutter, F.: Fixing weight decay regularization in adam (2018)
25. Glasser, M.F., et al.: A multi-modal parcellation of human cerebral cortex. Nature **536**, 171 (2016)

Multi-Feature Semi-Supervised Learning for COVID-19 Diagnosis from Chest X-Ray Images

Xiao Qi[1], David J. Foran[4], John L. Nosher[2], and Ilker Hacihaliloglu[2,3(✉)] 📍

[1] Department of Electrical and Computer Engineering, Rutgers University,
Piscataway, NJ, USA
[2] Department of Radiology, Rutgers Robert Wood Johnson Medical School,
New Brunswick, NJ, USA
ilker.hac@soe.rutgers.edu
[3] Department of Biomedical Engineering,
Rutgers University, Piscataway, NJ, USA
[4] Rutgers Cancer Institute of New Jersey, New Brunswick, NJ, USA

Abstract. Computed tomography (CT) and chest X-ray (CXR) have been the two dominant imaging modalities deployed for improved management of Coronavirus disease 2019 (COVID-19). Due to faster imaging, less radiation exposure, and being cost-effective CXR is preferred over CT. However, the interpretation of CXR images, compared to CT, is more challenging due to low image resolution and COVID-19 image features being similar to regular pneumonia. Computer-aided diagnosis via deep learning has been investigated to help mitigate these problems and help clinicians during the decision-making process. The requirement for a large amount of labeled data is one of the major problems of deep learning methods when deployed in the medical domain. To provide a solution to this, in this work, we propose a semi-supervised learning (SSL) approach using minimal data for training. We integrate local-phase CXR image features into a multi-feature convolutional neural network architecture where the training of SSL method is obtained with a teacher/student paradigm. Quantitative evaluation is performed on 8,851 normal (healthy), 6,045 pneumonia, and 3,795 COVID-19 CXR scans. By only using 7.06% labeled and 16.48% unlabeled data for training, 5.53% for validation, our method achieves 93.61% mean accuracy on a large-scale (70.93%) test data. We provide comparison results against fully supervised and SSL methods. The code and dataset will be made available after acceptance.

Keywords: Semi-supervised learning · Classification · COVID-19 · Chest X-ray

1 Introduction

Diagnostics is a key tool for improved management of Coronavirus disease 2019 (COVID-19), permitting healthcare workers to rapidly triage patients. Currently,

ⓒ Springer Nature Switzerland AG 2021
C. Lian et al. (Eds.): MLMI 2021, LNCS 12966, pp. 151–160, 2021.
https://doi.org/10.1007/978-3-030-87589-3_16

the gold standard diagnosis is based on reverse-transcription polymerase chain reaction (RT-PCR) tests. To improve the management of COVID-19, radiological assessment, based on computed tomography (CT) and chest X-ray (CXR), has also been incorporated into the decision-making process. Compared to CT, CXR provides additional advantages such as fast screening, being portable, and easy to setup (can be setup in isolation rooms). However, the interpretation of CXR images, compared to CT, by the expert radiologist is a difficult process as the visual cues for the disease can be subtle or similar to regular pneumonia. As such computer-aided diagnostic systems that can aid in the decision-making process have been investigated [20, 21, 26].

Computer-aided diagnosis via deep supervised learning has achieved strong performance when provided with a large labeled data set [20, 21, 26]. However, the requirement of expert knowledge for the labeling process is costly. The scarcity of available data, in particular with a new disease, also affects the success of the developed methods. In order to mitigate this problem methods based on semi-supervised learning (SSL) have been developed for diagnosing lung disease from CXR images.

The following papers provide a brief overview of prior work on lung disease classification from CXR data using SSL. Consistency regularization and pseudo labeling have been the two most dominant methods investigated for SSL. [25] proposed a multi-class abnormality detection from CXR data. Using 35% labeled and 35% unlabeled data as the training set an AUROC close to 0.82 was reported on 20% test data. In [10], disentangled stochastic latent space was used to improve self-ensembling for semi-supervised CXR classification. Using 3.12% labeled and 73.2% unlabeled data for training an AUROC value of 0.66 was obtained on 22% test data. In [9], the same group extended their prior work [10] by training SSL network on linear mixing of labeled and unlabeled data, at the input and latent space, to improve network regularization. 9.6% improvement of AUROC, compared to [10], was reported using same evaluation strategy as [10]. In [4], the authors proposed a graph-based label propagation method for semi-supervised CXR classification and achieved an AUROC of 0.78 with only using 20% of the labeled data. Unlabeled dataset size used for training and test dataset size was not reported in [4].

Deep learning methods can automatically learn the features of the data without the need for data preprocessing. Nonetheless, automation requires 1- the development of deeper and complex networks, and 2- annotated large training data. Preprocessing can provide solutions to some of these challenges. Most recently [21] local phase-based image processing was proposed for improved representation of lung CXR data. Quantitative results demonstrated the importance of local phase image features for improved diagnosis of COVID-19 disease from CXR scans. Motivated by this, in this work we propose a local phase-based SSL method for accurate diagnosis of COVID-19 from minimal training data. Our proposed method achieves similar accuracy as full supervised baseline architecture accuracy by only using 7.06% labeled and 16.48% unlabeled data for training, 5.53% for validation, on 70.93% testing data. Our contributions are as follows: 1- We introduce a novel multi-feature SSL model which outperforms baseline mono-feature-based SSL and fully supervised learning methods. 2- We perform ablation studies to show the effect of local-phase features for training and testing. 3- We evaluate our technique on 18,691 CXR data set of healthy (8,851), regular pneumonia (6,045), and COVID-19 (3,795) scans. This is the largest and first COVID-19 evaluation study reported for SSL. We provide a performance comparison of the presented method against the supervised baseline and several SSL methods.

2   Methods and Materials

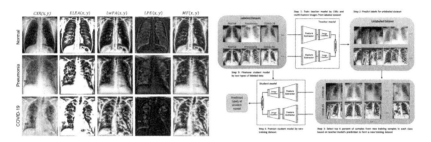

Fig. 1. Left:$CXR(x,y)$ from normal, bacterial pneumonia and COVID-19 lung and corresponding enhanced images. Green arrows point to consolidations. Right:Illustration of our proposed multi-feature guided SSL method. Step 1: Train a multi-feature teacher model on labeled dataset; Step 2: Run the trained multi-feature teacher model on unlabeled dataset to obtain the pseudo-labeled dataset; Step 3: Optimize the pseudo-labeled dataset by using hyperparameter K ; Step 4: Train a new student model on the optimized pseudo-labeled dataset; and Step 5: Finetune the trained student model on the labeled dataset.

Multi-feature Images. Enhancement of CXR images $(CXR(x,y))$ is based on the extraction of local phase image features using bandpass quadrature filters and L1 norm-based contextual regularization method [21]. Local phase image analysis (LPIA) has been extensively investigated for the processing of various medical data [1,2,21]. Since LPIA is intensity invariant [1,16], the enhancement results and subsequent image analysis methods are not affected by the intensity variations due to patient characteristics or machine acquisition settings. Monogenic filter, and α-scale space derivative quadrature filters (ASSD) are used as bandpass quadrature filters. Three different local phase $CXR(x,y)$ image features are extracted: 1- Local weighted mean phase angle $(LwPA(x,y))$, 2- $LwPA(x,y)$ weighted local phase energy $(LPE(x,y))$, and 3- Enhanced local energy attenuation image $(ELEA(x,y))$. During this work, we used the same filter parameters as explained in [21]. A multi-feature image, denoted as $MF(x,y)$, is created by combining these three types of local phase images as three-channel input. Since all enhanced images are grayscale, the combination results in an image with shape $(w*h*3)$, where w and h correspond to the width and height of the image. Qualitative results corresponding to the $MF(x, y)$ images are displayed in Fig. 1. Green arrows point to diffuse irregular patchy consolidations. The COVID-19 image shows bilateral peripherally distributed opacities (green arrows). Investigating Fig. 1 we can observe local phase and $MF(x, y)$ images of COVID-19 have improved opacity features related to COVID-19 compared to original $CXR(x, y)$ image. Although $LPE(x, y)$ image did not result in opacity detection improvement in the COVID -19 image, it could compliment the other two local phase enhanced CXR images. We can observe a similar enhancement of consolidations related to bacterial pneumonia image (middle row). The $MF(x, y)$ image and the $CXR(x, y)$ image are used as an input to train our proposed multi-feature teacher/student SSL model which is explained in the next section.

Semi-supervised Learning Pipeline. Our proposed method consists of five steps as illustrated in Fig. 1. The pipeline is an extension of [27], and the algorithm leverages the desirable property that learning is tolerant to a certain degree of noise [18]. Our goal here is to use a multi-feature convolutional neural network (CNN) architecture for the teacher model to limit the labeling noise for unlabeled dataset, thus forming a better teacher model without additional training samples. The same multi-feature CNN architecture is also used for the student model.

Our multi-feature CNN architecture consists of two mono-feature network streams for processing $CXR(x, y)$ images and the enhanced $MF(x, y)$ images respectively. In [2, 21] three different fusion strategies for the optimal fusion of features were investigated. The authors have shown that late-fusion outperformed early and mid-level fusion operations. Therefore we adopt the late-fusion strategy during this work. In our design, predictions are made based on high-level features from both network streams (Fig. 1-Step1 and 4). During this work we utilize ResNet50 [11] as the encoder network for both streams. Prior SSL methods used AlexNet [9] as their main architecture, however, in [21] ResNet50 outperformed the pre-trained AlexNet, for classifying COVID-19 from $CXR(x, y)$ images, and had slightly improved mean accuracy compared to SonoNet64 [5], XNet(Xception) [6], InceptionV4(Inception-Resnet-V2) [22] and EfficientNetB4 [23].

In our proposed work we denote x_{ld} as labeled data and y_{ld} the corresponding label. Unlabeled data is denoted as x_{uld} and the generated pseudo label is denoted as $\hat{y}_{uld}$. The teacher model, denoted as T, after finetuning with the labeled data, x_{ld}, performs a forward run on unlabeled images x_{uld} to obtain the class distribution $P(.|x_{uld}; \theta_T)$, where θ_T denotes the parameters of T. From this distribution, the trained teacher model predicts the pseudo-label $\hat{y}_{uld}$ for each image according to softmax prediction vector.

Once the teacher model, T, generates pseudo labels from the unlabeled data, the top K percent of images in each class, are retained as new positive training samples for each class. The ranking is done based on the corresponding label classification score. Optimization of hyperparameter K is performed using 10% of labeled and validation data. In terms of the selection of K, a small K

provides the simple and clean images without much labeling noise for each class. When K increases the images are less obvious and noisier introducing lots of false positives. Therefore, there is a significant trade-off on K. Based on our experiment, the classification accuracy goes down significantly when $K > 0.35$. And $K = 0.25$ gives us the best performance and it corresponds to 75% of the available unlabeled training dataset.

The paired $x_{uld}, \hat{y}_{uld}$ is then shown to the student model, denoted as S, to optimize its parameter θ_S. The optimization is based on the gradient calculated by back-propagation from the cross-entropy loss using SGD optimizer:

$$\theta_S^{(t+1)} := \theta_S^t - \eta * \left.\frac{\partial \ell(x_{uld}, \hat{y}_{uld}; \theta_S)}{\partial \theta_S}\right|_{\theta_S = \theta_S^{(t)}} = \theta_S^{(t)} - \eta * g_S^{(t)}, \qquad (1)$$

where η is the learning rate and $\ell(x, y; \theta)$ denotes the cross entropy loss calculated between input x_{uld} and label $\hat{y}_{uld}$ with parameter θ

After the student model optimizes its parameters, using Eq. 1, its parameters $\theta_S^{(t+1)}$ is further optimized on a labeled sample x_{ld}, y_{ld} using cross entropy again to minimize the loss:

$$\min_{\theta_S^{(t+1)}} \ell(x_{ld}, y_{ld}; \theta_S^{(t+1)}) \qquad (2)$$

Clearly, $\ell(x_{ld}, y_{ld}; \theta_S^{(t+1)})$ depends on $\theta_S^{(t+1)}$, which in turn depends on the pseudo label $\hat{y}_{uld}$

Dataset. We evaluate the performance of proposed method on following datasets: BIMCV [12], COVIDx [26], COVID-19-AR [8], and MIDRC-RICORD-1c [24]. The COVID-19-AR and MIDRC-RICORD-1c were downloaded from the Cancer Imaging Archive (TCIA) Public Access [7] thus denoted as TCIA dataset. The total data collection is consist of 18,691 CXR scans from 16,817 subjects. The detailed data distribution is shown in Fig. 2.

A subset containing 5,433 with a balanced distribution of classes consisting of 1- all COVID-19 images from COVIDx, 2- partial normal and pneumonia images from COVIDx, and 3- partial COVID-19 images from BIMCV were randomly selected as our evaluation dataset (Fig. 2). This data was split into a training dataset (4,400 scans), validation dataset (544 scans), and early stopping set (489 scans). We keep our validation set to a realistic size as heavy hyperparameter tuning on a large validation set may have limited real-world applicability [19]. All the random splits were repeated five times and average results are reported. During the random split, the evaluation and testing data did not include the same patient scans. We generate small labeled data by using 10%, 20%, and 30% of the training dataset and treat the rest of the training data as unlabeled examples. We would like to reiterate that 10%, 20%, and 30% of our training data corresponds to 2.35%, 4.7%, and 7.06% of the full labeled data (18,691 CXR scans).

Two test datasets were generated to show the robustness of our proposed approach. The remaining COVID-19 images in BIMCV plus the same amount of images in normal and pneumonia classes from COVIDx form our Test-1 dataset.

All remaining images in COVIDx and all images in the TCIA dataset were used to form our Test-2 dataset. The joined test data corresponds to 70.93% of the full dataset (18,691 CXR scans).

	Normal [17] (COVIDx)	Pneumonia [17] (COVIDx)	COVID-19 [17] (COVIDx)	COVID-19 [17] (BIMCV)	COVID-19 [17] (TCIA)
#images	8851	6045	400	2167	1228
#subjects	8851	6031	301	1183	451

		Normal	Pneumonia	COVID-19
Training & Stop & Validation	#images	1811	1811	1811
	#subjects	1811	1810	1038
Test Data 1	#images	756	756	756
	#subjects	756	756	446
Test Data 2	#images	6284	3478	1228
	#subjects	6284	3465	451

Fig. 2. Class distribution of evaluation dataset.

To evaluate the performance of our multi-feature SSL method to guide both teacher and student models, denoted as MF-TS, we compared it against three types of SSL methods: 1- SSL guided by $CXR(x,y)$ only, denoted as CXR-TS (this is also the baseline for Billion Scale SSL [27]); 2- SSL guided by $MF(x,y)$ only, denoted as Enh-TS; 3- SSL with the teacher model guided by both $CXR(x,y)$ and $MF(x,y)$ and the student model guided by $CXR(x,y)$ only, denoted as MF-T. We also compare our method against previously developed SSL methods: Temporal Ensembling [13], and Pseudo Labeling [15]. In addition, a comparison study between our proposed SSL method and supersized learning (SL) is also conducted.Three leading SL network architectures are evaluated: ResNet50 [11], trained by varying percent of labeled CXR samples (10%, 20%, 30%, and 100% of training data), XNet(Xception) [6] and InceptionV4(Inception-Resnet-V2) [22], trained on 100% of the labeled training data only. Finally, to investigate the effect of local phase features on the classification performance, we also trained the ResNet50 model using enhanced images (MF(x,y)) and multi-feature images (CXR(x,y)+MF(x,y)) [21].

During the comparative evaluation, we use the same amount of validation samples to tune hyperparameters for the investigated baselines. We evaluate the proposed and all the baseline methods by reporting the mean accuracy, precision, recall, and F1-score values.

3   Results

All proposed networks were trained for 50 epochs, using the early stopping technique [17] to avoid overfitting, a learning rate of 0.001 for the first epoch and a learning rate decay of 0.1 every 15 epochs with a mini-batches of size 32. All images were normalized to have zero mean and unit variance and resized to the suitable size for each network during training. All techniques were implemented in Python using the Pytorch framework

Firstly, we notice that all the investigated methods have better overall performance across all metrics for Test-1 dataset compared to Test-2 dataset (Table 1). Test-2 dataset is a more challenging dataset, which has a significantly larger number of samples compared with Test-1. The COVID-19 CXR images, in Test-2 dataset, are also collected from different populations and regions.

The proposed MF-TS model achieves an equivalent performance, by only using 30% labeled training data, in comparison with fully supervised ResNet50 [11] and outperforms the XNet [6] and the InceptionV4 [22] for both Test-1 and Test-2 data (Table 1). For Test-1 data, ResNet50 [11] accuracy drops down

Table 1. Quantitative results obtained from Test-1 and Test-2 data. Green shaded region corresponds to the highest scores obtained. Purple shaded region corresponds to the highest scores obtained when using 30% of the labeled training data. During this work 10%, 20%, and 30% of our labeled training data corresponds to 2.35%, 4.7%, and 7.06% of the full labeled data (18,691 CXR scans).

Method	Labeled Sample(%)	Precision		Recall		F1-Scores		Top-1(%)	
		Test-1	Test-2	Test-1	Test-2	Test-1	Test-2	Test-1	Test-2
XNet[6]	100	0.93	0.91	0.93	0.93	0.93	0.92	93.02	91.69
InceptionV4[22]	100	0.93	0.90	0.93	0.93	0.93	0.92	92.72	91.82
ResNet50[11] (CXR)	10	0.86	0.81	0.86	0.86	0.86	0.83	86.46	84.29
	20	0.90	0.84	0.90	0.89	0.90	0.86	89.61	87.29
	30	0.91	0.88	0.91	0.92	0.91	0.90	91.01	90.19
	100	0.94	0.92	0.94	0.94	0.94	0.93	93.54	92.41
ResNet50[11] (Enh)	10	0.89	0.81	0.89	0.86	0.89	0.83	89.11	80.97
	20	0.92	0.84	0.92	0.89	0.92	0.86	91.58	86.67
	30	0.92	0.88	0.92	0.92	0.92	0.90	91.84	88.77
	100	0.93	0.91	0.93	0.92	0.93	0.92	93.05	91.64
Fus-ResNet50[21] (CXR+Enh)	10	0.91	0.84	0.91	0.81	0.91	0.82	90.84	84.95
	20	0.93	0.89	0.93	0.92	0.93	0.90	92.63	89.90
	30	0.93	0.91	0.93	0.93	0.93	0.92	93.01	91.49
	100	0.94	0.93	0.94	0.95	0.94	0.94	94.17	93.26
Temporal Ensembling[14]	10	0.76	0.58	0.74	0.57	0.74	0.57	74.00	66.50
	20	0.83	0.71	0.82	0.67	0.82	0.68	81.85	77.79
	30	0.86	0.68	0.85	0.65	0.85	0.66	85.23	78.47
Pseudo-Labeling[3]	10	0.88	0.57	0.88	0.57	0.88	0.55	88.36	74.12
	20	0.92	0.84	0.92	0.77	0.92	0.79	92.06	83.72
	30	0.93	0.90	0.93	0.92	0.93	0.91	93.17	90.02
CXR-TS (Billion-Scale[27])	10	0.90	0.88	0.90	0.92	0.90	0.90	90.29	89.43
	20	0.92	0.90	0.92	0.93	0.92	0.91	91.59	90.69
	30	0.92	0.91	0.92	0.93	0.92	0.92	92.19	91.52
Enh-TS	10	0.91	0.85	0.91	0.81	0.91	0.82	90.99	85.98
	20	0.92	0.90	0.92	0.89	0.92	0.89	92.28	89.70
	30	0.92	0.90	0.92	0.90	0.92	0.90	92.47	90.17
MF-T	10	0.91	0.90	0.91	0.92	0.91	0.91	91.46	90.26
	20	0.92	0.90	0.92	0.93	0.92	0.91	92.25	91.01
	30	0.93	0.91	0.92	0.93	0.92	0.92	92.49	91.79
MF-TS	10	0.92	0.90	0.92	0.91	0.92	0.90	92.43	90.08
	20	0.93	0.92	0.93	0.93	0.93	0.93	93.41	91.84
	30	0.94	0.93	0.94	0.94	0.94	0.93	93.61	92.47
	50	0.94	0.93	0.94	0.94	0.94	0.94	93.94	93.18
	70	0.94	0.94	0.94	0.95	0.94	0.94	94.49	93.36

to 86.46% when trained on 10% labeled training data while the proposed MF-TS achieves a significantly higher mean accuracy of 92.43% ($p < 0.05$ with paired t-test). For Test-2 data we observe a similar significant ($p < 0.05$ with paired t-test) improvement between ResNet50 [11] and our proposed MF-TS network architecture (84.29% vs 90.08% mean accuracy). We can also observe that our proposed MF-TS model achieves significantly higher accuracy when using 10/20/30% of labeled data, compared to the fully supervised method proposed in [21] where both $MF(x,y)$ and $CXR(x,y)$, images are fused ($p < 0.05$ with paired t-test). Compared with the fully supervised method [21], trained with 100% labeled $MF(x,y)$ and $CXR(x,y)$ images, there is no significant difference ($p > 0.05$) when the proposed MF-TS model is trained on 50% of the labeled data. When the proposed method is trained on 50% and 70% labeled data, the accuracy continues to improve (Table 1).

We also compared our multi-feature guided method with three SSL techniques. From Table 1, we observe that MF-TS offers a substantial improvement over Temporal Ensembling [14] for all metrics at every different labeled sample. The largest differences are 18.43% and 23.58%, at 10% labeled training data, in terms of overall mean accuracy for Test-1 and Test-2 data respectively. Pseudo-labeling [3] provides close results for all metrics compared with the MF-TS for the Test-1 dataset. Our method, however, provides better stability in testing different datasets. In terms of overall accuracy, the average change, across different labeling samples between Test-1 and Test-2, of MF-TS is 1.69% against 8.58% of Pseudo-labeling [3]. Our MF-TS model, as an extension of CXR-TS (Billion-Scale [27]), improves the performance from all aspects. MF-TS model achieves significantly improved mean accuracy compared to CXR-TS model in Test-1 and Test-2 (paired t-test $p < 0.05$) except for 10% labeled training data results of Test-2 data (Table 1).

To further support the assertion that features from enhanced images are beneficial for SSL, we list two additional models, which are Enh-TS and MF-T. Enh-TS and MF-T slightly outperform CXR-TS as shown in Table 1. Investigating Table 1, only MF-T has a better performance in all metrics compared with CXR-TS. The improvement mainly comes from pseudo-label generation. Our fusion model has a more precise prediction for unlabeled samples, and thus it is beneficial for training a new student model.

4   Conclusions

We presented a novel multi-feature SSL method for classifying COVID-19 disease from CXR scans. To simulate a realistic scenario, which is often found in medical imaging applications where both obtaining data and labeling efforts are expensive, we opted to use a relatively small labeled dataset (16.48%) for training. We exhibit, on a small training (23.54%) and large testing dataset (> 70%), that the proposed multi-feature SSL provides improved classification results for diagnosing COVID-19 from CXR scans compared to prior SSL and SL methods. Our results suggest the feasibility of using local-phase CXR image features for improving the success rate of SSL methods and provide a strong foundation for future developments. Future work will include more extensive evaluation and investigation of the proposed method for classifying COVID-19 disease from CT and ultrasound data using SSL methods. We will also investigate the effect of each local phase image feature on the classification performance.

References

1. Alessandrini, M., Basarab, A., Liebgott, H., Bernard, O.: Myocardial motion estimation from medical images using the monogenic signal. IEEE Trans. Image Process. **22**(3), 1084–1095 (2012)
2. Alsinan, A.Z., Patel, V.M., Hacihaliloglu, I.: Automatic segmentation of bone surfaces from ultrasound using a filter-layer-guided CNN. Int. J. Comput. Assist. Radiol. Surg. **14**(5), 775–783 (2019)
3. Arazo, E., Ortego, D., Albert, P., O'Connor, N.E., McGuinness, K.: Pseudo-labeling and confirmation bias in deep semi-supervised learning (2020)
4. Aviles-Rivero, A.I., et al.: GraphX^{NET}: chest x-ray classification under extreme minimal supervision. In: Shen, D., et al. (eds.) MICCAI 2019. LNCS, vol. 11769, pp. 504–512. Springer, Cham (2019). https://doi.org/10.1007/978-3-030-32226-7_56
5. Baumgartner, C.F., et al.: Sononet: real-time detection and localisation of fetal standard scan planes in freehand ultrasound. IEEE Trans. Med. Imaging **36**(11), 2204–2215 (2017)
6. Chollet, F.: Xception: Deep learning with depthwise separable convolutions. In: Proceedings of the IEEE Conference on Computer Vision and Pattern Recognition, pp. 1251–1258 (2017)
7. Clark, K., et al.: The cancer imaging archive (TCIA): maintaining and operating a public information repository. J. Digit. Imaging **26**(6), 1045–1057 (2013). https://doi.org/10.1007/s10278-013-9622-7
8. Desai, S., et al.: Chest imaging representing a COVID-19 positive rural US population. Sci. Data **7**, 1–6 (2020). https://doi.org/10.1038/s41597-020-00741-6
9. Gyawali, P.K., Ghimire, S., Bajracharya, P., Li, Z., Wang, L.: Semi-supervised medical image classification with global latent mixing. In: Martel, A.L., et al. (eds.) MICCAI 2020. LNCS, vol. 12261, pp. 604–613. Springer, Cham (2020). https://doi.org/10.1007/978-3-030-59710-8_59
10. Gyawali, P.K., Li, Z., Ghimire, S., Wang, L.: Semi-supervised learning by disentangling and self-ensembling over stochastic latent space. In: Shen, D., et al. (eds.) MICCAI 2019. LNCS, vol. 11769, pp. 766–774. Springer, Cham (2019). https://doi.org/10.1007/978-3-030-32226-7_85
11. He, K., Zhang, X., Ren, S., Sun, J.: Deep residual learning for image recognition. In: Proceedings of the IEEE conference on computer vision and pattern recognition, pp. 770–778 (2016)
12. de la Iglesia Vayá, M., et al.: Bimcv COVID-19+: a large annotated dataset of RX and CT images from COVID-19 patients. arXiv preprint arXiv:2006.01174 (2020)
13. Laine, S., Aila, T.: Temporal ensembling for semi-supervised learning. arXiv preprint arXiv:1610.02242 (2016)
14. Laine, S., Aila, T.: Temporal ensembling for semi-supervised learning (2017)
15. Lee, D.H., et al.: Pseudo-label: the simple and efficient semi-supervised learning method for deep neural networks. In: Workshop on challenges in representation learning. In: ICML, vol. 3 (2013)

16. Li, Z., van Vliet, L.J., Stoker, J., Vos, F.M.: A hybrid optimization strategy for registering images with large local deformations and intensity variations. Int. J. Comput. Assist. Radiol. Surg. **13**(3), 343–351 (2017). https://doi.org/10.1007/s11548-017-1697-z
17. Mahsereci, M., Balles, L., Lassner, C., Hennig, P.: Early stopping without a validation set. CoRR abs/1703.09580 (2017). http://arxiv.org/abs/1703.09580
18. Natarajan, N., Dhillon, I.S., Ravikumar, P., Tewari, A.: Learning with noisy labels. In: Proceedings of the 26th International Conference on Neural Information Processing Systems, NIPS 2013, vol. 1. pp. 1196–1204. Curran Associates Inc., Red Hook (2013)
19. Oliver, A., Odena, A., Raffel, C., Cubuk, E.D., Goodfellow, I.J.: Realistic evaluation of deep semi-supervised learning algorithms. arXiv preprint arXiv:1804.09170 (2018)
20. Ozturk, T., Talo, M., Yildirim, E.A., Baloglu, U.B., Yildirim, O., Acharya, U.R.: Automated detection of COVID-19 cases using deep neural networks with X-ray images. Comput. Biol. Med. **121**, 103792 (2020)
21. Qi, X., Brown, L.G., Foran, D.J., Nosher, J., Hacihaliloglu, I.: Chest X-ray image phase features for improved diagnosis of COVID-19 using convolutional neural network. Int. J. Comput. Assist. Radiol. Surg. 1–10 (2020)
22. Szegedy, C., Ioffe, S., Vanhoucke, V., Alemi, A.: Inception-v4, inception-resnet and the impact of residual connections on learning (2016)
23. Tan, M., Le, Q.V.: Efficientnet: Rethinking model scaling for convolutional neural networks. arXiv preprint arXiv:1905.11946 (2019)
24. Tsai, E.B., et al.: The RSNA international COVID-19 open annotated radiology database (RICORD). Radiology 203957. https://doi.org/10.1148/radiol.2021203957, pMID: 33399506
25. Unnikrishnan, B., Nguyen, C.M., Balaram, S., Foo, C.S., Krishnaswamy, P.: Semi-supervised classification of diagnostic radiographs with noteacher: a teacher that is not mean. In: Martel, A.L., et al. (eds.) MICCAI 2020. LNCS, vol. 12261, pp. 624–634. Springer, Cham (2020). https://doi.org/10.1007/978-3-030-59710-8_61
26. Wang, L., Lin, Z.Q., Wong, A.: Covid-net: a tailored deep convolutional neural network design for detection of COVID-19 cases from chest x-ray images. Sci. Reports **10**(1), 1–12 (2020)
27. Yalniz, I.Z., Jégou, H., Chen, K., Paluri, M., Mahajan, D.: Billion-scale semi-supervised learning for image classification. CoRR abs/1905.00546 (2019). http://arxiv.org/abs/1905.00546

Transfer Learning with a Layer Dependent Regularization for Medical Image Segmentation

Nimrod Sagie[1], Hayit Greenspan[1], and Jacob Goldberger[2]([⊠])

[1] Tel-Aviv University, Tel-Aviv, Israel
[2] Bar-Ilan University, Ramat-Gan, Israel
Jacob.Goldberger@biu.ac.il

Abstract. Transfer learning is a machine learning technique where a model trained on one task is used to initialize the learning procedure of a second related task which has only a small amount of training data. Transfer learning can also be used as a regularization procedure by penalizing the learned parameters if they deviate too much from their initial values. In this study we show that the learned parameters move apart from the source task as the image processing progresses along the network layers. To cope with this behaviour we propose a transfer regularization method based on monotonically decreasing regularization coefficients. We demonstrate the power of the proposed regularized transfer learning scheme on COVID-19 opacity task. Specifically, we show that it can improve the segmentation of coronavirus lesions in chest CT scans.

Keywords: Transfer learning · Regularization · COVID-19 opacity

1 Introduction

Collecting annotated medical data is usually an expensive procedure that requires the collaboration of radiologists and researchers. One of the main differences between the medical imaging domain and computer vision is the need to cope with a limited amount of annotated samples [2,5,11,21]. Transfer learning is a popular strategy to overcome the difficulties posed by limited annotated training data. The goal of transfer learning is to transfer knowledge from a source task to a target task by using the parameter set of the source task in the process of learning the target task. Transfer learning utilizes models that are pre-trained on large datasets, that can either be scenery datasets such as ImageNet or medical datasets. There is a plethora of work on using transfer learning in different medical imaging applications (e.g. [3,22]). Due to the popularity of transfer learning in medical imaging, there has been also work analyzing its precise effects (see e.g. [13,15,19]).

This research was supported by the Ministry of Science & Technology, Israel.

C. Lian et al. (Eds.): MLMI 2021, LNCS 12966, pp. 161–170, 2021.
https://doi.org/10.1007/978-3-030-87589-3_17

A common procedure when using transfer learning is to start with a pre-trained model on the source task and to fine-tune the model, i.e. train it further, using a small set of data from the target task. Variants of transfer learning include fine-tuning of all network parameters, only the parameters of the last few layers, or simply just use the pre-trained model as a fixed feature extractor which is followed by a trained classifier. Injecting information into a network via parameter initialization is problematic since this information can be lost during the optimization procedure. Li et al. [9] recently proposed that, in addition to initialization, the pre-trained parameters can be also used as a regularization term. They implemented an L_2 penalty term to allow the fine-tuned network to have an explicit inductive bias towards the original pre-trained model.

In this study we show that the learned parameters move apart from the source task as the image processing progresses along the network layers, and that this occurs even if we regularize the learned parameters. To cope with this we propose a regularization method based on monotonically decreasing regularization coefficients that allows a gradually increasing distance of the learned parameters from the pre-trained model along the network layers. We applied this transfer learning regularization strategy to the task of COVID-19 opacity segmentation and show that it improves the segmentation of Coronavirus lesions in chest CT scans.

2 Transfer Learning via Gradual Regularization

Parameter regularization is a common technique for preventing overfitting to the training data. Let θ be the parameter set of a given neural network. The L_2 regularization modifies the loss function $\mathrm{Loss}(\theta)$ which we minimize by adding a regularization term that penalizes large weights:

$$\mathrm{Loss}(\theta) + \lambda\|\theta\|^2, \tag{1}$$

where λ is the regularization coefficient. Adding the L_2 term results in much smaller weights across the entire model, and for this reason it is known as weight decay. Network parameters are usually initialized by zero (with a small random perturbation to avoid trivial solutions) and the regularization term prevents the parameters from deviating too much from the initial zero values.

Transfer learning is a network training method where a model trained on a task with a large available annotated data, is reused as the starting point for a model on a second task. Several recent studies have suggested exploiting the full potential of the knowledge already acquired by the model on the source task, by penalizing the difference between the parameters of the source task and the parameters of the target task we aim to learn [9,10]. In transfer learning the target network is initialized by the source network parameters. Hence, a suitable L_2 regularized loss for transfer learning is:

$$\mathrm{Loss}(\theta) + \lambda\|\theta - \bar{\theta}\|^2, \tag{2}$$

where $\bar{\theta}$ is the parameter set of the source task model. The value for λ in the range of $(0, \infty)$ controls the amount of knowledge we want to transfer from the source task to the target task. In practice, λ is a hyper-parameter that can be tuned using cross-validation.

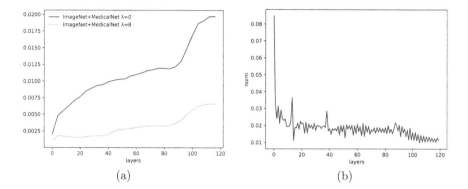

Fig. 1. (a) Average L_2 distance between the parameters of source and target networks at each layer, with ($\lambda = 8$) and without ($\lambda = 0$) regularization. (b) Average L_2 norm of the parameters of source network at each layer.

We next illustrate the tendency of the parameters of the target model to more deviate from the pre-trained values in deeper network layers. We used an image segmentation task implemented by a U-net architecture. The details of the source and target models are given below. We calculated the average L_2 distance between the original and the tuned parameters at each network layer. The distance between the target and the source values of each parameter is normalized by the norm of the source value. We examined two transfer learning cases, fine-tuning without regularization ($\lambda = 0$) and fine-tuning with a fixed regularization ($\lambda = 8$) (Eq. 2) that was found to be the optimal value for that setup. Figure 1a shows that at $\lambda = 0$, the distance of the tuned parameters from their original values increases along the network layers. For the case of $\lambda = 8$, as expected, the regularization reduces the distance between the pre-trained and the tuned model. However, the trend toward increased deviation along the network layers remains. Figure 1b shows the average parameter norms at each layer of the source network. We can see that, in contrast to transfer learning, in training from scratch there is no increased deviation from the near zero random starting point along the network layers.

Based on the analysis described above, in this study we propose to apply the transfer regularization gradually such that the transfer regularization coefficient λ decreased monotonically along the network layers. A larger value of λ results in a more aggressive knowledge transfer from the source to the target. The first network layers perform low-level processing that do not vary much between tasks applied to similar data types. As the data processing progresses along the network layers, the network is more focused on the target task which is different from the source task. Changing the parameters of a layer also modifies the input to the next layer, which causes the difference between the source and target tasks to accumulate along the network layers. Hence, it makes sense to gradually decrease the penalty of moving away from the pre-trained model as the data processing progresses along the network layers.

Denote the parameters of a target domain network by $\theta = (\theta_1, \theta_2, ..., \theta_k)$ such that θ_i are the parameter set of the i-th layer of the network and k is the number of layers in the network. In a similar way denote the parameters of the source network layers by $\bar{\theta} = (\bar{\theta}_1, \bar{\theta}_2, ..., \bar{\theta}_k)$.

The proposed regularized cost function for transfer learning is:

$$\text{Loss}(\theta) + \sum_{i=1}^{k} \lambda_i \|\theta_i - \bar{\theta}_i\|^2, \tag{3}$$

such that

$$\infty \geq \lambda_1 \geq \lambda_2 \geq \cdots \geq \lambda_k \geq 0.$$

Setting the transfer regularization hyper-parameter λ to ∞ results in freezing the regularized parameters. By setting the hyper-parameter λ to zero, we obtain standard transfer learning where the only way knowledge is transferred to the target task is via parameter initialization. In the case of the final layers that are learned from scratch, we can still initialize them with small random numbers and use standard L_2 regularization during training.

In this study we focus on U-net networks for image segmentation tasks. The U-net architecture [16] has become the state-of-the-art for medical image semantic segmentation. It is composed of two main pathways: a contraction path (the encoder) that captures the context by processing low-level information, and the expanding path (the decoder), which enables precise localization. The U-net encoder performs low and mid-level processing of the pixel map leading to a latent image representation. In contrast, the U-net decoder, generates the network's decisions based on the computed representation and is focused on a specific task accomplished by the network. The most common way of utilizing transfer learning with U-net is by initializing the encoder with pre-trained weights and then either freezing it, or allowing re-training, depending on the target's data size and computational power limitations. The decoder, which is task-dependent, is trained from scratch. We propose to exploit the full potential of the knowledge already acquired by the model on the source task, by enabling changes in weights, but under a certain constraint. The proposed cost function is:

$$\text{Loss}(\theta) + \sum_{i=1}^{k} \lambda_i \|\theta_{\text{encoder},i} - \bar{\theta}_{\text{encoder},i}\|^2 + \lambda' \|\theta_{\text{decoder}}\|^2 \tag{4}$$

s.t. $\bar{\theta} = (\bar{\theta}_{\text{encoder}}, \bar{\theta}_{\text{decoder}})$ and $\theta = (\theta_{\text{encoder}}, \theta_{\text{decoder}})$ are the parameters of the source and target networks, respectively and i goes over the encoder layers. In this scheme we refine the encoder regularization by setting a gradually decreasing regularization coefficients along the encoder layers as described in Eq. (3).

There are many ways to define a decreasing coefficient sequence. In this study we used slowly decreasing functions in the form of:

$$\lambda_i = \max(0, \lambda_0 - \alpha \cdot \log(i)) \qquad i = 1, ..., k \qquad (5)$$

such that λ_0 and α are hyper-parameters that can be tuned on a validation set using a grid search.

3   Network Implementation Details

We next describe the network architecture and pre-training used. We focused here on the task of COVID-19 opacity segmentation. We used a 2-D U-net [16] with a DenseNet121 [6] backbone. In our implementation, the decoder was composed of decoder blocks and a final segmentation head, which consists of a convolutional layer and softmax activation. Each decoder block consists of a transpose convolution layer, followed by two blocks of convolutional layers, batch normalization, and ReLU activation. For the cost function, we used weighted cross-entropy, where the weights were calculated using the class ratio in the dataset.

We investigated regularization in several different pre-training scenarios. We implemented three source tasks and used them to pre-train the encoder on the target task (the decoder was trained from scratch). The three source tasks were as follows:

– **Natural image pre-training network:** U-net with an encoder that was trained on ImageNet.
– **Medical image pre-training network:** U-net with encoder that was trained from scratch on several publicly available medical imaging segmentation tasks [20]. The network has a shared encoder for global feature extraction followed by several medical task-specific decoders [17]. We term this network "MedicalNet".
– **Combined natural and medical image pre-training network:** The U-net encoder was initialized with ImageNet weights and then trained on the medical datasets as above. We term this network "ImageNet+MedicalNet".

The overall system consisted of the trained model and a series of image processing techniques for both the pre, and the post-processing stages. For preprocessing, all the input slices were clipped and normalized to $[0, 1]$ using a window of $[-1000, 0]$ HU and then resized to a fixed spatial input size of 384×384. The trained network was applied to each slice separately. To construct the 3-D segmentation, we first concatenated the slice-level probabilities generated by the model, and then applied a post-processing pipeline that included morphological operations and removal of opacities outside the lungs.

4   Experiments and Results

We evaluated the system on the task of COVID-19 opacity segmentation using a small COVID-19 dataset [7] containing 29 non-contrast CT scans from three different distributions, from which 3,801 slices were extracted. Lungs and areas of infection were labeled by two radiologists and verified by an experienced radiologist. The given labels were of the lungs and infection. The train-validation-test split was: 21 cases (2446 slices) for training, 3 cases (442 slices) for validation, and 5 cases (913 slices) for testing, chosen at random. We compared two transfer regularization methods:

Table 1. Segmentation results for various source networks and transfer regularization schemes.

Pre-trained network	Regularization coefficient	Dice	Sensitivity	Precision
No pre-training	0	0.650	0.701	0.727
ImageNet	0	0.677	0.701	0.730
	3	0.638	0.706	0.632
	5	0.633	0.699	0.631
	8	0.627	0.674	0.638
MedicalNet	0	0.687	0.724	0.782
	3	0.711	0.835	0.734
	5	0.718	0.759	0.807
	8	0.710	0.832	0.704
	$15 - 1.5 \cdot \log(\text{layer-index})$	0.722	0.775	0.801
	$15 - 2.0 \cdot \log(\text{layer-index})$	0.776	0.772	0.828
	$20 - 1.5 \cdot \log(\text{layer-index})$	0.749	**0.847**	0.762
	$20 - 2.0 \cdot \log(\text{layer-index})$	0.767	0.766	0.824
ImageNet+MedicalNet	0	0.724	0.752	0.803
	3	0.743	0.836	0.778
	5	0.754	0.843	0.783
	8	0.764	0.789	0.825
	$15 - 1.5 \cdot \log(\text{layer-index})$	0.794	0.785	0.860
	$15 - 2.0 \cdot \log(\text{layer-index})$	0.784	0.781	0.851
	$20 - 1.5 \cdot \log(\text{layer-index})$	**0.799**	0.782	0.850
	$20 - 2.0 \cdot \log(\text{layer-index})$	0.792	0.765	**0.868**

- **Fixed regularization** [18]: Experiments performed with constant values of λ, starting with $\lambda = 0$; i.e., standard transfer learning via parameter initialization, up to $\lambda = 50$. A high penalty for deviation from the learned weights, which can be considered as basically freezing the encoder.

- **Layer-wize based regularization:** Experiments performed with a gradually decreased λ as a function of the U-net encoder layer's depth.

Given a 3-D chest CT scan, the system produced the correlated 3-D prediction mask for the lungs, as well as the COVID-19 related infections. Once the 3-D segmentation mask for the test set had been extracted, we compared it to the ground truth reference mask for the opacity class.

Table 1 summarizes the segmentation results for the three source tasks. The best segmentation results were attained with ImageNet+MedicalNet, for both the fixed and the monotonically decreasing regularizations. For the fixed regularization, $\lambda = 8$ was obtained as the optimal value, as the Dice score improved by 5.5% from 0.724 ($\lambda = 0$) to 0.764, with a p-value of 0.006. For the monotonically decreasing regularization, $\lambda = 20 - 1.5 \cdot \log(i)$ was found to be the optimal formula on a validation set. In this case the Dice score improved from the case of no regularization ($\lambda = 0$) by 10.3% with p-value < 0.0001.

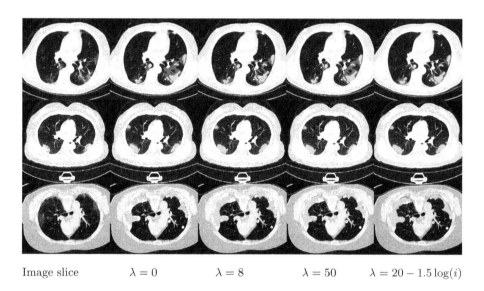

| Image slice | $\lambda = 0$ | $\lambda = 8$ | $\lambda = 50$ | $\lambda = 20 - 1.5 \log(i)$ |

Fig. 2. A qualitative comparison of COVID-19 opacity segmentation with different transfer learning regularizations. Three examples are shown. Green, red, and yellow represent TP, FP, and FN prediction, respectively. (Color figure online)

These results demonstrate that using an inductive bias towards the source parameters for transfer learning, overpowers initialization on its own, since the distributions of the source task and the target task are more similar. Thus, by using the regularization term, either as a function of the layer number or as a constant number, the segmentation results can be improved in cases where the transfer learning is from a source domain close to the target domain. In cases where the transfer learning comes from a source domain with a very different distribution than the target domain, as in the case from natural images to non-contrast chest CT images, it is better to allow deviation from the learned weights.

Qualitative results are shown in Fig. 2. For each input slice, the CT slice and the segmentation results are given for several values of λ, fixed or monotonically decreasing function, obtained by using the ImageNet+MedicalNet as source task. The given examples show the system prediction for slices from three different test cases with different disease demonstrates generalization capabilities of the proposed method in capturing ground-glass and consolidative opacities. It can be also seen that at $\lambda = 8$ and at $\lambda = 20 - 1.5 \cdot \log(l)$, the red and the yellow regions are demonstrably lessened compared to at $\lambda = 0$ and at $\lambda = 50$, which is indicative of improved results of the optimal regularization term.

Table 2. Classification results of the RSNA 2019 Brain CT Hemorrhage Challenge for various transfer regularization schemes.

Methods	Regularization coefficient	Accuracy
No regularization	0	0.583
Fixed regularization	4	0.699
Monotonically decreasing	$5 - 0.5 \cdot \log(\text{layer-index})$	**0.727**

There are several published results on the same COVID-19 dataset [7]. Wang et al. [23] suggested a Hybrid-encoder transfer learning approach. Laradji et al. [8] used a weakly supervised consistency-based strategy with point-level annotations. Muller et al. [12] implemented a 3-D U-Net and using a patch-based scheme. Paluru et al. [14] recently suggested an anamorphic depth embedding-based lightweight model. The reported Dice scores were 0.704 [23], 0.750 [8], 0.761 [12], 0.798 [14] and 0.698 [1]. Comparison here, however, is problematic due to different data-splits and different source tasks used for transfer learning. We note, however, that our transfer regularization approach is complementary to previous works and can be easily integrated into their training procedure.

To show that layerwise transfer learning regularization is a general concept we demonstrate it on another target task: The RSNA 2019 Brain CT Hemorrhage Challenge [4]. Detecting the hemorrhage, if present, is a critical step in treating the patient and there is a demand for computer-aided tools. The goal is to classify each single slice, to one of the following categories: normal, subarachnoid, intraventricular, subdural, epidural, and intraparenchymal hemorrhage. There is a large variability among images within the same class, making the classification task very challenging. We used the encoder described above, and we initialized it with the parameters of MedicalNet. On top of the encoder, we added two fully-connected layers for the classification task. By concatenating three instances of the same slice with different HU windowing (brain window, subdural window, and bone window) and a [0, 1] normalization, we formed a three channeled input. Since the dataset is highly imbalanced, we excluded most of the normal slices and slices with noisy labels, so eventually we were left with 23,031 images, that were split randomly into train (n = 13,819), validation (n = 4,606) and test (n = 4,606) sets. The parameters of the regularization term were tuned on the validation set using a grid search. Table 2 shows the classification results on the test set in terms of accuracy. The results demonstrate the added value of adding such regularization term, fixed or monotonically decreasing, to the standard classification loss.

To conclude, this study described a transfer learning regularization scheme based on using the parameters of the source task as a regularization term where the regularization coefficients decrease monotonically as a function of the layer depth. We concentrated on image segmentation problems handled by the U-net architecture where the encoder and the decoder need to be treated differently. We addressed the specific task of segmenting COVID-19 lesions in chest CT images and showed that adding a decreased regularization along the layer axis to the cost function, leads to improved segmentation results. The proposed transfer regularization method is general and can be incorporated in any situation where transfer learning from a source task to a target task is implemented.

References

1. Bressem, K.K., Niehues, S.M., Hamm, B., Makowski, M.R., Vahldiek, J.L., Adams, L.C.: 3D U-net for segmentation of COVID-19 associated pulmonary infiltrates using transfer learning: state-of-the-art results on affordable hardware. CoRR abs/2101.09976 (2021)
2. Cheplygina, V., de Bruijne, M., Pluim, J.P.: Not-so-supervised: a survey of semi-supervised, multi-instance, and transfer learning in medical image analysis. Med. Image Anal. **54**, 280–296 (2019)
3. De Fauw, J., et al.: Clinically applicable deep learning for diagnosis and referral in retinal disease. Nat. Med. **24**(9), 1342–1350 (2018)
4. Flanders, A.E., et al.: Construction of a machine learning dataset through collaboration: the RSNA 2019 brain CT hemorrhage challenge. Radiol. Artif. Intell. **2**(3), e190211 (2020)
5. Greenspan, H., van Ginneken, B., Summers, R.M.: Guest editorial deep learning in medical imaging: overview and future promise of an exciting new technique. IEEE Trans. Med. Imaging **35**(5), 1153–1159 (2016)
6. Huang, G., Liu, Z., Van Der Maaten, L., Weinberger, K.Q.: Densely connected convolutional networks. In: Proceedings of the IEEE Conference on Computer Vision and Pattern Recognition, pp. 4700–4708 (2017)
7. Jun, M., et al.: COVID-19 CT lung and infection segmentation dataset. Zenodo, 20 April 2020
8. Laradji, I., et al.: A weakly supervised consistency-based learning method for COVID-19 segmentation in CT images. In: Proceedings of the IEEE/CVF Winter Conference on Applications of Computer Vision, pp. 2453–2462 (2021)
9. Li, X., Grandvalet, Y., Davoine, F.: Explicit inductive bias for transfer learning with convolutional networks. In: International Conference on Machine Learning, pp. 2825–2834 (2018)
10. Li, X., Grandvalet, Y., Davoine, F.: A baseline regularization scheme for transfer learning with convolutional neural networks. Pattern Recogn. **98**, 107049 (2020)

11. Litjens, G., et al.: A survey on deep learning in medical image analysis. arXiv preprint arXiv:1702.05747 (2017)
12. Müller, D., Rey, I.S., Kramer, F.: Automated chest CT image segmentation of Covid-19 lung infection based on 3D U-net. arXiv preprint arXiv:2007.04774 (2020)
13. Neyshabur, B., Sedghi, H., Zhang, C.: What is being transferred in transfer learning? arXiv preprint arXiv:2008.11687 (2020)
14. Paluru, N., et al.: Anam-Net: anamorphic depth embedding-based lightweight CNN for segmentation of anomalies in COVID-19 chest CT images. IEEE Trans. Neural Netw. Learn. Syst. **32**(3), 932–946 (2021)
15. Raghu, M., Zhang, C., Kleinberg, J., Bengio, S.: Transfusion: understanding transfer learning for medical imaging. In: Advances in Neural Information Processing Systems (NIPS) (2019)
16. Ronneberger, O., Fischer, P., Brox, T.: U-net: convolutional networks for biomedical image segmentation. In: Navab, N., Hornegger, J., Wells, W.M., Frangi, A.F. (eds.) MICCAI 2015. LNCS, vol. 9351, pp. 234–241. Springer, Cham (2015). https://doi.org/10.1007/978-3-319-24574-4_28
17. Sagie, N., Almog, S., Talby, A., Greenspan, H.: COVID-19 opacity segmentation in chest CT via HydraNet: a joint learning multi-decoder network. In: Medical Imaging 2021: Computer-Aided Diagnosis, vol. 11597. SPIE (2021)
18. Sagie, N., Greenspan, H., Goldberger, J.: Transfer learning via parameter regularization for medical image segmentation. In: The European Signal Processing Conference (EUSIPCO) (2021)
19. Shirokikh, B., Zakazov, I., Chernyavskiy, A., Fedulova, I., Belyaev, M.: First U-Net layers contain more domain specific information than the last ones. In: Albarqouni, S., Bakas, S., Kamnitsas, K., Cardoso, M.J., Landman, B., Li, W., Milletari, F., Rieke, N., Roth, H., Xu, D., Xu, Z. (eds.) DART/DCL -2020. LNCS, vol. 12444, pp. 117–126. Springer, Cham (2020). https://doi.org/10.1007/978-3-030-60548-3_12
20. Simpson, A.L., et al.: A large annotated medical image dataset for the development and evaluation of segmentation algorithms. arXiv preprint arXiv:1902.09063 (2019)
21. Tajbakhsh, N., et al.: Convolutional neural networks for medical image analysis: full training or fine tuning? IEEE Trans. Med. Imaging **35**(5), 1299–1312 (2016)
22. Wang, X., Peng, Y., Lu, L., Lu, Z., Bagheri, M., Summers, R.M.: ChestX-ray8: hospital-scale chest x-ray database and benchmarks on weakly-supervised classification and localization of common thorax diseases. In: Proceedings of the IEEE Conference on Computer Vision and Pattern Recognition (2017)
23. Wang, Y., et al.: Does non-COVID19 lung lesion help? Investigating transferability in COVID-19 CT image segmentation. Comput. Methods Programs Biomed. **202**, 106004 (2021)

Multi-scale Self-supervised Learning for Multi-site Pediatric Brain MR Image Segmentation with Motion/Gibbs Artifacts

Yue Sun[1,2], Kun Gao[2], Weili Lin[2], Gang Li[2], Sijie Niu[1(✉)], and Li Wang[2(✉)]

[1] Department of Shandong Provincial Key Laboratory of Network Based Intelligent Computing, University of Jinan, Jinan 250022, China
[2] Department of Radiology and Biomedical Research Imaging Center, University of North Carolina at Chapel Hill, Chapel Hill, USA
li_wang@med.unc.edu

Abstract. Accurate tissue segmentation of large-scale pediatric brain MR images from multiple sites is essential to characterize early brain development. Due to imaging motion/Gibbs artifacts and multi-site issue (or domain shift issue), it remains a challenge to accurately segment brain tissues from multi-site pediatric MR images. In this paper, we present a multi-scale self-supervised learning (M-SSL) framework to accurately segment tissues for multi-site pediatric brain MR images with artifacts. Specifically, we first work on the downsampled images to estimate coarse tissue probabilities and build a global anatomic guidance. We then train another segmentation model based on the original images to estimate fine tissue probabilities, which are further integrated with the global anatomic guidance to refine the segmentation results. In the testing stage, to alleviate the multi-site issue, we propose an iterative self-supervised learning strategy to train a *site-specific* segmentation model based on a set of reliable training samples automatically generated for a to-be-segmented site. The experimental results on pediatric brain MR images with real artifacts and multi-site subjects from the iSeg-2019 challenge demonstrate that our M-SSL method achieves better performance compared with several state-of-the-art methods.

Keywords: Pediatric brain segmentation · Motion/Gibbs artifacts · Deep learning · Multi-site issue

1 Introduction

Accurate segmentation of the pediatric brain MR images into white matter (WM), gray matter (GM), and cerebrospinal fluid (CSF) is one of the most pivotal steps to characterize early brain development [1, 2]. However, compared with adult brain MR images, pediatric brain MR images exhibit low tissue contrast caused by the ongoing myelination and severe imaging artifacts caused by head motion, creating challenging tasks for tissue segmentation [3]. Therefore, existing tools developed for adult brains, e.g., BrainSuite

© Springer Nature Switzerland AG 2021
C. Lian et al. (Eds.): MLMI 2021, LNCS 12966, pp. 171–179, 2021.
https://doi.org/10.1007/978-3-030-87589-3_18

[4], FSL [5], FreeSurfer [6], and HCP pipeline [7], often perform poorly on the pediatric brain MR images.

Recently, many efforts based on convolutional neural networks have been devoted to the pediatric brain segmentation and achieved encouraging results. For instance, Wang et al. [8] designed an anatomy-guided densely-connected U-Net architecture to perform the infant brain segmentation task, with the anatomical prior as guidance to improve the segmentation accuracy. Nie et al. [9] proposed to train fully convolutional networks for each modality image, and then fuse their high-layer features together to obtain final segmentation. Zöllei et al. [10] proposed an automated segmentation and surface extraction pipeline (named as *Infant FreeSurfer*) for T1-weighted (T1w) neuroimaging data of infants aged 0–2 years. However, these previous works may not handle the motion/Gibbs artifacts. For example, as shown in Fig. 1, *Infant FreeSurfer* could achieve an accurate tissue segmentation result on an artifacts-free image (Fig. 1(a)), but fails on an image in presence of motion/Gibbs artifacts, as shown in Fig. 1(b).

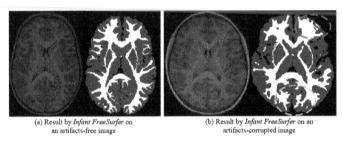

(a) Result by *Infant FreeSurfer* on an artifacts-free image   (b) Result by *Infant FreeSurfer* on an artifacts-corrupted image

Fig. 1. Impact of artifacts for tissue segmentation. Results by *Infant FreeSurfer* [10] on pediatric brain T1w MR images without/with motion/Gibbs artifacts at 24 months of age. The red and orange dashed ellipses indicate some unreliable results, which are mainly caused by motion/Gibbs artifacts as shown in (b).

Moreover, the collaborative use of multi-domain images (acquired from different imaging sites) is more prevalent recently, which makes the segmentation task more difficult. Generally, a model trained on a specific-site dataset often performs well on testing subjects from the same site, but poorly on subjects from other sites with different protocols/scanners. This is called the "multi-site issue" or "domain-shift issue" problem in medical image analysis. For instance, a MICCAI grand challenge on 6-month infant brain MRI segmentation from multiple sites (i.e., iSeg-2019, https://iseg2019.web.unc. edu/) reported and discussed this critical issue [1].

To accurately segment multi-site pediatric images with artifacts, we present a multi-scale self-supervised learning (M-SSL) framework in this paper. In the training stage, inspired by [8], we first train a segmentation model based on the downsampled images to estimate coarse tissue probabilities and build a global anatomic guidance. We then train another segmentation model based on the original images to estimate fine tissue probabilities. The global anatomic guidance and the fine tissue probabilities are integrated as inputs to train a final segmentation model. In the testing stage, to alleviate the multi-site issue, we propose an iterative self-supervised learning (SSL) strategy to train a *site-specific* segmentation model based on a set of reliable training samples, which are automatically generated and iteratively updated, for a to-be-segmented site. The main contributions of this paper are summarized as follows:

1. We propose a framework to accurately segment multi-site pediatric brain MR images with motion/Gibbs artifacts.
2. We leverage downsampled tissue segmentations to build a global anatomic guidance, which alleviates the motion/Gibbs artifacts.
3. We propose an iterative SSL strategy to train *site-specific* segmentation models to minimize the multi-site issue.

2   Dataset and Proposed M-SSL Method

Dataset and Preprocessing. T1w pediatric brain MR images used in this study for training were from the UNC/UMN Baby Connectome Project (BCP) [11]. They were acquired at around 24 months of age on Siemens Prisma scanners with 160 sagittal slices using parameters: TR/TE = 2400/2.2 ms and voxel resolution = $0.8 \times 0.8 \times 0.8$ mm^3. We randomly selected 5 subjects with manual labels as a training dataset. For validation, T1w MR images with real artifacts were from University of Houston, which were acquired with 160 sagittal slices using parameters: TR/TE = 1900/2.98 ms and voxel resolution = $1.0 \times 1.0 \times 1.0$ mm^3. For image preprocessing, the resolution of all images was resampled into $0.8 \times 0.8 \times 0.8$ mm^3, then in-house tools were used to perform skull stripping, intensity inhomogeneity correction, and cerebellum removal.

2.1   The Proposed Method

We propose a multi-scale self-supervised training (M-SST) framework to accurately segment tissues for multi-site pediatric MR images with motion/Gibbs artifacts, consisting of training and testing stages as shown in Fig. 2. We first elaborate on the training stage, consisting of training three segmentation models and a confidence model to detect reliability of automated segmentation results. Then, we design the testing stage to train a *site-specific* segmentation model based on a set of reliable training samples for the to-be-segmented site. Finally, we introduce the implementation details of the proposed method.

2.2   Training Stage

The architecture of a segmentation model can be chosen from U-Net [12], V-Net [13], U-Net++ [14], ADU-Net [8], and nnU-Net [15] et al. In this paper, we adopt ADU-Net as the segmentation architecture, which demonstrates outstanding performance on pediatric brain segmentation. As shown in Fig. 2, in the training stage, benefiting from downsampling and simulated motion/Gibbs artifacts, we first train two segmentation models (named as SegM-A and SegM-B) to generate global anatomic guidance and the fine tissue probabilities, which are integrated as inputs to train the third segmentation model (named as SegM-C) later. Next, an error map, defined as the differences between

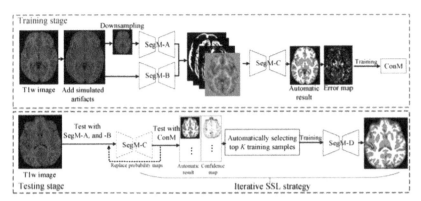

Fig. 2. Illustration of our M-SSL method for brain segmentation of pediatric MR images affected by motion/Gibbs artifacts. **Training stage:** Downsampled and original images with simulated artifacts, are input to train SegM-A to build a global anatomical guidance in the downsampled image space and SegM-B for 3 tissue probability maps in the original image space, respectively. Then a four-channel input (one signed distance map from SegM-A and three probability maps from SegM-B) are automatically generated for the training of SegM-C. Finally, a ConM is trained to evaluate the reliability of those automated segmentations at the voxel level. **Testing stage:** After inputting testing subjects into the trained SegM-A, -B and -C, we can obtain automated segmentation results. Then an iterative SSL strategy is proposed to train a *site-specific* segmentation model SegM-D for the to-be-segmented site.

ground truth and automated segmentations from SegM-C, is regarded as targets to train a confidence model (named as ConM) that is able to automatically detect reliability of automated segmentation results at the voxel level. Figure 3 presents the effectiveness of the confidence map evaluated on the automated segmentation of a testing subject, where the testing subject is not included in the training dataset of ConM. It can be seen that some gyral shapes (circled by a yellow dotted ellipse) in the right figure of Fig. 3(d) are not reasonable. The ConM can effectively detect these unreasonable regions, as indicated by the red color in Fig. 3(c), and also in the left figure of Fig. 3(d).

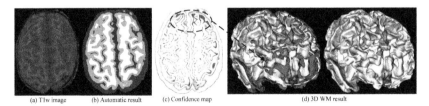

(a) T1w image   (b) Automatic result   (c) Confidence map   (d) 3D WM result

Fig. 3. The effectiveness of confidence map on a testing subject acquired at 24 months of age. (a) T1w testing image, (b) segmentation result by SegM-C, (c) the corresponding confidence map (generated by ConM, where some unreliable (darker) regions of WM are marked with red color), and (d) the 3D WM rendering result (the red region corresponds with the red region of (c)). Note that the testing subject is excluded in the training dataset of ConM.

Global Anatomic Guidance. Prior knowledge, e.g., the cortical thickness is within a certain range, could be employed as an anatomical guidance for the tissue segmentation [1, 8]. Considering the artifacts as high-frequency noise, we can alleviate the artifacts by simply downsampling the original images. Moreover, the downsampled images allow for a large receptive field during the network training. Therefore, instead of estimating the anatomical guidance from the original images with artifacts, we work on the downsampled images to train the SegM-A. Based on the trained model SegM-A, we then upsample the segmentation result into the original image space and construct a signed distance map with respect to the boundary of WM/GM to incorporate the cortical thickness as an anatomical guidance. In detail, after upsampling the result from SegM-A (see Fig. 4(a)), we can derive a label image (see Fig. 4(b)). Based on the label image, it is straightforward to construct a signed distance map with respect to the boundary of WM/GM, as shown in Fig. 4(c). Basically, the function value at each voxel is the shortest distance to its nearest point on the boundary of WM/GM, taking positive value for voxels inside of WM, and negative value for voxels outside of WM.

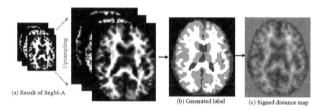

(a) Result of SegM-A

(b) Generated label   (c) Signed distance map

Fig. 4. (a) Shows the probability maps estimated by the SegM-A, then the maps are upsampled to the original size. (b) Is the generated label according to the upsampled probability maps. (c) Illustrates the signed distance map with respect to the WM/GM boundary.

2.3   Testing Stage

Due to the multi-site issue, the trained model in the source site cannot be directly applied to the to-be-segmented site. To alleviate the multi-site issue, we propose an iterative self-supervised learning (SSL) strategy to train a *site-specific* segmentation model for the to-be-segmented site. Based on the idea that the better the probability maps input to SegM-C, the better the outputs of SegM-C, we therefore replace the three tissue probability maps (part of the input of SegM-C) with the output of SegM-C. By iteratively updating the probability maps, the results of SegM-C are gradually refined (i.e., at Round N). Then, we apply the SSL method [2] to further refine the results at Round N, which can effectively refine the segmentation results of testing subjects from multiple sites. In detail, we utilize the SSL method to automatically generate a set of reliable training samples from the testing subjects, which are used to train a *site-specific* segmentation model (i.e., SegM-D). Finally, the testing subjects are directly input to the trained SegM-D to derive final results.

2.4 Implementation Details

In our experiment, we set 1.0 and 0.35 as the weight parameters for simulated motion and Gibbs artifacts [16] respectively, to preprocess MR images in the training stage of Fig. 2. Then, we randomly extract 1,000 patches (size: $32 \times 32 \times 32$) from each training subject. The loss for segmentation models (i.e., SegM-A, SegM-B, and SegM-C) is cross-entropy, and a spatially-weighted cross-entropy loss [2] is used for training SegM-D. The loss of ConM is multi-task cross-entropy. The kernels are initialized by Xavier [38], and we use SGD optimization strategy. The learning rate is 0.005 and multiplies by 0.1 after each epoch.

3 Experimental Results

To demonstrate the performance of our proposed M-SSL method, we first make ablation studies to verify the importance of each component, like downsampling, signed distance map, and the proposed iterative SSL strategy. Then, we validate our method on pediatric MR images with real motion/Gibbs artifacts. Finally, the method is applied on multi-site pediatric brain subjects from the iSeg-2019 challenge [1] to report quantitative analysis.

3.1 Ablation Study

Influence of Downsampling and the Signed Distance Map. To validate the effectiveness of downsampling and signed distance map, we make an ablation study to compare the results obtained by SegM-C trained without/with signed distance map (generated from original image space or downsampled image space) as shown in Fig. 5. Obviously, compared with the results in the second and third columns of Fig. 5, the gyrus of WM tissue (the fourth column in Fig. 5) are clearer and more reasonable with the guidance of the signed distance map (generated from downsampled image space).

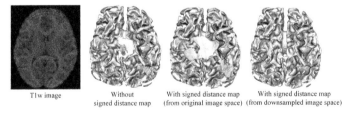

| T1w image | Without signed distance map | With signed distance map (from original image space) | With signed distance map (from downsampled image space) |

Fig. 5. The importance of downsampling and the signed distance map. From left to right: T1w image and the WM results obtained by SegM-C trained without/with the signed distance map (generated from original image space or downsampled image space).

Importance of Simulated Artifacts and Iterative SSL Strategy. In the training stage of Fig. 2, we use simulated motion/Gibbs artifacts to preprocess the intensity images to train SegM-B. Then, we propose an iterative SSL strategy to refine the segmentation results during the testing stage as discussed in Sect. 2.3. We are wondering whether these manners are helpful to improve the accuracy of testing results or not. Figure 6 shows the results of a testing subject with real motion/Gibbs artifacts generated by SegM-B and the iterative SSL strategy. First, we can see the SegM-B trained on images with simulated artifacts is more robust to deal with testing images with real artifacts (the second and third columns in Fig. 6). Second, by leveraging the proposed iterative SSL method, some ring-like tissue results caused by artifacts are gradually alleviated indicated by red arrows (see the fourth to sixth columns in Fig. 6). Therefore, the simulated artifacts added for training subjects and the proposed iterative SSL strategy are helpful to improve the accuracy of testing results.

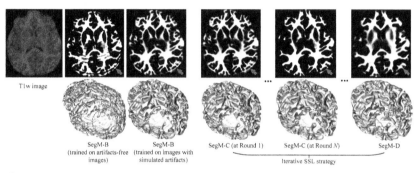

Fig. 6. Comparison of segmentation results for testing subjects with real artifacts. From left to right: T1w image, the results generated by SegM-B (trained on images without/with simulated artifacts), SegM-C and SegM-D (by iterative SSL strategy). Some regions are indicated by red arrows to show the difference, and the corresponding 3D rendering WM results are circled by red dotted circles.

3.2 Comparison Results on Pediatric Brain Images with Real Artifacts

We first verify the performance of our method on 9 brain T1w images with real motion and Gibbs artifacts at 24 months of age. In this experiment, we compare with three state-of-the-art pipelines/tools, including 1) *FreeSurfer* [6], 2) *Infant FreeSurfer* [10], and 3) *volBrain* [17]. Thanks to the freely releasing these pipelines/tools from pioneers, we can directly apply these pipelines/tools to derive tissue segmentation results according to their manuals. Figure 7 presents exemplary tissue segmentation results of one testing image with severe artifacts, obtained by three competing methods and our method. We can observe that most results generated by competing methods show ring-like tissue shapes circled by green dotted ellipses, which are mainly caused by artifacts. However, our results have smoother and more reasonable tissue segmentations as shown in the last column of Fig. 7. The qualitative comparison clearly demonstrates the advantage of the proposed method in terms of accuracy.

3.3 Comparisons on Multi-site Infant Subjects in the ISeg-2019 Challenge

To quantitatively validate of our proposed method, we test the multi-site brain images in the iSeg-2019 challenge. According to the review article [1], we choose one testing site

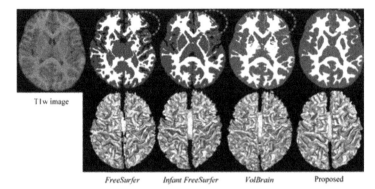

FreeSurfer   Infant FreeSurfer   VolBrain   Proposed

Fig. 7. Tissue segmentation comparison between *FreeSurfer* [6], *Infant FreeSurfer* [10], *volBrain* [17] and our proposed method on a pediatric brain MR image with severe artifacts. The first column shows pediatric T1w images with artifacts. The second to fifth columns are the corresponding segmentation results generated by three competing methods and our method, which are shown with 2D slices and 3D rendering WM tissues, respectively.

(i.e., Stanford University, exhibits different distribution in comparison of other sites) to test our method, which consists of five testing subjects, one of them are significantly affected by motion/Gibbs artifacts. We compare our method with top 3 methods (i.e., QL111111, Tao SMU, and FightAutism) in the challenge as reported in Table 1, in terms of CSF, GM and WM results. From Table 1, our method achieves the highest Dice ratio in terms of GM and WM results, with significant difference compared with others (p-value < 0.05).

Table 1. Dice ratio (%) of cross-site brain segmentation results on five testing subjects from the iSeg-2019 challenge. "+" indicates that our proposed method is significantly better than the top three methods with p-value < 0.05.

Method	QL111111	Tao_SMU	FightAutism	Proposed
CSF	82.77 ± 1.15	82.89 ± 1.18	82.28 ± 1.17	**82.48 ± 1.02**
GM	79.95 ± 2.01	79.08 ± 2.48	79.00 ± 1.78	**81.93 ± 1.27⁺**
WM	85.90 ± 2.75	84.13 ± 3.14	84.31 ± 2.45	**87.67 ± 1.69⁺**

4 Conclusion

To conclude, we propose a multi-scale self-supervised learning (M-SSL) framework to accurately segment tissues for multi-site pediatric brain MR images with motion/Gibbs artifacts. According to the above experiments, the M-SSL method can achieve encouraging results compared with several state-of-the-art methods. In future work, we will further improve our method and test on more multi-site subjects with artifacts.

Acknowledgements. This work was supported in part by National Institutes of Health grants MH109773 and MH117943, National Natural Science Foundation of China under Grant No. 61701192, 61872419, No.61873324, the Natural Science Foundation of Shandong Province, China, under Grant No. ZR2017QF004, No. ZR2019MF040, ZR2019MH106. This work utilizes approaches developed by an NIH grant (1U01MH110274) and the efforts of the UNC/UMN Baby Connectome Project Consortium.

References

1. Sun, Y., et al.: Multi-site infant brain segmentation algorithms: the iSeg-2019 challenge. IEEE Trans. Med. Imaging **40**(5), 1363–1376 (2021)
2. Sun, Y., Gao, K., Niu, S., Lin, W., Li, G., Wang, L.: Semi-supervised transfer learning for infant cerebellum tissue segmentation. In: Liu, M., Yan, P., Lian, C., Cao, X. (eds.) MLMI 2020. LNCS, vol. 12436, pp. 663–673. Springer, Cham (2020). https://doi.org/10.1007/978-3-030-59861-7_67
3. Wang, L., et al.: Benchmark on automatic 6-month-old infant brain segmentation algorithms: the iSeg-2017 challenge. IEEE Trans. Med. Imaging **38**(9), 2219–2230 (2019)
4. Shattuck, D.W., et al.: BrainSuite: an automated cortical surface identification tool. Med. Image Anal. **6**(2), 129–142 (2002)
5. Jenkinson, M., et al.: FSL. Neuroimage **62**(2), 782–790 (2012)
6. Fischl, B.: FreeSurfer. Neuroimage **62**(2), 774–781 (2012)
7. Glasser, M.F., et al.: The minimal preprocessing pipelines for the human connectome project. Neuroimage **80**, 105–124 (2013)
8. Wang, L., et al.: Volume-based analysis of 6-month-old infant brain MRI for autism biomarker identification and early diagnosis. In: Frangi, A.F., Schnabel, J.A., Davatzikos, C., Alberola-López, C., Fichtinger, G. (eds.) MICCAI 2018. LNCS, vol. 11072, pp. 411–419. Springer, Cham (2018). https://doi.org/10.1007/978-3-030-00931-1_47
9. Nie, D., et al.: Fully convolutional networks for multi-modality isointense infant brain image segmentation. In: 2016 IEEE 13th International Symposium on Biomedical Imaging (ISBI), pp. 1342–1345 (2016)
10. Zöllei, L., et al.: Infant FreeSurfer: an automated segmentation and surface extraction pipeline for T1-weighted neuroimaging data of infants 0–2 years. NeuroImage **218**, 116946 (2020)
11. Howell, B.R., et al.: The UNC/UMN Baby Connectome Project (BCP): an overview of the study design and protocol development. Neuroimage **185**, 891–905 (2019)
12. Milletari, F., et al.: V-net: fully convolutional neural networks for volumetric medical image segmentation. In: 2016 Fourth International Conference on 3D Vision (3DV), pp. 565–571 (2016)
13. Zhou, Z., Rahman Siddiquee, M.M., Tajbakhsh, N., Liang, J.: UNet++: a nested U-net architecture for medical image segmentation. In: Stoyanov, D., et al. (eds.) DLMIA/ML-CDS -2018. LNCS, vol. 11045, pp. 3–11. Springer, Cham (2018). https://doi.org/10.1007/978-3-030-00889-5_1
14. Isensee, F., et al.: nnU-Net: Breaking the Spell on Successful Medical Image Segmentation (2019). https://arxiv.org/abs/1904.08128
15. Czervionke, L.F., et al.: Characteristic features of MR truncation artifacts. Am. J. Neuroradiol. **151**(6), 1219–1228 (1988)
16. Manjón, J.V., et al.: volBrain: an online MRI brain volumetry system. (in English). Front. Neuroinf. **10**, 30 (2016)

Deep Active Learning for Dual-View Mammogram Analysis

Yutong Yan[1,2,3], Pierre-Henri Conze[2,3(✉)], Mathieu Lamard[1,2], Heng Zhang[4], Gwenolé Quellec[2], Béatrice Cochener[1,2,5], and Gouenou Coatrieux[2,3]

[1] University of Western Brittany, Brest, France
[2] Inserm, LaTIM UMR 1101, Brest, France
[3] IMT Atlantique, Brest, France
pierre-henri.conze@imt-atlantique.fr
[4] University of Rennes 1, IRISA UMR 6074, Rennes, France
[5] University Hospital of Brest, Brest, France

Abstract. Supervised deep learning on medical imaging requires massive manual annotations, which are expertise-needed and time-consuming to perform. Active learning aims at reducing annotation efforts by adaptively selecting the most informative samples for labeling. We propose in this paper a novel deep active learning approach for dual-view mammogram analysis, especially for breast mass segmentation and detection, where the necessity of labeling is estimated by exploiting the consistency of predictions arising from craniocaudal (CC) and mediolateral-oblique (MLO) views. Intuitively, if mass segmentation or detection is robustly performed, prediction results achieved on CC and MLO views should be consistent. Exploiting the inter-view consistency is hence a good way to guide the sampling mechanism which iteratively selects the next image pairs to be labeled by an oracle. Experiments on public DDSM-CBIS and INbreast datasets demonstrate that comparable performance with respect to fully-supervised models can be reached using only 6.83% (9.56%) of labeled data for segmentation (detection). This suggests that combining dual-view mammogram analysis and active learning can strongly contribute to the development of computer-aided diagnosis systems.

Keywords: Breast cancer · Mass segmentation · Mass detection · Dual-view mammogram analysis · Active learning · Computer-aided diagnosis

1 Introduction

Breast cancer is ranked as the leading cause of global cancer incidence among women in 2020, with an estimated 2.3 million new cases, representing about 25% of all cancers in women [1]. Digital X-ray mammography plays an essential role in diagnosing breast cancer at an early stage. In particular, masses are one of the most common and important type of targeted breast abnormalities. Conventional computer-aided diagnosis (CAD) systems usually use hand-crafted

© Springer Nature Switzerland AG 2021
C. Lian et al. (Eds.): MLMI 2021, LNCS 12966, pp. 180–189, 2021.
https://doi.org/10.1007/978-3-030-87589-3_19

features tailored for mass recognition. Recently, the rise of deep learning made the analysis of mammograms more automatic and accurate thanks to effective training methods, advances in hardware, and most importantly, large amounts of annotated training data [2]. Based on supervised learning using convolutional neural networks (CNN), recent studies have achieved impressive performance regarding mass segmentation [3–5] or detection [2,4,6–8]. Despite such success, supervised deep learning still faces obstacles, including data acquisition and high-quality manual annotations, which are expertise-needed and time-consuming.

Mammography screening involves two standard views acquired for left and right breasts: craniocaudal (CC) and mediolateral-oblique (MLO). In clinical routine, radiologists usually confirm the diagnosis through cross information arising from both views. Examining the CC/MLO correspondence and consistency between suspicious findings thus allows to improve clinical interpretations and subsequent decisions [9]. Computational analysis of dual-view mammograms [10–14] has been validated as an effective way to reduce false-positive cases and improves screening performance. Nevertheless, the labeling workload of radiologists is further increased. Therefore, it is greatly needed to develop an effective annotation suggestion algorithm to alleviate this issue.

Extensively studied in various fields, active learning (AL) aims at reducing human annotation efforts by adaptively selecting the most informative samples for labeling. As for medical imaging, AL has shown high potential in reducing the annotation cost [15]. Recent studies [16,17] proposed AL frameworks for breast cancer segmentation respectively on immunohistochemistry and biomedical images. However, AL methods have not been widely exploited in X-ray mammography analysis. Zhao et al. [18] first introduced AL into a mammography classification system based on a support vector machine (SVM) classifier. Shen et al. [19] proposed a mass detection framework that incorporates AL and self-paced learning (SPL) to improve the model generalization ability. These studies demonstrate great potential of AL in mammogram analysis. Contrary to existing studies based on the uncertainty and diversity of a single image, our goal is to score the dual-view mammograms according to their prediction consistency. Our work can be seen as a complement to existing methods, and proves that combining inter-view information can bring further improvements.

This paper provides the following contributions. First, we propose a novel approach of deep AL for dual-view mammogram analysis (including breast mass segmentation and detection), where the dual-view prediction consistency is integrated as selection criterion. Second, two task-specific neural networks are carefully designed for more effective mammogram mass segmentation and detection. Third, extensive experiments are conducted to reveal the relationship between dual-view consistency and mammogram informativeness.

2   Methods

To reduce the labeling efforts dealing with breast masses in mammograms, we propose a novel approach of deep active learning for dual-view mammogram analysis. Specifically, we consider two scenarios: mass segmentation and detection.

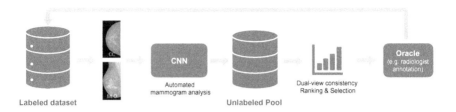

Fig. 1. Proposed deep active learning workflow. (Color figure online)

The key insight of our method is to use the consistency of mass segmentation or detection results arising from CC/MLO view-points as active learning criteria.

The proposed AL process starts by pre-training the model on a small labeled subset D_l. Then, we perform model inference on the unlabeled dataset D_u to select the most informative mammogram pairs according to the calculated dual-view prediction consistency. These selected pairs are then sent to radiologists for annotation and appended to D_l, where the model is consequently fine-tuned on. Such AL cycle is repeated several times to gradually improve the model performance, until the annotation budget is exhausted. The key feature of AL is the query algorithm for the informativeness ranking of unlabeled images, which in our work is the scoring function of the dual-view prediction consistency.

2.1   Proposed Network Architectures

Breast mass segmentation and detection are two main tasks in mammogram analysis. We take inspiration from recent advances of deep neural networks [20–22], and design simple and efficient networks for each of these tasks (Fig. 2).

Mass Segmentation Network (MSN). The architecture is composed of an encoder for feature extraction, a decoder for spatial detail reconstruction and several skip-connections between both branches to recover spatial information. Instead of using a standard symmetric encoder-decoder architecture [21,23], we apply an alternative asymmetric architecture where residual blocks are integrated into the encoder and 1×1 convolution layers are part of the decoder (Fig. 2(a)). The network complexity is greatly reduced while the performance stays unchanged. The optimization is supervised by the combination of binary cross-entropy (L_{bce}) and Dice (L_{dice}) losses following $L_{seg} = L_{dice} + \lambda_1 L_{bce}$ with:

$$L_{dice} = 1 - \frac{2|p \circ y|}{|p| + |y|} \tag{1}$$

$$L_{bce} = \begin{cases} -\log(p) & \text{if } y = 1 \\ -\log(1-p) & \text{otherwise} \end{cases} \tag{2}$$

where p and y represent the prediction mask and the ground truth mask respectively, $|.|$ and $\circ$ the pixel-wise sum and multiplication operations. The empirical factor λ_1 is set to 0.5 to prevent the combined loss from degenerating into L_{bce}.

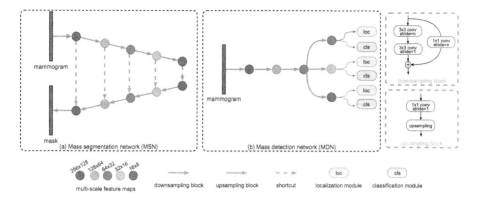

Fig. 2. Proposed network architectures for mass segmentation (a) and detection (b). A downsampling (upsampling) block is applied in each red (green) arrow. (Color figure online)

Mass Detection Network (MDN). We designed a single-stage mass detection network where a multi-scale prediction strategy is applied to detect masses of different scales. Three detection branches with different scales $\{64 \times 32, 32 \times 16, 16 \times 8\}$ are attached to a regular feature extraction network (Fig. 2(b)) consisting of 3 residual blocks. The multi-scale architecture allows the network to be more robust to lesions of different sizes, i.e. larger scale for smaller masses and vise-versa. Each branch consists of a localization module and a classification module, where the former is in charge of regressing the spatial transformation (4 coordinates offset) from predefined anchor boxes to ground truth boxes, and the latter predicts the mass presence probability for each anchor box. We use the focal loss (L_{focal}) to supervise classification modules and the balanced L1 loss (L_{bl1}) to supervise localization modules, following $L_{det} = L_{focal} + \lambda_2 L_{bl1}$ with:

$$L_{focal} = \begin{cases} -\alpha_1(1-p)^{\gamma_1}\log(p) & \text{if } y = 1 \\ -(1-\alpha_1)p^{\gamma_1}\log(1-p) & \text{otherwise} \end{cases} \qquad (3)$$

$$L_{bl1} = \begin{cases} \frac{\alpha_2}{\beta}(\beta|x|+1)\ln(\beta|x|+1) - \alpha_2|x| & \text{if } |x| < 1 \\ \gamma_2|x| + C & \text{otherwise} \end{cases} \qquad (4)$$

We use the default parameters of L_{focal} and L_{bl1} as respectively introduced in [22] and [24]: $\alpha_1 = 0.25, \gamma_1 = 2.0$ for L_{focal}, $\alpha_2 = 0.5, \gamma_2 = 1.5, \beta = 1.0$ for L_{bl1}. The final detection loss is the combination of L_{focal} and L_{bl1} with $\lambda_2 = 1$.

2.2 Dual-View Consistency

At the selection stage of each AL cycle, we aim at filtering the most informative mammograms in D_u through the analysis of dual-view consistency. Theoretically, given a pair of mammograms $\{I_{CC}, I_{MLO}\}$ from the same breast, the analysis results should be coherent. Many latent relationships can potentially be exploited

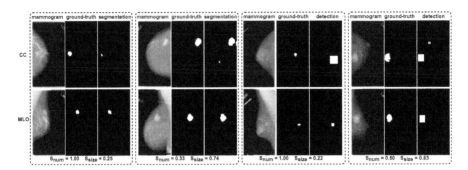

Fig. 3. Examples of mass segmentation (left half) and mass detection (right half) for CC/MLO pairs from DDSM-CBIS and corresponding dual-view consistency. Green delineations represent ground truth mass annotations. (Color figure online)

as query factors, such as the number of masses detected on both views, or the mass size, position, shape, texture... In our work, we consider the first two factors as consistency criteria since their correlation is more obvious. In particular, the number of identified masses from both views $\{N_{CC}, N_{MLO}\}$ should be identical and their sizes $\{S_{CC}, S_{MLO}\}$ (i.e. number of pixels) should be similar. We define two scores (S_{num} and S_{size}) as the measurements of the following factors:

$$S_{num} = \frac{min(N_{CC}, N_{MLO})}{max(N_{CC}, N_{MLO})}, S_{size} = \frac{min(S_{CC}, S_{MLO})}{max(S_{CC}, S_{MLO})} \tag{5}$$

where S_{num} and S_{size} varies from 0 (low consistency) and 1 (high consistency). Correct predictions should meet the above two conditions simultaneously, thus the final combined score is calculated as the minimum of S_{num} and S_{size}:

$$S = min(S_{num}, S_{size}) \tag{6}$$

The proposed consistency score S provides a rough estimation of the mass segmentation/detection prediction quality: mammogram pairs with higher S values are regarded as easy samples and vise-versa. Figure 3 shows mammogram pairs with different S values for both segmentation and detection tasks. When S is low, the prediction on at least one mammogram appears inaccurate. Considering the existence of labeling errors, verifying the number of found lesions from different views tends to avoid involving ambiguous or miss-annotated samples in the training set, towards better AL results. In this direction, our strategy selects mammogram pairs with consistent multi-view predictions such that the aforementioned examples are not taken into account in priority by the oracle.

2.3   Active Learning Strategies

The key of AL is to select the most informative samples to optimize a learnable model. However, the definition of informativeness is still an open question. In the common practice of AL, one considers examples with the most uncertainty or

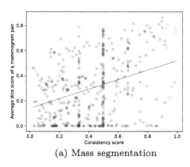 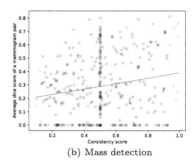

(a) Mass segmentation   (b) Mass detection

Fig. 4. Visualization of mammogram pairs selected by different AL strategies for mammogram segmentation (a) and detection (b) tasks. Red (green) points are picked by worstC (bestC) strategy. The straight line estimates the linear regression. (Color figure online)

examples that are most likely to be wrong as informative examples. However, we need to check if this paradigm remains valid for medical imaging. To this end, we implement three AL strategies: random (rand), best consistency (bestC) and worst consistency (worstC) selections. For each AL cycle, rand strategy randomly selects b mammogram pairs from unlabeled dataset D_u, while bestC (worstC) selects b pairs with the highest (lowest) consistency score S. We visualize Fig. 4 mammogram pairs selected by each AL strategy. Each point represents a CC/MLO pair. Red (green) points are b pairs selected by worstC (bestC). We estimate the linear regression between S and mass segmentation (Fig. 4(a)) or detection (Fig. 4(b)) accuracy. The consistency score appears as a reasonable reference of the prediction quality. Results were obtained during training (i.e. without full convergence) so some points fall in the area of low consistency scores.

3   Experiments

3.1   Implementation Details

We use two publicly-available datasets for our experiments: DDSM-CBIS (Digital Database for Screening Mammography) [25] and INbreast [26], with respectively 1514 and 107 cases containing ground truth mass delineations. For training AL cycles, 586 CC/MLO mammogram pairs are found from DDSM-CBIS and employed to compute the dual-view information consistency. These pairs are divided into a small labeled subset D_l and a simulated unlabeled pool D_u. For INbreast, all 107 images are employed as the test set since pair-wise data is not mandatory during inference. The original mammogram has a resolution of 4084×3328 or 3328×2560, which is computationally expensive. Therefore, we resize images to 512×256 for all experiments. Mammograms are normalized before feeding into neural networks. Random image rotation, cropping, padding, and flipping operations are applied during the training phase for data augmentation.

The proposed framework was implemented using PyTorch. We use SGD optimizer with a learning rate of 0.1 and a cosine annealing schedule. The proposed MSN (MDN) has 45,705 (80,202) learnable parameters in total and was trained for 2k (6k) iterations with a batch size of 32. Each experiment is repeated 5 times, and we report their average performance and the standard error. Following common practice, we adopt the Dice coefficient and the Average Precision (AP) score to respectively evaluate segmentation and detection performances. Dice coefficient is defined as $1 - L_{dice}$ (Eq. 1) whereas the AP score is calculated by taking the area under the precision-recall curve.

For each AL experiment, we start by training an initial model on a random labeled subset D_l containing b pairs. During each AL cycle, we adaptively select the next b pairs from DDSM-CBIS using three different AL strategies (rand, bestC or worstC) from unlabeled dataset D_u. These images are assigned with annotations and appended to D_l for fine-tuning at the next AL cycle. We fix an annotation budget B to end AL cycles. Concretely, we set b to 8 (16 images) for all experiments. Noting that the annotation cost for segmentation is much higher than for detection, we set the annotation budget B to 40 (80 images) for the mass segmentation task and 56 (112 images) for the detection task. In other words, we implement 4 (6) active cycles for segmentation (detection). Each cycle adds 1.37% of labeled data and the whole segmentation (detection) AL process takes 6.83% (9.56%) of labeled data in the training set.

3.2   Results

We conducted extensive experiments to evaluate the performance of rand, bestC and worstC AL strategies. Averaged results are shown in Fig. 5. It can be seen that the model performance is improved progressively cycle by cycle, and bestC ($Dice = 37.00\%$, $AP = 52.83\%$) is consistently better than the other strategies. bestC presents 1.62% Dice improvement and 4.02% AP gains relative to the rand baseline. Conversely, worstC ($Dice = 34.37\%$, $AP = 43.51\%$) is not superior to the baseline. From Fig. 5(b) and (d) we observe that both bestC and worstC reduce the performance instability of rand strategy to a certain extent. In particular, with only 6.83% (9.56%) labeling budget for mass segmentation (detection), bestC achieves performance comparable to the fully supervised model (37.00 vs 37.59% for segmentation, 52.83 vs 54.33% for detection), showing the great potential of our method in alleviating the annotation burden. Besides, we observe greater performance gaps for detection than segmentation. Since detection annotations only provide sparse box-level supervision, the detection task is more critical in terms of the amount of training images.

In the common practice of traditional AL, examples with high consistency scores provide better prediction quality, and could be seen as well-learned examples which are normally not included in AL cycles. Our results seem to contradict this practice, since pairs with higher consistency seem more useful than those with lower consistency. For these results, we propose some explanations: mammography analysis is actually more difficult than general natural image analysis tasks since it is difficult for humans without clinical knowledge to distinguish

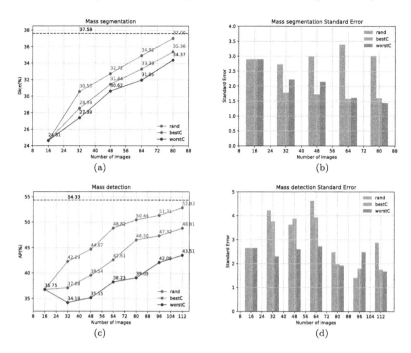

Fig. 5. Mass segmentation and detection performance with `rand` (green), `bestC` (red) and `worstC` (blue) AL strategies. Black dashed lines indicate results using the complete training set. We report average Dice score of mass segmentation (a), Dice score standard error (b), average AP score of mass detection (c) and AP score standard error (d). (Color figure online)

masses from surrounding healthy tissues. Medical imaging datasets can also be very biased due to different acquisition conditions. Learning with a small amount of medical images is challenging, especially for the first few AL cycles. For detection, Fig. 5(c) shows an AP drop for the first AL cycle of `worstC`, indicating that not all labeled data are beneficial when the model does not yet have a full understanding of what masses are. Picking examples with good prediction results helps to consolidate what has been learned while avoiding corner cases.

4   Conclusion

We propose a label-efficient deep learning approach that explores the prediction consistency arising from dual-view mammograms. The main novelty is the combination between multi-view mammogram analysis and active learning, which has not been studied in the field of medical imaging to our knowledge. Our contributions significantly alleviate the burden of manual labeling in breast mass segmentation and detection tasks, which is beneficial to the development of CAD tools. A future possible extension is to integrate existing single-view criteria into our current framework, towards a unified active learning system.

Acknowledgements. This work was partly funded by France Life Imaging (grant ANR-11-INBS-0006 from the French *Investissements d'Avenir* program).

References

1. Sung, H., et al.: Global cancer statistics 2020: GLOBOCAN estimates of incidence and mortality worldwide for 36 cancers in 185 countries. CA Cancer J. Clin. (2021)
2. Kooi, T., et al.: Large scale deep learning for computer aided detection of mammographic lesions. Med. Image Anal. **35**, 303–312 (2017)
3. Singh, V.K., et al.: Breast tumor segmentation and shape classification in mammograms using generative adversarial and convolutional neural network. Expert Syst. Appl. **139**, 112855 (2020)
4. Yan, Y., Conze, P.H., Quellec, G., Lamard, M., Cochener, B., Coatrieux, G.: Two-stage multi-scale mass segmentation from full mammograms. In: IEEE International Symposium on Biomedical Imaging (2021)
5. Yan, Y., et al.: Cascaded multi-scale convolutional encoder-decoders for breast mass segmentation in high-resolution mammograms. In: IEEE International Engineering in Medicine and Biology (2019)
6. Dhungel, N., Carneiro, G., Bradley, A.P.: A deep learning approach for the analysis of masses in mammograms with minimal user intervention. Med. Image Anal. **37**, 114–128 (2017)
7. Agarwal, R., Diaz, O., Lladó, X., Yap, M.H., Martí, R.: Automatic mass detection in mammograms using deep convolutional neural networks. J. Med. Imaging **6**(3), 1–9 (2019)
8. Ribli, D., Horváth, A., Unger, Z., Pollner, P., Csabai, I.: Detecting and classifying lesions in mammograms with deep learning. Sci. Rep. **8**(1), 1–7 (2018)
9. Vijayarajan, S.M., Jaganathan, P.: Breast cancer segmentation and detection using multi-view mammogram. Acad. J. Cancer Res. **7**(2), 131–140 (2014)
10. Yan, Y., Conze, P.H., Lamard, M., Quellec, G., Cochener, B., Coatrieux, G.: Multi-tasking siamese networks for breast mass detection using dual-view mammogram matching. In: International Workshop on Machine Learning in Medical Imaging, pp. 312–321 (2020)
11. Yan, Y., Conze, P.-H., Lamard, M., Quellec, G., Cochener, B., Coatrieux, G.: Towards improved breast mass detection using dual-view mammogram matching. Med. Image Anal. **71**, 102083 (2021)
12. Perek, S., Hazan, A., Barkan, E., Akselrod-Ballin, A.: Mammography dual view mass correspondence. arXiv preprint arXiv:1807.00637 (2018)
13. Ma, J., et al.: Cross-view relation networks for mammogram mass detection. arXiv preprint arXiv:1907.00528 (2019)
14. Gu, X., Shi, Z., Ma, J.: Multi-view learning for mammogram analysis: auto-diagnosis models for breast cancer. In: IEEE International Conference on Smart Internet of Things, pp. 149–153 (2018)
15. Budd, S., Robinson, E.C., Kainz, B.: A survey on active learning and human-in-the-loop deep learning for medical image analysis. arXiv preprint arXiv:1910.02923 (2019)
16. Shen, H., et al..: Deep active learning for breast cancer segmentation on immuno-histochemistry images. In: International Conference on Medical Image Computing and Computer-Assisted Intervention, pp. 509–518 (2020)

17. Li, H., Yin, Z.: Attention, suggestion and annotation: a deep active learning framework for biomedical image segmentation. In: International Conference on Medical Image Computing and Computer-Assisted Intervention, pp. 3–13 (2020)
18. Zhao, Yu., Chen, D., Xie, H., Zhang, S., Lixu, G.: Mammographic image classification system via active learning. J. Med. Biol. Eng. **39**(4), 569–582 (2019)
19. Shen, R., Yan, K., Tian, K., Jiang, C., Zhou, K.: Breast mass detection from the digitized X-ray mammograms based on the combination of deep active learning and self-paced learning. Future Gener. Comput. Syst. **101**, 668–679 (2019)
20. He, K., Zhang, X., Ren, S., Sun, J. :Deep residual learning for image recognition. In: IEEE Conference on Computer Vision and Pattern Recognition, pp. 770–778 (2016)
21. Ronneberger, O., Fischer, P., Brox, T.: U-net: convolutional networks for biomedical image segmentation. In International Conference on Medical Image Computing and Computer-Assisted Intervention, pp. 234–241 (2015)
22. Lin, T.Y., Goyal, P., Girshick, R., He, K., Dollár, P.: Focal loss for dense object detection. In IEEE International Conference on Computer Vision, pp. 2980–2988 (2017)
23. Conze, P.H., et al.: Abdominal multi-organ segmentation with cascaded convolutional and adversarial deep networks. Artif. Intell. Med. (2021)
24. Pang, J., Chen, K., Shi, J., Feng, H., Ouyang, W., Lin, D.: Libra R-CNN: towards balanced learning for object detection. In: IEEE Conference on Computer Vision and Pattern Recognition, pp. 821–830 (2019)
25. Lee, R., Gimenez, F., Hoogi, A., Miyake, K.K., Gorovoy, M., Rubin, D.: A curated mammography data set for use in computer-aided detection and diagnosis research. Sci. Data **4**, 170177 (2017)
26. Moreira, I.C., et al.: INbreast: toward a full-field digital mammographic database. Acad. Radiol. (2012)

Statistical Dependency Guided Contrastive Learning for Multiple Labeling in Prenatal Ultrasound

Shuangchi He[1,2,3], Zehui Lin[1,2,3], Xin Yang[1,2,3], Chaoyu Chen[1,2,3],
Jian Wang[1,2,3], Xue Shuang[1,2,3], Ziwei Deng[1,2,3], Qin Liu[1,2,3], Yan Cao[1,2,3],
Xiduo Lu[1,2,3], Ruobing Huang[1,2,3], Nishant Ravikumar[4,5],
Alejandro Frangi[1,4,5,6], Yuanji Zhang[7], Yi Xiong[7], and Dong Ni[1,2,3(✉)]

[1] National-Regional Key Technology Engineering Laboratory for Medical
Ultrasound, School of Biomedical Engineering, Health Science Center,
Shenzhen University, Shenzhen, China
nidong@szu.edu.cn
[2] Medical Ultrasound Image Computing (MUSIC) Lab,
Shenzhen University, Shenzhen, China
[3] Marshall Laboratory of Biomedical Engineering,
Shenzhen University, Shenzhen, China
[4] Centre for Computational Imaging and Simulation Technologies in Biomedicine
(CISTIB), University of Leeds, Leeds, UK
[5] Leeds Institute of Cardiovascular and Metabolic Medicine,
University of Leeds, Leeds, UK
[6] Medical Imaging Research Center (MIRC), KU Leuven, Leuven, Belgium
[7] Department of Ultrasound, Luohu People's Hospital, Shenzhen, China

Abstract. Standard plane recognition plays an important role in prenatal ultrasound (US) screening. Automatically recognizing the standard plane along with the corresponding anatomical structures in US image can not only facilitate US image interpretation but also improve diagnostic efficiency. In this study, we build a novel multi-label learning (MLL) scheme to identify multiple standard planes and corresponding anatomical structures of fetus simultaneously. Our contribution is three-fold. First, we represent the class correlation by word embeddings to capture the fine-grained semantic and latent statistical concurrency. Second, we equip the MLL with a graph convolutional network to explore the inner and outer relationship among categories. Third, we propose a novel cluster relabel-based contrastive learning algorithm to encourage the divergence among ambiguous classes. Extensive validation was performed on our large in-house dataset. Our approach reports the highest accuracy as 90.25% for standard planes labeling, 85.59% for planes and structures labeling and mAP as 94.63%. The proposed MLL scheme provides a novel perspective for standard plane recognition and can be easily extended to other medical image classification tasks.

S. He and Z. Lin—Contributed equally to this work.

Electronic supplementary material The online version of this chapter (https://doi.org/10.1007/978-3-030-87589-3_20) contains supplementary material, which is available to authorized users.

C. Lian et al. (Eds.): MLMI 2021, LNCS 12966, pp. 190–198, 2021.
https://doi.org/10.1007/978-3-030-87589-3_20

1   Introduction

Ultrasound (US) is widely used for the evaluation of fetal growth and congenital malformations in routine obstetric examinations [12]. During the scanning, US standard planes (SPs) that contain key anatomical structures (ASs) are selected and subsequent biometric measurements are performed [3]. For example, the abdominal circumference (AC) is measured on the transverse plane of the fetal abdomen with umbilical vein at the level of the portal sinus and stomach bubble visible (Fig. 1). The value of AC is then used to estimate the pre-birth weight of a fetus [3,12]. In clinical practice, the standard plane (SP) selection based on ASs identification is experience-dependent, cumbersome, and suffering from the inter-observer and intra-observer variability [1]. Hence, automatic recognition of SP is desired to improve the examinations.

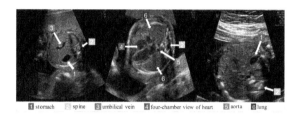

Fig. 1. Left: standard plane of fetal abdomen; Middle: standard plane of four chambers of fetal heart; Right: non-standard plane around fetal abdomen. All images are annotated with multiple anatomical structure labels.

In recent years, deep learning-based methods have witnessed significant growth in automated SP recognition. Chen et al. [3] proposed a composite neural network framework for the automatic recognition of three SPs. Burgos-Artizzu et al. [1] evaluated a large set of state-of-the-art convolutional neural networks for the classification of more than 6 maternal/fetal US planes. Cai et al. [2] presented a convolutional neural network (CNN) framework SonoEyeNet for the detection of SPs. They found that the eye movement tends to focus on the existence of ASs. These methods could distinguish the SPs from the non-standard ones directly with the plane-level labels. However, they did not explicitly incorporate the clues of key ASs, which limited the clinical interpretability and possible guidance for novice sonographers. Lin et al. [9] focused on the detection of key ASs, providing fine-grained information of SPs. However, as shown in Fig. 1, the presence of anatomical structure (AS) alone does not guarantee an accurate identification of the SP, as the SP is also defined by the global image appearance and subtle details [5]. Furthermore, the extensive annotations of each structure with bounding boxes are also labor-intensive and are difficult to obtain. Therefore, new frameworks and methods need to be devised to recognize SP and provide additional information on key ASs simultaneously.

In this paper, we build a novel multi-label learning (MLL) scheme to recognize multiple SPs and corresponding key ASs at the same time. Our contribution is three-fold. (*i*) Inspired by natural language processing techniques, the word embedding [7] is introduced to model the latent concurrency and statistical dependency among different classes, including SPs and ASs. These kinds of cues prove to be strong guidance for MLL prediction. (ii) To further capture the topological structures in the label space of the word embeddings, graph convolutional network (GCN) [4] is explored to propagate information between multiple classes to capture the inner and outter relationship among ASs and SPs. (iii) To tackle the high intra-class variation and low inter-class variation of different SPs and ASs (Fig. 1), we further devise a cluster relabel-based contrastive learning (CRC) to align the similarity and increase discrimination across different classes. We conduct extensive experiments on a large dataset which contains 9742 US images from 920 fetuses and 39 object classes (including 10 SPs and 29 ASs). Experiments prove that, the proposed MLL method can achieve promising results in classifying multiple SPs and identifying associated key ASs.

2   Methodology

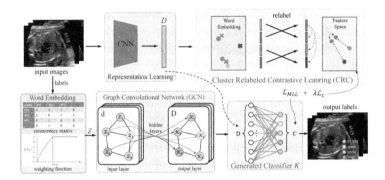

Fig. 2. Overall framework of the proposed MLL for US image recognition. The word embeddings $Z \in \mathbb{R}^{C \times d}$ are generated based on the concurrency matrix and weighting function. Stacked GCNs are learned over the graph to map these word embeddings into an inter-dependent object classifier, i.e., $K \in \mathbb{R}^{C \times D}$. CRC is used to improve the discrimination of the classifier. The classifier is then applied to the image representation from the input image via a CNN for MLL image recognition.

Figure 2 is the schematic view of our proposed method. We propose a MLL framework to recognize the multiple SPs and ASs simultaneously. To exploit the statistical dependency among classes, we firstly generate statistical word embeddings from label annotations. Then, we utilize GCN to model the hierarchical relationship among the classes. Further more, we propose CRC to align the high-level representation among samples of the same category. The MLL recognition output is obtained through representation learning and generated classifier.

2.1   Multi-label Learning with Word Embeddings

CNN is known for its ability in representation learning. As shown in Fig. 2, our MLL learning scheme is built upon a CNN to learn the feature of an image. In specific, we use ResNet [6] as the backbone model. Given an input image $\boldsymbol{I}$ with a size of 448×448 pixels, we can obtain an image-level feature $\boldsymbol{x}$:

$$\boldsymbol{x} = f_{cnn}(\boldsymbol{I}, \theta_{cnn}) \in \mathbb{R}^D, \tag{1}$$

where θ_{cnn} indicates model parameters and $D = 512$.

Inspired by the natural language processing techniques which aim to model the statistical dependency among words, phrases and sentences, we try to capture the fine-grained semantic dependency that exists among the SPs and ASs in the label space following the spirit of word embedding [7]. Since it is intractable to model the relationship among the SP and AS labels in prenatal US directly using the word embeddings pre-trained on natural languages, we build a corpus based on the labels from the trainng US dataset (An image sample is considered as a sentence, and the SP category and AS labels of the sample are considered as words.), and use the GloVe [11] to train the word embeddings. According to the label-based sentences, we construct a concurrency matrix X and use it as GloVe input. X_{ij} represents the number of times class i and class j appear together on the same sample in the dataset. Then, the relationship between word embeddings and the co-occurrence matrix is formulated as:

$$w_i^T \tilde{w}_j + b_i + \tilde{b}_j = log(X_{ij}), \tag{2}$$

where $w_i \in \mathbb{R}^d$ is word embedding and $\tilde{w}_j \in \mathbb{R}^d$ is separate context word embedding which reduces overfitting. b_i and $\tilde{b}_j$ are corresponding bias terms.

We can obtain the final word embedding output $z = w_i^T + \tilde{w}_j$ by optimizing the following loss function:

$$J = \sum_{i,j=1}^{C} f(X_{ij})(w_i^T \tilde{w}_j + b_i + \tilde{b}_j - log(X_{ij})), \tag{3}$$

where C is the size of the vocabulary (i.e. our class number, 39), f is the weighting function [11] that adjusts the frequency of concurrency in the corpus. Word embeddings matrix $Z = \{z_i\}_{i=1}^{C} \in \mathbb{R}^{C \times d}$ hence encodes the statistical dependency and distribution relationships among different labels and can be further explored in the following sections.

2.2 GCN for Class Dependency Learning

It is important to capture the internal relationships between ASs and SPs and leverage this relationship to improve the classification performance in multi-label US image recognition. In this paper, inspired by [4], we explore the GCN to model the class dependency in prenatal US images, which is an effective and flexible way to capture the topological structures in the word embeddings label space represented by Z. Specially, GCN is built to directly map the nodes (i.e. word embeddings Z) of the graph into an inter-dependent classifier (Fig. 2). The GCN based mapping function is defined as:

$$\boldsymbol{G}^{l+1} = h(\hat{\boldsymbol{B}} \boldsymbol{G}^l \boldsymbol{W}^l), \tag{4}$$

where $\boldsymbol{G}^l \in \mathbb{R}^{C \times d}$ are feature descriptions (C denotes the number of nodes and d indicates the dimensionality of node feature) and $\hat{\boldsymbol{B}} \in \mathbb{R}^{C \times C}$ is the normalized version of correlation matrix, and $h(\cdot)$ denotes a non-linear operation. In every back-propagation, the transformation matrix $\boldsymbol{W}^l \in \mathbb{R}^{d \times d'}$ will be updated.

As shown in Fig. 2, for the first layer of the stacked GCNs, the input is the word embeddings matrix Z. The output of the last GCN layer is $\boldsymbol{K} \in \mathbb{R}^{C \times D}$, which matches the dimensionality of the image representations extracted by the CNN. $\boldsymbol{K}$ contains the class dependency and hence regularizes the CNN prediction $\boldsymbol{x}$ as the final classifier. The multi-label prediction scores can be computed by applying the learned classifier to the image representation as follows:

$$\hat{\boldsymbol{y}} = \boldsymbol{K}\boldsymbol{x}, \tag{5}$$

where the ground truth labels of an image is represented as $\boldsymbol{y}$ with $y^i = \{0,1\}$ denoting whether label i appears in the image or not. The training of the whole network uses the traditional MLL classification loss as follows:

$$\mathcal{L}_{MLL} = \sum_{c=1}^{C} y^c log(\sigma(\hat{y}^c)) + (1 - y^c)log(1 - \sigma(\hat{y}^c)). \tag{6}$$

2.3   Cluster Relabeled Contrastive Learning

Borrowing the idea of supervised contrastive learning [8], we propose to use contrastive learning (CL) to further increase the discriminative ability of learning. In CL, the samples belonging to the same class are encouraged to be similar to each other, while that of the different classes are encouraged to be different in high dimensional feature space. However, this principle can not be directly applied to our multi-label circumstance. One sample may have labels overlapped with the other samples, thus it is difficult to define the positive and negative sample pairs. On the other hand, semantically related concepts in the word embeddings space are found to be naturally close to each other [10]. Therefore, we propose to assign every sample a new single label based on the cluster of the word embeddings and perform supervised contrastive learning.

Specially, as shown in Fig. 2, we perform the k-means clustering algorithm in the word embeddings label space $Z \in \mathbb{R}^{C \times d}$. We use C as the sample size and d as the dimensionality to generate N clusters. Each sample with original multi-label $\boldsymbol{y}$ is represented as a vector $\overline{z_p}$. It is calculated through the mean value of the $\{z_i \in \mathbb{R}^d | y_i = 1, i = 1, \cdots, C\}$. The new single label $\boldsymbol{y}^*$ with $y_i^* = \{0,1\}$, $i \in [1, N]$ is assigned to the sample according to the nearest distance among these N cluster centroids. For our multi-label task, N is set to 10.

The contrastive loss to drive the learning of relabeled samples is defined as:

$$\mathcal{L}_c = \sum_i \sum_{j,k \neq i} \alpha(1 - sim(\boldsymbol{x}_i, \boldsymbol{x}_j)) + \beta(1 + sim(\boldsymbol{x}_i, \boldsymbol{x}_k)), \tag{7}$$

where $\boldsymbol{x}$ is the image representation, i and j is the positive sample pair with same $\boldsymbol{y}^*$, i and k is the negative sample pairs with different $\boldsymbol{y}^*$. $sim(\boldsymbol{a}, \boldsymbol{b}) =$

$\frac{a^T b}{\|a\|\|b\|}$ is the cosine similarity between two vectors a and b. α and β are the hyperparameter to weight the similarity. Since there are fewer pairs of positive samples than negative samples, we empirically set α to 0.75 and β to 0.25 to balance the loss weights. The total loss of the proposed method is defined as the summation of MLL loss $\mathcal{L}_{MLL}$ and contrastive loss $\mathcal{L}_c$

$$\mathcal{L} = \mathcal{L}_{MLL} + \lambda \mathcal{L}_c, \tag{8}$$

where λ is the hyperparameter to weight the contrastive loss. λ is set to 0.1 based on the validation results.

3 Experimental Results

Implementation Details. Our dataset contains 9742 prenatal US images from 920 fetuses, including 10 types of SP and 29 types of AS. The gestational age ranges from 18 to 28 week. The image size was set to 448×448. An experienced sonographer provided the ground truth labels. The dataset was randomly split into 4331, 2643 and 2768 images in fetus level for training, validation and testing. There was no overlap of fetus among datasets. Adequate data augmentation were performed. The model was implemented in PyTorch with an RTX 2080Ti GPU. We used Adam optimizer (learning rate 0.001) to train GloVe for 256 epochs to obtain 512-dimensional word embeddings. SGD optimizer (learning rate 0.01) is used to train the model for 100 epochs to obtain the MLL classifier.

Quantitative and Qualitative Analysis. We evaluated the classification in terms of the average overall precision (OP), recall (OR), F1 (OF1) and the average per-class precision (CP), recall (CR), F1 (CF1). The mean average precision (mAP), Hamming loss (HL), the accuracy of the standard plane classification (SP_ACC) and the multi-label classification accuracy that exactly matches the categories of all targets on the image (MLL_ACC) were also taken into consideration. Table 1 illustrates the detailed evaluation results.

Ablation study was conducted to compare different methods, including MLL without GCN and CL (Single-MLL), MLL with contrastive learning (MLL-CL, the non-relabeled version of CRC), MLL with CRC (MLL-CRC), MLL with GCN (MLL-GCN) [4], MLL with GCN and vallina contrastive learning (MLL-GCN-CL) and the full model (MLL-GCN-CRC). We also compared with state-of-the-art methods, including CNN-RNN [13] and SRN [14]. All the above methods were pre-trained with ImageNet. The ResNet34 served as the network backbone for Single-MLL, MLL-CL, MLL-CRC, MLL-GCN, MLL-GCN-CL and MLL-GCN-CRC. We can draw the following conclusions from the Table 1:

(a) GCN significantly improves the model performance (4% in MLL_ACC) under both CL and CRC conditions (i.e., MLL-GCN-CL vs. MLL-CL and MLL-GCN-CRC vs. MLL-CRC). It is attributed to the informative class dependency extracted from the statistical word embeddings by the GCN. A similar conclusion can be deduced through the comparison between MLL-GCN and Single-MLL.

Table 1. Quantitative evaluation of multi-label classification methods (in %).

Method	SP_ACC	MLL_ACC	mAP	HL	OP	OR	OF1	CP	CR	CF1
CNN-RNN	80.07	76.45	83.15	3.83	–	–	–	–	–	–
SRN	86.95	66.17	91.74	2.15	90.13	89.63	89.88	86.81	88.40	87.60
Single-MLL	88.26	81.04	93.75	1.64	92.00	92.80	92.40	88.84	89.61	89.22
MLL-CL	88.37	81.37	93.67	1.65	92.02	92.65	92.33	88.84	89.43	89.13
MLL-CRC	88.37	81.37	93.73	1.63	92.22	92.63	92.42	89.09	89.57	89.33
MLL-GCN	89.27	84.83	94.30	1.51	92.64	93.31	92.98	89.64	90.29	89.97
MLL-GCN-CL	90.07	85.52	94.62	1.45	92.43	94.16	93.28	89.67	91.74	90.69
MLL-GCN-CRC	**90.25**	**85.59**	**94.63**	**1.40**	**92.68**	**94.42**	**93.54**	**89.87**	**92.14**	**90.99**

(b) Comparing the MLL-GCN, MLL-GCN-CL and MLL-GCN-CRC, we can draw the conclusion that, CL can increase the discriminative ability of our method by about 0.7% in MLL_ACC. Besides, we can observe that the CRC methods consistently give better perfomances than the CL methods. The relabeled operation in CRC incorporating the fine-grained semantic in word embeddings space further boosts the similarity alignment of CL.

(c) Among all the state-of-the-art methods (CNN-RNN lacks some result due to its design), the proposed full model MLL-GCN-CRC achieves the best results regarding both the SPs classification and ASs identification. The statistical knowledge via graph manner and similarity alignment in MLL contributes to the capture of class dependency.

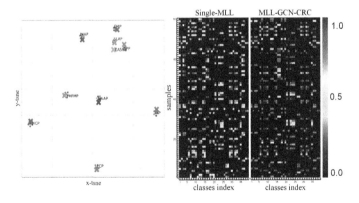

Fig. 3. Left: the t-SNE visualization of the word embeddings space of 39 classes. The dots of different colors represent categories. The cross represents the center of cluster. Right: two score matrices of Single-MLL and our proposed MLL-GCN-CRC. Each row in the matrix represents a sample class prediction. The little green points indicate the true class and the color of matrix element indicate the predicted class scores. (Color figure online)

In Fig. 3, we can observe that the related ASs and SPs embedding clustered together naturally, which builds a more semantic-reasonable label space. On the

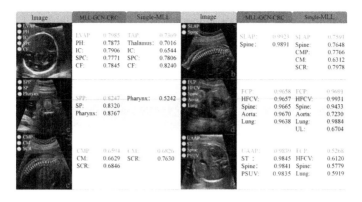

Fig. 4. Typical results of multi-label recognition on fetal US. Red box for ground truth. Blue circle for SPs and green circle for ASs. (Color figure online)

other hand, this result supports the feasibility of our CRC. The score matrices in Fig. 3 further illustrates the MLL prediction of some examples. More sample score matrices can be found in Fig. S1 in the Supplementary. It can be observed that the proposed method MLL-GCN-CRC obtains more matched cases (i.e. green point locates in the region with high score) than the Single-MLL does. This phenomenon reflects that the statistical knowledge encoded by GCN and the discriminative power enhanced by CRC are beneficial in promoting the class prediction and reducing the false positives.

Figure 4 shows the prediction comparisons of six samples. Among the six SPs of fetal LVAP, SPP, CMP, SLAP, FCP, and UAAP (see the detailed name list of SP and AS in the Table S1 of Supplementary) , our MLL-GCN-CRC obtaines high scores and correct predictions for most of the SPs and ASs (Fig. 4(a)(c)). More comparison results can be found in the Fig. S2 of Supplementary. On the contrary, the Single-MLL presents mis-classifications and false positives (Fig. 4(b)(f)).

4   Conclusion

In this paper, we propose a novel multi-label learning scheme (MLL-GCN-CRC) for multiple standard planes and corresponding anatomical structures recognition in prenatal ultrasound. Following the spirit of word embedding, the statistical concurrency knowledge is explored to capture the latent class dependency between standard planes and anatomical structures. A GCN is designed to further encode the dependency among the word embeddings. By performing relabeling based on the clusters in word embeddings space, the contrastive learning boosts the classification performance. Experiments on large dataset show that the proposed method obtains promising performances. Our proposed design is general and may inspire the community for multi-task labeling.

Acknowledgment. This work was supported by the SZU Top Ranking Project (No. 86000000210).

References

1. Burgos-Artizzu, X.P., et al.: Evaluation of deep convolutional neural networks for automatic classification of common maternal fetal ultrasound planes. Sci. Rep. 10(1), 1–12 (2020).
2. Cai, Y., Sharma, H., Chatelain, P., Noble, J.A.: SonoEyeNet: standardized fetal ultrasound plane detection informed by eye tracking. In: 2018 IEEE 15th International Symposium on Biomedical Imaging (ISBI 2018), pp. 1475–1478. IEEE (2018)
3. Chen, H., et al.: Ultrasound standard plane detection using a composite neural network framework. IEEE Trans. Cybern. 47(6), 1576–1586 (2017).
4. Chen, Z.M., Wei, X.S., Wang, P., Guo, Y.: Multi-label image recognition with graph convolutional networks. In: Proceedings of the IEEE/CVF Conference on Computer Vision and Pattern Recognition, pp. 5177–5186 (2019)
5. Dong, J., et al.: A generic quality control framework for fetal ultrasound cardiac four-chamber planes. IEEE J. Biomed. Health Inform. 24(4), 931–942 (2019)
6. He, K., Zhang, X., Ren, S., Sun, J.: Deep residual learning for image recognition. In: Proceedings of the IEEE Conference on Computer Vision and Pattern Recognition, pp. 770–778 (2016)
7. Hinton, G.E., et al.: Learning distributed representations of concepts. In: Proceedings of the Eighth Annual Conference of the Cognitive Science Society, Amherst, MA, vol. 1, p. 12 (1986)
8. Khosla, P., et al.: Supervised contrastive learning. arXiv preprint arXiv:2004.11362 (2020)
9. Lin, Z., et al.: Multi-task learning for quality assessment of fetal head ultrasound images. Med. Image Anal. 58, 101548 (2019)
10. Mikolov, T., Chen, K., Corrado, G., Dean, J.: Efficient estimation of word representations in vector space. arXiv preprint arXiv:1301.3781 (2013)
11. Pennington, J., Socher, R., Manning, C.D.: Glove: global vectors for word representation. In: Proceedings of the 2014 Conference on Empirical Methods in Natural Language Processing (EMNLP), pp. 1532–1543 (2014)
12. Salomon, L., et al.: Practice guidelines for performance of the routine mid-trimester fetal ultrasound scan. Ultrasound Obstet. Gynecol. 37(1), 116–126 (2011)
13. Wang, J., Yang, Y., Mao, J., Huang, Z., Huang, C., Xu, W.: CNN-RNN: a unified framework for multi-label image classification. In: Proceedings of the IEEE Conference on Computer Vision and Pattern Recognition, pp. 2285–2294 (2016)
14. Zhu, F., Li, H., Ouyang, W., Yu, N., Wang, X.: Learning spatial regularization with image-level supervisions for multi-label image classification. In: Proceedings of the IEEE Conference on Computer Vision and Pattern Recognition, pp. 5513–5522 (2017)

Semi-supervised Learning Regularized by Adversarial Perturbation and Diversity Maximization

Peng Liu and Guoyan Zheng[✉]

Institute of Medical Robotics, School of Biomedical Engineering, Shanghai Jiao Tong University, No. 800, Dongchuan Road, Shanghai 200240, China
`guoyan.zheng@sjtu.edu.cn`

Abstract. In many clinical settings, a lot of medical image datasets suffer from the imbalance problem, which makes the predictions of the trained models to be biased toward majority classes. Semi-supervised Learning (SSL) algorithms trained with such imbalanced datasets become more problematic since pseudo-labels of unlabeled data are generated from the model's biased predictions. Towards addressing this challenge, we propose a SSL framework which can effectively leverage unlabeled data for improving the performance of deep convolutional neural networks. It is a consistency-based method which exploits the unlabeled data by encouraging the prediction consistency of given input under adversarial perturbation and diversity maximization. We additionally propose to use uncertainty estimation to filter out low-quality consistency targets for the unlabeled data. We conduct comprehensive experiments to evaluate the performance of our method on two publicly available datasets, i.e., the ISIC 2018 challenge dataset for skin lesion classification and the ChestX-ray14 dataset for thorax disease classification. The experimental results demonstrated the efficacy of the present method.

Keywords: Semi-supervised learning · Adversarial perturbation · Diversity maximization · Uncertainty estimation

1 Introduction

It has been repeatedly reported that deep convolutional neural networks (DCNNs) can achieve human or even super-human level performance in various computer aided diagnosis tasks [1–3]. This success, however, crucially relies on the availability of large-scale labeled clinical datasets. For example, Rajpurkar et al. [2] trained DCNNs on the ChestX-ray14 dataset [4] which consists of hundreds of thousands of front-view chest X-ray images for 14 different thoracic diseases. Accurately annotating large amount of medical images, however, is often prohibitive due to time, effort and the requirement of expert annotators.

Electronic supplementary material The online version of this chapter (https:// doi.org/10.1007/978-3-030-87589-3_21) contains supplementary material, which is available to authorized users.

Semi-supervised learning (SSL) can mitigate this challenge by leveraging unlabeled data for improving the performance of DCNNs when only a small amount of labeled data are available [5].

Most of the SSL methods use consistency regularization, which is based on the smoothness assumption [5], i.e., if two samples are close in the input space, their labels should be the same. Accordingly, a good classification model should favor functions that give consistent output for similar data points. Several studies have proved that it is effective to make the model robust against random perturbation in SSL [6–9]. It was also found, however, that standard isotropic smoothing is inefficient in high dimensions as it often makes the model vulnerable to a small perturbation in a specific direction, i.e., the adversarial direction, which is the direction in the input space in which the labeled probability of the model is most sensitive [10,11]. Virtual Adversarial Training or VAT as introduced by Miyato et al. [8], searches for small perturbation that maximizes the change in the prediction model. Several recent studies have shown, however, that VAT can hurt generalization performance [12,13].

Another issue that has not been addressed by most of previous consistency regularization-based SSL methods is the assumption of a balanced class distribution for labeled and unlabeled data. However, in many realistic scenarios, the underlying class distribution of training data is highly imbalanced, which makes the predictions of the trained models to be biased toward majority classes [14]. For SSL algorithms, this issue becomes more problematic since pseudo-labels of unlabeled data are generated from the model's bias predictions. To alleviate this issue, different distribution alignment strategies [15,16] have been proposed. These methods either require prior knowledge on minority categories, which is difficult to obtain, or rely on prior knowledge estimated by pseudo-labels, which is usually less useful and straightforward towards more accurate prediction.

In this paper, we try to address the issue from a different perspective. Our method requires neither to know prior knowledge about the minority categories nor to estimate prior knowledge by pseudo-labels. It is a consistency-based method which exploits the unlabeled data by encouraging the prediction consistency of given input under adversarial perturbation and diversity maximization. Specifically, instead of aligning distributions, we propose a novel loss which explicitly enhances discriminability and diversity of the trained model. We name the new loss as Discriminability and Diversity Maximization (DDM) loss. We additionally propose to use uncertainty estimation to filter out low-quality consistency targets for the unlabeled data.

Our contributions can be summarized as follows: (1) we add adversarial perturbations to the unlabeled samples to regularize the model's predictions; (2) we minimize DDM loss in order to improve the model's generalization capability as well as the model's discriminability and diversity; (3) we introduce an effective uncertainty filtering to gradually enforce consistency regularization. By learning more meaningful knowledge, our method can mitigate the effect of confirmation bias in SSL [17], i.e., when the loss of inconsistency outweights that of misclassification, a model cannot learn any meaningful knowledge, and hence can stuck in a degenerated solution; (4) we conduct extensive experiments on two large-scale public datasets to evaluate the performance of the present method.

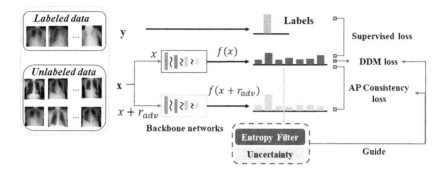

Fig. 1. A schematic illustration of the overall architecture of our method.

2   Method

Figure 1 shows the overall architecture of the proposed method. Our model is trained as follows: we first compute and add Adversarial Perturbations (AP) to the input samples and minimize the Kullback-Leibler (KL) divergence between the model's predictions from the original data and those from the data with the adversarial perturbations. Then for each batch we additionally minimize DDM loss to get better diversity. We further introduce uncertainty estimation to guide the model to learn more meaningful knowledge. We denote labeled dataset as $D_l = \{x_l^{(n)}, y_l^{(n)} | n = 1, ..., N_l\}$, and denote unlabeled dataset as $D_u = \{x_u^{(m)} | m = 1, ..., N_u\}$. We train our model with D_l and D_u..

Adversarial Perturbation-based SSL. Compared with other consistency-based methods which use a random perturbation, our method is based on adversarial perturbation as introduced in [8]. Specifically, we first compute the perturbation in the adversarial direction. Then we add this adversarial perturbation to the input data, and evaluate the consistency between the output of the original data and the output of the adversarially perturbed data.

We use $p(y|x, \theta)$ to denote the output distribution parametrized by θ. $\hat{\theta}$ denotes the model parameters at a specific training iteration. The model $p(y|x, \theta)$ is trained with D_l and D_u.

Our adversarial perturbation-based loss function can be described as follows:

$$L_{AP}(D_u, \theta) := D[p(y|x_u, \hat{\theta}), p(y|x_u + r_{adv}, \theta)]$$
$$where \quad r_{adv} = \underset{r; ||r|| \leq \epsilon}{\arg \max} \, D[p(y|x_u, \hat{\theta}), p(y|x_u + r, \theta)] \tag{1}$$

where $D[\cdot, \cdot]$ is the KL divergence between two distributions; r_{adv} is the adversarial perturbation, $p(y|x_u, \hat{\theta})$ is the current prediction of the unlabeled data. At each unlabeled data point x, $L_{AP}(D_u, \theta)$ can be regarded as a negative measure of the local smoothness. Therefore, with the reduction of $L_{AP}(D_u, \theta)$, the model would become smooth at each unlabeled data point. Please note that we only add adversarial perturbations to unlabeled data to calculate consistency loss while learning from labeled data is dominated by supervised loss.

There exists no closed form to calculate r_{adv}, but it can be approximated by:

$$r_{adv} = \frac{g}{||g||_2}, \ where \ g = \frac{\nabla_r D(r, x_u, \theta)|_{r=\zeta d}}{\zeta} \tag{2}$$

with $\zeta \neq 0$. For simplicity, we use $D(r, x_u, \theta)$ to denote $D[p(y|x_u, \widehat{\theta}), p(y|x_u + r, \theta)]$. The gradient vector g can get computed by back-propagation on the network. More details on this approximation can be found in [8].

Discriminability and Diversity Maximization (DDM). Inputting a randomly selected data batch **B** containing B unlabeled samples to the model, we will get a prediction output probability matrix $P \in \mathbb{R}^{B \times C}$, where C denotes the number of classes. The elements $\{p_{ic}\}$ of P satisfy:

$$\sum_{c=1}^{C} p_{ic} = 1, \forall i \in 1, ..., B, \ where \ p_{ic} > 0, \forall i \in 1, ..., B, c \in 1, ..., C \tag{3}$$

It has been shown previously that the model's discriminability can be enhanced by maximizing the sum of squared probabilities of the batch output [18]. Inspired by their work, we introduce a Discriminability and Diversity Maximization (DDM) loss which encodes discriminability and diversity in one equation:

$$L_{DDM}(D_u, \theta) = -\frac{\sum_{i=1}^{B} \sum_{c=1}^{C} (p_{ic})^2}{\sum_{i,j=1,i \neq j}^{B} \sqrt{\sum_{c=1}^{C} p_{ic} p_{jc} + \gamma \sum_{i=1}^{B} \sum_{c=1}^{C} (p_{ic})^2}} \tag{4}$$

where $\gamma > 0$ is a parameter to avoid gradient saturation near the optimal solution.

Proposition 1. Equation (4) is a robust version of following loss, named as Less Robust DDM (LR-DDM) Loss, done by replacing L_2-norms with $L_{2,1}$-norms [19] (Derivation omitted. See supplemental material for details.):

$$L_{LR-DDM}(D_u, \theta) = -\frac{1}{\frac{\sum_{c=1}^{C} (\sum_{i=1}^{B} p_{ic})^2}{\sum_{i=1}^{B} \sum_{c=1}^{C} (p_{ic})^2} + \gamma - 1} \tag{5}$$

Minimizing Eq. (5) has two-fold effects: maximizing $\sum_{i=1}^{B} \sum_{c=1}^{C} (p_{ic})^2$ and minimizing $\sum_{c=1}^{C} (\sum_{i=1}^{B} p_{ic})^2$. Maximizing $\sum_{i=1}^{B} \sum_{c=1}^{C} (p_{ic})^2$ encourages discriminate predictions as shown in [18]. Now let's have a look of $\sum_{i=1}^{B} p_{ic}$, which can be used to measure the size of c-th class under soft classification. We have:

Proposition 2. One necessary condition for the optimal solution to the minimization of $\sum_{c=1}^{C} (\sum_{i=1}^{B} p_{ic})^2$ is that it has balanced class sizes (Derivation omitted. See supplemental material for details.).

Thus, minimizing $\sum_{c=1}^{C} (\sum_{i=1}^{B} p_{ic})^2$ can enforce the class sizes to be balanced, i.e. regardless of class, the predictions contain the same class size, thus maximizing prediction diversity. We conclude that LR-DDM loss encourages discriminative predictions as well as prediction diversity. As Eq. (4) is a robust version of Eq. (5) to reduce the influence of outliers and noises, the DDM loss also can simultaneously enhance model's discriminability and diversity.

Proposition 3. The DDM loss can be extended to multi-label classification problems (Derivation omitted. See supplemental material for details.).

Certainty Guided Semi-supervised Learning. Existing consistency-based methods have a common issue that they enforce the outputs to be consistent under different perturbations regardless of the quality of the outputs, which may lead to confirmation bias [17]. Toward addressing this issue, we first estimate the uncertainty of the outputs, and then filter out less confident outputs to let the model learn more from more confident outputs, and less from less confident outputs. By doing this, the model will learn reliable and meaningful information, rather than some error prone information. Specifically, we use D_l and D_u to train our model parameters, aiming to minimize the supervised classification loss for the labeled dataset, along with a consistency loss for the unlabeled dataset:

$$L_{cls}(D_l, \theta) + \lambda(e)M(D_u)L_{cons}(D_u, \theta) \tag{6}$$

where $L_{cls}(D_l, \theta)$ is the standard cross-entropy loss which is only computed for labeled data, and $L_{cons}(D_u, \theta)$ is the consistency loss computed for unlabeled data in order to measure the distance between the prediction of the original input and that of the input with adversarial perturbations. $\lambda(e)$ is an ramp-up weighting function which depends on training epochs e, which controls the trade-off between the supervised loss and unsupervised loss. $M(D_u)$ is the certainty guided function.

In each iteration of our training process, it includes two stages: label prediction and uncertainty estimation. Specifically, there are two sets of label predictions: one from the original input and the other from the input with perturbations. We estimate the uncertainty of the latter predictions and dynamically select a subset of $k = \beta e$ low uncertainty predictions that are associated with more reliable targets to do the consistency regularization, where β is a fixed coefficient.

Overall Loss Function. Based on three components described above, we formulate the overall loss function of our method as:

$$L_{cls}(D_l, \theta) + \lambda(e)M(D_u)(\lambda_{DDM}L_{DDM}(D_u, \theta) + \lambda_{AP}L_{AP}(D_u, \theta)) \tag{7}$$

where $L_{cls}(D_l, \theta)$ is the cross entropy loss for labeled dataset, $\lambda(e)$ is a ramp-up weight. $M(D_u)$ is the uncertainty filtering mask. λ_{DDM} is the weight of DDM loss, λ_{AP} is the weight of Adversarial Perturbation (AP) loss.

Implementation Details. We used Python with PyTorch library to implement our method. We trained our model with 2 NVIDIA Tesla V100 GPUs. Adam optimizer [20] was used to train our model. The batch size was set to be 96, containing 24 labeled images and 72 unlabeled images. The learning rate was originally set as $1e - 4$ and decayed with a power of 0.9 after each epoch. For each task, We trained 300 epochs. We empirically set $\lambda_{DDM} = 0.1$, $\lambda_{AP} = 1$, $\beta = 0.024$, and $\gamma = 2.0$. Training a model takes about 10 h while testing is very quick.

3   Experiments and Results

To demonstrate the performance of the proposed method, we designed and conducted comprehensive experiments on two public datasets: the 2018 International Skin Imaging Collaboration (ISIC 2018) challenge dataset [21] and the ChestX-ray14 dataset [4]. Five macro-averaged metrics were adopted from [22] to evaluate performance of our method on the ISIC 2018 challenge dataset, including Area Under Curve (AUC), Accuracy, Sensitivity, Specificity and F1. For the ChestX-ray14 dataset, we adopted AUC metric for evaluation in order to compare with previous approaches.

Results on the ISIC 2018 Challenge Dataset. The ISIC 2018 challenge dataset [21] contains 10,015 skin lesion dermoscopy images, which are labeled into 7 different types of skin lesions. All images were resized into 224×224 for model training. In order to use the pre-trained model on ImageNet [23], we normalized each image with statistics calculated for ImageNet. The entire dataset was randomly split to 70% for training, 20% for testing and 10% for validation, since we don't have access to the ground truth labels of the official validation and testing set. DenseNet121 [24] pre-trained on ImageNet was adopted as the backbone network.

We compared our method with the state-of-the-art methods on skin lesion semi-supervised classification task, including Mean Teacher (MT [7] and Self-Relation Consistency Mean Teacher (SRC-MT) [22]. Table 1 shows the performance of these methods using 20% labeled data for model training. The upper-bound performance was obtained by training a model with 100% labeled data in a fully supervised manner. The baseline was obtained by training a model with 20% labeled data in a fully supervised manner. As shown in Table 1, the MT method achieved a AUC of 93.48%, which was better than the baseline. The SRC-MT method enforces the semantic relation consistency among different samples, achieving consistent improvements over the MT method, showing that the SRC paradigm is indeed helpful for more effective utilization of the unlabeled data. Notably, our method achieved consistently improved results in all metrics over the SRC-MT method, demonstrating effectiveness of our method in semi-supervised learning from imbalanced data.

Table 1. Comparison with state-of-the-art semi-supervised learning methods on the ISIC 2018 challenge dataset

Method	Percentage		Metrics				
	Labeled	Unlabeled	AUC	Sensitivity	Specificity	Accuracy	F1
Upper bound	100%	0	97.88	89.27	94.43	95.61	72.18
Baseline	20%	0	92.59	81.89	89.63	91.54	61.5
MT	20%	80%	93.48	80.86	89.73	91.71	65.89
SRC-MT	20%	80%	93.92	79.65	89.77	91.74	66.88
Ours	20%	80%	**95.69**	**84.35**	**91.72**	**93.95**	**69.63**

Table 2. Comparison with state-of-the-art semi-supervised methods on the ChestX-ray14 dataset.

Labeled percentage	2%	5%	10%	15%	20%	100%
GraphXNET	53	58	63	68	78	N/A
SRC-MT	66.95	72.29	75.28	77.76	79.23	81.75
Ours	**69.87**	**74.2**	**77.4**	**79.18**	**80.43**	81.75

Results on the ChestX-ray14 Dataset. We used the ChestX-ray14 dataset [4] to evaluate the performance of our semi-supervised method in multi-label thorax disease classification task. There are in total 112,120 chest x-rays in the dataset, which are labeled with one or multiple types of 14 thorax diseases. We resized all original images into 256×256 for model training. In order to use the pre-trained model on ImageNet, we normalized each image with statistics calculated for ImageNet [23]. The official data division of ChestX-ray14 [2] is adopted for a fair comparison with previous approaches [22,25]. The entire dataset was split into 70% for training, 20% for testing and 10% for validation. A deeper network, i.e., DenseNet169 [24] was adopted as our backbone network.

Table 2 presents the AUC achieved by all three methods under different percentages of labeled data. Although the $GraphX^{NET}$ [25] achieved an AUC of 78% when 20% labeled data was used, its performance rely heavily on the availability of the amount of the labeled data. When only 2% of labeled data was used, the AUC of the trained model decreased 25%. In comparison, the SRC-MT method [22] achieved an AUC of 66.95% even when 2% of labeled data was used, indicating the effectiveness of the SRC strategy for semi-supervised learning. By combining adversarial perturbation training with DDM training and by using certainty guided training process, our method achieved a further improvement of AUC to 69.87% when only 2% of labeled data was used. Moreover, regardless of the percentage of labeled data used, our method achieved consistently improved results over the other two SSL method. Specifically, using only 15% labeled data, our method achieved an equivalent performance to the SRC-MT method when 20% labeled data was used, demonstrating the advantage of our method in leveraging unlabeled data for improved performance of the underlying DCNNs.

Analytical Ablation Study. To analyze the effect of different losses used in our method, we conducted an ablation study using 20% labeled data of the ISIC 2018 challenge dataset. The results are shown in Table 3. We first compared the LR-DDM loss with the DDM loss. As shown in Table 3, by adding the DDM loss to the baseline model, the trained model achieved consistently improved results in almost all metrics over the one trained by adding the LR-DDM loss, indicating the effectiveness of the $L_{2,1}$-norms in the DDM loss in handling outliers and noises. By adding adversarial perturbation training to the baseline model, the trained model achieved better AUC and specificity over the model trained by adding the DDM loss but lower sensitivity, accuracy and F1 score, suggesting that adversarial perturbation training can boost the overall performance of semi-supervised learning but the trained model may be biased towards majority classes. The comparison results also suggest that the DDM training may help the model to handle imbalanced data but overall performance is still not optimal. By adding both the adversarial perturbation training and the DDM training to the baseline model, the trained model achieved a further improvement on almost all metrics, indicating that the DDM training and the adversarial perturbation training can benefit from each other. The best results were achieved when we further added uncertainty filtering, with a significant improvement on sensitivity and F1 score.

Table 3. Impact of different losses.

Method	Percentage		Metrics				
	Labeled	Unlabeled	AUC	Sensitivity	Specificity	Accuracy	F1
Baseline	20%	0	92.59	81.89	89.63	91.54	61.5
+LR-DDM	20%	80%	93.03	82.19	90.72	92.82	62.98
+DDM	20%	80%	93.25	82.16	91.07	93.10	64.63
+AP	20%	80%	94.71	81.56	91.62	92.92	60.74
+AP+DDM	20%	80%	95.40	82.70	91.98	93.87	65.39
Ours(+AP+DDM+Filtering)	20%	80%	**95.69**	**84.35**	**91.72**	**93.95**	**69.63**

4   Conclusion

In this paper, we introduced a certainty-driven SSL framework which could effectively leverage unlabeled data for improving the performance of DCNNs. Our method used consistency-based regularization which enforces the prediction consistency of given input under adversarial perturbation and diversity maximization for effective use of the unlabeled data. The experimental results demonstrated the efficacy of the present method.

Acknowledgments. This study was partially supported by the Natural Science Foundation of China via project U20A20199, and by Shanghai Municipal Science and Technology Commission via Project 20511105205 and 20DZ2220400.

References

1. Esteva, A., et al.: Dermatologist-level classification of skin cancer with deep neural networks. Nature **542**(7639), 115–118 (2017)
2. Rajpurkar, P., et al.: Deep learning for chest radiograph diagnosis: a retrospective comparison of the CheXNeXt algorithm to practicing radiologists. PLoS Med. **15**(11), e1002686 (2018)
3. Skrede, O.J., et al.: Deep learning for prediction of colorectal cancer outcome: a discovery and validation study. The Lancet **395**(10221), 350–360 (2020)
4. Wang, X., Peng, Y., Lu, L., Lu, Z., Bagheri, M., Summers, R.M.: ChestX-ray8: hospital-scale chest x-ray database and benchmarks on weakly-supervised classification and localization of common thorax diseases. In: Proceedings of the IEEE Conference on Computer Vision and Pattern Recognition, pp. 2097–2106 (2017)
5. Cheplygina, V., de Bruijne, M., Pluim, J.P.: Not-so-supervised: a survey of semi-supervised, multi-instance, and transfer learning in medical image analysis. Med. Image Anal. **54**, 280–296 (2019)
6. Laine, S., Aila, T.: Temporal ensembling for semi-supervised learning. arXiv preprint arXiv:1610.02242 (2016)
7. Tarvainen, A., Valpola, H.: Mean teachers are better role models: weight-averaged consistency targets improve semi-supervised deep learning results. In: Advances in Neural Information Processing Systems, pp. 1195–1204 (2017)
8. Miyato, T., Maeda, S.I., Koyama, M., Ishii, S.: Virtual adversarial training: a regularization method for supervised and semi-supervised learning. IEEE Trans. Pattern Anal. Mach. Intell. **41**(8), 1979–1993 (2018)
9. Xie, Q., Luong, M.T., Hovy, E., Le, Q.V.: Self-training with noisy student improves ImageNet classification. In: Proceedings of the IEEE/CVF Conference on Computer Vision and Pattern Recognition, pp. 10687–10698 (2020)
10. Szegedy, C., et al.: Intriguing properties of neural networks. arXiv preprint arXiv:1312.6199 (2014)
11. Goodfellow, I.J., Shlens, J., Szegedy, C.: Explaining and harnessing adversarial examples. arXiv preprint arXiv:1412.6572 (2014)
12. Nakkiran, P.: Adversarial robustness may be at odds with simplicity. arXiv preprint arXiv:1901.00532 (2019)
13. Tsipras, D., Santurkar, S., Engstrom, L., Turner, A., Madry, A.: Robustness may be at odds with accuracy. arXiv preprint arXiv:1805.12152 (2018)
14. Dong, Q., Gong, S., Zhu, X.: Imbalanced deep learning by minority class incremental rectification. IEEE Trans. Pattern Anal. Mach. Intell. **41**(6), 1367–1381 (2018)
15. Kim, J., Hur, Y., Park, S., Yang, E., Hwang, S.J., Shin, J.: Distribution aligning refinery of pseudo-label for imbalanced semi-supervised learning. In: Advances in Neural Information Processing Systems, vol. 33 (2020)
16. Berthelot, D., et al.: ReMixMatch: semi-supervised learning with distribution alignment and augmentation anchoring. arXiv preprint arXiv:1911.09785 (2019)
17. Arazo, E., Ortego, D., Albert, P., O'Connor, N.E., McGuinness, K.: Pseudo-labeling and confirmation bias in deep semi-supervised learning. In: 2020 International Joint Conference on Neural Networks (IJCNN), pp. 1–8. IEEE (2020)
18. Chen, M., Xue, H., Cai, D.: Domain adaptation for semantic segmentation with maximum squares loss. In: Proceedings of the IEEE/CVF International Conference on Computer Vision, pp. 2090–2099 (2019)

19. Nie, F., Huang, H., Cai, X., Ding, C.: Efficient and robust feature selection via joint 2, 1-norms minimization. In: Advances in Neural Information Processing Systems, vol. 23 (2010)
20. Kingma, D.P., Ba, J.: Adam: a method for stochastic optimization. arXiv preprint arXiv:1412.6980 (2014)
21. Codella, N., et al.: Skin lesion analysis toward melanoma detection 2018: a challenge hosted by the international skin imaging collaboration (ISIC). arXiv preprint arXiv:1902.03368 (2019)
22. Liu, Q., Yu, L., Luo, L., Dou, Q., Heng, P.A.: Semi-supervised medical image classification with relation-driven self-ensembling model. IEEE Trans. Med. imaging **39**(11), 3429–3440 (2020)
23. Russakovsky, O., et al.: ImageNet large scale visual recognition challenge. Int. J. Comput. Vis. **115**, 211–252 (2015)
24. Huang, G., Liu, Z., Van Der Maaten, L., Weinberger, K.Q.: Densely connected convolutional networks. In: Proceedings of the IEEE Conference on Computer Vision and Pattern Recognition, pp. 4700–4708 (2017)
25. Aviles-Rivero, A.I., et al.: Graph-net-chest x-ray classification under extreme minimal supervision. arXiv preprint arXiv:1907.10085 (2019)

TransforMesh: A Transformer Network for Longitudinal Modeling of Anatomical Meshes

Ignacio Sarasua$^{(\boxtimes)}$, Sebastian Pölsterl, Christian Wachinger,
and for the Alzheimer's Disease Neuroimaging

Artificial Intelligence in Medical Imaging (AI-Med), Department of Child and
Adolescent Psychiatry, Ludwig-Maximilians-Universität, Munich, Germany
ignacio@ai-med.de

Abstract. The longitudinal modeling of neuroanatomical changes related to Alzheimer's disease (AD) is crucial for studying the progression of the disease. To this end, we introduce TransforMesh, a spatio-temporal network based on transformers that models longitudinal shape changes on 3D anatomical meshes. While transformer and mesh networks have recently shown impressive performances in natural language processing and computer vision, their application to medical image analysis has been very limited. To the best of our knowledge, this is the first work that combines transformer and mesh networks. Our results show that TransforMesh can model shape trajectories better than other baseline architectures that do not capture temporal dependencies. Moreover, we also explore the capabilities of TransforMesh in detecting structural anomalies of the hippocampus in patients developing AD.

1 Introduction

Alzheimer's disease (AD) is a neurodegenerative disease characterized by progressive cognitive impairment due to neuronal loss (brain atrophy). To prevent or slow-down cognitive decline, current research suggests that it is critical to begin treatment before widespread brain damage occurs [21]. An emerging direction of research focuses on using machine learning to predict patient-specific biomarker trajectories to determine a patient's expected rate of cognitive decline, without relying on clinical diagnosis (e.g. [2,20]). Of particular interest are changes to the hippocampus, because it is among the first brain structures to show signs of atrophy [16]. If we could reliably predict neuroanatomical changes in the hippocampus, we could derive a powerful predictor for the expected rate of cognitive decline.

Since temporal modeling requires longitudinal patient data for training, a major challenge is how to deal with the data heterogeneity: patients enroll at different stages of the disease, drop-out at different time points, and might miss a number of intermediate follow-up visits before returning. Therefore, each patient's trajectory will differ in terms of length and time between recorded visits. To tackle these challenges and satisfy our first requirement of modeling neuroanatomical changes in the hippocampus, we seek a method that learns

C. Lian et al. (Eds.): MLMI 2021, LNCS 12966, pp. 209–218, 2021.
https://doi.org/10.1007/978-3-030-87589-3_22

from heterogeneous trajectories and predicts spatio-temporal changes in the hippocampus. In addition, we need to find a shape representation of the hippocampus that is sensitive to small changes and lends itself to deep learning.

In this paper, we propose the TransforMesh, a longitudinal mesh autoencoder that incorporates heterogeneous trajectories of varying lengths and missing visits via a transformer model, and predicts neuroanatomical changes by modifying a mesh representation of the patient's hippocampus. To the best of our knowledge, this is the first spatio-temporal model that uses transformers for longitudinal neuroanatomical shape analysis that can directly forecast changes to the 3D shape of the hippocampus. In our experiments, we show that our proposed method is able to predict future trajectories with lower reconstruction error than methods that do not include temporal information. In addition, we have also observed that our model is able to capture very fine changes on shapes belonging to patients developing AD.

Related Work. Previous approaches on longitudinal medical image analysis often use a convolutional neural network (CNN) to independently extract image descriptors in an encoding step, which are subsequently passed to a recurrent neural network (RNN) that processes the sequence to form a latent vector summarizing the entire image sequence. This vector can be used as a mean to aggregate the entire sequence [3,5,9,23,26,28] for the purpose of classifying the latest image in the sequence, or as input to another RNN to produce a sequence of predictions over time [14,29]. All these works use longitudinal image information for classification, except the work in [14] that uses a brain connectivity graph as input and produces another sequence of graphs as output.

While RNNs are used in all works on learning from longitudinal medical images, they have since been deprecated in natural language processing (NLP) by the breakthrough of Vaswani et al. [27] and their Transformer network. Transformers are now the de-facto standard in NLP, but it has only been recently that they have been applied to natural images [7]. Their application to medical image analysis is extremely challenging, because they require enormous amounts of training data [6,7,17].

Prior work on AD classification and shape generation has operated on point clouds [12]. However, point cloud representations can be limited for capturing subtle shape changes and therefore mesh representations have become more popular because they provide an efficient, non-uniform representation of a shape. Based on these benefits of meshes, several deep neural networks have recently been proposed in computer vision for learning on meshes [8,11,13,24]. Apart from deep learning, the advantage of meshes for identifying anatomical changes over a large population have earlier been noted [4].

2   Method

Let $\{M_t^i \mid t = 0, ..., T_i\}$ be the set of hippocampi meshes of subject i from the time of enrollment $t = 0$ to the last visit $t = T_i$. One of the biggest challenges

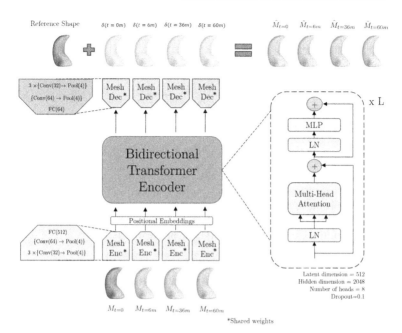

Fig. 1. TransforMesh overview: a shared mesh encoder block extracts the latent features from the input sequence of meshes (orange). The Bidirectional Transformer Encoder models the time trajectory and outputs the reconstructed set of latent vectors, which are fed into a shared decoder block that generates the deformations (yellow) that are applied to the reference mesh (gray). The model is trained end-to-end in an autoencoder manner. Details about the architectures are written next to each block. Further information about the parameters can be found in [11] and [27] (Color figure online)

of longitudinal studies is that not every subject remains in the study for the same amount of time, nor have all attended every follow-up visit. The goal of our method is to model longitudinal shape changes on this highly heterogeneous data. An overview of our proposed method can be found in Fig. 1. First, the *mesh encoder* extracts latent vectors from each mesh. Second, the latent vectors are fed into a *bidirectional transformer encoder*, which models the time dependency between the latent representations. Finally, the *mesh decoder* outputs the reconstructed meshes. The network is trained in an autoencoder manner similar to [17]. We provide more details about this training strategy in Sect. 2.3.

2.1   Mesh Network

A 3D mesh, $M = (V, E, A, F)$, is defined by a set of vertices, $V \in \mathbb{R}^{N \times 3}$, and edges, E, that connect the vertices. $A \in \{0, 1\}^{N \times N}$ is the adjacency matrix that indicates the connection between two vertices (e.g. $A_{i,j} = 1$ if vertex i is connected to vertex j and 0 otherwise) and F are the faces formed by a set of

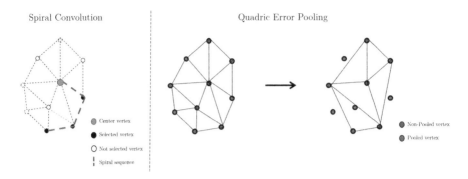

Fig. 2. Left: Definition of a spiral sequence given a vertex v (orange); Right: Quadric Pooling operation used by SpiralNet++. Mesh features are down-sampled by removing red vertices that minimize quadric error. (Color figure online)

edges (three in case of triangular meshes). As opposed to images, meshes are not represented on a regular grid of discrete values and therefore common CNN operations such as *convolution* and *pooling* are not explicitly defined anymore. In particular, defining local neighborhood for a given vertex becomes challenging.

Gong et al. [11] proposed SpiralNet++, a novel *message passing* approach to deal with irregular representations, such as meshes and it has achieved state of the art performance in computer vision tasks, such as reconstruction and classification. To the best of our knowledge, this work is the first application of SpiralNet++ for anatomical shape analysis. SpiralNet++ encoder and decoder blocks are formed by three main operations: spiral convolution, mesh pooling, and un-pooling.

Spiral Convolution: Due to the nature of triangular meshes, a spiral serialization of neighboring nodes is possible. Given a vertex in V, [11] defines its spiral sequence by choosing an arbitrary starting direction in counter-clockwise manner. An example of a spiral sequence can be found in Fig. 2. In comparison to SpiralNet [18], SpiralNet++ defines these sequences only once for a template shape and then applies them to the aligned samples in the dataset, highly increasing the efficiency of the method. The convolution operation in layer k for features $\mathbf{x}_i$ associated to the i-th vertex is therefore defined as: $\mathbf{x}_i^{(k)} = \gamma^{(k)} \left(\underset{j \in S(i,l)}{\|} \mathbf{x}_j^{(k-1)} \right)$, where γ denotes MLPs and $\|$ is the concatenation operation. $S(i,l)$ is an ordered set consisting of l vertices inside the spiral. More details about the method and its implementation can be found in [11].

Mesh Pooling and Un-pooling: The *pooling* is obtained by iteratively contracting vertex pairs that would minimize the quadratic error [10]. In Fig. 2, we illustrate this process. For efficiency, the coordinates of the vertices that must be pooled in each level are computed for the template and then applied to the samples in the dataset. The un-pooling operation is done by using the barycentric coordinates of the closest triangle in the downsampled mesh [25].

2.2 Transformer Network

Figure 1 shows the transformer encoder network, which is inspired by the one proposed in [7]. The Bidirectional Transformer Encoder is formed by concatenating L encoder blocks. Each encoder block is formed by a Multi-Head-Attention and an MLP, each preceded by a Layer Normalization (LN) block. In addition, residual connections are added after every block and GELU is used as non-linearity in the MLP block. Following [7], we also use learnable 1D position embeddings to retain positional information.

2.3 Training Procedure

As illustrated in Fig. 1, our model predicts the mesh deformation field $\delta(t)$ with respect to a *reference mesh*, which is a person's baseline scan $M_{t=0}^i$.

Missing Shapes: In order to account for variable lengths within the sequences, transformer networks use padding tokens. These are ignored during training by adding key masks and disregarded when computing the regression loss. For our application, we use the encoding of the reference shape, as produced by the mesh network, to generate the "missing value" embedding.

Data Augmentation: Since the goal of our method is to model longitudinal shape change in the presence of missing values, some shapes are randomly selected and substituted by the reference shape. Contrary to the real missing shapes, these are not ignored during training: the self-attention masks do take them into consideration and they are included when computing the final regression loss.

Loss Function: To account for the drop in the number of scans as the study goes on, we introduce an exponential factor that gives more weight to scans further from the reference. The training objective function is defined as: $\mathcal{L} = \sum_{i=0}^{P} \sum_{t=0}^{T_i} e^t ||M_{t=0} + \delta(t) - M_t|| - \alpha ||\delta(t)||$, where α (empirically set to 10^{-4}) is a regularization constant that prevents the network from copying the reference shape, and P is the number of patients in the set.

3 Experiments

For all our experiments, we use the shape of the left hippocampus, given its importance in AD pathology [16]. We use data from the Alzheimer's Disease Neuroimaging Initiative (ADNI) database (adni.loni.usc.edu) [15]. Structural scans are segmented with FIRST [22] from the FSL Software Library, which provides meshes for the segmented samples. FIRST segments the subcortical structures by registering them to a reference template, creating voxel-wise correspondences (and therefore, also vertex-wise) between the template and every sample in the dataset. We use this template as the reference for all the pre-computations of the mesh network in Sect. 2.1. We limit the number of follow-ups $S = 8$ (from baseline to 72 months) and the latent dimension for the Mesh Encoder $D = 512$. Our complete dataset consists of 1247 patients split 70/15/15 (train/validation/test) following a data stratification strategy that accounts for age, sex and diagnosis, so they are represented in the same proportion in each set. The average number of follow-up scans per patient is 3.25 and only a 3% of the patients have attended all the follow-up sessions.

3.1   Implementation Details

We define 5 network architectures to evaluate our experiments. First, we define three versions of our method: Tiny TransforMesh (TTM), Small TransforMesh (STM) and Base TransforMesh (BTM) with depths $L = 1$, $L = 3$ and $L = 12$, respectively. Further, we define a baseline and an upper-bound method.

Baseline. The main contribution of transformer networks is their capability of modeling the time dimension in the latent space. To measure the effect of this, we design a baseline method that does not take the time component into account. Inspired by the work in [12], after the mesh encoder path, the transformer network is substituted by a fully connected bottle neck (FCBN) formed by 3 MLP blocks (dimensions are $[S \times D, S, S \times D]$ respectively). The latent vectors obtained by the mesh encoder are concatenated and passed to the FCBN. The output is decoded as in our method. Notice that, as in [12], this method does not capture the temporal component (only the spatial). Contrary to our method, this method is not flexible to any sequence length (S), since the FCBN dimensions directly depend on this parameter.

Mesh Autoencoder. In order to put the reconstruction error into context, we train a mesh autoencoder, MeshAE, to reconstruct the shapes, which uses the same encoder and decoder architecture as our model. The main difference between MeshAE and our model is that MeshAE is only reconstructing the shape from its latent representation, while our model has never seen the shape. Since the mesh part of both networks are identical, we expect our network to be limited by this method. Details about the architecture can be found in Fig. 1.

3.2   Longitudinal Shape Modeling

Given the set of meshes belonging to a subject, $\{M_t^i \mid t = 0, ..., T_i\}$, the goal of our method is to predict missing shapes. We divide our experiments in three scenarios: shape interpolation, shape extrapolation and trajectory prediction.

Interpolation: In this first experiment, we aim to interpolate the mesh of an intermediate follow-up session, giving future and past shapes. From every patient in the test set, we remove from the sequence the shape in the middle $M_{t=\mu}^i$ with $\mu = \lfloor T_i/2 \rfloor$ and pass the sequence through the network. We compute the interpolation error as the Mean Absolute Error (MAE) between $M_{t=\mu}^i$ and $\hat{M}_{t=\mu}^i$, where $\hat{M}_{t=\mu}^i$ is the predicted mesh for $t = \mu$.

Table 1. The table reports the median error and median absolute deviation for the interpolation, extrapolation, and trajectory experiments. Errors were multiplied by 10^2 to facilitate presentation. We compare to the mesh autoencoder (MeshAE), the fully connected bottle neck (FCBN), and tiny, small, and base TransforMesh. Also reported are the model's number of parameters.

Method	Interpolation	Extrapolation	Trajectory	#params
MeshAE	22.2 ± 1.5	22.4 ± 1.7	22.2 ± 1.7	$0.9M$
FCBN	20.4 ± 6.0	26.0 ± 7.7	28.0 ± 8.9	$5.1M$
TTM (1 block)	20.2 ± 5.5	25.7 ± 7.5	27.7 ± 8.9	$4.1M$
STM (3 blocks)	20.3 ± 5.6	25.5 ± 7.3	27.6 ± 8.9	$10.4M$
BTM (12 blocks)	20.3 ± 5.6	25.4 ± 7.4	27.7 ± 8.9	$38.7M$

Extrapolation: In the second experiment, the goal is to predict the mesh of the last available shape, based on all the previous ones. Hence, from every patient in the test set, we remove from the sequence the shape $M_{t=T_i}^i$ and input the remaining shapes to the network. The extrapolation MAE is computed between $M_{t=T_i}^i$ and $\hat{M}_{t=T_i}^i$.

Trajectory Prediction: The third experiment is similar to the extrapolation experiment, but it predicts shapes that are more distant in time. Therefore, we only input into the network the shape belonging to the baseline scan $M_{t=0}$ and predict all the shapes that are at least 2 years apart. We define the future error as $FME = \mathbb{E}\left[\| \hat{M}_t - M_t \| \,|24m \leq t \leq T_i\right]$.

3.3 Results

Table 1 reports the median error and the median absolute deviation along all the patients and scans. We observe that all the Transformer models bring a significant improvement with respect to the baseline method, confirming our initial hypothesis regarding the importance of spatio-temporal models. It is also worth mentioning that the most shallow transformer network (TTM) is able to predict the missing shapes better than the baseline method while having 20% less trainable parameters.

For the extrapolation and trajectory experiments, MeshAE yielded better results, as expected. However, for the interpolation experiment, the temporal methods yielded a lower median error than MeshAE. These results indicate that it could be advantageous to have access to the longitudinal sequence even if the mesh from the actual time point T is not included, especially when T is close to 0. Yet, the increased deviation also indicates that the variation of the temporal models is higher.

3.4 Anomaly Visualization

A common application of Autoencoder Networks in medical image analysis is anomaly detection [1]. The idea is training the network on healthy controls and

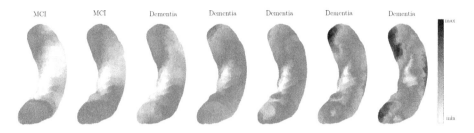

Fig. 3. Vertex-wise Absolute Error between the meshes of a patient converting from MCI to dementia and our method's prediction when trained on HC and MCI subjects.

passing patients with anomalies during inference time. Given an autoencoders' denoising behaviour, these anomalies can be detected in an unsupervised manner by computing the differences between the original and the reconstructed shapes. Those areas with higher reconstruction error are considered anomalies, since they are not part of the healthy training distribution.

We replicate this experiment in our longitudinal setting. To the best of our knowledge, this is the first work exploring longitudinal anomaly detection on anatomical shapes. We train our model on subjects that have not been diagnosed with dementia. During inference, we include AD patients and compute the error between the input and the output. In Fig. 3, we can observe an example belonging to a patient converting from MCI to AD. Even though we do not have an anomaly mask to compare to (like in other techniques [1]) we can observe a strong correlation between the areas with highest error (medial part of the body in the subiculumarea, the lateral part of the body in the CA1 area and the inferior part of the hippocampus head in the subiculum area) and those proven to be more affected my Alzheimer's disease [19].

4   Conclusion

In this work, we have proposed the first deep spatio-temporal model for longitudinal analysis of anatomical shapes. To the best of our knowledge, this is the first approach that combines mesh and transformer networks. Our experiments demonstrated that TransforMesh outperforms the baseline methods, which rely only on spatial features, even on its lightest implementation. This makes it ideal for medical applications where the amount of data is limited. We further illustrated that TransforMesh can be used to detect structural anomalies on patients converting from MCI to AD.

Acknowledgements. This research was supported by the Bavarian State Ministry of Science and the Arts and coordinated by the Bavarian Research Institute for Digital Transformation, and the Federal Ministry of Education and Research in the call for Computational Life Sciences (DeepMentia, 031L0200A).

References

1. Baur, C., Denner, S., Wiestler, B., Navab, N., Albarqouni, S.: Autoencoders for unsupervised anomaly segmentation in brain MR images: a comparative study. Med. Image Anal. **69**, 101952 (2021)
2. Bilgel, M., Prince, J.L., Wong, D.F., Resnick, S.M., Jedynak, B.M.: A multivariate nonlinear mixed effects model for longitudinal image analysis: application to amyloid imaging. Neuroimage **134**, 658–670 (2016)
3. Campanella, G., et al.: Clinical-grade computational pathology using weakly supervised deep learning on whole slide images. Nat. Med. **25**(8), 1301–1309 (2019)
4. Cong, S., et al.: Building a surface atlas of hippocampal subfields from MRI scans using Freesurfer, FIRST and SPHARM. In: 2014 IEEE 57th International Midwest Symposium on Circuits and Systems (MWSCAS), pp. 813–816. IEEE (2014)
5. Cui, R., Liu, M., Initiative, A.D.N., et al.: RNN-based longitudinal analysis for diagnosis of Alzheimer's disease. Comput. Med. Imaging Graph. **73**, 1–10 (2019)
6. Devlin, J., Chang, M.W., Lee, K., Toutanova, K.: BERT: pre-training of deep bidirectional transformers for language understanding. In: NAACL (2019)
7. Dosovitskiy, A., et al.: An image is worth 16x16 words: transformers for image recognition at scale. arXiv:2010.11929 (2020)
8. Feng, Y., Feng, Y., You, H., Zhao, X., Gao, Y.: MeshNet: mesh neural network for 3D shape representation. In: Proceedings of the AAAI Conference on Artificial Intelligence, vol. 33, pp. 8279–8286 (2019)
9. Gao, R., et al.: Distanced LSTM: time-distanced gates in long short-term memory models for lung cancer detection. In: Suk, H.I., Liu, M., Yan, P., Lian, C. (eds.) Machine Learning in Medical Imaging, vol. 11861, pp. 310–318. Springer, Cham (2019). https://doi.org/10.1007/978-3-030-32692-0_36
10. Garland, M., Heckbert, P.S.: Surface simplification using quadric error metrics. In: Proceedings of the 24th Annual Conference on Computer Graphics and Interactive Techniques, pp. 209–216 (1997)
11. Gong, S., Chen, L., Bronstein, M., Zafeiriou, S.: Spiralnet++: a fast and highly efficient mesh convolution operator. In: Proceedings of the IEEE/CVF International Conference on Computer Vision Workshops (2019)
12. Gutiérrez-Becker, B., Sarasua, I., Wachinger, C.: Discriminative and generative models for anatomical shape analysis on point clouds with deep neural networks. Med. Image Anal. **67**, 101852 (2021)
13. Hanocka, R., Hertz, A., Fish, N., Giryes, R., Fleishman, S., Cohen-Or, D.: MeshCNN: a network with an edge. ACM Trans. Graph. (TOG) **38**(4), 1–12 (2019)
14. Hwang, S.J., Mehta, R.R., Kim, H.J., Johnson, S.C., Singh, V.: Sampling-free uncertainty estimation in gated recurrent units with applications to normative modeling in neuroimaging. In: Proceedings of the 35th Uncertainty in Artificial Intelligence Conference, vol. 115, pp. 809–819 (2020)
15. Jack, C.R., et al.: The Alzheimer's disease neuroimaging initiative (ADNI): MRI methods. J. Magn. Reson. Imaging **27**(4), 685–691 (2008)
16. Jack, C.R., Holtzman, D.M.: Biomarker modeling of Alzheimer's disease. Neuron **80**(6), 1347–1358 (2013)
17. Lewis, M., et al.: BART: denoising sequence-to-sequence pre-training for natural language generation, translation, and comprehension. In: Proceedings of the 58th Annual Meeting of the Association for Computational Linguistics, pp. 7871–7880 (2020)

18. Lim, I., Dielen, A., Campen, M., Kobbelt, L.: A simple approach to intrinsic correspondence learning on unstructured 3D meshes. In: Proceedings of the European Conference on Computer Vision (ECCV) Workshops (2018)

19. Lindberg, O., et al.: Shape analysis of the hippocampus in Alzheimer's disease and subtypes of frontotemporal lobar degeneration. J. Alzheimer's Dis.: JAD **30**(2), 355 (2012)

20. Marinescu, R.V., et al.: Dive: a spatiotemporal progression model of brain pathology in neurodegenerative disorders. NeuroImage **192**, 166–177 (2019)

21. Mehta, D., Jackson, R., Paul, G., Shi, J., Sabbagh, M.: Why do trials for Alzheimer's disease drugs keep failing? A discontinued drug perspective for 2010–2015. Expert Opin. Investig. Drugs 26(6), 735–739 (2017)

22. Patenaude, B., Smith, S.M., Kennedy, D.N., Jenkinson, M.: A Bayesian model of shape and appearance for subcortical brain segmentation. NeuroImage **56**(3), 907–922 (2011)

23. Perek, S., Ness, L., Amit, M., Barkan, E., Amit, G.: Learning from longitudinal mammography studies. In: Shen, D. et al. (eds.) Medical Image Computing and Computer Assisted Intervention, vol. 11769, pp. 712–720. Springer, Cham (2019). https://doi.org/10.1007/978-3-030-32226-7_79

24. Ranjan, A., Bolkart, T., Sanyal, S., Black, M.J.: Generating 3D faces using convolutional mesh autoencoders. In: European Conference on Computer Vision (ECCV), pp. 704–720 (2018)

25. Ranjan, A., Bolkart, T., Sanyal, S., Black, M.J.: Generating 3D faces using convolutional mesh autoencoders. In: Proceedings of the European Conference on Computer Vision (ECCV), pp. 704–720 (2018)

26. Santeramo, R., Withey, S., Montana, G.: Longitudinal detection of radiological abnormalities with time-modulated LSTM. In: Stoyanov D. et al. (eds.) Deep Learning in Medical Image Analysis and Multimodal Learning for Clinical Decision Support, vol. 11045, pp. 326–333. Springer, Cham (2018). https://doi.org/10.1007/978-3-030-00889-5_37

27. Vaswani, A., et al.: Attention is all you need. In: Proceedings of the 31st International Conference on Neural Information Processing Systems, pp. 6000–6010 (2017)

28. Xu, Y., et al.: Deep learning predicts lung cancer treatment response from serial medical imaging. Clin. Cancer Res. **25**(11), 3266–3275 (2019)

29. Yang, D., et al.: Deep image-to-image recurrent network with shape basis learning for automatic vertebra labeling in large-scale 3D CT volumes. In: Descoteaux, M., Maier-Hein, L., Franz, A., Jannin, P., Collins, D., Duchesne, S. (eds.) Medical Image Computing and Computer Assisted Intervention, vol. 10435, pp. 498–506. Springer, Cham (2017). https://doi.org/10.1007/978-3-319-66179-7_57

A Recurrent Two-Stage Anatomy-Guided Network for Registration of Liver DCE-MRI

Wenjun Shen[1,2], Liyun Chen[1,2], Dongming Wei[1], Yuanfang Qiao[1], Yiqiang Zhan[2], Dinggang Shen[2], and Qian Wang[1(✉)]

[1] School of Biomedical Engineering, Shanghai Jiao Tong University, Shanghai, China
wang.qian@sjtu.edu.cn
[2] Shanghai United Imaging Intelligence Co., Ltd., Shanghai, China

Abstract. Registration of hepatic dynamic contrast-enhanced magnetic resonance imaging (DCE-MRI) series remains challenging, due to variable uptakes of the agent on different tissues or even the same tissues in the liver. The differences reflect on the intensity variations, which typically makes traditional intensity-based deformable registration methods fail to align small anatomical structures in liver such as vessels. Although deep-learning-based registration methods have become popular because of their superior efficiency for several years, registration of DCE-MRI series with dynamic intensity change is still under tackle. To solve this challenge, we present a two-stage registration network, in which the first stage aligns the whole liver and the second stage focuses on the registration of anatomical structures like vessels and tumors. Furthermore, we adopt a recurrent registration strategy for the deformation refinement. To evaluate our proposed method, we used clinical DCE-MRI series of 60 patients, and registered the arterial phase and the portal venous phase images onto the pre-contrast phases. Experimental results showed that the proposed method achieved a better registration performance than the traditional method (i.e., SyN) and the deep-learning-based method (i.e., VoxelMorph), especially in aligning anatomical structures such as vessel branches in liver.

Keywords: DCE-MRI · Deformable registration · Neural network

1 Introduction

Early diagnosis of liver cancer is particularly meaningful in reducing its high mortality. Dynamic contrast enhanced magnetic resonance imaging (DCE-MRI) typically involves a series of scanning, including the abdomen imaging before and after administration of a contrast agent. It is a noninvasive method for detecting and characterizing hepatic lesions, as it can provide the microcirculation and tissue characteristics in liver [1]. The longitudinal intensity changes in the DCE-MRI series indicate changes in the concentration of the contrast agent in the

© Springer Nature Switzerland AG 2021
C. Lian et al. (Eds.): MLMI 2021, LNCS 12966, pp. 219–227, 2021.
https://doi.org/10.1007/978-3-030-87589-3_23

tissue, which can be used to quantitatively analyze tumor perfusion and vascular anatomy [11].

During the acquisition process, although the patient is required to hold their breath to limit the breathing motion, the intensity inconsistency of DCE-MR series is inevitable due to patient motion (e.g., respiratory, heart motion, and bowel movement) [15]. Even the voxel-wise misalignment in adjacent time frames can interfere with the quantitative analysis of liver tissue characteristics in liver. Thus, the deformable registration of the DCE-MRI series is a crucial pre-processing step for the diagnosis of liver lesions. The goal of image registration is to compute a deformation field that transforms the moving image to be aligned with the fixed image. For decades, existing registration methods, such as SyN [2], diffeomorphic Demons [13], that compute the deformation field through iteratively maximizing the image similarity under some smoothness regularization. However, such methods are limited in the application stage by this time-consuming iterative process.

In recent years, deep-learning-based methods [9] have been proved to predict the accurate DVF in a short time. Some of these methods requires the ground-truth deformations to train the model in a supervised manner [4,5,12]. Since data annotation are tedious and laborious, unsupervised learning-based methods have been proposed. For example, VoxelMorph [4] uses unsupervised training framework with the help of a spatial transformer network (STN) [10] to regress the deformation based on the image similarity metric and deformation smoothness regularization without any need for annotation. However, such unsupervised method cannot be well applied in registering DCE-MRI series. It uses mean-square-error and cross-correlation as the image similarity metric, which cannot measure the similarity of DCE-MR series with dynamic intensity change. For solving the cross-modality registration, CK Guo et al. [8] involve the mutual information into the training loss. Although the similarity of images with different intensity distribution can be measured, the small structure (e.g., vessels in the liver) alignment is not well preserved as only global intensity information is considered.

This work mainly focuses on the registration between pre- and post-contrast phases of liver DCE-MRI. We proposed a two-stage network which incorporates anatomical constraints into the training process, and invoke the trained network in a recurrent way for the incremental refinement of deformation fields instead of using cascaded training [14]. Our contributions can be summarized as follows. (1) We aim to align different phases in DCE-MRI series using anatomical constraint to avoid the mis-alignment caused by intensity changes. The tumor region and vessel masks are introduced as the additional information to guide the registration, through which the network can focus its attention on the specific tissues. (2) The network is trained hierarchically through a two-stage framework, in which the first stage registers the global liver region while the second stage focuses on the local anatomical structures (e.g., vessel and tumor regions) corresponding with intensity changes. (3) The network is applied recurrently for incremental refinement, and the final deformation field is composed by all the incremental deformations. The experimental results show that our framework achieves a better performance on liver DCE-MRI series than baseline models, especially for vessel structures.

2 Method

2.1 Network Structures

Segmentation Network. We firstly segmented the liver from the whole abdomen as region of interest (ROI) through 3D U-Net [7] and the following work would be focus on the liver region. Furthermore, vessels are important structures in liver to locate tumor based on their relationships, but they are hard to align due to the variance intensity appearance between different time points in DCE-MRI series. To solve this issue, the masks of liver vessels play as the guidance for the alignment in our work. For this purpose, we trained a U-Net to segment vessels for the arterial phase and portal venous phase in a supervised manner. Note the segmentation network did not participate in the training of registration.

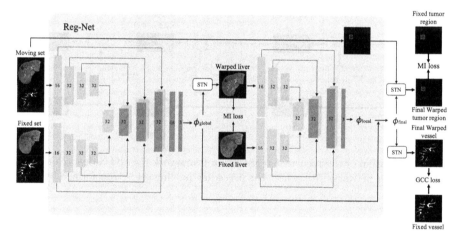

Fig. 1. Pipeline of the registration network (Reg-Net).

Alignment Network. The registration network, as shown in Fig. 1, includes two sub-networks, of which the first sub-network predicts a global deformation field ϕ_{global} to align the whole liver while the second sub-network focuses on the alignment of anatomical structures and predicts a deformation field ϕ_{local} based on the warped result ϕ_{global} for further local alignment. The input set includes two phase image I_m and I_f in liver DCE-MRI series and liver vessel segmentation masks V_m and V_f. In each sub-network, we use a dual-stream encoder architecture with shared parameters. The pair of the original image $(I_m$ or $I_f)$ and its segmentation mask $(V_m$ or $V_f)$ input into each encoder. Skip connections are applied for the combination of low-level features between encoder and decoder. Finally, we compose the ϕ_{global} that align global images and ϕ_{local} that align local anatomical structures together and obtain the final deformation field ϕ_{final} as follows,

$$\phi_{final} = \phi_{global} \circ \phi_{local}. \tag{1}$$

2.2   Recurrent Framework

Let S_m and S_f denote the input pairs that includes the moving set and the fixed set, respectively. For deformable registration, the pair of S_m and S_f is input into a pre-trained Reg-Net which shares the same weight as was trained in the registration network and predict a smooth deformation field ϕ_1 together with the warped moving set $S_m(\phi_1)$. Then the new input pairs $S_m(\phi_1)$ and S_f will be fed into the same Reg-Net to get the incremental deformation field ϕ_2. We recurrently perform above steps by n iterations, and the final deformation field can be given by $\phi = \phi_1 \circ \phi_2 \circ \cdots \phi_n$ (Fig. 2).

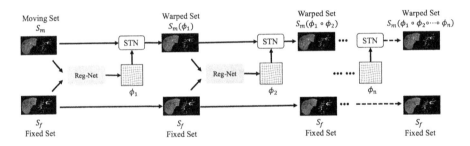

Fig. 2. Illustration of the recurrent registration framework [14]. STN denotes the spatial transformer networks [10], where the preceding moving set is transformed by deformation field ϕ.

2.3   The Training Strategy

The training is organized in two stages. First, we optimize the global registration sub-network using the loss function given by

$$\mathcal{L}_{global} = -\alpha_1 MI(I_m(\phi_{global}), I_f) + \alpha_2 R(\phi_{global}) + \alpha_3 J(\phi_{global}), \tag{2}$$

where $MI(\cdot)$ denotes the mutual Information (MI) loss implemented in [8], which approximated the common objective function mutual information in a differentiable way thus can be fused with the learning-based methods as loss function for handling the unlinear intensity relationship in DCE-MRI. $R(\cdot)$ is the L2-norm of the gradient of deformation field as defined in VoxelMorph. However, the gradient regularization of VoxelMorph is insufficient for avoid folding in recurrent registration. Therefore, we use Jacobian determinant $J(\cdot)$ as the anti-folding constraints to prevent folding in predicted transformation.

In the second stage, the second sub-network is trained to be focus on the alignment of local structures. The loss function is defined as

$$\mathcal{L}_{local} = -\beta_1 MI(T_m(\phi_{final}), T_f) + \beta_2 GCC(V_m(\phi_{final}), V_f) + \beta_3 R(\phi_{final}) + \beta_4 J(\phi_{final}). \tag{3}$$

Here, a manual labeled bounding boxes are used to acquire the tumor region. We take the union of the bounding boxes in moving and fixed images and dilate it, and the final region will be considered as the ROI of tumor. Then we compute and minimize the MI loss between ROI in the image pairs T_m and T_f and the results are represented as the first term in $\mathcal{L}_{local}$. For the vessel constraints, we apply a vessel-awareness gradient correlation coefficient (GCC) loss that measures the similarity of tubular-shaped tissue in liver as was implemented in [6]. The GCC loss calculate the horizontal and vertical normalized gradient correlation coefficient of two vessel masks which can describe the included angle between gradient vectors. In addition, α_1, α_2, α_3 in Eq. (2) and β_1, β_2, β_3, β_4 in Eq. (3) are hyper-parameters to balance the weight for each term.

3 Experiments

3.1 Dataset, Preprocessing and Implementation

The proposed method was evaluated on in-house 3D liver DCE-MRI scan series of 60 patients. Each DCE-MRI series contains pre-contrast T1-weighted, post-contrast T1-weighted scans of the arterial phase and portal venous phase. The preprocessing consists of: 1) cropping the liver from the origin image using liver mask, 2) applying affine registration based on the ANTs [3] package for linear registration, 3) normalizing the intensity to a range of $[0, 1]$, and 4)cropping and resampling all the images into size $256 \times 256 \times 96$ with $1.19 \times 1.19 \times 3$ mm^3 voxel resolution. The dataset was split into 48 for training and 12 for testing. Adam optimization was adopted with an initial learning rate of $1e-4$ and a decay rate of 0.01.

The proposed Reg-Net was implemented using Pytorch and trained on 2 NVIDIA Titan GPUs with batch size as 2. In this study, we focused on registering the post-contrast enhanced scans in the artery-phase and portal venous phase onto the pre-contrast scans.

3.2 Metrics for Evaluation

We quantified the performance through several metrics for different tissues. For the whole liver, we calculated the Dice Similarity Coefficient (DSC) based on the liver mask. For the vessel, Hausdorff distance and average symmetric surface distance (ASSD) were computed for evaluation. In addition, we manually annotated the tumor mask in the test dataset to quantify the registration of tumor via DSC and distance of tumor geometrical centers.

3.3   Comparison with Existing Methods

We performed the registration over the testing dataset and compared the result with SyN in the ANTs toolkit and VoxelMorph. For SyN and VoxelMorph, we used the cropped liver image as input and chose MI as objective and loss function.

The quantitative results for registration between different method are shown in Table 1. It can be observed that the metrics of the proposed anatomy-aware two-stage Reg-Net outperforms the VoxelMorph and SyN over each metric. After we recurrently performed Reg-Flow, the framework obtained significant improvement than the comparing methods, especially on the anatomical structures in liver such as vessels and tumors. It should be noted that the registration in the portal venous phase performed better than in the arterial phase, which is possibly caused by the fact that bright and dark vessels both appear in the arterial phase, which increases the difficulty of registration.

Table 1. The registration performance of SyN, VoxelMorph and the proposed method in the Arterial Phase and Portal Venous Phase. Iteration number of Resurrente Reg-Net is set to 7 according to our ablation studies.

Tissue	Metrics	Unregistered	Affine	SyN	VoxelMorph	Reg-Net	Recurrent Reg-Net
(a) Arterial phase							
Liver	DSC (%)	87.94 ± 7.94	94.68 ± 1.29	97.36 ± 0.56	97.50 ± 0.51	97.95 ± 0.39	$\mathbf{98.48 \pm 0.27}$
Vessel	Hausdorff (mm)	5.11 ± 1.79	4.37 ± 2.02	4.25 ± 1.90	4.26 ± 1.79	4.21 ± 1.92	$\mathbf{4.02 \pm 1.83}$
	ASSD (mm)	2.94 ± 1.57	1.94 ± 1.35	1.85 ± 1.30	1.83 ± 1.24	1.79 ± 0.62	$\mathbf{1.09 \pm 0.47}$
Tumor	DSC (%)	64.33 ± 25.43	78.10 ± 10.04	75.33 ± 10.56	78.18 ± 9.85	78.86 ± 8.85	$\mathbf{80.05 \pm 6.19}$
	Dist (mm)	4.58 ± 3.77	2.57 ± 1.97	2.54 ± 1.97	2.54 ± 2.12	2.34 ± 1.11	$\mathbf{2.03 \pm 0.85}$
(b) Portal venous phase							
Liver	DSC (%)	87.19 ± 6.69	94.37 ± 1.48	97.12 ± 0.56	97.37 ± 0.39	97.98 ± 0.33	$\mathbf{98.40 \pm 0.40}$
Vessel	Hausdorff (mm)	4.11 ± 1.34	2.69 ± 0.90	2.75 ± 0.75	2.77 ± 0.85	2.62 ± 0.84	$\mathbf{2.07 \pm 0.91}$
	ASSD (mm)	2.33 ± 1.15	1.33 ± 0.56	1.29 ± 0.53	1.31 ± 0.55	1.16 ± 0.50	$\mathbf{0.74 \pm 0.51}$
Tumor	DSC (%)	52.93 ± 24.29	78.10 ± 7.50	78.32 ± 6.65	79.24 ± 5.96	79.12 ± 4.50	$\mathbf{80.12 \pm 4.67}$
	Dist (mm)	5.07 ± 2.96	2.04 ± 0.87	1.99 ± 0.85	1.96 ± 1.06	1.92 ± 0.73	$\mathbf{1.65 \pm 0.69}$

The typical results are visually inspected in Fig. 3, confirming that the proposed framework obtains accurate alignment on anatomical structures compared with other methods.

We further evaluated the effects of the number of iterations in the recurrent framework. Table 2 shows that result varies with the growing of the iteration number. The performance achieved significantly gains before the 5th iteration and then tended to be stabilized. Figure 4 plots the result for better illustrating the trend. Note that for some metrics the performance decreased at 10th iteration such as the ASSD of vessel in the arterial phase, which may be caused by the amplified folding during recurrence. We chose $n = 10$ in arterial phase and $n = 7$ in portal venous phase for comparison since the performance is optimal at these points.

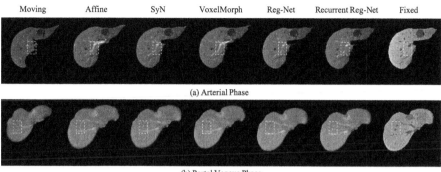

| Moving | Affine | SyN | VoxelMorph | Reg-Net | Recurrent Reg-Net | Fixed |

(a) Arterial Phase

(b) Portal Venous Phase

Fig. 3. Typical liver registration results of different methods for the arterial phase (a) and the portal venous phase (b). In (a) and (b), the first column is the post-contrast phase that was considered as the moving image while the last column is the pre-contrast phase that represented the fixed image.

Table 2. The comparison with different numbers of iterations in the recurrent framework.

Iteration numbers	Liver	Vessel		Tumor	
	DSC (%)	Hausdorff (mm)	ASSD (mm)	DSC (%)	Dist (mm)
(a) Arterial phase					
1	97.95 ± 0.39	4.21 ± 1.92	1.79 ± 0.62	78.86 ± 6.85	2.34 ± 1.08
2	98.29 ± 0.32	4.15 ± 1.92	1.53 ± 0.61	79.02 ± 7.21	2.19 ± 1.11
3	98.32 ± 0.25	4.11 ± 1.84	1.44 ± 0.52	79.16 ± 6.87	2.19 ± 1.11
5	98.36 ± 0.29	4.08 ± 1.76	1.18 ± 0.51	79.65 ± 6.64	2.08 ± 1.17
7	98.44 ± 0.23	4.05 ± 1.82	$\mathbf{1.07 \pm 0.50}$	79.94 ± 6.37	2.07 ± 0.97
10	$\mathbf{98.48 \pm 0.27}$	$\mathbf{4.02 \pm 1.83}$	1.09 ± 0.47	$\mathbf{80.05 \pm 6.19}$	$\mathbf{2.03 \pm 0.85}$
(b) Portal venous phase					
1	97.98 ± 0.33	2.62 ± 0.84	1.16 ± 0.50	79.12 ± 4.50	1.92 ± 0.73
2	98.07 ± 0.29	2.40 ± 0.96	1.03 ± 0.48	79.29 ± 4.79	1.88 ± 0.82
3	98.32 ± 0.27	2.16 ± 1.06	0.86 ± 0.45	79.54 ± 4.71	1.80 ± 0.75
5	98.39 ± 0.26	2.08 ± 1.09	0.78 ± 0.43	80.01 ± 4.03	1.68 ± 0.80
7	98.40 ± 0.40	$\mathbf{2.07 \pm 0.91}$	$\mathbf{0.74 \pm 0.51}$	$\mathbf{80.12 \pm 4.67}$	$\mathbf{1.65 \pm 0.69}$
10	$\mathbf{98.42 \pm 0.43}$	2.10 ± 0.88	0.75 ± 0.50	80.04 ± 4.21	1.67 ± 0.83

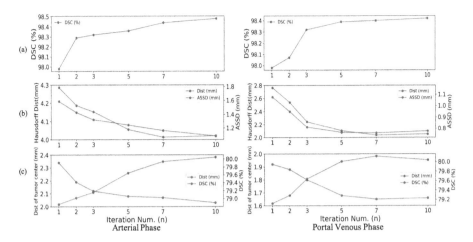

Fig. 4. Plot of the results with respect to the number of iterations, corresponding to the data in Table 2. (a) plots the DSC evaluated on the whole liver. (b) plots the Hausdorff distance and ASSD on liver vessels while (c) plots the center distance and DSC on tumors.

4   Conclusion

This paper presents a two-stage anatomy-guided learning-based registration network for liver DCE-MRI series. The first sub-network of the two-stage framework registers the global information and the second sub-network focuses on the vessel and tumor structures. Nevertheless, we applied a recurrent alignment framework for further refinement. Experimental results based on diverse evaluation metrics demonstrate that the proposed method achieves significant performance improvement over the baseline methods. The promised performance reveals the applicability of the learning-based registration method on the liver DCE-MRI series, and we expect that the proposed method can potentially be extended to the group-wise registration tasks.

Acknowledgement. This study has received funding by the National Natural Science Foundation of China (No. 91859107).

References

1. Aronhime, S., et al.: DCE-MRI of the liver: effect of linear and nonlinear conversions on hepatic perfusion quantification and reproducibility. J. Magn. Reson. Imaging **40**(1), 90–98 (2014)
2. Avants, B.B., Epstein, C.L., Grossman, M., Gee, J.C.: Symmetric diffeomorphic image registration with cross-correlation: evaluating automated labeling of elderly and neurodegenerative brain. Med. Image Anal. **12**(1), 26–41 (2008)

3. Avants, B.B., Tustison, N.J., Song, G., Cook, P.A., Klein, A., Gee, J.C.: A reproducible evaluation of ants similarity metric performance in brain image registration. Neuroimage **54**(3), 2033–2044 (2011)
4. Balakrishnan, G., Zhao, A., Sabuncu, M.R., Guttag, J., Dalca, A.V.: An unsupervised learning model for deformable medical image registration. In: Proceedings of the IEEE Conference on Computer Vision and Pattern Recognition, pp. 9252–9260 (2018)
5. Cao, X., et al.: Deformable image registration based on similarity-steered CNN regression. In: Descoteaux, M., Maier-Hein, L., Franz, A., Jannin, P., Collins, D., Duchesne, S. (eds.) Medical Image Computing and Computer Assisted Intervention, vol. 10433, pp. 300–308. Springer, Cham (2017). https://doi.org/10.1007/978-3-319-66182-7_35
6. Chen, L., et al.: Semantic hierarchy guided registration networks for intra-subject pulmonary CT image alignment. In: Martel, A.L. et al. (eds.) Medical Image Computing and Computer Assisted Intervention, vol. 12263, pp. 181–189. Springer, Cham (2020). https://doi.org/10.1007/978-3-030-59716-0_18
7. Çiçek, Ö., Abdulkadir, A., Lienkamp, S.S., Brox, T., Ronneberger, O.: 3D U-Net: learning dense volumetric segmentation from sparse annotation. In: Ourselin, S., Joskowicz, L., Sabuncu, M., Unal, G., Wells, W. (eds.) Medical Image Computing and Computer-Assisted Intervention, vol. 9901, pp. 424–432. Springer, Cham (2016). https://doi.org/10.1007/978-3-319-46723-8_49
8. Guo, C.K.: Multi-modal image registration with unsupervised deep learning. Ph.D. Thesis, Massachusetts Institute of Technology (2019)
9. Haskins, G., Kruger, U., Yan, P.: Deep learning in medical image registration: a survey. Mach. Vis. Appl. **31**(1), 1–18 (2020). https://doi.org/10.1007/s00138-020-01060-x
10. Jaderberg, M., Simonyan, K., Zisserman, A., Kavukcuoglu, K.: Spatial transformer networks. arXiv preprint arXiv:1506.02025 (2015)
11. Newatia, A., Khatri, G., Friedman, B., Hines, J.: Subtraction imaging: applications for nonvascular abdominal MRI. Am. J. Roentgenol. **188**(4), 1018–1025 (2007)
12. Rohé, M.M., Datar, M., Heimann, T., Sermesant, M., Pennec, X.: SVF-Net: learning deformable image registration using shape matching. In: Descoteaux, M., Maier-Hein, L., Franz, A., Jannin, P., Collins, D., Duchesne, S. (eds.) Medical Image Computing and Computer Assisted Intervention, vol. 10433, pp. 266–274. Springer, Cham (2017). https://doi.org/10.1007/978-3-319-66182-7_31
13. Vercauteren, T., Pennec, X., Perchant, A., Ayache, N.: Diffeomorphic demons: efficient non-parametric image registration. NeuroImage **45**(1), S61–S72 (2009)
14. Wei, D., et al.: An auto-context deformable registration network for infant brain MRI. arXiv preprint arXiv:2005.09230 (2020)
15. Wollny, G., Kellman, P., Santos, A., Ledesma-Carbayo, M.J.: Automatic motion compensation of free breathing acquired myocardial perfusion data by using independent component analysis. Med. Image Anal. **16**(5), 1015–1028 (2012)

Learning Infant Brain Developmental Connectivity for Cognitive Score Prediction

Yu Li[1], Jiale Cheng[1,2], Xin Zhang[1(✉)], Ruiyan Fang[1], Lufan Liao[1], Xinyao Ding[1], Hao Ni[3], Xiangmin Xu[1], Zhengwang Wu[2], Dan Hu[2], Weili Lin[2], Li Wang[2], John Gilmore[2], and Gang Li[2]

[1] South China University of Technology, Guangzhou, China
eexinzhang@scut.edu.cn
[2] University of North Carolina at Chapel Hill, Chapel Hill, USA
[3] University College London, London, UK

Abstract. During infancy, the human brain develops rapidly in terms of structure, function and cognition. The tight connection between cognitive skills and brain morphology motivates us to focus on individual level cognitive score prediction using longitudinal structural MRI data. In the early postnatal stage, the massive brain region connections contain some intrinsic topologies, such as small-worldness and modular organization. Accordingly, graph convolutional networks can be used to incorporate different region combinations to predict the infant cognitive scores. Nevertheless, the definition of the brain region connectivity remains a problem. In this work, we propose a crafted layer, the Inter-region Connectivity Module (ICM), to effectively build brain region connections in a data-driven manner. To further leverage the critical cues hidden in the development patterns, we choose path signature as the sequential data descriptor to extract the essential dynamic information of the region-wise growth trajectories. With these region-wise developmental features and the inter-region connectivity, a novel Cortical Developmental Connectivity Network (CDC-Net) is built. Experiments on a longitudinal infant dataset within 3 time points and hundreds of subjects show our superior performance, outperforming classical machine learning based methods and deep learning based algorithms.

Keywords: Infant cognition prediction · Brain region connectivity · Longitudinal analysis

1 Introduction

Infancy is an immense period to shape individuals' cognitive abilities [9]. Therefore, building the direct quantitative relationship between the longitudinal cor-

This work is supported by Science and Technology Program of Guangzhou (2018-1002-SF-0561) and Natural Science Foundation of Guangdong Province (2018A030313295) to X Z.

Y. Li and J. Cheng—Equal contribution.

C. Lian et al. (Eds.): MLMI 2021, LNCS 12966, pp. 228–237, 2021.
https://doi.org/10.1007/978-3-030-87589-3_24

tical structure and the cognitive scales is significant for us to better understand early brain development and related brain disorders. In this study, we aim to take longitudinal brain structural MRI scans as input and predict Mullen scales of early learning for each individual infant. Due to the challenges caused by the small sample size problem and the dynamic brain characteristics of infants, there are very few studies involving cognitive prediction [1,23].

Learning a set of compact representations, which effectively capture the spatial and temporal cortical developmental patterns, is one of the most important techniques to deal with the small sample size problem. However, existing methods usually simply flatten brain morphological feature vectors at every time point into a vector [1,23,24], which obviously neglect the potential connectivity among brain regions. It has been revealed that the massive brain region connections form elegant topologies, such as small-worldness and modular organization, which can be probed using the graph theoretical modeling method [25]. Hence, in recent works of functional MRI analysis [7,19], researchers build brain connectivity by Pearson's correlation. This kind of brain connectivity may also be constructed from structural MRI data and provide rich and useful information [6,18].

To explore the hidden dynamic information in growth trajectories, previous works utilize a set of learnable matrices to fuse the structural information of brain regions and get the predicted cognitive scales [1,23]. However, as brain regions develop hetergeneously during infancy [16], it may be beneficial to get a more effective temporal feature representation by designing a region-wise dynamic dependency descriptor for the developmental trajectory of each region. Path signature is a graded sequence of statistics to characterize streamed data. It has been applied to acquire effective features of the path [13,14,17,24], and might be helpful to extract the developmental information in brain regions.

In this work, our contributions are in three aspects. **1)** We innovatively propose a Inter-region Connectivity Module (ICM) to adaptively connect brain region pairs for constructing brain region developmental connectivity. **2)** Path signature is used to extract the developmental features to describe the growth trajectory of each brain region. **3)** Taking the learned connectivity as an adjacent matrix and the developmental path signature feature as the node feature, a Graph Convolutional Network (GCN) based predictor is leveraged to predict the cognitive scores. The whole structure is called the Cortical Developmental Connectivity Network (CDC-Net). Extensive experiments showed that our method achieves the state-of-the-art performance among various baselines.

2   Dataset and Feature Extraction

Longitudinal T1w and T2w brain MR images from 110 subjects at 0, 1 and 2 years of age were acquired. The T1w imaging parameters were TR/TE = 1900/4.38 ms, Flip Angle = $7°$, and isotropic 1 mm resolution; and the T2w imaging parameters were TR/TE = 7380/119 ms, Flip Angle = $150°$, and resolution = $1.25 \times 1.25 \times 1.95$ mm^3. All images were processed by an infant-tailored

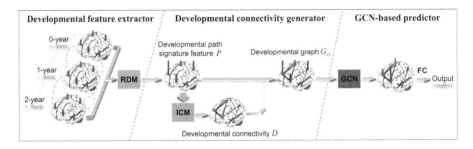

Fig. 1. Flowchart of our proposed method, which consists of three parts, including the developmental feature extractor, the developmental connectivity generator and a GCN-based predictor. Two modules: Region-wise Development Module (RDM) and Inter-region Connectivity Module (ICM), are proposed to extract developmental feature P and developmental connectivity D.

public computational pipeline[1]. For each hemisphere, inner and outer cortical surfaces were reconstructed [15] and 4 morphological features for each vertex were computed, including the cortical thickness, surface area, average convexity, and mean curvature. Then, we mapped the inner cortical surface onto a sphere using FreeSurfer [5] and aligned the spherical surface onto the 4D Infant Cortical Surface Atlas[2] to propagate the Desikan [4] parcellation with 70 regions from the atlas onto each indiviual surface. Finally, for each region, a 4-dimensional feature, i.e., the average cortical thickness, total surface area, average absolute convexity, and average absolute mean curvature were computed. Five Mullen cognitive scores of early learning are estimated for each participant at 2 years of age, i.e., Visual Receptive Scale (VRS), Fine Motor Scale (FMS), Receptive Language Scale (RLS), Expressive Language Scale (ELS) and Early Learning Composite (ELC).

3   CDC-Net

As illustrated in Fig. 1, our network consists of the development feature extractor, developmental connectivity generator and GCN-based predictor. Region-wise Development Module (RDM) is proposed to extract developmental features P of the longitudinal brain structural data. We select the developmental path signature features, which combines path signature theory and deep learning framework to generate *developmental path signature feature* for each brain region. Then, an Inter-region Connectivity Module (ICM) is proposed to learn the relationship between brain region pairs, and generate a developmental connectivity matrix D.

By considering brain regions as nodes, the developmental feature P can be seen as node features, and the developmental connectivity matrix D is actually

---

[1] http://www.ibeat.cloud.

[2] https://www.nitrc.org/projects/infantsurfatlas/.

an adjacent matrix of graph. By combining P and D, we acquire a complete developmental graph $G_D = (P, D)$ to represent the developmental brain network. Finally, a GCN-based score predictor which contains one 2-layer GCN and one fully connected (FC) layer is used to predict the cognitive score.

3.1   Region-Wise Development Module (RDM)

The region-wise development module is designed to extract the developmental features for each brain region, which is critical to build a meaningful developmental connectivity matrix D. We can summarize the function of RDM as follows:

$$P_i = f_{RDM}(X_i), \tag{1}$$

where $X \in \mathbb{R}^{70 \times 3 \times 4}$ represents our whole brain structural MRI data, $X_i \in \mathbb{R}^{3 \times 4}$ is longitudinal features of the ith region, and $P_i \in \mathbb{R}^{32}$ means the developmental feature of region i. The function $f_{RDM}(\cdot)$ can be any sequential feature extractor, which takes in the longitudinal feature of a brain region and returns its developmental features. In this work, we specifically introduce the path signature method as one of its instantiations.

Path Signature Preliminary. Suppose $\chi : [a, b] \rightarrow \mathbb{R}^d$ is a d-dimensional path defined on the time interval $[a, b]$. Conventionally, we regard the sequential data such as the region-wise growth trajectory, has a natural path-like structure. For any $t \in [a, b]$, $\chi_{(t)}$ can be written as $\{\chi_{(t)}^1, \chi_{(t)}^2, \cdots, \chi_{(t)}^n, \cdots, \chi_{(t)}^d\}$, where $\chi_{(t)}^n$ denotes the nth coordinate of $\chi_{(t)}$.

The signature of a path is a graded infinite series, which contains all the kth fold iterated integrals. Let $Sig_k(\chi)_{a,b}$ denoted the truncated signature of χ up to degree k as follows:

$$Sig_k(\chi)_{a,b} = (1, \ S_1(\chi)_{a,b}, \ S_2(\chi)_{a,b}, \cdots, \ S_k(\chi)_{a,b}), \tag{2}$$

with the kth fold iterated integral calculated as: $S_k(\chi)_{a,b}^{n_1, n_2, \cdots, n_k} = \frac{1}{k!} \prod_{j=1}^{k} (\chi_b^{n_j} - \chi_a^{n_j})$. $n_1, n_2, \cdots, n_k \in \{1, 2, \cdots, d\}$ are the indexes of coordinates.

It is noteworthy that the first fold iterated integral $S_1(\chi)_{a,b}$ equals the increment of path χ during a certain time period. The higher fold iterated integrals and path signature itself have many algebraic and analytic proprieties, which make it an effective feature set of the streamed data. More details about path signature can be found in [2].

Developmental Path Signature. In this work, we construct paths on the growth trajectories of brain regions to explore the geometrical properties of the developmental patterns. For each brain region, their structural feature of 0 year, 1 year and 2 years of age are constructed as a path of length 3 in chronological order. Inspired by [11] that path signature transformation can be integrated into the network as a layer, we propose our developmental path signature as,

$$P_i = \mathbf{W_3} \cdot [ReLU(Conv(\mathbf{W_1} \cdot X_i)) || (\mathbf{W_2} \cdot Sig_k(\mathbf{W_1} \cdot X_i))], \tag{3}$$

where $\mathbf{W_1}$, $\mathbf{W_2}$ and $\mathbf{W_3}$ are *learnable* matrices, which are introduced for a deep integration between path signature theory and machine learning. Sig_k computes the developmental path signature features as defined in equation (2).

3.2   Inter-region Connectivity Module (ICM)

Inspired by the fMRI-based connectivity, the relationship between pairs of brain regions may also benefit the sMRI data analyses. Considering the few sampling time points (only 3 here), the Pearson correlation coefficient cannot sufficiently describe the similarity between the growth trajectories of two brain regions. Thereby, with the developmental features extracted, we designed an Inter-region Connectivity Module (ICM) that learns a similarity coefficient for each pair of brain regions, as shown in Fig. 2.

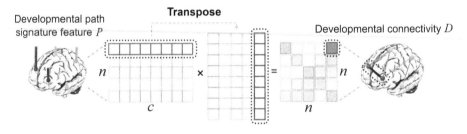

Fig. 2. Illustration of our ICM module. We calculate the cosine similarity of pairs of brain regions by multiplying feature matrices, and then construct a symmetric connectivity matrix D. Here n is the number of brain regions and c is the feature dimension.

With the developmental feature P from RDM, we evaluate the connectivity coefficient between the ith and j^{th} brain regions as follows:

$$f_{ICM}(P_i, P_j) = \frac{(\mathbf{W_{icm}} \cdot P_i) \cdot (\mathbf{W_{icm}} \cdot P_j)^\mathsf{T}}{||\mathbf{W_{icm}} \cdot P_i||_2 \cdot ||\mathbf{W_{icm}} \cdot P_j||_2}. \tag{4}$$

Here P_i and P_j means the developmental feature of region i and j. This equation essentially calculates the cosine similarity between the feature vector of $(\mathbf{W_{icm}} \cdot P_i)$ and $(\mathbf{W_{icm}} \cdot P_j)$. The *learnable* parameter $\mathbf{W_{icm}}$ is introduced to adaptively transform feature for the connectivity evaluation.

By calculating the similarity of all pairs of brain regions, we can get an adjacent matrix $D = \{d_{i,j} | d_{i,j} = f_{ICM}(P_i, P_j), d_{i,j} \in [-1, 1]\}$. It is based on developmental matrix P, so we call D the developmental connectivity. $f_{ICM}(P_i, P_j) = f_{ICM}(P_j, P_i)$ is obviously symmetric according to equation (4).

3.3   GCN-Based Score Predictor

Finally, we conduct the multi-layer graph convolution by using developmental connectivity $D \in \mathbb{R}^{70 \times 70}$ in Eq. (4) as the adjacent matrix, and developmental features $P \in \mathbb{R}^{70 \times 32}$ in Eq. (3) as node feature:

$$P^{l+1} = ReLU(D \cdot P^l \cdot \mathbf{W_{gcn}} + P^l), \tag{5}$$

where superscript l means the layer of feature, and $\mathbf{W_{gcn}}$ is a *learnable* matrix for GCN predictor. After a two-layer GCN, one fully connected layer is used to predict the cognitive score.

4   Experiments

As mentioned in Sect. 2, we have five cognitive Mullen Scales of early learning, including VRS, FMS, RLS, ELS and ELC. These five tasks are predicted together by the proposed method, and we used Root Mean Squared Error (RMSE) to evaluate the prediction error. L2 loss function is used, and all learnable parameters are initialized randomly and updated by BP algorithm. 5-fold cross validation was taken for comprehensive assessment of the performance.

4.1   Ablation Study

In this paper, our CDC-Net consists of two new modules: ICM to adaptively generate inter-region connectivity and RDM to extract the dynamic information of region-wise growth trajectories. Here, we will show the effectiveness of our ICM and developmental path signature by an ablation study.

From the GCN aspect, the connectivity matrix and node features both play important roles. As for connectivity matrix, we compared three types of matrices obtained by Pearson's correlation, graph attention layer (GAT) [22] and our ICM with raw data ($A = \{a_{i,j} | a_{i,j} = f_{ICM}(X_i, X_j)\}$) separately. As for node features, we used the raw feature X in Table 1. It can be observed that *learnable* matrices, including GAT and ICM work better than *static* Pearson's correlation matrix, confirming that our ICM can construct more meaningful brain region connectivity with few time points. Besides, benefited from the explicitly computation in Eq. (4), our ICM performs better than GAT. In Table 2, we compared different sequential data descriptor, including GRU [3], LSTM [8] and our Developmental Path Signature (DPS) to serve as the node features. GAT is applied to build the cross-region connectivity. By replacing the other sequential model with our developmental path signature, the overall performance of all five cognitive scores is improved, verifying the effectiveness of our DPS feature.

Table 1. Evaluation of adjacent matrices for GCN backbone.

Methods	RMSE					
	VRS	FMS	RLS	ELS	ELC	Average
Pearson	0.1855	0.1951	0.1680	0.1794	0.1757	0.1807 ± 0.0178
GAT [22]	0.1784	0.1946	0.1760	0.1771	0.1598	0.1757 ± 0.0203
ICM	**0.1728**	**0.1819**	**0.1451**	**0.1641**	**0.1586**	**0.1645 ± 0.0149**

Table 2. Evaluation of sequential data descriptor for node features.

Methods	RMSE					
	VRS	FMS	RLS	ELS	ELC	Average
LSTM [8]	**0.1726**	0.1867	0.1687	**0.1761**	0.1657	0.1740 ± 0.0193
GRU [3]	0.1734	0.1867	0.1687	0.1763	0.1686	0.1717 ± 0.0228
DPS	0.1727	**0.1832**	**0.1504**	0.1777	**0.1582**	**0.1684 ± 0.0192**

4.2 Comparison with State-of-the-Art Methods

In the past, some machine learning methods have been popular for analyzing the brain sMRI, e.g., KNN [10], SVR [20] and RF [21]. Recently, with the development of deep learning, more effective methods are available, e.g., LSTM [8], GRU [3], GCN [12] and GAT [22].

We compared our CDC-Net with other popular methods mentioned above. Table 3 shows the detailed RMSE of all methods and the ratio of performance improvement compared to KNN. The GCN-based methods perform better than the non-deep learning methods. With the assistance of sequential models to provide developmental information, the performance is further improved. Our proposed CDC-Net performs the best on almost all tasks by introducing the effective inter-region connectivity module and developmental path signature features. Considering the average performance of all tasks, our method reaches a state-of-the-art RMSE 0.1631. It is noteworthy that BrainPSNet [24] introduces the path signature in Eq. (2) to describe the development pattern of brain regions, but it is poor-performed for neglecting the graph structure among brain regions and the benefits may brought by the learnable matrices in Eq. (3).

Table 3. Comparison with state-of-the-art methods.

Methods	RMSE						
	VRS	FMS	RLS	ELS	ELC	Average	Ratio
KNN [10]	0.1878	0.1944	0.1680	0.1848	0.1767	0.1823 ± 0.0220	0%
SVR [20]	0.1865	0.1988	0.1692	0.1789	0.1814	0.1830 ± 0.0216	−0.4%
RF [21]	0.1892	0.1946	0.1666	0.1801	0.1824	0.1826 ± 0.0216	−0.2%
GCN [12]	0.1855	0.1951	0.1680	0.1794	0.1757	0.1807 ± 0.0178	+0.9%
GAT [22]	0.1784	0.1946	0.1760	0.1771	0.1598	0.1757 ± 0.0203	+3.7%
LSTM+GCN	**0.1765**	0.1805	0.1660	0.1799	0.1644	0.1735 ± 0.0193	+4.8%
GRU+GCN	0.1805	0.1895	0.1594	0.1683	0.1738	0.1743 ± 0.0247	+4.4%
BrainPSNet [24]	0.1829	0.1873	0.1587	0.1776	0.1680	0.1749 ± 0.0179	+4.2%
CDC-Net	0.1809	**0.1796**	**0.1451**	**0.1568**	**0.1529**	**0.1631 ± 0.0165**	**+11.5%**

4.3 Illustration of Brain Region Developmental Connectivity

By averaging the developmental connectivity matrices of all subjects, we can get the general brain region developmental connectivity, shown in Fig. 3. Figure 3(a) is the whole connectivity matrix, while Fig. 3(b) only shows the top 3% strong connections. Our 70 brain regions can be divided into 5 larger areas on each side of the brain, and details can be consulted in [4].

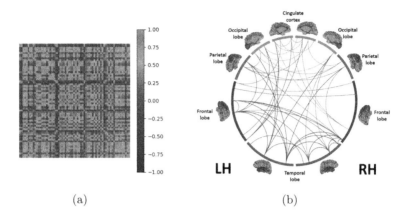

(a)   (b)

Fig. 3. Illustration of Brain Region Developmental Connectivity. (a) shows the whole connectivity matrix and (b) shows top 3% strong connections.

Figure 3(a) shows a phenomenon of modularization, indicating that brain regions in the same area have similar function. Considering the top 3% connections in Fig. 3(b), most of the connections are concentrated in temporal lobe and frontal lobe. Temporal lobe plays an important role in organizing language, while frontal lobe controls the execution of voluntary of muscle movement and other high-order functions. This is consistent with Mullen Scales assessment which mainly involves language and motor skills. The brain regions with the strongest connections incluce the precuneus cortex, transverse temporal cortex and cuneus cortex, which are all involved in vision or language.

5   Conclusion

In the task of infant cognitive score prediction based on longitudinal brain structure data, we proposed two novel modules: Region-wise Developmental Module (RDM) and Inter-region Connectivity Module (ICM). RDM is the feature extractor to explore the developmental information of brain regions, and ICM is the connectivity matrix builder to construct connections between brain regions. Our proposed method with RDM, ICM and 2-layer GCN backbone obtains the state-of-the-art prediction result. Furthermore, by visualizing the brain region developmental connectivity learned by RDM and ICM, we find that several brain regions associated with cognitive ability are connected, demonstrating the rationality of our proposed framework.

References

1. Adeli, E., Meng, Y., Li, G., et al.: Multi-task prediction of infant cognitive scores from longitudinal incomplete neuroimaging data. NeuroImage **185**, 783–792 (2019)
2. Chevyrev, I., Kormilitzin, A.: A primer on the signature method in machine learning. arXiv preprint arXiv:1603.03788 (2016)
3. Cho, K., Van Merriënboer, B., Gulcehre, C., et al.: Learning phrase representations using RNN encoder-decoder for statistical machine translation. arXiv preprint arXiv:1406.1078 (2014)
4. Desikan, R.S., Ségonne, F., Fischl, B., et al.: An automated labeling system for subdividing the human cerebral cortex on MRI scans into gyral based regions of interest. NeuroImage **31**(3), 968–980 (2006)
5. Fischl, B., Sereno, M.I., Dale, A.M.: Cortical surface-based analysis: II: inflation, flattening, and a surface-based coordinate system. NeuroImage **9**(2), 195–207 (1999)
6. Ghribi, O., Li, G., Lin, W., et al: Multi-regression based supervised sample selection for predicting baby connectome evolution trajectory from neonatal timepoint. Med. Image Anal. **68**, 101853 (2021)
7. Griffanti, L., Rolinski, M., Szewczyk-Krolikowski, K., et al.: Challenges in the reproducibility of clinical studies with resting state FMRI: an example in early Parkinson's disease. NeuroImage **124**, 704–713 (2016)
8. Hochreiter, S., Schmidhuber, J.: Long short-term memory. Neural Comput. **9**(8), 1735–1780 (1997)
9. Kagan, J., Herschkowitz, N.: A Young Mind in a Growing Brain. Psychology Press, Hove (2006)
10. Keller, J.M., Gray, M.R., Givens, J.A.: A fuzzy k-nearest neighbor algorithm. IEEE Trans. Syst. Man Cybern. **SMC-15**(4), 580–585 (1985)
11. Kidger, P., Bonnier, P., Perez Arribas, I., et al.: Deep signature transforms. Adv. Neural Inf. Process. Syst. **32**, 3105–3115 (2019)
12. Kipf, T.N., Welling, M.: Semi-supervised classification with graph convolutional networks. In: Proceedings of International Conference on Learning Representations (2017)
13. Lai, S., Zhu, Y., Jin, L.: Encoding Pathlet and SIFT features with bagged VLAD for historical writer identification. IEEE Trans. Inf. Forensics Secur. **15**, 3553–3566 (2020)
14. Li, C., Zhang, X., Liao, L., et al.: Skeleton-based gesture recognition using several fully connected layers with path signature features and temporal transformer module. In: Proceedings of the AAAI Conference on Artificial Intelligence, pp. 8585–8593 (2019)
15. Li, G., Nie, J., Wu, G., et al.: Consistent reconstruction of cortical surfaces from longitudinal brain MR images. NeuroImage **59**(4), 3805–3820 (2012)
16. Li, G., Wang, L., Shi, F., et al.: Mapping longitudinal development of local cortical gyrification in infants from birth to 2 years of age. J. Neurosci. **34**(12), 4228–4238 (2014)
17. Liao, L., Zhang, X., Li, C.: Multi-path convolutional neural network based on rectangular kernel with path signature features for gesture recognition. In: Proceedings of IEEE Visual Communications and Image Processing. IEEE (2019)
18. Seidlitz, J., Váša, F., Shinn, M., et al.: Morphometric similarity networks detect microscale cortical organization and predict inter-individual cognitive variation. Neuron **97**(1), 231–247 (2018)

19. Smith, S.M., Vidaurre, D., Beckmann, C.F., et al.: Functional connectomics from resting-state FMRI. Trends Cogn. Sci. **17**(12), 666–682 (2013)
20. Smola, A.J., Schölkopf, B.: A tutorial on support vector regression. Stat. Comput. **14**(3), 199–222 (2004). https://doi.org/10.1023/B:STCO.0000035301.49549.88
21. Svetnik, V., Liaw, A., Tong, C., et al.: Random forest: a classification and regression tool for compound classification and QSAR modeling. J. Chem. Inf. Comput. Sci. **43**(6), 1947–1958 (2003)
22. Veličković, P., Cucurull, G., Casanova, A., et al.: Graph attention networks. In: Proceedings of International Conference on Learning Representations (2018)
23. Zhang, C., Adeli, E., Wu, Z., et al.: Infant brain development prediction with latent partial multi-view representation learning. IEEE Trans. Med. Imaging **38**(4), 909–918 (2018)
24. Zhang, X., Cheng, J., Ni, H., et al.: Infant cognitive scores prediction with multistream attention-based temporal path signature features. In: Martel, A.L. et al. (eds.) Medical Image Computing and Computer Assisted Intervention, vol. 12267, pp. 134–144. Springer, Cham (2020). https://doi.org/10.1007/978-3-030-59728-3_14
25. Zhao, T., Xu, Y., He, Y.: Graph theoretical modeling of baby brain networks. NeuroImage **185**, 711–727 (2019)

Hierarchical 3D Feature Learning for Pancreas Segmentation

Federica Proietto Salanitri[1]([✉]), Giovanni Bellitto[1], Ismail Irmakci[2,3], Simone Palazzo[1], Ulas Bagci[3], and Concetto Spampinato[1]

[1] PeRCeiVe Lab, University of Catania, Catania, Italy
federica.proiettosalanitri@phd.unict.it
[2] CE, Ege University, Izmir, Turkey
[3] Department of Radiology and BME, Northwestern University, Chicago, IL, USA

Abstract. We propose a novel 3D fully convolutional deep network for auto-mated pancreas segmentation from both MRI and CT scans. More specifically, the proposed model consists of a 3D encoder that learns to extract volume features at different scales; features taken at different points of the encoder hierarchy are then sent to multiple 3D decoders that individually predict intermediate segmentation maps. Finally, all segmentation maps are combined to obtain a unique detailed segmentation mask. We test our model on both CT and MRI imaging data: the publicly available NIH Pancreas-CT dataset (consisting of 82 contrast-enhanced CTs) and a private MRI dataset (consisting of 40 MRI scans). Experimental results show that our model outperforms existing methods on CT pancreas segmentation, obtaining an average Dice score of about 88%, and yields promising segmentation performance on a very challenging MRI data set (average Dice score is about 77%). Additional control experiments demonstrate that the achieved performance is due to the combination of our 3D fully-convolutional deep network and the hierarchical representation decoding, thus substantiating our architectural design.

Keywords: CT and MRI pancreas segmentation · Fully convolutional neural networks · Hierarchical encoder-decoder architecture

1 Introduction

Pancreatic cancer is a growing public health concern worldwide. In 2021, an estimated 60,430 new cases of pancreatic cancer will be diagnosed in the US and 48,220 people will die from this disease [19]. Early detection of pancreas cancer [14] is very hard and options in treatment are very limited. Radiology imaging and automated image analysis play key roles in diagnosis, prognosis, treatment, and intervention of pancreatic diseases; thus, there is a strong, unmet, need for computer aided analysis tools supporting these tasks. The first step in such analysis is to automate the medical image segmentation procedures, since manual segmentation (current standard) is tedious, prone to error, and it is not practical in routine clinical evaluation of the diseases [20]. Beyond the known challenges of medical image segmentation problems, pancreas is one of the most difficult organs to segment despite the recent advances in deep segmentation models.

© Springer Nature Switzerland AG 2021
C. Lian et al. (Eds.): MLMI 2021, LNCS 12966, pp. 238–247, 2021.
https://doi.org/10.1007/978-3-030-87589-3_25

Computed tomography (CT) and magnetic resonance imaging (MRI) are the two most common modalities for pancreas imaging. CT is the modality of choice for pancreatic cancer at the moment, while MRI is mostly used for finding other pancreatic diseases including cysts and diabetes. Compared to CT, MRI has advantages such as the lack of ionizing radiation, better resolution and soft tissue contrast. However, MRI has other unique difficulties, including field inhomogeneity, non-standard intensity distributions due to variations in scanners, patients, field strengths, and high similarity in pancreas and non-pancreas tissue densities.

Image-based pancreas analysis is by itself a challenging task. Shapes and sizes greatly vary across different patients, making it difficult to use robust priors for improving the delineation procedures. Intensity similarities to non-pancreatic tissues, and smooth or invisible boundaries (due to resolution limitations of medical scanners) are other challenges that need to be addressed in a successful segmentation method. Moreover, in presence of a cyst, tumor, or other abnormalities in pancreases, segmentation algorithms may easily fail to delineate correct boundaries.

To address these challenges, in this work we propose a novel 3D fully convolutional encoder-decoder network with hierarchical multi-scale feature learning, for general, fully-automated pancreas segmentation applicable to CT and MRI scans. Major contributions of this study are the following:

- Our segmentation network is unique in the sense that it is volumetric, learns to extract 3D volume features at different scales, and decodes features hierarchically, leading to improved segmentation results;
- We show the efficacy of our work both on CT and MRI scans. Our architecture successfully extracts pancreases from CT and MRI with high accuracy, obtaining new state-of-the-art results on a publicly-available CT benchmark and first-ever volumetric pancreas segmentation from MRI in the literature.
- Our work on MRI pancreas segmentation is an important application contribution, due to the very limited published research on this task using MRI data with deep learning. It is our belief that our method provides a significant state-of-the-art baseline to be compared with for further MRI pancreas research.

2   Related Work

Following the success of deep learning methods applied in medical image segmentation, researchers have recently shown an increasing interest in pancreas segmentation, in order to support physicians in early stage diagnosis for pancreas cancer. Although this application field is still in its infancy—also due to variabilities in texture, size and imaging contrast—a line of promising approaches has been proposed in the literature, mainly on CT scans [2, 8, 10–12, 15–17, 21, 22, 24, 25]. We here describe the most significant ones which relate to our proposed model.

In [16], a two-stage cascaded approach for pancreas localization and pancreas segmentation is proposed. In the first stage, the method localizes the pancreas in the entire 3D CT scan, providing a reliable bounding box for a more refined segmentation step, based on an efficient application of holistically-nested convolutional networks (HNNs) on the three views of pancreas CT image. Per-pixel probability maps are then fused to produce a 3D bounding box of the pancreas. Projective adversarial networks [8] incorporate high-level 3D information through 2D projections and introduce an attention module that supports a selective integration of global information from the segmentation module to an adversarial network. More recently, [22] proposes a dual-input v-mesh fully-convolutional network, which receives original CT scans and images processed by contrast-specific graph-based visual saliency, in order to enhance the soft tissue contrast and highlight differences among local regions in abdominal CT scans.

All of the above works tackle the problem of pancreas segmentation on CT scans. However, as already mentioned, MRI acquisitions have several advantages over CT—most importantly, fewer risks to the patients. On the other hand, MRI pancreas segmentation presents additional challenges to automated visual analysis. For this reason and others (e.g., the lack of public benchmarks), very few works have addressed pancreas segmentation on MRI data: to the best of our knowledge, the major attempts are [1–3]. In [3], two CNN models are combined to perform, respectively, tissue detection and boundary detection; the results are provided as input to a conditional random field (CRF) for final segmentation. In [1], an algorithmic approach based on hand-crafted features is proposed, employing an ad-hoc multi-stage pipeline: contrast enhancement within coarsely detected pancreas regions is applied to differentiate between pancreatic and surrounding tissue; 3D segmentation and edge detection through max-flow and min-cuts approach and structured forest are performed; finally, non-pancreatic contours are removed via morphological operations on area, structure and connectivity.

3   Method

Our 3D fully-convolutional pancreas segmentation model—*PankNet*—is based on an encoder-decoder architecture; however, unlike standard encoder-decoder schemes with a single decoding path (see Fig. 1a), we have parallel decoders at different abstraction levels, generating multiple intermediate segmentation maps (Fig. 1c). Hierarchical decoding is also fundamentally different from using skip connections (Fig. 1b), since these have the purpose to ease gradient flow and forward low-level features for output reconstruction, while our multiple decoders aim to extract local and global dependencies. The detailed architecture is shown in Fig. 2: the input data (either CT or MRI volume) is first processed by the encoder stream of the model which aggregates volumetric features at different abstraction levels. These features are then given as input to

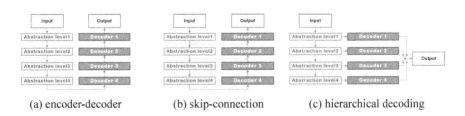

(a) encoder-decoder   (b) skip-connection   (c) hierarchical decoding

Fig. 1. A comparison between our proposed architecture and other types of networks used for segmentation: (a) standard encoder–decoder architecture; (b) encoder–decoder architecture with skip connections; (c) encoder–hierarchical decoder architecture (ours).

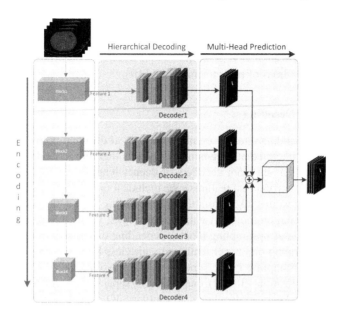

Fig. 2. *PanKNet* architecture: the encoding path extracts aggregated volumetric features, while the decoding path predicts four different intermediate segmentation masks (coarse to fine). Finally, intermediate segmentations are integrated into a detailed output mask. (Color figure online)

different decoder streams, each generating a segmentation mask volume. All intermediate masks are concatenated along the channel dimension and finally merged through a convolutional layer in order to predict the final segmentation mask for all input slices.

3.1 Volume Feature Encoding

The model's encoder performs aggregation of volumetric features from the input data. It is based on S3D [23], a network originally proposed for action recognition using 3D spatial and temporal separable 3D convolution layers, pretrained on the Kinetics Dataset [6]. We use the pretrained network, similarly to other works [3,8,9], to ease convergence given the limited training data we have from both CT and MRI datasets. Our encoder processes $D = 48$ slices from an input scan by progressively aggregating volumetric cues down to a more compact representation of size $1024 \times \frac{W}{8} \times \frac{H}{32} \times \frac{D}{32}$ (channels $\times$ width $\times$ height $\times$ depth). Features at the bottleneck and at the outputs of the second, third and fourth pooling layers are fed to separate decoders, described in the following section, to implement our hierarchical decoding strategy.

The proposed approach can be easily adapted to different encoder architectures. Thus, we additionally design a lightweight variant of our *PanKNet* network by replacing the S3D-based encoder with an encoder based on MobileNetV2 [18], where 2D convolutions are replaced with 3D ones through inflation. In particular, the 2D kernels are replicated along the third dimension, and the values of the weights are divided by the number of replications as proposed in [4]. In this case, as input to the decoders, we select the output of the second, third, fourth and sixth *bottleneck* blocks of MobileNetV2, providing a more compact feature map of size $160 \times \frac{W}{16} \times \frac{H}{32} \times \frac{D}{32}$. This lightweight variant has 10 times fewer parameters (2.5 millions of parameters, 9.33 MB) and than the S3D counterpart (25.6 millions of parameters, 97.88 MB).

3.2   Hierarchical Decoding

Our hierarchical decoding strategy employs features at different points of the encoder stream to generate intermediate segmentation masks that aim to capture and combine fine segmentation (derived from decoders of deeper features) to coarse segmentation (derived from decoders of initial features). We include four decoders: each one processes a set of volumetric features taken from the corresponding level in the encoder stack and performs segmentation on the input volume (see Fig. 2, yellow blocks). Each decoder consists of a cascade of upsampling blocks, depending on the size of the input feature map: decoders operating on deeper features require less blocks to recover the original input size. Each upsampling block contains a 3D convolutional block (convolutional layer + batch normalization + ReLU), one or two 3D separable convolutional blocks, and a trilinear upsample layer. As last layer, a pointwise 3D convolution outputs a volume with size $2 \times W \times H \times D$, where W, H and D are the same as the input volume.

3.3   Pancreas Segmentation

Intermediate segmentation maps predicted by each of the model's decoders are combined into a global mask. In particular, the four intermediate maps are concatenated into a $8 \times W \times H \times D$ tensor, which then goes through a last layer performing a voxel-wise convolution to generate a single segmentation map of size $2 \times W \times H \times D$.

The whole model (encoder, hierarchical decoders and output layer) is trained end-to-end using a hierarchical Dice loss [13] between ground-truth mask, intermediate generated masks and the output segmentation mask. Formally, given the predicted output segmentation masks $\mathbf{S}_v$ for the input volume, the four maps $\hat{\mathbf{S}}_{v^i}$ estimated by the decoders, and the ground-truth segmentation maps $\mathbf{G}_v$ for the input data, the *segmentation loss* $\mathcal{L}_s$ is:

$$\mathcal{L}_s\left(\mathbf{S}_v, \hat{\mathbf{S}}_{v^i}, \mathbf{G}_v\right) = \sum_{i=1}^{4} \frac{2\sum_j \hat{S}_{v^{i,j}} G_{v^j}}{\sum_j \hat{S}_{v^{i,j}}^2 + \sum_j G_{v^j}^2} + \frac{2\sum_j S_{v^j} G_{v^j}}{\sum_j S_{v^j}^2 + \sum_j G_{v^j}^2} \tag{1}$$

where index i iterates over the four intermediate maps and index j iterates over voxels.

4   Experiments

4.1   Dataset

We evaluate the accuracy of our proposed deep segmentation method in both CT and MRI modalities. For the former, we use the publicly available NIH Pancreas-CT dataset, which is the most used pancreas segmentation dataset for benchmarking [15]. This dataset includes 82 abdominal contrast-enhanced 3D CT scans. The resolution of the

CT scans is $512 \times 512 \times Z$, with Z (between 181 and 466) indicating the number of slices along the transverse axis. Voxel spacing ranges from 0.5 mm to 1 mm. More details on this dataset are available in [15].

In our experiments with MRI data, we use 40 in-house collected T2-weighted MRI scans from 40 patients, who have either IPMN (intraductal papillary mucinous neoplasm) cysts detected in their pancreases or invasive pancreatic ductal carcinoma. Two expert radiologists annotated pancreases manually and consensus segmentation masks were generated at the end of the ground-truth labeling procedure with agreement. MRI images were resized (in the transverse plane) to 256×256 pixels, with voxel spacing of varying from 0.468 mm to 1.406 mm. To minimize uncertainties in MRI scans, we applied a set of pre-processing steps: N4 bias field correction followed by an edge-preserving Gaussian smoothing, and intensity standardization procedure to standardize MRI scans across patients, scanners, and time.

4.2   Training and Evaluation Procedure

We apply the same training procedure for the two datasets, with the only difference regarding how model backbones are pre-trained. On the NIH Pancreas-CT dataset, we pre-train S3D on Kinetics [6] and MobileNetV2 on ImageNet [5] with weight inflation; on our MRI data, *Pancreas-MRI*, we employ the backbones pre-trained on the CT task.

Input CT and MRI scans are re-oriented using the RAS axes convention for consistency. We then perform voxel resampling through trilinear interpolation in order to have isotropic (1 mm) voxel spacing, and normalize the values of each scan between 0 and 1. During training, data augmentation is performed with random horizontal flipping, random 90° rotation and random crops of size $128 \times 128 \times 48$ (in RAS coordinates). We minimize our multi-part Dice loss with mini-batch gradient descent using the Adam optimizer (learning rate: 0.001) and batch size 8, for a total of 3000 epochs.

At inference time, we compute output segmentation masks by running a sliding window routine over an entire input scan, using $256 \times 256 \times 48$ windows overlapping by 25%. Voxel labels from overlapping segmentations are obtained by averaging the set of predictions. For evaluation, we carry out 4-fold cross-validation. At each iteration, the set of training folds is further split into the actual training set and a validation set, that is used to select the epoch at which Dice score on the test fold is reported. As metrics for quantitative evaluation, we employ: *Dice score coefficient* (DSC), *Positive Predictive Value* (PPV) and *Sensitivity*.

Experiments are performed on an NVIDIA Quadro P6000 GPU. The proposed approach was implemented in PyTorch and MONAI; all code will be publicly released.

4.3   Results

We first test our model (as well as its lightweight variant) on the NIH Pancreas-CT dataset and compare it to existing methods (which share our evaluation strategy with 4-fold cross-validation), namely, [2, 8, 10–12, 15–17, 21, 22, 24, 25]. Summarized in Table 1, our results indicate that *PanKNet* outperforms existing methods over different metrics. Note that PanKNet does not require any auxiliary regularization networks [8], nor additional inputs [22], nor upstream pancreas localization module [12]. Remarkably,

Table 1. Comparison of *PanKNet* against multiple state-of-the-art models for pancreas segmentation on NIH Pancreas-CT dataset using 4-fold cross-validation. **Best performance in bold,** *second best in italic.*

Method	DSC			PPV	SENS
	Avg	Max	Min		
Roth et al. [15]	71.42 ± 10.11	86.29	23.99	–	–
Roth et al. [16]	78.01 ± 8.20	88.65	34.11	–	–
Roth et al. [17]	81.27 ± 6.27	88.96	50.69	–	–
Zhou et al. [25]	82.37 ± 5.68	90.85	62.43	–	–
Cai et al. [2]	82.40 ± 6.70	90.10	60.00	–	–
Li et al. (2019) [10]	83.50 ± 6.20	–	–	84.50 ± 6.90	83.70 ± 10.40
Liu et al. (2020) [11]	84.10 ± 4.90	–	–	83.60 ± 5.90	85.30 ± 08.20
You et al. [24]	84.50 ± 4.97	91.02	62.81	–	–
Khosravan et al. [8]	85.53 ± 1.23	88.71	83.20	–	–
Wang et al. (2020) [21]	85.90 ± 3.40	–	–	–	–
Man et al. (2019) [12]	86.90 ± 4.90	–	–	–	–
Wang et al. [22]	87.04 ± 6.80	–	–	$\mathbf{89.50 \pm 5.80}$	87.70 ± 7.90
PanKNet$_{Light}$	*87.13 ± 4.58*	93.49	72.77	86.85 ± 6.52	*88.48 ± 5.12*
PanKNet	$\mathbf{88.01 \pm 4.74}$	93.84	70.62	*88.25 ± 5.45*	$\mathbf{88.69 \pm 5.99}$

even the lightweight variant of PanKNet yields accuracy comparable to the full model, while outperforming existing models, showing that the choice of the backbone is not as important as the overall employed hierarchical architecture. The best trade-off between accuracy and computational resources for CT pancreas segmentation is represented by PanKNet$_{Light}$, whose memory occupation is about 10 MB compared to about 100 MB of PanKNet, but with very similar performance.

We then test our model on pancreas segmentation from MRI data. In this case, we compare the 3D-UNet, proposed in [7], pre-trained on the NIH Pancreas-CT dataset and fine-tuned on our MRI dataset. Furthermore, we add to this evaluation some control experiments to show the effectiveness of the designed architecture. Consequently, we define as baseline our encoder-decoder architecture without hierarchical decoding strategy, decoding only the features at the model's bottleneck. Results in Table 2 indicate that both PanKNet variants outperform the state-of-the-art 3D U-Net model [7]. The baseline (with either backbones) also performs better than 3D U-Net model [7] demonstrating that even our 3D fully convolutional network, ablated from the hierarchical decoding, is effective for MRI pancreas segmentation. Adding hierarchical decoding leads to enhanced segmentation performance, especially on DSC and PPV. Different from CT segmentation and from baseline models, PanKNet largely outperforms its lightweight counterpart, demonstrating that MRI pancreas segmentation is far more complex and challenging than CT segmentation and calls for high-capacity networks to be solved.

Example segmentation masks, corresponding to the highest and lowest Dice scores reported in Tables 1 and 2 for CT and MRI pancreas segmentation, are illustrated in Fig. 3.

Table 2. Segmentation performance on Pancreas-MRI dataset (4-fold CV).

Method	DSC			PPV	SENS
	Avg	Max	Min		
3D-UNet [7]	65.05 ± 9.17	84.58	49.80	61.55 ± 7.55	74.42 ± 13.99
Baseline$_{Light}$	69.17 ± 8.10	83.86	49.92	64.64 ± 7.49	84.19 ± 11.72
Baseline	65.16 ± 9.11	84.00	49.49	61.92 ± 8.22	75.22 ± 12.46
PanKNet$_{Light}$	72.96 ± 10.33	88.54	49.90	71.39 ± 11.21	79.76 ± 11.53
PanKNet	**77.46 ± 08.62**	**89.07**	**52.30**	**76.63 ± 08.66**	**80.91 ± 10.51**

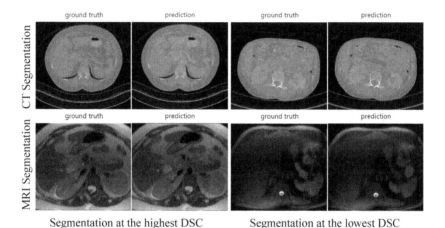

Fig. 3. Segmentation masks at the highest (left column) and lowest Dice score (right column) on NIH Pancreas-CT (first row) and Pancreas-MRI dataset (second row).

5   Conclusion

In this study, we propose a novel 3D fully-convolutional network for pancreas segmentation from MRI and CT scans. Our proposed deep network aims at learning and combining multi-scale features, namely a hierarchical decoding strategy, to generate intermediate segmentation masks for a coarse-to-fine segmentation process. The intermediate masks, capturing fine details, are derived from decoders of deeper features while coarse segmentation details are derived from decoders of initial features. We evaluated the efficacy of our method (a) on CT scans from the publicly available NIH CT-Pancreas benchmark, and obtained a new state of the art Dice score **88.01%**, outperforming all previous methods; and (b) on MRI scans, obtaining a Dice score of **77.46%**, which can be used as a baseline for future works on MRI pancreas segmentation. Noting that MRI pancreas segmentation methods are extremely limited due to the challenging nature of the problem, our study offers a fresh insight into MRI analysis of pancreas from a fully automated volumetric segmentation strategy. PanKNet is tested for pancreas segmentation, but its architecture is general and can be applied to any 3D object segmentation problem in medical domain.

References

1. Asaturyan, H., Gligorievski, A., Villarini, B.: Morphological and multi-level geometrical descriptor analysis in CT and MRI volumes for automatic pancreas segmentation. Comput. Med. Imaging Graph. **75**, 1–13 (2019)
2. Cai, J., Lu, L., Xie, Y., Xing, F., Yang, L.: Improving deep pancreas segmentation in CT and MRI images via recurrent neural contextual learning and direct loss function. arXiv preprint arXiv:1707.04912 (2017)
3. Cai, J., Lu, L., Xie, Y., Xing, F., Yang, L.: Pancreas segmentation in MRI using graph-based decision fusion on convolutional neural networks. In: Descoteaux, M., Maier-Hein, L., Franz, A., Jannin, P., Collins, D.L., Duchesne, S. (eds.) MICCAI 2017. LNCS, vol. 10435, pp. 674–682. Springer, Cham (2017). https://doi.org/10.1007/978-3-319-66179-7_77
4. Carreira, J., Zisserman, A.: Quo vadis, action recognition? A new model and the kinetics dataset. In: CVPR, pp. 6299–6308 (2017)
5. Deng, J., Dong, W., Socher, R., Li, L., Li, K., Li, F.: ImageNet: a large-scale hierarchical image database. In: Computer Society Conference on Computer Vision and Pattern Recognition, pp. 248–255 (2009). https://doi.org/10.1109/CVPR.2009.5206848
6. Kay, W., et al.: The kinetics human action video dataset. arXiv preprint arXiv:1705.06950 (2017)
7. Kerfoot, E., Clough, J., Oksuz, I., Lee, J., King, A.P., Schnabel, J.A.: Left-ventricle quantification using residual U-Net. In: Pop, M., et al. (eds.) STACOM 2018. LNCS, vol. 11395, pp. 371–380. Springer, Cham (2019). https://doi.org/10.1007/978-3-030-12029-0_40
8. Khosravan, N., Mortazi, A., Wallace, M., Bagci, U.: PAN: projective adversarial network for medical image segmentation. In: Shen, D., et al. (eds.) MICCAI 2019. LNCS, vol. 11769, pp. 68–76. Springer, Cham (2019). https://doi.org/10.1007/978-3-030-32226-7_8
9. LaLonde, R., et al.: INN: inflated neural networks for IPMN diagnosis. In: Shen, D., et al. (eds.) MICCAI 2019. LNCS, vol. 11768, pp. 101–109. Springer, Cham (2019). https://doi.org/10.1007/978-3-030-32254-0_12
10. Li, H., Lü, Q., Chen, G., Huang, T., Dong, Z.: Convergence of distributed accelerated algorithm over unbalanced directed networks. IEEE Trans. Syst. Man Cybern. Syst., 1–12 (2019). https://doi.org/10.1109/TSMC.2019.2946287
11. Liu, S., et al.: Automatic pancreas segmentation via coarse location and ensemble learning. IEEE Access **8**, 2906–2914 (2020). https://doi.org/10.1109/ACCESS.2019.2961125
12. Man, Y., Huang, Y., Feng, J., Li, X., Wu, F.: Deep Q learning driven CT pancreas segmentation with geometry-aware U-Net. IEEE Trans. Med. Imaging **38**(8), 1971–1980 (2019). https://doi.org/10.1109/TMI.2019.2911588
13. Milletari, F., Navab, N., Ahmadi, S.: V-Net: fully convolutional neural networks for volumetric medical image segmentation. In: 2016 Fourth International Conference on 3D Vision (3DV), pp. 565–571 (2016). https://doi.org/10.1109/3DV.2016.79
14. Oberstein, P.E., Olive, K.P.: Pancreatic cancer: why is it so hard to treat? Ther. Adv. Gastroenterol. **6**(4), 321–337 (2013)
15. Roth, H.R., et al.: DeepOrgan: multi-level deep convolutional networks for automated pancreas segmentation. In: Navab, N., Hornegger, J., Wells, W.M., Frangi, A.F. (eds.) MICCAI 2015. LNCS, vol. 9349, pp. 556–564. Springer, Cham (2015). https://doi.org/10.1007/978-3-319-24553-9_68
16. Roth, H.R., Lu, L., Farag, A., Sohn, A., Summers, R.M.: Spatial aggregation of holistically-nested networks for automated pancreas segmentation. In: Ourselin, S., Joskowicz, L., Sabuncu, M.R., Unal, G., Wells, W. (eds.) MICCAI 2016. LNCS, vol. 9901, pp. 451–459. Springer, Cham (2016). https://doi.org/10.1007/978-3-319-46723-8_52

17. Roth, H.R., et al.: Spatial aggregation of holistically-nested convolutional neural networks for automated pancreas localization and segmentation. Med. Image Anal. **45**, 94–107 (2018)
18. Sandler, M., Howard, A., Zhu, M., Zhmoginov, A., Chen, L.C.: MobileNetV 2: inverted residuals and linear bottlenecks. In: IEEE Conference on Computer Vision and Pattern Recognition, pp. 4510–4520 (2018)
19. American Cancer Society: Cancer Facts & Figures. American Cancer Society (2021)
20. European Society of Radiology (ESR) communications@myesr.org Emanuele Neri Nandita de Souza Adrian Brady Angel Alberich Bayarri Christoph D. Becker Francesca Coppola Jacob Visser, E.S.: What the radiologist should know about artificial intelligence-an esr white paper. Insights into imaging **10**, 1–8 (2019)
21. Wang, W., et al.: A fully 3D cascaded framework for pancreas segmentation. In: 2020 IEEE 17th International Symposium on Biomedical Imaging (ISBI), pp. 207–211 (2020). https://doi.org/10.1109/ISBI45749.2020.9098473
22. Wang, Y., et al.: Pancreas segmentation using a dual-input V-Mesh network. Med. Image Anal. **69**, 101958 (2021)
23. Xie, S., Sun, C., Huang, J., Tu, Z., Murphy, K.: Rethinking spatiotemporal feature learning: speed-accuracy trade-offs in video classification. In: Ferrari, V., Hebert, M., Sminchisescu, C., Weiss, Y. (eds.) ECCV 2018. LNCS, vol. 11219, pp. 318–335. Springer, Cham (2018). https://doi.org/10.1007/978-3-030-01267-0_19
24. Yu, Q., Xie, L., Wang, Y., Zhou, Y., Fishman, E.K., Yuille, A.L.: Recurrent saliency transformation network: incorporating multi-stage visual cues for small organ segmentation. In: 2018 IEEE/CVF Conference on Computer Vision and Pattern Recognition, pp. 8280–8289 (2018). https://doi.org/10.1109/CVPR.2018.00864
25. Zhou, Y., Xie, L., Shen, W., Wang, Y., Fishman, E.K., Yuille, A.L.: A fixed-point model for pancreas segmentation in abdominal CT scans. In: Descoteaux, M., Maier-Hein, L., Franz, A., Jannin, P., Collins, D.L., Duchesne, S. (eds.) MICCAI 2017. LNCS, vol. 10433, pp. 693–701. Springer, Cham (2017). https://doi.org/10.1007/978-3-319-66182-7_79

Voxel-Wise Cross-Volume Representation Learning for 3D Neuron Reconstruction

Heng Wang[1], Chaoyi Zhang[1], Jianhui Yu[1], Yang Song[2], Siqi Liu[3], Wojciech Chrzanowski[4,5], and Weidong Cai[1(✉)]

[1] School of Computer Science, University of Sydney, Sydney, Australia
tom.cai@sydney.edu.au
[2] School of Computer Science and Engineering, University of New South Wales, Sydney, Australia
[3] Paige AI, New York, NY, USA
[4] Sydney Pharmacy School, University of Sydney, Sydney, Australia
[5] Sydney Nano Institute, University of Sydney, Sydney, Australia

Abstract. Automatic 3D neuron reconstruction is critical for analysing the morphology and functionality of neurons in brain circuit activities. However, the performance of existing tracing algorithms is hinged by the low image quality. Recently, a series of deep learning based segmentation methods have been proposed to improve the quality of raw 3D optical image stacks by removing noises and restoring neuronal structures from low-contrast background. Due to the variety of neuron morphology and the lack of large neuron datasets, most of current neuron segmentation models rely on introducing complex and specially-designed submodules to a base architecture with the aim of encoding better feature representations. Though successful, extra burden would be put on computation during inference. Therefore, rather than modifying the base network, we shift our focus to the dataset itself. The encoder-decoder backbone used in most neuron segmentation models attends only intra-volume voxel points to learn structural features of neurons but neglect the shared intrinsic semantic features of voxels belonging to the same category among different volumes, which is also important for expressive representation learning. Hence, to better utilise the scarce dataset, we propose to explicitly exploit such intrinsic features of voxels through a novel voxel-level cross-volume representation learning paradigm on the basis of an encoder-decoder segmentation model. Our method introduces no extra cost during inference. Evaluated on 42 3D neuron images from BigNeuron project, our proposed method is demonstrated to improve the learning ability of the original segmentation model and further enhancing the reconstruction performance.

Keywords: Deep learning · Neuron reconstruction · 3D image segmentation · 3D optical microscopy

1 Introduction

3D neuron reconstruction is essential for analysis of brain circuit activities to understand how human brain works [12,13,21,25]. It traces neurons and recon-

© Springer Nature Switzerland AG 2021
C. Lian et al. (Eds.): MLMI 2021, LNCS 12966, pp. 248–257, 2021.
https://doi.org/10.1007/978-3-030-87589-3_26

structs their morphology from 3D light microscopy image stacks for neuroscientists to investigate the identity and functionality of neurons. Traditional tracing algorithms rely on hand-crafted features to capture neuronal structures but they are sensitive to the image quality. However, due to various imaging conditions, obtained neuron images suffer from different extent of noises and uneven labelling distribution. To attain better tracing performance for 3D neuron reconstruction, an accurate segmentation method to distinguish a neuron voxel from its surrounding low-contrast background is in high demand and necessary. However, due to the complexity of neuronal structures and various imaging artefacts, the precise restoration of neuronal voxels remains a challenging task.

A line of deep learning based segmentation models [9,10,17,19,20,22,26] has recently been proposed to demonstrate their advances in neuron segmentation studies. Since the neuronal structures range from long tree-like branches to blob-shape somas, [10] adopted inception networks [16] with various kernel sizes to better learn neuron representations from an enlarged receptive field. The invention of U-Net [15] gave rise to a line of encoder-decoder architectures and popularised them to be one of the de-facto structures in medical image segmentation tasks. Under the unsupervised setting, traditional tracing algorithm is combined with 3D U-Net [6] to progressively learn representative feature and the learned network, in turn, helps improve the tracing performance [26]. In the fully-supervised manner, modifications have been made based on 3D U-Net to extend the receptive field of kernels. In MKF-Net [19], a multi-scale spatial fusion convolutional block, where kernels with different size are processed in parallel and fused together, was proposed to replace some of the encoder blocks of 3D U-Net and achieved better segmentation results. Further, [20] introduces graph-based reasoning to learn longer-range connection for more complete features. The bottom layer of 3D U-Net encodes the richest semantic information but loses the spatial information. To alleviate the loss of spatial cues, [9] proposed to replace the bottom layer with a combination of dilated convolutions [3] and spatial pyramid pooling operations [8]. Although larger local regions have been aggregated after introducing these modifications and better segmentation results have been gained, additional overhead for the inference has also been introduced when using these proposed models. To avoid such overhead, [22] and [17] use additional teacher-student model and GAN-based data augmentation techniques, respectively. However, the focus of these segmentation models resides in the learning of local structural information of neurons within a single volume while the intrinsic semantic information of voxels among different volumes has been rarely touched.

As the semantic label is able to help obtain a better representation for image pattern recognition [23,24], such intrinsic features are beneficial to visual analysis. Siamese networks [1] based unsupervised representation learning methods [4,5,7] have recently gained impressive performance over image-level classification task by generating better latent representations through the comparison between two augmented views of the same image. Inspired by these work, we propose to encode the intrinsic semantic features into the representation of each voxel by maximising the similarity between two voxels belonging to the same class in a high-dimensional latent space. In our work, we follow the prevalent encoder-decoder architecture as the base segmentation model. Without the need of negative pairs [4] and momentum encoder [7], we design a class-aware voxel-wise Simple Siamese (SimSiam) [5] learning paradigm in a fully-supervised manner to maximise the similarity between two voxels with the same semantic meaning and encourage the base encoder to learn a better latent space for voxels of neuron images. Rather than being restricted within the same volume, to fully utilise the dataset, the voxel pairs can be sampled among different volumes. After training, only the original segmentation base model will be kept. Therefore, no extra cost is required to perform inference. Experimental results on 42 3D optical microscopy neuron images from the BigNeuron project [14] show that our proposed framework is able to achieve better segmentation results and further increase the accuracy of the tracing results.

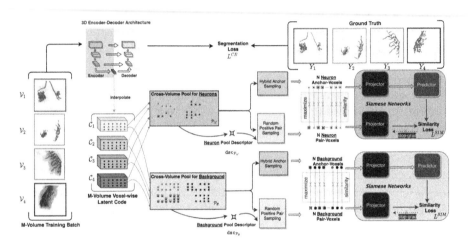

Fig. 1. Our proposed training paradigm.

2   Method

2.1   Supervised Encoder-Decoder Neuron Segmentation

We apply a 3D encoder-decoder architecture as the base model to perform neuron segmentation. As shown in Fig. 1, it consists of an encoder path, a decoder path, and skip connections linking in between, whose training progress is supervised via a binary cross-entropy loss that $L^{CE} = -log\left[Ylog(\widetilde{Y}) + (1-Y)log(\widetilde{Y})\right]$, where Y and $\widetilde{Y}$ represent the ground truth and the predictions of neuron segmentation masks, respectively.

L^{CE} provides semantic cues for each individual voxel but the relation learnt through the encoder is only within a local neighbourhood of each voxel, which ignores the correlation among voxels belonging to the same semantic class in a long distance, i.e., cross-volume. To decrease the distance of voxels with same category in the latent space, we introduce our proposed voxel-wise cross-volume SimSiam representation learning in next section.

2.2 VCV-RL: Voxel-Wise Cross-Volume SimSiam Representation Learning

As demonstrated in Fig. 1, for each volume $\mathcal{V}_i \in R^{D \times H \times W}$ where D, H, and W denote the depth, height, and width, respectively, we first obtain its corresponding downsized d-dimensional latent code through encoders, and then interpolate it back to same size as $\mathcal{V}_i$ to reach a voxel-wise embedding $\mathcal{C}_i$. Then, these M volumes of latent codes $\mathcal{C} \in \mathbb{R}^{M \times D \times H \times W \times d}$ are divided according to the ground truth labels Y_i, for the construction of two cross-volume point pools, one for the neurons voxels $\mathcal{P}_\mathcal{N}$ and one for the background voxels $\mathcal{P}_\mathcal{B}$. In other words, each pool contains the latent codes of voxels belonging to the same class from all the volumes within a training batch.

To enforce the voxel embeddings within the same class-aware pool to be closer in their latent space, we adopt Siamese networks [1] to perform the similarity comparison for each input voxel pair, and we denote the voxel to compare as the *anchor-voxel* and the voxel being compared as the *pair-voxel*. Following [5], we use a 3-layer MLP projector f as the channel mapping function to raise the latent code dimension from d to d_p, for both types of voxels. Another 2-layer MLP predictor h is employed to match the output of anchor-voxel to that of the pair-voxel. To prevent the Siamese networks from collapsing to constant solution, a stop-gradient operation [5] is also adopted as shown in Fig. 1.

Voxel Similarity. Following SimSiam [5], the symmetrised similarity loss between an anchor-voxel j and a pair-voxel k is defined with cos operator as:

$$L_{j,k}^{\cos} = \frac{1}{2}(1 - \cos[h(f(j)), f(k)]), \tag{1}$$

where $\cos[u,v] = \frac{u}{\|u\|_2} \cdot \frac{v}{\|v\|_2}$ denotes the cosine similarity between two embeddings. The total similarity loss comparing N voxel pairs is formulated as:

$$L^{SIM} = \sum_{* \in [\mathcal{N}, \mathcal{B}]} \frac{1}{N} \sum_{j \in \mathcal{P}_*, k \in \mathcal{P}_*}^{N} (\frac{1}{2}L_{j,k}^{cos} + \frac{1}{2}L_{k,j}^{cos}) \tag{2}$$

Together with loss L^{CE}, the encoded feature embedding for each voxel is expected to contain the intrinsic semantic cues shared among voxels of the same category. As the point-wise cross-entropy loss is complimentary to the representation learning loss [23], we keep the weight of these two losses the same. The segmentation loss is then formulated as $L^{seg} = L^{CE} + L^{SIM}$.

Anchor-Voxel Sampling Strategy. Given that latent codes of voxels being misclassified by the base segmentation model are more important for the representation learning [23], we design three different strategies to sample N anchor-voxels from each point pool $\mathcal{P}_*$:

- Random sampling ($\mathcal{AS}_{\mathbf{random}}$): Randomly sample N anchor-voxels from the whole point pool $\mathcal{P}_*$;
- Purely hard sampling ($\mathcal{AS}_{\mathbf{PH}}$): Randomly sample N anchor-voxels from a subset of point pool $\mathcal{P}_*$. The subset is the collection of voxels whose prediction is wrong;
- Hybrid combination sampling ($\mathcal{AS}_{\mathbf{hybrid}}$): $\frac{N}{2}$ anchor-voxels are randomly sampled from the whole point pool $\mathcal{P}_*$ while the rest $\frac{N}{2}$ anchor-voxels are randomly sampled from the subset stated in $\mathcal{AS}_{\mathbf{PH}}$.

Pair-Voxel Sampling Strategy. For each selected anchor-voxel, a pair-voxel is sampled from the same point pool to feed together into the Siamese networks. Apart from the candidate points in point pool $\mathcal{P}_*$, we propose a virtual point $\mathrm{dsc}_{\mathcal{P}_N}$ and $\mathrm{dsc}_{\mathcal{P}_B}$ as point descriptor for point pool $\mathcal{P}_N$ and $\mathcal{P}_B$, respectively to represent an aggregation of semantic feature for each class. We design two ways of computing such a virtual point:

- Relaxed: Average pooling of the entire point pool;
- Strict: Average pooling of the subset of correctly classified points from the entire point pool.

To avoid outliers and stabilise the learning, inspired by Momentum^2 Teacher [11], we propose to use a momentum update mechanism to keep the semantic information of past pool descriptors. Formally, the pool descriptor for each pool is defined as:

$$\mathrm{dsc}_{\mathcal{P}_*}^k = (1 - \alpha)\mathrm{dsc}_{\mathcal{P}_*}^k + \alpha\mathrm{dsc}_{\mathcal{P}_*}^{k-1}, \tag{3}$$

where k is the current iteration. α is the momentum coefficient and decreases from $\alpha_{base} = 1$ to 0 with cosine scheduling policy [11] defined as $\alpha = \alpha_{base}$ $(\cos(\frac{\pi k}{K}) + 1)/2$, where K is the total number of iterations.

3   Experiments and Results

3.1   Dataset and Implementation Details

Dataset. Our studies on 3D neuron reconstruction were conducted on the publicly available 42-volume Janelia dataset developed for the BigNeuron project [14], which was further divided into 35, 3, and 4 samples as training, validation, and testing set, respectively. We applied random horizontal and vertical flipping, rotation, and cropping, as data augmentation techniques, to amplify our training set with 3D patches of size $128 \times 128 \times 64$. Given that the number of neuron voxels is dramatically smaller than that of the background voxels, to ensure the existence of foreground neuron voxels, we re-choose the patch until the foreground voxels make up over 0.1% of the whole patch.

Network Setting and Implementation. All the models involved in the experiments were implemented in PyTorch 1.5 and trained from scratch for 222 epochs. A model was saved when it reached a better F1-score on the validation set. We use Adam as the optimizer with the learning rate of 1×10^{-3} and the weight decay of 1×10^{-4}. The batch size M is set as 4. We choose $\mathcal{AS}_{\mathbf{hybrid}}$ with the number of anchor-voxels N as 512 and strict pool descriptor with MoUp-date mechanism. The hidden layer dimension is 512 and 128 for the projector and predictor, respectively. d and d_p are set as 128 and 512, respectively. To make fair comparison, all the 3D U-Net based segmentation models including Li2019 [9] and MKF-Net [19] were implemented with feature dimensions of 16, 32, 64, and 128 for respective layer from top to bottom. As for 3D LinkNet [2], in addition, we replaced the head and final blocks with convolutional layer without spatial changes to keep the same 3 times downsampling of feature maps.

3.2 Results and Analysis

To quantitatively compare among different methods for 3D neuron segmentation, we reported F1, Precision, and Recall as the evaluation metrics. They measure the similarity between the prediction and the ground truth segmentation. Following [9], we employed three extra metrics for the reconstruction results: entire structure average (ESA), different structure average (DSA), and percentage of different structures (PDS) for the measurement between the traced neuron and the manually annotated neurons.

Segmentation Results. The quantitative segmentation result is presented in Table 1. Our proposed method achieves the best F1 score among all the other state-of-the-art segmentation methods and improves the performance of the base U-Net by 2.32%. It is also noticeable that our model includes no additional cost during inference. Our proposed VCV-RL module can be applied on the encoder output of encoder-decoder architecture easily. When applied on the 3D LinkNet, VCV-RL can enhance the segmentation performance by almost 2%.

Table 1. Segmentation performance on 3D neuron reconstruction.

Method	F1 (%)	Precision (%)	Recall (%)	#Params
Li2019 [9]	$52.69_{\pm 12.49}$	$46.50_{\pm 12.70}$	$61.06_{\pm 11.50}$	2.3M
3DMKF-Net [19]	$52.31_{\pm 12.14}$	$46.51_{\pm 11.82}$	$59.88_{\pm 12.32}$	1.5M
3D U-Net [6]	$51.26_{\pm 12.16}$	$45.26_{\pm 11.98}$	$59.35_{\pm 12.28}$	1.4M
+ **VCV-RL (proposed)**	$\mathbf{53.54}_{\pm 11.39}$	$\mathbf{46.68}_{\pm 11.58}$	$\mathbf{63.02}_{\pm 10.38}$	**1.4M**
3D LinkNet [2]	$50.74_{\pm 12.67}$	$42.93_{\pm 11.41}$	$62.22_{\pm 14.50}$	2.1M
+ **VCV-RL**	$52.66_{\pm 12.14}$	$46.10_{\pm 11.92}$	$61.55_{\pm 11.87}$	2.1M

Table 2. Tracing performance on 3D neuron reconstruction.

Method	ESA ↓	DSA ↓	PDS ↓	F1 (%)↑	Prec. (%)↑	Recall (%)↑
APP2 [25]	$3.62_{\pm 0.76}$	$6.80_{\pm 1.24}$	$0.34_{\pm 0.03}$	$55.84_{\pm 12.96}$	$57.48_{\pm 13.77}$	$\mathbf{64.94}_{\pm 23.10}$
+ U-Net [6]	$1.59_{\pm 0.19}$	$3.66_{\pm 0.81}$	$0.22_{\pm 0.02}$	$64.92_{\pm 15.33}$	$86.23_{\pm 7.76}$	$56.30_{\pm 19.39}$
+ MKF-Net [19]	$1.62_{\pm 0.21}$	$3.85_{\pm 0.76}$	$0.22_{\pm 0.03}$	$65.15_{\pm 14.93}$	$\mathbf{89.03}_{\pm 8.74}$	$55.07_{\pm 18.13}$
+ **Proposed**	$\mathbf{1.52}_{\pm 0.21}$	$\mathbf{3.48}_{\pm 0.29}$	$\mathbf{0.21}_{\pm 0.02}$	$\mathbf{66.17}_{\pm 14.39}$	$87.25_{\pm 7.41}$	$56.87_{\pm 17.92}$

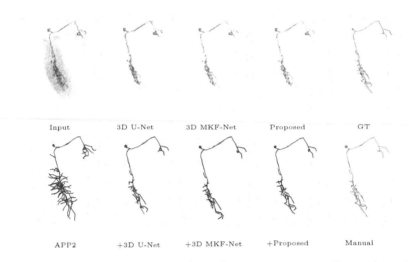

Fig. 2. Visualization of segmentation (above) and reconstruction (bottom) results for an example 3D neuron image. + refers performing APP2 on the segmented results from the segmentation model.

Neuron Reconstruction Results. To validate whether the proposed segmentation method can facilitate the tracing algorithm, we choose the state-of-the-art tracing algorithm APP2 [25] as the main tracer. As in [10], we perform the tracing algorithm on adjusted input volume using the probability map predicted from each segmentation model. As presented in Table 2, without extra overhead during inference, our proposed method combined with the tracer achieves the best quantitative tracing results on all the metrics among all the other deep learning based reconstruction methods except for the precision. The reason why plain APP2 reaches such high recall is that it overtraces the neuron structures, which is likely to include more real neuron points. Figure 2 displays the enhanced segmentation results after the image adjustment operation proposed in [10] in the first row. The second row presents the tracing results after applying APP2 [25] on the segmented images produced by different segmentation methods. Our proposed method combined with APP2 achieves competitive tracing result. We note that joint training of segmentation and tracing may further improve the 3D neuron reconstruction performance [18].

Ablation Study. As presented in Table 3, model B, C, and D.1 reach better F1 score than Model A, which demonstrates the effect of representation learning in improving learning ability of the base U-Net model and the superiority of hybrid anchor sampling strategy. The reason why model D.2 and D.3 outperform D.1 is because the existence of the proposed pool descriptor $dsc_\mathcal{P}$ can help improve the generality of the latent space. The strict way of computing $dsc_\mathcal{P}$ can further enhance the segmentation performance. We also try to remove the momentum update mechanism of $dsc_\mathcal{P}$ and the experiment result of model E demonstrates the importance of past information storage. In addition, we conducted experiments on the proposed method with different number of anchor-voxels N. The result is presented in Fig. 3. When N is 512, the F1-score is largest.

Table 3. Results of ablation studies.

ID	Method	F1 (%)
A	U-Net	51.26
B	$+\mathcal{AS}_{\text{random}}$	53.19
C	$+\mathcal{AS}_{\text{PH}}$	52.69
D	$+\mathcal{AS}_{\text{hybrid}}$	
D.1	w/o $dsc_\mathcal{P}$	53.20
D.2	w/ $dsc_\mathcal{P}$ (relaxed)	53.34
D.3	textbfw/ $dsc_\mathcal{P}$ (strict)	**53.54**
E	D.3 w/o MoUpdate	53.09

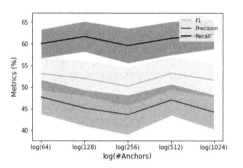

Fig. 3. $\mu \pm \frac{\sigma}{3}$ curves against $\log N$.

4   Conclusions

In this paper, we propose a novel voxel-wise cross-volume representation learning with the aim of facilitating the challenging 3D neuron reconstruction task. The segmentation of 3D neuron image is beneficial to improve the performance of tracing algorithms but is challenging due to various imaging artefacts and complex neuron morphology. Recent deep learning based methods rely on specially-designed structures in order to fully utilise the scarce 3D dataset. Though effective, extra computational costs have also been introduced. To make better use of the small dataset without sacrificing the efficiency during inference, we propose a novel training paradigm which consists of a SimSiam representation learning module to explicitly encode the semantic cues into the latent code of each voxel by comparing two voxels belonging to the same semantic category among different volumes. Compared to other methods, our proposed method learns a better latent space for the base model without modifying any part of it. And our proposed method shows superior performance in both the segmentation and reconstruction tasks.

References

1. Bromley, J., Guyon, I., LeCun, Y., Säckinger, E., Shah, R.: Signature verification using a "Siamese" time delay neural network. In: Advances in Neural Information Processing Systems (NeurIPS), p. 737 (1994)
2. Chaurasia, A., Culurciello, E.: LinkNet: exploiting encoder representations for efficient semantic segmentation. In: 2017 IEEE Visual Communications and Image Processing (VCIP), pp. 1–4. IEEE (2017)
3. Chen, L.C., Zhu, Y., Papandreou, G., Schroff, F., Adam, H.: Encoder-decoder with atrous separable convolution for semantic image segmentation. In: Proceedings of the European Conference on Computer Vision (ECCV), pp. 801–818 (2018)
4. Chen, T., Kornblith, S., Norouzi, M., Hinton, G.: A simple framework for contrastive learning of visual representations. In: International Conference on Machine Learning (ICML), pp. 1597–1607. PMLR (2020)
5. Chen, X., He, K.: Exploring simple Siamese representation learning. arXiv preprint arXiv:2011.10566 (2020)
6. Çiçek, Ö., Abdulkadir, A., Lienkamp, S.S., Brox, T., Ronneberger, O.: 3D U-Net: learning dense volumetric segmentation from sparse annotation. In: Ourselin, S., Joskowicz, L., Sabuncu, M.R., Unal, G., Wells, W. (eds.) MICCAI 2016. LNCS, vol. 9901, pp. 424–432. Springer, Cham (2016). https://doi.org/10.1007/978-3-319-46723-8_49
7. Grill, J.B., et al.: Bootstrap your own latent: a new approach to self-supervised learning. arXiv preprint arXiv:2006.07733 (2020)
8. He, K., Zhang, X., Ren, S., Sun, J.: Spatial pyramid pooling in deep convolutional networks for visual recognition. IEEE Trans. Pattern Anal. Mach. Intell. (TPAMI) **37**(9), 1904–1916 (2015)
9. Li, Q., Shen, L.: 3D neuron reconstruction in tangled neuronal image with deep networks. IEEE Trans. Med. Imaging (TMI) **39**(2), 425–435 (2019)
10. Li, R., Zeng, T., Peng, H., Ji, S.: Deep learning segmentation of optical microscopy images improves 3-D neuron reconstruction. IEEE Trans. Med. Imaging (TMI) **36**(7), 1533–1541 (2017)
11. Li, Z., Liu, S., Sun, J.: Momentum^{2}teacher: momentum teacher with momentum statistics for self-supervised learning. arXiv preprint arXiv:2101.07525 (2021)
12. Liu, S., Zhang, D., Liu, S., Feng, D., Peng, H., Cai, W.: Rivulet: 3D neuron morphology tracing with iterative back-tracking. Neuroinformatics **14**(4), 387–401 (2016)
13. Liu, S., Zhang, D., Song, Y., Peng, H., Cai, W.: Automated 3-D neuron tracing with precise branch erasing and confidence controlled back tracking. IEEE Trans. Med. Imaging (TMI) **37**(11), 2441–2452 (2018)
14. Peng, H., Hawrylycz, M., Roskams, J., Hill, S., Spruston, N., Meijering, E., Ascoli, G.A.: BigNeuron: large-scale 3D neuron reconstruction from optical microscopy images. Neuron **87**(2), 252–256 (2015)
15. Ronneberger, O., Fischer, P., Brox, T.: U-Net: convolutional networks for biomedical image segmentation. In: Navab, N., Hornegger, J., Wells, W.M., Frangi, A.F. (eds.) MICCAI 2015. LNCS, vol. 9351, pp. 234–241. Springer, Cham (2015). https://doi.org/10.1007/978-3-319-24574-4_28
16. Szegedy, C., et al.: Going deeper with convolutions. In: Proceedings of the IEEE Conference on Computer Vision and Pattern Recognition (CVPR), pp. 1–9 (2015)
17. Tang, Z., et al.: 3D conditional adversarial learning for synthesizing microscopic neuron image using skeleton-to-neuron translation. In: 2020 IEEE 17th International Symposium on Biomedical Imaging (ISBI), pp. 1775–1779. IEEE (2020)

18. Tregidgo, H.F.J., et al.: 3D reconstruction and segmentation of dissection photographs for MRI-free neuropathology. In: Martel, A.L., et al. (eds.) MICCAI 2020. LNCS, vol. 12265, pp. 204–214. Springer, Cham (2020). https://doi.org/10.1007/978-3-030-59722-1_20
19. Wang, H., et al.: Multiscale kernels for enhanced U-shaped network to improve 3D neuron tracing. In: Proceedings of the IEEE/CVF Conference on Computer Vision and Pattern Recognition (CVPR) Workshops, pp. 1105–1113 (2019)
20. Wang, H., et al.: Single neuron segmentation using graph-based global reasoning with auxiliary skeleton loss from 3D optical microscope images. In: 2021 IEEE 18th International Symposium on Biomedical Imaging (ISBI), pp. 934–938. IEEE (2021)
21. Wang, H., et al.: Memory and time efficient 3D neuron morphology tracing in large-scale images. In: 2018 Digital Image Computing: Techniques and Applications (DICTA), pp. 1–8. IEEE (2018)
22. Wang, H., et al.: Segmenting neuronal structure in 3D optical microscope images via knowledge distillation with teacher-student network. In: 2019 IEEE 16th International Symposium on Biomedical Imaging (ISBI), pp. 228–231. IEEE (2019)
23. Wang, W., Zhou, T., Yu, F., Dai, J., Konukoglu, E., Van Gool, L.: Exploring cross-image pixel contrast for semantic segmentation. arXiv preprint arXiv:2101.11939 (2021)
24. Wei, L., et al.: Can semantic labels assist self-supervised visual representation learning? arXiv preprint arXiv:2011.08621 (2020)
25. Xiao, H., Peng, H.: APP2: automatic tracing of 3D neuron morphology based on hierarchical pruning of a gray-weighted image distance-tree. Bioinformatics **29**(11), 1448–1454 (2013)
26. Zhao, J., et al.: Progressive learning for neuronal population reconstruction from optical microscopy images. In: Shen, D., et al. (eds.) MICCAI 2019. LNCS, vol. 11764, pp. 750–759. Springer, Cham (2019). https://doi.org/10.1007/978-3-030-32239-7_83

Diagnosis of Hippocampal Sclerosis from Clinical Routine Head MR Images Using Structure-constrained Super-Resolution Network

Zehong Cao[1,2], Feng Shi[2], Qiang Xu[6], Gaoping Liu[3], Tianyang Sun[2], Xiaodan Xing[2], Yichu He[2], Guangming Lu[3,4], Zhiqiang Zhang[3,4], and Dinggang Shen[2,5(✉)]

[1] School of Biomedical Engineering, Southern Medical University, Guangzhou, China
[2] Shanghai United Imaging Intelligence Co., Ltd., Shanghai, China
[3] Department of Radiology, Jinling Hospital, Nanjing Medical University, Nanjing, China
[4] State Key Laboratory of Analytical Chemistry for Life Science,
Nanjing University, Nanjing, China
[5] School of Biomedical Engineering, ShanghaiTech University, Shanghai, China
dgshen@shanghaitech.edu.cn
[6] Department of Radiology, Jinling Hospital, Nanjing University, Nanjing, China

Abstract. Medical images routinely acquired in clinical facilities are mostly low resolution (LR), in consideration of acquisition time and efficiency. This renders challenging for clinical diagnosis of hippocampal sclerosis where additional sequences for hippocampus need to be acquired. In contrast, high-resolution (HR) images provide more detailed information for disease investigation. Recently, image super-resolution (SR) methods were proposed to reconstruct HR images from LR inputs. However, current SR methods generally use simulated LR images and intensity constraints, which limit their applications in clinical practice. To solve this problem, we utilized real paired LR and HR images and trained a Structure-Constrained Super Resolution (SCSR) network. First, we proposed a single image super-resolution framework where mixed loss functions were introduced to enhance the reconstruction of brain tissue boundaries besides intensity constraints; Second, since the structure hippocampus is relatively small, we further proposed a weight map to enhance the reconstruction of subcortical regions. Experimental results using 642 real paired cases showed that the proposed method outperformed the the-state-of-the-art methods in terms of image quality with a PSNR of 27.0405 and an SSIM of 0.9958. Also, experiments using Radiomics features extracted from hippocampus on SR images obtained through the proposed method achieved the best accuracy of 95% for differentiating subjects with left and right hippocampal sclerosis from normal controls. The proposed method shows its potential for disease screening using clinical routine images.

1 Introduction

Due to the ability of better reflecting the soft tissue contrast than that of CT and other modalities and no radiation, MRI has been widely used in clinical facilities disease diagnosis. Normally plain MRI sequences are acquired for disease screening, where several

© Springer Nature Switzerland AG 2021
C. Lian et al. (Eds.): MLMI 2021, LNCS 12966, pp. 258–266, 2021.
https://doi.org/10.1007/978-3-030-87589-3_27

2D sequences are scanned such as T1, T2, and T2 FLAIR. These images generally have a fine in-plane resolution such as 1×1 mm^2 while the between slice thickness is large as 5 mm or even higher. Given the limitation of large slice thickness, these clinical routine images could not be used for precise quantification of small brain structures, especially for subtle disease related changes. High-resolution MRI image, e.g., 1 mm slice thickness, is highly desired in these scenarios while it suffers from the long acquisition time that is proportional to the slice thickness decrease. For example, hippocampal sclerosis (HS) is the most common pathological manifestation of intractable mesial temporal lobe epilepsy (mTLE) in adolescents and young adults, clinically accompanied with cognitive deficits in memory dysfunction, and manifests on imaging as atrophy of the hippocampal volume on conventional images, loss of internal structures, and atrophy of hippocampal temporal lobe structures [1]. Clinical routine images with large slice thickness and low resolution are unable to accurately capture the fine structures of the hippocampus, making it difficult for clinical screening of HS. In practice, specific high-resolution scanning sequence for hippocampus is often required for further diagnosis.

To fully utilize these clinical images, image processing methods were proposed to enhance the image resolution as complementary approaches. There are three main methods used for super-resolution tasks: 1) Interpolation-based methods; 2) Machine learning-based methods; and 3) Deep learning-based methods. The interpolation methods are the most fundamental and widely used to resample low-resolution images, but blurry boundaries and artifacts usually occur [2]. Machine learning methods were also proposed using dictionary learning and low rank [3], while the performance is still insufficient. Recently, deep learning methods were proposed for the SR tasks with superior performance than conventional methods [4]. For example, Dong et al. proposed a super-resolution convolutional neural network (SRCNN), which used several convolutional layers in an end-to-end manner to map LR to SR [5]. They also proposed an accelerated extension of SRCNN (FSRCNN), which used LR original image as the input and deconvolution layer to upsample SR images to HR space at the end of the network [6]. Li et al. presented Super-Resolution FeedBack Network (SRFBN), which mainly used the backhaul mechanism to improve the effect of super-resolution without introducing too many parameters [7]. However, the network is still too deep that require large GPU memory. The Enhanced Deep Residual Networks (EDSR) [8] removed the batch normalization from ResNet and introduced a constant scaling layer to use more layers for finer feature extraction to improve the performance of the network. However, these methods were intentionally designed for 2D natural images, and also their LR images were generally simulated by downsampling HR images, which limits their capacity for clinical usage.

In this paper, we proposed a novel framework named Structure-Constrained Super Resolution (SCSR) for HS diagnosis using clinical routine head MR images. We trained the SCSR with real paired LR-SR images. Different from a single super-resolution network, in SCSR, we combined super-resolution training and segmentation learning simultaneously, so that the resulting SR images achieved better ROI segmentation results. To learn SR network by segmentation knowledge, we designed to utilize a mixed loss including Dice loss in segmentations, MSE loss, and Mean Structure Similarity (MSSIM) loss [9] for SR quality in intensity images to improve the reconstruction performance. In summary, the method not only applies to the segmentation of the whole images, but also could focus on specific structures, such as the hippocampus and surrounding areas, to generate a more precise SR image with improvements on the local structures.

2 Method

2.1 Overview of the Method

In this work, we first upsampled the input LR images to the resolution of HR images using trilinear interpolation method, and then utilized neural networks for the improvement of image quality with high-frequency information. Briefly, we used thick slice brain MRI scans as inputs to a Deep Mapping Network (DMN) to reconstruct SR images. Here DMN mainly includes several resblocks. We further leveraged mixed loss functions between network outputs (SR images) and training targets (HR images) for better synthesis details, which consists of MSE loss, Dice loss, and MSSIM loss.

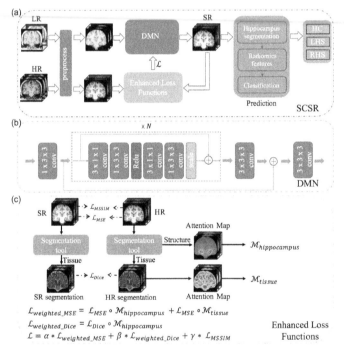

Fig. 1. The Framework of Structure-Constrained Super-Resolution Network. (a) The overview of our SCSR framework; (b) Detail structure of DMN backbone; and (c) the proposed Enhanced Loss Function Module. Note that the SR segmentation and hippocampus and tissue weight matrices were obtained per image per iteration to calculate the Loss Function.

2.2 Proposed Deep Mapping Network (DMN)

Image Preprocessing. Image preprocessing steps are as follows: 1) lr images were resampled to a standard resolution of $1 \times 1 \times 1$ mm^3; 2) skull stripping were performed on the resampled lr images; 3) hr images were linearly aligned to the lr images to maintain their consistency.

Super-Resolution Deep Mapping Network (DMN). We adopt edsr as the backbone of the deep mapping network. Edsr was previously proposed for natural 2D image super-resolution. Briefly, it includes several residual blocks at the low-resolution stage followed by an upsampling layer. We made a few modifications to the edsr structure for our task. First, the network was modified to 3D with image patches of size [32, 32, 32] as inputs. Second, we removed the upsampling module of the network since the lr images were already interpolated to the final resolution in the image preprocessing stage. Third, we applied an anisotropic convolution strategy, given that the main focus of the task is to improve the image fidelity between slices. We leveraged two consecutive convolution kernels with shapes $1 \times 3 \times 3$ and $3 \times 1 \times 1$ in place of the traditional $3 \times 3 \times 3$ kernel for lower computation costs during training.

Finally, the DMN includes 32 ResBlock Modules, and the channel number is 256. We normalized the intensity value of the inputs to $[-1, 1]$, as well as the corresponding training targets (patches cropped from the HR images).

2.3 Enhanced Loss Functions to Improve the SR Image Quality

To improve the performance of the reconstruction, we proposed to constrain the intensity similarity, brain tissue and structure similarities during the supervised learning process. To do this, we acquired tissue segmentation maps of the SR image and the corresponding HR ground truth using an in-house brain image analysis toolbox. Then we used an enhanced loss function consisting of MSE, MSSIM, and Dice losses between the SR and HR to update the weights of DMN. Note that the tissue segmentation of SR was updated every epoch to reflect the reconstruction changes.

Intensity Similarity and Brain Tissue Similarity. We first used mse loss for intensity similarity between sr and hr images. As a complementary method, SSIM index is also employed, which measures the difference of the image quality such as brightness, contrast, and structure between sr and hr. As widely used, we calculated the mean ssim (i.e., mssim) in the window size of 7 and 1.5 gaussian kernel to quantify the structural consistency.

For brain tissue similarity, we adopted Dice loss between tissue segmentation of both network outputs and training targets into our loss function, to promote clearer reconstruction results not only between foreground and background but also between anatomical boundaries.

Attention Maps for Structure Constraints. For the mse loss, we added a weighting matrix $\{\mathcal{M}_{tissue} | \mathcal{M}_{tissue} \in R^{xyz}, m_{i,j,k} \in \mathcal{M}_{tissue}\}$ Across the whole 3D image. The weighting matrix is designed to emphasize the boundaries across different tissues (gm, wm and csf), in order to recover the tissue distribution. Each element $m_{i,j,k}$ In the weighting matrix decreases with larger spatial distance to the tissue boundaries. We then smoothed the matrix via a gaussian kernel. The final loss function is a scaled average of mse loss through all 3D voxels weighted by $\mathcal{M}_{tissue}$.

Tissue constraints could help improve the GM/WM boundary clarity of SR outputs while leaving blurry subcortical regions that affect the precise analysis of the SR images. Although our focus in this paper is hippocampal dysfunction, we also involved parahippocampal gyrus, and entorhinal cortex as they are also highly related with neurodegenerative diseases. Thus, these structures acquired from the segmentation were used as attention map to generate the weighted matrix $\mathcal{M}_{hippocampus}$. Similarly, each element in the weighting matrix decreases with larger spatial distance to these structures. Finally, the loss function of the SCSR was defined as the combination of weighted MSE, MSSIM and weighted Dice losses (Fig. 1).

2.4 Diagnosis for HS on Clinical Routine MR Images

We propose to utilize MR routine images for the diagnosis of left HS (LHS) and right HS (RHS) out of normal controls (NC) according to the following procedure: 1) We obtain the hippocampus segmentation results by the proposed SCSR method as well as comparison methods; 2) Radiomics features [10] were extracted from the segmented hippocampus using the Pyradiomics package [11]. In detail, 215 Radiomics features include volume, shape, and texture matrices of Gray-level Co-occurrence Matrix (GLCM), Gray-Level Run-length Matrix (GLRLM), Local Binary Pattern (LBP), and Histogram of Oriented Gradient (HOG). Note that here we do not use sophisticated feature extraction methods since the main purpose is to evaluate the quality of image SR; 3) Least absolute shrinkage and selection operator (LASSO) [12] was used to select the most useful features. Support Vector Machine (SVM) classifier with default parameters was then employed to classify patients with hippocampal sclerosis (LHS, RHS, and NC).

3 Experiments

3.1 Data

We used the largest dataset till now, to our best knowledge, involving 642 subjects with sequentially acquisitions of clinical routine images and high-resolution images. The age ranges from 5 to 71 years with mean of 28 and standard devision of 11. In this dataset, 460 were clinical epilepsy patients, 129 were mTLE patients, and 53 were normal controls. All images included no focal lesion, and were reviewed by an experienced radiologist. All data were acquired on a 3T MR scanner (Siemens TimTrio, Erlangen, Germany). The 2D clinical routine thick-slice images were acquired about one and a half minutes per subject using spin-echo T1WI sequence: TR = 350 ms, TE = 2.46 ms, FOV = 24 cm × 24 cm, matrix = 320 × 320, thickness/gap = 4/0.4 mm, voxel size = 0.75 × 0.75 × 4.4 mm^3, number of slices = 30. Also, high resolution 3D-T1w images were acquired about seven and a half minutes per subject using MPRAGE sequence: TR = 2300 ms, TE = 2.98 ms, FOV = 256 × 256, matrix = 256 × 256, slice thickness = 1 mm, voxel size = 0.5 × 0.5 × 1 mm^3, number of slices = 176.

In this study, we used 230 epilepsy cases as training data. The remaining data were divided into two testing datasets. The first testing dataset included 230 epilepsy cases with normal image appearance. The second testing dataset contained 129 cases of mTLE patients with hippocampal sclerosis (L-mTLE 69 cases, R-mTLE 60 cases) and 53 cases of healthy control (HC). The diagnosis of the mTLE-HS was performed by experienced radiologists.

3.2 Experimental Setting

We used two comparison methods, known as FSRCNN and SRFBN. Hyperparameters were tuned during our experiment:$\alpha = 2$, $\beta = 1$, $\gamma = 0.2$. We considered three objective image quality metrics (IQMs) to evaluate the reconstruction quality of our SCSR Network, including root mean square error (RMSE), peak signal-to-noise ratio (PSNR), and structural similarity index measure (SSIM). Both RMSE and SSIM have a range of 0 to 1. Lower RMSE indicates better performance while 1 represents perfect for SSIM. PSNR ranges from 0 to infinity. In general, a high PSNR is expected. For the evaluation of diagnosis performance, we mainly used three metrics from classification: accuracy, sensitivity, and specificity.

3.3 Results

Evaluation of Image SR on the First Testing Data with 230 Cases. In the first test-ing dataset, we compared the reconstruction results of interpolation, FSRCNN, SRFBN, and SCSR with hr regarding anatomical information, shown in Fig. 2. Considering the clarity of the gyrus and sulci, as well as subcortical structures, the interpolation method lost most of the details while the FSRCNN outputs a slightly clearer gyrus and sulci but still lacking subcortical details. The SRFBN improved upon the previous two methods while the reconstruction quality was still less desirable compared to the HR targets. The proposed SCSR reconstructions were more satisfactory that provide information closest to the HR images. We zoomed in on the hippocampus areas in Fig. 2, showing that only SCSR reconstructs a clear boundary of hippocampus and surrounding regions among all methods. Figure 3 also demonstrates the consistent results with average difference maps especially on zoom-in areas for first testing data.

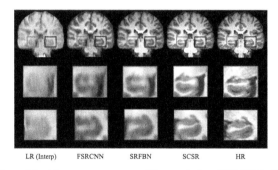

LR (Interp)   FSRCNN   SRFBN   SCSR   HR

Fig. 2. Visualization of the hippocampus areas of Interpolation, FSRCNN, SRFBN, SCSR, HR.

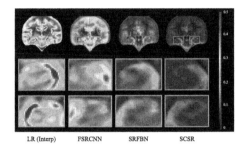

LR (Interp)   FSRCNN   SRFBN   SCSR

Fig. 3. The average difference map between each of Interpolation, FSRCNN, SRFBN, SCSR with HR in the first testing data.

We also quantitatively analyze the performance of the proposed method by comparing the other comparison methods illustrated in Table 1. Overall, SCSR outperforms all other methods regarding RMSE, PSNR, and SSIM.

Table 1. The mean metrics of four method results.

	Interpolation	FSRCNN	SRFBN	SCSR*	SCSR
Test1 (230)					
RMSE	0.1901 ± 0.016	0.1108 ±0.010	0.0657 ± 0.005	0.0488 ± 0.005	**0.0447 ± 0.005**
PSNR	14.50 ± 0.756	19.31 ± 0.814	24.18 ± 0.678	26.28 ± 0.882	**27.04 ± 0.932**
SSIM	0.9363 ± 0.007	0.9703 ± 0.003	0.9832 ± 0.001	0.9868 ± 0.001	**0.9958 ± 0.0009**
Test2 (182)					
RMSE	0.1893 ± 0.014	0.1082 ± 0.009	0.0659 ±0.005	0.0530 ± 0.005	**0.0486 ± 0.005**
PSNR	14.48 ± 0.629	19.35 ± 0.716	23.64 ± 0.631	25.55 ± 0.841	**26.31 ± 0.868**
SSIM	0.9430 ± 0.006	0.9778 ± 0.003	0.9908 ± 0.001	0.9941 ± 0.001	**0.9951 ± 0.001**

SCSR* represents the results from SCSR without hippocampus-weighted.

Evaluation of HS Diagnosis on Second Testing Data with 182 Cases. From Fig. 4, we can see that the hippocampus morphology of the three methods. The results of the SCSR, compared with the other two methods, gives a smoother surface and closer shape to HR images; the results of FSRCNN failed to recover the hippocampal boundaries in detail, and generate holes in the rendered 3D view; the results of srfbn were much better than those of fsrcnn, but still inferior to scsr in detail.

We calculate the asymmetry of hippocampus volume between left and right seen in Fig. 5. It shows that the results of the SCSR method are the closest to the distribution of hippocampal volume ratios in human brains of various populations: the value of HC is close to 0, LHS means the volume of the left hippocampus is smaller than the volume on the right.

Table 2. compared the results of the diagnosing HS using LR images, SR images, and HR images, and the metrics reported a very close performance using our reconstructed SR and the true HR images in this classification task, giving a much more desirable result than using the LR images. Moreover, the accuracy using SR outputs for detecting LHS and RHS exceeded that of HR images reporting 0.967 and 0.984 respectively.

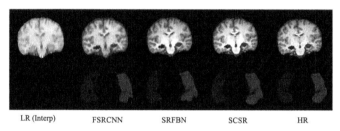

LR (Interp)   FSRCNN   SRFBN   SCSR   HR

Fig. 4. Visualization of the reconstruction results of the hippocampus in the second testing data. The brown ones represent the left hippocampus, the green ones are the right hippocampus.

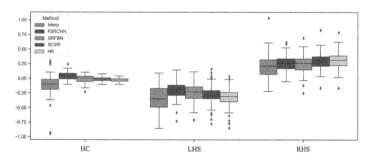

Fig. 5. The asymmetry of hippocampus volume between left and right in the second testing data. Asymmetry is defined as 2(LHV-RHV)/(LHV + RHV).

Table 2. The HS classification results with images of LR, SR, and HR in the second testing data.

	Sensitivity(LR/SR/HR)	Specificity(LR/SR/HR)	Accuracy(LR/SR/HR)
HC	0.906 / 0.981 / **1.00**	0.900 / **0.938** / **0.938**	0.920 / 0.951 / **0.956**
LHS	0.886 / **0.928** / 0.913	0.947 / 0.991 / **1.00**	0.923 / **0.967** / 0.951
RHS	0.850 / **0.950** / **0.950**	0.976 / **1.00** / 0.992	0.940 / **0.984** / 0.978

4 Conclusion

In this work, we presented a structure-constrained framework with a deep mapping network to improve the reconstruction of 3D images from clinical routine 2D images. Experiments show superior fidelity of the reconstructed images in comparison to other state-of-the-art methods. As aforementioned, the method has great potential in many applications such as the early screening of neurodegenerative disease patients where precise brain structure measurements are required.

References

1. Thom, M.: Hippocampal sclerosis in epilepsy: a neuropathology review. Neuropathol. Appl. Neurobiol. **40**(5), 520–543 (2014)
2. Siu, W.-C., Hung, K.-W.: Review of image interpolation and super-resolution. In: Proceedings of The 2012 Asia Pacific Signal and Information Processing Association Annual Summit and Conference. IEEE (2012)
3. Shi, F., et al.: LRTV: MR image super-resolution with low-rank and total variation regularizations. IEEE Trans. Med. Imaging **34**(12), 2459–2466 (2015)
4. Zhang, Y., et al.: Longitudinally guided super-resolution of neonatal brain magnetic resonance images. IEEE Trans. Cybern. **49**(2), 662–674 (2018)
5. Dong, C., et al.: Learning a Deep Convolutional Network for Image Super-Resolution. Springer International Publishing, Cham (2014)
6. Dong, C., Loy, C.C., Tang, X.: Accelerating the Super-Resolution Convolutional Neural Network. Springer International Publishing, Cham (2016)
7. Li, Z., et al.: Feedback network for image super-resolution. In: Proceedings of the IEEE Conference on Computer Vision and Pattern Recognition. (2019)
8. Lan, R., et al.: Cascading and enhanced residual networks for accurate single-image super-resolution. IEEE Trans. Cybern. **51**(1), 115–125 (2021)
9. Zhou, W., et al.: Image quality assessment: from error visibility to structural similarity. IEEE Trans. Image Process. **13**(4), 600–612 (2004)
10. Feng, Q., et al.: Hippocampus radiomic biomarkers for the diagnosis of amnestic mild cognitive impairment: a machine learning method. Front Aging Neurosci. **11**, 323 (2019)
11. van Griethuysen, J.J.M., et al.: Computational radiomics system to decode the radiographic phenotype. Can. Res. **77**(21), e104–e107 (2017)
12. Kukreja, S.L., Löfberg, J., Brenner, M.J.: A least absolute shrinkage and selection operator (LASSO) for nonlinear system identification. IFAC Proc. Volumes **39**(1), 814–819 (2006)

U-Net Transformer: Self and Cross Attention for Medical Image Segmentation

Olivier Petit[1,2(✉)], Nicolas Thome[1], Clement Rambour[1], Loic Themyr[1], Toby Collins[3], and Luc Soler[2]

[1] CEDRIC - Conservatoire National des Arts et Metiers, Paris, France
`olivier.petit@visiblepatient.com`
[2] Visible Patient SAS, Strasbourg, France
[3] IRCAD, Strasbourg, France

Abstract. Medical image segmentation remains particularly challenging for complex and low-contrast anatomical structures. In this paper, we introduce the U-Transformer network, which combines a U-shaped architecture for image segmentation with self- and cross-attention from Transformers. U-Transformer overcomes the inability of U-Nets to model long-range contextual interactions and spatial dependencies, which are arguably crucial for accurate segmentation in challenging contexts. To this end, attention mechanisms are incorporated at two main levels: a self-attention module leverages global interactions between encoder features, while cross-attention in the skip connections allows a fine spatial recovery in the U-Net decoder by filtering out non-semantic features. Experiments on two abdominal CT-image datasets show the large performance gain brought out by U-Transformer compared to U-Net and local Attention U-Nets. We also highlight the importance of using both self- and cross-attention, and the nice interpretability features brought out by U-Transformer.

Keywords: Medical image segmentation · Transformers · Self-attention · Cross-attention · Spatial layout · Global interactions

1 Introduction

Organ segmentation is of crucial importance in medical imaging and computed-aided diagnosis, *e.g.* for radiologists to assess physical changes in response to a treatment or for computer-assisted interventions.

Currently, state-of-the-art methods rely on Fully Convolutional Networks (FCNs), such as U-Net and variants [2,7,9,18]. U-Nets use an encoder-decoder architecture: the encoder extracts high-level semantic representations by using

Electronic supplementary material The online version of this chapter (https://doi.org/10.1007/978-3-030-87589-3_28) contains supplementary material, which is available to authorized users.

C. Lian et al. (Eds.): MLMI 2021, LNCS 12966, pp. 267–276, 2021.
https://doi.org/10.1007/978-3-030-87589-3_28

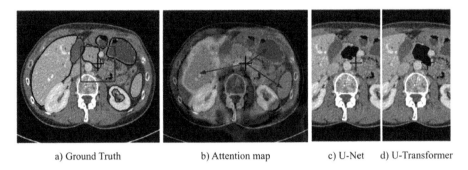

a) Ground Truth   b) Attention map   c) U-Net   d) U-Transformer

Fig. 1. Global context is crucial for complex organ segmentation but cannot be captured by vanilla U-Nets with a limited receptive field, *i.e.* blue cross region in a) with failed segmentation in c). The proposed U-Transformer network represents full image context by means of attention maps b), which leverage long-range interactions with other anatomical structures to properly segment the complex pancreas region in d). (Color figure online)

a cascade of convolutional layers, while the decoder leverages skip connections to re-use high-resolution feature maps from the encoder in order to recover lost spatial information from high-level representations.

Despite their outstanding performances, FCNs suffer from conceptual limitations in complex segmentation tasks, *e.g.* when dealing with local visual ambiguities and low contrast between organs. This is illustrated in Fig. 1a) for segmenting the blue cross region corresponding to the pancreas with U-Net: the limited Receptive Field (RF) framed in red does not capture sufficient contextual information, making the segmentation fail, see Fig. 1c).

In this paper, we introduce the U-Transformer network, which leverages the strong abilities of transformers [13] to model long-range interactions and spatial relationships between anatomical structures. U-Transformer keeps the inductive bias of convolution by using a U-shaped architecture, but introduces attention mechanisms at two main levels, which help to interpret the model decision. Firstly, a self-attention module leverages global interactions between semantic features at the end of the encoder to explicitly model full contextual information. Secondly, we introduce cross-attention in the skip connections to filter out non-semantic features, allowing a fine spatial recovery in the U-Net decoder.

Figure 1b) shows a cross-attention map induced by U-Transformer, which highlights the most important regions for segmenting the blue cross region in Fig. 1a): our model leverages the long-range interactions with respect to other organs (liver, stomach, spleen) and their positions to properly segment the whole pancreas region, see Fig. 1d). Quantitative experiments conducted on two abdominal CT-image datasets show the large performance gain brought out by U-Transformer compared to U-Net and to the local attention in [11].

Related Work. Attention mechanisms are a relatively recent problem in medical imaging [8,10–12,16]. Attention in segmentation is often based on multi-resolution features combined with a simple attention module [6,16]. These contributions however fail to incorporate long-range dependencies. Recent works successfully tackle this aspect through Dual attention networks [5,12] proving the importance of full range attention but to the cost of large parameter overhead and multiple concurrent loss functions.

Transformers [13] models also bring global attention and have witnessed increasing success in the last five years, started in natural language processing with text embeddings [3]. A pioneer use of transformers in computer vision is non-local networks [15], which combine self-attention with a convolutional backbone. Recent applications include object detection [1], semantic segmentation [14,17], and image classification [4].

U-Transformer combines the power of Transformers to grasp long-range dependencies and multi-resolution information processing through self- and cross-attention modules. Our cross-attention mechanism shares the high-level motivation of Attention U-Net [11] to help the recovery of fine spatial information from rich semantic features, with the noticeable difference that the U-Transformer's attention embraces all input features whereas Attention U-Net's attention uses each local feature independently.

2 The U-Transformer Network

As mentioned in Sect. 1, encoder-decoder U-shaped architectures lack global context information to handle complex medical image segmentation tasks. We introduce the U-Transformer network, which augments U-Nets with attention modules built from multi-head transformers. U-Transformer models long-range contextual interactions and spatial dependencies by using two types of attention modules (see Fig. 2): Multi-Head Self-Attention (MHSA) and Multi-Head Cross-Attention (MHCA). Both modules are designed to express a new representation of the input based on its self-attention in the first case (*cf.* Sect. 2.1) or on the attention paid to higher level features in the second (*cf.* Sect. 2.2).

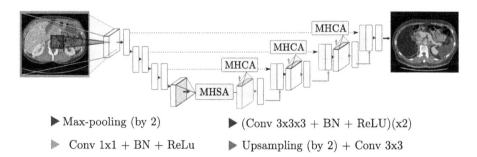

▶ Max-pooling (by 2)   ▶ (Conv 3x3x3 + BN + ReLU)(x2)

▷ Conv 1x1 + BN + ReLu   ▶ Upsampling (by 2) + Conv 3x3

Fig. 2. U-Transformer augments U-Nets with transformers to model long-range contextual interactions. The Multi-Head Self-Attention (MHSA) module at the end of the U-Net encoder gives access to a receptive field containing the whole image (shown in purple), in contrast to the limited U-Net receptive field (shown in blue). Multi-Head Cross-Attention (MHCA) modules are dedicated to combine the semantic richness in high level feature maps with the high resolution ones coming from the skip connections. (Color figure online)

2.1 Self-attention

The MHSA module is designed to extract long range structural information from the images. To this end, it is composed of multi-head self-attention functions as described in [13] positioned at the bottom of the U-Net as shown in Fig. 2. The main goal of MHSA is to connect every element in the highest feature map with each other, thus giving access to a receptive field including all the input image. The decision for one specific pixel can thus be influenced by any input pixel. The attention formulation is given in Eq. 1. A self-attention module takes three inputs, a matrix of queries $Q \in \mathbb{R}^{n \times d_k}$, a matrix of keys $K \in \mathbb{R}^{n \times d_k}$ and a matrix of values $V \in \mathbb{R}^{n \times d_k}$.

$$\text{Attention}(Q, K, V) = \text{softmax}(\frac{QK^T}{\sqrt{d_k}})V = AV \qquad (1)$$

A line of the attention matrix $A \in \mathbb{R}^{n \times n}$ corresponds to the similarity of a given element in Q with respect to all the elements in K. Then, the attention function performs a weighted average of the elements of the value V to account for all the interactions between the queries and the keys as illustrated in Fig. 3. In our segmentation task, Q, K and V share the same size and correspond to different learnt embedding of the highest level feature map denoted by X in Fig. 3. The embedding matrices are denoted as W_q, W_k and W_v. The attention is calculated separately in multiple heads before being combined through another embedding. Moreover, to account for absolute contextual information, a positional encoding is added to the input features. It is especially relevant for medical image segmentation, where the different anatomical structures follow a fixed spatial position. The positional encoding can thus be leveraged to capture absolute and relative position between organs in MHSA.

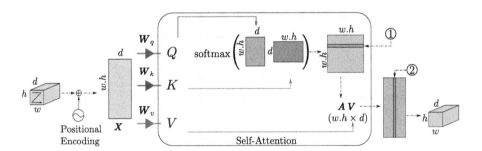

Fig. 3. MHSA module: the input tensor is embedded into a matrix of queries Q, keys K and values V. The attention matrix A in purple is computed based on Q and K. (1) A line of A corresponds to the attention given to all the elements in K with respect to one element in Q. (2) A column of the value V corresponds to a feature map weighted by the attention in A.

2.2 Cross-attention

The MHSA module allows to connect every element in the input with each other. Attention may also be used to increase the U-Net decoder efficiency and in particular enhance the lower level feature maps that are passed through the skip connections. Indeed, if these skip connections insure to keep a high resolution information they lack the semantic richness that can be found deeper in the network. The idea behind the MHCA module is to turn off irrelevant or noisy areas from the skip connection features and highlight regions that present a significant interest for the application. Figure 4 shows the cross-attention module. The MHCA block is designed as a gating operation of the skip connection S based on the attention given to a high level feature map Y. The computed weight values are then re-scaled between 0 and 1 through a sigmoid activation function. The resulting tensor, denoted Z in Fig. 4, is a filter where low magnitude elements indicate noisy or irrelevant areas to be reduced. A cleaned up version of S is then given by the Hadamard product $Z \odot S$. Finally, the result of this filtering operation is concatenated with the high level feature tensor Y. Here, the keys and queries are computed from the same source as we are designing a filtering operation whereas for NLP tasks, having homogeneous keys and values may be more meaningful. This configuration proved to be empirically more effective.

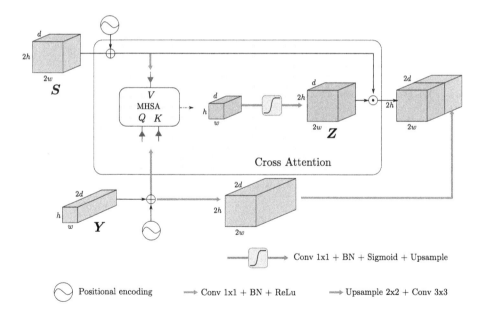

Fig. 4. MHCA module: the value of the attention function corresponds to the skip connection S weighted by the attention given to the high level feature map Y. This output is transformed into a filter Z and applied to the skip connection.

3 Experiments

We evaluate U-Transformer for abdominal organ segmentation on the TCIA pancreas public dataset, and an internal multi-organ dataset.

Accurate pancreas segmentation is particularly difficult, due to its small size, complex and variable shape, and because of the low contrast with the neighboring structures, see Fig. 1. In addition, the multi-organ setting assesses how U-transformer can leverage attention from multi-organ annotations.

Experimental Setup. The TCIA pancreas dataset[1] contains 82 CT-scans with pixel-level annotations. Each CT-scan has around $181 \sim 466$ slices of 512×512 pixels and a voxel spacing of $([0.66 \sim 0.98] \times [0.66 \sim 0.98] \times [0.5 \sim 1.0])$ mm^3.

We also experiment with an Internal Multi-Organ (IMO) dataset composed of 85 CT-scans annotated with 7 classes: liver, gallbladder, pancreas, spleen, right and left kidneys, and stomach. Each CT-scan has around $57 \sim 500$ slices of 512×512 pixels and a voxel spacing of $([0.42 \sim 0.98] \times [0.42 \sim 0.98] \times [0.63 \sim 4.00])$ mm^3.

All experiments follow a 5-fold cross validation, using 80% of images in training and 20% in test. We use the Tensorflow library to train the model, with Adam optimizer (10^{-4} learning rate, exponential decay scheduler).

We compare U-Transformer to the U-Net baseline [9] and Attention U-Net [11] with the same convolutional backbone for fair comparison. We also report performances with self-attention only (MHSA, Sect. 2.1), and the cross-attention only (MHCA, Sect. 2.2). U-Net has $\sim$30M parameters, the overhead from U-transformer is limited (MHSA $\sim$5M, each MHCA block $\sim$2.5M).

3.1 U-Transformer Performances

Table 1 reports the performances in Dice averaged over the 5 folds, and over organs for IMO. U-Transformer outperforms U-Net by 2.4pts on TCIA and 1.3pts for IMO, and Attention U-Net by 1.7pts for TCIA and 1.6pts for IMO. The gains are consistent on all folds, and paired t-tests show that the improvement is significant with p-values <3% for every experiment.

Table 1. Results for each method in Dice similarity coefficient (DSC, %)

Dataset	U-Net [9]	Attn U-Net [11]	MHSA	MHCA	U-Transformer
TCIA	76.13 ($\pm$0.94)	76.82 ($\pm$1.26)	77.71 ($\pm$1.31)	77.84 ($\pm$2.59)	**78.50** ($\pm$1.92)
IMO	86.78 ($\pm$1.72)	86.45 ($\pm$1.69)	87.29 ($\pm$1.34)	87.38 ($\pm$1.53)	**88.08** ($\pm$1.37)

Figure 5 provides qualitative segmentation comparison between U-Net, Attention U-Net and U-Transformer. We observe that U-Transformer performs

---

[1] https://wiki.cancerimagingarchive.net/display/Public/Pancreas-CT.

better on difficult cases, where the local structures are ambiguous. For example, in the second row, the pancreas has a complex shape which is missed by U-Net and Attention U-Net but U-Transformer successfully segments the organ.

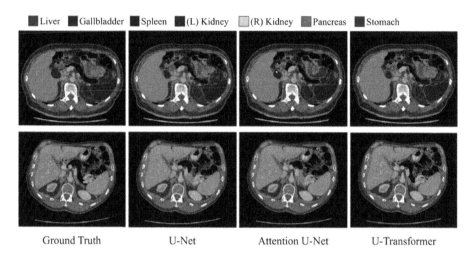

■Liver ■Gallbladder ■Spleen ■(L) Kidney □(R) Kidney ■Pancreas ■Stomach

Ground Truth   U-Net   Attention U-Net   U-Transformer

Fig. 5. Segmentation results for U-Net [9], Attention U-Net [11] and U-Transformer on the multi-organ IMO dataset (first row) and on TCIA pancreas (second row).

In Table 1, we can see that the self-attention (MHSA) and cross-attention (MHCA) alone already outperform U-Net and Attention U-Net on TCIA and IMO. Since MHCA and Attention U-Net apply attention mechanisms at the skip connection level, it highlights the superiority of modeling global interactions between anatomical structures and positional information instead of the simple local attention in [11]. Finally, the combination of MHSA and MHCA in U-Transformer shows that the two attention mechanisms are complementary and can collaborate to provide better segmentation predictions.

Table 2 details the results for each organ on the multi-organ IMO dataset. This further highlights the interest of U-Transformer, which significantly outperforms U-Net and Attention U-Net for the most challenging organs: pancreas: +3.4pts, gallbladder: +1.3pts and stomach: +2.2pts. This validates the capacity of U-Transformer to leverage multi-label annotations to drive the interactions between anatomical structures, and use easy organ predictions to improve the detection and delineation of more difficult ones. We can note that U-Transformer is better for every organ, even the liver which has a high score >95% with U-Net.

Table 2. Results on IMO in Dice similarity coefficient (DSC, %) detailed per organ.

Organ	U-Net [9]	Attn U-Net [11]	MHSA	MHCA	U-Transformer
Pancreas	69.71 (±3.74)	68.65 (±2.95)	71.64 (±3.01)	71.87 (±2.97)	**73.10** (±2.91)
Gallbladder	76.98 (±6.60)	76.14 (±6.98)	76.48 (±6.12)	77.36 (±6.22)	**78.32** (±6.12)
Stomach	83.51 (±4.49)	82.73 (±4.62)	84.83 (±3.79)	84.42 (±4.35)	**85.73** (±3.99)
Kidney(R)	92.36 (±0.45)	92.88 (±1.79)	92.91 (±1.84)	92.98 (±1.70)	**93.32** (±1.74)
Kidney(L)	93.06 (±1.68)	92.89 (±0.64)	92.95 (±1.30)	92.82 (±1.06)	**93.31** (±1.08)
Spleen	95.43 (±1.76)	95.46 (±1.95)	95.43 (±2.16)	95.41 (±2.21)	**95.74** (±2.07)
Liver	96.40 (±0.72)	96.41 (±0.52)	96.82 (±0.34)	96.79 (±0.29)	**97.03** (±0.31)

3.2   U-Transformer Analysis and Properties

Positional Encoding and Multi-level MHCA. The Positional Encoding
(PE) allows to leverage the absolute position of the objects in the image. Table 3
shows an analysis of its impact, on one fold on both datasets. For MHSA, the
PE improves the results by +0.7pt for TCIA and +0.6pt for IMO. For MHCA,
we evaluate a single level of attention with and without PE. We can observe an
improvement of +1.7pts for TCIA and +0.6pt for IMO between the two versions.

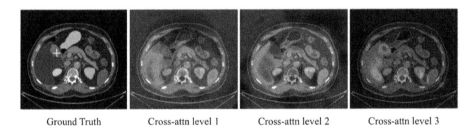

Ground Truth   Cross-attn level 1   Cross-attn level 2   Cross-attn level 3

Fig. 6. Cross-attention maps for the yellow-crossed pixel (left image). (Color figure
online)

Table 3 also shows the favorable impact of using multi *vs* single-level attention
for MHCA: +1.8pts for TCIA and +0.6pt for IMO. It is worth noting that
Attention U-Net uses multi-level attention but remains below MHCA with a
single level. Figure 6 shows attention maps at each level of U-Transformer: level
3 corresponds to high-resolution features maps, and tends to focus on more
specific regions compared to the first levels.

Table 3. Ablation study on the positional encoding and multi-level on one fold of
TCIA and IMO.

	U-Net	Attn U-Net	MHSA		MHCA		
			wo PE –	w PE	1 lvl wo PE –	1 lvl w PE –	multi-lvl w PE
TCIA	76.35	77.23	78.17	**78.90**	77.18	78.88	**80.65**
IMO	88.18	87.52	88.16	**88.76**	87.96	88.52	**89.13**

Further Analysis. To further analyse the behaviour of U-Transformer, we evaluate the impact of the number of attention heads for MHSA (supplementary, Fig. 1): more heads lead to better performances, but the biggest gain comes from the first head (*i.e.* U-Net to MHSA). Finally, the evaluation of U-Transformer with respect to the Hausdorff distance (supplementary, Table 1) follows the same trend than with Dice score. This highlights the capacity of U-Transformer to reduce prediction artefacts by means of self- and cross-attention.

4   Conclusion

This paper introduces the U-Transformer network, which augments a U-shaped FCN with Transformers. We propose to use self and cross-attention modules to model long-range interactions and spatial dependencies. We highlight the relevance of the approach for abdominal organ segmentation, especially for small and complex organs. Future works could include the study of U-Transformer in 3D networks, with other modalities such as MRI or US images, as well as for other medical image tasks.

References

1. Carion, N., Massa, F., Synnaeve, G., Usunier, N., Kirillov, A., Zagoruyko, S.: End-to-end object detection with transformers. In: Vedaldi, A., Bischof, H., Brox, T., Frahm, J.-M. (eds.) ECCV 2020. LNCS, vol. 12346, pp. 213–229. Springer, Cham (2020). https://doi.org/10.1007/978-3-030-58452-8_13
2. Çiçek, Ö., Abdulkadir, A., Lienkamp, S.S., Brox, T., Ronneberger, O.: 3D U-Net: learning dense volumetric segmentation from sparse annotation. In: Ourselin, S., Joskowicz, L., Sabuncu, M.R., Unal, G., Wells, W. (eds.) MICCAI 2016. LNCS, vol. 9901, pp. 424–432. Springer, Cham (2016). https://doi.org/10.1007/978-3-319-46723-8_49
3. Devlin, J., Chang, M., Lee, K., Toutanova, K.: BERT: pre-training of deep bidirectional transformers for language understanding. CoRR abs/1810.04805 (2018). arXiv:1810.04805
4. Dosovitskiy, A., et al.: An image is worth 16x16 words: transformers for image recognition at scale. In: International Conference on Learning Representations (2021)
5. Fu, J., et al.: Dual attention network for scene segmentation. In: Proceedings of the IEEE/CVF Conference on Computer Vision and Pattern Recognition (CVPR), June 2019
6. Li, C., et al.: Attention based hierarchical aggregation network for 3D left atrial segmentation. In: Pop, M., et al. (eds.) STACOM 2018. LNCS, vol. 11395, pp. 255–264. Springer, Cham (2019). https://doi.org/10.1007/978-3-030-12029-0_28
7. Milletari, F., Navab, N., Ahmadi, S.: V-Net: fully convolutional neural networks for volumetric medical image segmentation. In: 2016 Fourth International Conference on 3D Vision (3DV), pp. 565–571 (2016)
8. Nie, D., Gao, Y., Wang, L., Shen, D.: ASDNet: attention based semi-supervised deep networks for medical image segmentation. In: Frangi, A.F., Schnabel, J.A., Davatzikos, C., Alberola-López, C., Fichtinger, G. (eds.) MICCAI 2018. LNCS, vol. 11073, pp. 370–378. Springer, Cham (2018). https://doi.org/10.1007/978-3-030-00937-3_43

9. Ronneberger, O., Fischer, P., Brox, T.: U-Net: convolutional networks for biomedical image segmentation. In: Navab, N., Hornegger, J., Wells, W.M., Frangi, A.F. (eds.) MICCAI 2015. LNCS, vol. 9351, pp. 234–241. Springer, Cham (2015). https://doi.org/10.1007/978-3-319-24574-4_28

10. Roy, A.G., Navab, N., Wachinger, C.: Concurrent spatial and channel 'squeeze & excitation' in fully convolutional networks. In: Frangi, A.F., Schnabel, J.A., Davatzikos, C., Alberola-López, C., Fichtinger, G. (eds.) MICCAI 2018. LNCS, vol. 11070, pp. 421–429. Springer, Cham (2018). https://doi.org/10.1007/978-3-030-00928-1_48. arXiv:1803.02579

11. Schlemper, J., et al.: Attention gated networks: learning to leverage salient regions in medical images. Med. Image Anal. **53** (2019). https://doi.org/10.1016/j.media.2019.01.012

12. Sinha, A., Dolz, J.: Multi-scale self-guided attention for medical image segmentation. IEEE J. Biomed. Health Inform. 1 (2020)

13. Vaswani, A., et al.: Attention is all you need. In: NeurIPS, pp. 5998–6008 (2017)

14. Wang, H., Zhu, Y., Green, B., Adam, H., Yuille, A., Chen, L.-C.: Axial-DeepLab: stand-alone axial-attention for panoptic segmentation. In: Vedaldi, A., Bischof, H., Brox, T., Frahm, J.-M. (eds.) ECCV 2020. LNCS, vol. 12349, pp. 108–126. Springer, Cham (2020). https://doi.org/10.1007/978-3-030-58548-8_7

15. Wang, X., Girshick, R., Gupta, A., He, K.: Non-local neural networks. In: Proceedings of the IEEE Conference on Computer Vision and Pattern Recognition, pp. 7794–7803 (2018)

16. Wang, Y., et al.: Deep attentional features for prostate segmentation in ultrasound. In: Frangi, A.F., Schnabel, J.A., Davatzikos, C., Alberola-López, C., Fichtinger, G. (eds.) MICCAI 2018. LNCS, vol. 11073, pp. 523–530. Springer, Cham (2018). https://doi.org/10.1007/978-3-030-00937-3_60

17. Ye, L., Rochan, M., Liu, Z., Wang, Y.: Cross-modal self-attention network for referring image segmentation. In: Proceedings of the IEEE/CVF Conference on Computer Vision and Pattern Recognition, pp. 10502–10511 (2019)

18. Zhou, Z., Rahman Siddiquee, M.M., Tajbakhsh, N., Liang, J.: UNet++: a nested U-Net architecture for medical image segmentation. In: Stoyanov, D., et al. (eds.) DLMIA/ML-CDS -2018. LNCS, vol. 11045, pp. 3–11. Springer, Cham (2018). https://doi.org/10.1007/978-3-030-00889-5_1

Pre-biopsy Multi-class Classification of Breast Lesion Pathology in Mammograms

Tal Tlusty[1]([✉]), Michal Ozery-Flato[1], Vesna Barros[1,3], Ella Barkan[1],
Mika Amit[1], David Gruen[1], Michal Guindy[2], Tal Arazi[2], Mona Rozin[2],
Michal Rosen-Zvi[1,3], and Efrat Hexter[1]

[1] IBM Research, Haifa University, Mount Carmel 31905, Israel
ttlusty@ibm.com
[2] Department of Imaging, Assuta Medical Centers, Tel Aviv, Israel
[3] The Faculty of Medicine, The Hebrew University, Jerusalem, Israel

Abstract. Characterization of lesions by artificial intelligence (AI) has been the subject of extensive research. In recent years, many studies demonstrated the ability of convolution neural networks (CNNs) to successfully distinguish between malignant and benign breast lesions in mammography (MG) images. However, to date, no study has assessed the specific sub-type of lesions in MG images, as detailed in histolopathology reports. We present a method for finer classification of breast lesions in MG images into multiple pathology sub-types. Our approach works well with radiologists' diagnostic workflow, and uses data available in radiology reports. The proposed Dual-Radiology Dual-Resolution Network (Du-Rad Du-Res Net) receives dual input from the radiologist and dual image resolutions. The radiologist input includes annotation of the lesion area and semantic radiology features; the dual image resolutions comprise a low resolution of the entire mammogram and a high resolution of the lesion area. The network estimates the likelihood of malignancy, as well as the associated pathological sub-type. We show that the combined input of the lesion region of interest (ROI) and the entire mammogram is important for optimizing the model's performance. We tested the AI in a reader study on a dataset of 100 heldout cases. The AI outperformed three breast radiologists in the task of lesion histopathology sub-typing.

Keywords: Breast cancer · Mammography · Deep neural networks

1 Introduction

Breast cancer is the most common cancer among women, accounting for 30% of female cancers in the United States in 2020 [1]. Breast cancer detection commonly starts with a breast mammography (MG), which is currently the main

Electronic supplementary material The online version of this chapter (https:// doi.org/10.1007/978-3-030-87589-3_29) contains supplementary material, which is available to authorized users.

C. Lian et al. (Eds.): MLMI 2021, LNCS 12966, pp. 277–286, 2021.
https://doi.org/10.1007/978-3-030-87589-3_29

imaging modality used for breast cancer screening. Following an examination of MG images and additional clinical information by a radiologist, approximately 10% of screening patients are recalled for further imaging examinations [2–4]. Among the recalled patients, about 10% will undergo needle biopsy due to a suspicious abnormal findings [4]. However, the majority of these biopsies are ultimately classified as having a non-malignant pathology. Only 10 to 30% of breast biopsies were classified as malignant in the USA [5], thus there is significant opportunity to reduce those that turn out to have been unnecessary. In addition to malignant/non-malignant classification, suspicious lesions in histology images are further classified to more subtle sub-categories ranging from benign tumors to invasive cancers. In this study we focus on predicting a finer sub-typing of breast histopathology.

The correlation between the appearance of lesions in MG images and their histology sub-typing has been the subject of several works. Histology sub-categories often contain special or unique characteristics that can be found inside the MG images. For example, Holland and Hendriks [6] present a tight correlation between micro-calcification appearance and characterization in MG, and their appearance in histology images, for different ductal carcinoma in situ (DCIS) sub-types. Other works suggest a correlation between the tumor size, tumor margin, and tumor shape that appear in the MG to the specific histology sub-type [7]. A detailed review of additional works that study the connection between MG and histology images can be found in [8].

The proliferation in AI-based medical diagnostics has led to extensive research in the field of lesion classification in MG exams. The most common approach is to classify the entire mammogram image into two coarse categories of "malignant" versus "non-malignant" using a convolution neural network (CNN) [9–13]. Another approach is to identify "normal" images without any lesion, to reduce the screening workload of the radiologist [14–18]. To the best of our knowledge, there is no previous study on classifying lesions in MG images into finer histopathological sub-types. AI systems typically process MG images in lower resolution, since the actual size of a mammogram image is too large for the memory of standard computers. Several studies demonstrated improved accuracy for classifiers trained on full resolution image patches of ROIs extracted by lesion annotations [11–13]. In a recent work [19], Shen et al. present a novel method for using both the full image, and the suspicious lesions that are extracted in a weakly-supervised manner, to predict the presence or absence of benign and malignant lesions in a breast.

This paper presents a novel CNN multi-class architecture for the task of predicting lesion histopathology sub-types in MG images. Our study builds on information available in radiologists' reports: (i) annotations of suspicious areas, and (ii) semantic features, such as lesion size and type (e.g., mass, calcification). We evaluated our model on a heldout dataset and in a reader study conducted with three specialized breast-imaging radiologists.

2   Methods

2.1   Dataset

We used an in-house dataset of MG images with local annotations of visible suspected findings. This dataset was collected from 4 different sites and multiple imaging machines in the United States. In all, the dataset covers 2860 unique patients that were sent for biopsy. Each case includes 1 or more images from screening and diagnostic studies. All known pathology diseases can be grouped into categories according to their further management in clinical work-up: malignant, benign, and benign with risk. We further split the malignant category into DCIS and invasive sub-groups, and the benign with risk category into lesions with atypia (lifetime high-risk) and benign pathologies associated with pre-malignant tumors (e.g., papilloma). Given that, data labels including pathology-based sub-categories were grouped into five major categories: DCIS, invasive, papilloma, high risk and benign lesions. A table of diseases and classes is displayed in the Supplementary material. Images with more than one class were assigned multiple class labels.

We randomly split the data into training, validation, and test sets, while ensuring no patient overlap between the three. The random partition was made in a stratified manner to preserve the frequency of the five classes in each of the three data sets. Statistics on the number of images and lesions appear in Table 1. All images contain local annotations of the biopsied lesions, including contour around the lesion, pathology label, and semantic features extracted from the radiologist's report. These features were extracted using basic text processing techniques and include: lesion size, lesion type (i.e., mass, calcification), shape (oval, round, irregular), margins (circumscribed, spiculated), and appearance of suspicious structures (e.g., grouped microcalcification or architectural distortion).

For weights initialization, we used a large dataset of globally labelled images and pre-trained the backbone of the network on the related task of classifying an MG image into normal/benign/malignant. This globally annotated dataset, which totalled 6341 benign, 5341 malignant, and 10507 normal images, was assembled by augmenting our in-house dataset (after excluding validation and heldout sets) with data from the OPTIMAM [20] database and a second in-house dataset.

Table 1. Data statistics: number of images; number of lesions.

	DCIS	Invasive	High Risk	Papilloma	Benign	Total
Train	476; 539	1060; 1238	155; 192	228; 263	2398; 2735	4317; 4967
Validation	104; 110	227; 245	31; 41	50; 56	487; 574	899; 1026
Heldout	105; 113	234; 269	29; 39	45; 62	523; 586	936; 1069
Reader-study	22; 22	24; 25	7; 9	16; 20	29; 30	98; 106

2.2   Proposed Method

In this work, we used both the full image and patches of the lesion ROI. Thus, the algorithm received the context low-resolution (low-res) information that may affect the finding's classification, and the fine-detailed high-resolution (high-res) information from the lesion area. To extract the patches, we cropped the lesion area with margins of 60 pixels and resized all patches to the same size of 512×512. The alternative of cropping patches of the same size could possibly crop out information in larger lesions. The other alternative of cropping different patches from the lesion's ROI and aggregating the results would be very noisy. The full images were cropped around the breast area and resized to 2200×1200 while maintaining the aspect ratio. These images were then zero-padded to a final size of 2260×1260. As a final step, the obtained images were normalized to $[0;1]$.

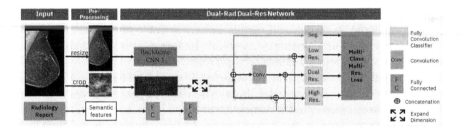

Fig. 1. Architecture of the Du-Rad Du-Res Network. The architecture contains low-res and high-res sub-networks, with Seg. branch for the former. Additionally it contains a third dual-res sub-network in which patch and full image representations from the two other sub-networks are concatenated to create a dual-res representation. A fully convolutional classification head is applied to each of the sub-networks to compute multi-class predictions. A fourth sub-network, uses two fully connected layers to learn a representation for the semantic radiology features, and is concatenated to the input for the sub-network classification heads.

Figure 1 depicts our Du-Rad Du-Res Net. Our proposed model architecture receives dual input from the radiologist: (i) annotation of the lesion area and (ii) semantic features describing the lesion's visual characterizations. The model also receives the dual resolution images: (i) a low-res full image and (ii) a high-res patch. These inputs are processed by 3 inter-connected networks, that do not share weights: full image network, patch network, and semantic features network. We used the well-known InceptionResnet-V2 (IRV2) architecture [21] as a baseline model and as a backbone for both the full image and patch networks. We optimized the model for small datasets by cutting the network off after 14 IRV2 blocks, before the global max pooling (GMP) layer. The semantic features network consisted of 2 (FC) layers with 64 and 384 neurons, respectively. Our model creates features that encapsulate information from both the full and patch images. First, the $[H, W, C]$ features from the patch network are concatenated to the $[H, W, C]$ features from the full image network. This is done after the features from the patch network were duplicated to fit the dimensions of the full image network. Then, a 3×3 convolution layer is applied to the resulting $[H, W, C]$ features, leading to $[H, W, C]$ features of mixed resolutions. The features corresponding to full image, patch, and the mixed resolutions are then concatenated with the output features from the semantic features network, producing features of size $[2H, W, C]$, where $H = 384, W = 139, C = 77$. Our model has 3 main classification branches corresponding to the 3 features types described above: low-res full image, high-res patch, and dual-res full-image-patch mixture. These branches share the same architecture, which is a GMP layer followed by 5 classification sub-branches, each having a single hidden layer of 256 neurons. Thus, overall, our model produces 15 prediction scores. In addition, we follow Ness et al. [22] and add a Seg. branch to the full image network before the global pooling, using a binary segmentation map of the lesion as ground truth. The final loss is a combination of cross-entropy for each of the 3 branches (multi-resolution, or MR) per class (multi-class, or MC) predicting a one vs. all probability and the segmentation branch loss (seg):

$$L_{\text{MC-MR}} = \sum_{i \in C_{1,\ldots,5}} \left(L_{\text{LR,i}} + L_{\text{HR,i}} + L_{\text{DR,i}} \right) + \lambda L_{\text{seg}} \qquad (1)$$

where λ was set to 2 after some hyper parameter tuning on the validation set, and $C_{1,\ldots,5}$ are the 5 sub-typing classes.

2.3   Model Training

We trained the models for 100 epochs using NVIDIA Tesla V100 GPUs, PyTorch, and FuseMedML framework [23]. Each minibatch consisted of 2 images, 1 from malignant classes and 1 from benign classes. We initialized the weights of the full image and patch branches using our pretained network as mentioned in 2.1. We then fine-tuned the entire network. For data augmentation, applied to both patches and full images, we used random rotation and translation, flipping, and zooming. In addition, we applied color augmentations of contrast, gamma, and intensity. We used Adam optimizer, with an initial learning rate of 10^{-4}.

3   Experiments and Results

We evaluated our Du-Res-Du-Rad model on the 1069 image-lesion pairs in our heldout dataset (see Table 1). We set the predicted class of the model as the class with the highest predicted score. Table 2 presents the accuracy of the model's predicted class and the ROC curve (AUC) per class. As shown, the invasive classification had the highest AUC (0.85 [95% CI 0.82, 0.87]), while accuracy was highest for the benign category (0.83 [95% CI 0.76, 0.83]). Note that benign and invasive are our largest classes, and also those found at the two extremes of the sub-categories' spectrum.

Ablation Study. To evaluate the contribution of the different components in our Du-Rad Du-Res Net model, we tested four incremental model variants: testing the contribution of the segmentation branch, dual-res architecture, model pre-training, and semantic features to our model performance. The results of this ablation study are presented in Table 2, where models with the Du-Res component are presented with three AUCs corresponding to the low-res, high-res, and dual-res classification heads. As shown, the addition of each tested component incrementally improves the overall performance of the classification heads. This indicates, the importance of these components to the overall performance of our model. For all models, adding the low-res and high-res heads, contributed to the model performance due to regularization effect in training. For variants with dual-res architecture, the model prediction is done by the dual-res head.

Table 2. Du-Rad Du-Res Net model performance: accuracy of predicted class, and a comparison of AUC with simpler model variants.

Model	Seg	Pre-trained	low-res	high-res	dual-res	DCIS	Invasive	High Risk	Papilloma	Benign
IRV2			✓			0.55	0.68	0.58	0.56	0.66
+Seg.	✓		✓			0.72	0.72	0.55	0.62	0.69
+Du-Res	✓		✓			0.71	0.72	0.53	0.62	0.69
	✓			✓		0.75	0.80	0.56	0.59	0.73
	✓				✓	0.77	0.79	0.60	0.69	0.74
+Pretrained	✓	✓	✓			0.72	0.80	0.59	0.62	0.77
	✓	✓		✓		0.76	0.81	0.58	0.55	0.76
	✓	✓			✓	0.73	0.83	0.62	0.69	0.78
+Du-Rad	✓	✓	✓			**0.80**	0.83	0.67	0.67	0.77
[Du-Rad	✓	✓		✓		**0.80**	0.84	0.65	0.62	0.76
Du-Res]	✓	✓			✓	0.78	**0.85**	**0.69**	**0.72**	**0.78**
Accuracy	✓	✓			✓	0.37	0.64	0.21	0.02	0.83

Malignant vs. Benign Classification. Using the softmax function, we transformed a multi-class classifier, for the finer histopathology sub-types into a binary classifier for malignant vs. benign. We computed the malignancy score as the sum of the predicted scores for the DCIS and invasive sub-types. The significance of the results, was tested with DeLong test. With this transformation, our Du-Rad Du-Res Net model achieved an AUC of 0.816 [95% CI 0.789, 0.842] in the task of malignant vs. benign classification. As a baseline classifier, we used our low-res head, modified for binary classification, and trained it with the coarse malignant vs. benign labels. Compared to our model, this baseline had a significantly lower AUC of 0.71 [95% CI 0.68, 0.74] ($P \ll 0.0001$). To evaluate the net contribution of finer sub-type labels to the binary classification of malignant vs. benign, we compared this baseline model (i.e., the low-res binary classification head) to the multi-class version of this model trained with the finer sub-types (i.e., the original low-res multi-class classification head). The AUC of the latter model showed an improved AUC of 0.73 [95% CI 0.69, 0.76] ($P < 0.05$). The results imply that malignant classification can be improved by incorporating additional information from pathology reports. Finally, we explored the ability of our model to reduce the number of unnecessary biopsies. This can be especially useful in other body organs where the biopsy procedure is risky. The results presented in

Table 3 show that the model could potentially assist in preventing 19% of the unnecessary biopsies, while missing 1% of the malignant cases.

Table 3. Du-Rad Du-Res Net model specificity at sensitivity levels for the task of malignant/benign.

Sensitivity	Specificity	95% CI for Specificity
0.99	0.19	[0.06,0.21]
0.98	0.24	[0.11,0.26]
0.95	0.38	[0.23,0.38]

Reader Study. The reader study was conducted by three specialists in breast-imaging radiology, each with 15 to 30 years of experience, all from different organizations. Each radiologist viewed 100 images of the heldout set. The images were selected randomly, with no more than a single image per study. Two of the cases were disqualified by the readers due to poor quality; (see Table 1). As a first step, each radiologist viewed the low-res full image; and was then asked to predict the probability of malignancy. In addition, the radiologist was asked to assign it to one of the 5 sub-classes. Next, the radiologist repeated the process after being given both the full image and the lesion's ROI. For the task of malignancy prediction, The AI outperformed the radiologists with an AUC of 0.75, compared to 0.68, 0.69 and 0.71 AUC for the readers. The significance of the results, was tested with Wilcoxon signed-rank test. The improved results were found significant only for Rad 1 ($P = 0.03$). The AI outperformed all 3 breast radiologists in the task of lesion histopathology sub-typing, with an accuracy of 45% compared to 28%,35% and 43% accuracy for the readers (detailed results in the Supplementary material). Here too, the improved results were found significant only when compared to Rad 1. Comparing the AI and radiologists' performance of low-res versus dual-res, similar trends can be observed. For the benign class, we found that exposure to high-res impairs the results, giving us an average accuracy of 48% versus 64% for the low-res ($P < 0.0003$). In contrast, when we compared the performance for the invasive class, the accuracy in the dual-res scenario improves, from 59% to 67% when moving from low to high resolution. We measured the agreement between the readers and the AI model, and between pairs of readers, by calculating Cohen's Kappa [24] scores. For the malignancy prediction, the average agreement between the readers and between the AI and the readers was low (Cohen's Kappa of 0.31 [0.28–0.34] and 0.17 [0.08–0.25]). For the subtyping task, the level of agreement between radiologists was moderate (0.47 [0.4–0.56]). The agreement with the AI system was similar (0.45 [0.40–0.5]). Comparing the agreement between the two types of tasks, the agreement of the latter is higher due to a greater increase in the number of cases for which both are incorrect. Most of the disagreement between the readers and the AI was in the benign classes, which contain the hardest cases to analyze.

4 Discussion

In the United States each year, more than 1 million women undergo breast biopsy, yet, on average, approximately 80% yield benign findings. When combined with the anxiety, discomfort, cost and potential complications of even minimally invasive fine needle biopsies, it is apparent that more accurate distinction of the lesions can have enormous clinical significance. Although biopsy is effective in cancer diagnosis, adding another modality to the diagnostic process could help radiologists offer more accurate diagnoses. In addition, feasibility in breast MG could pave the way for additional imaging modalities and other body organs, where biopsy may be more risky or impractical. A large number of works, presented impressive results on the task of malignant versus benign lesions classification in MG. However, to the best of our knowledge, this study is the first to use a dual-radiology dual-resolution network to classify lesions into pathologic sub-types, which more challenging due to the fine differences between the sub-classes. Although further work is needed, these encouraging results offer additional supporting evidence that AI can improve the accuracy, sensitivity, and specificity of breast cancer diagnosis.

Waugh et al. [26], show the feasibility of differentiating between breast cancer sub-types by texture features from MRI. Our work differs from their work in four aspects: (i) We use the MG, which is the most common screening modality so no other imaging procedures need be applied. (ii) We perform finer sub-typing classification of malignant lesions as well as benign lesions. (iii) We use features that are extracted from both the full image, the ROI and the clinical report. (iv) The relatively large dataset allows us to use deep learning. We note that there is a limitation associated with our reader study, since the radiologists are trained to predict malignancy and not the sub-typing. Hence, we compare AI and readers' performance in two tasks: malignancy prediction and sub-type classification.

The ground truth labels used in this study are based on pathological observations that may be noisy. Studies show that this categorization is often subjective and prone to observer variability [27]. Moreover, specifically for the DCIS category, a poor accuracy rate of only 65% was achieved in the core biopsy procedure [28]. In addition, a study published in JAMA in 2015 [25], showed that while pathologists have high concordance (96%) in diagnosing invasive carcinoma for needle biopsies, the agreement among pathologists was only 48% for atypical hyperplasia. It is possible that histology classification can also be improved by analyzing additional modalities, such as MG. This is a direction for future research work. In addition, we based our method on annotations given by the radiologists. Future work will focus in replacing this input by machine computed parameters, as previous works have shown large accuracy in predicting them [19].

References

1. https://acsjournals.onlinelibrary.wiley.com/doi/full/10.3322/caac.21588
2. Smith-Bindman, R., et al.: Comparing the performance of mammography screening in the USA and the UK. J. Med. Screening **12**(1), 50–54 (2005)

3. Schell, M.J., et al.: Evidence-based target recall rates for screening mammography. Radiology **243**(3), 681–689 (2007)
4. Neal, L., Tortorelli, C.L., Nassar, A.: Clinician's guide to imaging and pathologic findings in benign breast disease. Mayo Clin. Proc. **85**(3), 274–279 (2010). Elsevier
5. Kopans, D.B.: The positive predictive value of mammography. AJR Am. J. Roentgenol. **158**(3), 521–526 (1992)
6. Holland, R., Hendriks, J.H.: Microcalcifications associated with ductal carcinoma in situ: mammographic-pathologic correlation. Seminars Diagn. Pathol. **11**(3), 181–192 (1994)
7. Lamb, P.M., et al.: Correlation between ultrasound characteristics, mammographic findings and histological grade in patients with invasive ductal carcinoma of the breast. Clin. Radiol. **55**(1), 40–44 (2000)
8. Hamidinekoo, A., et al.: Deep learning in mammography and breast histology, an overview and future trends. Med. Image Anal. **47**, 45–67 (2018)
9. Cao, H., et al.: Multi-tasking U-shaped Network for benign and malignant classification of breast masses. IEEE Access **8**, 223396–223404 (2020)
10. Agnes, S.A., et al.: Classification of mammogram images using multiscale all convolutional neural network (MA-CNN). J. Med. Syst. **44**(1), 1–9 (2020)
11. Yi, D., et al.: Optimizing and visualizing deep learning for benign/malignant classification in breast tumors. arXiv preprint arXiv:1705.06362 (2017)
12. Li, H., et al.: Classification of breast mass in two' view mammograms via deep learning. IET Image Processing (2020)
13. Hamidinekoo, A., et al.: Comparing the performance of various deep networks for binary classification of breast tumours. In: 14th International Workshop on Breast Imaging (IWBI 2018), vol. 10718. International Society for Optics and Photonics (2018)
14. Chorev, M., et al.: The case of missed cancers: applying AI as a radiologist's safety net. In: Martel, A.L., et al. (eds.) MICCAI 2020. LNCS, vol. 12266, pp. 220–229. Springer, Cham (2020). https://doi.org/10.1007/978-3-030-59725-2_22
15. Akselrod-Ballin, A., et al.: Predicting breast cancer by applying deep learning to linked health records and mammograms. Radiology **292**(2), 331–342 (2019)
16. Yala, A., et al.: A deep learning model to triage screening mammograms: a simulation study. Radiology **293**(1), 38–46 (2019)
17. Rodriguez-Ruiz, A., et al.: One-view digital breast tomosynthesis as a stand-alone modality for breast cancer detection: do we need more? Eur. Radiol. **28**(5), 1938–1948 (2018)
18. Kyono, T., Gilbert, F.J., van der Schaar, M.: Improving workflow efficiency for mammography using machine learning. J. Am. Coll. Radiol. **17**(1), 56–63 (2020)
19. Shen, Y., et al.: An interpretable classifier for high-resolution breast cancer screening images utilizing weakly supervised localization. Med. Image Anal. **68**, 101908 (2021)
20. https://medphys.royalsurrey.nhs.uk/omidb/
21. Szegedy, C., et al.: Inception-v4, inception-resnet and the impact of residual connections on learning. In: Proceedings of the AAAI Conference on Artificial Intelligence, vol. 31(1) (2017)
22. Ness, L., Barkan, E., Ozery-Flato, M.: Improving the performance and explainability of mammogram classifiers with local annotations. In: Cardoso, J., et al. (eds.) IMIMIC/MIL3ID/LABELS -2020. LNCS, vol. 12446, pp. 33–42. Springer, Cham (2020). https://doi.org/10.1007/978-3-030-61166-8_4
23. IBM Research, H.: Fusemedml: https://github.com/ibm/fuse-med-ml (2021). https://doi.org/10.5281/ZENODO.5146491. https://zenodo.org/record/51464

24. Cohen, J.: A coefficient of agreement for nominal scales. Educ. Psychol. Measure. **20**(1), 37–46 (1960)
25. Elmore, J.G., et al.: Diagnostic concordance among pathologists interpreting breast biopsy specimens. Jama **313**(11), 1122–1132 (2015)
26. Waugh, S.A., et al.: Magnetic resonance imaging texture analysis classification of primary breast cancer. Eur. Radiol. **26**(2), 322–330 (2016)
27. Harrison, B.T., et al.: Quality assurance in breast pathology: lessons learned from a review of amended reports. Arch. Pathol. Lab. Med. **141**(2), 260–266 (2017)
28. Dillon, M.F., et al.: Diagnostic accuracy of core biopsy for ductal carcinoma in situ and its implications for surgical practice. J. Clin. Pathol. **59**(7), 740–743 (2006)

Co-segmentation of Multi-modality Spinal Image Using Channel and Spatial Attention

Yaocong Zou[1]([✉]) and Yonghong Shi[1,2,3]([✉])

[1] Shanghai Medical College, Fudan University, Shanghai 200032, China
{yczou17,yonghong.shi}@fudan.edu.cn
[2] Digital Medical Research Center, School of Basic Medical Sciences, Fudan University,
Shanghai 200032, China
[3] Shanghai Key Laboratory of Medical Imaging Computing and Computer Assisted
Intervention, Shanghai 200032, China

Abstract. Clinicians usually examine and diagnose patients with multimodality images such as CT and MRI because different modality data of the same anatomical structure are often complementary. This can provide doctors with a variety of information and help doctors to make accurate diagnoses. Inspired by this, the paper proposes a novel method of collaborative spinal segmentation based on spinal CT and MRI images. We use Siam network as architecture and ResNet50 as backbone network to extract high-level semantic features and low-level detail features of two modal images at the same time. Firstly, the high-level feature is enhanced by expanding the receptive field, and then it is input into the channel and spatial attention structure to achieve the optimal combination of high-level semantic information with the help of average pooling and maximum pooling, and learn the mutual information between different modal images. The learned high-level semantic correlation of different modalities will be combined with the up-sampled low-level features for maintaining the uniqueness of their respective modality, and finally the spinal segmentation results of the two modal images will be obtained at the same time. The experimental results show that the performance of multimodal co-segmentation is better than that of single-modal co-segmentation and ResNet50 segmentation. All codes and data described are available at: https://github.com/1shero/CoSeg_CSA_MLMI.

Keywords: Co-segmentation · Channel and spatial attention · Spinal image

1 Introduction

Spinal image segmentation is an important research field of computer-aided medicine. An good segmentation method can help doctors diagnose spinal lesions quickly and accurately and avoid long-term compression of spinal cord and nerve roots caused by further examination.

There are many spinal image segmentation methods in the literature. R. Windsor et al. [1] proposed that the learned vector field and probability graph can be used to obtain the overall information of the multi-vertebrae spinal MRI image. S. Zhao et al.

© Springer Nature Switzerland AG 2021
C. Lian et al. (Eds.): MLMI 2021, LNCS 12966, pp. 287–295, 2021.
https://doi.org/10.1007/978-3-030-87589-3_30

[2] proposed a discriminant dictionary embedding network, which achieved good results in image recognition, classification and tumor identification of the spinal MRI. H. Chang et al. [3] proposed a label attention network, which improved the performance of spine MR segmentation by multiplying the generated label probability vector with the output. N. Masuzawa et al. [4] used cascaded convolution neural network to integrate the segmentation, localization and recognition of spinal CT images into two stages and obtained good results.

However, these methods consider only a single modal image of the spine. Different modal images of the spine have their own advantages and disadvantages. CT shows the clearest imaging of bones, but it is less effective in imaging small coincident parts (such as nerve foramen) and soft tissue components (such as intervertebral discs). MRI has a good imaging effect on soft tissue such as intervertebral disc and nerve root, but the hydrophobic tissue such as bone structure is darker and poor contrast on MRI. There is an inherent relationship between different modal data, which is often ignored by the segmentation method based on single modal image. At present, the method of combining the complementary features of multimodal images has not been widely studied.

In the field of image segmentation, co-segmentation (or collaborative segmentation) was first proposed by C. Rother et al. [5], which means to segment two images containing the same or similar objects at the same time. Cui et al. [6] proposed a segmentation method by learning a single image and "transforming" the segmentation results to a group of related images. B. Li et al. [7] further proposed to use common attention recursive unit (Co-Attention Recurrent Unit, CARU) to represent the features of a group of images to achieve collaborative segmentation.

According to the application of co-segmentation in computer vision, this paper proposes a semantic collaborative spine segmentation model based on Siam network. We assume that the vertebrae of different modal images (i.e. CT and MRI) have similar characteristics in higher resolution which can be identified and utilized by neural network. Firstly, the different modal images will be input into the convolution neural network with 3D ResNet50 as the backbone at the same time, and the high-resolution semantic features and low-resolution detail features of each image will be extracted simultaneously. Then, the receptive field of the high-level feature is expanded, and it is input into the channel and spatial attention structure to realize the optimal combination of the high-level semantic features of the two modal images with the help of average pooling and maximum pooling, and learn the mutual information between the two images at the same time. The optimal combination features learned will be combined with the upsampled low-resolution features that keep the modal uniqueness. After the combination results are consolidated, the respective segmentation results are finally obtained at the same time. Experiments on the open datasets show that the performance of multi-modal collaborative segmentation is better than that of single-modal segmentation.

Contributions. We verified that dual-input co-segmentation has better performance than single-input ResNet backbone baseline. Meanwhile, we find that multi-modal dual-input co-segmentation can achieve better precision than single-model dual-input co-segmentation, which indicates that CSA structure has the potential to fuse the high dimensional features of CT and MR images.

2   Method

Figure 1A shows the proposed approach architecture based on the Siamese network. The spine CT and MRI images are simultaneously input into the encoding part of the respective backbone network, i.e., the 3D ResNet50, to extract different levels of features. In particular, the last level of feature is input to atrous spatial pyramid pooling (ASPP) to increase the receptive field. The results are input into channel and spatial attention (CSA) to fuse the channel and spatial information of the high-level semantic features of the two modal images. The fusion results are returned to their respective decoding parts and combined with the up-sampled low-level features that maintain their respective personalities. After the combination results are consolidated respectively, the corresponding segmentation results are obtained at the same time. Below, the four parts that make up the method, namely, the encoding part, the ASPP, the CSA and the decoding part, are described in detail.

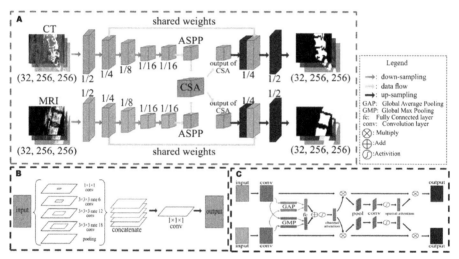

Fig. 1. Illustration of the proposed approach framework in (A), the atrous spatial pyramid pooling (ASPP) in (B), and the channel and spatial attention (CSA) in (C).

The Encoding Part. The proposed method is based on the Siamese network. The Siam network consists of two identical neural networks, each of which receives one of the two input images and calculates the corresponding high-level features and low-level features. Here, each neural network takes ResNet50 [8] as the backbone network. ResNet50 mainly calculates the features of two-dimensional images. In order to extract the features of three-dimensional spinal images, it is modified to process three-dimensional data, while keeping the structure and convolution kernel number of each stage unchanged. In this paper, the paired CT and MRI image blocks of fixed size ($32 \times 256 \times 256$) are input into the network, and they go through five stages of convolutional block processing (the strides are 1/2, 1/2, 1/2, 1/2, and 1, respectively), the extracted feature blocks with the size of 1/16 input image block are obtained. These features will be imported into the ASPP module for enhancement.

Atrous Spatial Pyramid Pooling. Atrous spatial pyramid pooling (ASPP) was proposed by L.C. Chen et al. in 2018 [9]. The basic structure of ASPP consists of atrous convolution (i.e., dilation convolution), which uses the dilation rate to upsample the original filter and introduces zeros between the filter values [10]. Its characteristic is that it can get a larger receptive field without changing the size of feature mapping and the amount of computation. We change its original two-dimensional form to the three-dimensional one so that it can process three-dimensional data. As shown in Fig. 1B, the ASPP consists of a convolution layer with the filter size of $1 \times 1 \times 1$ and 256 channels, 3 convolution layers each of which is composed of the filter size of $3 \times 3 \times 3$ and 256 channels (but each convolution layer has a dilation rate of 6, 12 and 18, respectively), and a global pooling layer. The parallel extracted features are successively concatenated and upsampled, and finally restored to the original image size. It is then input to the convolution layer with the filter size of $1 \times 1 \times 1$ and 256 channels to get output. When the input features are combined by multiple dilation convolution with different dilation rates, multi-scale information can be extracted, thus improving the performance of the network model in the segmentation task.

Channel and Spatial Attention. As shown in Fig. 1C, Channel and spatial attention (CSA) is composed of channel attention and spatial attention, which was proposed by J. Chen in 2020 [11]. In the channel attention part, the two inputs are pooled by global maximum pooling (GMP) and global average pooling (GAP), and two vectors with a length of 256 are obtained, respectively. Then the two vectors obtained by GMP and the two vectors obtained by GAP are concatenated and input to two full connection layer with 256 nodes, respectively, and the outputs of the two full connection layers are added together to get the channel attention (note that the number of nodes in the full connection layer is the same as the number of input channels, which keeps the length of the output vector unchanged, so that the resulting attention can be directly multiplied by the input). Since the two inputs share the same channel attention, multiply with it, the characteristics enhanced by the channel attention is gotten. Spatial attention separately maximizes and averages the depth-oriented pooling of the features enhanced by the channel attention. After the two pooling results are stacked in depth, it is processed by a single-channel convolution layer with the filter size of 3×3, and the output is obtained. Finally, the two channel-enhanced features generate two spatial attentions which multiply with the channel-enhanced features to get the two enhanced features, respectively.

The Decoding Part. Each enhanced feature output by the CSA needs to be upsampled to 4 times its size, making the number of channels 256. This will be combined with the low-level detail feature (with 256 channels) output in the second stage of the encoding part. In order to increase the accuracy of the segmentation result, instead of directly upsampling 4 times, a convolution layer with the filter size of $3 \times 3 \times 3$ and 64 channels is used to upsample the enhanced features output by the CSA twice, and finally make the number of channels of the output features reach 256. When combining with low-level features, in order to avoid confusion caused by too many low-level features, we first use a convolution layer with the filter size of $3 \times 3 \times 3$ and 48 channels to deal with the low-level features, and the results are concatenated with the enhanced features [9]. Then, the concatenated results will be consolidated successively using two convolution layers with the filter size of $3 \times 3 \times 3$ and 256 channels. Finally, the integration result will be upsampled to the size of the original input image, and processed by a single channel convolution layer with the filter size of $1 \times 1 \times 1$, and the segmentation result will be obtained.

Loss Function. Given the CT and MR images of any spine with the same vertebra distribution, the ResNet50 network uses the following formula to calculate the binary cross entropy loss (take CT as an example),

$$loss_{CT} = -\frac{1}{h \times w \times d} \sum_{i=1}^{h \times w \times d} y_i \cdot \log \hat{y}_i + (1 - y_i) \cdot \log(1 - \hat{y}_i), \qquad (1)$$

where y represents the predicted segmentation result of the CT (or MRI) image, and its height, width and depth are h, w and d, respectively. $\hat{y}$ is the label image. $\hat{y}_i$ is the label value of the ith pixel in the image, and y_i is the predicted value of the ith pixel in the image.

For our complete method, the spine CT and MRI images are input into the two ResNet50 networks and the segmentation results are obtained simultaneously. We take the mean of the binary cross entropy of the two networks as the final loss of the whole network, which is defined as follows,

$$loss = 0.5 \times loss_{CT} + 0.5 \times loss_{MRI}. \qquad (2)$$

3   Experimental Results

Dataset. In the experiment, we use the CT dataset from xVerSeg Challenge [12] and the MR dataset from [6]. The former has 15 CT images, mainly including part of the lumbar vertebrae of L1–L5 and the corresponding label. The latter also includes part of the lumbar spine of L1–L5, but there is no label. We select 5 images with T1 FLAIR parameters and label them manually. To evaluate the method properly and improve reproducibility, we use the k-fold cross-validation method in the experiment. We divide the data into 5 parts, one of which is used as the validation set for each training, and the rest is used as the training set. Precision and accuracy are used as evaluation measures for the experimental results. Precision measures the voxels that have been segmented correctly, and accuracy estimates how close the predicted results are to the real values as a whole.

Image Preprocessing. Due to cost and time, the original image is reduced to half of the original size. Then, we randomly crop the image into blocks with the size of $32 \times 256 \times 256$, and adjust window and level to show the vertebrae clearly. For CT, the window level is 1100 and the window width is 500; for MRI, the window level is 600 and the window width is 300. Finally, the data is normalized so that the average intensity is 0 and the standard deviation is 1.

Implementation Details. During the training process, the fitting batch size is 2, the Adam is used as the optimizer, and the initial learning rate is set to 0.001. If the loss value does not decrease on 3 consecutive epoch validation sets, the learning rate is multiplied by 0.1. The device used is NVIDIA Tesla V100 (video memory 32 GB, physical memory 64 GB, CUDA version 10.2), and the development environment is Huawei ModelArts. The training model and image processing program are implemented under TensorFlow2.3, Keras2.4.3 and Python3.6.

Results. In order to evaluate the performance of the proposed method, we use the ResNet50 network as the benchmark. In other words, any CT or MRI is input into the ResNet50 network and gone through the operation of encoding part, ASPP module and decoding part, and finally the segmentation result is output, which will be used to compare with our method. We performed 5 groups of experiments: single-input CT, single-input MR, double-input CT/CT, double-input MRI/MRI and double-input CT/MRI. 5-fold cross validation was carried out in each group, and a total of 25 experiments were carried out. The experimental results are shown in Table 1, and the data are the average of five experiments.

Table 1. Using the precision and accuracy to evaluate the performance of the proposed co-segmentation method by comparison with the ResNet50 network (mean ± standard deviation).

	CT/MRI-CT	CT/MRI-MRI	CT/CT	MRI/MRI	CT (ResNet)	MRI (ResNet)
Precision	0.740 ± 0.100	0.717 ± 0.208	0.678 ± 0.061	0.709 ± 0.241	0.673 ± 0.051	0.656 ± 0.221
Accuracy	0.950 ± 0.049	0.918 ± 0.032	0.983 ± 0.004	0.864 ± 0.134	0.982 ± 0.007	0.907 ± 0.045

As can be seen from Table 1, for spinal CT images, the average precisions of single-input CT, double-input CT/CT and CT/MRI-CT are 0.673, 0.678 and 0.740, respectively, and the average accuracies are 0.982, 0.983 and 0.950, respectively. For spinal MRI images, the average precisions of single-input MRI, double-input MRI/MRI and CT/MRI-MRI are 0.656, 0.709 and 0.717, respectively, and the average accuracies are 0.907, 0.864 and 0.918, respectively. These data show that the performance of co-segmentation of multi-modal images is slightly better than that of single image segmentation, and the precision of multi-modal co-segmentation is higher than that of single-modal co-segmentation. The performance improvement of collaborative segmentation may be related to the complementary characteristics of the two data input from two networks simultaneously. The precision of multi-modal co-segmentation is better than that of single-modal co-segmentation, probably because the features extracted from CT and MRI are synthesized by CSA module, thus the performance of both is improved. In addition, in the same input mode (i.e., double-input CT/CT or MRI/MRI), the standard deviation of MRI is higher than that of CT, which indicates that the segmentation performance of our method is unstable when segmenting MRI data, which may be related to the lack of MRI data or the influence of random clipping on MRI data.

After applying the threshold of 0.4 to the segmentation result, the segmentation result is visually displayed in Figs. 2 and 3. Figure 2 shows the segmentation results of any CT image in sagittal, axial and coronal view. The original CT image and its segmentation ground truth, as well as the CT segmentation results of single-input CT, double-input CT/CT and double-input CT/MRI are displayed from top to bottom, respectively. Figure 3 shows the segmentation results of any MRI image in sagittal, axial and coronal view. The original MRI image and its segmentation ground truth, as well as the MRI segmentation results of single-input MRI, double-input MRI/MRI and double-input CT/MRI are displayed from top to bottom, respectively. As can be seen from the figures, because the CSA module enhances the high-level semantic features of multi-modal images, the ability of the network to deal with some regions with low gray values and easy to be confused increases, which makes the prediction results of our method closer to the real values. In our method, the main reasons for the errors in the segmentation of spinal MRI may be that the high gray value of spinal MRI interferes with the segmentation, and the distinguishing ability of low gray value leads to false positive or false negative value.

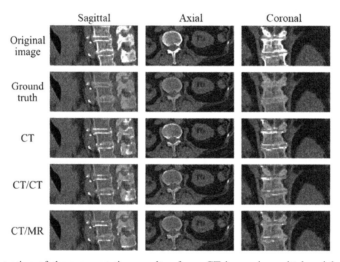

Fig. 2. Illustration of the segmentation results of any CT image in sagittal, axial and coronal planes. The original CT image and its segmentation ground truth, as well as the CT segmentation results of single-input CT, double-input CT/CT and double-input CT/MRI are displayed from top to bottom, respectively.

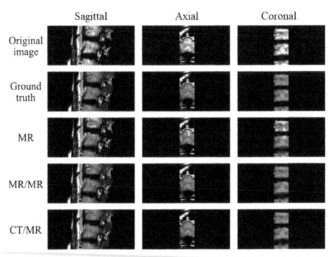

Fig. 3. Illustration of the segmentation results of any MRI image in sagittal, axial and coronal planes. The original MRI image and its segmentation ground truth, as well as the MRI segmentation results of single-input MRI, double-input MRI/MRI and double-input CT/MRI are displayed from top to bottom, respectively.

4 Conclusion

This paper proposed a co-segmentation method based on Siamese network for segmenting multimodal spinal images. The 3D ResNet50 was used as the backbone network. ASPP was used to enhance the extracted features, and the enhanced features were further input into the CSA module for feature enhancement and cross-correlation learning to get enhanced high-level features. The low-level features extracted by backbone and the high-level features enhanced by CSA were combined, and the segmentation results were obtained after upsampling and integration. The experimental results showed that co-segmentation has higher accuracy than single image segmentation, and multi-modal co-segmentation has higher accuracy than single-modal co-segmentation. In the future, our method can be improved from the following three aspects: Firstly, to find a better backbone network that matches the co-segmentation network. Secondly, expand the spinal MRI data set, or use public datasets of other anatomical structures. Thirdly, use images with higher resolution to obtain finer segmentation results.

Acknowledgement. This study is supported by the Medical-Industrial Integration Project of Fudan University.

References

1. Windsor, R., Jamaludin, A., Kadir, T., Zisserman, A.: A convolutional approach to vertebrae detection and labelling in whole spine MRI. In: Martel, A.L., et al. (eds.) MICCAI 2020. LNCS, vol. 12266, pp. 712–722. Springer, Cham (2020). https://doi.org/10.1007/978-3-030-59725-2_69

2. Zhao, S., Chen, B., Chang, H., Wu, X., Li, S.: Discriminative dictionary-embedded network for comprehensive vertebrae tumor diagnosis. In: Martel, A.L., et al. (eds.) MICCAI 2020. LNCS, vol. 12266, pp. 691–701. Springer, Cham (2020). https://doi.org/10.1007/978-3-030-59725-2_67

3. Chang, H., Zhao, S., Zheng, H., Chen, Y., Li, S.: Multi-vertebrae segmentation from arbitrary spine MR images under global view. In: Martel, A.L., et al. (eds.) MICCAI 2020. LNCS, vol. 12266, pp. 702–711. Springer, Cham (2020). https://doi.org/10.1007/978-3-030-59725-2_68

4. Masuzawa, N., Kitamura, Y., Nakamura, K., Iizuka, S., Simo-Serra, E.: Automatic segmentation, localization, and identification of vertebrae in 3D CT images using cascaded convolutional neural networks. In: Martel, A.L., et al. (eds.) MICCAI 2020. LNCS, vol. 12266, pp. 681–690. Springer, Cham (2020). https://doi.org/10.1007/978-3-030-59725-2_66

5. Rother, C., et al.: Cosegmentation of image pairs by histogram matching-incorporating a global constraint into MRFS. In: 2006 IEEE Computer Society Conference on Computer Vision and Pattern Recognition (CVPR 2006), vol. 1, pp. 993–1000 (2006)

6. Cai, Y., et al.: Multi-modality vertebra recognition in arbitrary views using 3D deformable hierarchical model. IEEE Trans. Med. Imaging 34(8), 1676–1693 (2015)

7. Li, B., et al.: Group-wise deep object co-segmentation with co-attention recurrent neural network. In: Proceedings of the IEEE/CVF International Conference on Computer Vision 2019, pp. 8519–852 (2019)

8. He, K., et al.: Deep residual learning for image recognition. In: Proceedings of the IEEE Conference on Computer Vision and Pattern Recognition 2016, pp. 770–778 (2016)

9. Chen, L.-C., Zhu, Y., Papandreou, G., Schroff, F., Adam, H.: Encoder-decoder with atrous separable convolution for semantic image segmentation. In: Ferrari, V., Hebert, M., Sminchisescu, C., Weiss, Y. (eds.) ECCV 2018. LNCS, vol. 11211, pp. 833–851. Springer, Cham (2018). https://doi.org/10.1007/978-3-030-01234-2_49

10. Chen, L.-C., et al.: Deeplab: Semantic image segmentation with deep convolutional nets, atrous convolution, and fully connected CRFs. IEEE Trans. Pattern Anal. Mach. Intell. 40(4), 834–848 (2017)

11. Chen, J., et al.: Channel and spatial attention based deep object co-segmentation. Knowl. Based Syst. 211, 106550 (2021)

12. The xVertSeg Challenge. http://lit.fe.uni-lj.si/xVertSeg/. Accessed 03 Mar 2020

Hetero-Modal Learning and Expansive Consistency Constraints for Semi-supervised Detection from Multi-sequence Data

Bolin Lai[1]([✉]), Yuhsuan Wu[1], Xiao-Yun Zhou[2], Peng Wang[4], Le Lu[2], Lingyun Huang[1], Mei Han[3], Jing Xiao[1], Heping Hu[4], and Adam P. Harrison[2]

[1] Ping An Technology, Shanghai, China
[2] PAII Inc., Bethesda, MD, USA
[3] PAII Inc., Palo Alto, CA, USA
[4] Eastern Hepatobiliary Surgery Hospital, Shanghai, China

Abstract. Lesion detection serves a critical role in early diagnosis and has been well explored in recent years due to methodological advances and increased data availability. However, the high costs of annotations hinder the collection of large and completely labeled datasets, motivating semi-supervised detection approaches. In this paper, we introduce mean teacher hetero-modal detection (MTHD), which addresses two important gaps in current semi-supervised detection. *First*, it is not obvious how to enforce unlabeled consistency constraints across the very different outputs of various detectors, which has resulted in various compromises being used in the state of the art. Using an anchor-free framework, MTHD formulates a mean teacher approach without such compromises, enforcing consistency on the soft-output of object centers and size. *Second*, multi-sequence data is often critical, *e.g.*, for abdominal lesion detection, but unlabeled data is often missing sequences. To deal with this, MTHD incorporates hetero-modal learning in its framework. Unlike prior art, MTHD is able to incorporate an expansive set of consistency constraints that include geometric transforms and random sequence combinations. We train and evaluate MTHD on liver lesion detection using the largest MR lesion dataset to date (1099 patients with > 5000 volumes). MTHD surpasses the best fully-supervised and semi-supervised competitors by 10.1% and 3.5%, respectively, in average sensitivity.

Keywords: Semi-supervised detection · Mean teacher · Hetero-modal learning

1 Introduction

Lesion detection is a fundamental task in medical imaging, as an end goal [6] or as a critical step for computer-aided diagnosis (CAD) [17]. This puts great

Electronic supplementary material The online version of this chapter (https://doi.org/10.1007/978-3-030-87589-3_31) contains supplementary material, which is available to authorized users.

C. Lian et al. (Eds.): MLMI 2021, LNCS 12966, pp. 296–305, 2021.
https://doi.org/10.1007/978-3-030-87589-3_31

impetus on developing powerful lesion detectors and there are many successful deep-learning efforts [8,10,12,22,24]. Because of the data-driven nature of deep learning, they rely on a large number of manually annotated images. But annotating bounding boxes (bboxes) is labor-intensive and time-consuming, requiring roughly 15 minutes per study [8]. A highly promising alternative is to use localization annotations found within hospital picture archiving and communication systems (PACSs) [25], but such annotations are not typically recorded and are incomplete even when present [5]. Thus, much like other medical imaging applications [18], lesion detection requires effective techniques for imperfect labels, particularly ones that can best handle unlabeled and heterogeneous PACS data, which is often the only realistic source of large-scale data.

Most semi-supervised learning (SSL) methods are designed for classification [2,11,19] or segmentation [15,28], and SSL detection is understudied. Unique to detection, its outputs are not easily made consistent, where even the number of predicted bboxes can differ. This has prevented a general framework for enforcing detector consistency. STAC [16] and unbiased teacher [14] offer two good examples, both of which generate pseudo bbox labels to get around this issue. However, pseudo-labels are generated by hard thresholding detection confidences, which can amplify noise near the margin. Also detection confidence does not necessarily filter good bbox size predictions, which is why unbiased teacher avoids using bbox size consistency [14]. Finally, because bboxes are restricted to have horizontal and vertical edges, geometric transforms change their widths and heights, meaning it is actually incorrect to enforce SSL consistency after such transformations. CSD [9] avoids it by only enforcing consistency on horizontally flipped counter-parts, but this is a highly limited augmentation scheme. There are successful domain specific techniques, *e.g.*, for lung nodules [20] or fracture regression [21], but these are not generally applicable for lesion detection.

Another challenge is dealing with multi-sequence data, which is particularly important in abdominal studies [4]. Early fusion (EF) is the most straightforward approach, which concatenates sequences together as multi-channel input. However, PACS data is frequently heterogeneous, *i.e.*, missing sequences/contrast phases [15]. Given the expense of annotations, studies with all sequences available are typically chosen for annotation and missing-sequences studies remain unlabeled. EF would abandon such studies, which wastes valuable medical data.

We introduce mean teacher hetero-modal detection (MTHD)—an SSL and multi-sequence detector that addresses the above gaps. We build MTHD off of the anchor-free CenterNet framework [27], a leading solution for lesion detection [5,23]. MTHD incorporates hetero-modal learning [7], which naturally can handle challenging unlabeled PACS data with missing sequences. Moreover, exploiting CenterNet's heatmap based predictions, MTHD uses a semi-supervised formulation that enforces *valid* consistency across "centered-ness" and bbox size, using an *expansive and rich* set of SSL consistency transforms that include intensity, geometric, and hetero-modal sequence combinations. We integrate these constraints within a mean teacher framework [19]. Experiments employ a multi-sequence dataset of 1099 liver magnetic resonance (MR) studies, 70% of which are unlabeled and heterogeneous. We evaluate on 430 labeled

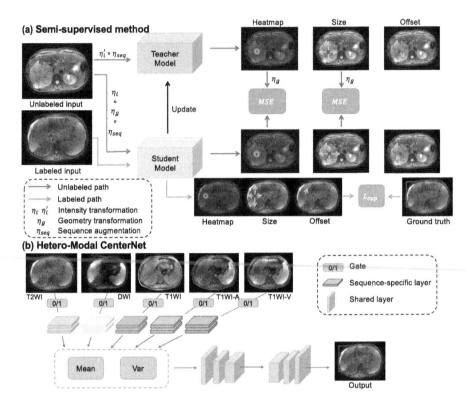

Fig. 1. The proposed (a) semi-supervised framework and (b) Hetero-Modal CenterNet.

studies via cross-validation. When tested on labeled data, MTHD improves the average sensitivity by 10.1% compared to the best fully-supervised alternative. MTHD also outperforms leading SSL approaches [11,14,16] by significant margins, *e.g.*, 3.5% over the best competitor [14].

2   Method

Figure 1 illustrates the MTHD approach. We assume we are given a multi-sequence dataset consisting of M labeled patients $\mathcal{S} = \{\mathcal{X}_i, \mathcal{Y}_i\}_{i=1}^{M}$ and N unlabeled patients $\mathcal{U} = \{\mathcal{X}_i\}_{i=1}^{N}$, where $\mathcal{X}_i$ and $\mathcal{Y}_i$ denote sets of sequences and bbox annotations, respectively. We do not assume that every study has all sequences. This is particularly likely for unlabeled data, which we can divide up into complete and incomplete studies: $\mathcal{U} = \mathcal{U}_{\text{cplt.}} \bigcup \mathcal{U}_{\text{incplt.}}$.

2.1   Hetero-Modal CenterNet

Like many SSL approaches, MTHD is built off of a fully-supervised foundation. For reasons elaborated in Sect. 2.2, we opt for an anchor-free fully-convolutional network (FCN) detector. While there are several viable options, we choose CenterNet [27], which is a leading lesion detector [5,23]. Very briefly, instead of anchors, CenterNet encodes ground-truth locations by splatting a Gaussian kernel at the center of each object onto a heatmap, $Y \in \mathcal{R}^{H/R \times W/R}$ where H and W are height and width of input, and R is the FCN downsampling factor. A chosen FCN is used to generate a predicted heatmap, $\hat{Y}$, which is encouraged to match Y using a focal-style loss. In addition to this "centeredness" heatmap, the FCN also outputs bbox size, $\hat{D} \in \mathcal{R}^{H/R \times W/R \times 2}$, and offset, $\hat{O} \in \mathcal{R}^{H/R \times W/R \times 2}$, activation maps, which are trained using an L1 regression loss, but only on ground-truth center locations. See Zhou et al. [27] for more details.

To fully exploit all unlabeled data, especially those found within PACSs, a detector must be flexible enough to handle studies with missing sequences. To do this, we adapt the CenterNet architecture for hetero-modal learning [7]. As Fig. 1(b) illustrates, if we have k sequences, we construct a set of sequence-specific layers early in the FCN backbone, producing a set of sequence-specific intermediate activation maps $\mathcal{A} = \{a_{ijk}^1, \ldots a_{ijk}^k\}$. Here we assume 2D inputs to create 3D activations. Since some sequences may be missing, this set of activation maps should be aggregated with any operator that can accept a variable number of inputs. As argued by Havaei et al. [7], the mean and variance are excellent choices, since their expected values should be constant across different numbers of inputs. If the output activation is denoted b_{ijk}, this can be expressed as

$$b_{ijk} = \mathrm{mean}(\mathcal{A}) \oplus \mathrm{var}(\mathcal{A}), \tag{1}$$

where $\oplus$ as channel-wise concatenation. After (1), a shared set of layers processes b_{ijk} just like any other standard FCN. During training, random combinations of sequences are chosen, while in inference all available sequences are used to produce the most confident prediction. In addition to being able to handle heterogeneous learning, hetero-modal learning can act as a powerful form of data augmentation. As we show later, this can even boost performance in fully-supervised setups when all sequences are available. Additionally, hetero-modal learning can provide another form of SSL consistency transformation.

2.2 Semi-supervised Detection

As shown in Fig. 1(a), the proposed semi-supervised detector is based off of the popular mean teacher framework [19]. The labeled images, $\mathcal{S}$, are first used to train a hetero-modal CenterNet, denoted $f_\theta(\mathcal{X}_i)$. The resulting network weights are used to initialize teacher and student parameters, θ_t and θ_s, respectively. The student model is then trained by transforming unlabeled data via strong augmentations, denoted η, and penalizing any inconsistencies with the teacher model:

$$\mathcal{L} = \sum_{\{\mathcal{X}_i, \mathcal{Y}_i\} \in \mathcal{S}} \ell_{sup}(f_{\theta_s}(\mathcal{X}_i), \mathcal{Y}_i) + \sum_{\mathcal{X}_i \in \mathcal{U}} \ell_{cons}(f_{\theta_s}(\mathcal{X}_i, \eta), f_{\theta_t}(\mathcal{X}_i, \eta')), \tag{2}$$

where $\ell_{sup}(.)$ is the standard CenterNet loss [27], $\ell_{cons}(.)$ is some consistency loss, and $f_\theta(.)$ has also been modified to accept the augmentations. Only the student model is updated by backpropagation. On the other hand, the "mean teacher" parameters are updated using a moving average of the current student parameters:

$$\theta_t^k = \alpha\theta_t^{k-1} + (1-\alpha)\theta_s^k, \tag{3}$$

where α denotes the moving average rate.

The challenge with SSL detection is how to construct the strong data augmentations and a consistency loss. To avoid difficulties with enforcing consistency across detection outputs, prior student/teacher approaches have relied on using teacher pseudo labels [14,16], but this only allows for hard labels and requires filtering out spurious teacher outputs using a confidence threshold, which does not necessarily filter out poor bbox size predictions [14]. Moreover, geometric augmentations are not easily applied to bboxes, since a rotation changes their shape. Liu *et al.* [14] avoid geometric transforms in their mean teacher approach, but given the importance of well-chosen data augmentations [19], this is critical limitation. Here is where the adoption of an anchor-free FCN-style framework, like CenterNet [27], provides important benefits. For one, unlike anchor-based frameworks, consistency can be straightforwardly enforced on FCN outputs, avoiding the complexities of generating pseudo-labels and allowing for soft constraints, similar to what is done in SSL classification and segmentation. Moreover, CenterNet's Gaussian kernel-based heatmap, $\hat{Y}$, is equivariant with rotations, meaning such geometric transforms are not problematic. This simple insight can allow for important performance gains.

Additionally, because MTHD is hetero-modal, it can apply missing sequence augmentations, η_{seq}, in addition to intensity and geometric ones, η_i and η_g. Like Liu *et al.* [14], we only apply "weak" transformations to the teacher, which in our case foregoes geometric transformations. Instead we transform the *output* of the teacher model with the same geometric transform used on the *input* of the student model. This scheme avoids the danger of a geometric transform occluding a bbox in the teacher input that is visible to the student, which would produce misleading supervisory signals. Formulaically, this is expressed as

$$\hat{Y}_s, \hat{D}_s, \hat{O}_s = f_{\theta_s}(\eta_i \circ \eta_g \circ \eta_{seq} \circ \mathcal{X}_i), \tag{4}$$

$$\hat{Y}_t, \hat{D}_t, \hat{O}_t = f_{\theta_t}(\eta_i' \circ \eta_{seq} \circ \mathcal{X}_i), \tag{5}$$

$$\ell_{cons} = MSE(\hat{Y}_s, \eta_g \circ \hat{Y}_t) + \lambda\, MSE(\hat{D}_s, \eta_g \circ \hat{D}_t), \tag{6}$$

where consistency is enforced using the mean-squared error (MSE). $\hat{Y}$, $\hat{D}$ and $\hat{O}$ are heatmap, bbox size and offset, respectively, as mentioned in Sect. 2.1. Here, the student and teacher use the same set of random sequences, η_{seq}, because we found training was unstable otherwise. Note that no consistency is used on offset, $\hat{O}$, because it is a local refinement, which is quite different before and after geometric transforms, so it is not correct to enforce consistency across those outputs.

3   Results

Data and Setup. We tested MTHD on liver lesion detection from multi-sequence MR studies. We collected 1099 studies from Eastern Hepatobiliary Surgery Hospital. Each *ideally* comprises T1-weighted imaging (T1WI), T2-weighted imaging (T2WI), venous-phase T1WI (T1WI-V), arterial-phase T1WI (T1WI-A), and diffusion-weighted imaging (DWI) MR sequences, but some studies are missing at least one sequence. The use of multi-sequence data is critical for clinical detection and characterization [1]. Any patients with both radiological and pathological reports between 2006 and 2019 were included in the dataset, 430 of which, with all five sequences available, were annotated with 2D bboxes on each slice under the supervision of a hepatic physician with > 10 years experience. Of the remaining 669 unlabeled studies, 495 have all five sequences ($\mathcal{U}_{\text{cplt.}}$) and 174 were missing at least one sequence ($\mathcal{U}_{\text{incplt.}}$). Labeled data was split using five-fold cross validation with 70%, 10% and 20% as training, validation, and test sets, respectively. The unlabeled data was used to augment the training data for each fold. Thus, the training set contains 969 studies (300 labeled and 669 unlabeled studies) in each fold, where labeled data accounts for $\approx 30\%$. However, as per prior work [2,9,14,16], SSL is typically developed and evaluated with a more severe ratio of labeled data (usually $<= 10\%$) because it is a more realistic environment and better reveals differences between competitor approaches. Hence, we have three experimental settings: (1) training models with a 10% label ratio, where we randomly treat a portion of labeled data as unlabeled, *i.e.*, 100 labeled and 869 unlabeled studies, to make a more illuminating comparison with baselines; (2) training models using all available 30% labeled data to show the best performance MTHD can achieve on our dataset; and (3) experiments on different ratios of labeled studies to show the generalizability of MTHD across different SSL settings. Due to page limitations, results under setting (1) are demonstrated in the main body and comparisons under setting (2) and (3) are shown in supplementary. The conclusions for (2) and (3) match (1), in that MTHD outperforms all SSL competitors and fully-supervised alternatives by large margins.

In implementation, gamma transformation was used as the intensity transformation (η_i), and random rotations, resizing, and shifting were adopted as geometric transformations (η_g). The official implementation and hyper-parameters of 2D CenterNet [27] were used in all experiments. Since up to five sequences are inputted to the model at the same time, 3D models consume too much GPU memory for practical use. The hyperparameters can be found in the supplementary. For SSL, α and λ were empirically set as 0.999 and 0.02, respectively. We measured performance using the average sensitivity of different numbers of false-positives (FPs) *per patient*, which as Cai *et al.* argue [5], is a more meaningful and challenging metric than the *per-slice* numbers often used in DeepLesion works. A prediction bounding box is seen as true positive when IoU>0.5.

Comparison with State of the Art. We compare MTHD with state of the art in Table 1. First, we compare Hetero-Modal CenterNet with three early fusion

Table 1. Comparison against state of the art. In all experiments, the results are average of five-fold cross validation on test set with 10% training data labeled.

Methods	FPs per patient					Avg.
	1	2	4	8	16	
Fully-supervised models						
3DCE [22]	17.3	22.6	29.3	36.8	43.0	29.8
CenterNet [27]	23.2	32.1	42.0	52.0	59.3	41.9
ATSS [26]	25.4	31.0	37.6	44.6	50.2	37.7
Hetero-Modal CenterNet	25.8	35.6	45.8	55.6	62.9	45.1
Semi-supervised models						
Co-training [15]	17.3	22.3	28.8	36.1	43.2	29.5
Π-Model [11]	29.4	38.4	47.5	56.4	64.6	47.3
CSD [9]	29.8	37.6	45.7	54.4	62.7	46.1
STAC [16]	30.5	40.5	50.0	57.6	63.0	48.3
Unbiased Teacher [14]	32.1	42.8	53.3	62.0	68.3	51.7
MTHD	**37.0**	**47.6**	**56.3**	**64.5**	**70.8**	**55.2**

(EF) detectors. Even though 3DCE [22] and ATSS [26] are strong baselines for DeepLesion [25] and COCO [13], respectively, CenterNet [27] outperforms them by a large margin, which validates its reported effectiveness on lesion detection [5,23]. Hetero-Modal CenterNet can boost performance even further, producing an average sensitivity of 45.1%, surpassing its EF counter-part by 3.2%. This occurs even though inference is always performed using all five MR sequences. We postulate that by randomly selecting sequences during training, hetero-modal learning can provide a form of data augmentation that benefits even non-heterogeneous inference.

We compare against several SSL methods, most of which follow a student/teacher framework [9,11,14,16]. Because we have multi-sequence data, we also compare against a co-training [3] implementation that follows Raju *et al.*'s approach [15]. *To conduct a fair comparison, we use Hetero-Modal CenterNet as the detector framework for all options.* Among prior work, unbiased teacher [14] achieves the best performance. This is not surprising because, like MTHD, it uses the mean teacher framework, which is a highly effective SSL approach. However, unbiased teacher relies on pseudo labels that are generated by hard thresholding detection confidences. This means that any inaccuracies in the teacher output near the threshold will be amplified by polarizing the labels to hard positives or negatives. Moreover, unbiased teacher does not apply a consistency loss to the bbox size regression because thresholding the detection confidence does not filter for good width and height predictions [14]. In contrast, MTHD uses a consistency loss to penalize differences between student and teacher outputs on both center locations and bounding box sizes. This can be seen as a kind of soft pseudo label, which does

Table 2. Ablation study of MTHD. Results correspond to the average of validation results across the five folds using the 10% labeled training data setting.

Methods	Intensity Trans.	Geometric Trans.	FPs per study					Avg.
			1	2	4	8	16	
2.5D Hetero-Modal CenterNet	N/A	N/A	26.3	33.0	41.0	47.8	53.9	40.4
2D Hetero-Modal CenterNet	N/A	N/A	24.9	34.8	45.0	53.4	60.5	43.7
2D+center consistency	√	×	32.9	39.2	46.0	53.1	58.9	46.0
2D+center consistency	√	√	38.8	46.1	54.1	62.1	67.7	53.8
MTHD	√	√	**39.0**	**47.1**	**56.0**	**63.0**	**69.1**	**54.9**
MTHD w/o $\mathcal{U}_{\text{incplt.}}$	√	√	34.9	41.0	47.4	53.5	58.6	47.1

not amplify noise in the teacher outputs. The significant improvement of 3.5% validates its superiority. Visual examples, seen in the supplementary, provides further qualitative validation.

Ablation Study. Table 2 presents an ablation study. In terms of fully-supervised backbone, we also tried a 2.5D variant, seen in other works [5], but its performance was inferior to the 2D version. This may be due to the large slice thickness (8 mm) of raw MR data. Once unlabeled data is included, it can be seen that enforcing consistency on the centeredness heatmap, $\hat{Y}$, improves the average sensitivity by 2.3% when only using intensity transformations. Geometric transformations further boost the score from 46.0% to 53.8%, highlighting how critical strong data augmentations are to mean teacher frameworks. Adding consistency on bounding-box sizes then completes MTHD, which adds another 1.1% improvement. Finally, to demonstrate the importance of using all unlabeled data, we train MTHD without studies having missing sequences ($\mathcal{U}_{\text{incplt.}}$), which results in a considerable decrease in performance. This underscores that using all available data, even if imperfect, can be critical for performance. As such, this further validates our use of hetero-modal learning as a foundation for SSL detection on multi-sequence data.

4   Conclusion

Lesion detection greatly benefits from having large-scale data. Nonetheless, given the cost of manual annotations, SSL is a crucial strategy to best exploit clinical data archives. Towards this end, we articulate a hetero-modal SSL detection technique, called MTHD, that (1) uses hetero-modal learning to handle missing sequences and to exploit all available unlabeled data; and (2) employs expansive consistency constraints for much more effective SSL. Using the largest MR liver detection dataset to date, MTHD improves the average sensitivity by up to 10% compared to fully-supervised counter-parts and can outperform leading SSL alternatives by 3.5% to 9.1%. Consequently, the concepts outline here can help push forward the state of SSL lesion detection.

References

1. Aubé, C., et al.: EASL and AASLD recommendations for the diagnosis of HCC to the test of daily practice. Liver Int. **37**(10), 1515–1525 (2017)
2. Berthelot, D., Carlini, N., Goodfellow, I., Papernot, N., Oliver, A., Raffel, C.: Mixmatch: a holistic approach to semi-supervised learning. In: Advances in Neural Information Processing Systems (2019)
3. Blum, A., Mitchell, T.: Combining labeled and unlabeled data with co-training. In: Proceedings of the Eleventh Annual Conference on Computational Learning Theory, pp. 92–100. COLT' 98, Association for Computing Machinery, New York, NY, USA (1998)
4. Burrowes, D.P., Medellin, A., Harris, A.C., Milot, L., Wilson, S.R.: Contrast-enhanced us approach to the diagnosis of focal liver masses. RadioGraphics **37**(5), 1388–1400 (2017). pMID: 28898188
5. Cai, J., et al.: Lesion-harvester: iteratively mining unlabeled lesions and hard-negative examples at scale. IEEE Trans. Med. Imaging **40**(1), 59–70 (2020)
6. Castellino, R.A.: Computer aided detection (cad): an overview. Cancer Imaging Official Publ. Int. Cancer Imaging Soc. **5**(1), 17 (2005)
7. Havaei, M., Guizard, N., Chapados, N., Bengio, Y.: HeMIS: hetero-modal image segmentation. In: Ourselin, S., Joskowicz, L., Sabuncu, M.R., Unal, G., Wells, W. (eds.) MICCAI 2016. LNCS, vol. 9901, pp. 469–477. Springer, Cham (2016). https://doi.org/10.1007/978-3-319-46723-8_54
8. Huo, Y., et al.: Harvesting, detecting, and characterizing liver lesions from large-scale multi-phase ct data via deep dynamic texture learning. arXiv preprint arXiv:2006.15691 (2020)
9. Jeong, J., Lee, S., Kim, J., Kwak, N.: Consistency-based semi-supervised learning for object detection. In: Advances in Neural Information Processing Systems (2019)
10. Jiang, C., Wang, S., Liang, X., Xu, H., Xiao, N.: Elixirnet: relation-aware network architecture adaptation for medical lesion detection. In: Proceedings of the AAAI Conference on Artificial Intelligence, vol. 34, pp. 11093–11100 (2020)
11. Laine, S., Aila, T.: Temporal ensembling for semi-supervised learning. In: Proceedings of the International Conference on Learning Representations (2017)
12. Li, Z., Zhang, S., Zhang, J., Huang, K., Wang, Y., Yu, Y.: MVP-Net: multi-view FPN with position-aware attention for deep universal lesion detection. In: Shen, D., et al. (eds.) MICCAI 2019. LNCS, vol. 11769, pp. 13–21. Springer, Cham (2019). https://doi.org/10.1007/978-3-030-32226-7_2
13. Lin, T.Y., et al.: Microsoft COCO: common objects in context. In: Fleet, D., Pajdla, T., Schiele, B., Tuytelaars, T. (eds.) ECCV 2014. LNCS, vol. 8693, pp. 740–755. Springer, Cham (2014). https://doi.org/10.1007/978-3-319-10602-1_48
14. Liu, Y.C., et al.: Unbiased teacher for semi-supervised object detection. In: Proceedings of the International Conference on Learning Representations (2021)
15. Raju, A., et al.: Co-heterogeneous and adaptive segmentation from multi-source and multi-phase CT imaging data: a study on pathological liver and lesion segmentation. In: Vedaldi, A., Bischof, H., Brox, T., Frahm, J.-M. (eds.) ECCV 2020. LNCS, vol. 12368, pp. 448–465. Springer, Cham (2020). https://doi.org/10.1007/978-3-030-58592-1_27
16. Sohn, K., Zhang, Z., Li, C.L., Zhang, H., Lee, C.Y., Pfister, T.: A simple semi-supervised learning framework for object detection. arXiv preprint arXiv:2005.04757 (2020)

17. Suzuki, K.: A review of computer-aided diagnosis in thoracic and colonic imaging. Quant. Imaging Med. Surg. **2**(3), 163–176 (2012)
18. Tajbakhsh, N., Jeyaseelan, L., Li, Q., Chiang, J.N., Wu, Z., Ding, X.: Embracing imperfect datasets: a review of deep learning solutions for medical image segmentation. Med. Image Anal. **63**, 101693 (2020)
19. Tarvainen, A., Valpola, H.: Mean teachers are better role models: weight-averaged consistency targets improve semi-supervised deep learning results. In: Advances in Neural Information Processing Systems, pp. 1195–1204 (2017)
20. Wang, D., Zhang, Y., Zhang, K., Wang, L.: Focalmix: semi-supervised learning for 3D medical image detection. In: Proceedings of the IEEE/CVF Conference on Computer Vision and Pattern Recognition, pp. 3951–3960 (2020)
21. Wang, Y., et al.: Knowledge distillation with adaptive asymmetric label sharpening for semi-supervised fracture detection in chest x-rays. In: Information Processing in Medical Imaging (2020)
22. Yan, K., Bagheri, M., Summers, R.M.: 3D context enhanced region-based convolutional neural network for end-to-end lesion detection. In: Frangi, A.F., Schnabel, J.A., Davatzikos, C., Alberola-López, C., Fichtinger, G. (eds.) MICCAI 2018. LNCS, vol. 11070, pp. 511–519. Springer, Cham (2018). https://doi.org/10.1007/978-3-030-00928-1_58
23. Yan, K., et al.: Learning from multiple datasets with heterogeneous and partial labels for universal lesion detection in ct. In: IEEE Transactions on Medical Imaging (2020)
24. Yan, K., et al.: MULAN: multitask universal lesion analysis network for joint lesion detection, tagging, and segmentation. In: Shen, D., et al. (eds.) MICCAI 2019. LNCS, vol. 11769, pp. 194–202. Springer, Cham (2019). https://doi.org/10.1007/978-3-030-32226-7_22
25. Yan, K., Wang, X., Lu, L., Summers, R.M.: Deeplesion: automated mining of large-scale lesion annotations and universal lesion detection with deep learning. J. Med. Imaging **5**(3), 036501 (2018)
26. Zhang, S., Chi, C., Yao, Y., Lei, Z., Li, S.Z.: Bridging the gap between anchor-based and anchor-free detection via adaptive training sample selection. In: Proceedings of the IEEE/CVF Conference on Computer Vision and Pattern Recognition, pp. 9759–9768 (2020)
27. Zhou, X., Wang, D., Krähenbühl, P.: Objects as points. arXiv preprint arXiv:1904.07850 (2019)
28. Zhou, Y., et al.: Collaborative learning of semi-supervised segmentation and classification for medical images. In: Proceedings of the IEEE/CVF Conference on Computer Vision and Pattern Recognition, pp. 2079–2088 (2019)

STRUDEL: Self-training with Uncertainty Dependent Label Refinement Across Domains

Fabian Gröger[1,2], Anne-Marie Rickmann[1(✉)], and Christian Wachinger[1]

[1] Artificial Intelligence in Medical Imaging (AI -Med), KJP, LMU München,
Munich, Germany
arickman@med.lmu.de
[2] Computer Aided Medical Procedures, Technische Universität München,
Munich, Germany

Abstract. We propose an unsupervised domain adaptation (UDA) approach for white matter hyperintensity (WMH) segmentation, which
uses Self-TRaining with Uncertainty DEpendent Label refinement
(STRUDEL). Self-training has recently been introduced as a highly effective method for UDA, which is based on self-generated pseudo labels.
However, pseudo labels can be very noisy and therefore deteriorate
model performance. We propose to predict the uncertainty of pseudo
labels and integrate it in the training process with an uncertainty-guided
loss function to highlight labels with high certainty. STRUDEL is further improved by incorporating the segmentation output of an existing
method in the pseudo label generation that showed high robustness for
WMH segmentation. In our experiments, we evaluate STRUDEL with a
standard U-Net and a modified network with a higher receptive field. Our
results on WMH segmentation across datasets demonstrate the significant improvement of STRUDEL with respect to standard self-training.

1 Introduction

Dementia presents a highly relevant societal challenge due to the ever-aging population. Research shows that aging-related structural and functional changes
in the brain may manifest as cerebral small vessel disease (SVD), which is a
major contributor to the risk of developing dementia [16]. A promising neuroimaging biomarker for SVD are white matter hyperintensities (WMHs) of presumed vascular origin. WMHs are visible in fluid-attenuated inversion recovery
(FLAIR) magnetic resonance imaging (MRI) as diffuse regions of brighter intensity than surrounding white matter [14]. Convolutional neural networks (CNNs)
have achieved remarkable performances for WMH segmentation [11]. However,
CNNs are highly dependent on the training set and the performance can steeply

Electronic supplementary material The online version of this chapter (https://
doi.org/10.1007/978-3-030-87589-3_32) contains supplementary material, which is
available to authorized users.

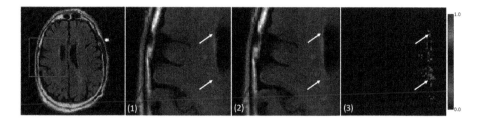

Fig. 1. Illustration of a FLAIR scan with (1) ground truth WMH, (2) noisy pseudo labels, and (3) corresponding uncertainty map. White arrows point to false positive predictions with higher uncertainty values (brighter pixels).

decrease on a target domain with a large domain shift. In a recent comparison, this resulted in traditional segmentation software producing higher quality WMH labels than CNNs [24]. Unsupervised domain adaptation (UDA) attempts to overcome the problem of domain shift without using target data annotations. Not relying on target annotations is a key benefit, as medical annotations are typically scarce and therefore will not cover the wide range of acquisition protocols, scanner types, artifacts, or patient statistics that make up domain differences in MRI. Self-training is a recent approach for UDA, where a segmentation model is first trained on annotated source data and then applied on target data to infer pseudo labels. These self-generated pseudo labels are then integrated into the network training to achieve the domain adaptation. However, pseudo labels tend to be noisy, as illustrated in Fig. 1, which necessitates estimating the reliability of pseudo labels to avoid propagating label errors.

In this work, we propose STRUDEL, a Self-TRaining approach with Uncertainty DEpendent Label refinement. It is motivated by earlier work on brain lesion segmentation [12], which demonstrated that uncertainty measures are an indicator for erroneous pixel-wise predictions. Following a Bayesian segmentation approach, we estimate the uncertainty for pseudo labels, see Fig. 1, and then integrate it in successive model refinements with an uncertainty-guided loss function. To further improve the initial pseudo label generation in STRUDEL, we propose to integrate the output of the lesion prediction algorithm (LPA) [19], which was reported to achieve robust results across domains [24]. In our experiments, we evaluate STRUDEL with a U-Net as the backbone and a modified network with a higher receptive field. Our results on WMH segmentation across datasets demonstrate the necessity for domain adaptation and further the significant improvement for integrating uncertainty and LPA in the training process.

1.1 Related Work

White Matter Hyperintensity Segmentation methods have recently been assessed in the WMH challenge [11]. All top-ranking methods were deep-learning-based and some have achieved superior performance to human observers. Specific inter-scanner robustness experiments showed there is still a need for improving the robustness of these methods which coincides with the findings in [24].

Unsupervised Domain Adaptation (UDA) approaches transfer a model from a source domain without direct supervision on the target domain and are commonly based on adversarial learning. These types of methods attempt to learn domain invariant features by minimizing the discrepancy between source and target domain or convert images from one domain to the other. Different approaches have demonstrated their effectiveness for medical applications [9,10]. However, the training process for adversarial networks can be a multi-faceted and complex endeavor [1,15].

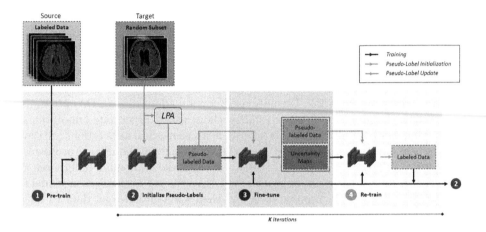

Fig. 2. Illustration of our Self-training pipeline for domain adaptation.

Self-training presents an alternative approach to UDA, which has recently been shown to be highly efficient [28]. It follows the principle that predictions generated in previous steps are used as pseudo labels for the next stage of network learning. The literature has addressed self-training for semantic segmentation in non-medical and medical applications and showed state-of-the-art performance on benchmark datasets [13,20,25,26,29,30]. A further categorization of methods for handling limited dataset annotations is available in the review by Tajbakhsh et al. [21]. The potential of integrating uncertainty guidance in self-training was recently demonstrated for segmenting sparsely annotated micro-CT scans [27].

2   Methods

2.1   Problem Definition

Given a labeled dataset from the source domain S with samples $X^S = \{x_i^S\}_{i=1}^N$ and labels $Y^S = \{y_i^S\}_{i=1}^N$, and an unlabeled dataset from the target domain T with samples $X^T = \{x_i^T\}_{i=1}^M$, the goal is to predict labels in the target domain. We want to achieve this goal by incorporating the large number of unlabeled target samples in the network training, where typically $M > N$. In self-training, this is achieved by inferring pseudo target labels $\tilde{y}^T$ and updating them iteratively to improve the label quality and consequently the learning process.

2.2   STRUDEL: Self-training with Uncertainty

Figure 2 provides a graphical overview of our proposed Self-TRaining with Uncertainty DEpendent Label refinement, with the pseudo-code in Algorithm 1. First, we pre-train a base model on the source dataset (X^S, Y^S) with standard supervised learning. The base model is then applied to a random subset (drawn without replacement) of the target sample, $r(X^T)$, of size P to infer pseudo labels. However, these pseudo labels will initially not be of high quality due to the domain gap. Consequently, we propose to leverage existing WMH segmentation software, where we use LPA, to increase the quality of pseudo labels. To this end, we apply a pixel-wise OR operator between base model predictions and LPA prediction to obtain the pseudo target labels $\tilde{Y}^T = \{\tilde{y}_i^T\}_{i=1}^P$.

Algorithm 1: Self-Training with Uncertainty on Noisy Labels

input : Source data X^S, Source labels Y^S, Target data X^T
output: Output model $\mathcal{M}_K$

1 $r() \leftarrow$ random sampler;
2 $\mathcal{M}_0 \leftarrow$ train base model with (X^S, Y^S);
3 $D_{\text{fix}} \leftarrow (X^S, Y^S)$;   // initialize fixed training set
4 **for** $k \leftarrow 1$ **to** K **do**
5 $\quad X_k^T \leftarrow r(X^T)$;   // sample random subset
6 $\quad \tilde{Y}_k^T \leftarrow pixel - wise\_or(\mathcal{M}_{k-1}(X_k^T), LPA(X_k^T))$;   // init. pseudo labels
7 $\quad D_k \leftarrow D_{\text{fix}} \cup (X_k^T, \tilde{Y}_k^T)$;   // merge training data
8 $\quad \mathcal{M}_{k-1} \leftarrow$ fine-tune $\mathcal{M}_{k-1}$ with D_k;
9 $\quad \tilde{Y}_k^T, U_k \leftarrow \mathcal{M}_{k-1}(X_k^T)$;   // update labels and get uncertainty
10 $\quad D_k' \leftarrow D_{\text{fix}} \cup (X_k^T, \tilde{Y}_k^T)$;   // merge training data
11 $\quad \mathcal{M}_k \leftarrow$ re-train model with D_k' and uncertainty U_k;
12 $\quad D_{\text{fix}} \leftarrow D_{\text{fix}} \cup (X_k^T, \mathcal{M}_k(X_k^T))$;   // update fixed training set
13 **end**
14 **return** $\mathcal{M}_K$;

In a third step, the base model is fine-tuned with $\tilde{Y}^T$. With this model, we segment again the same random sample drawn earlier, $r(X^T)$, producing pseudo labels of higher quality. Here, we assume that a model can generate better predictions than its noisy training labels [6]. Next to the labels, we also infer the segmentation uncertainty U at this stage. Finally, a new model is trained from scratch, where we add the updated pseudo labels $\tilde{Y}^T$ with the corresponding uncertainty U to the training set. Training a new model at this point has advantages over fine-tuning as also noted in [28]. The labels inferred from this model on the random subset are then added to the fixed training set, which initially only consists of the annotated source data. We then continue with the next iteration in step 2, where this model serves as a base model, another random subset is sampled and pseudo labels are inferred.

2.3   Uncertainty-Guided Pseudo Labels

Inferring pseudo target labels usually has the disadvantage of label noise. To increase the robustness of our method against label noise, we propose an uncertainty-guidance that strengthens regions of low uncertainty and penalizes regions of high uncertainty. To estimate the uncertainty, we follow a Bayesian machine learning approach with an estimation of Monte Carlo (MC) samples by dropout [5]. Accordingly, we train the backbone segmentation network with dropout layers and perform C stochastic forward passes at test time to obtain Monte Carlo samples. The expectation over the MC samples $\mathbb{E}(\hat{y})$ provides us a more robust label prediction, which we use to update the pseudo-label. Further, computing the variance across C MC samples gives us a pixel-wise measure of uncertainty of the predicted segmentation:

$$U(\hat{y}) = \{\sigma_1, ..., \sigma_{H \times W}\} = \frac{1}{C} \sum_{i=1}^{C} (\hat{y}_i - \mathbb{E}(\hat{y})))^2 , \qquad (1)$$

where $U(\hat{y})$ denotes the uncertainty map, σ_i the pixel-wise variance, H, W the images height and width, and $\hat{y}_i$ the model prediction from the ith MC sample. Anticipating small values for the uncertainties, we rescale the values into the range $[0, 1]$. We integrate the uncertainty into network training by the definition of an uncertainty-aware binary cross entropy (UBCE) loss:

$$\mathcal{L}_{\text{UBCE}} = -\frac{1}{H \times W} \sum_{n=1}^{H \times W} (1 - \sigma_n) \left[\tilde{y}_n \cdot log\,(\hat{y}_n) + (1 - \tilde{y}_n) \cdot log\,(1 - \hat{y}_n) \right]. \qquad (2)$$

Note that the uncertainty-aware cross entropy $\mathcal{L}_{\text{UBCE}}$ is only applied to pseudo-labeled data within the re-training step (see Algorithm 1 line 11), whereas the standard cross entropy $\mathcal{L}_{\text{BCE}}$ is applied to fixed data samples. The combined loss function is defined as:

$$\mathcal{L} = \mathcal{L}_{\text{Dice}} + \mathcal{L}_{\text{BCE}} + \mathcal{L}_{\text{UBCE}}. \qquad (3)$$

2.4   Segmentation Backbone Architectures

For the segmentation model M, we evaluate two neural network architectures. First, the U-Net [17], which has proven its performance in difficult segmentation tasks and has been adapted and improved for various applications since then. Second, to explore the effects of a superior model architecture within the framework, we include a novel network architecture, the OctSE-Net, which introduces two modifications to the U-Net. The first modification is to replace all convolution layers with octave convolutions [4], which factorize feature maps by their frequencies. Octave convolutions can increase segmentation performance, by offering a wider context, while reducing memory consumption. The second modification is to integrate squeeze & excitation (SE) blocks [8], more precisely the channel and spatial SE block [18], which can boost accuracy by re-calibrating feature maps. Both modifications increase the receptive field without substantially increasing model parameters. This can improve the segmentation accuracy without encouraging overfitting and therefore help the generalization across domains. For uncertainty estimation, we insert dropout layers after each convolutional block in both architectures.

3   Experiments and Results

3.1   Datasets

As source dataset, we use data from the WMH challenge (https://wmh.isi.uu.nl), and, as target dataset, we use data from the Alzheimer's Disease Neuroimaging Initiative (ADNI) (http://adni.loni.usc.edu). The WMH segmentation challenge dataset provides manual annotations for 60 subjects from 3 sites. For each subject, co-registered 3D T1-weighted and 2D multi-slice FLAIR scans are available, where we work with bias-field corrected T1 scans and original FLAIR scans as suggested by [7,22]. The large ADNI-2 dataset [3] with over 3,000 scans from 58 sites serves as our multi-domain target data. For each subject, T1-weighted and 2D FLAIR scans are available, which we have linearly aligned within a session with ANTs [2]. T1 scans have further been bias field corrected with N4 normalization [23], using the ANTs implementation. To quantitatively evaluate our methods on the ADNI dataset, we extracted a subset of 30 subjects based on scanner type and WMH lesion load for manual annotation[1]. 21 of these annotated scans serve solely as a test set and 9 are used to train alternative segmentation approaches, described in the next section.

3.2   Implementation Details

We implemented the proposed framework and all baseline experiments in Pytorch v.1.6.0 (see footnote 1). Experiments were performed employing the Adam optimizer with default parameters (betas = (0.9, 0.999), eps = 1e-08), learning rate 1e-4 and batch size 4. Image intensities were normalized to zero-mean and unit-variance and axial slices center cropped to a consistent size of 192×192 pixels across all datasets. Standard spatial augmentation techniques (flipping, rotation, scaling, and elastic transformation) were used during training for regularization. In self-training, the size of the random subset per iteration is set to $P = 35$. The thresholds to obtain binary segmentation maps for the creation of pseudo-labels were set to 0.5 and 0.75 for the network prediction and LPA, respectively. The high threshold for LPA mitigates hypersensitive responses. We use 80 epochs for training from scratch and 20 epochs for fine-tuning. We set the number of stochastic forward passes to $C = 10$. We found that increasing C further does not improve the segmentation performance. The drop-out rate was set to 0.2. No explicit post-processing was performed in any experiment. A Geforce Titan RTX GPU was predominantly used for training and testing.

---

[1] Code and manual segmentations are available under: https://github.com/ai-med/STRUDEL.

3.3 Experiments

We perform several experiments to evaluate the domain transfer performance of different approaches. First, as **Base Model**, we directly apply a model trained on the source data on the target domain data without any adaptation. Next, we evaluate two approaches that use a small labeled subset of the target domain during training. The **Joint Model** combines the target and source training data, and the **Fine-Tuning** model uses the labeled target data to fine-tune the Base Model. Finally, we evaluate two UDA approaches with pseudo labels: **Self-Training** without uncertainty guidance and the proposed **STRUDEL** with uncertainty guidance, both using LPA labels. We also report results for STRUDEL without LPA and for LPA itself, where we set the threshold parameter to 0.45, as suggested in [24] for ADNI. We evaluate the segmentation accuracy for all experiments by following evaluation metrics suggested by the WMH segmentation challenge [11]: (1) Dice Similarity Coefficient (DSC), (2) modified Hausdorff distance (95th percentile; H95), (3) absolute log-transformed volume difference (lAVD), (4) sensitivity for detecting individual lesions (Recall), and (5) F1-score for individual lesions (F1).

Table 1. Comparison of segmentation methods, network architectures, type of used data (S: Source manual labels, T: Target manual labels, P: target pseudo labels), and their mean performance ± standard deviation on the metrics: Dice Coefficient (DSC), 95th Percentile Hausdorff Distance (H95), log transformed absolute volume difference (lAVD), lesion Recall and F1.

Methods	S	T	P	DSC ↑	H95 [mm] ↓	lAVD ↓	Recall ↑	F1 ↑
LPA	✗	✗	✗	0.57 ± 0.16	23.1 ± 23.4	0.71 ± 0.49	0.81 ± 0.16	0.39 ± 0.18
U-Net								
Base Model	✓	✗	✗	0.45 ± 0.28	27.1 ± 37.5	1.09 ± 1.70	0.67 ± 0.32	0.48 ± 0.21
Joint Model	✓	✓	✗	0.64 ± 0.19	17.2 ± 25.0	0.60 ± 0.52	0.74 ± 0.29	0.52 ± 0.15
Fine-Tuning	✓	✓	✗	**0.73 ± 0.16**	**11.2 ± 23.0**	0.36 ± 0.41	**0.75 ± 0.22**	**0.65 ± 0.14**
Self-Training	✓	✗	✓	0.64 ± 0.20	17.8 ± 28.8	0.51 ± 0.68	0.51 ± 0.27	0.50 ± 0.23
STRUDEL	✓	✗	✓	0.69 ± 0.18	**11.2 ± 14.5**	**0.30 ± 0.32**	0.58 ± 0.27	0.64 ± 0.22
OctSE-Net								
Base Model	✓	✗	✗	0.60 ± 0.23	19.7 ± 29.5	0.77 ± 1.12	0.80 ± 0.26	0.61 ± 0.19
Joint Model	✓	✓	✗	0.73 ± 0.15	11.8 ± 24.7	0.34 ± 0.37	**0.89 ± 0.10**	0.59 ± 0.14
Fine-Tuning	✓	✓	✗	0.73 ± 0.15	11.4 ± 23.4	0.41 ± 0.38	0.77 ± 0.18	0.64 ± 0.17
Self-Training	✓	✗	✓	0.73 ± 0.13	14.7 ± 18.2	**0.25 ± 0.27**	0.56 ± 0.21	0.63 ± 0.17
STRUDEL	✓	✗	✓	**0.78 ± 0.10**	**7.79 ± 8.52**	0.27 ± 0.23	0.77 ± 0.16	**0.70 ± 0.15**
↪ w/o LPA	✓	✗	✓	0.67 ± 0.20	12.9 ± 13.4	0.63 ± 0.58	0.58 ± 0.23	0.66 ± 0.18

3.4   Results and Discussion

Table 1 reports the quantitative segmentation results on the ADNI target domain. Figure 3 shows the DSC in more details as boxplots. As a reference, the DSC of the Base Models on the source dataset are 0.73 for U-Net and 0.76 for OctSE-Net. We observe that the direct transfer of the Base Model on the ADNI dataset performs poorly, regardless of the backbone architecture. LPA beats the baseline U-Net by a large margin, which is in accordance with the results described in [24]. OctSE-Net outperforms LPA, which confirms our assumption that OctSE-Net is a more robust architecture. However, the more detailed results in Fig. 3 show that both base models produce some predictions with zero DSC, whereas LPA does not. These outliers can lead to poorly initialized pseudo labels, which is the reason for including LPA in pseudo-label initialization, confirmed by the bad results for STRUDEL w/o LPA. We report results for all self-training-based experiments after 5 iterations, as no further improvement was observed afterwards. STRUDEL outperforms all other methods in DSC and H95, and is best or second best in lAVD and lesion F1. LPA and the joint model perform best in terms of lesion recall. Both of these methods have relatively poor performance in lesion F1, which is the result of a high number of false positive predictions. STRUDEL performs strongly in both metrics, which we believe is due to the uncertainty capturing false positives reliably. Results of a Wilcoxon signed-rank test on DSC show that the improvement of STRUDEL with respect to Self-Training is significant for OctSE-Net ($p < 0.005$) and also that the improvement of OctSE-Net with respect to U-Net is significant for STRUDEL ($p < 0.001$).

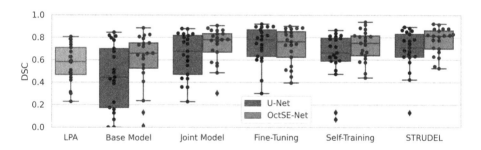

Fig. 3. Boxplot of Dice Similarity Coefficient for the different methods. Points outside the whiskers are determined as outliers based on the inter-quartile range.

4   Conclusion

Self-Training is a simple and effective approach for UDA, however noisy pseudo labels can limit its effectiveness. In this work, we proposed STRUDEL, an uncertainty-guided self-training method for unsupervised domain adaptation. We found that introducing uncertainty into the objective function can efficiently guide the learning process in the presence of noisy labels. We further demonstrated that leveraging an existing algorithm (LPA) for pseudo label initialization can additionally boost performance. Our experimental results showed that Self-Training with uncertainty guidance is a strong approach for UDA, which in combination with a strong and robust network architecture, can even outperform supervised methods.

Acknowledgments. This research was partially supported by the Bavarian State Ministry of Science and the Arts and coordinated by the bidt, and the BMBF (DeepMentia,031L0200A).

References

1. Arjovsky, M., Bottou, L.: Towards principled methods for training generative adversarial networks. arXiv preprint arXiv:1701.04862 (2017)
2. Avants, B.B., Tustison, N., Song, G.: Advanced normalization tools (ants). Insight J. **2**(365), 1–35 (2009)
3. Beckett, L., Donohue, M., Wang, C., Aisen, P., Harvey, D., Saito, N.: The alzheimer's disease neuroimaging initiative phase 2: increasing the length, breadth, and depth of our understanding. Alzheimer's Dementia J. Alzheimer's Assoc. **11**, 823–31 (2015)
4. Chen, Y., et al.: Drop an octave: reducing spatial redundancy in convolutional neural networks with octave convolution, pp. 3434–3443 (2019)
5. Gal, Y., Ghahramani, Z.: Dropout as a bayesian approximation: representing model uncertainty in deep learning. In: Proceedings of The 33rd International Conference on Machine Learning (2015)
6. Guan, M., Gulshan, V., Dai, A., Hinton, G.: Who said what: modeling individual labelers improves classification. In: Proceedings of the AAAI Conference on Artificial Intelligence, vol. 32 (2018)
7. Hernández, M.D.C.V., et al.: On the computational assessment of white matter hyperintensity progression: difficulties in method selection and bias field correction performance on images with significant white matter pathology. Neuroradiology **58**(5), 475–485 (2016)
8. Hu, J., Shen, L., Sun, G.: Squeeze-and-excitation networks. In: The IEEE Conference on Computer Vision and Pattern Recognition (CVPR), June 2018
9. Huo, Y., Xu, Z., Bao, S., Assad, A., Abramson, R.G., Landman, B.A.: Adversarial synthesis learning enables segmentation without target modality ground truth. In: 2018 IEEE 15th International Symposium on Biomedical Imaging (ISBI 2018), pp. 1217–1220. IEEE (2018)
10. Kamnitsas, K., et al.: Unsupervised domain adaptation in brain lesion segmentation with adversarial networks, pp. 597–609 (2017)
11. Kuijf, H.J., et al.: Standardized assessment of automatic segmentation of white matter hyperintensities and results of the wmh segmentation challenge. IEEE Trans. Med. Imaging **38**(11), 2556–2568 (2019)
12. Nair, T., Precup, D., Arnold, D., Arbel, T.: Exploring uncertainty measures in deep networks for multiple sclerosis lesion detection and segmentation, pp. 655–663 (2018)

13. Nie, D., Gao, Y., Wang, L., Shen, D.: ASDNet: attention based semi-supervised deep networks for medical image segmentation. In: Frangi, A.F., Schnabel, J.A., Davatzikos, C., Alberola-López, C., Fichtinger, G. (eds.) MICCAI 2018. LNCS, vol. 11073, pp. 370–378. Springer, Cham (2018). https://doi.org/10.1007/978-3-030-00937-3_43

14. Prins, N., Scheltens, P.: White matter hyperintensities, cognitive impairment and dementia: an update. Nature reviews. Neurology **11**, 157–165 (2015)

15. Radford, A., Metz, L., Chintala, S.: Unsupervised representation learning with deep convolutional generative adversarial networks. arXiv preprint arXiv:1511.06434 (2015)

16. Raz, L., Knoefel, J., Bhaskar, K.: The neuropathology and cerebrovascular mechanisms of dementia. J. Cerebral Blood Flow Metabol. Official J. Int. Soc. Cerebral Blood Flow Metabol. **36**, 172–186 (2015)

17. Ronneberger, O., Fischer, P., Brox, T.: U-Net: convolutional networks for biomedical image segmentation. In: Navab, N., Hornegger, J., Wells, W.M., Frangi, A.F. (eds.) MICCAI 2015. LNCS, vol. 9351, pp. 234–241. Springer, Cham (2015). https://doi.org/10.1007/978-3-319-24574-4_28

18. Roy, A.G., Navab, N., Wachinger, C.: Recalibrating fully convolutional networks with spatial and channel "squeeze and excitation" blocks. IEEE TMI **38**(2), 540–549 (2019)

19. Schmidt, P.: Bayesian inference for structured additive regression models for large-scale problems with applications to medical imaging. Ph.D. thesis, Ludwig-Maximilians-Universität München (2017)

20. Shin, I., Woo, S., Pan, F., Kweon, I.S.: Two-phase pseudo label densification for self-training based domain adaptation. In: Vedaldi, A., Bischof, H., Brox, T., Frahm, J.-M. (eds.) ECCV 2020. LNCS, vol. 12358, pp. 532–548. Springer, Cham (2020). https://doi.org/10.1007/978-3-030-58601-0_32

21. Tajbakhsh, N., Jeyaseelan, L., Li, Q., Chiang, J.N., Wu, Z., Ding, X.: Embracing imperfect datasets: a review of deep learning solutions for medical image segmentation. Med. Image Anal. **63**, 101693 (2020)

22. Tubi, M.A., et al.: White matter hyperintensities and their relationship to cognition: effects of segmentation algorithm. NeuroImage **206**, 116327 (2020)

23. Tustison, N.J., et al.: N4itk: improved n3 bias correction. IEEE Trans. Med. Imaging **29**(6), 1310–1320 (2010)

24. Vanderbecq, Q., et al.: Comparison and validation of seven white matter hyperintensities segmentation software in elderly patients. NeuroImage Clin. **27**, 102357 (2020)

25. Xia, Y., et al.: 3D semi-supervised learning with uncertainty-aware multi-view co-training. In: Proceedings of the IEEE/CVF Winter Conference on Applications of Computer Vision, pp. 3646–3655 (2020)

26. Yu, L., Wang, S., Li, X., Fu, C.-W., Heng, P.-A.: Uncertainty-aware self-ensembling model for semi-supervised 3D left atrium segmentation. In: Shen, D., et al. (eds.) MICCAI 2019. LNCS, vol. 11765, pp. 605–613. Springer, Cham (2019). https://doi.org/10.1007/978-3-030-32245-8_67

27. Zheng, H., et al.: Cartilage segmentation in high-resolution 3D Micro-CT images via uncertainty-guided self-training with very sparse annotation. In: Martel, A.L., et al. (eds.) MICCAI 2020. LNCS, vol. 12261, pp. 802–812. Springer, Cham (2020). https://doi.org/10.1007/978-3-030-59710-8_78

28. Zoph, B., et al.: Rethinking pre-training and self-training. In: Advances in Neural Information Processing Systems, vol. 33 (2020)

29. Zou, Y., Yu, Z., Kumar, B.V., Wang, J.: Unsupervised domain adaptation for semantic segmentation via class-balanced self-training. In: Proceedings of the European Conference on Computer Vision (ECCV), pp. 289–305 (2018)
30. Zou, Y., Yu, Z., Liu, X., Kumar, B.V., Wang, J.: Confidence regularized self-training. In: The IEEE International Conference on Computer Vision (ICCV), October 2019

Deep Reinforcement Learning for L3 Slice Localization in Sarcopenia Assessment

Othmane Laousy[1,2,3]([✉]), Guillaume Chassagnon[2]([✉]), Edouard Oyallon[4]([✉]),
Nikos Paragios[5]([✉]), Marie-Pierre Revel[2]([✉]), and Maria Vakalopoulou[1,3]([✉])

[1] CentraleSupélec, Paris-Saclay University, Mathématiques et Informatique pour la Complexité et les Systèmes, Gif-sur-Yvette, France
{othmane.laousy,maria.vakalopoulou}@centralesupelec.fr
[2] Radiology Department, Hôpital Cochin AP-HP, Paris, France
{guillaume.chassagnon,marie-pierre.revel}@aphp.fr
[3] Inria Saclay, Gif-sur-Yvette, France
[4] Centre National de Recherche Scientifique, LIP6, Paris, France
edouard.oyallon@lip6.fr
[5] Therapanacea, Paris, France
n.paragios@therapanacea.eu

Abstract. Sarcopenia is a medical condition characterized by a reduction in muscle mass and function. A quantitative diagnosis technique consists of localizing the CT slice passing through the middle of the third lumbar area (L3) and segmenting muscles at this level. In this paper, we propose a deep reinforcement learning method for accurate localization of the L3 CT slice. Our method trains a reinforcement learning agent by incentivizing it to discover the right position. Specifically, a Deep Q-Network is trained to find the best policy to follow for this problem. Visualizing the training process shows that the agent mimics the scrolling of an experienced radiologist. Extensive experiments against other state-of-the-art deep learning based methods for L3 localization prove the superiority of our technique which performs well even with a limited amount of data and annotations.

Keywords: L3 slice · CT slice localization · Deep reinforcement learning · Sarcopenia

1 Introduction

Sarcopenia corresponds to muscle atrophy which may be due to ageing, inactivity, or disease. The decrease of skeletal muscle is a good indicator of the overall health state of a patient [11]. In oncology, it has been shown that sarcopenia is linked to outcome in patients treated by chemotherapy [5,14], immunotherapy [20], or surgery [9]. There are multiple definitions of sarcopenia [7,23] and consequently multiple ways of assessing it. On CT imaging, the method used is based on muscle mass quantification. Muscle mass is most commonly assessed at a level passing through the middle of the third lumbar vertebra area (L3),

© Springer Nature Switzerland AG 2021
C. Lian et al. (Eds.): MLMI 2021, LNCS 12966, pp. 317–326, 2021.
https://doi.org/10.1007/978-3-030-87589-3_33

which has been found to be representative of the body composition [29]. After manual selection of the correct CT slice at the L3 level, segmentation of muscles is performed to calculate the skeletal muscle area [8]. In practice, the evaluation is tedious, time-consuming, and rarely done by radiologists, highlighting the need for an automatic diagnosis tool that could be integrated into clinical practice. Such automated measurement of muscle mass could be of great help for introducing sarcopenia assessment in daily clinical practice.

Muscle segmentation and quantification on a single slice have been thoroughly addressed in multiple works using simple 2D U-Net like architectures [4,6]. Few works, however, focus on L3 slice detection. The main challenges for solving this task rely on the inherent diversity in patient's anatomy, the strong resemblance between vertebrae, the variability of CT fields of view as well as their acquisition and reconstruction protocols.

The most straightforward approach to address L3 localization is by investigating methods for multiple vertebrae labeling in 3D images using detection [25] or even segmentation algorithms [22]. Such methods require a substantial volume of annotations and are computationally inefficient when dealing with the entire 3D CT scan. In fact, even if our input is 3D, a one-dimensional output as the z-coordinate of the slice is sufficient to solve the L3 localization problem.

In terms of L3 slice detection, the closest methods leverage deep learning [3, 13] and focus on training simple convolutional neural networks (CNN). These techniques use maximal intensity projection (MIP), where the objective is to project voxels with maximal intensity values into a 2D plane. Frontal view MIP projections contain enough information towards the body and vertebra's bone structure differentiation. On the sagittal view, restricted MIP projections are used to focus solely on the spinal area. In [3] the authors tackle this problem through regression, training the CNN with parts of the MIP that contain the L3 vertebra only. More recently, in [13] a UNet-like architecture (L3UNet-2D) is proposed to draw a 2D confidence map over the position of the L3 slice.

In this paper, we propose a reinforcement learning algorithm for accurate detection of the L3 slice in CT scans, automatizing the process of sarcopenia assessment. The main contribution of our paper is a novel formulation for the problem of L3 localization, exploiting different deep reinforcement learning (DRL) schemes that boost the state of the art for this challenging task, even on scarce data settings. Moreover, in this paper we demonstrate that the use of 2D approaches for vertebrae detection provides state of the art results compared to 3D landmark detection methods, simplifying the problem, reducing the search space and the amount of annotations needed. To the best of our knowledge, this is the first time a reinforcement learning algorithm is explored on vertebrae slice localization, reporting performances similar to medical experts and opening new directions for this challenging task.

2   Background

Reinforcement Learning is a fundamental tool of machine learning which allows dealing efficiently with the exploration/exploitation trade-off [24]. Given state-reward pairs, a reinforcement learning agent can pick actions to reach unexplored states or increase its accumulated future reward. Those principles are appealing for medical applications because they imitate a practitioner's behavior and self-learn from experience based on ground-truth. One of the main issues of this class of algorithm is its sample complexity: a large amount of interaction with its environment is needed before obtaining an agent close to an optimal state [26]. However, those techniques were recently combined with deep learning approaches, which efficiently addressed this issue [17] by incorporating priors based on neural networks. In the context of highly-dimensional computer vision applications, this approach allowed RL algorithms to obtain outstanding accuracy [12] in a variety of tasks and applications.

In medical imaging, model-free reinforcement learning algorithms are highly used for landmark detection [10] as well as localization tasks [16]. In [1], a Deep Q-Network (DQN) that automates the view planning process on brain and cardiac MRI was proposed. This framework takes as an input a single plane and updates its angle and position during the training process until convergence. Moreover, in [27] the authors again present a DQN framework for the localization of different anatomical landmarks introducing multiple agents that act and learn simultaneously. DRL has also been studied for object or lesion localization. More recently, in [19] the authors propose a DQN framework for the localization of 6 different organs from CT scans achieving a performance comparable to supervised CNNs. This framework uses a 3D volume as input with 11 different actions to generate bounding boxes for these organs. Our work is the first to explore and validate a RL scheme on MIP representations for a single slice detection using the discrete and 2D nature of the problem.

3   Reinforcement Learning Strategy

In this paper, we formulate the slice localization problem as a *Markov Decision Process* (MDP), which contains a set of states S, actions A, and rewards R.

States S: For our formulation, the environment $\mathcal{E}$ that we explore and exploit is a 2D image representing the frontal MIP projection of the 3D CT scans. This projection allows us to reduce our problem's dimensionality from a volume of size $512 \times 512 \times N$ (N being the varying heights of the volumes) to an image of size $512 \times N$. The reinforcement learning agent is self-taught by interacting with this environment, executing a set of actions, and receiving a reward linked to the action taken. An input example is shown in Fig. 1. We define a state $s \in S$ as an image of size 512×200 in $\mathcal{E}$. We consider the middle of the image to be the slice's current position on a z-axis. To highlight this, we assign a line of maximum intensity pixel value to the middle of each image provided as input to our DQN.

Actions A: We define a set of discrete actions $\mathcal{A} = \{t_z^+, t_z^-\} \in \mathbb{R}^2$. t_z^+ corresponds to a positive translation (going up by one slice) and t_z^- corresponds to a negative translation (going down by one slice). These two actions allow us to explore the entirety of our environment $\mathcal{E}$.

Rewards R: In reinforcement learning designing a good reward function is crucial in learning the goal to achieve. To measure the quality of taking an action $a \in A$ we use the distance over z between the current slice and the annotated slice g. The reward for non-terminating states is computed with:

$$R_a(s, s') = sign(\mathcal{D}(p, g) - \mathcal{D}(p', g)) \tag{1}$$

where we denote as s and s' the current and next state and g the ground truth annotation. Moreover, our positions p and p' are the z-coordinates of the current and next state respectively. $\mathcal{D}$ is the Euclidean distance between both coordinates over the z-axis. The reward is non-sparse and binary $r \in \{-1, +1\}$ and helps the agent differentiate between good and bad actions. A good action being when the agent gets closer to the correct slice. For a terminating state, we assign a reward of $r = 0.5$.

Starting States: An episode starts by randomly sampling a slice over the z-axis and ends when the agent has achieved its goal of finding the right slice. The agent then executes a set of actions and collects rewards until the episode terminates. When reaching the upper or lower borders of an image, the current state is assigned to the next state (i.e., the agent does not move), and a reward of $r = -1$ is appointed to this action.

Final States: During training, a terminal state is defined as a state in which the agent has reached the right slice. A reward of $r = 0.5$ is assigned in this case, and an episode is terminated. During testing, the termination of an episode happens when oscillations occur. We adopted the same approach as [1], and chose actions with the lowest Q-value, which have been found to be closest to the right slice since the DQN outputs higher Q-values to actions when the current slice is far from the ground truth.

3.1 Deep Q-Learning

To find the optimal policy π^* of the MDP, a state-action value function $Q(s, a)$ can be learned. In Q-Learning, the expected value of the accumulated discounted future rewards can be estimated recursively using the Bellman optimality equation:

$$Q_{i+1}(s, a) = \mathbb{E}[r + \gamma \max_{a'} Q_i(s', a') \mid s, a] \tag{2}$$

In practice, since the state S is not easily exploitable, we can take advantage of neural networks as universal function approximators to approximate $Q(s, a)$ [18]. We utilize an experience replay technique that consists in storing the agent's experience $e_t = (s, a, r, s', a')$ at each time step in a replay memory M. To break the correlation between consecutive samples, we will uniformly batch a set of experiences from M. The Deep Q-Network (DQN) will iteratively optimize its parameters θ by minimizing the following loss function:

$$L_i(\theta_i) = \mathbb{E}_{(s,a,r,s',a')} \left[\left(r + \gamma \max_{a'} Q_{target}(s', a'; \theta_i^-) - Q_{policy}(s, a; \theta_i) \right)^2 \right] \tag{3}$$

with θ_i and θ_i^- being the parameters of the policy and the target network respectively. To stabilize rapid policy changes due to the distribution of the data and the variations in Q-values, the DQN uses $Q_{target}(\theta_i^-)$, a fixed version of $Q_{policy}(\theta_i)$ that is updated periodically. For our experiments, we update θ_i^- every 50 iterations.

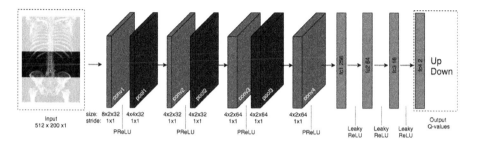

Fig. 1. The implemented Deep Q-Network architecture for L3 slice localization. The network takes as input an image of size 512×200 with a single channel. The output is the q-values corresponding to each of the two actions.

3.2 Network Architecture

Our Deep Q-Network takes as input the state s and passes it through a convolutional network. The network contains four convolution layers separated by parametric ReLU in order to break the linearity of the network, and four linear layers with LeakyReLU. Contrary to [1], we chose not to add the history of previously visited states in our case. We opted for this approach since there is a single path that leads to the right slice. This approach allows us to simplify our problem even more. Ideally, our agent should learn, just by looking at the current state, whether to go up or down when the current slice is respectively below or above the L3 slice. An overview of our framework is presented in Fig. 1.

We also explore dueling DQNs from [28]. Dueling DQNs rely on the concept of an advantage which calculates the benefit that each action can provide. The advantage is defined as $A(s, a) = Q(s, a) - V(s)$ with $V(s)$ being our state value function. This algorithm will use the advantage of the Q-values to distinguish between actions from the state's baseline values. Dueling DQNs were shown to provide more robust agents that are wiser in choosing the next best action. For our dueling DQN, we use the same architecture as the one in Fig. 1 but change the second to last fully connected layer to compute state values on one side, and action values on the other.

3.3 Training

Since our agent is unaware of the possible states and rewards in $\mathcal{E}$, the exploration step is implemented first. After a few iterations, our agent can start exploiting what it has learned on $\mathcal{E}$. In order to balance between exploration and exploitation, we use an ϵ-greedy strategy. This strategy consists of defining an exploration rate ϵ, which is initialized to 1 with a decay of 0.1, allowing the agent to become greedy and exploit the environment. A batch size of 48 and an experience replay of 17×10^3 are used. The entire framework was developed in Pytorch [21] library using an NVIDIA GTX 1080Ti GPU. We trained our model for 10^5 episodes, requiring approximately 20-24 hours. Our source code is available on GitHub: https://git.io/JRyYw.

4   Experiments and Results

4.1   Dataset

A diverse dataset of 1000 CT scans has been retrospectively collected for this study. CT scans were acquired on four different CT models from three manufacturers (Revolution HD from GE Healthcare, Milwaukee, WI; Brillance 16 from Philips Healthcare, Best, Netherlands; and Somatom AS+ & Somatom Edge from Siemens Healthineer, Erlangen, Germany). Exams were either abdominal, thoracoabdominal, or thoraco-abdominopelvic CT scans acquired with or without contrast media injection. Images were reconstructed using abdominal kernel with either filtered back-projection or iterative reconstruction. Slice thickness ranged from 0.625 to 3 mm, and the number of slices varied from 121 to 1407. The heterogeneity of our dataset highlights the challenges of the problem from a clinical perspective.

Experienced radiologists manually annotated the dataset, indicating the position of the middle of the L3 slice. Before computing the MIP, all of the CT scans are normalized to 1 mm over the z-axis. This normalisation step harmonises our network's input, especially since the agent performs actions along the z-axis. After the MIP, we apply a threshold of 100 HU (Hounsfield Unit) to 1500 HU allowing us to eliminate artifacts and foreign metal bodies while keeping the skeleton structure. The MIP are finally normalized to [0,1]. From the entire dataset, we randomly selected 100 patients for testing and the rest 900 for training and validation. For the testing cohort, annotations of L3 from a second experienced radiologist have been provided to measure the interobserver performance.

4.2   Results and Discussion

Our method is compared with other techniques from the literature. The error is calculated as the distance in millimeters (mm) between the predicted L3 slice and the one annotated by the experts. In particular, we performed experiments with the L3UNet-2D [13] approach and the winning SC-Net [22] method of the Verse2020[1] challenge. Even if SC-Net is trained on more accurate annotations with 3D landmarks as well as vertebrae segmentations, and addresses a different problem, we applied it to our testing cohort. The comparison of the different methods is summarised in Table 1. SC-Net reports 12 CT scans with an error

---

[1] https://verse2020.grand-challenge.org/.

higher than 10 mm. Moreover, L3UNet-2D [13] reports a mean error of 4.24 mm ± 6.97 mm when the method is trained on the entire training set, giving only 7 scans with an error higher than 10 mm for the L3 detection. Our proposed method gives the lowest errors with a mean error of 3.77 mm ± 4.71 mm, proving its superiority. Finally, we evaluated our technique's performance with a Duel DQN strategy, reporting higher errors than the proposed one. This observation could be linked to the small action space that is designed for this study. Duel DQNs were proven to be powerful in cases with higher action spaces and in which the computation of the advantage function makes a difference.

Table 1. Quantitative evaluation of the different methods using different number of training samples (metrics in mm).

Method	# of samples	Mean	Std	Median	Max	Error > 10 mm
Interobserver	–	2.04	4.36	1.30	43.19	1
SC-Net [22]	–	6.78	13.96	**1.77**	46.98	12
L3UNet-2D [13]	900	4.24	6.97	2.19	40	**7**
Ours (Duel-DQN)	900	4.30	5.59	3	38	8
Ours	900	**3.77**	**4.71**	2.0	**24**	9
L3UNet-2D [13]	100	145.37	161.91	32.8	493	68
Ours	100	**5.65**	**5.83**	4	**26**	**19**
L3UNet-2D [13]	50	108.7	97.33	87.35	392.02	86
Ours	50	**6.88**	**5.79**	**6.5**	**26**	**11**
L3UNet-2D [13]	10	242.85	73.07	240.5	462	99
Ours	10	**8.97**	**8.72**	7	**56**	**33**

For the proposed reinforcement learning framework, trained on the whole training set, 9 CTs had a detection error of more than 10 mm. These scans were analysed by a medical expert who indicated that 2 of them have a lumbosacral transitional vertebrae (LSTV) anomaly [15]. Transitional vertebrae cases are common and observed in 15–35% of the population [2] highlighting once again the challenges of this task. For both cases, the localization of the L3 vertebra for sarcopenia assessment is ambiguous for radiologists and consequently for the reinforcement learning agent. In fact, the only error higher than 10 mm in the interobserver comparison corresponds to an LSTV case where each radiologist chose a different vertebrae as a basis for sarcopenia assessment. Even if the interobserver performance is better than the one reported by the algorithms, our method reports the lowest errors, proving its potential.

Qualitative results are displayed in Fig. 2. The yellow line represents the medical expert's annotation and the blue one the prediction of the different employed models. One can notice that all of the different methods converge to the correct L3 region with our method reporting great performance. It is important to note that for sarcopenia assessment, an automatic system does not

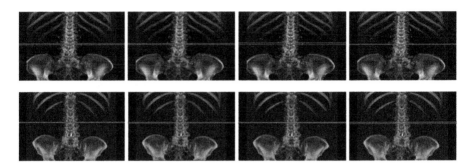

Fig. 2. Qualitative comparison of different localization methods for two patients. First left to right represents: interobserver (4 mm/2 mm), SC-Net (4 mm/2 mm), L3UNet-2D (8 mm/1 mm), Ours (1 mm/4 mm). In the parenthesis we present the reported errors for the first and second row respectively. The yellow line represents the ground truth and the blue one the prediction.

need to be at the exact middle of the slice; a few millimeters around will not skew the end result since muscle mass in the L3 zone does not change significantly. Concerning prediction times for the RL agent, they depend on the initial slice that is randomly sampled on the MIP. Computed inference time for a single step is approximately 0.03 s.

To highlight the robustness of our network on a low number of annotated samples, we performed different experiments using 100, 50, and 10 CTs corresponding respectively to 10%, 5% and 1% of our dataset. We tested those 3 agents on the same 100 patients test set and report results in Table 1. Our experiments prove the robustness of reinforcement learning algorithms compared to traditional CNN based ones [13] in the case of small annotated datasets. One can observe that the traditional methods fail to be trained properly with a small number of annotations, reporting errors higher than 100 mm for all three experiments. In our case, decreasing the dataset size does not significantly affect the performance. In fact, trained on only 10 CTs with the same number of iterations and memory size, our agent was able to learn a correct policy and achieve a mean error of 8.97 mm ± 8.72 mm. Learning a valid policy from a low number of annotations is one of the strengths of reinforcement learning. Traditional deep learning techniques rely on pairs of images and annotations in order to build a robust generalization. Thus, each pair is exploited only once by the learning algorithm. Reinforcement learning, however, relies on experiences, each experience $e_t = (s, a, r, s', a')$ being a tuple of state, action, reward, next state and next action. Therefore, a single CT scan can provide multiple experiences to the self-learning agent, making our method ideal for slice localization problems using datasets with a limited amount of annotations.

5   Conclusion

In this paper, we propose a novel direction to address the problem of CT slice localization. Our experiments empirically prove that reinforcement learning schemes work very well on small datasets and boost performance compared to classical convolutional architectures. One limitation of our work lies in the fact that our agent is always moving 1 mm independently of the location, slowing down the process. In the future, we aim to explore different ways to adapt the action taken depending on the current location, with one possibility being to incentivize actions with higher increments. Future work also includes the use of reinforcement learning in multiple vertebrae detection with competitive or collaborative agents.

References

1. Alansary, A., et al.: Automatic view planning with multi-scale deep reinforcement learning agents. In: Frangi, A.F., Schnabel, J.A., Davatzikos, C., Alberola-López, C., Fichtinger, G. (eds.) MICCAI 2018. LNCS, vol. 11070, pp. 277–285. Springer, Cham (2018). https://doi.org/10.1007/978-3-030-00928-1_32
2. Apazidis, A., Ricart, P.A., Diefenbach, C.M., Spivak, J.M.: The prevalence of transitional vertebrae in the lumbar spine. Spine J. **11**(9), 858–862 (2011)
3. Belharbi, S., et al.: Spotting l3 slice in ct scans using deep convolutional network and transfer learning. Comput. Biol. Med. **87**, 95–103 (2017)
4. Blanc-Durand, P., et al.: Abdominal musculature segmentation and surface prediction from ct using deep learning for sarcopenia assessment. Diagnost. Intervent. Imaging **101**(12), 789–794 (2020)
5. Bozzetti, F.: Forcing the vicious circle: sarcopenia increases toxicity, decreases response to chemotherapy and worsens with chemotherapy. Ann. Oncol. **28**(9), 2107–2118 (2017)
6. Castiglione, J., Somasundaram, E., Gilligan, L.A., Trout, A.T., Brady, S.: Automated segmentation of abdominal skeletal muscle in pediatric ct scans using deep learning. Radiol. Artif. Intell. **3**(2), e200130 (2021)
7. Cruz-Jentoft, A.J., et al.: Sarcopenia: revised european consensus on definition and diagnosis. Age Ageing **48**(1), 16–31 (2019)
8. Derstine, B.A., et al.: Quantifying sarcopenia reference values using lumbar and thoracic muscle areas in a healthy population. J. Nutr. Health Aging **21**(10), 180–185 (2017)
9. Du, Y., Karvellas, C.J., Baracos, V., Williams, D.C., Khadaroo, R.G.: Sarcopenia is a predictor of outcomes in very elderly patients undergoing emergency surgery. Surgery **156**(3), 521–527 (2014)
10. Ghesu, F.C., Georgescu, B., Grbic, S., Maier, A., Hornegger, J., Comaniciu, D.: Towards intelligent robust detection of anatomical structures in incomplete volumetric data. Med. Image Anal. **48**, 203–213 (2018)
11. Gilligan, L.A., Towbin, A.J., Dillman, J.R., Somasundaram, E., Trout, A.T.: Quantification of skeletal muscle mass: sarcopenia as a marker of overall health in children and adults. Pediatric Radiol. **50**(4), 455–464 (2020)
12. Grill, J.B., et al.: Bootstrap your own latent: A new approach to self-supervised learning. arXiv preprint arXiv:2006.07733 (2020)

13. Kanavati, F., Islam, S., Aboagye, E.O., Rockall, A.: Automatic l3 slice detection in 3D ct images using fully-convolutional networks (2018)
14. Lee, J., et al.: Skeletal muscle loss is an imaging biomarker of outcome after definitive chemoradiotherapy for locally advanced cervical cancer. Clin. Cancer Res. **24**(20), 5028–5036 (2018)
15. Lian, J., Levine, N., Cho, W.: A review of lumbosacral transitional vertebrae and associated vertebral numeration. Eur. Spine J. **27**(5), 995–1004 (2018)
16. Maicas, G., Carneiro, G., Bradley, A.P., Nascimento, J.C., Reid, I.: Deep reinforcement learning for active breast lesion detection from DCE-MRI. In: Descoteaux, M., Maier-Hein, L., Franz, A., Jannin, P., Collins, D.L., Duchesne, S. (eds.) MICCAI 2017. LNCS, vol. 10435, pp. 665–673. Springer, Cham (2017). https://doi.org/10.1007/978-3-319-66179-7_76
17. Mnih, V., et al.: Playing atari with deep reinforcement learning. arXiv preprint arXiv:1312.5602 (2013)
18. Mnih, V., et al.: Human-level control through deep reinforcement learning. Nature **518**(7540), 529–533 (2015)
19. Navarro, F., Sekuboyina, A., Waldmannstetter, D., Peeken, J.C., Combs, S.E., Menze, B.H.: Deep reinforcement learning for organ localization in ct. In: Medical Imaging with Deep Learning, pp. 544–554. PMLR (2020)
20. Nishioka, N., et al.: Association of sarcopenia with and efficacy of anti-pd-1/pd-l1 therapy in non-small-cell lung cancer. J. Clin. Med. **8**(4), 450 (2019)
21. Paszke, A., et al.: Automatic differentiation in PyTorch. In: NIPS Autodiff Workshop (2017)
22. Payer, C., Štern, D., Bischof, H., Urschler, M.: Coarse to fine vertebrae localization and segmentation with spatialconfiguration-net and u-net. In: Proceedings of the 15th International Joint Conference on Computer Vision, Imaging and Computer Graphics Theory and Applications - Volume 5: VISAPP, vol. 5, pp. 124–133 (2020). https://doi.org/10.5220/0008975201240133
23. Santilli, V., Bernetti, A., Mangone, M., Paoloni, M.: Clinical definition of sarcopenia. Clin. Miner. Bone Metab. **11**(3), 177 (2014)
24. Sutton, R.S., Barto, A.G.: Reinforcement Learning: An introduction. MIT Press, Cambridge (2018)
25. Suzani, A., Seitel, A., Liu, Y., Fels, S., Rohling, R.N., Abolmaesumi, Purang: Fast automatic vertebrae detection and localization in pathological CT scans - a deep learning approach. In: Navab, N., Hornegger, J., Wells, W.M., Frangi, A.F. (eds.) MICCAI 2015. LNCS, vol. 9351, pp. 678–686. Springer, Cham (2015). https://doi.org/10.1007/978-3-319-24574-4_81
26. Tarbouriech, J., Garcelon, E., Valko, M., Pirotta, M., Lazaric, A.: No-regret exploration in goal-oriented reinforcement learning. In: International Conference on Machine Learning, pp. 9428–9437. PMLR (2020)
27. Vlontzos, A., Alansary, A., Kamnitsas, K., Rueckert, D., Kainz, B.: Multiple landmark detection using multi-agent reinforcement learning. In: Shen, D., et al. (eds.) MICCAI 2019. LNCS, vol. 11767, pp. 262–270. Springer, Cham (2019). https://doi.org/10.1007/978-3-030-32251-9_29
28. Wang, Z., de Freitas, N., Lanctot, M.: Dueling network architectures for deep reinforcement learning. CoRR abs/1511.06581 (2015). http://arxiv.org/abs/1511.06581
29. Zopfs, D., et al.: Single-slice ct measurements allow for accurate assessment of sarcopenia and body composition. Eur. Radiol. **30**, 1701–1708 (2019)

MIST GAN: Modality Imputation Using Style Transfer for MRI

Jaya Chandra Raju[1,2(✉)], Kompella Subha Gayatri[1,2], Keerthi Ram[2],
Rajeswaran Rangasami[3], Rajoo Ramachandran[3],
and Mohanasankar Sivaprakasam[1,2]

[1] Indian Institute of Technology Madras, Chennai, India
[2] Healthcare Technology Innovation Centre, Chennai, India
[3] Sri Rama Chandra Institute of Higher Education and Research, Chennai, India

Abstract. MRI entails a great amount of cost, time and effort for generation of all the modalities that are recommended for efficient diagnosis and treatment planning. Recent advancements in deep learning research show that generative models have achieved substantial improvement in the aspects of style transfer and image synthesis. In this work, we formulate generating the missing MR modality from existing MR modalities as an imputation problem using style transfer. With a multiple-to-one mapping, we model a network that accommodates domain specific styles in generating the target image. We analyse the style diversity both within and across MR modalities. Our model is tested on the BraTS'18 dataset and the results obtained are observed to be on par with the state-of-the-art in terms of visual metrics, SSIM and PSNR. After being evaluated by two expert radiologists, we show that our model is efficient, extendable, and suitable for clinical applications.

Keywords: MRI · Image imputation · Deep learning · Image synthesis

1 Introduction

MRI is a versatile imaging technique that can image in multiple sequences each highlighting specific tissue type based on the TR (Repetition time) and TE (Time to Echo) parameters. The most commonly generated sequences are T1 (with optional gadolinium enhancement), T2 and FLAIR (Fluid-Attenuated Inversion Recovery), which are primarily used for diagnostic purposes in a clinical setting. It is stated in [5] that in brain MRI, MR images with T1, T2, or FLAIR contrast, are all required for accurate diagnosis and segmentation of cancer margin. At the same time, acquiring all the sequences from a patient is also an extremely tedious task, especially given the long scanning time and the added cost. Even if all the sequences are acquired, there could be faulty scans with distorted contrast which often are a result of motion artifacts or noise. T1c requires injecting a dye which may induce side effects and is not recommended for pregnant women or patients with kidney or heart issues. Considering these clinical

© Springer Nature Switzerland AG 2021
C. Lian et al. (Eds.): MLMI 2021, LNCS 12966, pp. 327–336, 2021.
https://doi.org/10.1007/978-3-030-87589-3_34

limitations and feasibility, it is therefore necessary to explore the potential of cross modality synthesis for efficient data infilling with respect to MRI.

Generating a new MRI modality from the existing modalities can be formulated as an image-to-image translation problem, whose goal is to learn a mapping between two different visual domains [9]. Here, domain implies a set of images that can be grouped as a visually distinctive category and each image has a unique appearance which is the style. For example, in MR images, each sequence can be considered as a domain and different variants in each sequence can be considered as different styles. In recent years, Generative Adversarial Networks (GANs) [6] have been extensively studied for image-to-image translation applications and showed promising outcomes [2,11,22]. Lately, cross domain image translation using GANs have been considered for MRI. Due to limited scalability, i.e. requiring N(N-1) generators for N-domain image transfer, many networks have been restricted for one-to-one synthesis [4,20]. The next formulation of significance is multiple input-to-multiple target synthesis of which models such as MM-GAN [18], AutoSyncoder [17], latent representation [1], have performed efficiently, but these networks remain either bulky or in need for a target based training. However, in medical image synthesis, multiple-to-one mapping models are more relevant as learning proprietary information of individual modalities can be more synergistic [13,14]. They affirm that a more accurate target image is generated when information from all the input modalities is used for training.

Though the aforementioned recent works successfully addressed the scalability issue for multi-domain MRI synthesis/imputation, none of the models consider the style diversity within the domain. Various MRI imaging centres follow various imaging protocols giving multiple possible combinations of sequence parameters each of which can be considered as a different style. Learning these styles and being able to generate a target image of the required style would greatly improve the comparability between two scans. For example, in multiple sclerosis (MS) patients, treatment planning and decisions are majorly based on longitudinal comparisons of MRI studies. Also, existing lesion quantification tools require completely identical modalities scanned at multiple time points [14]. It is therefore important to extract and reflect this style information. For this, a very recent framework, StarGAN v2 [3] introduced the concept of style code to represent diverse styles of a specific domain and it is learnt using two additional modules, style encoder and mapping network. Another recent work ReMIC [19], used style based representation disentanglement for image completion and segmentation in MRI.

Contribution: 1) We propose MIST GAN for MRI, which is a unified domain scalable architecture that learns the multiple-to-one cross-modality mapping with just a single generator and discriminator. 2) We establish that our model is capable of learning and reflecting the diverse styles present within a domain. We also prove that, contrary to the popular assumption, an MRI modality is not just a single style, but a spectrum of styles.

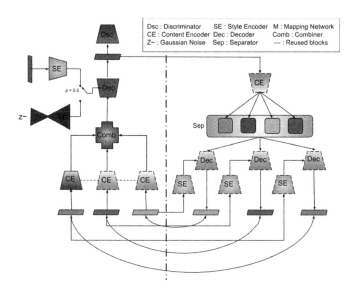

Fig. 1. MIST GAN overview. Left: Generates a target image from the given three input images which is of the style specified by *SE*. Right: The cycle to reconstruct the input images from the generated image with the help of *Sep* and reusable *CE*, *SE* and *Dec*.

2   Proposed Framework

Let $D = \{p, q, r, ..\}$ be the set of domains and x_d be the image drawn from domain $d \in D$. Given the target domain, t that is randomly sampled from D, the set of input domains will be $I = D - \{t\}$. When set of input images, $X_d = \{x_d | d \in I\}$ and a target image x_t is given, the goal of the network is to learn the mappings from multiple input modalities to the target modality while also learning varied styles within the domain. The following modules are included in our network to realize the same:

Generator (G):

Style Encoder (SE): extracts the fine style features present in a given reference image. *SE* takes the image, x_d and the domain information, d, as input to generate the domain specific style code $s_d = SE(x_d, d)$.

Mapping Network (M): is a fully connected network that learns to map random Gaussian noise (z), to latent style code distribution, $s_d = M(z, d)$. In applications where the target image is not given for the style encoding, a random target domain specific style code can be generated by using this mapping network which is guided by the discriminator(Dsc) during the training.

Content Encoder (CE): is expected to take in an image, X, and generate its content, $c = CE(x)$, in latent representation which is modality invariant. Instance normalization (IN) is used for the down-sampling blocks.

Decoder (Dec): Given the content(c) generated by CE our decoder tries to generate the target image, $\hat{x}_t = Dec(c, s_t)$, reflecting the provided target style code s_t. Here, we use Adaptive Instance Normalization(AdaIN) [8] for the up-sampling blocks and add the skip connections through a high-pass filter.

Combiner and Separator (Comb, Sep): The combiner generates a combined content, $cc = Comb(concatenate(c_p, c_q, c_r))$ by integrating the content generated by CE for different domain images, c_p, c_q, c_r where $p, q, r \in I$. cc stores the useful key features extracted from all the input modalities. The target image $\hat{x}_t$ is generated by passing combined content, cc, through the Dec. Once we have generated the target image, $\hat{x}_t$, it is encoded again using the CE in order to get the new combined content, $\hat{cc}$. Now, the separator takes in this content ($\hat{cc}$) and the modality information (d) of each input and generates domain specific content, $c_d = Sep(\hat{cc}, d)$. This is given to the Dec that generates back the original input images (Fig. 1).

Discriminator (Dsc) used in this network is a multi-task discriminator with multiple output branches where each branch learns a binary classification determining whether an image is a real image (x_d), or a decoder outputted fake image, ($\hat{x}_t$). It takes in an image(x_d) and its domain information(d), and predicts the probability($p_{r/f}$)= $Dsc(x_d/\hat{x}_t; d/t)$ of how real the given image is.

Training Objectives:

Cyclic Loss: To ensure that the network properly preserves the domain invariant characteristics of the given inputs, we enforce the Dec to reconstruct the input images from the target image by incorporating a cyclic loss $(CL = CSL + CCL)$ which consists of cyclic style loss(CSL) and cyclic content loss(CCL). CSL is the loss defined between the style information encoded from the generated fake image and the style information encoded from the target image sampled. To synthesize the input images from the imputed target image, $\hat{x}_t$, CE, Sep are used to get the corresponding modality specific content and SE, Dec are reused to generate the respective images, $\hat{x}_p$, $\hat{x}_q$ and $\hat{x}_r$. Now, CCL is defined as the summation of domain-wise $L1$ loss between translated images and source images.

$$CSL = ||SE(x_t) - SE(\hat{x}_t)||_1 \tag{1}$$

$$CCL = \sum_{d=p,q,r} ||\hat{x}_d - x_d||_1 \tag{2}$$

Adversarial Loss: The adversarial loss is defined to impose the network to produce images that are indistinguishable from the true distribution. The adversarial loss components are different for the generator part (G_{adv}) and the Dsc (Dsc_{adv}) as shown below:

$$G_{adv} = 1 - Dsc(\hat{x}_t, t) \tag{3}$$

$$Dsc_{adv} = (1 - Dsc(x_t, t)) + (0 - Dsc(\hat{x}_t, t)) \tag{4}$$

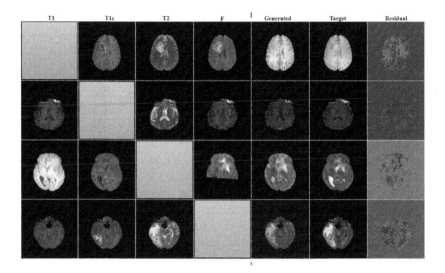

Fig. 2. Qualitative results achieved by the proposed model on the HGG BraTS'18 dataset. A sample of the domain-wise generated images along with their respective ground truths are shown. The blue squares represent the missing/target modality. (Color figure online)

Style Diversification Loss (SDL): To enable the network to produce diverse styled images, we regularize it with the diversity sensitive loss [3]. Two reference images are randomly sampled from the target domain $(x_{t1}; x_{t2})$ to get $s_{t1}; s_{t2}$ estimated by SE. We generate images using these two styles and maximize the difference between them which forces the network to learn meaningful style features. The total objective can be summarized for generator and discriminator as follows:

$$\min_{G}(CL + G_{adv} - SDL) \quad \text{and} \quad \min_{Dsc}(Dsc_{adv}) \tag{5}$$

3   Results and Discussion

Dataset and Implementation: The performance of the proposed architecture is evaluated on BraTS'18 dataset [15,16] that consists of 75 LGG and 210 HGG patient records of T1, T1c, T2 and Flair modalities all of which are axially acquired, co-registered and skull stripped. The dataset is split for training, validation and testing in the ratio 3:1:1. All the slices are padded to a uniform size of 256×256 and the intensity values are linearly normalized between (0,1). For the training, the batch size is set to 2 and the model is trained for 200k iterations. The training duration is approximately two days on a single NVIDIA GEFORCE GTX 1080 Ti GPU, occupying about 10 GB. Adam [12] optimizer is adopted for training and the learning rate is set to 1e-4 for all modules except mapping network for which it is set as 1e-6. All weights are initialized using [10].

Table 1. Quantitative performance of the model on BraTS'18 HGG and LGG cohort.

	Modality	Metrics	Stargan-v2	MM-GAN	Proposed
LGG	T1	SSIM	0.90143 ± 0.01685	**0.9288 ± 0.0137**	0.90193 ± 0.02157
		PSNR	22.96876 ± 3.28355	**25.9841 ± 2.1926**	24.68469 ± 3.87952
	T1c	SSIM	0.89411 ± 0.01774	**0.9086 ± 0.0253**	0.89597 ± 0.02294
		PSNR	24.61375 ± 2.66390	22.4980 ± 1.1684	**25.07036 ± 3.43035**
	T2	SSIM	0.88846 ± 0.02120	**0.9030 ± 0.0303**	0.88411 ± 0.02094
		PSNR	24.60764 ± 2.88763	**25.6005 ± 2.7909**	24.761391 ± 2.57905
	F	SSIM	0.86251 ± 0.02914	0.8692 ± 0.0224	**0.87498 ± 0.02930**
		PSNR	22.67723 ± 2.65003	23.0852 ± 1.5142	**23.72480 ± 2.86680**
HGG	T1	SSIM↑	0.88832 ± 0.01787	**0.9228 ± 0.0190**	0.88226 ± 0.02094
		PSNR	23.99222 ± 2.58524	24.173 ± 3.2754	**25.62511 ± 1.94889**
	T1c	SSIM	0.87734 ± 0.02066	**0.9239 ± 0.0375**	0.87303 ± 0.02411
		PSNR	24.09959 ± 3.23823	24.372 ± 2.2792	**26.42274 ± 1.80974**
	T2	SSIM	0.86777 ± 0.01812	**0.9349 ± 0.0262**	0.87160 ± 0.02162
		PSNR	23.45378 ± 2.26071	**28.678 ± 2.3290**	24.88888 ± 1.33794
	F	SSIM	0.84534 ± 0.02543	**0.9150 ± 0.0275**	0.85406 ± 0.02861
		PSNR	21.95010 ± 2.74631	**26.397 ± 1.9733**	24.79925 ± 2.10912

Qualitative and Quantitative Analysis: We generate a target modality that not only learns relevant information from multiple input modalities, but also reflects the user specified style. Any 3 of 4 MRI modalities, irrespective of their ordering are taken as inputs and the missing modality is imputed using style transfer. We ensure domain scalability and limit the network complexity by reusing the core blocks (CE, SE, Dec). The model is studied for both HGG and LGG datasets of BraTS'18 and quantitatively evaluated using PSNR (Peak Signal-to-Noise Ratio) and SSIM (Structural Similarity Index) metrics. The results for the same are shown in Table 1. Based on these metrics it is observed that our synthesized T1c is closer to the real T1c for both HGG and LGG, thereby opening the possibility of generating a contrast enhanced image without actually injecting the contrast agent, an application which is of high clinical relevance.

We compare the achieved results with the results of a baseline model, MM-GAN, which incorporates a target based training with supervised loss. Despite being blind to the target image and solely guided by the discriminator, our model performed on par, better for few modalities (LGG - T1c, F; HGG - T1, T1c) in terms of PSNR (Table 1). We have also implemented StarGAN v2, a style based model for MRI and recorded the metrics for comparison. We trained the model as is, but while testing, instead of just giving one image, we reused the encoder of the generator for multiple inputs and averaged the content in the latent space and passed it to the decoder to get the generated image. It can be seen from Table 1 that the proposed model clearly outperformed StarGAN v2. A sample of our modality-wise generated output is shown in Fig. 2 and it can be observed

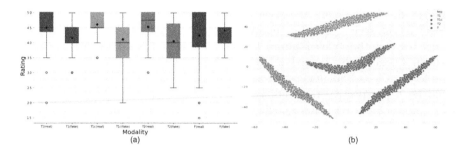

Fig. 3. (a): Validation scores assigned by expert radiologists for all the modalities when blinded to the ground truth label of the image. (b): t-SNE plot showing disengagement of style codes across four domains and a spectrum of styles within each domain.

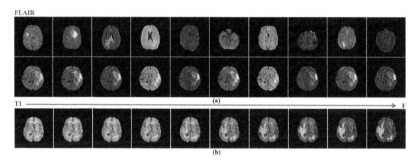

Fig. 4. (a) Representation of style forms: the second row shows the Flair generated images when its corresponding first row image is given as a reference image to the style encoder. (b) Style interpolation: T1 to F transition, achieved by linearly interpolating the styles between mean style of T1 and mean style of Flair in steps of 0.1.

that, generated T1 and T1c are enhanced and more sharp than their respective ground truths, a proper T2 is generated even though one of its source images is faulty and generated Flair is quite close to its original image.

Style Analysis: In order to study the efficiency of our model in learning, extracting and reflecting varied styles, we attempted generating different style forms of a single image in a given domain. We fed the network with any arbitrary style image and made it generate the same target image in the specified random style. Figure 4(a) shows the style forms for Flair domain: top row shows the given reference style images and the bottom row shows the corresponding generated target images. Clear variations can be noticed in the generated images indicating that any modality is not just a single style but a spectrum of styles and our model succeeded in capturing them. The same is evident from the t-SNE(t-distributed stochastic neighbor embedding) plot Fig. 3(b) of the latent style codes, in which it is also seen that style information is properly disentangled between the four domains. With the acquired domain specific mean style value, we observed the transition of a generated image between two domains, while gradually changing the style (simple weighted average/linear interpolation of latent style), thereby enabling operator, the flexibility to choose and change between the styles. Figure 4(b) shows one such smooth transition achieved between T1 and Flair. Indeed an unseen intermediate modality which is neither completely T1 or Flair has surfaced in this interpolation whose clinical utility could be a prospective research. However, some droplet artifacts are noticed when interpolated with intermediate styles which is probably due to a harsh style diversification loss. For quantitative analysis, mean FID (Frechét inception distance) [7] and mean LPIPS (Learned perceptual image patch similarity) [21] are extracted for all modalities. The mean FID (between the generated images and the true images) and LPIPS (across the generated images) scores for T1, T1c, T2 and Flair are [49.37, 70.52, 49.66, 74.25] and [0.0159, 0.0261, 0.0092, 0.0277] respectively when assessed for BraTS'18 dataset.

Visual Evaluations by Radiologists: For qualitative assessment of the achi--eved results, we consulted two expert radiologists with 10+ years of professional experience. They each evaluated the output image on a scale of 5 (1, being very poor and 5, being excellent). In each trial, they were provided with four images, the left three were the source images and on the right, a corresponding target image was shown, which is either a real image or a generated one. Radiologists are blinded to the nature (real or fake) of the image so as to avoid any bias and to collect a more objective feedback. Each of them are presented with 400 such random trials for all the four modalities. The rating was given based on the image quality, capturing of small sub-structures and efficiency in mapping the source images to the target image. The ratings of the two radiologists are averaged and the boxplot of the results are shown in Fig. 3(a). From the plot it is noticed that for T1, T1c and T2, the mean rating of synthesized images lags the mean rating of real images by about 0.5. But for the Flair modality, the generated images were given a better rating when compared to the real images. Both the experts acknowledged that the synthetic Flair images are as good as the original flair. The radiologists' opinion was also taken on the performed style analysis of the output images and they both affirmed that the model captured multiple styles within each domain and that such work would help in longitudinal MR comparability studies of patients with progressive neurological conditions.

4 Conclusion

Motivated by the need to create an efficient and novel method for MRI modality synthesis, we have derived a style transfer formulation for imputation and realized an efficient GAN based network that generates an enhanced version of the missing MR modality from given three input modalities. We show that our network is not only scalable to any number of input modalities, but also capable of picking up style variations within each modality. Evaluation on the BraTS'18 multimodal brain MRI dataset suggests that the method is promising and opens new avenues for further research.

References

1. Chartsias, A., Joyce, T., Giuffrida, M.V., Tsaftaris, S.A.: Multimodal MR synthesis via modality-invariant latent representation. IEEE Trans. Med. Imaging **37**(3), 803–814 (2018)
2. Choi, Y., Choi, M., Kim, M., Ha, J.W., Kim, S., Choo, J.: StarGAN: unified generative adversarial networks for multi-domain image-to-image translation, pp. 8789–8797 (2018)
3. Choi, Y., Uh, Y., Yoo, J., Ha, J.W.: Stargan v2: diverse image synthesis for multiple domains. In: Proceedings of the IEEE Conference on Computer Vision and Pattern Recognition (2020)
4. Dar, S.U., et al.: Image synthesis in multi-contrast mri with conditional generative adversarial networks. In: IEEE Transactions on Medical Imaging (2019)
5. Drevelegas, A., Papanikolaou, N.: Imaging modalities in brain tumors. In: Drevelegas, A. (ed.) Imaging of Brain Tumors with Histological Correlations, pp. 13–33. Springer, Heidelberg (2011). https://doi.org/10.1007/978-3-540-87650-2_2
6. Goodfellow, I.J., et al.: Generative adversarial networks. Commun. ACM **63**(11), 139–144 (2014)
7. Heusel, M., Ramsauer, H., Unterthiner, T., Nessler, B., Hochreiter, S.: Gans trained by a two time-scale update rule converge to a local nash equilibrium. Asian J. Appl. Sci. Eng. **8**, 25–34 (2018)
8. Huang, X., Belongie, S.: Arbitrary style transfer in real-time with adaptive instance normalization (2017)
9. Isola, P., Zhu, J.Y., Zhou, T., Efros, A.A.: Image-to-image translation with conditional adversarial networks (2018)
10. He, K., Zhang, X., Ren, S., Sun, J.: Delving deep into rectifiers: surpassing human-level performance on imagenet classification. In: ICCV (2015)
11. Kim, T., Cha, M., Kim, H., Lee, J.K., Kim, J.: Learning to discover cross-domain relations with generative adversarial networks (2017)
12. Kingma, D.P., Ba., J.: Adam: a method for stochastic optimization. In: ICLR (2015)
13. Lee, D., Moon, W.J., Ye, J.C.: Assessing the importance of magnetic resonance contrasts using collaborative generative adversarial networks. Nat. Mach. Intell. **2**(1), 34–42 (2020)
14. Li, H., et al.: DiamondGAN: unified multi-modal generative adversarial networks for MRI sequences synthesis. In: MICCAI 2019
15. Menze, B., et al.: The multimodal brain tumor image segmentation benchmark (BRATS). IEEE Trans. Med. Imaging, **34**(10), 1993–2024 (2014)
16. MenzeBjoernHetal: The multimodal brain tumor image segmentation benchmark (brats). IEEE Trans. Med. Imaging **34**(10), 1993–2024 (2015). https://doi.org/10.1109/TMI.2014.2377694
17. Raju, J., Murugesan, B., Ram, K., Sivaprakasam, M.: AutoSyncoder: an adversarial autoencoder framework for multimodal MRI synthesis. In: Deeba, F., Johnson, P., Würfl, T., Ye, J.C. (eds.) MLMIR 2020. LNCS, vol. 12450, pp. 102–110. Springer, Cham (2020). https://doi.org/10.1007/978-3-030-61598-7_10
18. Sharma, A., Hamarneh, G.: Missing MRI pulse sequence synthesis using multimodal generative adversarial network. IEEE Trans. Med. Imaging **39**(4), 1170–1183 (2020)
19. Shen, L., et al.: Multi-domain image completion for random missing input data. IEEE Trans. Med. Imaging **40**(4), 1113–1122 (2021). https://doi.org/10.1109/TMI.2020.3046444

20. Yang, Q., L.N.Z.Z., et al.: Mri cross-modality image-to-image translation. Sci. Rep. **10**, 3753 (2020). https://doi.org/10.1038/s41598-020-60520-6
21. Zhang, R., Isola, P., Efros, A.A., Shechtman, E., Wang, O.: The unreasonable effectiveness of deep features as a perceptual metric. In: CVPR (2018)
22. Zhu, J.Y., Park, T., Isola, P., Efros, A.A.: Unpaired image-to-image translation using cycle-consistent adversarial networks (2020)

Biased Extrapolation in Latent Space for Imbalanced Deep Learning

Suhyeon Jeong and Seungkyu Lee[✉]

Kyunghee University, Yongin, Republic of Korea
{jun9suhyun,seungkyu}@khu.ac.kr

Abstract. Addressing class data imbalance to improve generalization on minor classes is critical in medical applications. Traditional approaches including re-weighing and re-sampling have shown the potential of the generalization but ignore statistical characteristics of classes. We study the potential and effectiveness of data extrapolation in latent space of deep learning networks to address data imbalance. We propose biased normal sample selection and latent space sample extrapolation methods for imbalanced deep learning. Two types of biases in the extrapolations are sample bias and extrapolation bias. Experimental evaluation is performed for ulcer classification in endoscopy images and Cardiomegaly detection from CXR. We show that new abnormal samples extrapolated asymmetrically from biased normal samples of low probability improve the separation between normal and abnormal classes.

Keywords: Imbalanced learning · Extrapolation · Biased learning

1 Introduction

Traditionally, minor sample re-weighing or interpolation for re-sampling have been used for class imbalanced learning. Chawla et al. [2] select observations of a minor class and then create new minor samples among neighbor samples using K-nearest neighbors method. Chen et al. [12] present RAMOBoost which adaptively determines the ranking of minor classes according to data distribution in an ensemble learning system. Traditional oversampling methods generate minority samples in some conditions, making learning more difficult. To resolve the problem, Barua et al. [11] implement MWMOTE using a clustering method and weighting according to Euclidean distance. Mullick et al. [10] propose GAMO to reduce misclassification by oversampling minority classes and to handle the class imbalance problem. Bae and Yoon [3] propose a data sampling-based boosting framework for detecting polyps with techniques such as least square analysis. However, this method has difficulty in obtaining distant new samples from given abnormal samples. Lin et al. [4] implement a general imbalanced classification model on deep reinforcement learning by using a deep Q learning algorithm and formulating their classification task as a sequential decision-making process. Xu et al. [5] apply automatic data augmentation to medical image segmentation tasks using a stochastic natural gradient method which is a time-consuming

© Springer Nature Switzerland AG 2021
C. Lian et al. (Eds.): MLMI 2021, LNCS 12966, pp. 337–346, 2021.
https://doi.org/10.1007/978-3-030-87589-3_35

process. All of such interpolation approaches create new minor samples inside the given minor class region of feature space showing limited performance in expanding the distribution of minor class.

On the other hand, extrapolation methods find new minor samples from the outskirts of a given minor class region. Han et al. [6] add minor samples that are difficult to classify with the current classifier by creating samples on the borderline with other classes, rather than just sampling within the minor classes. Lee et al. [7] conduct feature space extrapolation for ulcer data augmentation. However, the method shows limited robustness to out-liers because the extrapolation is performed based on Euclidean distance in relatively high dimensional feature space. To compensate for the limitation, Lee et al. [8] improve imbalanced learning proposing decision boundary re-sampling (DBR) that finds new ulcer samples on the decision boundary in the latent space of deep convolutional neural network. They extrapolate abnormal samples with good normal samples of high probability. When the number of training samples is very small, we need to create new samples from distant locations not simply from the neighbor of given minor samples. In the processing of an under-represented class, Li et al. [9] propose to move the activation distribution towards under-represented class and across the decision boundary, while the samples of an over-represented class remain stable. They propose a margin tuning method by using asymmetric large margin loss and mix-up. Cao et al. [13] propose label-distribution-aware margin (LDAM) loss that gives a larger margin to minor classes for the learning with imbalanced data sets.

In most of the prior approaches, highly confident samples are chosen for both interpolation and extrapolation in order to let new synthesized samples contain good features of expected class. These schemes mostly produce good but less effective samples in imbalanced deep learning, because they are able to neither create unseen new samples nor expand the distribution of minor classes significantly. Our insight is that synthesized minor samples do not have to contain representative good feature values for the class in the current training stage. Preferably, we expect synthesized samples are located at struggling locations of the feature space for the current classification such as very around the decision boundary. Then the synthesized samples promote deep learning networks to further investigate the separation of normal and abnormal samples located at the challenging location of the feature space that was difficult to be clearly classified in the current training step. With such extrapolated samples, deep learning network is induced to draw improved decision boundary at the local area and clarify the separation of those weak samples around the decision boundary of initial classification. In other words, multiplied challenging samples reinforce the classification providing fine-grained descriptions of the classes.

In this work, we propose biased training sample extrapolation for imbalanced deep learning based on decision boundary resampling scheme [8]. We propose two types of biases in our extrapolations: sample bias and extrapolation bias. For sample bias, sample probability or density is used to assign a bias in the selection of existing normal samples for decision boundary resampling. For extrapolation bias, asymmetric probability bias is used to adjust the location of abnormal instance sampling. In imbalanced learning, classifiers tend to be over-fitted on major(normal) class shifting decision boundary toward minor(abnormal) class

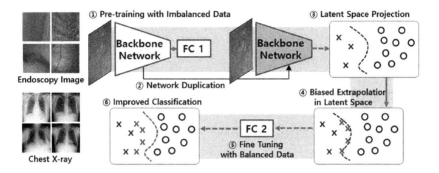

Fig. 1. Biased sample extrapolation: (Step 4 to Step 6) is performed repeatedly adding new abnormal samples of fixed proportion until class imbalance is resolved. The initial parameters of FC2 are set to pre-trained FC1 parameters.

region [9, 13]. Therefore, abnormal sample extrapolation on such decision boundary has improved classification accuracy [8]. Our asymmetry probability bias assigns normal preferred bias that shifts the location of abnormal sample extrapolation from decision boundary toward the normal class region in the feature space. We demonstrate the effectiveness of our biased extrapolation in imbalanced deep learning with two classification tasks: ulcer classification in endoscopy images of our own data set and cardiomegaly detection from chest x-ray (CXR) on both our own and NIH CXR data sets [1].

2   Biased Extrapolation

Figure 1 illustrates the procedure of proposed biased extrapolation. First, a backbone network is trained with imbalanced data set building an initial classifier. And then representations of all training data in a latent space are acquired (Latent space projection). In order to minimize the dimension of feature space for efficient extrapolation, we project training data onto the latent space right before the fully connected layer. Biased extrapolation is performed in the latent space manipulating feature vectors of normal and abnormal training samples. More specifically, all minor(abnormal) instances are paired to the closest major(normal) instances. And new minor(abnormal) samples are extrapolated in the feature space from the normal-abnormal pairs. If we choose a sample of 0.5 probability for both normal and abnormal classes, we are adding a new sample that is located on the current decision boundary (decision boundary sampling). This extrapolation process is performed repeatedly adding new abnormal samples of fixed proportion until the class imbalance is resolved. In our extrapolation, we use two types of sample bias (probability and probability density) and one type of extrapolation bias (asymmetry bias). Sample bias decides which normal instance is going to be used for building normal-abnormal pair and corresponding decision boundary sampling. Biased selection of major class instances (normal in usual medical tasks) for minor class sample extrapolation gives improved control of minor sample synthesis. On the other hand, extrapolation bias decides in

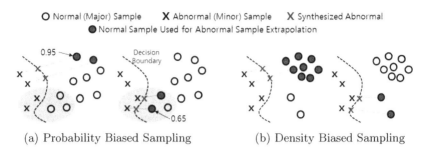

(a) Probability Biased Sampling   (b) Density Biased Sampling

Fig. 2. Extrapolation with sample bias: (a) Abnormal sample extrapolation with low (or high) probability normal instances (b) Abnormal sample extrapolation with low (or high) density normal instances

which location in the latent space we make the new minor sample. Asymmetry bias finds new abnormal samples from one-sided location toward either normal or abnormal class region.

2.1   Probability Biased Sampling

Figure 2(a) illustrates how abnormal samples extrapolated with relatively low probability (biased) normal instances affect the improvement of classification accuracy. Extrapolated abnormal samples from good normal instances (of high probability like 0.95) do not make significant changes in the decision because good normal instances are already very well classified in the current step. On the other hand, extrapolated abnormal samples from biased normal instances (of low probability like 0.65) affect more on the improvement of classification accuracy providing a finer scale of class separation in such struggling local regions of feature space. Therefore, we choose normal instances of lower probability for abnormal sample extrapolation.

2.2   Density Biased Sampling

Probability bias only considers separate characteristics of very local region around a training instance. On the other hand, in the condition that we have enough amount of training instances, the density of samples reveals the characteristic of the broader area of the distribution. In this regard, we assign sample density bias to normal samples. Parzen windowing is applied on given samples of each class to obtain the sample density of each normal and abnormal sample. Similar to probability bias as illustrated in Fig. 2(b), abnormal samples are extrapolated with the low (or high) density of normal samples.

2.3   Asymmetry Probability Biased Sampling

Asymmetry probability bias is a type of extrapolation bias. The location of the new abnormal sample is adjusted based on the sample density d_{normal} of

Fig. 3. Asymmetry probability biased sampling: the location of new abnormal sample extrapolation is moved towards normal side in proportion to corresponding normal sample density.

normal of corresponding normal-abnormal pair. Following equations are basic formulations for the decision of the location of new abnormal samples.

$$P_{Asym} = 0.5 + \alpha \cdot P_{max} \cdot d_{normal} \tag{1}$$

$$P_{Asym} = 0.5 + \beta \cdot P_{max} \cdot \ln(d_{normal}) \tag{2}$$

where P_{Asym} indicates the abnormal probability of extrapolated abnormal sample in the current feature space and P_{max} is upper bound of asymmetry probability. If α or β is zero, $P_{Asym} = 0.5$ means that we extrapolate new abnormal samples exactly on decision boundary. With the positive values of the weights α or β, the location of new abnormal sample extrapolation is moved towards the abnormal side in proportion to corresponding normal sample density d_{normal} (abnormal probability bias). With the negative values of the weights α or β, the location of new abnormal sample extrapolation is moved towards the normal side in proportion to corresponding normal sample density d_{normal} as illustrated in Fig. 3 (normal probability bias). Equation (1) or (2) apply identical asymmetry probability bias at each extrapolation iteration (step 4–6 in Fig. 1).

$$P_{Asym} = 0.5 + \gamma \cdot P_{max} \cdot (max(1, (d_{normal})/(d_{abnormal})) - 1) \tag{3}$$

On the other hand, Eq. (3) considers the density ratio of normal and abnormal samples. Therefore, asymmetry probability bias decreases as abnormal sample density $d_{abnormal}$ increases.

3   Experimental Evaluation

We evaluate our biased extrapolation in imbalanced learning for two classification tasks: ulcer classification in endoscopy images of our own data set and cardiomegaly detection from chest x-ray (CXR) on both our and NIH data sets.

3.1   Ulcer Classification in Endoscopy Image

In ulcer image classification, total number of normal and abnormal(ulcer) samples are 4500 and 1000 respectively. VGG19 is used as our backbone network that is initially trained (epoch = 25). Extrapolation is repeated 19 times adding new ulcer samples up to 10% for each iteration (epoch = 10). Empirical value of P_{max} is 0.3 and multiple α, β, and γ values are tested. For probability and density bias, 30% of biased normal samples are selected for all tests. Table 1 summarizes all results of our three tests.

Table 1. Ulcer classification tests: precision-recall AUC

Bias	Type	PR AUC
	No extrapolation	0.1760
	Random [8]	0.1883
Test 1: Probability	Top 30%	0.1819
	Bottom 30%	**0.1995**
Test 2: Density	Top 30%	0.1699
	Bottom 30%	0.1679
Test 3: Asymmetry	$\alpha = -1.0$	0.2064
	$\beta = -1.0$	0.2062
	$\gamma = -0.4$	**0.2073**

Test 1: Probability Bias. Figure 4(a) compares precision-recall curves of four cases: No extrapolation, Extrapolation with randomly chosen normal samples (Random) [8], Extrapolation with top 30% of normal sample probability (Top), and Extrapolation with bottom 30% of normal sample probability (Bottom). 30% is the maximum ratio of normal sample selection which is chosen empirically to allow enough number of normal samples in the iterative extrapolation steps. Experimental result shows that extrapolating ulcer samples with normal samples of low probability (Bottom 30%) outperforms other cases. Compared to Top 30% test result, it shows that our deep learning network comes to extract improved decision boundary with new ulcer training samples clarifying the separation with weak normal samples around the decision boundary of initial classification.

Test 2: Density Bias. Similar to test 1, we have extrapolated ulcer samples with top 30% and bottom 30% of normal sample density. However, in this test, simple biased sampling with the sample density decreases classification performance. With limited amount of training data, sample density itself may vary along the composition of the training samples. Therefore, sample density does not reveal any critical characteristic of latent class distribution.

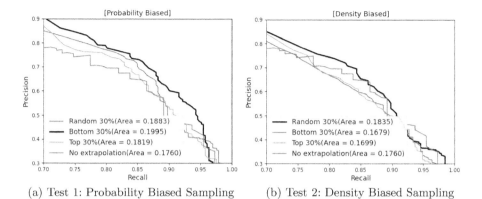

(a) Test 1: Probability Biased Sampling   (b) Test 2: Density Biased Sampling

Fig. 4. Ulcer classification tests: precision-recall curves for Test 1 and 2

Table 2. PR AUC with varying positive and negative α, β values

α	AUC_α	α	AUC_α	β	AUC_β	β	AUC_β
−1.0	0.2064	0.2	0.1968	−1.0	0.2062	0.2	0.1962
−0.8	0.2055	0.4	0.2032	−0.8	0.2043	0.4	0.1982
−0.6	0.1984	0.6	0.1949	−0.6	0.2053	0.6	0.1938
−0.4	0.1988	0.8	0.1967	−0.4	0.2041	0.8	0.1973
−0.2	0.1971	1.0	0.1955	−0.2	0.2016	1.0	0.2010

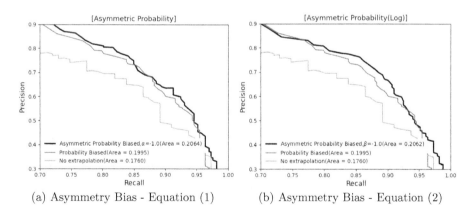

(a) Asymmetry Bias - Equation (1)   (b) Asymmetry Bias - Equation (2)

Fig. 5. Ulcer classification tests: precision-recall curves for Test 3 compared to the best result of Test 1

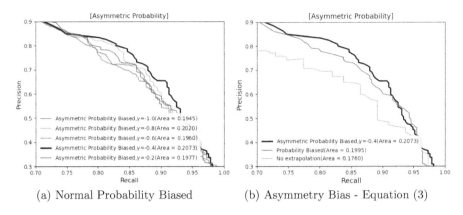

(a) Normal Probability Biased   (b) Asymmetry Bias - Equation (3)

Fig. 6. Ulcer classification tests: precision-recall curves for Test 3 with varying negative γ values

Test 3: Asymmetry Probability Bias. Asymmetry probability bias is implemented on the probability biased sampling of test 1. Asymmetry extrapolation based on the Eqs. (1), (2), and (3) are evaluated. Table 2 compares test results with varying positive and negative α, β values of the Eqs. (1) and (2). As we have expected, for both cases, normal probability bias with negative α and β show best performance. Figure 5 compares asymmetry probability bias results with the result of test 1 showing that asymmetry probability improves classification performance further over simple probability bias (Table 1). Figure 6 compares test results with varying negative γ of Eq. (3) showing best classification improvement over all other cases.

3.2   Cardiomegaly Detection from CXR

In the cardiomegaly detection test on chest X-ray (CXR), two data sets are used: our own cardiomegaly CXR dataset and NIH Chest X-ray dataset [1]. In our cardiomegaly CXR dataset, total number of normal and abnormal(cardiomegaly) samples are 1753 and 816 respectively. VGG19 is used as a backbone network that is initially trained (epoch = 15). In the NIH Chest X-ray dataset, total number of normal and abnormal(cardiomegaly) samples are 2000 and 1094 respectively. ResNet [14] is used as a backbone network that is initially trained (epoch = 15). Figure 7(a) shows our own cardiomegaly CXR results. In this test, high probability biased (Top 30%) or random extrapolation results are better than low probability biased (Bottom 30%) result. However, asymmetry probability bias improves the probability biased result showing best PR AUC. Figure 7(b) shows NIH cardiomegaly CXR results. Probability biased extrapolation (bottom 30%) improves the result of random extrapolation. Furthermore, asymmetry probability bias improves the probability biased result.

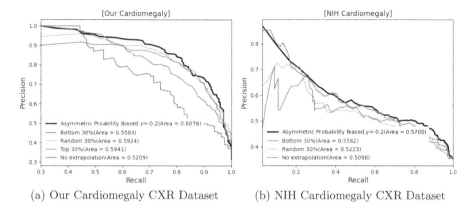

(a) Our Cardiomegaly CXR Dataset   (b) NIH Cardiomegaly CXR Dataset

Fig. 7. Cardiomegaly detection tests: precision-recall curves

4   Conclusion

We propose biased normal sample selection and latent space sample extrapolation method for class imbalanced deep learning. Proposed probability bias with asymmetry bias has achieved the best abnormal training sample extrapolation and classification result over prior work. Our effective but simple method can be applied to any imbalanced deep learning task in medical applications.

References

1. Wang, X., Peng, Y., Lu, L., Lu, Z., Bagheri, M., Summers, R.: ChestX-ray8: hospital-scale chest x-ray database and benchmarks on weakly-supervised classification and localization of common thorax diseases. In: IEEE CVPR, pp. 3462–3471 (2017)
2. Chawla, N.V., Bowyer, K.W., Hall, L.O., Kegelmeyer, W.P.: Smote: synthetic minority over-sampling technique. J. Artif. Intell. Res. **16**, 321–357 (2002)
3. Bae, S., Yoon, K.: Polyp detection via imbalanced learning and discriminative feature learning. IEEE Trans. Med. Imaging **34**(11), 2379–2393 (2015)
4. Lin, E., Chen, Q., Qi, X.: Deep reinforcement learning for imbalanced classification. Appl. Intell. **50**(8), 2488–2502 (2020). https://doi.org/10.1007/s10489-020-01637-z
5. Xu, J., Li, M., Zhu, Z.: Automatic data augmentation for 3D medical image segmentation. In: International Conference on Medical Image Computing and Computer-Assisted Intervention, pp. 378–387 (2020)
6. Han, H., Wang, W.Y., Mao, B.H.: Borderline-smote: a new over-sampling method in imbalanced data sets learning. In: Proceedings of International Conference on Intelligent Computing, pp. 878–887 (2005)
7. Lee, C., Min, J., Cha, J., Lee, S.: Feature space extrapolation for ulcer classification in wireless capsule endoscopy images. In: The IEEE International Symposium on Biomedical Imaging (ISBI), pp. 100–103 (2019)
8. Lee, C., Min, J., Cha, J., Lee, S.: Decision boundary re-sampling in imbalanced learning for ulcer detection. IEEE Access **8**, 186274–186278 (2020)

9. Li, Z., Kamnitsas, K., Glocker, B.: Overfitting of neural nets under class imbalance: analysis and improvements for segmentation. In: International Conference on Medical Image Computing and Computer-Assisted Intervention, pp. 402–410 (2019)
10. Mullick, S., Data, S., Das, S.: Generative adversarial minority oversampling. In: Proceedings of the IEEE/CVF International Conference on Computer Vision, pp. 1695–1704 (2019)
11. Barua, S., Islam, M.M., Yao, X., Murase, K.: MWMOTE-majority weighted minority oversampling technique for imbalanced data set learning. IEEE Trans. Knowl. Data Eng. 26(2), 405–425 (2012)
12. Chen, S., He, H., Garcia, E.A.: RAMOBoost: ranked minority oversampling in boosting. IEEE Trans. Neural Netw. 21(10), 1624–1642 (2010)
13. Cao, K., Wei, C., Gaidon, A., Arechiga, N., Ma, T.: Learning imbalanced datasets with label-distribution-aware margin loss. In: Conference on Neural Information Processing Systems (NeurIPS) (2019)
14. He, K., Zhang, X., Ren, S., Sun, J.: Deep residual learning for image recognition. In: IEEE Conference on Computer Vision and Pattern Recognition (CVPR) (2016)

3DMeT: 3D Medical Image Transformer for Knee Cartilage Defect Assessment

Sheng Wang[1,4], Zixu Zhuang[1,4], Kai Xuan[1], Dahong Qian[1], Zhong Xue[4],
Jia Xu[2], Ying Liu[3], Yiming Chai[2], Lichi Zhang[1], Qian Wang[1(✉)],
and Dinggang Shen[4(✉)]

[1] School of Biomedical Engineering, Shanghai Jiao Tong University, Shanghai, China
wang.qian@sjtu.edu.cn
[2] Shanghai Jiao Tong University Affiliated Sixth People's Hospital, Shanghai, China
[3] Pekinng University Third Hospital, Beijing, China
[4] Shanghai United Imaging Intelligence Co., Ltd., Shanghai, China
dgshen@shanghaitech.edu.cn

Abstract. While convolutional neural networks (CNNs) are dominating the area of computer-aided 3D medical image diagnosis, they are incapable of capturing global information due to the intrinsic locality of convolution. Transformers, another type of neural network empowered with self-attention mechanism, are good at representing global relations, yet computationally expensive and do not generalize well on small datasets. Applying Transformers on 3D medical images has two major problems: 1) medical 3D volumes are bigger in size than natural images which makes training process computationally impractical, 2) and 3D medical image datasets are usually smaller than natural image datasets since medical images are expensive to collect. In this paper, we propose the 3D Medical image Transformer (3DMeT) to address these two issues. 3DMeT introduces 3D convolutional layers to perform block embedding instead of the original linear embedding to cut the computational cost. Additionally, we propose a teacher-student training strategy to address the data-hungry issue by adapting convolutional layers' weights from a CNN teacher. We conduct experiments on knee images, results demonstrate that the 3DMeT (70.2) confidently outperforms the 3DCNNs (65.3) and Vision Transformer (58.7).

Keywords: Transformer · 3D medical image

1 Introduction

3D medical image, e.g. magnetic resonance imaging (MRI) image and computed tomography (CT) image, is frequently employed for the clinical diagnosis. Deep Convolutional Neural Network (CNN) based computer-aided diagnosis

---

S. Wang and Z. Zhuang—Contributed equally.

This work was supported by the National Key Research and Development Program of China (2018YFC0116400), National Natural Science Foundation of China (NSFC) grants (62001292), Shanghai Pujiang Program(19PJ1406800), and Interdisciplinary Program of Shanghai Jiao Tong University.

© Springer Nature Switzerland AG 2021
C. Lian et al. (Eds.): MLMI 2021, LNCS 12966, pp. 347–355, 2021.
https://doi.org/10.1007/978-3-030-87589-3_36

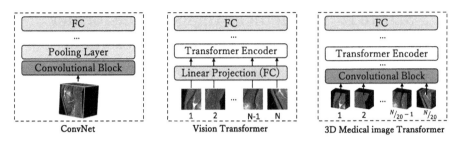

Fig. 1. Framework of conventional convolutional neural networks, Vision Transformer and our proposed 3D Medical image Transformer (3DMeT)

recently have become dominant on 3D medical image diagnosis [12]. Existing deep learning algorithms can be roughly summarized into two categories: slice-wise and subject-wise. Slice-wise methods [8] use 2DCNN to get slice-wise classification results and summarize it, which only pay attention to features and contexts within each single slice of a 3D volume. Subject-level methods [3,9] tackle slice-wise methods' problem by using 3DCNN. In spite of CNNs' success, these approaches generally exhibit limitations for modeling explicit long-range dependency, due to the intrinsic locality of convolution operations [15]. To better model the long-range dependency in natural images, researchers increase the model depth to enlarge the receptive field [6,11]. However, due to the size of 3D medical images, deeper models usually suffer from massive computational cost and become unfeasible.

On the other hand, Transformer [14], a type of deep neural network mainly based on self-attention mechanism, demonstrates its capability at modeling global relation. Inspired by the strong representation ability of Transformer, researchers propose to extend Transformers for computer vision (CV) tasks. By processing the image as sequence data, these methods show huge potential in CV-NLP joint tasks [10] and excellent performance on a wide range of visual tasks [2]. However, to our best knowledge, there has been no works incorporating the Transformer into 3D medical image diagnosis. It is because Transformers lack some inductive biases like translation invariance and locality, and therefore do not generalize well when trained on medical image datasets (usually many orders of magnitude difference with natural image datasets). In addition to data-starving, a 3D medical image input is much larger than a 2D natural image input thus has a much longer input sequence and the computational complexity of Transformers are quadratic in respect to the sequence length. Current Transformer structures are simply too expensive to compute (Fig. 1).

In this work, we propose 3D Medical image Transformer (3DMeT) to process 3D medical images. We demonstrate that Transformer could generalize well on a small datasets and being computationally practical on big 3D images. In 3DMeT, a 3D image is partitioned into 3D blocks and embedded by convolutional kernels instead of the a fully connected layer used in ViT. Table 2 compares the computational costs (float point operations, FLOPs) per image and

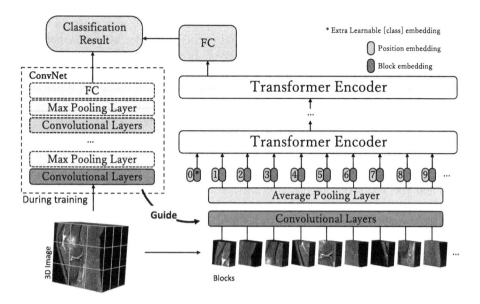

Fig. 2. The framework of the 3DMeT and ConvNet teacher mechanism. During training, we share several low-level convolutional layers of ConvNet to the 3DMeT for better encoded feature representation. The dotted-line parts of the ConvNet do not share weights with 3DMeT during training. In evaluation stage, the ConvNet is not used.

accuracy of different methods. Our methods are able to achieve the comparable performance using approximately 1/8 computational resources comparing with existing 3D CNNs. In addition, we propose ConvNet teacher mechanism to train 3DMeT with insufficient amount of medical image data. Since 3DMeT share the same components with CNNs (convolutional layers), ConvNet teacher mechanism takes trained convolution kernels from the first one/few layer(s) and uses in the 3DMeT (denoted as 3DMeT^C). This mechanism brings CNNs' spatial inductive biases such as translation invariance and locality to Transformers. Results demonstrate ConvNet teacher mechanism brings a considerable performance improvement on small datasets.

2 Method

In this section, we first introduce the Transformer and self-attention mechanism [14] followed by the Vision Transformer. Then we introduce our proposed 3D Medical image Transformer and the ConvNet teacher mechanism optimization strategy which is more data-efficient.

2.1 Transformer and ViT

Transformer. Transformer was firstly applied on the machine translation task in neural language processing [14]. It consists of an encoder module a decoder module. Encoder is composed of a self-attention layer [2] and a feed-forward neural network. Decoder has an additional encoder-decoder attention between the self-attention layer and feed-forward layer. In self-attention layer, the input vector is firstly transformed into three different vectors: q for query vector, k for key vector, and v for value vector. All three vectors have the same dimension d. Vectors derived from sequence inputs are then packed together into three matrices: Q, K and V. After that, the attention function can be calculated by:

$$\text{Attention}(Q, K, V) = \text{softmax}(\frac{Q \cdot K^T}{\sqrt{d}}) \cdot V \tag{1}$$

Then the self-attention layer is further improved by adding a mechanism called multi-head attention in order to boost the performance of the vanilla self-attention layer. Specifically, considering h attention heads: h different Q, K and V matrices are used for these different heads. Above process is irrelevant to the order of sequence input, thus the self-attention layer lacks the ability of capturing the positional information of inputs. To address this, a positional encoding with dimension d is added to the original input to get the final input vector.

Vision Transformer. Recently, [2] proposed vision Transformer (ViT), an image classifier which directly use image patches as sequence input. ViT only employs the encoder of the standard Transformer and a fully-connected layer is followed by the output of the Transformer encoder. ViT performs well on image classification tasks, specifically, when pretrained at sufficient scale (JFT- 300M) dataset, ViT approaches or beats state-of-the-art on multiple image recognition benchmarks, reaching accuracy of 88.36% on ImageNet. However, the author concluded that Transformers *"do not generalize well when trained on insufficient amounts of data"*, and the training of these models involved extensive computing resources.

2.2   Medical Image Transformer

Medical image Transformer (3DMeT) is a classifier Transformer architecture that process 3D medical image efficiently. Following the ViT [2], here we introduce how to handle 3D volume images in a similar scheme. As shown in Algorithm. 1, the image $x \in \mathbb{R}^{W \times H \times Z}$ is reshaped into a sequence of flattened 3D blocks $x_p \in \mathbb{R}^{(\frac{W \cdot H \cdot Z}{w \cdot h \cdot z}, w \cdot h \cdot z)}$. (W, H, Z) is the resolution of the original 3D volume and (w, h, z) is the size of each block. $N = (\frac{W \cdot H \cdot Z}{w \cdot h \cdot z})$ is then the sequence length for the Transformer. Then, a trainable linear projection maps each vectorized block to block embedding. As illustrated in Fig. 2, similar to BERT's [class] token [1], a learnable embedding is concatenate to the sequence of block embedding. In addition, 1D position embedding are added to the patch embedding to retain positional information. The joint embedding severs as input to the Transformer encoder. Finally, a fully-connected layer (FC) is followed by the output of the last Transformer encoder to give classification result.

Above straightforward solution could have huge computational cost, which could make the network training process unpractical. In Table 2.2, we list two

Algorithm 1. Process 3D volume data with transformer

Input: A 3D medical image volume X, a 3D tensor which have shape $[W, H, Z]$. Block size $BlockSize = [w, h, z]$, vector of 3 dimension.

Output: Classification result Y.

1: **if** linear embedding **then**
2:     reshape X from $[W, H, Z]$ to $[\frac{W \cdot H \cdot Z}{w \cdot h \cdot z}, w \cdot h \cdot z]$;
3:     $BlockEmbedding = \text{FC}(X)$;
4: **end if**
5: **if** convolutional embedding **then**
6:     **for** each 3DConv $\in$ 3DConvolutionalLayers **do**
7:         $X = $ 3DConv(X);
8:         $X = \text{ReLU}(X)$;
9:     **end for**
10:     $BlockEmbedding = \text{AvgPool3D}(X)$;
11:     $BlockEmbedding = \text{FC}(BlockEmbedding)$;
12:     reshape $BlockEmbedding$ from $[\frac{W}{w}, \frac{H}{h}, \frac{Z}{z}, d]$ to $[\frac{W \cdot H \cdot Z}{w \cdot h \cdot z}, d]$;
13: **end if**
14: $BlockEmbedding = \text{Concatnate}(ClsToken, X)$;
15: $BlockEmbedding = BlockEmbedding + PosEmbedding$;
16: $ImgRepresentation = \text{Transformer}(BlockEmbedding)$;
17: $Y = \text{FC}(ImgRepresentation)$;
18: **return** Y;

---

very common resolution setting of medical images and the memory needed for training respectively. In this table, the we choose the smallest ViT architecture setting in original paper [2] and set the batch size to 1. As shown in the Table 1, if we follow the setting of ViT, the input sequence length for the Transformer $(\frac{W \cdot H \cdot Z}{w \cdot h \cdot z})$ can be too huge for 3D medical image volume. We could use larger block size, e.g. 16×16×10 or 32×32×20 to reduce the massive computational cost. However, empirically, the linear projection would be incapable to extract representative feature from such a large volume (FC didn't work well on CIFAR 10 which is only 32×32×3). Naturally, we thought of the convolutional kernel when our target is extracting representative features. We introduce convolutional embedding, as shown in the Algorithm 1 line 6 to line 9, the image $x \in \mathbb{R}^{W \times H \times Z}$ first pass few sets of 3D convolutional layers and activation layers. Then, an average pool layer is added to the pipeline to further downsample the data to

Table 1. ViT-Base [2] training memory consumption with different block sizes when batch size is set to 1.

Block Size	$224 \times 224 \times 3$ (natural image)	$256 \times 256 \times 30$ (MRI image)	$512 \times 512 \times 500$ (CT image)
$16 \times 16 \times 1$	0.4G	34.2G	2620.2G
$32 \times 32 \times 1$	0.1G	8.6G	655.1G

$[\frac{W}{w}, \frac{H}{h}, \frac{Z}{z}, d]$, where d is the dimension of the block embedding vector as well as the channel number of the last convolutional layer. Finally, the representative block embedding is then reshaped and processed as aforementioned.

2.3   ConvNet Teacher Mechanism

Transformers lack some inductive biases inherent to CNNs, such as translation invariance and locality, and therefore do not generalize well when trained on insufficient amounts of data [4]. The author [2] concluded that ViT "do not generalize well when trained on insufficient amounts of data". We address this issue by introducing a training strategy: ConvNet teacher mechanism.

As mentioned before, 3DMeT replaces the linear embedding of the ViT with convolutional embedding. The structure of convolutional embedding can be find on CNNs, thus, can be trained jointly with CNNs which have same layers. Like the mean teacher model [13], convnet teacher mechanism contains two models (e.g. 3DMeT student model and CNN teacher model) with same convolutional layers. The 3DMeT student model's convolutional embedding layers are updated by exponential moving average [7]. The weight of 3DMeT student's convolutional embedding θ_i^{3DMeT} at training step i is updated by

$$\theta_i^{3DMeT} = \alpha\theta_{i-1}^{3DMeT} + (1 - \alpha)\theta_i^{ConvNet} \tag{2}$$

where α is a hyper-parameter that controls the learning factor and $\theta^{ConvNet}$ is the weight of CNN teacher's convolutional layers. In ConvNet teacher mechanism, only first few layers of a ConvNet are used as shown in Fig. 2.

3   Experiment and Result

3.1   Dataset and Metrics

We evaluate our methods on a knee cartilage defect assessment dataset based on MRI images. The input is a $448 \times 448 \times 20$ volume data and the output includes 3 classes: grade 0 (no defect), grade 1 (mild defect), grade 2 (severe defect). The whole dataset is split into two sets: training and testing (denoted as train and test) with 607 and 260 images, respectively. We compare our approaches with other methods on test set and use accuracy, recall and F1-score as metrics. Also, since output class digit and severity have positive correlation, we also compare the mean absolute error (MAE).

3.2   Efficiency and Performance: A Comparative Study with CNNs

Experiment Setting. We compare our results with other existing method, including 2D CNNs, 3D CNNs, and ViT [2]. To explain the validity of the our method better, we do not make changes to the widely-used training scheme of the image classification. In addition, we don't use tricks such as using label smoothing and data augmentation (e.g. Mixup). Specifically, for 3D convnets, we use 3D ResNet [5]. For 2D convnets, we use ResNet [6] as slice level classifier followed by a FC layer to summarize outputs from all layers. And for the ViT, we partition the 3D volume into 3D blocks and process them using a standard transformer like ViT. ViT has a block size of (10, 32, 32), embedding dimension $d = 512$, 6 layer transformer encoder with 16 heads. For the 3DMeT, we have the same transformer architecture as ViT3D while set the block size = (20, 32, 32) to have a similar number of operations with ViT3D, while the convolutional embedding (we use the first residual block of 3D ResNet18) replaced the linear embedding. $3DMeT^C$ denote the same 3DMeT trained by ConvNet teacher mechanism.

Table 2. Comparison with existing methods.

	FLOPs	Accuracy	Recall	F1-score	Error
3D-ResNet18 [5]	346.57G	0.645	0.594	0.684	0.404
3D-ResNet34 [5]	634.57G	0.653	0.630	0.622	0.456
2D-ResNet18 [6]	145.82G	0.664	0.666	**0.661**	0.421
2D-ResNet50 [6]	329.65G	0.672	0.663	0.654	0.400
ViT [2]	43.62G	0.617	0.546	0.548	0.513
3DMeT	**41.12G**(−88.1%)	0.664	0.624	0.617	0.371
$3DMeT^C$	**41.12G**(−88.1%)	**0.702**	**0.682**	0.655	**0.313**

Experiment Result. Table 2 illustrates the performance of each method on the different metrics and their computational cost in floating-point operations (FLOPs). We can see that original ViT with linear embedding is worse than CNNs since its poor generalization capability on small dataset. Also, the linear embedding is not good at encoding the information in a large block. Our 3DMeT have better performance comparing with both 2D CNNs and 3D CNNs in most cases while have significantly less computation (88.1% less comparing to 3D-ResNet18). With ConvNet teacher mechanism for training process, the performance is be further improved.

3.3 Ablation Study

In this section, we perform a controlled study of different architectures choice by evaluating accuracy. We explore how number of multi-head attention, number of transformer encoder layers and transformer dimension influence the performance.

Experiment Setting. The default settings in this experiment are: block size is set to $20 \times 32 \times 32$, number of head is 16, number of transformer layer is 6, transformer dimension is 1024 and 3 convolutional layers in the convolutional embedding. In the number of multi-head experiment, we have the number of head to be 2, 4, 8, 16, 32. In the number of transformer layers experiments, we set it to be 2, 4, 6, 8, 12, 16. In the transformer dimension experiments, we try 128, 256, 512, 768,1024 and 2048.

Experiment Result. Figure 3 illustrates the accuracy of each architecture setting. We can see that the optimal number of head is around 16 which is similar to ViT in natural images (ViT-Large and ViT-Huge). However, the optimal layers setting is around 6, not as large as ViT in natural images which is usually more than 20. Dimension are also smaller than ViT on natural images. Also we find the more convolutional layers brings better results as well as being more computational costly.

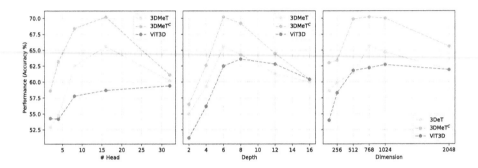

Fig. 3. Effect of head number, depth, and dimension of the Transformer.

In 3D medical image classification task, we find bigger model do not always bring better performance due to insufficiency of training sample. We find 3DMeT outperform ViT in most cases which further verify the effectiveness of the convolutional embedding.

4   Conclusion

In this paper, we first present the 3DMeT, a Transformer-based computer-aided diagnosis framework for 3D medical image. With Transformer's inherent ability to model long-range relation in big medical images and convolutional embedding, 3DMeT demonstrated comparable performance with CNNs while using significant less computational resources. And the performance were further boosted with the proposed ConvNet teacher mechanism by co-training with 3D CNNs.

Acknowledgement. This work was supported by the National Key Research and Development Program of China (2018YFC0116400), National Natural Science Foundation of China (NSFC) grants (62001292), Shanghai Pujiang Program(19PJ1406800), and Interdisciplinary Program of Shanghai Jiao Tong University.

References

1. Devlin, J., Chang, M.W., Lee, K., Toutanova, K.: BERT: pre-training of deep bidirectional transformers for language understanding. arXiv preprint arXiv:1810.04805 (2018)
2. Dosovitskiy, A., et al.: An image is worth 16×16 words: transformers for image recognition at scale. arXiv preprint arXiv:2010.11929 (2020)
3. Feng, C., et al.: Deep learning framework for Alzheimer's disease diagnosis via 3D-CNN and FSBI-LSTM. IEEE Access **7**, 63605–63618 (2019)
4. Han, K., et al.: A survey on visual transformer. arXiv preprint arXiv:2012.12556 (2020)
5. Hara, K., Kataoka, H., Satoh, Y.: Can spatiotemporal 3D cnns retrace the history of 2D CNNs and ImageNet? In: Proceedings of the IEEE Conference on Computer Vision and Pattern Recognition (CVPR), pp. 6546–6555 (2018)
6. He, K., Zhang, X., Ren, S., Sun, J.: Deep residual learning for image recognition. In: Proceedings of the IEEE Conference on Computer Vision and Pattern Recognition, pp. 770–778 (2016)
7. Laine, S., Aila, T.: Temporal ensembling for semi-supervised learning. arXiv preprint arXiv:1610.02242 (2016)
8. Liu, F., et al.: Fully automated diagnosis of anterior cruciate ligament tears on knee MR images by using deep learning. Radiol. Artif. Intell. **1**(3), 180091 (2019)
9. Pedoia, V., Norman, B., Mehany, S.N., Bucknor, M.D., Link, T.M., Majumdar, S.: 3D convolutional neural networks for detection and severity staging of meniscus and PFJ cartilage morphological degenerative changes in osteoarthritis and anterior cruciate ligament subjects. J. Magn. Reson. Imaging **49**(2), 400–410 (2019)
10. Ramesh, A., et al.: Zero-shot text-to-image generation. Proc. Mach. Learn. Res. **139**, 8821–8831 (2021)
11. Simonyan, K., Zisserman, A.: Very deep convolutional networks for large-scale image recognition. arXiv preprint arXiv:1409.1556 (2014)
12. Singh, S.P., Wang, L., Gupta, S., Goli, H., Padmanabhan, P., Gulyás, B.: 3D deep learning on medical images: a review. Sensors **20**(18), 5097 (2020)
13. Tarvainen, A., Valpola, H.: Mean teachers are better role models: weight-averaged consistency targets improve semi-supervised deep learning results. arXiv preprint arXiv:1703.01780 (2017)
14. Vaswani, A., et al.: Attention is all you need. arXiv preprint arXiv:1706.03762 (2017)
15. Wang, X., Girshick, R., Gupta, A., He, K.: Non-local neural networks. In: Proceedings of the IEEE Conference on Computer Vision and Pattern Recognition, pp. 7794–7803 (2018)

A Gaussian Process Model for Unsupervised Analysis of High Dimensional Shape Data

Wenzheng Tao[2(✉)], Riddhish Bhalodia[1,2], and Ross Whitaker[1,2]

[1] Scientific Computing and Imaging Institute, University of Utah, Salt Lake City, USA
[2] School of Computing, University of Utah, Salt Lake City, USA
wztao@cs.utah.edu

Abstract. Applications of medical image analysis are often faced with the challenge of modelling high-dimensional data with relatively few samples. In many settings, normal or healthy samples are prevalent while pathological samples are rarer, highly diverse, and/or difficult to model. In such cases, a robust model of the normal population in the high-dimensional space can be useful for characterizing pathologies. In this context, there is utility in hybrid models, such as probabilistic PCA, which learns a low-dimensional model, commensurates with the available data, and combines it with a generic, isotropic noise model for the remaining dimensions. However, the isotropic noise model ignores the inherent correlations that are evident in so many high-dimensional data sets associated with images and shapes in medicine. This paper describes a method for estimating a Gaussian model for collections of images or shapes that exhibit underlying correlations, e.g., in the form of smoothness. The proposed method incorporates a Gaussian-process noise model within a generative formulation. For optimization, we derive a novel expectation maximization (EM) algorithm. We demonstrate the efficacy of the method on synthetic examples and on anatomical shape data.

Keywords: Computational anatomy and physiology · Gaussian process

1 Introduction

High-dimensional data with limited samples is a common predicament in medical image analysis and computational anatomy, where practitioners frequently rely on statistical methods and/or domain-specific context for models with appropriate generalizability. In many cases normal examples are readily available, whereas pathological samples are not available, extremely rare, or highly diverse, making them difficult to model. This is especially true in shape analysis/morphology applications, where pathology is recognized in how it compares to the statistics of normal populations. This kind

---

Supported by the National Institutes of Health under R21-EB026061, NIBIB-U24EB029011, NIAMS-R01AR076120, NHLBI-R01HL135568, NIBIB-R01EB016701, and NIGMS-P41GM103545. Authors thank Erin Anstadt, MD and Jesse A Goldstein, MD for providing cranial shapes. The content is solely the responsibility of the authors and does not necessarily represent the official views of the National Institutes of Health.

Electronic supplementary material The online version of this chapter (https://doi.org/10.1007/978-3-030-87589-3_37) contains supplementary material, which is available to authorized users.

The original version of this chapter was revised: the acknowledgement has been corrected. The correction to this chapter is available at https://doi.org/10.1007/978-3-030-87589-3_73

of shape characterization has been used, for example, in orthopedics [1] and neurology [5].

A typical strategy to model limited-data, high-dimension normal populations is to learn a low dimensional subspace, presumably without over fitting to the samples. Principle component analysis (PCA) has been used in the context of shape [1,2], and images [27], and there are generalizations to curved manifolds [6,23]. In the particular context of deformation-based descriptions of images [21], low-dimensional models of displacement fields or momentum fields (in the case of diffeomorphisms) can characterize the statistics of a population. In the case of nonlinear models, autoencoders have been used to characterize the variability of shapes [18].

However, such low dimensional models fall short when the goal is to learn a normative model that can characterize previously unseen pathologies. *The reason is that pathological cases often differ from the normal population in ways outside the normal variability.* Thus, a projection onto a subspace learned from a normal population may (and often does) hide the differences that quantify pathology. A typical way to address this problem is to include a statistical model of the population behavior away from the learned subspace—i.e., null space. Given small sample size and high dimensionality, any model of the null space will need to be relatively simple (e.g., few degrees of freedom). Previously, the data in the null space is assumed to be isotropic-Gaussian— formulated in the form of a generative model, leads to probabilistic PCA (PPCA) [25].

In many applications in medical imaging, the high-dimensional data (points, vectors, pixels, correspondences) demonstrates substantial correlation, which is often observed as continuity or smoothness. Thus, a *noise model* in the null space that assumes independence is prone to incorrectly estimate the probabilities of samples that differ from the learned subspace in ways that are inherently correlated and reflect the underlying spatial statistics of the shape or image. In this paper, we present a novel formulation of PCA that accounts for the underlying correlations among the dimensions in data samples to construct statistical models that can be used to estimate the log likelihoods of unseen samples that may differ from the training data. We focus on cases where the correlations are described by a Gaussian process model, and thus we call the method Gaussian process PCA (GP-PCA). We formulate the model, which is a combination of linear processes, within an expectation-maximization (EM) framework. While PPCA typically learns a low-dimensional subspace and applies an isotropic noise to the null space, here we augment that with a correlated noise model, from a Gaussian process that is learned from the data. We present methods for simultaneously estimating the subspace and associated hyperparameters within a unified EM framework. We demonstrate the efficacy on simulated and clinical data, relative to state-of-the-art linear methods that do not account for the correlated structure of the data.

2   Related Works

The proposed method builds on using EM to estimate linear, latent models set by Roweis [20], as well as PPCA, proposed by Tipping and Bishop [25]. The isotropic noise model results in equal contributions from the null space dimensions, which limits efficacy when the data space exhibits inherent correlations between points in the shapes or images. The proposed method explicitly models these correlations by introducing a correlated noise model, the solution of which requires a novel, EM formulation.

The modeling of high-dimensional data is especially prevalent in the field of statistical shape modeling and deformable registration. Correspondence based models [3,22] model image/surface population with sets of geometrically consistent points. Dense correspondences or displacement fields [9,21] often exhibit a low-dimensional structure, and dimensionality reduction is a prerequisite to perform shape analysis and other downstream tasks [2,24]. Mahalanobis distance (log likelihood) is used for quantifying the deviation of normal shape statistics from unseen/pathological variations [15].

Also relevant is the work in statistical models that incorporate Gaussian processes (GPs). In [12] they apply GPs to the mappings from the embedded space of PCA, forming a nonlinear model that closely linked to kernel-based methods. In [8] they model latent factors using GPs in a form of nonlinear regression. Here we need to carefully distinguish these and related methods from this paper. These other works apply GPs *among samples*, typically based on some a-priori, parameterized kernel, and they introduce isotropic (uncorrelated) noise to penalize the fit. Here we introduce GPs so that each individual sample is generated from a nonisotropic, linear model that reflects the smoothness of the fields (e.g., shapes, images) from which these samples are obtained, and we learn the different components of that model using an EM optimization strategy.

Alternatively, in medical/biological applications, GPs have been applied for image segmentation tasks, where the smoothness of labels maps is used as a prior [13]. GP modeling is also used for image denoising [16], in CT reconstruction [19] and, in image registration [7]. Lüthi et al. [17] have proposed *Gaussian process morphable models*, which is a generalization of statistical shape models and incorporate GPs into the low-dimensional shape variations that are observed in the data. They combine learned linear subspaces (e.g., PCA) with GPs in order to obtain more effective low-dimensional representations of deformations. Here we extend the basic model of learned linear subspaces and GPs in several distinct directions. First, whereas [17] works towards learning a better low-dimensional statistical shape model, we model the GPs to model the entire high-dimensional space of variations. We also present a unified EM framework that marginalizes over the unknown latent variables to estimate the model parameters. Finally, while the GP morphable models aim to improve/regularize image registration or segmentation, the proposed method is designed to improve estimates of likelihoods on unseen data for the purposes of characterizing pathology in an unsupervised setting.

3   Methods

Here we describe the GP-PCA formulation and the associated EM algorithm. Although the formulation applies to any high-dimensional data samples, the discussion will refer to data as set of correspondence points on 3D surfaces, whose positions are correlated by their proximity on the underlying surface that they represent.

3.1   Gaussian Process Model

The correspondences of the ith shape are $\mathcal{Y}_i \in \mathbb{R}^{c \times p}$, where p is the number of points and c is the ambient dimension (e.g., 2D or 3D), and are assumed to be generated (approximately) by a multidimensional GP. The sample mean is denoted as $\bar{\mathcal{Y}}$, which we subtract to get the centered observations $\tilde{\mathcal{Y}}_i$ with a vector/flattened version $Y_i \in \mathbb{R}^{c \cdot p}$.

The generative model on the collection of shapes is:

$$Y_i \sim \mathcal{N}(0, \Sigma) \text{ where } \Sigma = S + \sigma_K^2 K_\theta + \sigma_I^2 I, \tag{1}$$

where S is a low-rank (of rank l) representation of the specific population, K_θ is a kernel matrix representing a stationary Gaussian process, with parameter(s), θ (e.g., the bandwidth), and I is the identity matrix representing uncorrelated noise. Thus, the covariance for the GP consists of three different *layers* with varying dependence on the input data. The uncorrelated noise requires the estimation of only a single parameter, σ_I, and the structure of I includes no knowledge of the structure of the input data. The correlated noise $\sigma_K^2 K_\theta$ represents the correlations between points/data within each example, with the assumption that proximal points are correlated, which results in some degree of regularity (e.g., smoothness). In this model, the correlation between two points on a surface $\bar{x}_j \in \mathbb{R}^c$ and $\bar{x}_k \in \mathbb{R}^c$, is stationary, so that $(K_\theta)_{jk} = k(d(\bar{x}_j, \bar{x}_k)/\theta)$, where $d(\cdot)$ is the distance function between points (e.g., Euclidean or geodesic) and θ is the bandwidth that controls the degree of smoothness. Typically choices of $k(\cdot)$ include negative exponential functions of the distance to some power (e.g., $\exp(-d^\alpha)$). For this work we treat the individual coordinates in the ambient space $\mathbb{R}^c$ as independent of one another, and thus K_θ has a block structure that depends only on the distance between correspondences on the base shape or mean. The low-rank matrix S, as in PCA, describes the global variations among shapes, and is learned from the n examples Y_is.

3.2 Optimization

The parameters are estimated by maximizing the log likelihood of the observations, assuming, by the design of the model, that the parameters are fewer relative to samples. The log-likelihood is $\mathcal{L} = n \log |\Sigma| + \text{Tr}(Y^T \Sigma^{-1} Y)$. Notice, the layered structure of Σ does not lend itself to a closed form solution of Θ as it does with PPCA [25], because the parameters interact with each other, as well as the learned subspace S.

Therefore, we propose a new formulation. Here we use the notation X to refer to the latent variables and W to refer to the linear transformation that maps those variables into a particular part of the model. We let Q denote the covariance of the *noise*, which is the part of the data not explained by the latent variables.

In this general case the form of the generative model with latent variables is $Y_i = WX_i + \epsilon_i$, where $\epsilon_i \sim \mathcal{N}(0, Q)$ and $X_i \sim \mathcal{N}(0, I)$.

Marginalizing the log-likelihood $\mathcal{L} = \sum_i \log (p(X_i, Y_i))$ over the latent variables:

$$\mathbb{E}(\mathcal{L}) = n \log |Q| + \text{Tr}(Y^T Q^{-1} Y - 2\mathbb{E}[X]^T W^T Q^{-1} Y + W^T Q^{-1} W \mathbb{E}[XX^T] + \mathbb{E}[X^T X]) \tag{2}$$

The expectations of the latent variables (*E-step*), with M as $(W^T Q^{-1} W + I)^{-1}$, are:

$$\mathbb{E}[X] = MWQ^{-1}Y, \mathbb{E}[XX^T] = nM + \mathbb{E}[X]\mathbb{E}[X]^T, \text{Tr}(\mathbb{E}[X^T X]) = n \text{Tr}(M) \tag{3}$$

The *M-step*, for a particular parameter, will set the derivative of (2) to zero and solve for that parameter. Here we use the () superscript to represent the iterations.

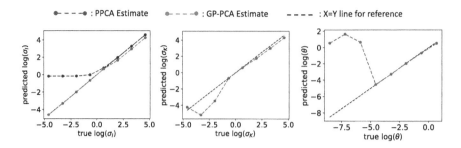

Fig. 1. Parameter estimates for PPCA and GP-PCA for Legendre data. In each case the X axis represent the true value, and the y axis represent the predicted, $x = y$ line is shown for reference.

Update for S: This follows more directly the form in [25], so that $Q = \sigma_I^2 I + \sigma_K^2 K_\theta$, $S = W^T W$, and $X_i \sim \mathcal{N}(0, I) \in \mathbb{R}^l$. We get the following subspace update:

$$W^{(i+1)} = f_S(Y, \Theta^{(i)}) \equiv Y\mathbb{E}[X^T]^{(i)}(\mathbb{E}[XX^T]^{(i)})^{-1} \text{ and } S^{(i+1)} = W^{(i+1)}(W^{(i+1)})^T \tag{4}$$

Update for θ: The bandwidth θ has no similar closed-form update, and thus we propose a (local) gradient descent optimization for this parameter. Alternatives, such as a line search, would also be effective.

Update for σ_I: Using $X_i \sim \mathcal{N}(0, I) \in \mathbb{R}^{c \cdot p}$, with substitution from a decomposition (e.g., SVD) $WW^T = S + \sigma_K^2 K_\theta$, and $Q = \sigma_I^2 I$, we get the following update:

$$\sigma_I^{(i+1)} = f_I(Y, \Theta^{(i)}) \equiv \{\frac{1}{npc} \text{Tr}(Y^T Y - 2(\mathbb{E}[X^T])^{(i)} W^{(i)T} Y + (\mathbb{E}[XX^T])^{(i)} W^{(i)T} W^{(i)})\}^{\frac{1}{2}} \tag{5}$$

Update for σ_K: Using $X_i \sim \mathcal{N}(0, I) \in \mathbb{R}^{c \cdot p}$ as the latent variable, with the substitutions $WW^T = \sigma_K^2 K_\theta$, and $Q = S + \sigma_I^2 I$, we get the following update:

$$\sigma_K^{(i+1)} = f_K(Y, \Theta^{(i)}) \equiv \frac{\sigma_K^{(i)} \text{Tr}((\mathbb{E}[X]^{(i)})^T W^{(i)T} Q^{(i)-1} Y)}{\text{Tr}(W^{(i)T} Q^{(i)-1} W^{(i)} (\mathbb{E}[XX^T])^{(i)})}, \tag{6}$$

A detailed mathematical derivation of the optimization methodology and update rules, as well as the initialization strategy are provided in the supplementary material.

Convergence and Model Selection: Parameters convergence via EM is based on the relative change of $\Theta^{(i)}$ except W per iteration, if it's less than 1% we stop the optimization. The model, specifically the dimension l of subspace S, is chosen using a cross-validation strategy. We have a held-out validation set and progressively increase l, for each model we perform the EM optimization and evaluate the validation loss via likelihood $\mathcal{L}(Y_{val}|\Theta^*)$. Finally, we choose the model with the best validation loss. For θ we use an Adam [11] descent scheme, implemented in PyTorch.

4   Results

4.1   Legendre Data

We construct a synthetic dataset using the Legendre basis functions representing the low-dimension modes of variations. The Legendre basis functions are orthogonal polynomials in range $[-1, 1]$ expressed via 500 uniform point samples in range. Legendre polynomials of degrees $[3, 4, 30, 39]$ with standard deviations of $[25, 30, 20, 20]$ are used to form W. Each sample is generated from W combined with the isotropic and Gaussian process components. We use a Radial Basis Function (RBF) covariance for the Gaussian process. To validate the parameter estimation, we generate samples using various parameter sets (σ_I, σ_K and θ) and report the estimation results. Test set (acts as pathological samples) is generated using twice the standard deviations. Training and test sets each has 200 samples. Correlation between the true and estimated Mahalanobis distance quantify model's ability to characterize both normal and pathological variations.

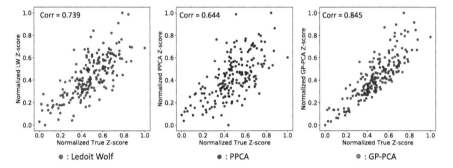

Fig. 2. Scatter plots showing comparison between Z-scores from different models with the true scores generated from synthetic Legendre data.

Table 1. Correlation of the predicted Z-scores with expert scores of testing scans.

Method	Pearson Correlation Coefficient	Spearman Correlation Coefficient	AUC
GP-PCA	0.828	**0.835**	**0.953**
Ledoit-Wolf [14]	0.815	0.812	0.910
OAS [4]	0.815	0.812	0.910
PPCA(EM)	**0.844**	0.809	0.925

In Fig. 1 we see that GP-PCA is able to estimate the three parameters very close to the true values, verifying the ability of the estimation routine. There are large offsets in GP-PCA estimation for small bandwidths, while the overall estimated covariance matrices are still correct. The reason is that $\sigma_K{}^2 K_\theta$ becomes isotropic with small bandwidth. For the 3 smallest bandwidth, GP-PCA estimated $\sigma_K = [0.0147, 0.0120, 0.0395]$ and

$\sigma_I = [1.3981, 1.3974, 1.3949]$ respectively. As the true $\sigma_K = \sigma_I = 1$ here, GP-PCA estimates the correct isotropic variance in total with estimated $\sigma_I{}^2 + \sigma_K{}^2 \approx 2$.

We calculated the average Pearson Correlation Coefficients for these 24 estimations. It is 0.822 for GP-PCA, 0.69 for PPCA and 0.715 for Ledoit-Wolf [14], which is a competing methodology. Percentage of correctly estimated subspace dimensions, l, is 0.792 for GP-PCA and 0.375 for PPCA. The results show that GP-PCA estimates unknown parameters better leading to better quantification of abnormality represented by Mahalanobis distance/Z-scores. Z-score is equal to negative log-likelihood here as it is a linear transformation of the later. Scatter plots comparing the true Z-score versus the predicted (for a single parameter set) on the test data are given in Fig. 2.

Sensitivity Analysis: To investigate the robustness of the estimation to the initialization, we perform sensitivity analysis on the choice of EM initialization of parameters. We use the initialization strategy detailed in the supplementary materials. In Fig. 3 we show that with a wide range of initialization choices for each of the parameters θ, σ_I and σ_K, the estimation of the proposed method is consistent.

4.2 Craniosynostosis Data

Craniosynostosis is a morphological disorder occurring in infants, where sutures in the brain fuses prematurely. For instance, in metopic craniosynostosis the metopic suture is fused early leading to a triangular forehead and compensatory expansion at the back. Current medical practice has experts/surgeons subjectively gauge the severity of the condition to make a decision to engage in a risky operative procedure or not. This subjectivity can at times lead to severe cases remaining untreated or mild/moderate cases to be operated upon, additionally, different institutions can incur an operational bias. Such subjective practice presents a need for an objective severity metric.

Data Description: We have 74 CT scans of normal subjects, as the control set. We use *ShapeWorks* [3], an automatic correspondence generation software to express each with 2048 3D points, this set is used for training. Additionally, we have 20 held out normal, 30 metopic and 20 sagittal scans as the test set.

Quantitative Validation: We obtained expert severity ratings from 36 physicians for the held normal and metopic scans through an online portal and aggregated scores using

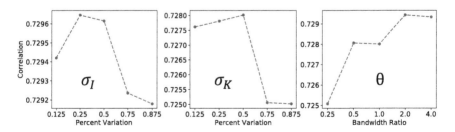

Fig. 3. Sensitivity analysis on initialization: Correlations are measured by Pearson coefficients. For σ_I and σ_K, the x-axis is the percentage of sample variance used to initialize. For bandwidth the x-axis is the ratio of bandwidth used to initialize versus the average nearest neighbor distance.

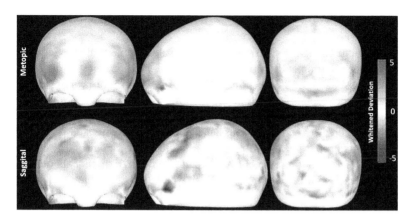

Fig. 4. Localized Z-score for metopic (top row) and sagittal (bottom row). The variation is the deviation in whitened space averaged by pathological samples, the positive values (red) represent outward variation and negative values (blue) represent inward variation. (Color figure online)

Latent trait theory [26]. We correlate the Z-scores coming from different methods with the expert ratings and diagnosis labels (shown in Table 1). We do model selection for l in the range between 5 to 23 and GP-PCA selects 13. We can see that GP-PCA outperforms the other null space characterization methods, demonstrating its superiority in characterizing the pathology. We focus more on Spearman Correlation here, as Pearson Correlation can be limited by the linearity assumption. We also compare the detection task (normal versus pathological) for both metopic as well as sagittal with Area Under Curve (AUC) scores. We have 20 sagittal scans, and AUC score of GP-PCA is 0.972. However, we stress that the advantage of GP-PCA is not in a detection task (neural networks can perform better) but rather in characterization of the pathology.

Localized Z-score: By discovering an analytical covariance matrix one can compute individual Mahalanobis distance (Z-score) at each correspondence point after transforming to the whitened space. This is computed via the following equation: $\hat{Y}_i = US^{-\frac{1}{2}}U^T Y_i$, where USV is the SVD of the estimated covariance of GP-PCA. Taking a dot product of this score with the vertices normal, we visualize this point wise whitened deviation using colored maps. The mean of such deviations the pathological scans will represent *localized pathological deformation*. Such a characterizationi is important for clinicians to make decisions, hence, we showcase it as a bipolar-heatmap projected on a mean normal head shape in Fig. 4. The metopic shape clearly shows an average depression in the forehead, one of the major symptoms and degree of which has been related to severity [10], it also showcases the slight bulging in the back which compensates for the forehead constriction. The localized Z-score heatmaps for the sagittal shows narrow sides and elongated back, which are the most significant features.

5   Conclusions

Analysis of structured high-dimensional data is prevalent in medical imaging and shape analysis. In order to efficiently model such data and capture both "in-data" (in low-dimensional subspace) and "unseen" (null space) variations we propose a Gaussian Process model that account for such inherent correlations. It also tackles the challenge of good generalization when provided with limited observations. The proposed method extends PPCA model by adding a Gaussian Process based correlated noise model in addition to the isotropic noise. We propose a unified EM framework to estimate all the model parameters, and derive all the update functions. Experiments on a synthesized curve dataset with Legendre polynomials validate the accuracy and stability of our proposed method. Additionally, we apply the model on metopic and sagittal craniosynostosis data for pathology characterization, and achieve best performance among the competing methods.

References

1. Atkins, P.R., et al.: Quantitative comparison of cortical bone thickness using correspondence-based shape modeling in patients with cam femoroacetabular impingement. J. Orthopaedic Res. **35**(8), 1743–1753 (2017)
2. Bhalodia, R., Dvoracek, L.A., Ayyash, A.M., Kavan, L., Whitaker, R., Goldstein, J.A.: Quantifying the severity of metopic craniosynostosis: a pilot study application of machine learning in craniofacial surgery. J. Craniofac. Surg. **31**(3), 697–701 (2020)
3. Cates, J., Fletcher, P.T., Styner, M., Shenton, M., Whitaker, R.: Shape modeling and analysis with entropy-based particle systems. In: Karssemeijer, N., Lelieveldt, B. (eds.) IPMI 2007. LNCS, vol. 4584, pp. 333–345. Springer, Heidelberg (2007). https://doi.org/10.1007/978-3-540-73273-0_28
4. Chen, Y., Wiesel, A., Eldar, Y.C., Hero, A.O.: Shrinkage algorithms for MMSE covariance estimation. IEEE Trans. Signal Process. **58**(10), 5016–5029 (2010)
5. Doan, N.T., van Lew, B., Lelieveldt, B., van Buchem, M.A., Reiber, J.H.C., Milles, J.: Deformation texture-based features for classification in Alzheimer's disease. In: Medical Imaging 2013: Image Processing, vol. 8669, pp. 601–607. International Society for Optics and Photonics, SPIE (2013)
6. Fletcher, P.T., Lu, C., Pizer, S.M., Joshi, S.: Principal geodesic analysis for the study of nonlinear statistics of shape. IEEE Trans. Med. Imaging **23**(8), 995–1005 (2004)
7. Gerig, T., Shahim, K., Reyes, M., Vetter, T., Lüthi, M.: Spatially varying registration using Gaussian processes. In: Golland, P., Hata, N., Barillot, C., Hornegger, J., Howe, R. (eds.) MICCAI 2014. LNCS, vol. 8674, pp. 413–420. Springer, Cham (2014). https://doi.org/10.1007/978-3-319-10470-6_52
8. Gu, M., Shen, W.: Generalized probabilistic principal component analysis of correlated data. J. Mach. Learn. Res. **21**(13), 1–41 (2020)
9. Joshi, S.C., Miller, M.I., Grenander, U.: On the geometry and shape of brain sub-manifolds. Int. J. Pattern Recogn. Artif. Intell. **11**(08), 1317–1343 (1997)
10. Kellogg, R., Allori, A.C., Rogers, G.F., Marcus, J.R.: Interfrontal angle for characterization of trigonocephaly: part 1: development and validation of a tool for diagnosis of metopic synostosis. J. Craniofac. Surg. **23**(3), 799–804 (2012)
11. Kingma, D.P., Ba, J.: Adam: a method for stochastic optimization. arXiv preprint arXiv:1412.6980 (2014)

12. Lawrence, N., Hyvärinen, A.: Probabilistic non-linear principal component analysis with Gaussian process latent variable models. J. Mach. Learn. Res. **6**(11), 1783–1816 (2005)
13. Lê, M., Unkelbach, J., Ayache, N., Delingette, H.: GPSSI: Gaussian process for sampling segmentations of images. In: Navab, N., Hornegger, J., Wells, W.M., Frangi, A.F. (eds.) MICCAI 2015. LNCS, vol. 9351, pp. 38–46. Springer, Cham (2015). https://doi.org/10.1007/978-3-319-24574-4_5
14. Ledoit, O., Wolf, M.: A well-conditioned estimator for large-dimensional covariance matrices. J. Multivariate Anal. **88**(2), 365–411 (2004)
15. Leys, C., Klein, O., Dominicy, Y., Ley, C.: Detecting multivariate outliers: use a robust variant of the Mahalanobis distance. J. Exp. Soc. Psychol. **74**, 150–156 (2018)
16. Liu, P.J.: Using Gaussian process regression to denoise images and remove artefacts from microarray data. University of Toronto (2007)
17. Lüthi, M., Gerig, T., Jud, C., Vetter, T.: Gaussian process morphable models. IEEE Trans. Pattern Anal. Mach. Intell. **40**(8), 1860–1873 (2017)
18. Oktay, O., et al.: Anatomically constrained neural networks (ACNNs): application to cardiac image enhancement and segmentation. IEEE Trans. Med. Imaging **37**(2), 384–395 (2018). https://doi.org/10.1109/TMI.2017.2743464
19. Purisha, Z., Jidling, C., Wahlström, N., Schön, T.B., Särkkä, S.: Probabilistic approach to limited-data computed tomography reconstruction. Inverse Probl. **35**(10), 105004 (2019)
20. Roweis, S.: EM algorithms for PCA and SPCA. In: Proceedings of the 10th International Conference on Neural Information Processing Systems, pp. 626–632, NIPS 1997. MIT Press, Cambridge (1997)
21. Rueckert, D., Frangi, A.F., Schnabel, J.A.: Automatic construction of 3-D statistical deformation models of the brain using nonrigid registration. IEEE Trans. Med. Imaging **22**(8), 1014–1025 (2003)
22. Styner, M., et al.: Framework for the statistical shape analysis of brain structures using SPHARM-PDM. Insight J. **1071**, 242–250 (2006)
23. Styner, M., Gerig, G.: Medial models incorporating object variability for 3D shape analysis. In: Insana, M.F., Leahy, R.M. (eds.) IPMI 2001. LNCS, vol. 2082, pp. 502–516. Springer, Heidelberg (2001). https://doi.org/10.1007/3-540-45729-1_53
24. Beiging, E.T., Morris, A., Wilson, B.D., McGann, C.J., Marrouche, N.F., Cates, J.: Left atrial shape predicts recurrence after atrial fibrillation catheter ablation. J. Cardiovasc. Electrophysiol. **29**(7), 966–972 (2018). https://doi.org/10.1111/jce.13641
25. Tipping, M.E., Bishop, C.M.: Probabilistic principal component analysis. J. R. Stat. Soc. Ser. B **61**(3), 611–622 (1999)
26. Uebersax, J.S., Grove, W.M.: A latent trait finite mixture model for the analysis of rating agreement. Biometrics **49**(3), 823–835 (1993)
27. Wood, B.C., et al.: What's in a name? Accurately diagnosing metopic craniosynostosis using a computational approach. Plastic Reconstr. Surg. **137**(1), 205–213 (2016)

Standardized Analysis of Kidney Ultrasound Images for the Prediction of Pediatric Hydronephrosis Severity

Pooneh Roshanitabrizi[1]([✉]), Jonathan Zember[2], Bruce Michael Sprague[3], Steven Hoefer[2], Ramon Sanchez-Jacob[2], James Jago[4], Dorothy Bulas[2], Hans G. Pohl[3], and Marius George Linguraru[1,5]

[1] Sheikh Zayed Institute for Pediatric Surgical Innovation, Children's National Hospital, Washington, DC, USA
proshnani2@childrensnational.org
[2] Radiology Department, Children's National Hospital, Washington, DC, USA
[3] Urology Division, Children's National Hospital, Washington, DC, USA
[4] Philips Healthcare, Bothell, WA, USA
[5] Departments of Radiology and Pediatrics, School of Medicine and Health Sciences, George Washington University, Washington, DC, USA

Abstract. Congenital hydronephrosis, which is the dilatation of the renal collecting system, is common in children, resolving in most and treatable in the remaining 25%. Sonography is routinely used for hydronephrosis detection and longitudinal evaluation but lacks standardization in acquisition and provides no information about kidney function. These facts make the visual assessment of hydronephrosis from ultrasound subjective and variable. In this paper, we present an automatic method to standardize the analysis of the kidney regions in sonograms for the quantification of hydronephrosis severity as well as the prediction of obstruction. First, the field-of-view in images is standardized by segmenting the kidney regions using convolutional neural networks and reorienting them along their longest axes in the coronal view. Then, the core areas of the kidney containing the pelvis and calyces are identified by correlation analysis. Each standardized kidney image slice is evaluated using a deep learning-based approach to predict the obstruction severity, and the slice-based predictive scores are fused based on a weighted-voting technique to determine the final risk score. The performance of the method was evaluated on 54 hydronephrotic kidneys with known clinical outcome. Results show that our method could automatically predict the obstruction severity with an average accuracy of 0.83, a significant improvement over the common clinical approach (p-value < 0.001). Our method has the potential to predict kidney function from routine ultrasound evaluation.

Keywords: Classification · Deep neural networks · Kidney · Pediatric hydronephrosis · Standardization · Ultrasound

© Springer Nature Switzerland AG 2021
C. Lian et al. (Eds.): MLMI 2021, LNCS 12966, pp. 366–375, 2021.
https://doi.org/10.1007/978-3-030-87589-3_38

1   Introduction

Hydronephrosis (HN) refers to dilatation of the upper urinary tract (renal pelvis and ureters) and often presents as a congenital finding on screening prenatal sonography but may also be found during an evaluation for urinary tract infection, abdominal pain and voiding abnormalities [1]. Ultrasound (US) imaging is commonly used for HN detection in pediatrics due to its non-invasive nature, cost effectiveness, and high-speed property [2]. While most cases of hydronephrosis do not represent urinary obstruction, if obstruction is unrecognized and left untreated, permanent loss of kidney function can occur [1]. To visually assess HN from US images, the Society of Fetal Ultrasound (SFU) has developed a widely accepted grading system to classify HN into four categories from mild to severe (grades 1–4) [3]. While the clinical detection of HN with US is accurate, the SFU grading system is subjective and correlates poorly with the kidney function [4]. Thus, children with moderate or severe HN (SFU grade 3 or 4) are often further evaluated by nuclear medicine examination such as diuresis renography (DR) [1]. DR evaluates kidney function by measuring the time to drain half of the radiotracer from its peak (T½) [1]. However, DR is an invasive procedure requiring ionizing radiation [5]. Improving the ability of US to correlate with a reduction in renal function would be important to safely limit the number of pediatric DR exams.

Previous methods have been proposed to estimate the HN outcome from US, including: (1) ratios of collecting system area to kidney area (hydronephrosis index) [6, 7] and renal parenchyma to hydronephrosis area [8], (2) automatic SFU grading using deep neural networks [9, 10], and (3) morphological analysis of renal units [11]. These methods are impeded by: (1) intense user interaction [6, 11], (2) low inter-observer variability in measurements [8], (3) dependency on accurate kidney and collecting system segmentation, which is challenging for pathological kidneys [6, 7, 11], and (4) lack of correlation with kidney function [9, 10]. To address these limitations, we propose a deep learning-based approach to automatically predict the obstruction severity—correlated to the result of the DR exam—by standardizing the view in the US images followed by robust predictive analysis.

A variety of methods have been proposed to segment the kidney in US images [2, 12–18]. Shape model-based approach for kidney segmentation [2, 12–16] are limited by: (1) user dependency for parameter initialization or slice selection [2, 12–14] and (2) lack of validation for diseased pediatric cases [15]. In [2, 14–16], kidney regions were segmented in 3D US images. Although 3D US imaging could provide important information for the comprehensive analysis of the kidney appearance and morphology, 2D US is commonly used in pediatric urology due to its cost effectiveness. Convolutional neural networks (CNNs) have also been applied to segment kidney regions in US images [17, 18]. In [17], kidneys were segmented using U-net and shape prior based on a classification problem. The method proposed in [18] incorporated boundary distance regression and pixel classification networks for clinical 2D US images. However, this method assumed the kidney texture at boundary is homogenous across the images, which is not valid in clinical US images.

Our aim is to improve the clinical interpretation of pediatric HN and kidney function from clinical US images, and thus eliminate as much as possible the need for DR exams in children. A preliminary result of this work was presented in [19]. In this paper, we present an automatic framework for the prediction of DR-based renal function from US images. We standardize the field-of-view (FOV) of the US images by detecting the kidney regions and reorienting them along their longest axes in the coronal view. We use the kidney middle slice with its correlated adjacent slices—to include the pelvis and calyces—for a comprehensive analysis of the presentation of the renal units using a deep learning-based approach. Then, the slice-based predictive scores are fused using a weighted-voting technique to provide the final prediction risk score of obstruction severity.

2 Materials

We used 54 retrospective US images (IRB approved) from 49 patients (33 left and 21 right kidneys) who were diagnosed with differing degrees of HN based on SFU (mean age 19 $\pm$ 28 months; age range 0 to 154 months) and evaluated with DR. All patients underwent both US and DR imaging within a time range of $[-180{:}90]$ days. Volumetric US images were acquired with iU22 (29 images) or Epiq (25 images) scanner (Philips, Amsterdam, Netherlands), using an X6-1 xMATRIX array transducer. The average US volume size was $476 \times 405 \times 244$ voxels and the voxel size ranged from 0.15 mm to 0.63 mm, which we resampled to the same size of $0.28 \times 0.22 \times 0.44$ mm in the pre-processing step. All US images included the full renal capsule in the coronal view. The reference kidney regions were semi-automatically localized for validation by an imaging expert under the supervision of a pediatric radiologist using the interactive method in [14]. DR exams were accomplished using 1.0 mCi 99mTechnetium MAG3 on a Siemens e-Cam Signature series (Siemens, Erlangen, Germany). We considered a safety threshold value of 20 min for T½ (T½ > 20 min) to define critical ureteropelvic junction obstruction indicative of severe HN. Among 54 cases, 13 kidneys were recognized with severe obstruction.

3 Methods

In Fig. 1, we present the flowchart of the proposed method, including: (1) US image standardization, and (2) obstruction severity prediction. US images were standardized by the FOV of the kidneys along their longest axes in the coronal view. This step also allowed our analysis to be focused only on the relevant regions of the kidney. Hence, obstruction severity was predicted by analyzing the dilated kidney in the middle slice and its adjacent correlated slices to capture the appearance of the pelvis and calyces. The details of each step are described below.

3.1 US Image Standardization

US image standardization was performed to reduce the non-discriminative variability between datasets acquired during routine clinical evaluations. US is particularly vulnerable to the orientation of the probe and the proposed level of standardization improves the ability of models to learn image-based information, shown in Fig. 2.

Kidney Region Segmentation. We aimed to investigate the region inside the kidney that was relevant to characterizing HN. To localize the kidney region on each US image, we compared the performance of two recent extensions of the U-net architecture: (1) LinkNet [20, 21] and (2) UNet++ [22, 23]. LinkNet was proposed to increase the segmentation accuracy without significant increasement in the number of parameters. UNet++ included a nested structure with improved skip connections between the encoding and decoding path. For both networks, we applied VGG16 as the pretrained encoder on ImageNet datasets. The input to the network was an US image of size 128 × 128 pixels to decrease the memory size and increase the process time. To provide the input of the network, we resized all images and padded them, if necessary, with zeros to provide the same size of 128 × 128 pixels. We employed a 1 × 1 convolution filter to update the input shape dimension to the pre-trained VGG16. The model was trained with a batch size of eight to minimize the negative value of the Dice similarity coefficient (DSC) as a loss function with Adam optimization algorithm, 100 epochs, and an adaptive learning rate ($l_r = \left(Unit(epoch - 50) \times 10^{-4} \times e^{-0.1}\right) + \left(Unit(50 - epoch) \times 10^{-4}\right)$; $Unit$ represents the Heaviside step function).

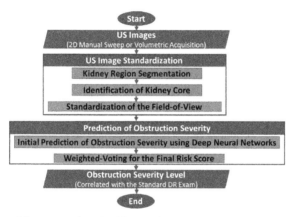

Fig. 1. Flowchart of the proposed method for prediction of obstruction severity using US images

Identification of Kidney Core. The kidney middle slice was selected using the maximum cross-sectional area in the coronal view. Then, we calculated the Pearson correlation coefficient [24] to identify the adjacent slices correlated to the kidney middle slice. We considered the correlation value >0.8 which provided the number of selected slices in the range [18, 89]. We then found that the correlation value that provided the maximum accuracy (see *Results*). These slices of the kidney cover the anatomical regions most relevant to the assessment of obstruction.

Standardization of the Field-of-View. Orientation of the kidney region was corrected along the longest axis in the coronal view (Fig. 2). Also, the pixel values inside the kidney region were normalized in the range [0, 1] and padded images with zero to standardize all images to the same size of 256 × 256 pixels to improve the process time.

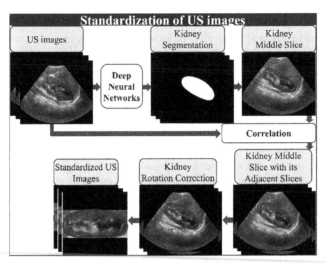

Fig. 2. One sample of the standardized kidney region

3.2 Prediction of Obstruction Severity

To assess the obstruction severity, we first evaluated each selected and standardized US slice. Then, the final risk score was computed by concatenation of all slice-based scores using a new weighted-voting technique.

Initial Prediction of Obstruction Severity. We employed a deep learning-based model with a structure proposed in [9, 10]. This model was successfully used for SFU grade classification. We used it to predict the obstruction severity. The network contained five convolutional layers, a fully connected layer of 512 units, and a final output layer of 1 unit to compute relevant image features to the obstruction. We used standardized US images of the kidney of size 256×256 pixels as the input of the network. We employed a sigmoid probability function in the last layer to predict kidney function. In the last fully connected layer, we applied a dropout layer with a keep rate of 0.3, selected empirically, to reduce overfitting. We trained the prediction model using the binary cross-entropy loss function, a stochastic gradient descent with momentum (Adam with learning rate l_r), batch size of 4, and 100 epochs. To increase the number of training datasets, we applied geometric augmentation, including vertical flip and rotation in the range $[-10°:5°:10°]$. These parameters were selected based on the required memory for processing.

Final Prediction of Obstruction Severity. We fused the slice-based predictive scores using a weighted-voting technique. To assign a weight to each US slice, we first ranked all slices using the Pearson correlation coefficient (r) between each slice and the kidney middle slice. Then, we used a multiquadric function $\emptyset(r) = \mathrm{sqrt}\left(1 + (\varepsilon r)^2\right)$ to calculate the weights. ε is the scale parameter to tune the shape of $\emptyset(r)$ which was considered in the range $[0.001:0.001:0.01]$. Finally, we fused all weighted predictive values and applied a threshold value in the range $[0.1:0.1:0.9]$ to obtain the final predictive score of a T½ value higher than 20 min. We selected the best ε and threshold values based on the provided maximum accuracy (see *Results*). We compared this method with the clinically-used SFU grading system [3]. We used the Wilcoxon signed-rank method with a significance level of 0.05 to compare the results.

4 Experimental Results

The proposed framework was implemented using Keras (version 2.2.4) on NVIDIA GeForce GTX TITAN X. To evaluate it, we applied a 13-fold cross validation due to 13 obstructed cases. Each fold contained 12 obstructed and 12 randomly selected, non-obstructed cases for training (on average 1,050 and 10,500 images before and after data augmentation, respectively) and the remaining obstructed case and 29 non-obstructed cases for testing (on average 1,450 and 14,500 images before and after data augmentation, respectively). The training of the method (including the segmentation and prediction tasks) based on the LinkNet and UNet++ structures took a total of 7 h and 12 h, respectively. Our method used 4 GB of memory for the segmentation/prediction task.

Table 1 presents a summary of quantitative results (mean and standard deviation) obtained for kidney segmentation and middle slice selection using the LinkNet and UNet++ structures. In Table 2, the influence of different parameters of the weighted-voting technique on the prediction accuracy is investigated. Based on these results, the multiquadric radial basis function with a ε value of 0.001 and a threshold value of 0.9 along with a correlation coefficient value of 0.83 provided the best prediction accuracy. These parameters determined the best number of selected correlated slices as an indicator of dilation size of the collecting system, especially for severe hydronephrosis cases.

Table 1. Quantitative results obtained for kidney segmentation and middle slice detection.

Method	Middle slice difference	DSC of kidney segmentation	HD (mm) of kidney Segmentation
LinkNet	8.56 ± 8.22	$0.84 \pm 0.17^*$	2.92 ± 3.18
UNet++	7.39 ± 6.6	$0.82 \pm 0.2^*$	3.13 ± 3.94

DSC – Dice similarity coefficient; HD – Hausdorff distance; *p-value < 0.001.

Table 3 reports quantitative prediction results for obstruction severity. Our method significantly outperforms the SFU grading system by reducing the variability in image interpretation (p-value < 0.001). Integrating the prediction scores using the weighted-voting technique further significantly increased the prediction accuracy due to the comprehensive analysis of the renal units (p-value < 0.001 for both LinkNet and UNet++). Compared to the equal voting, results showed that the weighted-voting technique improved the accuracy of 0.52 to 0.83 (p-value < 0.001). To select the deep learning model structure, we assessed our method based on LeNet [25] with great success in handwritten character recognition. Results showed that our model is more capable to extract the relevant obstructive information from ultrasound images (p-value < 0.001).

Table 2. Parameters of the weighted-voting technique on the prediction accuracy.

$\emptyset(r)$	Multiquadric						Gaussian*	Inverse multiquadric**
ε	1e − 3	**1e − 3**	1e − 3	1e − 3	1e − 3	2e − 3	1e − 3	1e − 3
Threshold value	0.9	**0.9**	0.9	0.8	0.7	0.9	0.9	0.9
Correlation coefficient	0.82	**0.83**	0.84	0.83	0.83	0.83	0.83	0.83
Prediction accuracy	0.8	**0.83**	0.78	0.78	0.63	0.81	0.81	0.80

*Gaussian: $\emptyset(r) = \exp\left(-(\varepsilon r)^2\right)$

**Inverse multiquadric: $\emptyset(r) = 1/\text{sqrt}\left(1 + (\varepsilon r)^2\right)$

Table 3. Quantitative prediction results using different methods

Method	Accuracy	Sensitivity	Specificity	p-Value
LinkNet	**0.83 ± 0.38**	**0.46 ± 0.52**	**0.95 ± 0.22**	–
REF	0.83 ± 0.39	0.62 ± 0.51	0.9 ± 0.3	0.26
LinkNet (Before weighted-voting technique)	0.56 ± 0.33	0.72 ± 0.25	0.51 ± 0.34	<0.001
LinkNet (Equal voting)	0.52 ± 0.5	0.77 ± 0.52	0.44 ± 0.22	<0.001
LinkNet (based on LeNet [25])	0.74 ± 0.44	0.15 ± 0.38	0.93 ± 0.26	<0.001
UNet++	0.76 ± 0.43*	0.38 ± 0.51	0.88 ± 0.33	<0.001
UNet++ (Before weighted-voting technique)	0.51 ± 0.3*	0.73 ± 0.26	0.44 ± 0.28	<0.001
SFU (≥3) [3]	0.5 ± 0.5	0.69 ± 0.48	0.44 ± 0.5	<0.001

REF: reference-based approach using the semi-automatic segmentation method [14]; p-values were computed against our results from the LinkNet; *p-value < 0.001.

Compared to UNet++, LinkNet provided higher prediction accuracy of 0.83 vs. 0.76 (p-value < 0.001). This result is similar to the reference-based approach (REF) using the semi-automatic segmentation method [14] (p-value of 0.26); however, our method does not need the user knowledge for kidney middle slice selection and kidney segmentation. In addition, it is more reproducible due to its automatic nature, which is better suited for clinical application. The high specificity value of our approach shows that most cases that were examined by DR in our dataset did not have severe obstruction, and could be safely monitored in the clinic by only US imaging.

5 Discussion

HN is a common congenital condition that is often identified by US imaging. Children with this condition are also followed up by US exams to monitor evolution. Severe cases undergo surgical intervention to prevent renal damage and permanent loss of function. However, the clinical interpretation of US does not accurately reveal if kidney function is compromised and children at potential risk are evaluated by DR. In this paper, we proposed an accurate automated framework that predicted obstruction severity from US imaging, in agreement with the T½ value from the standard DR exam. We evaluated the obstruction severity based on a deep learning-based approach with a new weighted-voting technique for fusing information from adjacent, correlated image slices. The inclusion of the adjacent slices provided 3D image information for the comprehensive analysis of the renal appearance and morphology.

Application of CNN for US-based kidney analysis is challenging due to the large variability in acquisition, image quality and kidney appearance, especially when the appearance of the kidney is modified as for children with HN. To deal with these challenges, our method standardized the US image FOV by identifying the kidney region and reorienting it along its longest axis in the coronal view. By reducing the variability in image interpretation, our method increased the reproducibility and accuracy of the US-based analysis of the renal unit, even as it was modified by the accumulation urine in its collecting system. In addition, our approach can avoid future investigation by DR for non-severe cases, which is invasive and harmful especially for children.

A limitation of this study is its reliance on the accuracy of kidney segmentation. Kidney segmentation is challenging for cases with high values of noise and shadows. However, results showed that our automatic method could predict obstruction severity with an accuracy similar to the reference-based approach. This feasibility study is also limited by the low number of cases, which we plan to increase in the future for large scale validation. We will also investigate how additional clinical information can improve the image-based prediction of obstruction severity.

6 Conclusion

We presented an automatic approach to improve the accuracy of the interpretation of ultrasound images for obstruction assessment in pediatric hydronephrosis using deep neural networks. First, kidney regions were localized using convolutional neural networks. Then, the field of view in the ultrasound image was standardized by reorienting the kidney regions along their longest axes in the coronal view. This allowed us to select the kidney core area that contained the pelvis and calyces by correlation analysis for the comprehensive analysis of the renal units. We next used a deep learning-based approach to predict the obstruction severity from each ultrasound image slice. The final severity score was computed by integrating the slice-based scores using a weighted-voting technique. Results showed a significant improvement over the common clinical approach. Our approach has the potential to improve the accuracy of US-based assessment of renal function and reduce the number of invasive imaging techniques for vulnerable pediatric populations.

References

1. Peters, C., Chevalier, R.L.: Congenital urinary obstruction: pathophysiology and clinical evaluation. In: Campbell-Walsh Textbook of Urology, 10th edn. Elsevier, Philadelphia (2012)
2. Cerrolaza, J.J., Safdar, N., Biggs, E., Jago, J., Peters, C.A., Linguraru, M.G.: Renal segmentation from 3D ultrasound via fuzzy appearance models and patient-specific alpha shapes. IEEE Trans. Med. Imaging **35**(11), 2393–2402 (2016)
3. Fernbach, S.K., Maizels, M., Conway, J.J.: Ultrasound grading of hydronephrosis: introduction to the system used by the Society for Fetal Urology. Pediatr. Radiol. **23**(6), 478–480 (1993)
4. Keays, M.A., et al.: Reliability assessment of society for fetal urology ultrasound grading system for hydronephrosis. J. Urol. **180**(4), 1680–1683 (2008)
5. Koizumi, K., et al.: Japanese consensus guidelines for pediatric nuclear medicine: part 1: pediatric radiopharmaceutical administered doses (JSNM pediatric dosage card). Part 2: technical considerations for pediatric nuclear medicine imaging procedures. Ann. Nucl. Med. **28**(5), 498–503 (2014)
6. Shapiro, S.R., Wahl, E.F., Silberstein, M.J., Steinhardt, G.: Hydronephrosis index: a new method to track patients with hydronephrosis quantitatively. Urology **72**(3), 536–538 (2008)
7. Tabrizi, P.R., et al.: Automatic segmentation of the renal collecting system in 3D pediatric ultrasound to assess the severity of hydronephrosis. In: International Symposium on Biomedical Imaging, Venice, Italy, pp. 1717–1720. IEEE (2019)
8. Rickard, M., Lorenzo, A.J., Braga, L.H.: Renal parenchyma to hydronephrosis area ratio (PHAR) as a predictor of future surgical intervention for infants with high-grade prenatal hydronephrosis. Urology **101**, 85–89 (2017)
9. Dhindsa, K., Smail, L.C., McGrath, M., Braga, L.H., Becker, S., Sonnadara, R.R.: Grading prenatal hydronephrosis from ultrasound imaging using deep convolutional neural networks. In: 15th Conference on Computer and Robot Vision, CRV, Toronto, ON, Canada, pp. 80–87. IEEE (2018)
10. Smail, L.C., Dhindsa, K., Braga, L.H., Becker, S., Sonnadara, R.R.: Using deep learning algorithms to grade hydronephrosis severity: toward a clinical adjunct. Front. Pediatr. **8**(1), 1–8 (2020)
11. Cerrolaza, J.J., Peters, C.A., Martin, A.D., Myers, E., Safdar, N., Linguraru, M.G.: Quantitative ultrasound for measuring obstructive severity in children with hydronephrosis. J. Urol. **195**(4), 1093–1099 (2016)
12. Xie, J., Jiang, Y., Tsui, H.T.: Segmentation of kidney from ultrasound images based on texture and shape priors. IEEE Trans. Med. Imaging **24**(1), 45–57 (2005)
13. Mendoza, C.S., Kang, X., Safdar, N., Myers, E., Peters, C.A., Linguraru, M.G.: Kidney segmentation in ultrasound via genetic initialization and active shape models with rotation correction. In: International Symposium on Biomedical Imaging, San Francisco, CA, USA, pp. 69–72 (2013)
14. Ardon, R., Cuingnet, R., Bachuwar, K., Auvray, V.: Fast kidney detection and segmentation with learned kernel convolution and model deformation in 3D ultrasound images. In: International Symposium on Biomedical Imaging, New York, NY, USA, pp. 267–271. IEEE (2015)
15. Marsousi, M., Plataniotis, K.N., Stergiopoulos, S.: An automated approach for kidney segmentation in three-dimensional ultrasound images. IEEE J. Biomed. Heal. Informatics **21**(4), 1079–1094 (2017)
16. Tabrizi, P.R., Mansoor, A., Cerrolaza, J.J., Jago, J., Linguraru, M.G.: Automatic kidney segmentation in 3D pediatric ultrasound images using deep neural networks and weighted fuzzy active shape model. In: International Symposium on Biomedical Imaging, Washington, DC, USA, pp. 1170–1173. IEEE (2018)

17. Ravishankar, H., Venkataramani, R., Thiruvenkadam, S., Sudhakar, P., Vaidya, V.: Learning and incorporating shape models for semantic segmentation. In: Descoteaux, M., Maier-Hein, L., Franz, A., Pierre Jannin, D., Collins, L., Duchesne, S. (eds.) MICCAI 2017. LNCS, vol. 10433, pp. 203–211. Springer, Cham (2017). https://doi.org/10.1007/978-3-319-66182-7_24

18. Yin, S., et al.: Automatic kidney segmentation in ultrasound images using subsequent boundary distance regression and pixelwise classification networks. Med. Image Anal. **60**, 101602 (2020)

19. Roshanitabrizi, P., et al.: Pediatric hydronephrosis severity assessment using convolutional neural networks with standardized ultrasound images. In: International Symposium on Biomedical Imaging, Nice, Acropolis, France. pp. 1803–1806. IEEE (2021)

20. Chaurasia, A., Culurciello, E.: LinkNet: exploiting encoder representations for efficient semantic segmentation. In: Visual Communications and Image Processing, St. Petersburg, FL, USA, pp. 1–5. IEEE (2017)

21. Yakubovskiy, P.: Segmentation Models. GitHub Repository. GitHub (2019)

22. Zhou, Z., Rahman Siddiquee, M.M., Tajbakhsh, N., Liang, J.: UNet++: redesigning skip connections to exploit multiscale features in image segmentation. IEEE Trans. Med. Imaging **39**(6), 1856–1867 (2020)

23. Zhou, Z., Rahman Siddiquee, M.M., Tajbakhsh, N., Liang, J.: Unet++: a nested U-net architecture for medical image segmentation. In: Stoyanov, D., et al. (eds.) DLMIA/ML-CDS -2018. LNCS, vol. 11045, pp. 3–11. Springer, Cham (2018). https://doi.org/10.1007/978-3-030-00889-5_1

24. Pearson, K.: Mathematical contributions to the theory of evolution. III. Regression, heredity, and panmixia. Philos. Trans. R. Soc. London **187**, 253–318 (1896)

25. LeCun, Y., Bottou, L., Bengio, Y., Haffner, P.: Gradient-based learning applied to document recognition. Proc. IEEE. **86**(11), 2278–2323 (1998)

Automated Deep Learning-Based Detection of Osteoporotic Fractures in CT Images

Eren Bora Yilmaz[1(✉)], Christian Buerger[2], Tobias Fricke[3],
Md Motiur Rahman Sagar[4], Jaime Peña[1], Cristian Lorenz[2],
Claus-Christian Glüer[1], and Carsten Meyer[4,5]

[1] Section Biomedical Imaging, Department of Radiology and Neuroradiology,
University Hospital Schleswig-Holstein (UKSH), Campus Kiel, Kiel, Germany
eren.yilmaz@rad.uni-kiel.de
[2] Philips Research, Hamburg, Germany
[3] Department of Radiology and Neuroradiology, UKSH, Campus Kiel, Kiel, Germany
[4] Department of Computer Science, Ostfalia University of Applied Sciences,
Wolfenbüttel, Germany
[5] Department of Computer Science, Faculty of Engineering,
Kiel University, Kiel, Germany

Abstract. Automating opportunistic screening of osteoporotic fractures in computed tomography (CT) images could reduce the underdiagnosis of vertebral fractures. In this work, we present and evaluate an end-to-end pipeline for the detection of osteoporotic compression fractures of the vertebral body in CT images. The approach works in 2 steps: First, a hierarchical neural network detects and identifies all vertebrae that are visible in the field of view. Second, a feed-forward convolutional neural network is applied to patches containing single vertebrae to decide if an osteoporotic fracture is present or not. The maximum of the classifier's output scores then allows to classify if there is at least one fractured vertebra in the image. On a per-patient basis our pipeline classifies 145 CT images—annotated by an experienced musculoskeletal radiologist—with a sensitivity of 0.949 and a specificity of 0.815 regarding the presence of osteoporotic fractures. The fracture classifier even distinguishes grade 1 deformities from grade 1 osteoporotic fractures with an area under the ROC-curve of 0.742, a task potentially challenging even for human experts. Our approach demonstrates robust and accurate diagnostic performance and thus could be applied to opportunistic screening.

Keywords: Deep learning · Osteoporosis · CT · Automatic vertebral fracture detection

1 Introduction

In a hospital's daily routine, computed tomography (CT) images are taken for a variety of purposes, including diagnosis of bone disorders, lung diseases or

C. Lian et al. (Eds.): MLMI 2021, LNCS 12966, pp. 376–385, 2021.
https://doi.org/10.1007/978-3-030-87589-3_39

tumors. Diagnosis of vertebral fractures on such images is possible but time-consuming and rarely done, thus an automated system could reduce the under-diagnosis of vertebral fractures.

In the past few years, the popularity of convolutional neural networks (CNNs) for automated detection of vertebral fractures has grown. Tomita et al. [20] proposed an automated pipeline consisting of a CNN and a Long Short-Term Memory (LSTM) network, requiring the CT image to include the whole thoracolumbar spine in the center. Nicolaes et al. [16] derived segmentation masks from ground-truth coordinates and trained a 3D CNN for voxelwise classification. Husseini et al. [10] proposed new pre-training methods for this task, but excluded mild fractures from their dataset, which are particularly difficult to diagnose. Chettrit et al. [6] evaluated sequential classification techniques for fracture detection and proposed a comparably slow ensemble of models (60s on CPU). The ground truth used by these approaches [6,10,16,20] relies on the semiquantitative Genant grading [8], ignoring whether a deformity is degenerative or constitutes an osteoporotic fracture. Pisov et al. [19] even directly computed a variant of the Genant grade from heights measured between points in the image. Recently, Yilmaz et al. [21] developed a tool that classifies osteoporotic fractures on vertebra level but still needs vertebra coordinates for automation (Fracture Classification Network, *fNet*).

In the context of finding vertebral fractures, an important first step is to automatically detect vertebrae in the CT scans. Multiple segmentation-based methods have been presented in the past, such as Klinder et al. [13] or Kelm et al. [11]. Recently presented deep learning-based techniques include a combination of random forest classifiers with a neural network [5], a patch-based iterative instance-by-instance segmentation [2], coupling CNN's with deformable models [14], and CNN based keypoint localizers [15]. Similar to the coarse to fine approach as presented in [18], Buerger et al. [4] presented a Hierarchical Convolutional Neural Network (*hNet*) to identify all vertebrae from CT images.

Here, we propose a fully automated pipeline for localizing and classifying osteoporotic compression fractures of the vertebral body in CT images. The pipeline consists of the *hNet* [4] to localize all visible vertebrae in the CT image and the *fNet* [21] that classifies vertebrae regarding the presence of osteoporotic fractures. A final step aggregates the outputs of *fNet* to a single value per CT image and applies a classification threshold (operating point) to decide if there is at least one osteoporotic fracture present.

2   Methods

Given a 3D CT image the task of our proposed pipeline is to decide if any visible vertebra has an osteoporotic fracture or not (binary classification). In contrast to most other publications [6,10,16,19,20], this decision treats degenerative deformities as non-fractured even if the Genant grade [8] is above 0, as this distinction is important for therapeutic decisions. We split this task in three sub-tasks (see Fig. 1): Vertebra localization (Sect. 2.1), vertebra classification (Sect. 2.2), and an aggregation step (Sect. 2.3).

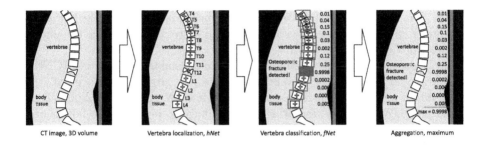

Fig. 1. Illustration of the overall pipeline working in 3 steps. (1) Vertebra localization: *hNet* detects vertebra body centers as labeled landmarks from CT. (2) Vertebra classification: *fNet* classifies single-vertebra patches to determine if a vertebra shows an osteoporotic fracture or not. (3) Patient level aggregation to a single fracture score per image.

2.1   *hNet*: Vertebra Localization and Identification

The first step is responsible for detecting and identifying vertebra landmarks in the visible field of view (FOV). The method has previously been published in [4] and follows a hierarchical Convolutional Neural Network based approach, hence the name *hNet*. A coarse to fine network inference is executed: The coarse single-class 3D model (working with sliding window patches on an isotropic 5 mm Cartesian grid) separates the spine from the background. A subsequent fine multi-class 3D model (working with sliding patches that are placed along the detected coarse spine segmentation on an isotropic 2 mm grid) segments several key vertebrae and a landmark class representing the vertebra body center. The combination of these key vertebrae (for which the label is known) and the landmark segmentations (for which the label is not known) can finally be used to derive identifications of vertebra-body landmarks over the complete FOV.

To increase the robustness of *hNet* with respect to generalization to unseen test data, an extensive data augmentation scheme was applied during training of both coarse and fine models. Next to common geometric (random rigid transformations) and intensity augmentations (random noise and Gaussian blur), we also employed FOV augmentations by mixing in images tightly cropped to the spine (where each cropped FOV is defined by the bounding box around the respective whole-spine ground truth annotation) for robustness against FOV variations of the input data during inference.

2.2   *fNet*: Fracture Classification

The second stage model is responsible for classifying whether sub-volumes ("patches") extracted from the CT images and containing a single vertebra show an osteoporotic fracture or not. As sketched in Fig. 2, the patches have size $40 \times 50 \times 60$ mm^3 = $40 \times 50 \times 20$ voxels (vx) at a resolution of $1 \times 1 \times 3$ $\frac{\text{mm}^3}{\text{vx}}$ (longitudinal $\times$ anteroposterior $\times$ lateral) and are centered on each single vertebra. For this step we employ a simple feedforward CNN (*fNet*) as used by [21].

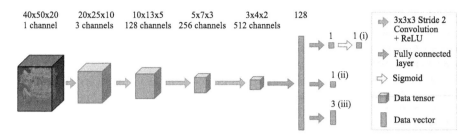

Fig. 2. *fNet.* (i)–(iii) denote outputs referring to fracture classification (used here), grade estimation and deformity estimation, respectively. Image source (modified): Yilmaz et al. [21].

Out of the three outputs described in [21] (see Fig. 2) we are interested in output (i) that does a binary classification of the osteoporotic fracture status, but we use all outputs for training. Like [21], we train the vertebra classifier using the Adam optimizer [12] combined with Stochastic Weight Averaging [17].

As preprocessing, Hounsfield-values are divided by 2048. To improve diversity in training data, various data augmentation techniques are applied. First, the vertebra patches are randomly rotated by up to 18° along the left-right-axis. To simulate missing values at the border of the CT image, we randomly crop the images, replacing cropped areas by 0-valued voxels. Next the images are randomly mirrored on the sagittal plane and translated by up to 4 vx in each dimension. Finally, with equal probability one of the following three settings is chosen for further data augmentation:

– Additive (σ_a) and multiplicative (σ_m) Gaussian noises are applied, each in two settings: (1) with a separate random value for each voxel—simulating low image quality—and (2) with a single value that is applied to all voxels—simulating miscalibrated scanners. Based on earlier experiments, we chose $\sigma_{a,1} = 100$ HU, $\sigma_{a,2} = 50$ HU, $\sigma_{m,1} = 0$ % and $\sigma_{m,2} = 16$ %.
– With 50 % probability the image is blurred with a Gaussian kernel with $\sigma = 1.5 \times 1.5 \times 1.5$ vx, simulating different reconstruction kernels used in the CT reconstruction.
– No augmentation is applied.

2.3   Patient Level Aggregation

After one score per vertebra is obtained from the outputs of the *fNet*, these scores are aggregated to estimate if there is at least one osteoporotic fracture in the image or not. Previous work [3,6] uses LSTM or BiLSTM networks to include information on neighboring vertebrae. We did not obtain significantly better results with LSTMs and thus chose the simpler maximum function for the aggregation step. Note that only vertebrae labeled T4–L4 by the *hNet* were included in the aggregation step, because the other vertebrae rarely contain osteoporotic fractures and are not annotated in our dataset.

3   Experiments

Two in-house datasets are used in this study, which we refer to as "Dataset 1" and "Dataset 2".

Dataset 1 (same dataset as in [4]) contains CT images from a general trauma setting (targeting rib fractures, only minor appearances of vertebral fractures) of 147 patients and was used to train and validate hNet (both coarse and fine models) for vertebra body detection and identification. This data showed quite some variation in terms of FOV and varied from thoracic (always covering T1 - T12) to whole-body images, with an axial FOV range of 431.17 ± 62.95 mm and a feet-head FOV range of 518.14 ± 279.55 mm. An average axial resolution of 0.84 ± 0.12 mm and a slice thickness of 1.80 ± 0.88 mm were measured.

Dataset 2 contains low-dose (120 mA, 80 kV peak) CT images of 159 patients that were acquired in 7 centers in the course of the Diagnostik Bilanz study of the BioAsset project [9]. Using SpineAnalyzer^{TM} [1] each vertebra from T4 to L4 was annotated with the fracture grade, vertebra centroids and a differential diagnosis indicating if the vertebra shows either a "deformity" (1019 cases), an "osteoporotic fracture" (128), is "unevaluable" due to noise (5) or "normal" (802). The goal is to distinguish "osteoporotic fractures" from "deformity" plus "normal" (binary classification, "unevaluable" vertebrae were excluded).

Dataset 1 is used to train the hNet, but contains no fracture annotation, so training and testing of the fNet was only possible on Dataset 2. Additionally, Dataset 2 is used to evaluate the trained hNet. Although it was originally trained on the whole-body CT images in Dataset 1, the evaluation of the hNet on Dataset 2 is restricted to T4–L4, because (a) osteoporotic fractures are most common in this area and (b) Dataset 2 only contains annotations for vertebrae in this range. Additionally we excluded one complete hospital (14 cases, 145 remaining) from the evaluation of the hNet and the whole pipeline due to unrealistically low image quality because of the choice of the reconstruction kernel, even for low-dose images (The whole Dataset 2 is used for training and evaluating the fNet on vertebra level).

3.1   hNet: Localization evaluation

For the quantitative evaluation of the hNet we used the metrics defined by Mader et al. [15]. They count true negatives (TN, vertebra center not in the image and not detected by the localizer) and true positives (TP, localizer output within 10 mm of the ground truth position for the corresponding vertebra level). Analogously they count false positives (FP, vertebra center missing, but detected), false negative (FN, vertebra center present, but not detected), and mislocalizations (distance to ground truth above 10 mm). The *success rate* [15] is then defined as $(TP + TN)/(N \cdot K_{test})$ where K_{test} is the number of images in the test set and N is the maximum number of vertebrae. In our case $N = 13$ since at most T4–L4 were annotated in the images (some images contain 1 or 2 fewer vertebrae). Note that this metrics puts emphasis on the vertebra identification—for example if each vertebra has the label of the vertebra above it, the success rate

Table 1. *hNet* results by fracture grade. Note that this only evaluates the localizer. TP, FN, Mislocalizations, TN, and FP are defined as in Sect. 3.1. TN (56) and FP (58) are not listed in this table as there are no corresponding annotated vertebrae.

1765 annotated Vertebrae			Grade 0		Grade 1		Grade 2		Grade 3	
			1611		61		63		30	
TP	1458	82.6%	1324	82.2%	55	90.2%	55	87.3%	24	80.0%
FN	30	1.7%	29	1.8%	0	0.0%	0	0.0%	1	3.3%
Misloc.	277	15.7%	258	16.0%	6	9.8%	8	12.7%	5	16.7%

would go to zero—although it does not make a difference for the level-agnostic fracture classifier (it plays a small role in the aggregation, see Sect. 2.3).

On Dataset 2 the *hNet* achieved a success rate of 0.813. The most common issues were misidentification by one vertebra level (20 of 145 spines), not finding the key vertebrae in the coarse segmentation step (2 of 145 spines), and partially visible vertebrae that were found by the localizer but not annotated by the radiologist (53 vertebrae). The latter two are problematic for the whole pipeline, because any missed vertebra could be a fractured one and any additional vertebrae may be misclassified as fractured. However, misidentification has no effect on the fracture pipeline, except for the exclusions described in Sect. 2.3. As it is important for the fracture detection task to first be able to localize the *fractured* vertebrae, Table 1 breaks down the *hNet*'s results: Although the fraction of TP was slightly smaller for grade 2 and 3 as compared to grade 1 (more mislocalizations), we did not see evidence for a reduced detection and identification performance on fractured vertebrae on our dataset.

3.2 *fNet*: Fracture Classifier Evaluation

The fracture classifier was trained using the patches based on the ground truth coordinates. Dataset 2 was split into four subsets defining a 4-fold cross-validation (CV) setup, by assigning each patient randomly to one of the folds. In each run, two data subsets were used as training data, one as validation data (for early stopping and choosing the operating point) and one subset as test data. To address the class imbalance, for training the *fNet*, each vertebra in each mini-batch was chosen randomly, such that with 50% probability a fractured one was chosen.

First, the fracture classifier was evaluated on vertebra level—step (2) in Fig. 1—independently of the localization tool by using the ground truth coordinates for patch extraction. Operating points were chosen to maximize the product of sensitivity and specificity. The *fNet* was highly accurate in classifying individual vertebrae with an area under the ROC curve (ROC-AUC), specificity and sensitivity of 0.986, 0.905, and 0.960, respectively. Figure 3 provides an overview of the results on the dataset, illustrating the highly accurate classification.

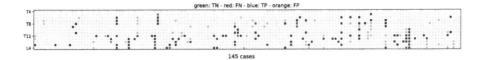

Fig. 3. Evaluation of the *fNet* on Dataset 2 using ground truth vertebra coordinates and the operating point chosen on vertebra level.

The Largest Difficulty is the Distinction of Osteoporotic Fractures from Degenerative Deformities. If we restrict the test set to grade 1 vertebrae (66) in the evaluation of the *fNet*, the ROC-AUC, specificity and sensitivity drop to 0.742, 0.694, and 0.635, respectively. Note that this restriction implies that there are only degenerative deformities (38) and mild osteoporotic fractures (28) in the dataset, which are difficult to distinguish even for human experts. For comparison, if the test set is restricted to grade 0 and 3 vertebrae, sensitivity and ROC-AUC are 1.

3.3   Patient Level Evaluation

The full pipeline (*hNet* + *fNet* + max), when classifying whole CT images, achieved an ROC-AUC of 0.945 (see Table 2). Using the operating points determined on the respective validation set of the vertebra classifier, the sensitivity and specificity on the test set were 0.815 and 0.949, respectively. Note that these operating points need to be tuned individually for different localizers (i.e. *hNet* vs. ground truth) and differ from the vertebra level operating points.

The Patient-Level Performance only Slightly Decreases When Using Automatic Vertebra Localization. As seen in Table 2, row 1, when using ground truth vertebral body center positions in the pipeline the ROC-AUC was 0.979. Note that the *fNet* is vertebra-agnostic (robust to misidentification) and was trained with random translations as data augmentation, so mislocalizations of a few mm should not be a problem. Out of the 20 spines where the *hNet* misidentified the vertebrae by one vertebra level, 2 were misclassified on patient level. Localizations of partially visible vertebrae resulted in 2 false positives and not finding the spine resulted in 1 false negative patient level classification.

Table 2. Experimental results for osteoporotic compression fracture classification (see Sect. 3.3) on patient level. Note that GT and *hNet* use different operating points (76.5 and 0.874, respectively, averaged across 4 CV folds) chosen to maximize the product of specificity and sensitivity on validation data. GT: ground truth vertebra center positions. max: maximum aggregation. σ: standard deviation across CV-folds.

Method	ROC-AUC	Specificity	Sensitivity
GT + *fNet* + max	0.979, $\sigma = 0.0042$	0.931, $\sigma = 0.045$	0.877, $\sigma = 0.089$
hNet + *fNet* + max	0.945, $\sigma = 0.023$	0.815, $\sigma = 0.16$	0.949, $\sigma = 0.030$

Source/Method	ROC-AUC	Specificity	Sensitivity
Chettrit et al. [6]	0.955	0.951	0.822
Husseini et al. [10]	-	0.985	0.769
Nicolaes et al. [16]	0.95	0.938	0.905
Pisov et al. [19]	0.93	0.68,	0.94,
Tomita et al. [20]	0.982	0.952	0.852
hNet + fNet + max	**0.945**	**0.815**	**0.949**

Fig. 4. Literature comparison. Note that authors work with different datasets on different tasks, so rows are not directly comparable. For example, we classify osteoporotic fractures while other authors classify based on Genant grading. **bold**: our pipeline. Right: Correct localization but FN classification at a mildly fractured T9 vertebra.

Data Augmentation is Crucial for Successfully Training the Models. Without noise, blur and cropping in the training of the *hNet*, success rate dropped to 0.747. When omitting all data augmentation in the training of the *fNet*, overfitting prevented any improvement beyond random guessing (test loss at output (i) after each epoch above $\ln(2)$). We leave it for future work to assess the importance of different kinds of data augmentation in more detail.

4 Discussion and Conclusions

In this paper we present and evaluate a pipeline that first localizes and then classifies individual vertebrae to finally decide if the patient's CT image contains at least one vertebra with an osteoporotic fracture. As seen in Tables 1 and 4, our end-to-end (whole pipeline) results on data from a multi-center osteoporosis study document strong performance, both in terms of sensitivity (0.95—most relevant for opportunistic screening) and specificity (0.81). Published data [6,10,16,19,20], shows similar performance levels but a direct comparison is not possible since different datasets and definitions of the classification task have been used. Unlike some other studies that excluded mild fractures that are more difficult to detect [10], our approach showed robust performance even when including mild fractures. Actually, we were even able to discriminate (ROC-AUC of 0.742) mild osteoporotic fractures that enhance the risk of future fracture from mild degenerative deformities that have a similar shape but do not increase future fracture risk [7], a task often considered challenging even for radiologists. Our model performed robustly despite the challenge of data acquired in a multi-center setting including 7 different CT scanners. Warranting further successful evaluations on larger datasets, our approach could be turned into an AI assistant that would alert the radiologist about vertebral fractures and reduce the diagnostic and treatment gap in osteoporosis.

Acknowledgement. This work has been financially supported by the Federal Ministry of Education and Research (grants 01EC1005, 01EC1908C) and the Federal Ministry for Economic Affairs and Energy (KI-RAD). We thank Alexander Oliver Mader for a fruitful collaboration.

References

1. SpineAnalyzer. Optasia Medical Ltd., Cheadle Hulme, UK (2013)
2. Nikolas, L., van Ginneken, B., de Jong, P.A., Isgum, I.: Iterative fully convolutional neural networks for automatic vertebra segmentation and identification. Med. Image Anal. **53**, 142–155 (2019)
3. Bar, A., Wolf, L., Amitai, O.B., Toledano, E., Elnekave, E.: Compression fractures detection on CT. In: Proceedings of SPIE, vol. 10134, pp. 301–308. SPIE, Orlando, Florida, USA (2017)
4. Buerger, C., von Berg, J., Franz, A., Klinder, T., Lorenz, C., Lenga, M.: Combining deep learning and model-based segmentation for labeled spine CT segmentation. In: Medical Imaging 2020: Image Processing, vol. 11313, pp. 307–314. International Society for Optics and Photonics, SPIE (2020)
5. Chen, H., et al.: Automatic localization and identification of vertebrae in Spine CT via a joint learning model with deep neural networks. In: Navab, N., Hornegger, J., Wells, W.M., Frangi, A.F. (eds.) MICCAI 2015. LNCS, vol. 9349, pp. 515–522. Springer, Cham (2015). https://doi.org/10.1007/978-3-319-24553-9_63
6. Chettrit, D., et al.: 3D convolutional sequence to sequence model for vertebral compression fractures identification in CT. In: Martel, A.L., et al. (eds.) MICCAI 2020. LNCS, vol. 12266, pp. 743–752. Springer, Cham (2020). https://doi.org/10.1007/978-3-030-59725-2_72
7. Ferrar, L., Roux, C., Felsenberg, D., Glüer, C.C., Eastell, R.: Association between incident and baseline vertebral fractures in European women: vertebral fracture assessment in the Osteoporosis and Ultrasound Study (OPUS). Osteoporos. Int. **23**(1), 59–65 (2012)
8. Genant, H.K., Wu, C.Y., van Kuijk, C., Nevitt, M.C.: Vertebral fracture assessment using a semiquantitative technique. J. Bone Min. Res. Official J. Am. Soc. Bone Min. Res. **8**(9), 1137–1148 (1993)
9. Glüer, C.C.: New horizons for the in vivo assessment of major aspects of bone quality Microstructure and material properties assessed by quantitative computed tomography and quantitative ultrasound methods developed by the BioAsset consortium. Osteologie **22**, 223–233 (2013)
10. Husseini, M., Sekuboyina, A., Loeffler, M., Navarro, F., Menze, B.H., Kirschke, J.S.: Grading Loss: A Fracture Grade-based Metric Loss for Vertebral Fracture Detection (2020)
11. Kelm, B.M., et al.: Spine detection in CT and MR using iterated marginal space learning. Med. Image Anal. **17**(8), 1283–1292 (2013)
12. Kingma, D.P., Ba, J.L.: Adam: a method for stochastic optimization. In: 3rd International Conference on Learning Representations, San Diego (2015)
13. Klinder, T., Ostermann, J., Ehm, M., Franz, A., Kneser, R., Lorenz, C.: Automated model-based vertebra detection, identification, and segmentation in CT images. Med. Image Anal. **13**(3), 471–482 (2009)
14. Korez, R., Likar, B., Pernuš, F., Vrtovec, T.: Model-based segmentation of vertebral bodies from MR images with 3D CNNs. In: Ourselin, S., Joskowicz, L., Sabuncu, M.R., Unal, G., Wells, W. (eds.) MICCAI 2016. LNCS, vol. 9901, pp. 433–441. Springer, Cham (2016). https://doi.org/10.1007/978-3-319-46723-8_50
15. Mader, A.O., Lorenz, C., Bergtholdt, M., von Berg, J., Schramm, H., Modersitzki, J., Meyer, C.: Detection and localization of spatially correlated point landmarks in medical images using an automatically learned conditional random field. Comput. Vis. Image Underst. **176–177**, 45–53 (2018)

16. Nicolaes, J., Raeymaeckers, S., Wilms, G., Libanati, C., Debois, M.: Detection of vertebral fractures in CT using 3D Convolutional Neural Networks (2019)
17. Izmailov, P., Podoprikhin, D., Garipov, T., Vetrov, D., Wilson, A.G.: Averaging weights leads to wider optima and better generalization. In: Uncertain Artificial Intelligence, pp. 876–885. AUAI Press, Corvallis, Oregon, Monterey, California (2018)
18. Payer, C., Štern, D., Bischof, H., Urschler, M.: Coarse to fine vertebrae localization and segmentation with SpatialConfiguration-Net and U-Net. In: Proceedings of the 15th International Joint Conference on Computer Vision, Imaging and Computer Graphics Theory and Applications, pp. 124–133. SCITEPRESS - Science and Technology Publications, Valletta, Malta (2020)
19. Pisov, M., et al.: Keypoints Localization for Joint Vertebra Detection and Fracture Severity Quantification. arXiv:2005.11960 [cs, eess] (2020)
20. Tomita, N., Cheung, Y.Y., Hassanpour, S.: Deep neural networks for automatic detection of osteoporotic vertebral fractures on CT scans. Comput. Biol. Med. **98**, 8–15 (2018)
21. Yilmaz, E.B., Mader, A.O., Fricke, T., Peña, J., Glüer, C.-C., Meyer, C.: Assessing attribution maps for explaining CNN-based vertebral fracture classifiers. In: Cardoso, J., et al. (eds.) IMIMIC/MIL3ID/LABELS -2020. LNCS, vol. 12446, pp. 3–12. Springer, Cham (2020). https://doi.org/10.1007/978-3-030-61166-8_1

GT U-Net: A U-Net Like Group Transformer Network for Tooth Root Segmentation

Yunxiang Li[1], Shuai Wang[2], Jun Wang[3], Guodong Zeng[4], Wenjun Liu[1], Qianni Zhang[5], Qun Jin[6], and Yaqi Wang[7]([✉])

[1] Microelectronics CAD Center, Hangzhou Dianzi University, Hangzhou, China
[2] School of Mechanical, Electrical and Information Engineering, Shandong University, Weihai, China
[3] School of Biomedical Engineering, Shanghai Jiao Tong University, Shanghai, China
[4] sitem Center for Translational Medicine and Biomedical Entrepreneurship, University of Bern, Bern, Switzerland
[5] School of Electronic Engineering and Computer Science, Queen Mary University of London, London, UK
[6] Department of Human Informatics and Cognitive Sciences, Faculty of Human Sciences, Waseda University, Tokyo, Japan
[7] College of Media Engineering, Communication University of Zhejiang, Hangzhou, China
wangyaqi@cuz.edu.cn

Abstract. To achieve an accurate assessment of root canal therapy, a fundamental step is to perform tooth root segmentation on oral X-ray images, in that the position of tooth root boundary is significant anatomy information in root canal therapy evaluation. However, the fuzzy boundary makes the tooth root segmentation very challenging. In this paper, we propose a novel end-to-end U-Net like Group Transformer Network (GT U-Net) for the tooth root segmentation. The proposed network retains the essential structure of U-Net but each of the encoders and decoders is replaced by a group Transformer, which significantly reduces the computational cost of traditional Transformer architectures by using the grouping structure and the bottleneck structure. In addition, the proposed GT U-Net is composed of a hybrid structure of convolution and Transformer, which makes it independent of pre-training weights. For optimization, we also propose a shape-sensitive Fourier Descriptor (FD) loss function to make use of shape prior knowledge. Experimental results show that our proposed network achieves the state-of-the-art performance on our collected tooth root segmentation dataset and the public retina dataset DRIVE. Code has been released at https://github.com/Kent0n-Li/GT-U-Net.

Keywords: Image segmentation · Shape-sensitive loss · Group transformer · Root canal therapy

© Springer Nature Switzerland AG 2021
C. Lian et al. (Eds.): MLMI 2021, LNCS 12966, pp. 386–395, 2021.
https://doi.org/10.1007/978-3-030-87589-3_40

1   Introduction

In worldwide, approximately 743 million people are affected by severe periodontitis, which is considered the sixth most common health disorder [1]. Nowadays, root canal therapy is a routine periodontal treatment for periodontitis and an incorrect evaluation of the treatment result will impede timely follow-up [2,3]. Since the tooth root boundary is the significant anatomy feature for carrying out the evaluation, tooth root segmentation becomes the most important step of root canal therapy automatic evaluation. Unfortunately, performing an accurate tooth root segmentation is a very challenging task due to the following reasons: 1) the tooth root boundaries are blurry and some tissues around the teeth have similar intensities to the teeth, as shown in Fig. 1 (a); 2) other bones and tissues may overlap with the tooth root in the oral X-ray images, as shown in Fig. 1 (b); 3) the quality of the X-ray image may be very poor such as overexposed or underexposed, as shown in Fig. 1 (c).

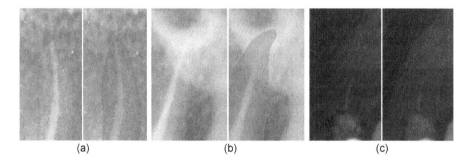

(a)   (b)   (c)

Fig. 1. Three examples of root canal therapy X-ray images. The left image in each group is the original image, and the right one depicts the tooth root and other tissues by red and cyan, respectively.

To address the problems mentioned above, Zhao et al. [4] provided a two-stage attention segmentation network, which can effectively alleviate the inhomogeneous intensity distribution problem by focusing on automatically catching the real tooth region. Lee et al. [5] adopted a fine-tuned mask R-CNN [6] algorithm to achieve the tooth segmentation. Nevertheless, these methods do not effectively solve the segmentation problem of fuzzy boundaries, and the performance improvement is mostly incremental. Chen et al. [7] proposed a novel MSLPNet with multi-scale structural similarity (MS-SSIM) loss, enhancing tooth segmentation that has fuzzy root boundaries. Cheng et al. [8] proposed U-Net+DFM to learn a direction field. It characterizes the directional relationship between pixels and implicitly restricts the shape of the segmentation result. Although these methods all achieve decent results in segmentation tasks, they are still limited by the intrinsic locality of convolutional neural networks (CNNs) and can not process global features very well. To relieve this problem, long-range dependencies via non-local operations are highly desired, and Transformer [9] provides a modeling pipeline to achieve that. Chen et al. proposed TransUNet [10], a Transformer-based encoder operating for segmentation, which adopts ViT [11] with 12 Transformer layers as the encoder. However, ViT relies on pre-trained weights obtained by a huge image corpus, which results in undesirable performance on insufficient datasets. To solve that, Aravind et al. [12] presented BoT-Net, an effective instance segmentation backbone, by combining transformer and convolution. Due to the high computational complexity of Transformer, BoTNet has only replaced a part of convolution in the last few layers of ResNet with Transformer.

In order to mitigate the problems existing in the present approaches, our network GT U-Net employs both the combination of convolution and Transformer without pre-training weights and the grouping structure and bottleneck structure that significantly reduces the amount of computation. Besides, FD loss also solves the problem of fuzzy boundary segmentation by making full use of shape prior knowledge. The main contributions of this paper are listed as follows:

(1) Our network retains the advantages of the general U-Net framework and introduces Transformer into the medical image segmentation application to solve the limitation of convolution.
(2) We design a grouping structure and a bottleneck structure, which greatly reduces the computation load of Transformer and makes it feasible in image segmentation.
(3) For the root segmentation task, we propose a shape-sensitive Fourier Descriptor loss function to deal with the problem of fuzzy boundary segmentation.

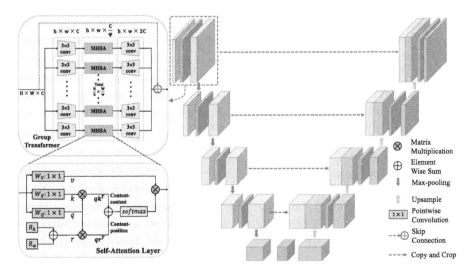

Fig. 2. The architecture of the proposed GT U-Net, which is composed of a U-shape general framework and group Transformer. Self-attention layer is the base structure of Multi-head self-attention (MHSA) where our MHSA has 4 heads, and we do not illustrate them on the figure for simplicity.

2   Method

GT U-Net follows the overall U-shaped structure, where both the encoders and decoders consist of group Transformer. It works with a shape-sensitive Fourier Description (FD) loss function for tooth root segmentation. The proposed method is described in full detail below.

2.1   U-Net Like Group Transformer Network

Overall Architecture: The architecture of GT U-Net is illustrated in Fig. 2. Specifically, our group Transformer is composed of a skip connection, a grouping module, 3×3 convolutions, Multi-head self-attention (MHSA) modules, and a merging module. Among them, the skip connection is used to solve the problem of gradient vanishing and keep the low-level information. Considering that MHSA requires $O(n^2 d)$ memory and computation when performing globally across n entities, the grouping module and 3×3 convolutions are designed to reduce the amount of MHSA computation.

Grouping Structure and Bottleneck Structure: Transformer was first proposed for natural language processing (NLP) tasks and has become a hot topic in many computer vision tasks as a non-local operation with long-range dependence. However, it is difficult to apply Transformer directly to medical image tasks on account of the discrepancy between natural language and image. The number of words in natural language is limited, but the number of pixels increases quadratic with the increase of image size. Because of this, we designed group Transformer to solve the problem of too much computation due to the image characteristics in medical image segmentation. Assuming that the original feature block size is $H \times W \times C$, the computation of MHSA will be greatly reduced through the grouping structure and bottleneck structure. The calculation amount of MHSA before improvement [13] and that of our group Transformer (GT) are given in Eq. (1) and (2), respectively:

$$\Omega(MHSA) = 4HWC^2 + 2(HW)^2 C \tag{1}$$

$$\Omega(GT) = 4hw(\frac{C}{\varphi})^2 + 2(hw)^2 \frac{C}{\varphi} = \frac{4hw}{\varphi^2 HW} HWC^2 + \frac{2h^2 w^2}{\varphi H^2 W^2}(HW)^2 C \tag{2}$$

where φ is the channel scaling factor of the bottleneck structure, and the size of each group Transformer unit is set to $h \times w$, which is determined by the input image size and task characteristics. The proposed method achieves local long-range dependencies at higher encoder and decoder layers. With the deepening of the network, the field of vision is gradually expanded to extract the junction features between different GT units and to achieve the global long-range dependencies.

Multi-head Self-attention: MHSA [9] is a type of attention mechanism which is more focused on the internal structure. The detailed architecture of MHSA is illustrated in Fig. 2, where the position encoding method is relative-distance-aware position encoding [14–16] with R_h and R_w for height and width. The attention logit is $qk^T + qr^T$, and q, k, v, r denote query, key, value, and position encodings, respectively.

Hybrid Structure of Convolution and Transformer: Transformer has the ability to extract global contexts, but it also has the limitation of lacking translation invariance while locality and translational equivariance are the basic property of convolution. Therefore, 3×3 convolution in our group Transformer is not only for the bottleneck structure, but it also helps construct a hybrid architecture where the convolution is responsible for feature extraction and Transformer is constructed to model the long-range dependencies.

2.2   Shape-Sensitive Fourier Descriptor Loss Function

From the anatomical prior knowledge, it is known that the tooth roots share a similar shape. Adding the shape information into the loss function helps better guide the model to segment tooth roots. (x_m, y_m) is one of the coordinates on a tooth root boundary that contains N pixels, and the boundary shape can be formed as a complex number $z(m) = x_m + jy_m$. The Fourier Descriptor [17] of this shape is defined as $DFT(z(m))$, that is Z(k) in Eq. (3):

$$Z(k) = DFT(z(m)) = \frac{1}{N} \sum_{m=0}^{N-1} z(m)e^{-j2\pi mk/N} \quad (k = 0, \cdots, N-1) \quad (3)$$

Fourier Descriptor is a quantitative representation of closed shapes independent of their starting point, scaling, location, and rotation. Therefore, the shape difference between the predicted boundary A and the manually labeled boundary B can be quantified, and it can be calculated based on Eq. (4). The original loss function is binary cross-entropy (BCE). In the proposed method, we add Fourier Descriptors to build a new shape-sensitive loss function. Considering their order of magnitude in the whole training process, we calculate the final Fourier Descriptor (FD) loss by Eq. (5):

$$\Delta Z(k) = |Z_A(k) - Z_B(k)|. \quad (4)$$

$$FD\ loss = BCE(A, B) \times \frac{1}{1 + e^{-\beta \times \Delta Z(k)}} \quad (5)$$

The novel FD loss pays equal attention to shape loss and BCE loss regardless of their order of magnitude. β is determined by the order of magnitude of $\Delta Z(k)$, and it is set to 10 in this paper.

3   Experiment

3.1   Tooth Root Segmentation Dataset

We have built a new root canal therapy X-ray image dataset with patients' consent. The tooth root segmentation dataset contains 248 root canal therapy X-ray images from different patients in total. And three experienced stomatologists helped to complete the tooth root annotation. Specifically, to minimize inter-observer variability, the final annotation results needed the agreement of all three stomatologists.

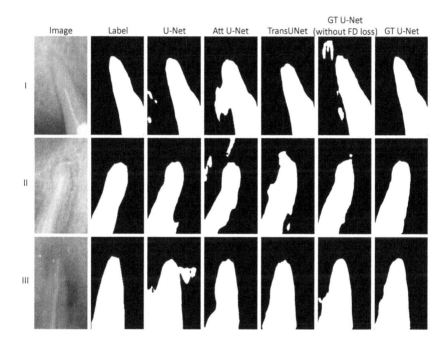

Fig. 3. Representative segmentation results achieved by different methods.

3.2   Implementation Details

In our implementation, three state-of-the-art methods are compared, U-Net [18], Attention U-Net [19], and TransUNet [10]. All compared methods and our method were implemented using PyTorch and trained on four RTX 2080Ti GPUs. Random center crop, random rotation, and axial flipping are adopted for the data augmentation to improve robustness and avoid overfitting. In the training, Adam optimizer is adopted and the initial learning rate and momentum are set to 2×10^{-4} and 0.5, respectively. The total number of training epochs is 200, and the batch size is 12. All the images are resized to 256×256 for input. $h \times w$ of the group Transformer unit is set to 8×8, and φ is set to 2.

To evaluate the segmentation performance, we compare the results with six metrics for evaluation, including Accuracy (ACC), Sensitivity (SE), Specificity (SP), Jaccard Similarity (JS), and Dice Coefficient (DICE). Besides, we adopt 3-fold cross-validation for testing the segmentation performance on the tooth root segmentation dataset.

Table 1. The evaluation performance of all methods on the Tooth Root Segmentation Dataset.

	U-Net	Att U-Net	TransUNet	GT U-Net (without FD loss)	GT U-Net
ACC (%)	92.81 ± 0.90	93.03 ± 0.56	93.41 ± 0.25	93.80 ± 0.32	**93.98 ± 0.11**
SE (%)	92.37 ± 0.68	89.36 ± 1.30	91.81 ± 0.74	**93.98 ± 0.38**	93.74 ± 1.18
SP (%)	93.74 ± 0.99	**95.60 ± 0.28**	95.05 ± 0.43	94.26 ± 0.35	94.79 ± 1.14
JS (%)	84.62 ± 1.30	83.85 ± 1.29	85.75 ± 0.65	86.63 ± 0.62	**86.87 ± 0.13**
DICE (%)	91.19 ± 1.01	91.15 ± 0.76	91.96 ± 0.40	92.32 ± 0.49	**92.54 ± 0.25**

3.3   Experimental Results

The average results with standard deviations of all methods are reported in Table 1. From this table, it can be summarized that our proposed GT U-Net achieves the best accuracy, sensitivity, Jaccard Similarity, and Dice Coefficient. In Fig. 3, we qualitatively compare the segmentation results on three oral X-ray images. It can be evidently seen that our method obtains the best segmentation results on the blurred boundaries. To verify the effectiveness of FD loss, we implement GT U-Net without FD loss. Based on the comparison in Table 1 and Fig. 3, FD loss can be effectively applied to the segmentation task with a fuzzy boundary but a similar shape by implicitly restricting the shape of the segmentation result with the anatomical prior knowledge.

Table 2. The segmentation performance achieved by all methods on the public DRIVE dataset.

	Year	ACC (%)	SE (%)	SP (%)	F1 (%)	AUC (%)
U-Net [18, 20]	2015	95.55	78.22	98.08	81.74	97.52
MS-NFN [21]	2018	95.67	78.44	98.19	–	98.07
Residual U-Net [22, 23]	2018	95.53	77.26	98.20	81.49	97.79
Recurrent U-Net [22, 23]	2018	95.56	77.51	98.16	81.55	97.82
R2U-Net [22, 23]	2018	95.56	77.92	98.13	81.71	97.84
DenseBlock-UNet [24]	2018	95.41	79.28	97.76	81.46	97.56
DUNet [20]	2019	95.58	78.63	98.05	81.90	97.78
ACE-Net [25]	2019	95.69	77.25	**98.42**	–	97.42
IterNet [23]	2020	95.73	77.35	98.38	82.05	**98.16**
GT U-Net	2021	**96.31**	**82.54**	98.24	**84.58**	98.02

3.4 Performance on the Public DRIVE Dataset

To further evaluate the performance of our GT U-Net, we applied it on the widely used retinal dataset DRIVE [26]. The dataset consists of 40 randomly selected color fundus retinal images of size 565×584. Officially, DRIVE is split into two equal groups for training and testing. The group Transformer unit parameters h and w are all set to 4, and φ is set to 2. Since FD loss is proposed for medical image segmentation with a similar shape, like tooth root segmentation, we did not implement it on the DRIVE dataset. To efficiently train our GT U-Net, we cropped image patches from the original images, and 64×64 is adopted as the sample patch size, which is widely adopted on this dataset.

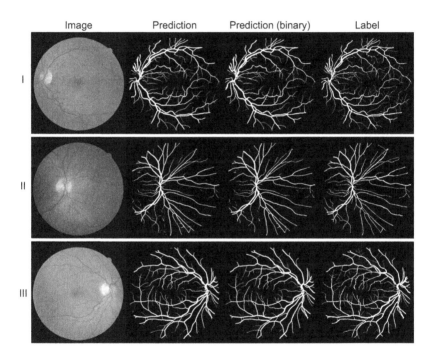

Fig. 4. Representative segmentation results of the public DRIVE dataset.

The comparison results are reported in Table 2. It can be observed that our GT U-Net outperforms other methods by at least 3.2% on SE, over 2.5% on F1, and over 0.5% on ACC. In terms of SP and AUC, our GT U-Net achieved competitive performance in comparison with the best methods, limiting the difference from them within 0.2%. Meanwhile, the visualization of some representative segmentation results is given in Fig. 4. It can be clearly observed that the vascular part of the retina can be segmented very well by our method, except for some very tiny areas.

4   Conclusion

In this paper, a novel end-to-end U-Net like group Transformer network (GT U-Net) for medical image segmentation is presented. GT U-Net uses the hybrid structure of convolution and Transformer to eliminate the need for pre-trained weights. Moreover, it significantly reduces the computational complexity of Transformer using a bottleneck structure and a grouping structure, and FD loss can be effectively applied to the segmentation task with a fuzzy boundary but a similar shape. Extensive experiments are conducted on tooth root and retinal vessel segmentation tasks, respectively, and the results show that the proposed GT U-Net has excellent segmentation performance and a great application perspective in a wide range of medical image segmentation tasks.

Acknowledgements. This work was supported by the National Key Research and Development Program of China (Grant No. 2019YFC0118404) and Public Projects of Zhejiang Province (Grant No. LGG20F020001).

References

1. Peres, M.A., et al.: Oral diseases: a global public health challenge. Lancet **394**(10194), 249–260 (2019)
2. Estrela, C., Holland, R., Estrela, C.R.D.A., Alencar, A.H.G., Sousa-Neto, M.D., Pécora, J.D.: Characterization of successful root canal treatment. Braz. Dent. J. **25**(1), 3–11 (2014)
3. Kaplan, T., Sezgin, G.P., Sönmez-Kaplan, S.: A survey study, Dental students' perception of difficulties concerning root canal therapy (2019)
4. Zhao, Y., et al.: TSASNet: tooth segmentation on dental panoramic x-ray images by two-stage attention segmentation network. Knowl.-Based Syst. **206**, 106338 (2020)
5. Lee, J.H., Han, S.S., Kim, Y.H., Lee, C., Kim, I.: Application of a fully deep convolutional neural network to the automation of tooth segmentation on panoramic radiographs. Oral Surg. Oral Med. Oral Pathol. Oral Radiol. **129**(6), 635–642 (2020)
6. He, K., Gkioxari, G., Dollár, P., Girshick, R.: Mask R-CNN. In: Proceedings of the IEEE International Conference on Computer Vision, pp. 2961–2969 (2017)
7. Chen, Q., et al.: MSLPNet: multi-scale location perception network for dental panoramic X-ray image segmentation. Neural Comput. Appl. **33**(16), 10277–10291 (2021). https://doi.org/10.1007/s00521-021-05790-5
8. Cheng, F., et al.: Learning directional feature maps for cardiac MRI segmentation. In: Martel, A.L., et al. (eds.) MICCAI 2020. LNCS, vol. 12264, pp. 108–117. Springer, Cham (2020). https://doi.org/10.1007/978-3-030-59719-1_11
9. Vaswani, A., et al.: Attention is all you need. In: NIPS, pp. 6000–6010 (2017)
10. Chen, J.: TransUNet: transformers make strong encoders for medical image segmentation. arXiv preprint arXiv:2102.04306 (2021)
11. Dosovitskiy, A., et al.: An image is worth 16 × 16 words: transformers for image recognition at scale. arXiv preprint arXiv:2010.11929 (2020)

12. Srinivas, A., Lin, T.Y., Parmar, N., Shlens, J., Abbeel, P., Vaswani, A.: Bottleneck transformers for visual recognition. arXiv preprint arXiv:2101.11605 (2021)
13. Liu, Z., et al. Swin transformer: hierarchical vision transformer using shifted windows. arXiv preprint arXiv:2103.14030 (2021)
14. Shaw, P., Uszkoreit, J., Vaswani, A.: Self-attention with relative position representations. In Proceedings of the 2018 Conference of the North American Chapter of the Association for Computational Linguistics: Human Language Technologies, Volume 2 (Short Papers), pp. 464–468 (2018)
15. Ramachandran, P., Parmar, N., Vaswani, A., Bello, I., Levskaya, A., Shlens, J.: Stand-alone self-attention in vision models. arXiv preprint arXiv:1906.05909 (2019)
16. Bello, I., Zoph, B., Vaswani, A., Shlens, J., Le, Q.V.: Attention augmented convolutional networks. In: Proceedings of the IEEE/CVF International Conference on Computer Vision (ICCV), October 2019
17. Zahn, C.T., Roskies, R.Z.: Fourier descriptors for plane closed curves. IEEE Trans. Comput. **100**(3), 269–281 (1972)
18. Ronneberger, O., Fischer, P., Brox, T.: U-Net: convolutional networks for biomedical image segmentation. In: Navab, N., Hornegger, J., Wells, W.M., Frangi, A.F. (eds.) MICCAI 2015. LNCS, vol. 9351, pp. 234–241. Springer, Cham (2015). https://doi.org/10.1007/978-3-319-24574-4_28
19. Oktay, O., et al.: Attention U-Net: learning where to look for the pancreas (2018)
20. Jin, Q., Meng, Z., Pham, T.D., Chen, Q., Wei, L., Su, R.: DUNet: a deformable network for retinal vessel segmentation. Knowl.-Based Syst. **178**, 149–162 (2019)
21. Wu, Y., Xia, Y., Song, Y., Zhang, Y., Cai, W.: Multiscale network followed network model for retinal vessel segmentation. In: Frangi, A.F., Schnabel, J.A., Davatzikos, C., Alberola-López, C., Fichtinger, G. (eds.) MICCAI 2018. LNCS, vol. 11071, pp. 119–126. Springer, Cham (2018). https://doi.org/10.1007/978-3-030-00934-2_14
22. Alom, M.Z., Hasan, M., Yakopcic, C., Taha, T.M., Asari, V.K.: Recurrent residual convolutional neural network based on U-Net (R2U-Net) for medical image segmentation. arXiv preprint arXiv:1802.06955 (2018)
23. Li, L., Verma, M., Nakashima, Y., Nagahara, H., Kawasaki, R.: IterNet: retinal image segmentation utilizing structural redundancy in vessel networks. In: Proceedings of the IEEE/CVF Winter Conference on Applications of Computer Vision, pp. 3656–3665 (2020)
24. Li, X., Chen, H., Qi, X., Dou, Q., Chi-Wing, F., Heng, P.-A.: H-DenseUNet: hybrid densely connected UNet for liver and tumor segmentation from CT volumes. IEEE Trans. Med. Imaging **37**(12), 2663–2674 (2018)
25. Zhu, Y., Chen, Z., Zhao, S., Xie, H., Guo, W., Zhang, Y.: ACE-Net: biomedical image segmentation with augmented contracting and expansive paths. In: Shen, D., et al. (eds.) MICCAI 2019. LNCS, vol. 11764, pp. 712–720. Springer, Cham (2019). https://doi.org/10.1007/978-3-030-32239-7_79
26. Staal, J., Abràmoff, M.D., Niemeijer, M., Viergever, M.A., Van Ginneken, B.: Ridge-based vessel segmentation in color images of the retina. IEEE Trans. Med. Imaging **23**(4), 501–509 (2004)

Information Bottleneck Attribution for Visual Explanations of Diagnosis and Prognosis

Ugur Demir[1]([✉]), Ismail Irmakci[1,2], Elif Keles[1], Ahmet Topcu[3], Ziyue Xu[4], Concetto Spampinato[5], Sachin Jambawalikar[6], Evrim Turkbey[7], Baris Turkbey[7], and Ulas Bagci[1]

[1] Department of Radiology and BME, Northwestern University, Chicago, IL, USA
ugurdemir2023@u.northwestern.edu
[2] ECE, Ege University, Izmir, Turkey
[3] Tokat State Hospital, Tokat, Turkey
[4] NVIDIA, Bethesda, MD, USA
[5] University of Catania, Catania, Italy
[6] Columbia University Medical Center, New York, NY, USA
[7] National Cancer Institute, National Institutes of Health, Bethesda, MD, USA

Abstract. Visual explanation methods have an important role in the prognosis of the patients where the annotated data is limited or unavailable. There have been several attempts to use gradient-based attribution methods to localize pathology from medical scans without using segmentation labels. This research direction has been impeded by the lack of robustness and reliability. These methods are highly sensitive to the network parameters. In this study, we introduce a robust visual explanation method to address this problem for medical applications. We provide an innovative visual explanation algorithm for general purpose and as an example application we demonstrate its effectiveness for quantifying lesions in the lungs caused by the Covid-19 with high accuracy and robustness without using dense segmentation labels. This approach overcomes the drawbacks of commonly used Grad-CAM and its extended versions. The premise behind our proposed strategy is that the information flow is minimized while ensuring the classifier prediction stays similar. Our findings indicate that the bottleneck condition provides a more stable severity estimation than the similar attribution methods. The source code will be publicly available upon publication.

Keywords: Visual explanations · Covid-19 · Weakly supervised · Information bottleneck attribution

1 Introduction

The role of visual explanation methods in deep learning received increased attention especially in high-risk applications. Determining the impacts of each pixel in the input on the classifier decision has huge importance to understand the behavior

© Springer Nature Switzerland AG 2021
C. Lian et al. (Eds.): MLMI 2021, LNCS 12966, pp. 396–405, 2021.
https://doi.org/10.1007/978-3-030-87589-3_41

of the classifiers [1]. Results from earlier studies demonstrate that these methods can be used to estimate coarse *heatmaps* (responses) of a deep network to a specific input [4]. Grad-CAM [13] and its variants have become a dominant approach for visual explanation and they have been used to highlight areas that provide evidence in favor of, and against choosing a certain class. Furthermore, locating objects and even weakly segmenting the object of interest are some of the other applications that these methods have been applied to [4,8]. Despite its practicality and easy-to-use nature, Grad-CAM has significant drawbacks of (i) tuning its parameters, (ii) high number of false positives, and (iii) overestimating the response regions [5,17]. Hence, the adoption of Grad-CAM and its variants is not completed and more trustable alternatives are needed to be developed, particularly for high-risk domains such as medicine. In this study, we develop a new visual attribution method that is superior to the commonly used Grad-CAM method. Although the proposed method is generic and can be applied to any classification-based applications, we demonstrate its efficacy in two important tasks in the medical imaging domain: diagnosis and prognosis. Due to the emergent need for helping the healthcare system to combat Covid-19, we focus on a Covid-19 problem where computed tomography scans are used for patient management.

Covid-19 Diagnosis and Prognosis. When the pandemic started, the researchers put lots of effort to develop automated diagnostic systems to fight against Covid-19. The main problem was the lack of knowledge about the disease and there was not enough resource to annotate dense and large datasets. Despite rapid and reliable advances in diagnostic tool, determining the severity of the Covid-19 positive patient remains a challenging goal, a crucial step for hospitalization planning and after Covid-19 clinics. Segmentation can be used to find morphometry and volumetry of lesions caused by Covid-19 but this requires labor-intensive segmentation labels from expert radiologists. More recent studies start to address the problem of prognosis (severity estimation) of Covid-19 patients from CT scans and Grad-CAM is still the most used visual explanation method therein [6]. Lesions caused by Covid-19 have been identified by Grad-CAM without using segmentation ground truths [6,11]. Since Grad-CAM-based approaches are not robust, they overestimate the related regions in the images, and often difficult to tune its parameters to find reliable visualization, there is a strong need to develop an alternative visualization method addressing the current drawbacks of the Grad-CAM.

Information Bottleneck. There is an increase in visual explanation methods that generate the heatmap in a more constrained way to ensure it focuses on the relevant region. In [12], a new method, Information Bottleneck Attribution (IBA), is introduced for an alternative visual explanation method based on the information bottleneck concept. Basically, IBA injects a noise to a specified intermediate layer of a network to mask out uninformative regions. The noise matrix is learned by an optimization process that tries to minimize the mutual information between the original features and the masked features. To prevent the trivial solution, it enforces the network to keep its prediction constant. This optimization promises to learn the mask that keeps the minimal enough information necessary for the prediction. Inspired by the IBA, we propose a conceptual theoretical framework based on information bottleneck that can both predict Covid-19 presence and estimate its severity from CT scans. We show that IBA results are more robust and reliable than the Grad-CAM. The visual results are inspected by four expert physicians to measure the quality of pathology localization as well as comparisons between the proposed IBA and the baseline Grad-CAM. Our experiments show that the results align with the lesions without having ground-truth masks.

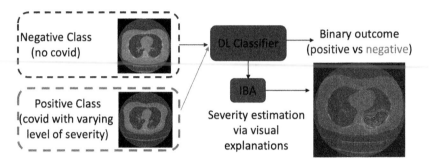

Fig. 1. Overview of information bottleneck based weak lesion localization.

2   Related Work

Visual Explanation Methods. There has been a significant number of studies that focus on the CNN visualization [3,12,13]. In [13], convolutional features and their gradient are used to estimate an importance map for the input image. These heatmaps indicate image regions that have discriminative features. However, the gradients can be unstable for certain inputs that will result in irrelevant outputs. In [12], information bottleneck [16] approach was utilized for visual explanation. It learns a mask that controls the information flow in the intermediate layers of a neural network to filter out irrelevant regions. It provides a theoretical guarantee that the masked out regions are unnecessary for the prediction; this is crucial for more robust visual explanation results.

Covid-19 Severity Estimation. Instead of diagnostic applications, recent studies focus on the severity of the COVID-19 cases where the spread of the lesions is estimated [2,6,7,9,11,14,15]. In [15], lungs are segmented and variety of features are extracted from the CT scans to predict the severity of the patient. Well-curated pipelines for feature extraction and segmentation were challenges. Using Grad-CAM for visual explanation on Covid-19 was also another popular approach to localize important regions [6,11]. However, it is known that Grad-CAM-based approaches are sensitive to many parameters in the network, and they are not robust Here, we introduce a better visual explanation method to localize Covid-19 related lesions. To the best of our knowledge, this is the first study where Information Bottleneck based attribution is adopted for both diagnosis and prognosis tasks.

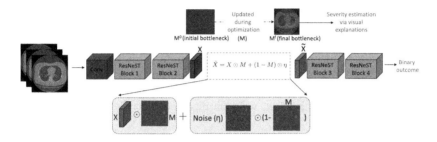

Fig. 2. Overall architecture. IBA method is applied to the pretrained ResNeSt.

3 Method

We used Information Bottleneck based approach that finds the critical input regions for the decision-making process. Empirical results show that those important regions align with Covid-19 related lesions. During the training, input scans are only considered as positive and negative samples instead of well and full labels. Figure 1 shows the general structure of our approach. We utilize recently introduced ResNeSt architecture [18] as a binary classifier on 2D slices since it uses channel-wise attention and multi-paths to increase the representation capacity. Other CNN architectures can be used as backbones instead of ResNeSt, we do not impose any restriction on the choice of networks. We employ IBA to the trained binary Covid-19 classifier to obtain the heatmaps as shown in Fig. 2.

3.1 Information Bottleneck Attribution

Information bottleneck is a technique that compresses the information provided by the input associated with the label [16]. Let X and Y be random variables associated with the input of the network and the prediction (output), respectively. In the standard classification setup, $p(y|x)$, the network uses the whole information from input X to predict Y. Information bottleneck introduces another random variable $\tilde{X}$ that is obtained by compressing the X. The $\tilde{X}$ is calculated by an optimization process that forces to use of only relevant information from the input related to Y. To find the optimal solution, the mutual information between X and $\tilde{X}$ is minimized while the mutual information between $\tilde{X}$ and Y is maximized for the $p(\tilde{X}|X)$;

$$\min_{p(\tilde{X}|X)} I(X;\tilde{X}) - \beta I(\tilde{X};Y) \tag{1}$$

where $I(\cdot; \cdot)$ is the mutual information and *beta* is the Lagrange multiplier that controls the amount of relevant information to flow.

In [12], information bottleneck is placed into a pre-trained network's intermediate layer that still has spatial information. Information flow is reduced by injecting noise into the representation. Similar to that, X and $\tilde{X}$ in our study express the original and the distorted representation, respectively. A matrix M that has the same shape with X controls where the noise will be injected spatially. The calculation is operated by

$$\tilde{X} = X \odot M + (1 - M) \odot \eta \tag{2}$$

where $\odot$ is Hadamard product and η is a random noise drawn from a certain distribution. The random noise is sampled from a normal distribution parametrized by the statistics of the original representation X; $\eta \sim \mathcal{N}(\mu_X, \sigma_X^2)$. For the Information Bottleneck optimization, we minimize the mutual information between X and $\tilde{X}$ by

$$I(X; \tilde{X}) = \mathbb{E}_X[D_{KL}(P_{\tilde{X}|X} || P_{\tilde{X}})] \tag{3}$$

where P is a probability distribution and D_{KL} is Kullback–Leibler divergence. Since calculating $P_{\tilde{X}}$ is intractable, it is replaced with the variation approximation $Q_Z \sim \mathcal{N}(\mu_X, \sigma_X^2)$. Then Eq. 3 becomes

$$I(X; \tilde{X}) = \mathbb{E}_X[D_{KL}(P_{\tilde{X}|X} || Q_{\tilde{X}})] - D_{KL}(P_{\tilde{X}} || Q_{\tilde{X}}). \tag{4}$$

In Eq. 4, calculating the second is not trivial because of $P_{\tilde{X}}$. Removing it can cause overestimation of the actual mutual information,

$$\tilde{I}(X; \tilde{X}) = \mathbb{E}_X[D_{KL}(P_{\tilde{X}|X} || Q_{\tilde{X}})], \tag{5}$$

but this will still guarantee that if the value calculated by $\tilde{I}(X; \tilde{X})$ is zero, the information from that region will not be necessary for the prediction. Finally, if we put $\tilde{I}(X; \tilde{X})$ into Eq. 1, we can optimize the M to find out important regions. We resized M to the input size to use it as a visual explanation.

4   Datasets

We used MOSMEDDATA [10] that contains lung CT scans from Covid-19 positive cases along with healthy lung scans. The dataset was collected in 2020 between March and April from municipal hospitals in Moscow, Russia. There were 1110 studies aged from 18 to 97 years old. It has 42% male, 56% female, and 2% other/unknown subjects. MOSMED-DATA contains subjects from 5 categories from healthy to the critical

Severity	Class	GGO	N
Zero	CT-0	None (Healthy)	254
Mild	CT-1	$\leq 25\%$	684
Moderate	CT-2	25–50%	125
Severe	CT-3	50–75%	45
Critical	CT-4	$\geq 75\%$	2

Table 1. Ground glass opacity (GGO) distributions in each severity category with class labels and number of samples

level. Table 1 shows the definition of categories and the number of samples per each class. During the training, we re-categorized the data as healthy and unhealthy for our binary classification task. The dataset was then split into training (70%) and testing (30%) sets without changing the class distribution. Also, we used lung segmentation maps to mask attribution results for both approaches.

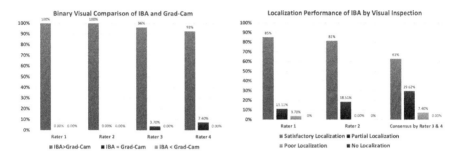

Fig. 3. Visual Evaluation results. Left: Visual inspection scores for method comparison. Right: localization performance based on the following scores: (1) satisfactory localization, (2) partial localization, (3) poor localization, and (4) no localization/unacceptable.

5   Experiments and Results

Our pre-trained ResNeSt classifier achieved 73.87% accuracy, 74.71% sensitivity and 71.05% specificity on the test set. This model was used both for IBA and Grad-CAM evaluations. As long as the same network architecture is used for backbone classification, comparisons of IBA and Grad-CAM will be fair. Any improvement in classification will improve localization ability of both methods.

Visual Evaluations. We have conducted two visual scoring experiments with four participating physicians who have never seen the data set that we have used in our study. We have selected representative CT scans from each severity class (30 CT scans) and fused both IBA and Grad-CAM visual results with original CT scans separately and used the same contrast/brightness windows for the visual inspection. In the first experiments, we asked four participating physicians to evaluate whether IBA or Grad-CAM is performing better, or tie. Figure 3 (left) shows visual inspection results for this part of the experiments: IBA wins these experiments with a large margin, and found to be equal to Grad-CAM only in a few cases. In the majority of the cases, IBA was found to be most successful. In the second experiment, we asked our participating physicians to evaluate the localization performance of the IBA method with the following scoring system: (1) satisfactory localization, (2) partial localization, (3) poor localization, and (4) no localization/unacceptable. Figure 3 (right) summarizes the visual scoring for this experiment: the majority of the cases were found to be successfully locating pathology regions with the IBA method, and only a few cases were found to be either partially or poorly localized.

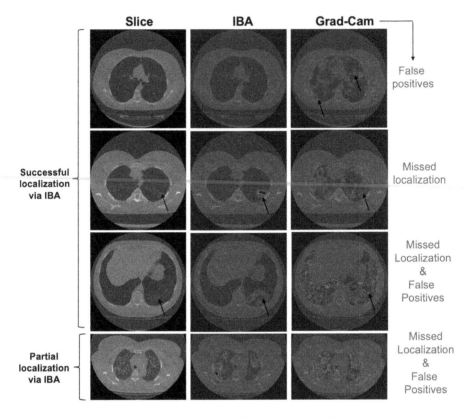

Fig. 4. Example heatmap predictions from IBA and Grad-CAM methods. In most of the examples IBA produces more precise localization results compared to Grad-CAM. Grad-CAM tends to over-predict localization.

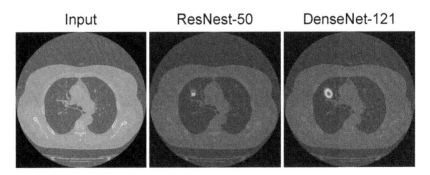

Fig. 5. IBA results from different network architectures.

Qualitative Results. In Fig. 4, we demonstrated sample outputs from the IBA and Grad-CAM approaches. IBA finds sharp and finer regions, mostly tries to focus on only the necessary lesion as we expected. On the other hand, Grad-CAM overestimates the focused regions, and there are a large number of false-positive findings. Interestingly, even for negative findings (first row in Fig. 4) are considered to have Covid-19. Since Grad-CAM does not have any mechanism to enforce networks to use information from a certain region, these results are not surprising. Another drawback of the Grad-CAM is that it is sensitive to hyper-parameters due to the gradient operation. Figure 4 exemplifies failure cases of Grad-CAM in varying levels of Covid-19 severity and shows the superiority of IBA. In the majority of the CT scans, as mentioned earlier, there is satisfactory localization with IBA, and only with a few cases, IBA is partially overlapping.

To show the effectiveness of IBA approach, alternatively, we used also a different network architecture, DenseNet, as a classifier model. Figure 5 shows IBA results from ResNeSt and DenseNet architectures. In both examples, IBA locates similar regions on the input image.

Lesion Detection. After obtaining the IBA heatmaps, we applied connected component analysis to detect the location of lesions. Figure 6 demonstrates the detection results for different inputs. These results show that lesions of varying sizes and multiple locations can be successfully located from IBA heatmaps.

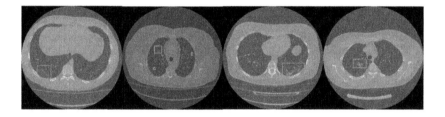

Fig. 6. Lesion localization by using IBA heatmaps.

6   Discussion and Concluding Remarks

This study sets out to critically examine the visual explanation methods for medical imaging. Our findings indicate that the information bottleneck-based approach, IBA, outperforms the Grad-CAM method both quantitatively and qualitatively. It should be noted that our proposed method has certain limitations that we plan to address in our extended study. For instance, IBA is highly sensitive in localizing the Covid-19 related pathologies; however, IBA tends to underestimate the region of interest contributing to the diagnosis and prognosis. This is understandable because information is minimized to increase most relevant features to be captured. Similar equations can be set up for background identification to handle this discrepancy. While we choose an application with an emerging need globally, IBA is general method that we plan to demonstrate its effectiveness in various different medical diagnosis tasks in our future work.

References

1. Bagci, U., et al.: A computational pipeline for quantification of pulmonary infections in small animal models using serial PET-CT imaging. EJNMMI Res. **3**(1), 55 (2013)
2. Chassagnon, G., et al.: Ai-driven quantification, staging and outcome prediction of COVID-19 pneumonia. Med. Image Anal. **67**, 101860 (2021)
3. Chattopadhay, A., Sarkar, A., Howlader, P., Balasubramanian, V.N.: Grad-CAM++: generalized gradient-based visual explanations for deep convolutional networks. In: 2018 IEEE Winter Conference on Applications of Computer Vision (WACV), pp. 839–847 (2018). https://doi.org/10.1109/WACV.2018.00097
4. Dubost, F., et al.: Weakly supervised object detection with 2D and 3D regression neural networks. Med. Image Anal. **65**, 101767 (2020)
5. Eitel, F., Ritter, K.: Testing the robustness of attribution methods for convolutional neural networks in MRI-based Alzheimer's disease classification. In: Suzuki, K., et al. (eds.) ML-CDS/IMIMIC -2019. LNCS, vol. 11797, pp. 3–11. Springer, Cham (2019). https://doi.org/10.1007/978-3-030-33850-3_1
6. Harmon, S.A., et al.: Artificial intelligence for the detection of COVID-19 pneumonia on chest CT using multinational datasets. Nat. Commun. **11**(1), 4080 (2020)
7. Li, K., et al.: CT image visual quantitative evaluation and clinical classification of coronavirus disease (COVID-19). Eur. Radiol. **30**(8), 4407–4416 (2020)
8. Li, K., Wu, Z., Peng, K.C., Ernst, J., Fu, Y.: Tell me where to look: guided attention inference network. In: Proceedings of the IEEE Conference on Computer Vision and Pattern Recognition (CVPR), June 2018
9. Li, Z., et al.: A novel multiple instance learning framework for COVID-19 severity assessment via data augmentation and self-supervised learning. Med. Image Anal. **69**, 101978 (2021)
10. Morozov, S.P., et al.: MosMedData: chest CT scans with COVID-19 related findings dataset (2020)
11. Panwar, H., Gupta, P., Siddiqui, M.K., Morales-Menendez, R., Bhardwaj, P., Singh, V.: A deep learning and grad-cam based color visualization approach for fast detection of COVID-19 cases using chest x-ray and CT-scan images. Chaos, Solitons Fractals **140**, 110190 (2020)
12. Schulz, K., Sixt, L., Tombari, F., Landgraf, T.: Restricting the flow: information bottlenecks for attribution. In: International Conference on Learning Representations (2020). https://openreview.net/forum?id=S1xWh1rYwB
13. Selvaraju, R.R., Cogswell, M., Das, A., Vedantam, R., Parikh, D., Batra, D.: Grad-CAM: visual explanations from deep networks via gradient-based localization. In: Proceedings of the IEEE International Conference on Computer Vision (ICCV), October 2017
14. Shan, F., et al.: Abnormal lung quantification in chest CT images of COVID-19 patients with deep learning and its application to severity prediction. Med. Phys. **48**(4), 1633–1645 (2021). https://doi.org/10.1002/mp.14609
15. Tang, Z., et al.: Severity assessment of COVID-19 using CT image features and laboratory indices. Phys. Med. Biol. **66**(3), 035015 (2021)

16. Tishby, N., Pereira, F.C., Bialek, W.: The information bottleneck method, pp. 368–377 (1999)
17. Young, K., Booth, G., Simpson, B., Dutton, R., Shrapnel, S.: Deep neural network or dermatologist? In: Suzuki, K., et al. (eds.) ML-CDS/IMIMIC -2019. LNCS, vol. 11797, pp. 48–55. Springer, Cham (2019). https://doi.org/10.1007/978-3-030-33850-3_6
18. Zhang, H., et al.: ResNeSt: split-attention networks. arXiv preprint arXiv:2004.08955 (2020)

Stacked Hourglass Network with a Multi-level Attention Mechanism: Where to Look for Intervertebral Disc Labeling

Reza Azad[1], Lucas Rouhier[2], and Julien Cohen-Adad[2,3,4]($\boxtimes$)

[1] Sharif University of Technology, Tehran, Iran
[2] NeuroPoly Lab, Institute of Biomedical Engineering,
Polytechnique Montreal, Canada
jcohen@polymtl.ca
[3] Mila, Quebec AI Institute, Montreal, Canada
[4] Functional Neuroimaging Unit, CRIUGM, University of Montreal,
Montreal, Canada

Abstract. Labeling vertebral discs from MRI scans is important for the proper diagnosis of spinal related diseases, including multiple sclerosis, amyotrophic lateral sclerosis, degenerative cervical myelopathy and cancer. Automatic labeling of the vertebral discs in MRI data is a difficult task because of the similarity between discs and bone area, the variability in the geometry of the spine and surrounding tissues across individuals, and the variability across scans (manufacturers, pulse sequence, image contrast, resolution and artefacts). In previous studies, vertebral disc labeling is often done after a disc detection step and mostly fails when the localization algorithm misses discs or has false positive detection. In this work, we aim to mitigate this problem by reformulating the semantic vertebral disc labeling using the pose estimation technique. To do so, we propose a stacked hourglass network with multi-level attention mechanism to jointly learn intervertebral disc position and their skeleton structure. The proposed deep learning model takes into account the strength of semantic segmentation and pose estimation technique to handle the missing area and false positive detection. To further improve the performance of the proposed method, we propose a skeleton-based search space to reduce false positive detection. The proposed method evaluated on spine generic public multi-center dataset and demonstrated better performance comparing to previous work, on both T1w and T2w contrasts. The method is implemented in ivadomed (https://ivadomed.org).

Keywords: Intervertebral disc labeling · Spine generic database · Pose estimation · Deep learning

1 Introduction

The human spine consists of four connected regions namely: cervical, thoracic, lumbar, and sacral vertebrae (Fig. 1). The vertebrae in each region perform

© Springer Nature Switzerland AG 2021
C. Lian et al. (Eds.): MLMI 2021, LNCS 12966, pp. 406–415, 2021.
https://doi.org/10.1007/978-3-030-87589-3_42

unique functionality, including protection of the spinal cord, load breathing, and etc. As shown in Fig. 1, intervertebral discs lie between adjacent vertebrae and connect the vertebrae column. Any injury in the vertebral disc may result in back pain or sensation in different parts of the human body. Vertebral injury usually comes from excessive strain or trauma to the spine. Thus, analyzing the intervertebral disc location and its shape is an important part of diagnosis. The analysis starts with the detection of intervertebal discs, which is a tedious job. To automate the intervertebal disc detection process, several approaches have been proposed in previous studies. A local descriptor based approach has been utilized by [11] to detect the intervertebral disc C2/C3. This approach uses the local descriptor to find the mutual information between the image from the patient and the template by looking at region that has the minimum distance to the spine template. Even though this handcrafted approach produces good results in general, its performances are greatly reduced when the images differs too much from the template. To overcome the limitation of handcrafted approaches, deep learning based models are used to perform robust interverte-bral disc labeling. In [6] the author proposed a 3D CNN model to perform the 3D segmentation on MRI data and retrieve the vertebral disc location. Cai et al. [4] also utilized a 3D Deformable Hierarchical Model to extract the vertebral disc location on 3D space. In an another example, the Count-ception model is trained on 2D MRI sagittal slices to perform vertebral discs detection [16]. The exact location of the vertebral discs in 2D space are extracted using local maximum technique on top of the prediction masks. The mains drawback of the deep segmentation approaches is the false positive rate. Therefore, it requires extensive post-processing and it often fails to retrieve the exact location of the disc. In this work, we aim to overcome the limitations of literature work by taking into account the importance of skeleton structure in intervertebral disc labeling. The main idea is to include the structural information of the intervertebral discs to increase True Positive (TP) rate while reducing the False Negative (FN) detection. We re-formulate the problem using pose estimation technique, to take into account the skeleton structure between intervertebral discs in loss function. To the best of our knowledge, this is the first attempt to label intervertebral discs using pose estimation technique. Our contribution can be summarized as follow:

- Adapting the pose estimation approach for intervertebral disc labeling.
- Scaling the representation space by multi-level attention mechanism.
- Skeleton based post-processing approach to reduce the false-positive rate.
- State-of-the-art results on the public data set.
- Publicly-available implementation source code [1]

2   Literature Review

Automatic intervertebral disc labeling is a crucial task in medical image analysis. The spine can be considered as an anatomical landmark that runs along the length of the neck and trunk of the human upper body. In recent decades, many approaches have been proposed in this field of research. Like other tasks

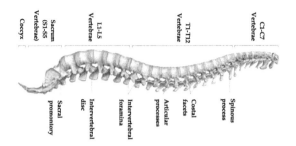

Fig. 1. Human vertebral column [2].

in computer vision, we can divide these approaches in two groups: handcrafted-based methods and deep learning-based approaches. Early studies in intervertebral discs labeling were based on handcrafted-based features. Glocker et al. [10] employed regression forests and hidden Markov model (HMM) to detect and localize intervertebral discs from CT scan images. Kim et al. [13] proposed a Sphere Surface Expansion method and iterative optimization framework by incorporating local appearance features with global translational symmetry and local reflection symmetry features. Zhang et al. [21] utilized a local articulated model to effectively model the spatial relations across vertebrae and discs. The authors derived disk locations from a cloud of responses from disc detectors which is robust to sporadic voxel-level errors.

Ullmann et al. [18] proposed a 3D-based analysis method, which supports both T1-weighted and T2-weighted contrasts. A template is utilized for detecting the intervertebral disc location. This template was computed from a collection of vertebral distances along the cervical, thoracic, and lumbar spine in adult humans. A method for detecting the spinal cord centerline on MRI volumes was proposed by Gros et al. [11]. The optimization procedure of that method aimed at striking a balance between a probabilistic localization map of the spinal cord center point and the overall spatial consistency of the spinal cord centerline. The authors also employed a post-processing step to split brain and spine region.

The handcrafted-based features suffers from overfitting problem due to the over-designed framework. Like other field of research in computer vision, in recent years, deep learning-based approaches have been proposed for detection of spine vertebrae. Rouhier et al. [16] combine a Fully Convolutional Network (FCN) with inception modules to localize intervertebral discs from MRI data. An FCN was also utilized by Suzani et al. [17] to predict the relative distance from the voxel to each vertebral centroid. Forsberg et al. [9] employed two separate detection and labeling pipelines, for the lumbar and the cervical cases. Each pipeline consists of two convolutional neural networks (CNNs) for detection of potential vertebrae (one general lumbar/cervical vertebra detector and one specific S1/C2 vertebra detector, respectively), followed by a parts-based graphical model for false positive removal and subsequent labeling.

A deep convolutional neural network was exploited by Chen et al. [5] to identify the vertebrae type based on the centroid proposals generated from the random forest classifier. Transformed Deep Convolution Network (TDCN) was introduced by Cai et al. [3] for multi-modal intervertebral disc recognition. The image features from different modalities are fused unsupervisely, and the pose of vertebra is then automatically rectified. Windsor et al. [19] proposed a CNN model for detection and localization of vertebrae from MRI images. The model is uses the image-to-image translation technique to label the intervertebral disc.

Lu et al. [14], utilized a natural-language-processing approach to extract level-by-level ground-truth labels from free-text radiology reports, and then employed a U-Net architecture combined with a spine-curve fitting method for intervertebral segmentation and disc-level localization. A multi-input, multi-task, and multi-class CNN is then exploited for central canal and foraminal stenosis grading. A U-Net like architecture is proposed by Yang et al. [20] to directly model the vertebrae centroids as a 27-class segmentation-like problem (26 vertebrae types and background) An FCN-based approach was employed by Chen et al. [6] for vertebrae localization. The authors then propose a post-processing method to reduce the dimension of the localization, and constrain the search space for the vertebrae centroids with a hidden Markov model. The reviewed deep learning-based methods does not consider the geometrical information in the learning process, hence, suffers from FP and FN detection. We aim to overcome this limitation using the pose estimation technique.

3   Proposed Method

General diagram of the proposed method is shown in Fig. 2. The first step in the proposed method is to pre-process the input data for the model. The model utilizes the pose estimation method with attention mechanism to learn intervertebral disc position. In the next subsections we will discuss these steps.

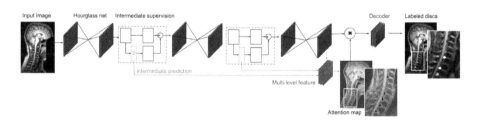

Fig. 2. Stacked hourglass network with an attention mechanism. The model considers the loss function between each hourglass prediction and ground truth mask (intermediate supervision). Further it feds the intermediate representation into an attention layer to produce the attention map. The attention map guides the decoder layer to focus on the intervertebral disc.

3.1   Pre-processing

As done in previous studies [16], we extract the average of 6 sagittal slices (centered in the middle slice) as a data sample for each subject. We normalize each image to be in range $[0, 1]$ to reduce the effect of data variation. In order to prepare the ground truth data for the training process, first, we extract the intervertebral disc position (single pixel) from the ground truth data then we convolve the image with a Gaussian kernel to generate a smooth ground truth with increased target size (radius 10). We repeat this process for each intervertebral disc separately to produce V channel ground truth, where V is the number of intervertebral discs. Since the Spine Generic dataset consists of samples with variable number of intervertebral discs (between 7–11), we extract 11 intervertebral discs for each subject. For any missing intervertebal disc we consider unknown position and eliminate its effect on the training process by simply filtering out with the visibility flag on the loss function.

3.2   Proposed Model

As shown in Fig. 2, the stacked hourglass network [15] learns the object pose using (N-1) intermediate (shown in 2 as intermediate prediction) prediction and one final prediction. Thus, it takes into account the multi-level representation in terms of the N stacked hourglass network. To further improve the power of representation space, we propose to use a multi-level attention mechanism. To this end, an intermediate representation generated by each hourglass network (shown in 2 with p) is concatenated to form a multi-level representation. This representation can be seen as a collective knowledge that is extracted from a different level of the network with various scales, thus, using this collective knowledge as a supervisory signal to calibrate the final representation can result in better representation. To include this supervisor signal, we stack all the intermediate representations. This stacked representation is fed to the attention block (series of point-wise convolution with sigmoid activation) to generate a single channel attention mechanism. We multiply this attention channel with the final representation to re-calibrate the representation space and teach the model to pay more attention to the disc location. This attention mechanism reduces the FP rates. We train the model using the sum of MSE loss (Eq. 1) between the predicted mask y' and the ground truth mask y. In Eq. 1, N shows the number of pixels in the ground truth mask.

$$\text{loss} = \frac{1}{V \times N} \sum_{i=1}^{V} \sum_{j=1}^{N} (y_j - \hat{y}_j)^2 \tag{1}$$

3.3   Post-processing

Even though the proposed network learns the intervertebral discs with high accuracy, the predicted mask needs to be post-processed to further reduce the FP

rate. In order to fine-tune the predicted result we propose a skeleton-based approach. In this approach, we create a general skeleton model based on the training
set. To this end, for all subjects in the training set, we extract the intervertebral disc location, then we calculate the distance from each intervertebral disc
(v^i) to the first intervertebral (v^1) disc. Which shows the relational structure
between intervertebral discs. To normalize this representation, we shift the v^1
to the world coordinate $(0,0)$ then we normalize all the relational distances by
dividing them to the distance from v^1 to v^5. Figure 3 (a) shows the structure of
the extracted skeleton from the training set. Each intervertebral disc v^i (shown
by triangle) is calculated based on the average of all subject's intervertebral disc
location v^i. Using the generated skeleton, we define the skeleton model S as 2:

$$S = \text{set} \left\{ v^i \right\}, i = 1, 2, \ldots V, v^i = (x, y) \tag{2}$$

Skeleton S consists of V intervertebral discs with relational 2D position (x,
y). On the test time, for each predicted mask, we create the search tree, where
each node in this tree shows one possible combination of ordered intervertebral
disc location, as illustrated in Fig. 3 (b). For each path of tree (from leaf to node
which forms a S') we calculate the error function between the general skeleton S
and the predicted skeleton S' using Eq. 3. Please note that for each intervertebral
disc (v_i) we have several candidates, thus, (v_i, c) shows the c_{th} candidate of v_i.
In our equation S' with minimum error is the solution. We use flag δ in Eq. 3
to represent that candidate availability. Hence, in case the algorithm misses any
intervertebral disc, it will not affect the error function.

$$\text{error}\,(S, S') = \sum_{j=1}^{N} d\left(v^j, v^{j'}\right), \quad d\left(v^j, v^{j'}\right) = \sqrt{\sum_{i=1}^{n} \delta \left(v_i^j - v_i'^j\right)^2} \tag{3}$$

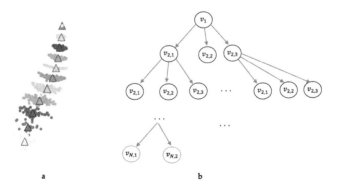

a   b

Fig. 3. (a): Vertebral disc skeleton structure (b): Skeleton search tree, where each level
of the tree contains candidates for invertebrate v_i. Path from root to leaf shows one
possible combination of detected candidates to form a S'.

4   Experimental Results

To evaluate the performance of the proposed method we consider the Spine Generic Dataset [7]. This dataset contains both T1w and T2w contrasts for each subject. Images were acquired in 42 different centers world wide. The dataset contains large sample variation in term of image quality, scale and imaging devices, hence, exhibits a challenging benchmark for intervertebral disc labeling.

4.1   Metrics

To demonstrate the performance of the proposed method and compare with the literature work, we consider different comparison metrics. We include the L2 distance between each predicted intervertebral disc position and the ground truth along the superior-inferior axis to calculate the preciseness of the prediction. We further use the False Positive Rate (FPR) and False Negative Rate (FNR) metrics. The FPR counts the number of prediction which are at least 5 mm away from the ground truth positions. Similarly the FNR counts the number of prediction where the ground truth has at least 5mm distance from the predicted intervertebral disc's real position.

4.2   Comparison Results

We train the proposed model for 150 epochs using Adam optimization with learning rate 0.00025 and batch size 4. In our experimental results we achieved best results on the validation set using 2 stacks. The implementation and model training was done in ivadomed [12] and the method can readily be used via the Spinal Cord Toolbox [8]. Comparison results of the proposed method with the literature work on the test set is provided in Table 1. We use the same setting as explained in [16] to compare our method with the literature work. As shown in Table 1, the proposed method outperformed literature work in both T1w and T2w modalities. On T1w modality, the proposed method has better results than the template matching technique on all metrics. Although the counting method on T1w modality has a slightly lower average distance to the target, our proposed method has a lower standard deviation value and statistically it is more reliable for retrieving intervertebral disc location. The efficiency of the proposed method is clearer on the T2 contrast, where the proposed method outperforms the Countception and template matching methods.

Table 1. Intervertebral disc labeling results on the spine generic public dataset.

Method	T1			T2		
	Distance to target (mm)	FNR (%)	FPR (%)	Distance to target (mm)	FNR (%)	FPR (%)
Template matching	1.97 (±4.08)	8.1	2.53	2.05 (±3.21)	11.1	2.11
Counting model [16]	1.03 (±2.81)	4.24	0.9	1.78 (±2.64)	3.88	1.5
Proposed method (without attention)	1.39 (±2.38)	1.3	0.0	1.31 (±3.34)	1.5	0.0
Proposed method	**1.32 (±1.33)**	**0.32**	0.0	**1.31 (±2.79)**	**1.2**	0.6

It is worth mentioning that the counting method [16] applies image straightening technique on the pre-processing step to extract the spinal cord area. This technique crops the spinal region from the input MRI image ad simplifies the data sample. However, in our proposed method we eliminated this step to learn the intervertebral disc labeling without such pre-processing requirement. In addition, the counting method doesn't consider the relation between intervertebral discs and simply applies the region based segmentation technique to extract the intervertebral disc location. Thus, it has high false negative rate (FNR). On the other hand, the proposed method takes into account the skeleton information to eliminate the FNR samples. In our post-processing method, to eliminate the FN samples and reduce the search burden we provide the number of desired intervertebral disc as a meta data. Thus, this metadata helps the search tree to create a less depth and retrieve the intervertebral disc location fast.

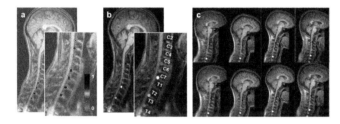

Fig. 4. (a): Attention map visualization. (b): Corresponding ground truth mask. (c): Prediction results in four representative T1w images. The top row shows the ground truths, the bottom row shows the predictions.

4.3   Effect of Attention Mechanism

The proposed method utilizes the attention mechanism to re-calibrate the representation space and guide the model to focus on target location. To analyse the effect of the attention mechanism we trained the model with and without attention mechanism. The comparison results demonstrated in Table 1, in which the attention mechanism in both T1w and T2w modalities increases the model performance. To visualize the effect of attention mechanism inside the network, we depicted a sample attention maps on the input images, Fig. 4(b).

5   Conclusion

In this work we formulated the intervertebral disc labeling using pose estimation technique. The proposed method learns the structural information between intervertebal discs to recover the TP locations. The proposed method uses the strength of attention mechanism to re-calibrate the representation space in way to focus more on interverterbral disc area. We further proposed a skeleton-based post processing approach to eliminate the FP and FN detection.

References

1. The source code used in this paper is available at: https://github.com/rezazad68/ Deep-Intervertebral-Disc-Labeling
2. Vertebral column: Overview. http://www.thieme.com/media/samples/pubid-455606490.pdf. Accessed 1 Jan 2021
3. Cai, Y., Landis, M., Laidley, D.T., Kornecki, A., Lum, A., Li, S.: Multi-modal vertebrae recognition using transformed deep convolution network. Comput. Med. Imaging Graph. **51**, 11–19 (2016)
4. Cai, Y., Osman, S., Sharma, M., Landis, M., Li, S.: Multi-modality vertebra recognition in arbitrary views using 3D deformable hierarchical model. IEEE Trans. Med. Imaging **34**(8), 1676–1693 (2015)
5. Chen, H., et al.: Automatic localization and identification of vertebrae in spine CT via a joint learning model with deep neural networks. In: Navab, N., Hornegger, J., Wells, W.M., Frangi, A.F. (eds.) MICCAI 2015. LNCS, vol. 9349, pp. 515–522. Springer, Cham (2015). https://doi.org/10.1007/978-3-319-24553-9_63
6. Chen, Y., Gao, Y., Li, K., Zhao, L., Zhao, J.: vertebrae identification and localization utilizing fully convolutional networks and a hidden Markov model. IEEE Trans. Med. Imaging **39**(2), 387–399 (2019)
7. Cohen-Adad, J., et al.: Open-access quantitative MRI data of the spinal cord and reproducibility across participants, sites and manufacturers. Sci. Data **8**, 1–17 (2021). https://doi.org/10.1038/s41596-021-00588-0
8. De Leener, B., et al.: SCT: Spinal cord toolbox, an open-source software for processing spinal cord MRI data. Neuroimage **145**, 24–43 (2017)
9. Forsberg, D., Sjöblom, E., Sunshine, J.L.: Detection and labeling of vertebrae in MR images using deep learning with clinical annotations as training data. J. Digit. Imaging **30**(4), 406–412 (2017)
10. Glocker, B., Feulner, J., Criminisi, A., Haynor, D.R., Konukoglu, E.: Automatic localization and identification of vertebrae in arbitrary field-of-view CT scans. In: Ayache, N., Delingette, H., Golland, P., Mori, K. (eds.) MICCAI 2012. LNCS, vol. 7512, pp. 590–598. Springer, Heidelberg (2012). https://doi.org/10.1007/978-3-642-33454-2_73
11. Gros, C., et al.: Automatic spinal cord localization, robust to MRI contrasts using global curve optimization. Med. Image Anal. **44**, 215–227 (2018)
12. Gros, C., et al.: ivadomed: a medical imaging deep learning toolbox. J. Open Source Softw. **6**(58), 2868 (2021). https://doi.org/10.21105/joss.02868
13. Kim, K., Lee, S.: Vertebrae localization in CT using both local and global symmetry features. Comput. Med. Imaging Graph. **58**, 45–55 (2017)
14. Lu, J.T., et al.: DeepSPINE: automated lumbar vertebral segmentation, disc-level designation, and spinal stenosis grading using deep learning. arXiv preprint arXiv:1807.10215 (2018)
15. Newell, A., Yang, K., Deng, J.: Stacked hourglass networks for human pose estimation. In: Leibe, B., Matas, J., Sebe, N., Welling, M. (eds.) ECCV 2016. LNCS, vol. 9912, pp. 483–499. Springer, Cham (2016). https://doi.org/10.1007/978-3-319-46484-8_29
16. Rouhier, L., Romero, F.P., Cohen, J.P., Cohen-Adad, J.: Spine intervertebral disc labeling using a fully convolutional redundant counting model. arXiv preprint arXiv:2003.04387 (2020)

17. Suzani, A., Seitel, A., Liu, Y., Fels, S., Rohling, R.N., Abolmaesumi, P.: Fast automatic vertebrae detection and localization in pathological CT scans - a deep learning approach. In: Navab, N., Hornegger, J., Wells, W.M., Frangi, A.F. (eds.) MICCAI 2015. LNCS, vol. 9351, pp. 678–686. Springer, Cham (2015). https://doi.org/10.1007/978-3-319-24574-4_81
18. Ullmann, E., Pelletier Paquette, J.F., Thong, W.E., Cohen-Adad, J.: Automatic labeling of vertebral levels using a robust template-based approach. Int. J. Biomed. Imaging **2014**, 719520 (2014)
19. Windsor, R., Jamaludin, A., Kadir, T., Zisserman, A.: A convolutional approach to vertebrae detection and labelling in whole spine MRI. In: Martel, A.L. (ed.) MICCAI 2020. LNCS, vol. 12266, pp. 712–722. Springer, Cham (2020). https://doi.org/10.1007/978-3-030-59725-2_69
20. Yang, D., et al.: Automatic vertebra labeling in large-scale 3D CT using deep image-to-image network with message passing and sparsity regularization. In: Niethammer, M., et al. (eds.) IPMI 2017. LNCS, vol. 10265, pp. 633–644. Springer, Cham (2017). https://doi.org/10.1007/978-3-319-59050-9_50
21. Zhan, Y., Maneesh, D., Harder, M., Zhou, X.S.: Robust MR spine detection using hierarchical learning and local articulated model. In: Ayache, N., Delingette, H., Golland, P., Mori, K. (eds.) MICCAI 2012. LNCS, vol. 7510, pp. 141–148. Springer, Heidelberg (2012). https://doi.org/10.1007/978-3-642-33415-3_18

TED-Net: Convolution-Free T2T Vision Transformer-Based Encoder-Decoder Dilation Network for Low-Dose CT Denoising

Dayang Wang, Zhan Wu, and Hengyong Yu$^{(\boxtimes)}$

Department of Electrical and Computer Engineering, University of Massachusetts
Lowell, Lowell, MA 01854, USA

Abstract. Low dose computed tomography (CT) is a mainstream for clinical applications. However, compared to normal dose CT, in the low dose CT (LDCT) images, there are stronger noise and more artifacts which are obstacles for practical applications. In the last few years, convolution-based end-to-end deep learning methods have been widely used for LDCT image denoising. Recently, transformer has shown superior performance over convolution with more feature interactions. Yet its applications in LDCT denoising have not been fully cultivated. Here, we propose a convolution-free T2T vision transformer-based Encoder-decoder Dilation Network (TED-Net) to enrich the family of LDCT denoising algorithms. The model is free of convolution blocks and consists of a symmetric encoder-decoder block with sole transformer. Our model (Codes are available at https://github.com/wdayang/TED-Net) is evaluated on the AAPM-Mayo clinic LDCT Grand Challenge dataset, and results show outperformance over the state-of-the-art denoising methods.

Keywords: Low-dose CT · Transformer · Token-to-Token · Encoder-decoder · Dilation

1 Introduction

In recent years, low dose computed tomography (LDCT) has become the mainstream in the clinical applications of medical imaging. However, the low quality of LDCT image has always been a barrier since it compromises the diagnosis value. To overcome this issue, traditional methods (e.g. iterative methods) manage to suppress the artifact and noise by using the physical model and/or prior information. For example, Compressive Sensing (CS) has been widely used for ill-posed inverse problems by learning sparse representations [1], and the representative total variation (TV)-based models assume that the clean image is piecewise constant and its gradient transform is sparse [2–5]. Xu *et al.* combined dictionary learning and statistic iterative reconstruction (SIR) [6] for LDCT denoising. Tan *et al.* proposed a tensor-based dictionary learning model for spectral and

© Springer Nature Switzerland AG 2021
C. Lian et al. (Eds.): MLMI 2021, LNCS 12966, pp. 416–425, 2021.
https://doi.org/10.1007/978-3-030-87589-3_43

dynamic CT [7]. Ma *et al.* designed a Non-Local Mean (NLM) method to utilize the redundancy of information across the whole image rather than local operations on neighboring image voxels [8]. Nonetheless, none of these algorithms is adopted in commercial CT scanners because of the hardware limitations and high computational cost [9].

In last few years, convolution-based methods attracted more attention in CT denoising and achieved the state-of-the-art performance. Chen *et al.* combined the auto-encoder, deconvolution network, and shortcut connections into a residual encoder-decoder convolution neural network (CNN) for CT imaging [10]. Yang *et al.* used WGAN-VGG in the denoising base network and adopted perceptual loss to evaluate the reconstructed image quality [11]. Fan *et al.* constructed a quadratic neuron-based autoencoder with more robustness and utility for model efficiency in contradiction of other CT denoising methods [12]. However, these convolution-based methods have limited ability to capture contextual information with long spatial dependence in image or feature maps.

Very recently, transformer [13] has gradually become the dominant method in the natural language processing (NLP) [14–17] and computer vision (CV) fields [18–27]. Transformer has achieved a great performance in high level tasks, such as classification, object detection, image segmentation, *etc.*. Dosovitskiy *et al.* first proposed vision transformer (ViT) in the CV field by mapping an image into 16 × 16 sequence words [21]. To overcome the simple tokenization in ViT, Yuan *et al.* further proposed a Token-to-Token method to enrich the tokenization process [28]. Moreover, Liu *et al.* designed a swin transformer to include patch fusion and cyclic shift to enlarge the perception of contextual information in tokens [29]. Researchers also explored the transformer for low vision task [20,22,23]. However, transformer in LDCT denoising has not been well explored. Zhang *et al.* designed a TransCT-net to utilize transformer in the high frequency (HF) and low frequency (LF) composite inference [30]. Nevertheless, Zhang's work includes a lot of convolutions in the HF/LF tokenization and Piecewise Reconstruction module. So far, there is no convolution-free model for LDCT denoising. In this paper, for the first time, we propose a convolution-free Token-to-Token (T2T) vision Transformer-based Encoder-decoder Dilation (TED-Net) model and evaluate its performance compared with other state-of-the-art models.

The rest of this paper is organized as follows. Section 2 introduces the key methods used in our proposed model. Section 3 reports the experiment details and results. Section 4 discusses some related issues and make a conclusion.

2   Methods

In this paper, we propose a convolution-free T2T vision transformer-based Encoder-decoder Dilation Network (TED-Net). As shown in Fig. 1, in the encode part, the model includes Tokenization block, Transformer Block (TB), Cyclic Shift Block (CSB), Token-to-Token block with Dilation (T2TD) and without dilation (T2T). The decoder part includes T2T, T2TD, Inverse Cyclic Shift Block (ICSB) and Detokenization Block.

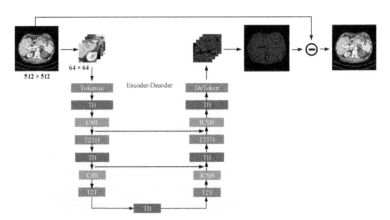

Fig. 1. The pipeline of our proposed TED-Net. Tokenize block uses unfold to extract tokens from image patches while DeToken block applies Fold to convert tokens back to image. TB includes a typical transformer block. CSB uses cyclic shift operation and ICSB employs inverse cyclic shift. T2T incorporates Token-to-Token block to enhance tokenization while T2TD includes Dilation in the T2T tokenization process. The final image is obtained by subtracting the model residual output from the noisy input image.

2.1   Noise Model

The LDCT denoising task attempts to recover a clean NDCT image $y \in \mathbb{R}^{N \times N}$ from a matching noise LDCT image $x \in \mathbb{R}^{N \times N}$ by using a general denoising model $D : \mathbb{R}^{N \times N} \to \mathbb{R}^{N \times N}$, and the Mean-Square Error (MSE) loss $\mathcal{L}_{MSE}$ is defined as follows:

$$\mathcal{L}_{MSE} = \|D(x) - y\|_2. \tag{1}$$

In this paper, we propose a Transformer-based model $T : \mathbb{R}^{N \times N} \to \mathbb{R}^{N \times N}$ to learn the deep features and capture the noise pattern of the input image. Then we recover the clean image y by removing the estimated residual image $T(x)$ from the original noisy image x,

$$y = x - T(x). \tag{2}$$

2.2   Transformer Block

In the Transformer Block (TB), we utilize a traditional transformer in the encoder and decoder stage between two T2T blocks which contain Multiple head Self-Attention (MSA), Multiple Layer Perceptron (MLP) and residual connection to promote the expressive power of this module. The output of TB $T' \in \mathbb{R}^{b \times n \times d}$ has the same size as the input tokens $T \in \mathbb{R}^{b \times n \times d}$. Here b is batch size, n is the number of tokens, and d is the token embedding dimension,

$$T' = MLP(MSA(T)). \tag{3}$$

2.3   Token-to-Token Dilation Block

Token-to-Token (T2T) block is recently utilized to overcome the simple tokenization of image in vision transformer. Traditional tokenization only includes one tokenization process using either reshape or convolution, while the T2T block adopts a cascade tokenization procedure. We further use dilation in the tokenization process to refine the contextual information fusion and seek relation across larger regions. Figure 2(a) illustrates the structure of T2TD block which consists of reshape and soft split with dilation.

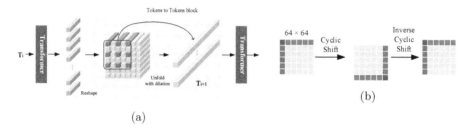

(a)   (b)

Fig. 2. (a)The architecture of T2TD block which includes reshape and unfold with dilation. (b)The structures of cyclic shift and inverse cyclic shift operations to enrich the tokenization process by fusion different kernel area.

Reshape. Given tokens T from last stage, they are first transposed to $T^T \in \mathbb{R}^{b \times n \times d}$ and then reshaped into $I \in \mathbb{R}^{b \times c \times h \times w}$,

$$I = Reshape(T), \tag{4}$$

where $c = d$ and $h = w = \sqrt{n}$ are the channel, height and width of feature map, respectively.

Soft Split with Dilation. With the feature maps from the reshape stage, the soft split stage will retokenize the reshaped feature map using unfold operation. In this stage the four dimension feature maps $I \in \mathbb{R}^{b \times c \times h \times w}$ are converted back to three dimensional tokens $T'' \in \mathbb{R}^{b \times n' \times d'}$. Through this operation, the number of tokens is reduced by combining several neighboring tokens into one unit though the embedding dimension is increased accordingly with several tokens concatenated together,

$$T'' = SoftSplit(I). \tag{5}$$

As demonstrated in Fig. 2(a), dilation is also used in the unfold process to capture the contextual information with longer dependence. After the soft split with dilation, the input feature maps $I \in \mathbb{R}^{b \times c \times h \times w}$ become $T''' \in \mathbb{R}^{b \times n'' \times d''}$

where $d'' = c \times \prod kernel$ and the total number of tokens n'' after the soft split operation is calculated as:

$$n'' = \left\lfloor \frac{h - dilation \times (kernel - 1) - 1}{stride} + 1 \right\rfloor, \qquad (6)$$

where $dilation, kernel, stride$ are related parameters in the Unfold operation.

Cyclic Shift. After the reshape process in the encoder network, we employ the cyclic shift to modify the shaped feature map. The pixel values in the feature map are shifted in a cyclic way that will add more information integration in the model. Then, an inverse cyclic shift is performed in the symmetric decoder network to avoid any pixel shifts in the final denoising results. Figure 2(b) exhibits the cyclic shift module and inverse cyclic shift module,

$$I' = CyclicShift(I). \qquad (7)$$

3   Experiments and Results

In this part, the data preparation, experiment settings and comparison results are presented. Our model is trained and evaluated on a public dataset, and the results show that our model outperforms other state-of-the-art models.

Dataset. A publicly released dataset from *2016 NIH-AAPM-Mayo Clinic LDCT Grand Challenge*[1] [31] is used for model training and testing. We employ the patient L506 data for evaluation and the other nine patients for model training. The pairs of quarter-dose LDCT and normal-dose CT (NDCT) images are used to train the model. Data augmentation is also applied to enlarge the dataset where we keep a copy of original image and then randomly apply image rotation (90, 180 or 270°) and flipping (up/down, left/right) to the original one.

Experiment Settings. The experiments are running on Ubuntu 18.04.5 LTS, with Intel(R) Core (TM) i9-9920X CPU @ 3.50 GHz using PyTorch 1.5.0 and CUDA 10.2.0. The models are trained with 2 NVIDIA 2080TI 11G GPUs. Here are the details of our experiment setting: in the encoder block, our model consists of three soft split stages, two transformer layers, and two cyclic shift layers, while in the decoder block it includes three inverse soft split stages with fold operations, two transformer blocks, and two corresponding inverse cyclic shift operations. One additional transformer layer is between the encoder and decoder parts to further incorporate feature inference. The kernel size for the three unfold/fold operations are $\{7 \times 7, 3 \times 3, 3 \times 3\}$ with a stride of $\{2, 1, 1\}$ and a dilation of $\{1, 2, 1\}$, respectively. Moreover, the shift quantity is 2 for the two cyclic shift layers and we use a patch number of 4 for training through 4000 epochs. Adam

---

[1] https://www.aapm.org/GrandChallenge/LowDoseCT/.

is adopted to minimize the MSE loss with an initial learning rate of 1e−5. In the evaluation stage, we segment the 512 × 512 image into overlapped 64 × 64 patches, and only crop out the center part of the model output to aggregate to the final whole predictions to overcome the boundary artifacts.

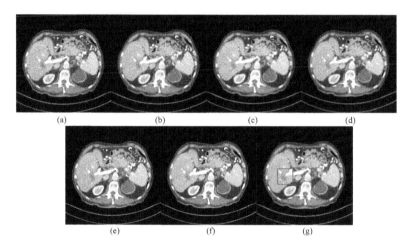

Fig. 3. The denoising results of different networks on L506 with lesion No. 575. (a) LDCT, (b) RED-CNN, (c) WGAN-VGG, (d) MAP-NN, (e) AD-NET, (f) proposed TED-Net, and (g) NDCT. The display window is [−160, 240] HU.

Comparison Results. SSIM and RMSE are adopted to quantitatively measure the quality of the denoised image. Our model is compared with state-of-the-art baseline algorithms: RED-CNN [10], WGAN-VGG [11], MAP-NN [32], and AD-NET [33]. RED-CNN, MAP-NN, and WGAN-VGG are popular low dose CT denoising models while AD-NET has high performance on gray image denoising. We retrain AD-NET on AAPM dataset with the same setting as other methods and obtain a comparison result. Figure 3 shows the results of different networks on L506 with lesion No.575. Figure 4 demonstrates the amplified ROIs from the rectangular area marked in Fig. 3. Figure 3 and Fig. 4 show that our TED-Net has a better performance in removing the noise/artifact and maintaining high-level spatial smoothness while keeping the details of the target image. However, other methods have more blotchy noisy textures. Additionally, quantitative results from Table 1(a) also confirm that our model outperforms other models.

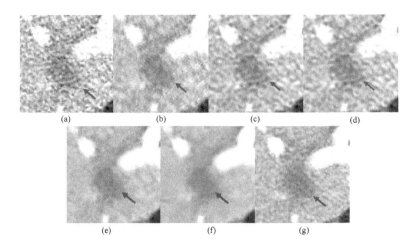

Fig. 4. The amplified ROIs of different network outputs in the rectangle marked in Fig. 3. (a) LDCT, (b) RED-CNN, (c) WGAN-VGG, (d) MAP-NN, (e) AD-NET, (f) proposed TED-Net, and (g) NDCT.

Table 1. (a): Quantitative results of different methods on L506, our method is marked by a star, 'no cs' means no cyclic shift blocks and bold numbers indicate best results. (b): Quantitative results of models with different token embedding size.

(a)

Method	SSIM	RMSE
LDCT	0.8759	14.2416
RED-CNN	0.8952	11.5926
WGAN-VGG	0.9008	11.6370
MAP-NN	0.8941	11.5848
AD-NET	0.9041	9.7166
TED-Net*(no cs)	0.9126	8.9518
TED-Net*	**0.9144**	**8.7681**

(b)

Method	Dim	#Param.	SSIM	RMSE
TED-T	64	1.45M	0.9090	9.1520
TED-S	256	2.35M	0.9115	9.0192
TED-B	1024	18.88M	0.9144	8.7681

3.1   Ablation Study

On the Influence of Cyclic Shift: In this work, cyclic shift is performed in the encoder and decoder blocks to enhance the perceptual fields of our model. To investigate the effectiveness of the this block, we run comparative test by removing the cyclic shift blocks from the original model. Results from Fig. 5 shows that TED-Net with cyclic shift enjoys more spatial smoothness compared to TED-Net without cyclic shift. Quantitative results from Table 1(b) also confirm the effectiveness of cyclic shift in improving the performance of our model.

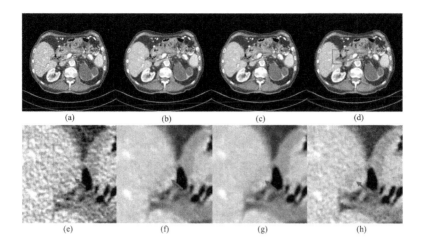

Fig. 5. The performance of TED-Net on case L506 with lesion No. 576. (a) LDCT, (b) TED-Net without cyclic shift, (c) TED-Net, and (d) NDCT. (e)–(h) are the co responding magnified ROIs from (a)–(d).

On the Influence of Model Size. With respect to the embedding size of the transformer block between the encoder and decoder parts, we provide several model variants of TED-T, TED-S and TED-B with embedding size of 64, 256, and 1024, respectively. Results from Table 1(b) shows that TED-B has best performance with the compromise of large model size.

4 Conclusion

In this paper, a novel pure transformer-based convolution-free LDCT denoising algorithm is developed for clinical applications. In contrast, the most state-of-art models are based on CNN. To the best of our knowledge, this is the first research to apply pure transformer for LDCT denoising. Our contributions are mainly three-folds: (1) A convolution-free T2T-vit-based denoising transformer model is developed. (2) The dilation is used in the T2T stage to enlarge the receptive field to obtain more contextual information from the feature maps. (3) A cyclic shift is used to furthermore refine the mode of image tokenization. Experimental results show our model outperforms other state-of-the-art models with the highest SSIM value and smallest RMSE value. In the future, this model can be further slimmed with a more powerful tokenization without downgrading of images.

References

1. Yu, H., Wang, G.: Compressed sensing based interior tomography. Phys. Med. Biol. **54**(9), 2791 (2009)
2. Liu, Y., Ma, J., Fan, Y., Liang, Z.: Adaptive-weighted total variation minimization for sparse data toward low-dose x-ray computed tomography image reconstruction. Phys. Med. Biol. **57**(23), 7923 (2012)
3. Sidky, E.Y., Pan, X.: Image reconstruction in circular cone-beam computed tomography by constrained, total-variation minimization. Phys. Med. Biol. **53**(17), 4777 (2008)
4. Tian, Z., Jia, X., Yuan, K., Pan, T., Jiang, S.B.: Low-dose CT reconstruction via edge-preserving total variation regularization. Phys. Med. Biol. **56**(18), 5949 (2011)
5. Zhang, Y., Zhang, W., Lei, Y., Zhou, J.: Few-view image reconstruction with fractional-order total variation. JOSA A **31**(5), 981–995 (2014)
6. Xu, Q., Yu, H., Mou, X., Zhang, L., Hsieh, J., Wang, G.: Low-dose x-ray CT reconstruction via dictionary learning. IEEE Transa. Med. Imaging **31**(9), 1682–1697 (2012)
7. Tan, S., et al.: Tensor-based dictionary learning for dynamic tomographic reconstruction. Phys. Med. Biol. **60**(7), 2803 (2015)
8. Ma, J., et al.: Low-dose computed tomography image restoration using previous normal-dose scan. Med. Phys. **38**(10), 5713–5731 (2011)
9. Yin, X., et al.: Domain progressive 3D residual convolution network to improve low-dose CT imaging. IEEE Trans. Med. Imaging **38**(12), 2903–2913 (2019)
10. Chen, H., et al.: Low-dose CT with a residual encoder-decoder convolutional neural network. IEEE Trans. Med. Imaging **36**(12), 2524–2535 (2017)
11. Yang, Q., et al.: Low-dose CT image denoising using a generative adversarial network with Wasserstein distance and perceptual loss. IEEE Trans. Med. Imaging **37**(6), 1348–1357 (2018)
12. Fan, F., et al.: Quadratic autoencoder (Q-AE) for low-dose CT denoising. IEEE Trans. Med. Imaging **39**(6), 2035–2050 (2019)
13. Vaswani, A., et al.: Attention is all you need. arXiv preprint arXiv:1706.03762 (2017)
14. Devlin, J., Chang, M.-W., Lee, K., Toutanova, K.: BERT: pre-training of deep bidirectional transformers for language understanding. arXiv preprint arXiv:1810.04805 (2018)
15. Xin, J., Tang, R., Lee, J., Yu, Y., Lin, J.: DeeBERT: dynamic early exiting for accelerating BERT inference. arXiv preprint arXiv:2004.12993 (2020)
16. Zhang, Y., et al.: DialoGPT: large-scale generative pre-training for conversational response generation. arXiv preprint arXiv:1911.00536 (2019)
17. Brown, T.B., et al.: Language models are few-shot learners. arXiv preprint arXiv:2005.14165 (2020)
18. Wu, H., et al.: CvT: introducing convolutions to vision transformers. arXiv preprint arXiv:2103.15808 (2021)
19. Chu, X., Zhang, B., Tian, Z., Wei, X., Xia, H.: Do we really need explicit position encodings for vision transformers? arXiv e-prints arXiv: 2102.10882 (2021)
20. Chen, M., et al.: Generative pretraining from pixels. In: International Conference on Machine Learning, pp. 1691–1703. PMLR (2020)
21. Dosovitskiy, A., et al.: An image is worth 16×16 words: transformers for image recognition at scale. arXiv preprint arXiv:2010.11929 (2020)

22. Yang, F., Yang, H., Fu, J., Lu, H., Guo, B.: Learning texture transformer network for image super-resolution. In: Proceedings of the IEEE/CVF Conference on Computer Vision and Pattern Recognition, pp. 5791–5800 (2020)
23. Chen, H., et al.: Pre-trained image processing transformer. arXiv preprint arXiv:2012.00364 (2020)
24. Choromanski, K., et al.: Rethinking attention with performers. arXiv preprint arXiv:2009.14794 (2020)
25. Cao, H., et al.: Swin-Unet: Unet-like pure transformer for medical image segmentation. arXiv preprint arXiv:2105.05537 (2021)
26. Touvron, H., Cord, M., Douze, M., Massa, F., Sablayrolles, A., Jégou, H.: Training data-efficient image transformers & distillation through attention. arXiv preprint arXiv:2012.12877 (2020)
27. Han, K., Xiao, A., Wu, E., Guo, J., Xu, C., Wang, Y.: Transformer in transformer. arXiv preprint arXiv:2103.00112 (2021)
28. Yuan, L., et al.: Tokens-to-token ViT: training vision transformers from scratch on ImageNet. arXiv preprint arXiv:2101.11986 (2021)
29. Liu, Z., et al.: Swin transformer: hierarchical vision transformer using shifted windows. arXiv preprint arXiv:2103.14030 (2021)
30. Zhang, Z., Yu, L., Liang, X., Zhao, W., Xing, L.: TransCT: dual-path transformer for low dose computed tomography. arXiv preprint arXiv:2103.00634 (2021)
31. McCollough, C.H., et al.: Low-dose CT for the detection and classification of metastatic liver lesions: results of the 2016 low dose CT grand challenge. Med. Phys. 44(10), e339–e352 (2017)
32. Shan, H., et al.: Competitive performance of a modularized deep neural network compared to commercial algorithms for low-dose CT image reconstruction. Nat. Mach. Intell. 1(6), 269–276 (2019)
33. Tian, C., Xu, Y., Li, Z., Zuo, W., Fei, L., Liu, H.: Attention-guided CNN for image denoising. Neural Netw. 124, 117–129 (2020)

Self-supervised Mean Teacher for Semi-supervised Chest X-Ray Classification

Fengbei Liu[1(✉)], Yu Tian[1,4], Filipe R. Cordeiro[2], Vasileios Belagiannis[3], Ian Reid[1], and Gustavo Carneiro[1]

[1] Australian Institute for Machine Learning, University of Adelaide, Adelaide, Australia
`fengbei.liu@adelaide.edu.au`
[2] Universidade Federal Rural de Pernambuco, Recife, Brazil
[3] Universität Ulm, Ulm, Germany
[4] South Australian Health and Medical Research Institute, Adelaide, Australia

Abstract. The training of deep learning models generally requires a large amount of annotated data for effective convergence and generalisation. However, obtaining high-quality annotations is a laboursome and expensive process due to the need of expert radiologists for the labelling task. The study of semi-supervised learning in medical image analysis is then of crucial importance given that it is much less expensive to obtain unlabelled images than to acquire images labelled by expert radiologists. Essentially, semi-supervised methods leverage large sets of unlabelled data to enable better training convergence and generalisation than using only the small set of labelled images. In this paper, we propose Self-supervised Mean Teacher for Semi-supervised (S^2MTS^2) learning that combines self-supervised mean-teacher pre-training with semi-supervised fine-tuning. The main innovation of S^2MTS^2 is the self-supervised mean-teacher pre-training based on the joint contrastive learning, which uses an infinite number of pairs of positive query and key features to improve the mean-teacher representation. The model is then fine-tuned using the exponential moving average teacher framework trained with semi-supervised learning. We validate S^2MTS^2 on the multi-label classification problems from Chest X-ray14 and CheXpert, and the multi-class classification from ISIC2018, where we show that it outperforms the previous SOTA semi-supervised learning methods by a large margin. Our code will be available upon paper acceptance.

Keywords: Semi-supervised learning · Chest X-ray · Self-supervised learning · Multi-label classification

1 Introduction

Deep learning has shown outstanding results in medical image analysis problems [16,19–21,28,29]. However, this performance usually depends on the avail-

---

F. Liu and Y. Tian—First two authors contributed equally to this work.

C. Lian et al. (Eds.): MLMI 2021, LNCS 12966, pp. 426–436, 2021.
https://doi.org/10.1007/978-3-030-87589-3_44

ability of labelled datasets, which is expensive to obtain given that the labelling process requires expert doctors. This limitation motivates the study of semi-supervised learning (SSL) methods that train models with a small set of labelled data and a large set of unlabelled data.

The current state-of-the-art (SOTA) SSL is based on pseudo-labelling methods [18,26], consistency-enforcing approaches [2,17,27], self-supervised and semi-supervised learning (S^{4}L) [5,34], and graph-based label propagation [1]. Pseudo-labelling is an intuitive SSL technique, where confident predictions from the model are transformed into pseudo-labels for the unlabelled data, which are then used to re-train the model [18]. Consistency-enforcing regularisation is based on training for a consistent output given model [22,27] or input data [2,17] perturbations. S^{4}L methods are based on self-supervised pre-training [4,12], followed by supervised fine-tuning using few labelled samples [5,34]. Graph-based methods rely on label propagation on graphs [1]. Recently, Yang et al. [33] suggested that self-supervision pre-training provides better feature representations than consistency-enforcing approaches in SSL. However, previous S^{4}L approaches use only the labelled data in the fine-tuning stage, missing useful training information present in the unlabelled data. Furthermore, self-supervised pre-training [4,12] tends to use limited amount of samples to represent each class, but recently, Cai et al. [3] showed that better representation can be obtained with an infinite amount of samples. Also, recent research [26] suggests that the student-teacher framework, such as the mean-teacher [27], works better in multi-label semi-supervised tasks than other SSL methods. We speculate that this is because other methods are usually designed to work with softmax activation that only works in multi-class problems, while mean-teacher [27] does not have this constraint and can work in multi-label problems.

In this paper, we propose a self-supervised mean-teacher for semi-supervised (S^{2}MTS^2) learning approach that combines S^{4}L [5,34] with consistency-enforcing learning based on the mean-teacher algorithm [27]. The main contribution of our method is the self-supervised mean-teacher pre-training with the joint contrastive learning [3]. To the best of our knowledge, this is the first approach, in our field, to train the mean teacher model with self-supervised learning. This model is then fine-tuned with semi-supervised learning using the exponential moving average teacher framework [27]. We evaluate our proposed method on the thorax disease multi-label datasets ChestX-ray 14 [32] and CheXpert [15], and on the multi-class skin condition dataset ISIC2018 [8,30]. We show that our method outperforms the SOTA on semi-supervised learning [1,11,22,31]. Moreover, we investigate each component of our framework for their contribution to the overall model in the ablation study.

2   Related Works

SSL is a research topic that is gaining attention from the medical image analysis community due to the expensive image annotation process [7] and the growing number of large-scale datasets available in the field [32]. The current SOTA

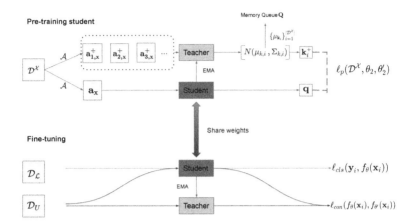

Fig. 1. Description of the proposed self-supervised mean-teacher for semi-supervised (S^2MTS^2) learning. The main contribution of the paper resides in the top part of the figure, with the self-supervised mean-teacher pre-training based on joint contrastive learning, which uses an infinite number of pairs of positive query and key features sampled from the unlabelled images to minimise $\ell_p(.)$ in (1). This model is then fine-tuned with the exponential moving average teacher in a semi-supervised learning framework that uses both labelled and unlabelled sets to minimise $\ell_{cls}(.)$ and $\ell_{con}(.)$ in (2).

SSL methods are based on consistency-enforcing approaches that leverage the unlabelled data to regularise the model prediction consistency [17,27]. Other related papers [9] extend the mean teacher [27] to encourage consistency between the prediction by the student and teacher models for atrium and brain lesion segmentation. The SOTA SSL method on Chest X-ray images [22] exploits the consistency in the relations between labelled and unlabelled data. None of these methods explores a self-supervised consistency-enforcing method to pre-train an SSL model, as we propose.

Self-supervised learning methods [4,12] are also being widely investigated in SSL because they can provide good representations [5,34]. However, these methods ignore the large amount of unlabelled data to be used during SSL, which may lead to unsatisfactory generalisation process. An important point in self-supervised learning is on how to define the classes to be learned. In general, each class is composed of a single pair of augmented images from the same image, and many pairs of augmentations from different images [3–5,12]. The use of a single pair of images to form a class has been criticised by Cai et al. [3], who propose the joint contrastive learning (JCL), which is an efficient way to form a class with an infinite number of augmented images from the same image to leverage the statistical dependency between different augmentations.

3   Method

In this section, we introduce our two-stage learning framework in detail (see Fig. 1). We assume that we have a small labelled dataset, denoted by $\mathcal{D}_L = \{(\mathbf{x}_i, \mathbf{y}_i)\}_{i=1}^{|\mathcal{D}_L|}$, where the image is represented by $\mathbf{x} \in \mathcal{X} \subset \mathbb{R}^{H \times W \times C}$, and class $\mathbf{y} \in \{0,1\}^{|\mathcal{Y}|}$, where $\mathcal{Y}$ represents the label set. We consider a multi-label problem and thus $\sum_{c=1}^{|\mathcal{Y}|} \mathbf{y}_i(c) \in [0, |\mathcal{Y}|]$. The unlabelled dataset is defined by $\mathcal{D}_U = \{\mathbf{x}_i\}_{i=1}^{|\mathcal{D}_U|}$ with $|\mathcal{D}_L| << |\mathcal{D}_U|$.

Our model consists of a student and a teacher model [27], denoted by parameters $\theta, \theta' \in \Theta$, respectively, which parameterize the classifier $f_\theta : \mathcal{X} \to [0,1]^{|\mathcal{Y}|}$. This classifier can be decomposed as $f_\theta = h_{\theta_1} \circ g_{\theta_2}$, with $g_{\theta_2} : \mathcal{X} \to \mathcal{Z}$ and $h_{\theta_1} : \mathcal{Z} \to [0,1]^{|\mathcal{Y}|}$. The first stage (top of Fig. 1) of the training consists of a self-supervised learning that uses the images from $\mathcal{D}_L$ and $\mathcal{D}_U$, denoted by $\mathcal{D}^{\mathcal{X}} = \{\mathbf{x}_i | \mathbf{x}_i \in \mathcal{D}_L^{\mathcal{X}} \bigcup \mathcal{D}_U\}_{i=1}^{|\mathcal{D}_{\mathcal{X}}|}$, with $\mathcal{D}_L^{\mathcal{X}}$ representing the images from the set $\mathcal{D}_L$, where our method minimises the joint contrastive learning loss [3], defined in (1). This means that during this first stage, we only learn the parameters for g_{θ_2}. The second stage (bottom of Fig. 1) fine-tunes this pre-trained student-teacher model using the semi supervised consistency loss defined in (2). Below we provide details on the losses and training.

3.1   Joint Contrastive Learning to Self-supervise the Mean-Teacher Pre-training

The self-supervised pre-training of the mean-teacher using joint contrastive learning (JCL) [3], presented in this section, is the main technical contribution of this paper. The teacher and student process an input image to return the keys $\mathbf{k} \in \mathcal{Z}$ and the queries $\mathbf{q} \in \mathcal{Z}$ with $\mathbf{k} = g_{\theta_2'}(\mathbf{x})$ and $\mathbf{q} = g_{\theta_2}(\mathbf{x})$. We also assume that we have a set of augmentation functions, i.e., random crop and resize, rotation and Gaussian blur, denoted by $\mathcal{A} = \{a_l : \mathcal{X} \to \mathcal{X}\}_{l=1}^{|\mathcal{A}|}$. Then JCL minimises the following loss [3]:

$$\ell_p(\mathcal{D}^{\mathcal{X}}, \theta_2, \theta_2') = -\frac{1}{|\mathcal{D}^{\mathcal{X}}|} \frac{1}{M} \sum_{i=1}^{|\mathcal{D}^{\mathcal{X}}|} \sum_{m=1}^{M} \left[\log \frac{\exp\left[\frac{1}{\tau} \mathbf{q}_i^\top \mathbf{k}_{i,m}^+\right]}{\exp\left[\frac{1}{\tau} \mathbf{q}_i^\top \mathbf{k}_{i,m}^+\right] + \sum_{j=1}^{K} \exp\left[\frac{1}{\tau} \mathbf{q}_i^\top \mathbf{k}_{i,j}^-\right]} \right],$$

$$(1)$$

where τ is the temperature hyper-parameter, the query $\mathbf{q}_i = g_{\theta_2}(a(\mathbf{x}_i))$, with $a \in \mathcal{A}$. the positive key $\mathbf{k}_{i,m}^+ \sim p(\mathbf{k}_i^+)$, with $p(\mathbf{k}_i^+) = \mathcal{N}(\mu_{\mathbf{k}_i}, \Sigma_{\mathbf{k}_i})$ and $\mathbf{k}_i = g_{\theta_2'}(a(\mathbf{x}_i))$ (i.e., a sample from the data augmentation distribution for $\mathbf{x}$), and the negative keys $\mathbf{k}_{i,j}^- \in \{\mu_{\mathbf{k}_j}\}_{i,j \in \{1,...,|\mathcal{D}^{\mathcal{X}}|\}, i \neq j}$ represents a negative key for query $\mathbf{q}_i$. In (1), M denotes the number of positive keys, and Cai et al. [3] describe a loss that minimises a bound to (1) for $M \to \infty$ – below, the minimisation of $\ell_p(.)$ in (1) is realised by the minimisation of this bound. As defined above, the generative model $p(\mathbf{k}_i^+)$ is denoted by the Gaussian $\mathcal{N}(\mu_{\mathbf{k}_i}, \Sigma_{\mathbf{k}_i})$, where the mean $\mu_{\mathbf{k}_i}$ and covariance $\Sigma_{\mathbf{k}_i}$ are estimated from a set of keys $\{\mathbf{k}_{i,l}^+ = g_{\theta_2'}(a_l(\mathbf{x}_i))\}_{a_l \in \mathcal{A}}$ formed by different views of $\mathbf{x}_i$. The set of negative keys $\{\mu_{\mathbf{k}_j}\}_{i,j \in \{1,...,|\mathcal{D}^{\mathcal{X}}|\}, i \neq j}$ is stored in a memory queue [12] that is updated in a first-in-first-out way, where the mean of the keys in $\{\mu_{\mathbf{k}_i}\}_{i=1}^{|\mathcal{D}^{\mathcal{X}}|}$ are inserted to the memory queue to replace the oldest key means from previous training iterations. The memory queue has been designed to increase the number of negative samples without sacrificing computation efficiency.

The training of the student-teacher model [12,27,35] is achieved by updating the student parameter using the loss in (1), as in $\theta_2(t) = \theta_2(t-1) - \nabla_{\theta_2}\ell_p(\mathcal{D}^{\mathcal{X}}, \theta_2, \theta_2')$, where t is the training iteration. The teacher model parameter is updated with exponential moving average (EMA) with $\theta_2'(t) = \alpha\theta_2'(t-1) + (1-\alpha)\theta_2(t)$, with $\alpha \in [0,1]$. For this pre-training stage, we notice that training for more epochs always improve the model regularisation given that it is difficult to overfit the training set with the loss in (1). Hence, we select the last epoch student model $g_{\theta_2}(.)$ to initialise the fine-tuning stage, defined below in Sec. 3.2.

3.2   Fine-Tuning the Mean Teacher

To fine tune the mean teacher, we follow the approach in [12,27] using the following loss to train the student model:

$$\ell_t(\mathcal{D}_L, \mathcal{D}_U, \theta, \theta') = \frac{1}{|\mathcal{D}_L|} \sum_{(\mathbf{x}_i, \mathbf{y}_i) \in \mathcal{D}_L} \ell_{cls}(\mathbf{y}_i, f_\theta(\mathbf{x}_i)) + \frac{1}{|\mathcal{D}|} \sum_{\mathbf{x}_i \in \mathcal{D}} \ell_{con}(f_\theta(\mathbf{x}_i), f_{\theta'}(\mathbf{x}_i)),$$
(2)

where $\ell_{cls}(\mathbf{y}_i, f_\theta(\mathbf{x}_i)) = -\mathbf{y}_i^\top \log(f_\theta(\mathbf{x}_i))$, $\ell_{con}(f_\theta(\mathbf{x}_i), f_{\theta'}(\mathbf{x}_i)) = \|f_\theta(\mathbf{x}_i) - f_{\theta'}(\mathbf{x}_i)\|^2$, and $\mathcal{D} = \mathcal{D}_U \bigcup \mathcal{D}_L^{\mathcal{X}}$. The training of the student-teacher model [12, 27,35] is achieved by updating the student parameter using the loss in (2), as in $\theta(t) = \theta(t-1) - \nabla_\theta\ell_t(\mathcal{D}_L, \mathcal{D}_U, \theta, \theta')$, where t is the training iteration. The teacher model parameter is updated with exponential moving average (EMA) with $\theta'(t) = \alpha\theta'(t-1) + (1-\alpha)\theta(t)$, with $\alpha \in [0,1]$. After finishing the fine-tuning stage, we select the teacher model $f_{\theta'}(.)$ to estimate the multi-label classification for test images.

4   Experiment

4.1   Dataset Setup

We use Chest X-ray14 [32], CheXpert [15] and ISIC2018 [8,30] datasets to evaluate our method.

Chest X-ray14 contains 112,120 chest x-ray images from 30,805 different patients. There are 14 different labels (each label represents a disease) in the dataset, where each patient can have multiple diseases at the same time, forming a multi-label classification problem. To compare with previous papers [1,22], we adopt the official train/test data split. For the self-supervised pre-training of the mean teacher, we used all the unlabelled images (86k samples) from the training set. For the semi-supervised fine-tuning of the mean teacher, we follow the papers [1,22] and experiment with training sets containing different proportions of labelled data (2%,5%,10%,15%,20%). We report the classification result on the official test set (26,000 samples) using area under the ROC curve (AUC).

CheXpert contains around 220,000 images with 14 different diseases, and similarly to Chest X-ray14, each patient can have multiple diseases at the same time. For pre-processing, we remove all lateral view images and treat uncertain label as negative labels. We follow the semi-supervised setup from [11], and experiment with 100/200/300/400/500 labelled samples per class. We report results on the official test set using AUC.

ISIC2018 is a multi-class skin condition dataset that contains 10,015 images with seven different labels. Each image is associated with one of the seven labels, forming a multi-class classification problem. We follow [22] train/test split for fair comparison, where the training contains 20% of the samples labelled, and the remaining 80% unlabelled. We report the AUC, Sensitivity, and F1 score results to compare with baselines.

4.2 Implementation Details

For all datasets, we use DenseNet121 [14] as our backbone model. For self-supervised pre-training, we follow [5] and replace the two-layer multi-layer perceptron (MLP) projection head by a three-layer MLP. For dataset pre-processing, we resized Chest X-ray14 images to 512×512 for faster processing and CheXpert and ISIC2018 to 224×224 for fair comparison with baselines. We use the data augmentation proposed in [4], consisting of random resize and crop, random rotation, random horizontal flipping, except for random grayscale because X-ray images are originally in grayscale. The batch size is 128 for Chest X-ray14 and 256 for CheXpert and ISIC2018, and learning rate is 0.05. For the fine-tuning stage, we use batch size 32 with 16 labelled and 16 unlabelled. The fine-tuning takes 30 epochs with learning rate decayed by 0.1 at 15 and 25 epochs for all datasets. We use the SGD optimiser with 0.9 momentum for the pre-training stage, and Adam optimiser in fine-tuning stage. The code is written in Pytorch [24]. We use 4 Nvidia Volta-100 for the self-supervised stage and 1 Nvidia RTX 2080ti for fine-tuning.

4.3 Experimental Results

We evaluate our approach on the official test set of ChestX-ray14 using different percentage of labelled training data (i.e., 2%, 5%, 10%, 15%, 20%), as shown in Table 1. The set of labelled data used for each percentage above follows the same strategy of previous works [1,22]. Our S^4L achieves the SOTA AUC results on all different percentages of labels. Our model surpasses the previous SOTA SRC-MT [22] by a large margin of 8.7% and 6.8% AUC for the 2% and 5% labelled set cases, respectively, where we use a backbone architecture of lower complexity (Densenet121 instead of the DenseNet169 of [22]). Using the same Densenet121 backbone, GraphXnet [1] fails to classify precisely for the 2% and 5% labelled set cases. Our method surpasses GraphXnet by more than 20% AUC in both cases.

Furthermore, we achieve the SOTA results of the field for the 10%, 15% and 20% labelled set cases, outperforming all previous semi-supervised methods [1, 22]. It is worth noting that our model trained with 5% of the labelled set achieves better results than SRC-MT with 15% of labelled. We also compare with a recently proposed self-supervised pre-training methods, MoCo V2 [6], adapted to our semi-supervised task, followed by the fine-tuning stage using different percentages of labelled data. Our method outperforms MoCo V2 by almost 10% AUC when using 2% of labelled set, and almost 3% AUC for 10% of labelled set. Our result for 20% labelled set achieves comparable 81.06% AUC performance as the supervised learning approaches – 81.20% from MoCo V2 (Densenet 121) and 81.75% from SRC-MT (Densenet 169) using 100% of the labelled samples. Such result indicates the effectiveness of our proposed S^2MTS^2 in SSL benchmark problems.

We also show the class-level performance using 20% of the labelled data and compare with other SOTA methods in Table 2. We compare with the previous baselines, namely original mean teacher (MT) with Densenet169, SRC-MT with Densenet169, MoCo V2, and GraphXNet with Densenet121. We also train a baseline Densenet121 model with 20% labelled data using Imagenet pre-trained model. Our method achieves the best results on nine classes, surpassing the original MT [27] and its extension SRC-MT [22] by a large margin, demonstrating the effectiveness of our self-supervised learning.

Furthermore, we compare our approach on the fully-supervised Chest X-ray14 benchmark in Table 3. To the best of our knowledge, Hermoza et al. [13] has the SOTA supervised classification method containing a complex structure (relying on the weakly-supervised localisation of lesions) with a mean AUC of 82.1% (over the 14 classes), while ours reports a mean AUC of 82.5%. Hence, our model, using the whole labelled set, achieves the SOTA performance on 8 classes and an average that surpasses the previous supervised methods by a minimum of 0.4% and a maximum of 8% AUC.

The results on CheXpert and ISIC2018 datasets are shown in Tables 4 and 5, respectively. In particular, for CheXpert in Table 4, we compare our method with LatentMixing [11] and our result is better in all cases. For ISIC2018 on Table 5, using the test set from SRC-MT [22], our method outperforms all baselines (Supervised, MT, and SRC-MT) for all measures.

4.4   Ablation Study

We study the impact of different components of our proposed S^2MTS^2 in Table 6 using Chest X-Ray14. Using the proposed self-supervised learning with just the student model, our model achieves at least 72.95% mean AUC on various percentages of labelled training data. Adding the JCL component improves the baseline by around 1% mean AUC on each training percentage. Adding the mean teacher boosts the result by 1.5% to 2% mean AUC on each training percentage. The combination of all our proposed three components achieves SOTA performance on semi-supervised task.

Table 1. Mean AUC result over the 14 disease classes of Chest X-Ray14 for different label set training percentages. * indicates the methods that use Densenet169 as backbone architecture.

Label Percentage	2%	5%	10%	15%	20%	100%
Graph XNet* [1]	53.00	58.00	63.00	68.00	78.00	N/A
SRC-MT* [22]	66.95	72.29	75.28	77.76	79.23	81.75
NoTeacher [31]	72.60	77.04	77.61	N/A	79.49	N/A
MOCO V2 [6]	65.97	73.84	77.07	79.37	80.17	81.20
Ours	**74.69**	**78.96**	**79.90**	**80.31**	**81.06**	**82.51**

Table 2. Class-level AUC comparison between our S^2MTS^2 and other semi-supervised SOTA approaches trained with **20% of labelled data** on Chest X-Ray14. * denotes the methods that use Densenet-169 as backbone.

Method	Densenet-121	GraphXNet [1]	MOCO V2 [6]	MT [22] *	SRC-MT [22] *	Ours
Atelectasis	75.75	71.89	77.21	75.12	75.38	**78.57**
Cardiomegaly	80.71	87.99	85.84	87.37	87.7	**88.08**
Effusion	79.87	79.2	81.62	80.81	81.58	**82.87**
Infiltration	69.16	**72.05**	70.91	70.67	70.4	70.68
Mass	78.40	80.9	81.71	77.72	78.03	**82.57**
Nodule	74.49	71.13	**76.72**	73.27	73.64	76.60
Pneumonia	69.55	**76.64**	71.08	69.17	69.27	72.25
Pneumothorax	84.70	83.7	85.92	85.63	86.12	**86.55**
Consolidation	71.85	73.36	74.47	72.51	73.11	**75.47**
Edema	81.61	80.2	83.57	82.72	82.94	**84.83**
Emphysema	89.75	84.07	91.10	88.16	88.98	**91.88**
Fibrosis	79.30	80.34	80.96	78.24	79.22	**81.73**
Pleural Thicken	73.46	75.7	75.65	74.43	75.63	**76.86**
Hernia	86.05	87.22	85.62	**87.74**	87.27	85.98
Mean	78.19	78.88	80.17	78.83	79.23	**81.06**

Table 3. Class-level AUC comparison between our S^2MTS^2 and other supervised SOTA approaches trained with **100% of labelled data** on Chest X-Ray14.

Method	Wang et al. [32]	Li et al. [20]	CheXNet [25]	CRAL [10]	Ma et al. [23]	Hermoza et al. [13]	Ours
Atelectasis	70	72.9	75.5	78.1	77.7	77.5	**78.7**
Cardiomegaly	81	84.6	86.7	88.3	**89.4**	88.1	87.4
Effusion	75.9	78.1	81.5	83.1	82.9	83.1	**83.8**
Infiltration	66.1	67.3	69.4	69.7	69.6	69.5	**70.9**
Mass	69.3	74.3	80.2	83	**83.8**	82.6	83.3
Nodule	66.9	75.8	73.5	76.4	77.1	78.9	**79.9**
Pneumonia	65.8	63.3	69.8	72.5	72.2	**74.1**	73.9
Pneumothorax	79.9	79.3	82.8	86.6	86.2	**87.9**	87.1
Consolidation	70.3	72	72.2	75.8	75	74.7	**75.9**
Edema	80.5	71	83.5	85.3	84.6	**84.6**	84.5
Emphysema	83.3	75.1	85.6	91.1	90.8	93.6	**93.7**
Fibrosis	78.6	76.1	80.3	82.6	82.7	83.3	**83.4**
Pleural Thicken	68.4	73	74.9	78	77.9	79.3	**79.3**
Hernia	87.2	66.8	89.4	91.8	**93.4**	91.7	93.3
Mean	74.5	73.9	78.9	81.6	81.7	82.1	**82.5**

Table 4. Mean AUC result (over the 14 disease classes) on CheXpert for different number of training samples per class.

Labelled	100	200	300	400	500
LatentMixing [11]	65.12	66.41	67.39	67.96	68.47
Ours	**66.15**	**67.85**	**70.83**	**71.37**	**71.58**

Table 5. AUC, Sensitivity and F1 result on ISIC2018 using 20% of labelled training samples.

Method	AUC	Sensitivity	F1
Supervised	90.15	65.50	52.03
MT	92.96	69.75	59.10
SRC-MT [22]	93.58	71.47	60.68
Ours	**94.71**	**72.14**	**62.67**

Table 6. Ablation studies of our method with different components on Chest X-Ray14. "Self-supervised" indicates the traditional self-supervised learning with contrastive loss [12]. "JCL" replaces contrastive loss with (1), "MT" stands for fine-tuned with student-teacher learning instead only fine-tuned on only labelled samples.

Self-supervised	JCL	MT	AUC (2%)	AUC (5%)	AUC (10%)	AUC (15%)	AUC (20%)
✓			72.95	76.82	78.54	79.28	80.14
✓	✓		73.60	77.46	79.18	79.83	80.62
✓		✓	74.80	77.66	79.08	79.70	80.57
✓	✓	✓	75.69	78.96	79.90	80.31	81.06

5 Conclusion

In this paper, we presented a novel semi-supervised framework, the Self-supervised Mean Teacher for Semi-supervised (S^2MTS^2) learning. The main contribution of S^2MTS^2 is the self-supervised mean teacher pre-trained based on joint contrastive learning [3], using an infinite number of pairs of positive query and key features. This model is then fine-tuned with the exponential moving average teacher framework. S^2MTS^2 is validated on the thorax disease multi-label classification problem from the datasets Chest X-ray14 [32] and CheXpert [15], and the multi-class classification from the skin condition dataset ISIC2018 [8,30]. The experiments show that our method outperforms the previous SOTA semi-supervised learning methods by a large margin in all benchmarks containing a varying percentage of labelled data. We also show that the method holds the SOTA results on Chest X-ray14 [32] even for the fully-supervised problem. The ablation study shows the importance of three main components of the method, namely self-supervised learning, JCL, and the mean-teacher model. We will investigate the performance of our method on other semi-supervised medical imaging benchmarks in the future.

Acknowledgement. This work was supported by Australian Research Council through grants DP180103232 and FT190100525.

References

1. Aviles-Rivero, A.I., et al.: GraphXNET - chest X-ray classification under extreme minimal supervision. arXiv preprint arXiv:1907.10085 (2019)
2. Berthelot, D., et al.: ReMixMatch: semi-supervised learning with distribution alignment and augmentation anchoring. arXiv preprint arXiv:1911.09785 (2019)
3. Cai, Q., Wang, Y., Pan, Y., Yao, T., Mei, T.: Joint contrastive learning with infinite possibilities. arXiv preprint arXiv:2009.14776 (2020)
4. Chen, T., Kornblith, S., Norouzi, M., Hinton, G.: A simple framework for contrastive learning of visual representations. In: International Conference on Machine Learning, pp. 1597–1607. PMLR (2020)
5. Chen, T., Kornblith, S., Swersky, K., Norouzi, M., Hinton, G.: Big self-supervised models are strong semi-supervised learners. arXiv preprint arXiv:2006.10029 (2020)
6. Chen, X., Fan, H., Girshick, R., He, K.: Improved baselines with momentum contrastive learning. arXiv preprint arXiv:2003.04297 (2020)
7. Cheplygina, V., de Bruijne, M., Pluim, J.P.: Not-so-supervised: a survey of semi-supervised, multi-instance, and transfer learning in medical image analysis. Med. Image Anal. **54**, 280–296 (2019)
8. Codella, N., et al.: Skin lesion analysis toward melanoma detection 2018: a challenge hosted by the international skin imaging collaboration (ISIC). arXiv preprint arXiv:1902.03368 (2019)
9. Cui, W., et al.: Semi-supervised brain lesion segmentation with an adapted mean teacher model. In: Chung, A.C.S., Gee, J.C., Yushkevich, P.A., Bao, S. (eds.) IPMI 2019. LNCS, vol. 11492, pp. 554–565. Springer, Cham (2019). https://doi.org/10. 1007/978-3-030-20351-1_43
10. Guan, Q., Huang, Y.: Multi-label chest X-ray image classification via category-wise residual attention learning. Pattern Recogn. Lett. **130**, 259–266 (2020)
11. Gyawali, P.K., Ghimire, S., Bajracharya, P., Li, Z., Wang, L.: Semi-supervised medical image classification with global latent mixing. In: Martel, A.L., et al. (eds.) MICCAI 2020. LNCS, vol. 12261, pp. 604–613. Springer, Cham (2020). https:// doi.org/10.1007/978-3-030-59710-8_59
12. He, K., Fan, H., Wu, Y., Xie, S., Girshick, R.: Momentum contrast for unsupervised visual representation learning. In: Proceedings of the IEEE/CVF Conference on Computer Vision and Pattern Recognition, pp. 9729–9738 (2020)
13. Hermoza, R., Maicas, G., Nascimento, J.C., Carneiro, G.: Region proposals for saliency map refinement for weakly-supervised disease localisation and classification. In: Martel, A.L., et al. (eds.) MICCAI 2020. LNCS, vol. 12266, pp. 539–549. Springer, Cham (2020). https://doi.org/10.1007/978-3-030-59725-2_52
14. Huang, G., Liu, Z., Van Der Maaten, L., Weinberger, K.Q.: Densely connected convolutional networks. In: Proceedings of the IEEE Conference on Computer Vision and Pattern Recognition (2017)
15. Irvin, J., et al.: CheXpert: a large chest radiograph dataset with uncertainty labels and expert comparison. In: Proceedings of the AAAI Conference on Artificial Intelligence, vol. 33, pp. 590–597 (2019)
16. Jonmohamadi, Y., et al.: Automatic segmentation of multiple structures in knee arthroscopy using deep learning. IEEE Access **8**, 51853–51861 (2020)
17. Laine, S., Aila, T.: Temporal ensembling for semi-supervised learning. arXiv preprint arXiv:1610.02242 (2016)
18. Lee, D.H., et al.: Pseudo-label: the simple and efficient semi-supervised learning method for deep neural networks. In: Workshop on Challenges in Representation Learning, ICML, vol. 3 (2013)

19. Lee, J.G., et al.: Deep learning in medical imaging: general overview. Korean J. Radiol. **18**(4), 570 (2017)
20. Li, Z., et al.: Thoracic disease identification and localization with limited supervision. In: Proceedings of the IEEE Conference on Computer Vision and Pattern Recognition, pp. 8290–8299 (2018)
21. Liu, F., Jonmohamadi, Y., Maicas, G., Pandey, A.K., Carneiro, G.: Self-supervised depth estimation to regularise semantic segmentation in knee arthroscopy. In: Martel, A.L., et al. (eds.) MICCAI 2020. LNCS, vol. 12261, pp. 594–603. Springer, Cham (2020). https://doi.org/10.1007/978-3-030-59710-8_58
22. Liu, Q., et al.: Semi-supervised medical image classification with relation-driven self-ensembling model. IEEE Trans. Med. Imaging **39**(11), 3429–3440 (2020)
23. Ma, C., Wang, H., Hoi, S.C.H.: Multi-label thoracic disease image classification with cross-attention networks. In: Shen, D., et al. (eds.) MICCAI 2019. LNCS, vol. 11769, pp. 730–738. Springer, Cham (2019). https://doi.org/10.1007/978-3-030-32226-7_81
24. Paszke, A., et al.: Pytorch: an imperative style, high-performance deep learning library. arXiv preprint arXiv:1912.01703 (2019)
25. Rajpurkar, P., et al.: ChexNet: radiologist-level pneumonia detection on chest x-rays with deep learning. arXiv preprint arXiv:1711.05225 (2017)
26. Rizve, M.N., Duarte, K., Rawat, Y.S., Shah, M.: In defense of pseudo-labeling: an uncertainty-aware pseudo-label selection framework for semi-supervised learning. arXiv preprint arXiv:2101.06329 (2021)
27. Tarvainen, A., Valpola, H.: Mean teachers are better role models: weight-averaged consistency targets improve semi-supervised deep learning results. arXiv preprint arXiv:1703.01780 (2017)
28. Tian, Yu., Maicas, G., Pu, L.Z.C.T., Singh, R., Verjans, J.W., Carneiro, G.: Few-shot anomaly detection for polyp frames from colonoscopy. In: Martel, A.L., et al. (eds.) MICCAI 2020. LNCS, vol. 12266, pp. 274–284. Springer, Cham (2020). https://doi.org/10.1007/978-3-030-59725-2_27
29. Tian, Y., Pu, L.Z., Singh, R., Burt, A.D., Carneiro, G.: One-stage five-class polyp detection and classification. In: 2019 IEEE 16th International Symposium on Biomedical Imaging (ISBI 2019), pp. 70–73. IEEE (2019)
30. Tschandl, P., Rosendahl, C., Kittler, H.: The HAM10000 dataset, a large collection of multi-source dermatoscopic images of common pigmented skin lesions. Sci. Data **5**(1), 1–9 (2018)
31. Unnikrishnan, B., Nguyen, C.M., Balaram, S., Foo, C.S., Krishnaswamy, P.: Semi-supervised classification of diagnostic radiographs with NoTeacher: a teacher that is not mean. In: Martel, A.L., et al. (eds.) MICCAI 2020. LNCS, vol. 12261, pp. 624–634. Springer, Cham (2020). https://doi.org/10.1007/978-3-030-59710-8_61
32. Wang, X., Peng, Y.a.: ChestX-ray8: Hospital-scale chest x-ray database and benchmarks on weakly-supervised classification and localization of common thorax diseases. In: CVPR (2017)
33. Yang, Y., Xu, Z.: Rethinking the value of labels for improving class-imbalanced learning. arXiv preprint arXiv:2006.07529 (2020)
34. Zhai, X., Oliver, A., Kolesnikov, A., Beyer, L.: S4L: self-supervised semi-supervised learning. In: ICCV, pp. 1476–1485 (2019)
35. Zhou, H.-Y., Yu, S., Bian, C., Hu, Y., Ma, K., Zheng, Y.: Comparing to learn: surpassing ImageNet pretraining on radiographs by comparing image representations. In: Martel, A.L., et al. (eds.) MICCAI 2020. LNCS, vol. 12261, pp. 398–407. Springer, Cham (2020). https://doi.org/10.1007/978-3-030-59710-8_39

VoxelEmbed: 3D Instance Segmentation and Tracking with Voxel Embedding based Deep Learning

Mengyang Zhao[1], Quan Liu[2], Aadarsh Jha[2], Ruining Deng[2], Tianyuan Yao[2],

Anita Mahadevan-Jansen[2], Matthew J. Tyska[2], Bryan A. Millis[2], and Yuankai Huo[2(✉)]

[1] Dartmouth College, Hanover, NH 03755, USA
[2] Vanderbilt University, Nashville, TN 37215, USA
yuankai.huo@vanderbilt.edu

Abstract. Recent advances in bioimaging have provided scientists a superior high spatial-temporal resolution to observe dynamics of living cells as 3D volumetric videos. Unfortunately, the 3D biomedical video analysis is lagging, impeded by resource insensitive human curation using off-the-shelf 3D analytic tools. Herein, biologists often need to discard a considerable amount of rich 3D spatial information by compromising on 2D analysis via maximum intensity projection. Recently, pixel embedding based cell instance segmentation and tracking provided a neat and generalizable computing paradigm for understanding cellular dynamics. In this work, we propose a novel spatial-temporal voxel-embedding (VoxelEmbed) based learning method to perform simultaneous cell instance segmenting and tracking on 3D volumetric video sequences. Our contribution is in four-fold: (1) The proposed voxel embedding generalizes the pixel embedding with 3D context information; (2) Present a simple multi-stream learning approach that allows effective spatial-temporal embedding; (3) Accomplished an end-to-end framework for one-stage 3D cell instance segmentation and tracking without heavy parameter tuning; (4) The proposed 3D quantification is memory efficient via a single GPU with 12 GB memory. We evaluate our VoxelEmbed method on four 3D datasets (with different cell types) from the ISBI Cell Tracking Challenge. The proposed VoxelEmbed method achieved consistent superior overall performance (OP) on two densely annotated datasets. The performance is also competitive on two sparsely annotated cohorts with 20.6% and 2% of data-set having segmentation annotations. The results demonstrate that the VoxelEmbed method is a generalizable and memory-efficient solution.

Keywords: Tracking · Segmentation · Instances · 3-D · Embedding

1 Introduction

Characterizing cellular dynamics, the behaviors of the fundamental units of life, is indispensable in translational biological research, such as organogenesis [4], immune response [24], drug development [31], and cancer metastasis [7]. With a tenet of "seeing is believing", recent advances in bioimaging have provided scientists unprecedented high spatial-temporal resolution to observe three dimensional dynamics (3D volumes

© Springer Nature Switzerland AG 2021
C. Lian et al. (Eds.): MLMI 2021, LNCS 12966, pp. 437–446, 2021.
https://doi.org/10.1007/978-3-030-87589-3_45

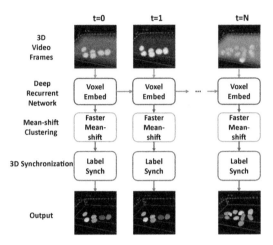

Fig. 1. The overall framework. The workflow of the proposed voxel embedding based deep learning framework is presented, for 3D cell instance segmentation and tracking.

+ time) of living cells [16]. Yet, large-scale bioimage data are concurrent with imaging innovations, causing fundamental computational challenges for quantifying cellular dynamics in translational biological research, where "quantifying is deciding" [19]. For example, a single lattice light-sheet microscope [6] (LLS) produce TB level rich spatial-temporal dynamic 3D volumetric videos [27]. Unfortunately, the 3D biomedical video analysis is lagging, impeded by resource insensitive human curation via off-the-shelf 3D analytic tools [5]. Recent deep learning techniques have achieved remarkable success in computer vision and biomedical image analysis. However, the large-scale quantification of "3D+time" cellular dynamics with deep learning (e.g., dense instance segmentation and tracking) is still hindered by the high dimensionality and heterogeneity of the dynamics. [11] Herein, biologists often need to discard a considerable amount of rich 3D spatial information by projecting the 3D videos to 2D space for downstream analyses.

When quantifying cellular dynamics, the "segment-then-track" two-stage paradigm [7, 8, 13, 28, 29] is a prevalent design for both conventional model based methods and deep learning approaches. Such a paradigm first segments instance objects across frames and then links the instance objects via association algorithms. However, the spatial and temporal information from the same individual object cannot be learned simultaneously with the two-stage design. To integrate and handle these two tasks simultaneously, Payer et al. [25] proposed a cosine pixel embedding based recurrent stacked hourglass network (RSHN) for simultaneous instance cell segmentation and tracking. The pixel embedding approach tackled instance segmentation and tracking within an uniformed "single-stage" framework. The key idea is to loosen the constraint of having each cell requiring a globally unique embedding, to just allowing the cell to have a different embedding relative to the nearby four cells (based on the Four Color Map theorem [1, 25]). Herein, the number of embeddings does not have to strictly increase with the number of cells, providing a scalable learning strategy. Based on the merits, it is appealing to adapt such strategy from 2D to 3D settings. However, the direct 3D adaptation (e.g., change a 2D network to a 3D version) requires higher computational resources, more training samples, and a considerably larger embedding space to differentiate each cell to all of its neighbors in 3D.

In this study, we propose a novel spatial-temporal voxel-embedding (VoxelEmbed) deep learning based method to perform 3D cell instance segmentation and tracking. As opposed to deploying a 3D network, we develop a simple multi-stream learning approach to learn spatial, temporal, and 3D context information simultaneously via a 2D network design. To aggregate 3D information, we introduce a 3D synchronization algorithm to build volumetric masks inspired by recent advances in slice propagation [3]. Briefly, the innovations of the proposed approach is in four-folds:

(1) The proposed VoxelEmbed approach generalizes the pixel embedding to a voxel embedding with 3D context information.
(2) A simple multi-stream learning approach is presented that allows effective spatial-temporal embedding.
(3) To our knowledge, this is the first embedding based deep learning approach for simultaneous 3D cell instance segmentation and tracking.
(4) The proposed method is memory efficient to a single GPU with 12 GB memory.

2   Methods

The principle of our proposed VoxelEmbed framework is presented Fig. 1. In VoxelEmbed, we extend spatial-temporal embedding by adding 3D information. The extra embedding information is obtained from the zigzag multi-stream straining.

2.1   Cosine Embedding Based Instance Segmentation and Tracking

Payer et al. [25] proposed a cosine embedding based recurrent stacked hourglass network (RSHN) that, for the first time, integrated the instance segmentation and tracking algorithms into a one-stage holistic learning framework with pixel-wise cosine embedding, which achieves instance cell segmentation and tracking in a one-stage framework. The entire pixel embedding based learning consist of two major stages: embedding encoding and feature clustering.

The RSHN network, combining convolutional GRUs [2, 12] (ConvGRUs) and the stacked hourglass network [21], is employed as the backbone network for our voxel embedding. In [25], each pixel in a 2D video sequence was encoded to a high-dimensional embedding vector with the intuition that all pixels from the same cell, across spatial and temporal, should have the same feature representation (embedding). The renown cosine similarity [14, 15] approach is commonly used to measure the similarities of any two pixels. For instance, $\mathbf{A}$ was the embedding vector of pixel a, and $\mathbf{B}$ was the embedding vector of pixel b. The cosine similarity of $\mathbf{A}$ and $\mathbf{B}$ is defined as:

$$cos\left(\mathbf{A}, \mathbf{B}\right) = \frac{\mathbf{A} \cdot \mathbf{B}}{\|\mathbf{A}\| \ \|\mathbf{B}\|} \tag{1}$$

which ranged from -1 to 1, where 1 indicates that two vectors have the same direction, 0 indicates orthogonal, and -1 indicates the opposite. The cosine similarities is used as a loss function to force the pixels from the same cell to have $cos\left(\mathbf{A}, \mathbf{B}\right)$ towards 1, while the pixels from different cells to be 0.

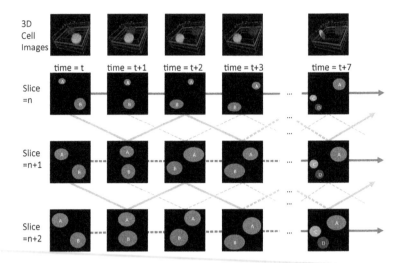

Fig. 2. VoxelEmbed at the training stage. Our VoxelEmbed model is trained by both blue and yellow streams, to ensure the same voxel embedding of the same object in terms of spatial-temporal domain and 3D context.

2.2 Voxel Embedding (Training Stage)

To encode the 3D context information for 3D cell instance segmentation and tracking, we generalize the pixel embedding principle to a voxel embedding scheme. Briefly, the VoxelEmbed approach learns additional 3D context features, and concatenates those features to original pixel embedding features. The additional 3D context features are learned from the new zigzag training path (the golden stream in Fig. 2) beyond the temporal embedding path (the blue stream in Fig. 2).

The sections with the same z-axis location from all 3D volumes across the temporal direction are used as the temporal embedding path, to ensure the same embedding of each cell (at the same z-axis location in 2D) has the same embedding along with migration and time.

Then, the sections at nearby z-axis locations are selected in a zigzag training path to enforce the embedding similarity along with the migration and 3D context, when learning all zigzag training paths converge the entire 3D space. The zigzag training paths are designed as an interleaved "W" shape. Using multi-stream training, the voxels from the same cell in a 3D video will be forced to have the same embedding along with migration, time, and 3D context.

2.3 Voxel Embedding (Testing Stage)

In implementation, T frames from time t to $t + T$ are used as a single training sample. The temporal embedding feature, a 14-dimensional feature vector, is encoded for each voxel:

$$f_{time} = [a_1, a_2, ..., a_{14}] \tag{2}$$

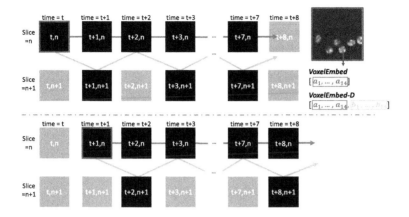

Fig. 3. VoxelEmbed at the testing stage. The blue and yellow streams cover two ways of forming testing samples (e.g., T = 8 in this study) for obtaining embedding of the first frame (red boarder) at the testing stage. With the embedded feature maps, the 3D cell instance segmentation and tracking results are obtained from the 3D synchronization (see Algorithm 1).

With the same principle, T frames from time t to $t + T$ with zigzag frames are used as another single training example. The 3D context embedding feature, another 14-dimensional feature vector, is encoded for each voxel:

$$f_{threeD} = [b_1, b_2, ..., b_{14}] \tag{3}$$

Then, the f_{time} and f_{threeD} are concatenated into a 28 dimension feature for each voxel as our final voxel embedding feature:

$$f_{spatial-temporal} = f_{time} \oplus f_{threeD} = [a_1, ..., a_{14}, b_1, ..., b_{14}] \tag{4}$$

where $\oplus$ is the concatenation operator.

In the testing stage, the encoded embedding images will be clustered to each individual cells. The standard Mean-shift clustering algorithm was used in Payer et al. [25]. In this study, we utilize a GPU accelerated Faster Mean-shift algorithm [30] to accelerate the embedding clustering. From the unsupervised clustering, a unique label will be assigned to each cell as different instances, which achieve the instance segmentation and tracking simultaneously.

2.4   3D Synchronization

In the original pixel-embedding approach, the temporal synchronization step was introduced to stick different short "video clips" to the a complete full cell video, with consistent instance numbers. Inspired by this approach, we extend the synchronization from temporal to spatial as well. To do so, we propose a 3-D synchronization Algorithm 1 to improve label synchronization on the z-axis direction.

Algorithm 1. 3D Synchronization Algorithm

Require: $InMasks$: Unsynchronized mask-set; $ReMasks$: Reference mask-set;
Ensure: $OutMasks$: Synchronized mask-set;
1: **for** $mask \in InMasks$ **do**
2:     **for** $rmask \in ReMasks$ **do**
3:         $S_{mask} \leftarrow J(rmask, mask)$ # Caluate the Jaccard similarity: $J(R, S) = \frac{|R \cap S|}{|R \cup S|}$
4:         **if** S_{mask} is largest in the layer of $mask$ **then**
5:             $List_{rmask}$ append $mask$ # Get the largest similarity instance in each layers
6:         **end if**
7:     **end for**
8: **end for**
9: **for** $list \in List_{rmask}$ **do**
10:     Find the most common label $l \in list$
11:     **for** $mask \in list$ **do**
12:         $mask \leftarrow l$ # Synchronize the mask
13:     **end for**
14: **end for**

In Algorithm 1, the input reference mask-set $ReMasks$ is the layer with the largest foreground ratio. In Jaccard Similarity [22], R is the set of pixels belonging to reference mask and S is the set of pixels belonging to matching mask.

3   Data and Implementation Details

In this study, we conduct an empirical validation via the ISBI Cell Tracking Challenge [26] dataset to evaluate the accuracy performance of our proposed framework. Specifically, we used 3-D microscope video sequences from the ISBI Cell Tracking Challenge, which is independent with training data. The following four video sequence datasets of different sizes, shapes, and textures cells were adopted to evaluate the performance: (1) Chinese Hamster Ovarian nuclei(Fluo-N3DH-CHO), (2) C.elegans developing embryo (Fluo-N3DH-CE), (3) Simulated nuclei of HL60 cells(Fluo-N3DH-SIM+), and (4) Simulated GFP-actin-stained A549 Lung Cancer cells (Fluo-C3DH-A549-SIM). Note that the labels of the official testing data have not been released. Since each cohort has two training videos, we used one as training while another as testing. The source code of benchmarks (LEID-NL [9], KTH-SE [17], and RSHN [25]) was also deployed on such data directly. Therefore, the reported results could be different from the online leader board [10].

All computation and training were performed via a standard NC6 [20] virtual machine platform at the Microsoft Azure cloud. The virtual machine includes half an NVIDIA Tesla K80 accelerator [23] card (12 GB accessible) and six Intel Xeon E5-2690 v3 (Haswell) processor. The multi-stream training and 3-D Synchronization Algorithm was implemented with tensorflow and Python3. All the models in this study were trained with 20,000 iterations. During model training, the learning rate was initially set to 0.0001, and decreases to 0.00001 after 10,000 iterations. The bandwidth hyper-parameter in mean-shift clustering is set to 0.1.

Table 1. Quantitative result of empirical validation

Method	Data-set					
	Fluo-N3DH-SIM+			Fluo-C3DH-A549-SIM		
	SEG	TRA	OP	SEG	TRA	OP
LEID-NL [9]	$0.643^{(3)}$	$\mathbf{0.971}^{(1)}$	$0.807^{(3)}$	$0.827^{(2)}$	1.000	$0.914^{(2)}$
KTH-SE [17]	$0.774^{(1)}$	$0.958^{(2)}$	$0.866^{(1)}$	$0.842^{(1)}$	1.000	$0.921^{(1)}$
RSHN [25]	0.748	0.909	0.829	0.798	1.000	0.899
VoxelEmbed (ours)	**0.818**	0.933	0.876	**0.852**	1.000	**0.926**
VoxelEmbed-D (ours)	0.806	0.952	**0.879**	0.838	1.000	0.919

* "(1), (2), (3)" indicates the current ranking in the Cell Tracking Challenge leader board [10].

Table 2. Quantitative result of empirical validation

Method	Data-set					
	Fluo-N3DH-CHO (20.6% frames have labels)			Fluo-N3DH-CE (2% frames have labels)		
	SEG	TRA	OP	SEG	TRA	OP
LEID-NL [9]	$0.901^{(4)}$	$0.923^{(5)}$	$0.912^{(5)}$	–	–	–
KTH-SE [17]	$\mathbf{0.907}^{(2)}$	$0.953^{(1)}$	$\mathbf{0.930}^{(1)}$	$0.667^{(3)}$	$\mathbf{0.945}^{(1)}$	$0.806^{(2)}$
RSHN [25]	0.837	0.946	0.892	0.673	0.875	0.774
VoxelEmbed (ours)	0.862	0.958	0.910	**0.717**	0.897	0.807
VoxelEmbed-D (ours)	0.860	**0.959**	0.910	0.709	0.925	**0.817**

* "(1), (2), (3)" indicates the current ranking of such approach in the Cell Tracking Challenge leader board [10]. LEID did not provide N3DH-CE, the performance is not available (–).

4 Results

The qualitative results are shown in Fig. 4, while the comparison of quantitative results are presented in Table 1 and 2. In Table 1 and 2, the Cell Tracking Challenge's official tool of measuring the performance of tracking (TRA) and segmentation (SEG) are employed. TRA is computed by the normalized Acyclic Oriented Graph Matching measure (AOGM) [18] and is used as the tracking accuracy metric. The SEG is computed by Jaccard index. Following the ISBI Cell Tracking Challenge's Benchmark, we also compute the overall performance (OP), which is the average of TRA and SEG.

Based on quantitative and qualitative results, our methods achieved a competitive accuracy performance, using the same network without heavy parameter tuning. For Fluo-N3DH-SIM+ and Fluo-C3DH-A549-SIM data-set, since the manual annotations are available for all video frames, our VoxelEmbed achieved the best SEG and OP. The tracking performance is also competitive. For Fluo-N3DH-CHO and Fluo-N3DH-CE, only sparse manual annotations are provided (Fig. 4 (b)). As a result, the OP of Fluo-N3DH-CHO is inferior compared with leading approaches.

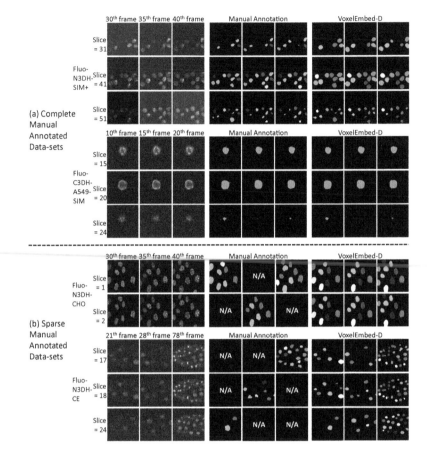

Fig. 4. The qualitative results of our VoxelEmbed framework. The 3D ISBI Cell Tracking Challenge dataset, (a) Complete and (b) Sparse manual annotations, as well as the VoxelEmbed results are presented.

5   Conclusion

In this paper, we introduce VoxelEmbed, an novel embedding based deep learning approach for 3D cell instance segmentation and tracking. The proposed detection method introduced a simple multi-stream training strategy to allow the embedding encoder to learn the spatial-temporal consistent voxel embedding across 3D context. The results show that the VoxelEmbed achieves decent performance compared with the leading method in four Cell Tracking Challenge datasets, with a uniformed learning framework. This study shows the promises of performing 3D cell instance segmentation and tracking with embedding based deep learning with a single GPU.

References

1. Appel, K., Haken, W., et al.: Every planar map is four colorable. Bull. Am. Math. Soc. **82**(5), 711–712 (1976)
2. Ballas, N., Yao, L., Pal, C., Courville, A.: Delving deeper into convolutional networks for learning video representations. arXiv preprint arXiv:1511.06432 (2015)
3. Cai, J., et al.: Accurate weakly-supervised deep lesion segmentation using large-scale clinical annotations: slice-propagated 3D mask generation from 2D RECIST. In: Frangi, A.F., Schnabel, J.A., Davatzikos, C., Alberola-López, C., Fichtinger, G. (eds.) MICCAI 2018. LNCS, vol. 11073, pp. 396–404. Springer, Cham (2018). https://doi.org/10.1007/978-3-030-00937-3_46
4. Cao, M., et al.: The single-cell transcriptional landscape of mammalian organogenesis. Nature **566**(7745), 496–502 (2019)
5. von Chamier, L., Laine, R.F., Henriques, R.: Artificial intelligence for microscopy: what you should know. Biochem. Soc. Trans. **47**(4), 1029–1040 (2019)
6. Chen, B.C., et al.: Lattice light-sheet microscopy: imaging molecules to embryos at high spatiotemporal resolution. Science **346**(6208), 1257998 (2014)
7. Condeelis, J., Pollard, J.W.: Macrophages: obligate partners for tumor cell migration, invasion, and metastasis. Cell **124**(2), 263–266 (2006)
8. Debeir, O., Van Ham, P., Kiss, R., Decaestecker, C.: Tracking of migrating cells under phase-contrast video microscopy with combined mean-shift processes. IEEE Trans. Med. imaging **24**(6), 697–711 (2005)
9. Meijering, E., Dzyubachyk, O., Smal, I.: Methods for cell and particle tracking. Methods Enzym. **504**, 183–200 (2012)
10. ISBI: Isbi cell tracking challenge benchmark leader boarder (2021). http://celltrackingchallenge.net/latest-ctb-results/
11. Jiang, C., Tsai, Y.J.: Enhanced crack segmentation algorithm using 3D pavement data. J. Comput. Civil Eng. **30**(3), 04015050 (2016)
12. Jiang, R., Gouvea, J., Hammer, D., Aeron, S.: Automatic coding of students' writing via contrastive representation learning in the wasserstein space. arXiv preprint arXiv:2011.13384 (2020)
13. Jin, B., Cruz, L., Goncalves, N.: Deep facial diagnosis: deep transfer learning from face recognition to facial diagnosis. IEEE Access **8**, 123649–123661 (2020)
14. Li, M., Chen, X., Li, X., Ma, B., Vitányi, P.M.: The similarity metric. IEEE Trans. Inform. Theory **50**(12), 3250–3264 (2004)
15. Liu, Q., et al.: Towards annotation-free instance segmentation and tracking with adversarial simulations. arXiv preprint arXiv:2101.00567 (2021)
16. Liu, T.L., et al.: Observing the cell in its native state: imaging subcellular dynamics in multicellular organisms. Science **360**(6386), eaaq1392 (2018)
17. Magnusson, K.E.: Segmentation and tracking of cells and particles in time-lapse microscopy. Ph.D. thesis, KTH Royal Institute of Technology (2016)
18. Matula, P., Maška, M., Sorokin, D.V., Matula, P., Ortiz-de Solórzano, C., Kozubek, M.: Cell tracking accuracy measurement based on comparison of acyclic oriented graphs. PloS ONE **10**(12), e0144959 (2015)
19. Meijering, E.: A bird's-eye view of deep learning in bioimage analysis. Comput. Struct. Biotech. J. **18**, 2312 (2020)
20. Microsoft: Azure NC-series (2020). https://docs.microsoft.com/en-us/azure/virtual-machines/nc-series
21. Newell, A., Yang, K., Deng, J.: Stacked hourglass networks for human pose estimation, pp. 483–499 (2016)

22. Niwattanakul, S., Singthongchai, J., Naenudorn, E., Wanapu, S.: Using of Jaccard coefficient for keywords similarity. In: Proceedings of the International Multiconference of Engineers and Computer Scientists, vol. 1, pp. 380–384 (2013)
23. NVIDIA: Nvidia, V. (2013). tesla k20 gpu accelerator board specification (2015). https://www.nvidia.com/content/PDF/kepler/tesla-k20-active-bd-06499-001-v03.pdf
24. Ong, E.Z., et al.: A dynamic immune response shapes Covid-19 progression. Cell Host Microbe 27(6), 879–882 (2020)
25. Payer, C., Štern, D., Feiner, M., Bischof, H., Urschler, M.: Segmenting and tracking cell instances with cosine embeddings and recurrent hourglass networks. Med. Image Anal. 57, 106–119 (2019)
26. Ulman, V., et al.: An objective comparison of cell-tracking algorithms. Nat. Methods 14(12), 1141–1152 (2017)
27. Wan, Y., McDole, K., Keller, P.J.: Light-sheet microscopy and its potential for understanding developmental processes. Annu. Rev. Cell Dev. Biol. 35, 655–681 (2019)
28. Yuan, W., Xu, W.: Neighborloss: a loss function considering spatial correlation for semantic segmentation of remote sensing image. IEEE Access 9, 75641–75649 (2021)
29. Zhao, M., Chang, C.H., Xie, W., Xie, Z., Hu, J.: Cloud shape classification system based on multi-channel CNN and improved FDM. IEEE Access 8, 44111–44124 (2020)
30. Zhao, M., et al.: Faster mean-shift: GPU-accelerated clustering for cosine embedding-based cell segmentation and tracking. Med. Image Anal. 71, 102048 (2021)
31. Zhou, X., Wong, S.T.: High content cellular imaging for drug development. IEEE Signal Process. Mag. 23(2), 170–174 (2006)

Using Spatio-Temporal Correlation Based Hybrid Plug-and-Play Priors (SEABUS) for Accelerated Dynamic Cardiac Cine MRI

Qingyong Zhu[1] and Dong Liang[1,2(✉)]

[1] Medical AI Research Centre, Shenzhen Institutes of Advanced Technology, Chinese Academy of Science, Shenzhen, Guangdong, China
dong.liang@siat.ac.cn
[2] Paul C. Lauterbur Research Centre for Biomedical Imaging, Shenzhen Institutes of Advanced Technology, Chinese Academy of Science, Shenzhen, Guangdong, China

Abstract. The plug-and-play prior (P^3) is known as denoising prior which has been successfully applied to various imaging problems. In this work for accelerated dynamic cardiac cine magnetic resonance imaging (Dcc-MRI), we introduce a Spatio-tEmporal correlAtion based hyBrid plUg-and-play priorS (SEABUS) integrating local P^3 and nonlocal P^3, which further help both suppress aliasing artifacts and capture dynamic features. Specifically, the local P^3 enforces the pixel-wise edge-orientation consistency by reference frame guided multi-scale orientation projection (MSOP) in a subset of few adjacent frames. The nonlocal P^3 constrains the cube-wise anatomic-structure similarity by cube matching and 4D filtering (CM4D) in all frames. By composite splitting algorithm (CSA), the SEABUS is coupled into fast iterative shrinkage-thresholding algorithm (FISTA) and then a new Dcc-MRI approach that is named as SEABUS-FCSA is proposed. The experimental results on the in-vivo cardiac MR datasets demonstrated the efficiency and potential of the proposed SEABUS-FCSA approach.

Keywords: Dynamic cardiac cine MRI · Local P^3 · Nonlocal P^3 · FISTA

1 Introduction

Dynamic cardiac cine MR imaging (Dcc-MRI) can produce a set of cardiac-structure images with periodic changes, which is great benefit to computer-aided diagnosis and vision-guided surgery. However, the slow MR data-acquisition process leads to the resolution limitation in spatio-temporal domain.

Both the compressed sensing (CS) [1,2] and matrix completion (MC) [3,4] provide the systematic theoretical frameworks to accelerate MRI by prior regularization from a small number of incoherently sampled k-space data. Currently, the prior regularization based Dcc-MRI methods usually take use of some properties of spatio-temporal domain, including sparsity [5,6], low rank [7] and the

© Springer Nature Switzerland AG 2021
C. Lian et al. (Eds.): MLMI 2021, LNCS 12966, pp. 447–456, 2021.
https://doi.org/10.1007/978-3-030-87589-3_46

combinations [8–11] enforcing both sparsity and low rank. For example, Lingala et al. [12] developed a blind CS (BCS) based imaging method that utilizes spatio-temporal multiplicative-factorization and constrains the sparsity of coefficients on temporal basis functions. Chen et al. [6] proposed a dynamic total variation (dTV) regularization that enforces the difference-domain sparsity between the target frame and the first frame. Otazo et al. [8] exploited an accelerated Dcc-MRI method using low-rank plus sparsity (L+S) model based on spatio-temporal additive-decomposition. Nevertheless, these methods seem to be less effective for aliasing-artifacts suppressing and dynamic-features capturing when it comes to large changes between frames and high reduction factors [13].

Recently, the plug-and-play prior (P^3) framework has attracted lots of attention in various imaging problems [14–16], which aims to replaces the proximal operator (regarded as a denoising step) in the family of proximal gradient algorithms (PGA) [17–19] with a generic denoiser such as nonlocal means (NLM) [20] and BM3D [21]. The earliest work was P^3-ADMM proposed by Venkatakrishnan et al. [14], then Kamilov et al. [15] proposed a P^3-FISTA, which is superior to the original ADMM-based formulation is that there is no need to perform an inversion on the forward model. However, the conventional framework based on NLM or BM3D is independent to imaging scene and fails to model scene-based denoiser to help improve imaging quality especially in terms of edge details.

In this work, we propose a new accelerated Dcc-MRI approach that utilizes a Spatio-tEmporal correlAtion based hyBrid plUg-and-play priorS (SEABUS) to further promote dynamic-features capturing as well as aliasing-artifacts suppressing. Specifically, the SEABUS integrates a local P^3 that enforces the pixel-wise edge-orientation consistency by reference frame guided multi-scale orientation projection (MSOP) in a subset of few adjacent frames and a nonlocal P^3 that constrains the cube-wise anatomic-structure similarity by cube matching and 4D filtering (CM4D) in all frames. By composite splitting algorithm (CSA) [22], SEABUS is coupled into FISTA and a new reconstruction approach named as SEABUS-FCSA for Dcc-MRI is proposed. Through experiments on the in-vivo cardiac MR datasets, the proposed SEABUS-FCSA approach yields reconstructions that have the best performance in terms of aliasing-artifacts suppressing and dynamic-features capturing even in high reduction factors compared with several state-of-the-art Dcc-MRI technologies.

2   Method

2.1   Accelerated Dcc-MRI: FISTA Framework

The observation forward model for accelerated Dcc-MRI can be formulated as a discretized linear system,

$$y = Ax + \zeta \tag{1}$$

where $x = [x_1, x_2, ..., x_T] \in \mathbb{C}^{N \times T}$ denotes the T frames to be reconstructed. Each column of which is certain frame with N pixels. The observed k-space data is given as $y = [y_1, y_2, ..., y_T] \in \mathbb{C}^{M \times T}$. The linear operator $A : \mathbb{C}^{N \times T} \to \mathbb{C}^{M \times T}$ formulates the observation forward model which performs a multiplication by coil sensitivities followed by an undersampled Fourier transform in each column. $\zeta \sim \mathcal{N}(0, \sigma^2)$ is usually assumed to be additive white Gaussian noise (AWGN). The backward program of reconstructing x is an ill-posed problem by the reason that the problem is under-determined with $M \leq N$. By adding the signal prior, the variational formulation for this inverse problem can be written as:

$$x^* \in \arg \min_{x \in \mathbb{C}^{N \times T}} \|Ax - y\|_2^2 / 2 + \lambda \mathcal{R}(x) \tag{2}$$

where the $\mathcal{R}(x)$ usually acts as a regularization term encoding sparsity or low rankness. The nonnegative scalar $\lambda \in R_+$ is as the regularization parameter. Due to nondifferentiability existing on $\mathcal{R}(x)$, a family of proximal gradient algorithms (PGAs) are in common usage. A PGA with Nestrerov momentum named as FISTA [17] for Dcc-MRI can be summarized as follows,

$$\begin{cases} z^t \leftarrow s^{t-1} + \tau A^H \left(y - As^{t-1} \right) & \text{(3a)} \\ x^t \leftarrow prox_{\mathcal{R}}(\lambda; z^t) & \text{(3b)} \\ s^t \leftarrow x^t + \dfrac{q_{t-1} - 1}{q_t}(x^t - x^{t-1}) & \text{(3c)} \end{cases}$$

where (3a) is gradient descent step on $\|Ax - y\|_2^2$ at $x = s^{t-1}$ with step-size $\tau \in R_+$. The proximal operator is defined as $prox_{\mathcal{R}}(\lambda; z^t) \triangleq \arg \min_x \|x - z^t\|_2^2 / 2 + \lambda \mathcal{R}(x)$. The proximal operator $prox_{\mathcal{R}}(\lambda; z^t)$ can be regarded as a denoiser of z^t under AWGN variance $\lambda \in R_+$. (3c) is Nestrerov acceleration step using $q_t = \frac{1 + \sqrt{1 + 4q_{t-1}^2}}{2}$ and $q_0 = 1$.

2.2 The Hybrid P^3 Based FISTA Framework: SEABUS-FCSA

The key idea of P^3 algorithms developed in recent years is to consider the proximal operator $prox_{\mathcal{R}}(\lambda; z^t)$ as a family of generic denoisers $\mathcal{D}(\lambda; z^t)$, for example, NLM and BM3D, which are not originated from any explicit regularizers. Ahmad's work [23] has pointed out that a state-of-the-art denoiser can help recover richer image structure. However, the denoisers in existing P^3 algorithms are usually chosen traditionally as scene-independent, failing to model scene based denoiser to help improve algorithm performance. Here we propose a spatio-temporal correlation based hybrid P^3-FISTA, named as SEABUS-FCSA, which integrates local P^3 and nonlocal P^3 by composite splitting algorithm (CSA) [22]. Specifically, the SEABUS-FCSA uses the iterative updates,

$$\begin{cases} z^t \leftarrow s^{t-1} + \tau A^H \left(y - As^{t-1} \right) & \text{(4a)} \\ x_1^t \leftarrow \overbrace{\mathcal{D}_{\text{MSOP}}(\eta; z^t, x_r)}^{\text{local P}^3} & \text{(4b)} \\ x_2^t \leftarrow \overbrace{\mathcal{D}_{\text{CM4D}}(\gamma; z^t)}^{\text{nonlocal P}^3} & \text{(4c)} \\ x^t \leftarrow \alpha x_1^t + (1 - \alpha) x_2^t & \text{(4d)} \\ s^t \leftarrow x^t + \dfrac{q_{t-1} - 1}{q_t} \left(x^t - x^{t-1} \right) & \text{(4e)} \end{cases}$$

where $\eta \in R_+$ and $\gamma \in R_+$ represent the denoising strengths of local and non-local P³, respectively. $\alpha \in R_+$ is a weighting parameter that balances local and nonlocal P³.

Multi-scale Orientation Projection (MSOP). A multi-scale orientation projection in our previous work [24] is utilized to enhance suppressing artifacts and capturing dynamic features based on edge orientation consistency of adjacent frames. Here we first estimate adaptively a reference frame x_r from a subset consisting of every four adjacent frames. By measurements averaging, the reference frame is with more k-space data and can be effectively reconstructed by TV with sensitivity encoding,

$$x_r^* \in \arg \min_{x_r \in \mathbb{C}^{N \times 1}} \|A_r x_r - \frac{\sum_{i=1}^4 y_i}{\sum_{i=1}^4 m_i}\|_2^2 / 2 + \mu \mathrm{TV}(x_r)$$

where $A_r : \mathbb{C}^{N_r \times 1} \to \mathbb{C}^{M_r \times 1}$ is the observation forward model for the reference frame. y_i and m_i are the k-space data and sampling mask for the i-th frame in the corresponding subset.

Each intermediate frame z_i will be effectively guided filtered by x_r. Specifically, let $\mathcal{G}x_i$ be a multiscale direction vector-field of x_i and $\mathcal{G}^{tangent}x_i = [\mathcal{M}_1 x_i, \mathcal{M}_2 x_i]$, $\mathcal{G}^{normal}x_i = [\mathcal{M}_2 x_i, -\mathcal{M}_1 x_i]$. We call

$$\mathrm{MSOP} : \mathbb{C}^N \to \mathbb{R}; \quad \mathrm{MSOP}(x_i, x_r) := \|< \mathcal{G}^{tangent}x_i, \frac{\mathcal{G}^{normal}x_r}{|\mathcal{G}^{normal}x_r| + \epsilon} >\|_{w,2}^2 \tag{5}$$

the multiscale directional projection. Here $\mathcal{G}^{tangent}x_i$ stands for the tangent vectors of image x_i at spatial locations, $\mathcal{G}^{normal}x_i$ represents the normal vectors with normalization perpendicular to the corresponding tangent vectors. $\langle \cdot, \cdot \rangle$ denotes vector inner product. We normalized the normal vectors of the reference frame to solely preserve orientation information. Since the anatomy of image x_i should be well aligned with x_r, the angle between edge orientation $\mathcal{G}^{tangent}x_i$ and $\mathcal{G}^{normal}x_r$ at same location should approach 90, that is to say, the edge orientations of x_i are orthogonal to the edge normal orientations of x_r, we have $\mathrm{MSOP}(x_i, x_r) \approx 0$. $\| \cdot \|_{w,2}^2$ represents a weighted l_2-norm with $w = 1/\|[\mathcal{M}_2 x_r, -\mathcal{M}_1 x_r]\|$. $\epsilon \in R_+$ is small constant to avoid dividing by zero. The multi-scale gradient operators $\mathcal{M}_1$ and $\mathcal{M}_2$ in horzional and vertical coordinates are defined as,

$$\mathcal{M}_1 x_i := \frac{1}{J} \sum_{j=1}^J \frac{x_i(u+j, v) - x_i(u, v)}{\sqrt{j}}; \quad \mathcal{M}_2 x_i := \frac{1}{J} \sum_{j=1}^J \frac{x_i(u, v+j) - x_i(u, v)}{\sqrt{j}}$$

where J is the scale number. u and v are the row and column indexes of image matrix, respectively. When J is small, the gradient responses on weak edges are small by the reason that the values of adjacent pixels on weak edges are close, which will fail to capture the orientation information of fine details; When J

is large, the orientation information on weak edges can be effectively captured by averaging gradient responses based on different difference-intervals. The captured orientation information on both sharp and weak edges of the reference frame could further help recover the target frames.

Then, considering to avoid the trivial solution, a Euclidean-distance constraint is added and the total objective function of $\mathcal{D}_{\text{local}}(\eta; z^t, x_r)$ can be expressed as,

$$
\mathcal{D}_{\text{local}}(\eta; z^t, x_r) \triangleq \arg\min_{x_i} \sum_{i=1}^{T} \| x_i - z_i^t \|_2^2 / 2\eta \\
+ \| < \mathcal{G}^{tangent} x_i, \mathcal{G}^{normal} x_r > \|_{w,2}^2
\tag{6}
$$

where $\eta \in R_+$ is the parameter controlling the denoising strength. The nonlinear conjugate gradient (NLCG) [25] optimization is used to solving the problem (6).

Cube Matching and 4D Filtering (CM4D). The CM4D [26] is developed by nonlocal structure-similarity to extend BM3D [21] to Dcc-MRI and is done by two cascading denoising-stages concluding hard-thresholding (HT) and wiener filtering (WF). First, HT based denoising-stage can be expressed as,

$$
\text{CM4D}_{\text{HT}} := Aggregation \left(\mathcal{T}_{4D}^{-1} \left(\mathcal{H}_\gamma \left(\mathcal{T}_{4D} \left(G_{z^t} \right) \right) \right) \right)
\tag{7}
$$

where G_{z^t} is a 4D array that groups the 3D similar shapes from 3D images frames z^t by nonlocal cube matching based on similarity metric of the corresponding intensities of the two input cubes. $\mathcal{H}_\gamma$ is hard-thresholding operator. $\mathcal{T}_{4D}(\cdot)$ and $\mathcal{T}_{4D}^{-1}(\cdot)$ are the forward and inverse 4D transform, which are separately applied to every dimension of the group.

Then, the WF is completed to restore more image details and is expressed as,

$$
\text{CM4D}_{\text{WF}} := Aggregation \left(\mathcal{T}_{4D}^{-1} \left(\mathcal{W} \cdot \left(\mathcal{T}_{4D} \left(G_{\text{CM4D}_{\text{HT}}} \right) \right) \right) \right)
\tag{8}
$$

with

$$
\mathcal{W} = \frac{\left| \mathcal{T}_{4D} \left(G_{\text{CM4D}_{\text{HT}}} \right) \right|^2}{\left| \mathcal{T}_{4D} \left(G_{\text{CM4D}_{\text{HT}}} \right) \right|^2 + \delta^2}
\tag{9}
$$

where δ denotes the standard deviation of the noise.

Finally, a complex-valued CM4D based $\mathcal{D}_{\text{nonlocal}}(\gamma; z^t)$ can be defined as,

$$
\mathcal{D}_{\text{nonlocal}}(\gamma; z^t) \triangleq \text{CM4D}(\gamma; re(z^t)) + i * \text{CM4D}(\gamma; im(z^t))
\tag{10}
$$

where $re(z^t) \in \mathbb{R}^N$ and $im(z^t) \in \mathbb{R}^N$ are the real part and imaginary part of z^t.

3   Numerical Results

3.1   Experiment Setting

The two groups of in-vivo breath-hold cardiac complex-valued datasets (Fig. 1) are used to evaluate reconstruction performance of SEABUS-FCSA, which have

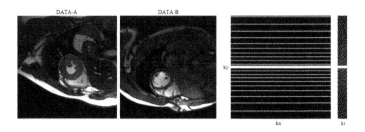

Fig. 1. Cardiac datasets (DATA-A and DATA-B) and Sampling Pattern.

respectively dimensions of $192 \times 192 \times 20 \times 18$ and $256 \times 264 \times 20 \times 22$ with the first two dimensions representing the number of pixels and the last two dimensions representing the number of coils and frames and both are acquired by a 3T scanner (SIEMENS MAGNETOM Trio) with a balanced steady-state free precession (bSSFP) sequence. In all experiments, variable density incoherent spatio-temporal acquisition (VISTA) [27] with full sampling in central region is adopted to retrospectively generate sampled k-space data. Three state-of-the-art dynamic MRI methods including the L+S [8], BCS [12] and dTV-FISTA were conducted for comparison. For dTV-FISTA, an average frame from all sampled frames is used as the fixed reference frame, which regards dTV as a denoiser and coupled into P^3-FISTA. The relative l_2-norm error (RLNE) and high frequency normalized error norm (HFEN) are used to quantitatively evaluate reconstruction performance of all methods. Specifically, the RLNE and HFEN are defined as,

$$ \text{RLNE} = \frac{\|\hat{x} - x\|_2}{\|x\|_2}; \qquad \text{HFEN} = \frac{1}{NT} \sum_{i=1}^{T} \sum_{j=1}^{N} \frac{\|\text{LoG}(\hat{x}_{i,j}) - \text{LoG}(x_{i,j})\|_2^2}{\|\text{LoG}(x_{i,j})\|_2^2} $$

The capped characters imply reconstructed objects. The LoG is a Laplacian of Gaussian filter that capture edges. In Gaussian filtering setting, the kernel size is 15×15 pixels, with a standard deviation of 1.5 pixels. In addition, the model parameters in all methods were tuned for fair comparison.

The central regions with size of 64×128 in k-space of the reference frame is used to estimate coil sensitivities. In our proposed method, the numbers of iteration for both DATA-A and DATA-B experiments are set to 60. The weighting parameter α is fixed as 0.3. All reconstructions were performed in Matlab R2019b on a standard laptop (Windows 10, 64 bit operation system, Intel(R) Core(TM) i7-9700 CPU, 3 GHz, 32 GB RAMS).

3.2   Reconstruction Performance

The Fig. 2 shows the reconstruction comparison on DATA-A (Top, R = 8) and DATA-B (Bottom, R = 10). Observed from the ROI images and error maps of $7th$ (Systole) frame and $18th$ (Diastole) frame on DATA-A, it is clearly visible

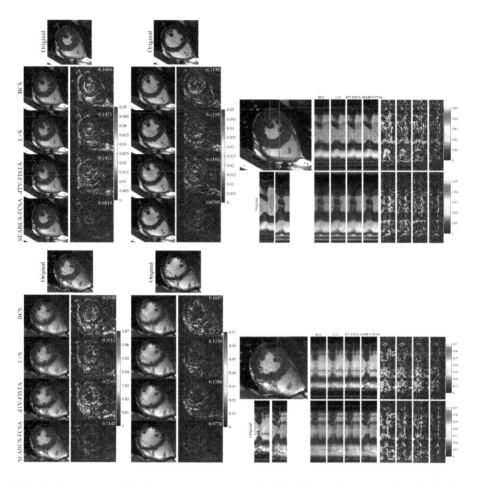

Fig. 2. Reconstructed ROI and error maps on DATA-A (Top-left, R = 8) and DATA-B (Bottom-left, R = 10). The values on the top-left corners of error maps are the RLNEs. The right-part of figure is the corresponding temporal profiles in x-t plane on fixed regions for DATA-A (Top-right) and DATA-B (Bottom-right). (Color figure online)

that lots of artifacts can be seen on the images reconstructed by BCS, L+S and dTV-FISTA, especially in terms of systolic-frame reconstructions. Our SEABUS-FCSA method provides the best balance in terms of suppressing artifacts and preserving details for both systolic and diastolic frames. Furthermore, for two fixed regions labeled by red solid lines, the reconstructions of all methods in x-t plane are shown (Top-right) and the all comparison methods except SEABUS-FCSA smooth out or distort the dynamic features. The experiment on DATA-B is used to further demonstrate the efficiency of the SEABUS-FCSA method. Specifically, it can be observed from ROI images and error maps of $6th$ (Systole) frame and $20th$ (Diastole) frame that the BCS method provides the worst

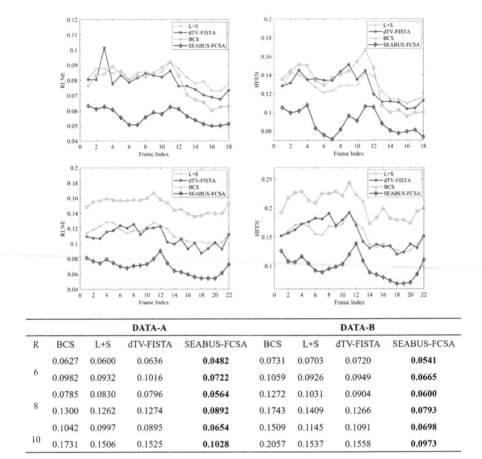

		DATA-A				DATA-B	
R	BCS	L+S	dTV-FISTA	SEABUS-FCSA	BCS	L+S	dTV-FISTA	SEABUS-FCSA
	0.0627	0.0600	0.0636	**0.0482**	0.0731	0.0703	0.0720	**0.0541**
6	0.0982	0.0932	0.1016	**0.0722**	0.1059	0.0926	0.0949	**0.0665**
	0.0785	0.0830	0.0796	**0.0564**	0.1272	0.1031	0.0904	**0.0600**
8	0.1300	0.1262	0.1274	**0.0892**	0.1743	0.1409	0.1266	**0.0793**
	0.1042	0.0997	0.0895	**0.0654**	0.1509	0.1145	0.1091	**0.0698**
10	0.1731	0.1506	0.1525	**0.1028**	0.2057	0.1537	0.1558	**0.0973**

Fig. 3. The quantitative comparisons of the all competing methods on DATA-A and DATA-B. Top: The RLNE and HFEN curves; Bottom: The data table of average RLNE and average HFEN with various reduction factors (R = 6, 8, 10).

reconstruction. The L+S and BCS methods return images that still contain some unacceptable aliasing and blurs. By contrast, the proposed SEABUS-FCSA method provides the images closer to original images than other comparison methods. The corresponding x-t profiles (Bottom-right) of two fixed regions also show that the proposed method outperforms other methods in terms of capturing the temporal variation. Figure 3 gives the RLNE and HFEN curves of all frames reconstructed by all methods. For the DATA-A (Top, R = 8), the proposed SEABUS-FCSA keeps the superiority to other methods especially in terms of heart systole frames (4th frame to 12th frame). For the DATA-B (Bottom, R = 10), the proposed SEABUS-FCSA method still has lower RLNE and HFEN values than other methods. In addition, the Fig. 3 summarizes the average RLNE and average HFEN values versus varying reduction factors (R = 6, 8, 10) on the

DATA-A and DATA-B datasets. We note that the proposed SEABUS-FCSA method evidently outperforms the L+S, dTV-FISTA and BCS methods.

The main limitation of the SEABUS-FCSA method is that the running speed is slower than other methods except BCS, which is because that the nonlocal cube-matching in CM4D filtering is time-consuming. There should be a trade-off between the algorithm speed and reconstruction quality for our proposed method. In practical, the speed can be further improved by other efficient implementations such as C++ and CUDA.

4   Conclusion

We proposed a new accelerated Dcc-MRI reconstruction approach based on hybrid P^3 exploiting both local P^3 and nonlocal P^3. Experimental results have shown that the proposed SEABUS-FCSA approach exhibits prominent performance over the state-of-the-art methods. Our method can be easily extended to other MRI applications such as multi-contrast imaging.

References

1. Donoho, D.L.: Compressed sensing. IEEE Trans. Inf. Theory **52**(4), 1289–1306 (2006)
2. Candes, E.J., Romberg, J., Tao, T.: Robust uncertainty principles: exact signal reconstruction from highly incomplete frequency information. IEEE Trans. Inf. Theory **52**(2), 489–509 (2006)
3. Wright, J., Ganesh, A., Rao, S., Ma, Y.: Robust principal component analysis: exact recovery of corrupted low-rank matrices (2009). https://arxiv.org/abs/0905.0233
4. Candès, E., Recht, B.: Exact matrix completion via convex optimization. Found. Comput. Math. **9**(6), 717 (2009)
5. Angshul, M., Ward, R.K., Tyseer, A.: Compressed sensing based real-time dynamic MRI reconstruction. IEEE Trans. Med. Imaging **31**(12), 2253–2266 (2012)
6. Chen, C., Li, Y.Q., Axel, L., Huang, J.Z.: Real time dynamic MRI by exploiting spatial and temporal sparsity. Magn. Reson. Imaging **34**(4), 473–482 (2016)
7. Fan, L.A., Dl, A., Xj, A., Wq, A., Qi, X.B., Bs, A.: Dynamic cardiac MRI reconstruction using motion aligned locally low rank tensor (mallrt) - sciencedirect. Magn. Reson. Imaging **66**, 104–115 (2020)
8. Otazo, R., Candès, E., Sodickson, D.K.: "Low" rank plus sparse matrix decomposition for accelerated dynamic MRI with separation of background and dynamic components. In: Magnetic Resonance in Medicine, vol. 73, no. 3, pp. 1125–1136 (2015)
9. Miao, X., et al.: Accelerated cardiac cine MRI using locally low rank and finite difference constraints. Magn. Reson. Imaging **34**(6), 707–714 (2016)
10. Ravishankar, S., Moore, B.E., Nadakuditi, R.R., Fessler, J.A.: Low-rank and adaptive sparse signal (LASSI) models for highly accelerated dynamic imaging. IEEE Trans. Med. Imaging **36**(5), 1116–1128 (2017)
11. Liu, Y., Liu, T., Liu, J., Zhu, C.: Smooth robust tensor principal component analysis for compressed sensing of dynamic MRI. Pattern Recogn. **102**(6), 107252 (2020)

12. Lingala, S.G., Jacob, M.: Blind compressive sensing dynamic MRI. IEEE Trans. Med. Imaging **32**(6), 1132–1145 (2013)
13. Shetty, G.N., Slavakis, K., Bose, A., Nakarmi, U., Scutari, G.: Bi-linear modeling of data manifolds for dynamic-MRI recovery. IEEE Trans. Med. Imaging **39**(3), 688–702 (2019)
14. Venkatakrishnan, S.V., Bouman, C.A., Wohlberg, B.: Plug-and-play priors for model based reconstruction. In: IEEE Global Conference on Signal and Information Processing, pp. 945–948 (2013)
15. Kamilov, U.S., Mansour, H., Wohlberg, B.: A plug-and-play priors approach for solving nonlinear imaging inverse problems. IEEE Signal Process. Lett. **24**(12), 1872–1876 (2017)
16. Eksioglu, E.M., Tanc, A.K.: Denoising AMP for MRI reconstruction: BM3D-AMP-MRI. SIAM J. Imaging Sci. **11**(3), 2090–2109 (2018)
17. Beck, A., Teboulle, M.: A fast iterative shrinkage-thresholding algorithm for linear inverse problems. SIAM J. Imaging Sci. **2**(1), 183–202 (2009)
18. Song, C., Yoon, S., Pavlovic, V.: Fast admm algorithm for distributed optimization with adaptive penalty. arXiv preprint arXiv:1506.08928 (2015)
19. Salim, A., Korba, A. , Luise, G.: The wasserstein proximal gradient algorithm (2020). https://arxiv.org/abs/2002.03035v3
20. Coll, B.: A review of image denoising algorithms, with a new one. Multiscale Model Simul. **4**(2), 490–530 (2005)
21. Dabov, K., Foi, A., Katkovnik, V., Karen, E.: Image denoising by sparse 3-D transform-domain collaborative filtering. IEEE Trans. Image Process. **16**(8), 2080–2095 (2007)
22. Huang, J., Zhang, S., Li, H., Metaxas, D.: Composite splitting algorithms for convex optimization. Comput. Vis. Image Underst. **115**(12), 1610–1622 (2011)
23. Ahmad, R., Bouman, C.A., Buzzard, G.T., Chan, S., Schniter, P.: Plug-and-play methods for magnetic resonance imaging: using denoisers for image recovery. IEEE Signal Process. Mag. **37**(1), 105–116 (2020)
24. Zhu, Q.Y., Wang, W., Cheng, J., Peng, X.: Incorporating reference guided priors into calibrationless parallel imaging reconstruction. Magn. Reson. Imaging **57**, 347–358 (2019)
25. Lustig, M., Donoho, D.L., Pauly, J.M.: Sparse MRI: the application of compressed sensing for rapid MR imaging. Magn. Reson. Med. **58**(6), 1182–1195 (2007)
26. Maggioni, M., Katkovnik, V., Egiazarian, K., Foi, A.: Nonlocal transform-domain filter for volumetric data denoising and reconstruction. IEEE Trans. Image Process. **22**(1), 119–133 (2013)
27. Ahmad, R., Hui, X., Giri, S., Yu, D., Simonetti, O.P.: Variable density incoherent spatiotemporal acquisition (VISTA) for highly accelerated cardiac MRI. Magn. Reson. Med. **74**(5), 1266–1278 (2015)

Window-Level Is a Strong Denoising Surrogate

Ayaan Haque[1,2(✉)], Adam Wang[2], and Abdullah-Al-Zubaer Imran[2]

[1] Saratoga High School, Saratoga, CA, USA
[2] Stanford University Department of Radiology, Stanford, CA, USA

Abstract. CT image quality is heavily reliant on radiation dose, which causes a trade-off between radiation dose and image quality that affects the subsequent image-based diagnostic performance. However, high radiation can be harmful to both patients and operators. Several (deep learning-based) approaches have been attempted to denoise low dose images. However, those approaches require access to large training sets, specifically the full dose CT images for reference, which can often be difficult to obtain. Self-supervised learning is an emerging alternative for lowering the reference data requirement facilitating unsupervised learning. Currently available self-supervised CT denoising works are either dependent on foreign domains or pretexts that are not very task-relevant. To tackle the aforementioned challenges, we propose a novel self-supervised learning approach, namely Self-Supervised Window-Leveling for Image DeNoising (SSWL-IDN), leveraging an innovative, task-relevant, simple, yet effective surrogate—prediction of the window-leveled equivalent. SSWL-IDN leverages residual learning and a hybrid loss combining perceptual loss and MSE, all incorporated in a VAE framework. Our extensive (in- and cross-domain) experimentation demonstrates the effectiveness of SSWL-IDN in aggressive denoising of CT (abdomen and chest) images acquired at 5% dose level only (Code available at https://github.com/ayaanzhaque/SSWL-IDN).

Keywords: Computed tomography · Image denoising · Self-supervised learning · Window-leveling · VAEs

1 Introduction

Computed Tomography (CT) imaging is one of the fundamental imaging modalities in medical practice. However, X-ray radiation is a clinical concern, as high radiation can be harmful to patients [3]. Low dose CT (LDCT) images could be acquired to reduce radiation dose as an alternative to full dose CT (FDCT). However, lowering radiation dose introduces higher noise and various imaging artifacts, resulting in degraded diagnostic and other image-based performance. To address this tradeoff, deep learning-based *denoising* methods have been investigated to improve and enhance CT imaging. Conventionally, denoising models map noisy LDCT (input) to cleaner FDCT (target). CT denoising is a popular

© Springer Nature Switzerland AG 2021
C. Lian et al. (Eds.): MLMI 2021, LNCS 12966, pp. 457–466, 2021.
https://doi.org/10.1007/978-3-030-87589-3_47

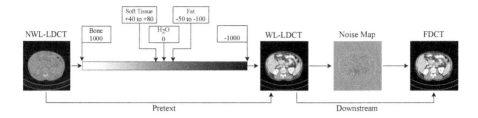

Fig. 1. Window-Leveling is the process of using CT numbers to adjust the contrast and brightness of the image. This image modification as a pretext learns important representations of the data, improving downstream denoising of predicting FDCT from LDCT by removing noise.

field of research because of its clinical importance, as being able to denoise, and thus use, low dose CT provides improved patient safety and diagnostic performance. Approaches include new architectures [1,5,7,21] and training procedures [6,22].

Acquiring reference images is challenging due to the harmful nature of radiation as well as the difficulty of performing two identical scans at different radiation doses. Thus, it is desirable to train denoising models with limited reference data. Self-Supervised Learning (SSL) has emerged as a promising alternative to fully-supervised learning in order to utilize large unlabeled training examples. In an SSL scheme, synthetic labels can be generated from the data itself, for both labeled and unlabeled data. Similar to transfer learning, SSL pre-trains a model on a surrogate task, but on the same dataset instead of one from a foreign domain, and then fine-tunes the pretrained model on a downstream, or main evaluation, task [17]. This SSL is not to be confused with other methods that are also called self-supervised which use no reference scans. Common surrogates available in literature include rotation prediction, colorization/restoration, and patch prediction. In general-purpose SSL denoising, popular works include [14,16,19,20]. In SSL CT denoising, popular works include [13,18,24]. However, these specific methods do not use any reference scans or have a downstream task, meaning they are more unsupervised than self-supervised. Thus, as argued by these papers, a method using FDCT references, like ours, is not comparable.

Variational autoencoders (VAEs) [12] are an extension of autoencoders (AEs), which use encoders and decoders to deconstruct inputs to low-dimensional representations and then reconstruct it. VAEs are generative as they use the *reparameterization trick* to inject randomized noise into the latent code. For denoising, VAEs have not been extensively used [10,25]. Additionally, residual learning [9] has gained interest in deep learning, so Residual (ResNet) VAEs have been proposed for image generation [11]. In medical image denoising, to our best knowledge, there is little literature using VAEs [2], and the use of RVAEs are even more scarce. Additionally, recent works support the use of Perceptual Loss because it optimizes high-level feature learning [1,21] as opposed to Mean-Squared Error (MSE) which optimizes on a pixel-wise scale for precise noise removal. Only a few CT denoising methods have used hybrid losses [8,15].

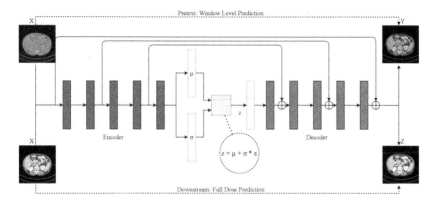

Fig. 2. Schematic of the proposed SSWL-IDN model. As a pre-text, the model predicts window-leveled images from non-window-leveled images. The model architecture uses residuals between the encoder and decoder with a VAE bottleneck.

In this paper, we use SSL to improve performance of deep denoising models with limited reference FDCT. We propose a novel denoising surrogate to predict *window-leveled* CT images from non-window-leveled images. Window-leveling in CT is the process of modifying the grayscale of an image, using the CT numbers, to highlight, brighten, and contrast important structures. Unlike many other existing self-supervised learning methods, our proposed self-supervised window-leveling (SSWL) is a task-relevant surrogate, as it is directly related to the downstream task, prioritizing similar feature learning. Furthermore, we limit all our experiments to 5% dose level potentially towards an aggressive dose reduction mechanism to demonstrate effectiveness even at such low dose settings.

Our primary contributions can be summarized as follows:

- A novel and task-relevant self-supervised window-level prediction CT denoising surrogate which is related to the downstream task.
- An innovative residual-based VAE architecture coupled with a hybrid loss function to simultaneously penalize the model pixel-wise and perceptually.
- Extensive experimentation with varied quantities of labeled data on different proposed components on in- and cross-domain data demonstrating improved and effective denoising even from extremely low dose (5%) CT images.

2   Methods

2.1   Denoising

To formulate the problem, we assume unknown data distribution $p(X, Y)$ over LDCT X and FDCT Y. We also assume access to labeled training set $\mathcal{D}_l$ sampled i.i.d. from $p(X, Y)$ and unlabeled training set $\mathcal{D}_u$ sampled i.i.d. from $p(X)$ after marginalizing out Y. In CT denoising, the input images are LDCT and the reference images are FDCT. This relationship can be represented by the equation

$$X = Y + n \tag{1}$$

where n is the resultant noise due to lowering dose. The deep denoising model is trained to remove n by encoding the input LDCT and recovering the FDCT.

A similar relationship can be found in the CT window-leveling task—non-window-leveled (NWL) scans as inputs and window-leveled (WL) scans for references. This relationship can be similarly represented as,

$$Z = aX + b, \tag{2}$$

where X is NWL, Z is WL, and a and b are window-leveling parameters (determined by the DICOM metadata). Figure 1 illustrates the process of window-leveling and its relation to denoising. Inspired from the relatedness of the two tasks, we leverage the first task to help learn or improve the second. The window-leveling labels enable us to train a deep denoising model as if it is fully-supervised. Specifically, formulating it as a pretext to the downstream denoising task is more appropriate when obtaining full dose reference images is difficult. Window-leveling is similar to denoising as both tasks from a computational view are image modifications. Since the task is domain-specific, it allows for more relevant feature learning than foreign or arbitrary surrogates.

Therefore, our proposed self-supervised learning training comprises of two steps: fully-supervised pre-training on the window-leveling task followed by fine-tuning on the small labeled denoising task. For pre-training, we prepare both a NWL and WL version of each LDCT scan for both labeled and unlabeled data. Loss is optimized for predicting the WL LDCT from input NWL LDCT. Our surrogate is end-to-end as opposed to many other methods which do not use related tasks, as no architectural or loss changes are required between tasks. After pre-training, we fine-tune the pre-trained network on the standard denoising task only with the limited pairs of LDCT and FDCT without freezing parameters.

2.2   Model Architecture and Training

For the model architecture, illustrated in Fig. 2 along with the SSL algorithm, we propose a Residual Variational Autoencoder (RVAE), which is a combination of [5] and [12]. While Residual-based VAEs have been proposed [11], they use residuals in the encoder and decoder separately (ResNet as encoder and Transposed ResNet as decoder) instead of using residual connections *between* the encoder and decoder like [5].

We use the base architecture of [5] and add a bottleneck component with Global Average Pooling and Linear Layers. This downsamples the input to a latent representation, where we use the reparameterization trick on the latent code, z. This improves FD predictions, as adding randomized noise ϵ, which is tunable through learnable parameters μ (mean) and σ (standard deviation), in the bottleneck can decrease overfitting, improve generalization, and act as a regularizer, which is experimentally verified through cross-domain evaluation. To reparameterize, we use the calculation $z = \mu + \sigma * \epsilon$. A generative model can allow for better FD predictions, as the noise in LDCT may hide important details and features which can be more easily recovered through generative models, as denoising tends to oversmooth and remove subtle information. As opposed to traditional VAEs, we use constant convolutional filters of 96 instead of downsampling convolutional layers. We use residuals from feature maps in the encoder and add them after corresponding layers in the decoder phase. We use standard convolutional layers in the encoder and transpose convolutional layers in the decoder.

For our loss, we use a hybrid loss combining MSE and perceptual loss. Perceptual loss encourages high-level visual feature learning/matching, while MSE optimizes precise, pixel-wise noise removal. Thus, we combine both into one loss to obtain both benefits. We use [27]'s perceptual loss where the model prediction and target are passed to an Image-Net pre-trained VGG-19 and the extracted features from hidden convolutional layers are used to calculate perceptual distance between the features. The perceptual loss can be defined as

$$\mathcal{L}_{perceptual} = \frac{1}{M} \sum_{i=1}^{M} \|\mathcal{F}(\hat{y}_i) - \mathcal{F}(y_i)\|^2, \qquad (3)$$

where M is the mini-batch size, $\hat{y}$ are the model predictions, y are the labels, and $\mathcal{F}$ is a feature extractor.

Our final loss function can be represented by

$$L(y, \hat{y}, \mu, \sigma) = L_{MSE}(\hat{y}, y) + \beta \mathcal{L}_{perceptual}(\mathcal{F}(\hat{y}), \mathcal{F}(y)) + \alpha \mathcal{L}_{KL}(\mu, \sigma), \qquad (4)$$

where L_{MSE} is standard MSE loss, the $\mathcal{L}_{perceptual}$ is perceptual loss, and β is $\mathcal{L}_{perceptual}$ weight. For the VAE, $\mathcal{L}_{KL}$ represents the KL divergence loss, and α is the $\mathcal{L}_{KL}$ weight. μ is the mean term, and σ is the standard deviation term, both from the latent space. The KL divergence attempts to reduce divergence of μ and σ in the training distribution from those of the target distribution. Both the surrogate and downstream task are trained with the same loss and architecture.

3 Empirical Evaluation

3.1 Data

We primarily collect abdomen scans from the publicly available Mayo CT data [4,23]. The dataset includes CT scans originally acquired at routine dose level (full dose), so simulated quarter dose images are reconstructed through inserting Poisson noise into each projection dataset. For thorough denoising evaluation, we generate the CT scans at 5% dose level using the full dose and quarter dose data (scaling the zero-mean independent noise from 25% to 5% dose level). While from a clinical perspective it is more ideal to have a well-denoised quarter dose as opposed to a lower quality denoised 5% dose, from a computational perspective, showing the ability to remove high volumes of noise can more appropriately evaluate the model's full potential to accurately remove noise. We use 15 full dose abdomen CT and the corresponding quarter (25%) dose CT scans: 10 scans (1,533 slices, 15% for validation) for training and 5 (633 slices) for testing. 5

chest scans (1,061 slices) are selected from the same library for cross-domain evaluation.[1]

3.2   Implementation Details

Baselines: For architectural baselines, we used a VAE [12], DnCNN [26], and RED-CNN [5]. For SSL surrogate task baselines, we use simple reconstruction (Rec), a recent general method Noisy-as-Clean (NAC) [20] as it uses a surrogate to downstream training like us, and Noise2Void [13], even though it is a no reference method, meaning a comparison is not fitting. **Training:** Models were trained at varying levels of supervision in terms of the number of labeled data (250, 500, 1000, full). Each experiment was repeated 5 times and mean scores were reported. All inputs were normalized and resized to $256 \times 256 \times 1$. **Hyperparameters:** The code was written in Python and PyTorch (newest versions) and trained on an NVIDIA K80 GPU with 12 GB RAM. We used the Adam optimizer with learning rates of 1e−5, momentum of 0.1 per 8 epochs, and a minibatch size 10. For loss, $\beta = 0.6$ and $\alpha = 1.0$ (as per VAEs). Based on tuning experiments, choice of hyperparameters do not noticeably affect performance. All parameters are identical for pre-text and downstream training. **Evaluation:** For evaluation, we used Peak Signal-to-Noise Ratio (PSNR), Structural Similarity Index Measure (SSIM), Mean-Squared Error (MSE), and Normalized RMSE (NRMSE). To evaluate statistical significance, we performed two-sample t-tests.

3.3   Results and Discussion

Table 1. Comparison of our RVAE architecture against other conventional and SoTA architectures on in- and cross-domain evaluation. The best fully-supervised scores are bolded, and best semi-supervised scores are underlined.

Model	$|\mathcal{D}_l|$	In-Domain						Cross-Domain					
		LDCT	U-Net	VAE	DnCNN	RED-CNN	RVAE	LDCT	U-Net	VAE	DnCNN	RED-CNN	RVAE
PSNR	250	20.179	17.325	17.278	19.814	23.414	24.829	16.876	16.434	16.318	16.861	18.625	19.033
	500	—	18.445	19.241	20.430	23.612	26.139	—	17.385	17.329	17.192	18.919	19.193
	1000	—	20.930	20.771	21.693	24.097	<u>26.286</u>	—	18.458	18.094	18.702	19.005	<u>19.259</u>
	Full	—	21.900	21.547	21.186	24.115	**26.574**	—	18.843	18.364	18.925	19.005	**19.288**
SSIM	250	0.7554	0.5685	0.4656	0.7031	0.8433	0.8483	0.6725	0.4781	0.3741	0.6132	0.6971	0.7438
	500	—	0.6606	0.5849	0.7372	0.8522	0.8538	—	0.6094	0.4859	0.6554	0.7264	0.7467
	1000	—	0.7697	0.6873	0.7460	0.8538	<u>0.8626</u>	—	0.6883	0.5634	0.6343	0.7327	<u>0.7497</u>
	Full	—	0.7918	0.7510	0.7907	0.8616	**0.8646**	—	0.7095	0.5971	0.6584	0.7467	**0.7490**
MSE	250	0.0107	0.0201	0.0192	0.0112	0.0054	0.0053	0.0207	0.0234	0.0238	0.0207	0.0139	0.0131
	500	—	0.0144	0.0121	0.0101	0.0051	0.0050	—	0.0187	0.0188	0.0202	0.0132	0.0127
	1000	—	0.0083	0.0086	0.0071	0.0052	<u>0.0046</u>	—	0.0147	0.0148	0.0157	0.0129	<u>0.0125</u>
	Full	—	0.0067	0.0073	0.0088	0.0047	**0.0045**	—	0.0135	0.0145	0.0136	0.0124	**0.0122**
NRMSE	250	0.3082	0.4204	0.4173	0.3171	0.2155	0.1889	0.4832	0.5186	0.5245	0.4777	0.4319	0.3861
	500	—	0.3629	0.3324	0.2981	0.2078	0.1889	—	0.4647	0.4671	0.4740	0.4019	0.3789
	1000	—	0.2746	0.2802	0.2533	0.1990	<u>0.1878</u>	—	0.4110	0.4266	0.4217	0.3819	<u>0.3763</u>
	Full	—	0.2470	0.2573	0.2749	0.1979	**0.1817**	—	0.3933	0.4091	0.3662	0.3745	**0.3690**

---

[1] For chest scans, the 5% dose level is simulated from routine and 10% dose level scans available in the Mayo data library.

Table 1 shows our proposed RVAE alone is able to outperform all baselines, including two state-of-the-art denoising architectures in DnCNN [26] and RED-CNN [5], and even with minimum data, we are able to match the fully-supervised metrics of those two architectures. While DnCNN and RED-CNN are older methods, as architectures alone, they are still SoTA. Our RVAE outperforms RED-CNN with statistical significance (p-value of 0.032 for PSNR), and also significantly outperforms the other models (p < 0.05). For the cross-domain chest dataset evaluation, similar improvements are shown from RVAE compared to the baselines and the state-of-the-art, proving the generalizatibility of the RVAE.

Table 2 displays the performance of RED-CNN and our RVAE with various SSL tasks (trained on MSE). SSL Reconstruction simply uses the LDCT as the input and reference, allowing the model to become familiar with the data. For the Noisy-As-Clean surrogate task [20], we follow their implementation where their surrogate task uses LDCT with additional injected Gaussian noise as the input and the plain LDCT as the reference, and subsequently perform denoising. As shown in the tables, SSWL has significantly improved performance over no pre-training and basic reconstruction (p-value of 0.021 and 0.039 respectively). Compared to NAC, the SoTA, we still have the best performance, confirming the importance and ability of our proposed SSL task. Against N2V, a SoTA no-reference method, we greatly outperform them (p-value < 0.5), proving the effectiveness of our new method. Similar to in-domain evaluations, the cross-domain metrics further confirm the superior performance and generalization of SSWL, showing the importance of a task-relevant surrogate.

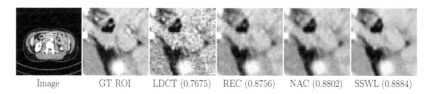

Image   GT ROI   LDCT (0.7675)   REC (0.8756)   NAC (0.8802)   SSWL (0.8884)

Fig. 3. ROI predictions with RVAE ($|\mathcal{D}_l| = 500$) and the 3 SSL tasks show higher visual noise removal and SSIM from SSWL.

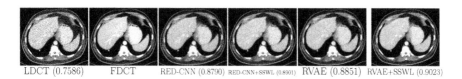

LDCT (0.7586)   FDCT   RED-CNN (0.8790) RED-CNN+SSWL (0.8901)   RVAE (0.8851)   RVAE+SSWL (0.9023)

Fig. 4. FDCT predictions from RED-CNN and RVAE ($|\mathcal{D}_l| = 500$) with and without SSWL show strong visual improvements and improved SSIM from our proposed models.

Table 2. SSWL is compared to Rec, NAC, and N2V trained on RED-CNN and RVAE architectures. SSWL is shown to be the most effective. Follows same key as Table 1.

Model	$|\mathcal{D}_l|$	In-domain				Cross-domain			
		PSNR	SSIM	MSE	NRMSE	PSNR	SSIM	MSE	NRMSE
RED-CNN + REC	250	23.414	0.8433	0.0054	0.2155	19.351	0.7551	0.0122	0.3726
	500	23.612	0.8522	0.0051	0.2078	19.430	0.7572	0.0120	0.3691
	1000	24.097	0.8538	0.0052	0.1990	19.499	0.7523	0.0117	0.3644
	Full	24.115	0.8616	0.0047	0.1979	19.546	0.7555	0.0120	0.3678
RVAE + REC	250	23.795	0.8522	0.0051	0.2120	19.417	0.7583	0.0120	0.3694
	500	24.088	0.8598	0.0050	0.2067	19.544	0.7602	0.0117	0.3640
	1000	24.071	0.8650	0.0048	0.2007	19.570	0.7598	0.0116	0.3632
	Full	24.301	0.8669	0.0047	0.1958	19.609	0.7613	0.0115	0.3614
RED-CNN + NAC	250	23.989	0.8375	0.0050	0.2050	19.122	0.7527	0.0124	0.3750
	500	24.033	0.8541	0.0048	0.2015	19.401	0.7565	0.0127	0.3703
	1000	24.219	0.8549	0.0045	0.1961	19.494	0.7577	0.0118	0.3660
	Full	24.422	0.8611	0.0045	0.1958	19.579	0.7588	0.0116	0.3625
RVAE + NAC	250	23.890	0.8612	0.0052	0.2029	19.313	0.7545	0.0123	0.3739
	500	24.168	0.8579	0.0047	0.1996	19.338	0.7532	0.0123	0.3726
	1000	24.019	0.8647	0.0047	0.1980	19.439	0.7580	0.0122	0.3686
	Full	24.186	0.8649	0.0046	0.1956	19.520	0.7596	0.0118	0.3650
RED-CNN + N2V	250	23.145	0.7899	0.008	0.2505	18.310	0.7170	0.0148	0.4098
	500	23.743	0.8000	0.0075	0.2390	18.075	0.7212	0.0157	0.4206
	1000	24.020	0.8032	0.0073	0.2342	18.169	0.7101	0.0153	0.4162
	Full	24.116	0.8083	0.0072	0.2318	18.417	0.7215	0.0145	0.4047
RVAE + N2V	250	23.928	0.8057	0.0073	0.2352	18.241	0.7177	0.0151	0.4129
	500	24.139	0.8107	0.0070	0.2297	18.307	0.7193	0.0149	0.4010
	1000	24.072	0.8111	0.0070	0.2301	18.313	0.7186	0.0148	0.4099
	Full	24.321	0.8113	0.0069	0.2269	18.174	0.7155	0.0153	0.4162
RED-CNN + SSWL	250	26.119	0.8509	0.0050	0.1890	19.550	0.7611	0.0117	0.3635
	500	26.300	0.8542	0.0050	0.1865	19.460	0.7614	0.0120	0.3679
	1000	26.710	0.8627	0.0046	0.1786	19.520	0.7617	0.0120	0.3652
	Full	26.747	0.8626	0.0045	0.1764	19.547	0.7619	0.0117	0.3642
RVAE + SSWL	250	26.150	0.8612	0.0051	0.1900	19.566	0.7632	0.0116	0.3630
	500	26.464	0.8659	0.0048	0.1820	19.619	0.7634	0.0115	0.3607
	1000	26.799	0.8669	<u>0.0043</u>	0.1781	19.549	0.7619	0.0117	<u>0.3505</u>
	Full	26.844	0.8701	0.0044	0.1774	19.617	0.7624	0.0115	0.3692
SSWL-IDN	250	26.581	0.8649	0.0046	0.1793	19.660	0.7645	0.0109	0.3556
	500	26.778	0.8723	0.0045	0.1783	19.854	0.7680	0.0107	0.3530
	1000	<u>27.018</u>	<u>0.8744</u>	<u>0.0043</u>	<u>0.1732</u>	<u>20.016</u>	<u>0.7706</u>	<u>0.0105</u>	<u>0.3505</u>
	Full	**27.800**	**0.8815**	**0.0042**	**0.1701**	**20.178**	**0.7739**	**0.0104**	**0.3458**

The improvements from our hybrid loss are shown in our final model, SSWL-IDN, in Table 2. When compared to RVAE + SSWL, which is trained on MSE, we see improved performance of up to 1 to 2 scores higher for both PSNR and SSIM. A full ablation for our hybrid loss is available in the supplemental. Figures 3 and 4 demonstrate precise removal of noise from whole scans as well as specific regions of interest (ROIs), proving the effectiveness of our model over baseline architectures and other self-supervised tasks. While certain structural details are lost, this is due to the ultra low dose, and our method recovers details best.

4   Conclusions

We present SSWL-IDN, a self-supervised denoising model with a novel, task-relevant, and efficient surrogate task of window-level prediction. We also propose a Residual-VAE specialized for denoising, as well as a hybrid loss leveraging benefits of both perceptual and pixel-wise optimization. We confirm each component of our method outperforms baselines on difficult 5% dose denoising for both in- and cross-domain evaluations, and when combined, the model significantly outperforms state-of-the-art methods. Improved denoising with limited reference data is of clinical significance to reduce harms to patients. Our future work will focus on developing cascaded and joint surrogate and downstream learning as well as 3D architectures to utilize information in the z-dimension.

References

1. Ataei, S., Alirezaie, J., Babyn, P.: Cascaded convolutional neural networks with perceptual loss for low dose CT denoising. In: 2020 International Joint Conference on Neural Networks (IJCNN), pp. 1–5. IEEE (2020)
2. Biswas, B., Ghosh, S.K., Ghosh, A.: DVAE: deep variational auto-encoders for denoising retinal fundus image. In: Bhattacharyya, S., Konar, D., Platos, J., Kar, C., Sharma, K. (eds.) Hybrid Machine Intelligence for Medical Image Analysis. SCI, vol. 841, pp. 257–273. Springer, Singapore (2020). https://doi.org/10.1007/978-981-13-8930-6_10
3. Brenner, D.J., Hall, E.J.: Computed tomography - an increasing source of radiation exposure. N. Engl. J. Med. **357**(22), 2277–2284 (2007). https://doi.org/10.1056/NEJMra072149. pMID: 18046031
4. Chen, B., Duan, X., Yu, Z., Leng, S., Yu, L., McCollough, C.: Development and validation of an open data format for CT projection data. Med. Phys. **42**(12), 6964–6972 (2015)
5. Chen, H., et al.: Low-dose CT with a residual encoder-decoder convolutional neural network. IEEE Trans. Med. Imaging **36**(12), 2524–2535 (2017). https://doi.org/10.1109/TMI.2017.2715284
6. Chen, H., Zhang, Y., Zhang, W., Liao, P., Li, K., Zhou, J., Wang, G.: Low-dose CT denoising with convolutional neural network. In: 2017 IEEE 14th International Symposium on Biomedical Imaging (ISBI 2017), pp. 143–146. IEEE (2017b)
7. Diwakar, M., Kumar, M.: A review on CT image noise and its denoising. Biomed. Signal Process. Control **42**, 73–88 (2018)
8. Gholizadeh-Ansari, M., Alirezaie, J., Babyn, P.: Deep learning for low-dose CT denoising using perceptual loss and edge detection layer. J. Digit. Imaging, 1–12 (2019)
9. He, K., Zhang, X., Ren, S., Sun, J.: Deep residual learning for image recognition. In: Proceedings of the IEEE Conference on Computer Vision and Pattern Recognition, pp. 770–778 (2016)

10. Im Im, D., Ahn, S., Memisevic, R., Bengio, Y.: Denoising criterion for variational auto-encoding framework. In: Proceedings of the AAAI Conference on Artificial Intelligence, vol. 31 (2017)
11. Kingma, D.P., Salimans, T., Jozefowicz, R., Chen, X., Sutskever, I., Welling, M.: Improving variational inference with inverse autoregressive flow. arXiv preprint arXiv:1606.04934 (2016)
12. Kingma, D.P., Welling, M.: Auto-encoding variational bayes. arXiv preprint arXiv:1312.6114 (2013)
13. Krull, A., Buchholz, T.O., Jug, F.: Noise2Void-learning denoising from single noisy images. In: Proceedings of the IEEE/CVF Conference on Computer Vision and Pattern Recognition, pp. 2129–2137 (2019)
14. Laine, S., Karras, T., Lehtinen, J., Aila, T.: High-quality self-supervised deep image denoising. In: Advances in Neural Information Processing Systems, vol. 32, pp. 6970–6980 (2019)
15. Ma, Y., Wei, B., Feng, P., He, P., Guo, X., Wang, G.: Low-dose CT image denoising using a generative adversarial network with a hybrid loss function for noise learning. IEEE Access **8**, 67519–67529 (2020)
16. Quan, Y., Chen, M., Pang, T., Ji, H.: Self2Self with dropout: learning self-supervised denoising from single image. In: Proceedings of the IEEE/CVF Conference on Computer Vision and Pattern Recognition (CVPR), June 2020
17. de Sa, V.R.: Learning classification with unlabeled data. In: Advances in Neural Information Processing Systems, pp. 112–119. Citeseer (1994)
18. Wu, D., Ren, H., Li, Q.: Self-supervised dynamic CT perfusion image denoising with deep neural networks. IEEE Trans. Radiat. Plasma Med. Sci. (2020)
19. Xie, Y., Wang, Z., Ji, S.: Noise2Same: optimizing a self-supervised bound for image denoising. In: Advances in Neural Information Processing Systems, vol. 33 (2020)
20. Xu, J., et al.: Noisy-As-Clean: learning self-supervised denoising from corrupted image. IEEE Trans. Image Process. **29**, 9316–9329 (2020)
21. Yang, Q., et al.: Low-dose CT image denoising using a generative adversarial network with wasserstein distance and perceptual loss. IEEE Trans. Med. Imaging **37**(6), 1348–1357 (2018). http://dx.doi.org/10.1109/TMI.2018.2827462
22. Yi, X., Babyn, P.: Sharpness-aware low-dose CT denoising using conditional generative adversarial network. J. Digital Imaging **31**(5), 655–669 (2018)
23. Yu, L., Shiung, M., Jondal, D., McCollough, C.H.: Development and validation of a practical lower-dose-simulation tool for optimizing computed tomography scan protocols. J. Comput. Assist. Tomogr. **36**(4), 477–487 (2012)
24. Yuan, N., Zhou, J., Qi, J.: Half2Half: deep neural network based CT image denoising without independent reference data. Phys. Med. Biol. **65**(21), 215020 (2020)
25. Yue, Z., Yong, H., Zhao, Q., Zhang, L., Meng, D.: Variational denoising network: Toward blind noise modeling and removal. In: The Thirty-third Annual Conference on Neural Information Processing Systems (2019)
26. Zhang, K., Zuo, W., Chen, Y., Meng, D., Zhang, L.: Beyond a gaussian denoiser: residual learning of deep CNN for image denoising. IEEE Trans. Image Process. **26**(7), 3142–3155 (2017)
27. Zhang, R., Isola, P., Efros, A.A., Shechtman, E., Wang, O.: The unreasonable effectiveness of deep features as a perceptual metric. In: CVPR (2018)

Cardiovascular Disease Risk Improves COVID-19 Patient Outcome Prediction

Diego Machado Reyes[1], Hanqing Chao[1], Fatemeh Homayounieh[2],
Juergen Hahn[1], Mannudeep K. Kalra[2], and Pingkun Yan[1(✉)]

[1] Department of Biomedical Engineering, Rensselaer Polytechnic Institute Troy,
New York, USA
{machad,chaoh,hahnj,yanp2}@rpi.edu
[2] Department of Radiology, Massachusetts General Hospital, Boston, MA, USA
mkalra@mgh.harvard.edu

Abstract. The pandemic of coronavirus disease 2019 (COVID-19) has
severely impacted the world. Several studies suggest an increased risk
for COVID-19 patients with underlying cardiovascular diseases (CVD).
However, it is challenging to quantify such risk factors and integrate
them into patient condition evaluation. This paper presents machine
learning methods to assess CVD risk scores from chest computed tomography together with laboratory data, demographics, and deep learning
extracted lung imaging features to increase the outcome prediction accuracy for COVID-19 patients. The experimental results demonstrate an
overall increase in prediction performance when the CVD severity score
was added to the feature set. The machine learning methods obtained
their best performance when all categories of the features were used for
the patient outcome prediction. With the best attained area under the
curve of 0.888, the presented research may assist physicians in clinical
decision-making process on managing COVID-19 patients.

Keywords: COVID-19 · Machine learning · Cardiovascular disease ·
Chest CT · Severity score

1 Introduction

Coronavirus disease 2019 (COVID-19), caused by severe acute respiratory syndrome coronavirus 2 (SARS-CoV-2), has had a global impact like no other pandemic or disease in modern times. COVID-19 patients present a clinical picture
similar to the other two coronaviruses that in previous years have caused pandemic diseases (SARS-CoV, MERS-CoV) [10]. Patients initially manifest a respiratory infection that can lead to viral pneumonia, which can culminate into
acute respiratory distress syndrome (ARDS) [15,20]. Early prediction of disease

This work was partially supported by National Institute of Biomedical Imaging and
Bioengineering (NIBIB) under award R21EB028001 and National Heart, Lung, and
Blood Institute (NHLBI) under award R56HL145172.

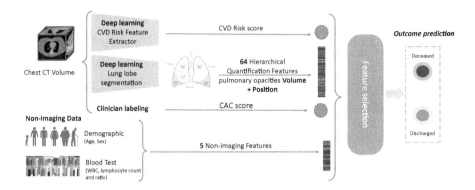

Fig. 1. The proposed framework utilizing both imaging and non-imaging features including lung severity and CVD risk for COVID-19 outcome prediction.

progression and severity assessment in at-risk patients can help determine and plan healthcare needs related to hospital admission, mechanical ventilation, and intensive care monitoring.

Several approaches have been taken to implement machine learning methods for COVID-19 patient outcome prediction [14]. For example, Chao et al. [2] developed a framework integrating various features extracted from chest CT scans with patient demographic information, vital signs, and laboratory blood exams to assess disease severity and predicting intensive care unit (ICU) admission. In Tang et al. [18], a random forest model was trained using quantitative features from the lungs to assess the severity of the disease. Said features were obtained from a deep learning-based chest CT analysis tool. Other approaches have focused on developing similar deep learning-based tools to extract relevant information from chest CT-scans and then use the obtained features for disease severity assessment. Fang et al. [4] employed a deep learning approach to segment the lung lobes and pulmonary opacities and computed severity scores from those. Afterward, random forest, SVM, and logistic regression models were trained with the severity scores to predict the patient outcome. However, the existing works mainly focus on the imaging features from the lungs. COVID-19 severity prognosis has been associated with multiple comorbidities through systematic reviews and meta-analyses and retrospective cohorts [6–8,12,17,19]. Among the comorbidities, cardiovascular disease (CVD) was found to have a risk ratio ranging from 2.25 to 3.15 [6,12,17] and a prevalence ranging from 5.8% to 25% [7,8,19]. It is thus important to take CVD risk into account when assessing patient prognosis outcomes.

In this work, we combine features of CVD analysis, chest CT image, and non-imaging data to predict the outcome of COVID-19 patients through machine learning, as shown in Fig. 1. By leveraging the correlation between CVD and COVID-19 severity, our model can more accurately separate patients with high mortality risks. The rest of this paper is organized as follows. First, the datasets

Table 1. Demographic statistics of the dataset.

Features	Deceased	Discharged
Gender (M:F)	55% : 45%	55% : 45%
Age (years)	72.9 ± 12.5	64.5 ± 18
WBC	2902 ± 3940	4148 ± 2799
Lym	341 ± 450.6	849.4 ± 676.8
Lym ratio (%)	14.6 ± 12.9	19.2 ± 9.9
CAC score (No:Mild:Moderate:Severe)	23% 38% 5% 34%	43% 28% 9% 20%
CVD score	0.511 ± 0.13	0.424 ± 0.15

employed are presented, then an exploratory analysis of the variables is introduced. Next, the feature selection and machine learning methods used for outcome prediction are discussed. Then the experimental results are presented, which show the performance improvement with including the CVD severity score. Finally, the findings and conclusions of this paper are discussed.

2 Materials and Methods

2.1 Datasets

The data employed in this work was obtained from two hospitals, Firoozgar Hospital (Tehran, Iran), and Massachusetts General Hospital (MGH) (Boston, MA, USA). These datasets were comprised of patients' demographics (age, sex), laboratory blood tests (white blood cell count, lymphocyte count, lymphocyte ratio), coronary artery calcification (CAC) score, and the outcome of the patients (discharged, deceased). Moreover, non-contrast chest CT scans without intravenous contrast injection were provided. The Firoozgar Hospital dataset included data from 113 patients, while the MGH dataset was comprised of 125 patients. Both datasets were combined into one larger dataset comprised of 238 patients. CT scans were manually inspected for lines, tubes, and imaging artifacts, the ones including such artifacts were removed from the dataset. A resulting total of 208 patients were available for outcome prediction. From the combined dataset, 108 patients were discharged and the remaining 100 were deceased. Other relevant demographic statistics of the patient population are presented in Table 1.

2.2 Outcome Prediction Methods

In order to perform the outcome prediction, both the imaging and metadata features were divided into three categories. 1) First, all the non-imaging features from demographic, vital signs, and blood examination (DVB) were placed together. These features being age, sex, white blood cell count, lymphocyte count, and lymphocyte ratio. 2) The second category consisted of 64 hierarchical lobe-wise quantification features (HLQ) calculated with the approach introduced in [2]. The five lung lobes and pulmonary opacities are first segmented by a deep neural network [5]. Then, in 8 regions of interest (ROIs) (whole lung, left & right lung, and 5 lung lobes), the volumes of pulmonary opacities grouped under 4 Hounsfield unit (HU) ranges and their ratios to the corresponding ROIs are calculated ($8 \times 4 \times 2 = 64$). 3) Furthermore, a third category was the collection of CVD severity scores that were calculated from the chest CT-scans.

In this paper, we studied two kinds of CVD scores. One is the deep learning-based CVD risk score automatically calculated based on the model introduced in [3]. The other one is the CAC score manually assigned by radiologists. A radiologist from MGH categorized all 208 CT images into 4 CAC levels, i.e., normal: no calcification, minimal: calcification less than $1/3$ of the length of coronary arterial length, moderate: calcification over $1/3$ to $2/3$ of the coronary arteries, and heavy: calcification over $2/3$ of the arterial length.

As a first step, an exploratory analysis was performed to examine the relevance of the CVD risk and CAC scores towards the differentiability of the patients. Next, the sets of non-imaging (DVB) and imaging (HLQ) features were used as inputs for machine learning models to predict the patient outcome. In further detail, experiments were run with the non-imaging features (DVB) and lung imaging features (HLQ) individually and together to establish a baseline. Afterward, the CVD or the CAC scores were added in the following experiments to check for performance improvement. Three well-known machine learning algorithms – random forest classifier, logistic regression, and support vector machine (SVM) – were used to perform the outcome prediction. Several configurations of their parameters were tested and the one with the best performance was chosen.

2.3   Exploratory Statistical Analysis

While CVDs have been widely associated with increased risk ratios and prevalence among the patients with fatal outcomes [6–8,12,17,19], a key initial step was to test if the deep learning-based CVD risk scores and the CAC scores showed significant differences between the patients with fatal outcome and the discharged patients. First, the normality of the distribution of the CVD risk scores and CAC scores of each group were tested through a Shapiro-Wilk test ($\alpha = 0.05$). Due to the non-normal distribution of the CVD scores, Mann-Whitney U tests were employed to determine significance between the CVD risk scores and CAC scores of two groups ($\alpha = 0.05$). Second, to enhance the explainability of the developed model, an exploratory factor analysis [16] was performed to examine the relevance of the features and further granularities in the data. First, the loadings for each factor were calculated to determine relations between the features present in the datasets. Second, groups of features were obtained for each factor; thus providing an improved understanding of which features share the most similar information between each other, and especially to the features of interest – the CVD risk score and the CAC scores. The factor analysis was implemented through the FactorAnalyzer python library, using a varimax rotation matrix. Bartlett's sphericity test [1] and Kaiser-Meyer-Olkin test [9] were used to test for adequacy of applying factor analysis for the datasets.

2.4   Feature Selection

As the ratio of available features to patients was not favorable, two feature reduction strategies were implemented depending on the machine learning method. For the SVM and logistic regression, a leave-one-out strategy was implemented to select the best set of features. The second feature reduction approach ranks features based on their Gini importance [11]. Because of the randomness of Random Forest, the Gini importance of each feature might vary in different runs. To alleviate such randomness, we calculate the features' Gini importance 100 times with different random seeds. Each time the ranks of each feature were recorded. The final rank was determined by sorting the total summed rank of each feature. Based on the final ranking, the model was trained and tested multiple times on the top $K \in [1,100]$ features and its performances were recorded. The combination of features that rendered the highest AUC was found. Furthermore, due to the small sample size a traditional train-test split was not ideal for independent assessment; therefore, a 5-fold cross-validation was implemented for all experiments to evaluate the methods performances differences. Finally, significance in performance increase was tested using Mann–Whitney U tests.

3   Results

3.1   Exploratory Analysis

The Mann-Whitney U tests revealed a significant difference between the CVD risk scores from the outcome groups in the Iran, MGH and combined datasets; while it revealed no significant difference between the CAC scores in any of the datasets. Therefore, for the individual datasets and the combined version of these, it can be understood that the CVD risk score is a valuable feature to differentiate patients' possible outcomes.

The factor analysis revealed granularities in the dataset with especial interest in the CVD risk and CAC scores. For the Iran data set, the factor analysis indicated that the CVD risk score – labeled as softmax_pred on the right end of the first subplot of Fig. 2 – has a strong correlation with age, and a mild correlation with features such as white blood cell count and the infection volume and ratios in lobes 2 and 3. These results are expected, as patients of older age have a higher prevalence of cardiovascular disorders [13]. The relation between the CVD risk score and the lung features could be explained in part due to the closeness of the heart and lobe 2 (middle lobe of the right lung). In the low dose CT-scans employed for the HLQ and CVD risk score features calculation parts of the lung and heart, in their respective cases, could appear in the field of view of the neural networks producing the respective quantitative measurements. On the other hand, for the MGH data set relationships were found between the CVD risk score and the infection volumes and ratios of multiple sections from the lobes 3 and 4. This can be seen on the right plot of Fig. 2. Similar to the Iran dataset factor analysis results, lobe 4 (superior lobe of the left lung) shares boundaries with the left atrium and aorta, which are major areas of interest

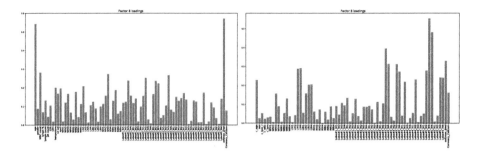

Fig. 2. Factor loadings plots for Iran (left) and MGH (right) datasets

for the CVD risk estimator network as calcification in the aorta is a key driver of this score. Furthermore, the factor loadings also showed a mild correlation between the CVD risk score and the age following a similar pattern as the Iran dataset results.

3.2   Inclusion of CVD Risk Score in Patient Classification

As mentioned before, multiple parameter configurations were tested for the implemented machine learning methods. All three machine learning methods used Scikit-Learn package implementation. For the parameters used, in the case of the SVM, the best results were accomplished when using a RBF kernel. As for the logistic regression, the best performance was achieved by using a stochastic average gradient descent solver. Furthermore, the random forest classifier obtained its best results when setting the number of estimators to 300.

Table 2 reports the results from the 5-fold cross-validation using each best-performing method and features used as input for prediction. Most of the experiments showed an increased performance, reflected on the AUC after the CVD severity score was included on the feature set. Nevertheless, after testing for significance with a Mann–Whitney U test, 5 were found to be significant, mainly improving when all features were introduced. Moreover, while the inclusion of the CAC scores resulted in increased performance in some of the experiments, mainly in the HLQ+CAC category, none of the improvements were found to be significant by the Mann–Whitney U test. The lack of significant improvement when including the CAC score as a feature concurs with the results from the exploratory analysis in which no significant differences were found in the distributions of the CAC scores between the deceased and discharged patients.

Table 2 and Fig. 3 show consistently increased performance for all the three models when the CVD severity score was added to the DVB+HLQ configuration, with the increase on the random forest and the logistic regression being validated as significant by the Mann–Whitney U test. Moreover, this feature configuration also achieves the best performances on each method, with the highest performing experiment being accomplished by the logistic regression using all 3 sets of features. Furthermore, similar increases in performance can be seen for the

Table 2. Comparison of 5-fold cross-validation results of RF, SVM and LR using various configurations of features. Sensitivity was obtained after setting specificity = 70%, results with p-value > 0.05 are marked with *.

Features	Random forest		SVM		Logistic regression	
	AUC	Sensitivity	AUC	Sensitivity	AUC	Sensitivity
DVB	0.790	0.685	0.751	0.666	0.734	0.704
DVB + CAC	0.757	0.759	0.746	0.599	0.741	0.629
DVB + CVD	0.822*	0.752	0.768*	0.676	0.743	0.685
HLQ	0.770	0.552	0.723	0.714	0.886	0.866
HLQ + CAC	0.820	0.600	0.733	0.704	0.886	0.866
HLQ + CVD	**0.862***	0.638	0.719	0.704	0.883	0.866
DVB + HLQ	0.822	0.686	0.823	**0.819**	0.845	**0.876**
DVB + HLQ + CAC	0.820	0.771	0.813	**0.819**	0.844	**0.876**
DVB + HLQ + CVD	**0.862***	**0.790**	**0.829**	0.819	**0.888***	0.829

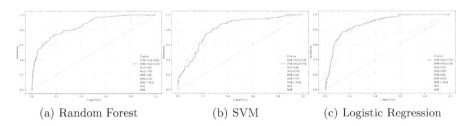

(a) Random Forest   (b) SVM   (c) Logistic Regression

Fig. 3. ROC curves for Random Forest, SVM, and Logistic regression.

DVB + CVD when compared to DVB alone. The improvements in AUC were found to be significant for the random forest and the SVM models. On the other hand, while SVM and logistic regression struggled to achieve an increased performance when the CVD severity score was added to the HLQ features, the random forest classifier was able to achieve this goal. Additionally, the random forest model was able to achieve significant performance increases when the CVD severity score was included in all 3 feature combinations (DVB+CVD, HLQ+CVD, DVB+HLQ+CVD); this could be in part due to the feature selection strategy and the robustness of the random forest due to its ensemble method design. However, it is interesting to notice that the best performance of all methods and feature combinations (0.888) was achieved by the logistic regression model when the CVD severity score was included in the DVB+HLQ configuration. Emphasizing the relevance of including the CVD severity score, as this best performing method had a significant increase in performance compared to the classification based only on the DVB and HLQ features.

4   Discussion and Conclusion

Although the majority of patients with COVID-19 infection are either asymptomatic or present minor symptoms, most healthcare systems from the most to the least developed countries in the world were overwhelmed, because the pandemic nature of COVID-19 pneumonia results in the substantial volume of patients with COVID-19 infection. Patients present with rapid deterioration requiring hospital admission and mechanical ventilation. Therefore, understanding and predicting patients at risk of severe and potentially life-threatening infection is key to resource planning and prognosis prediction. One commonly recognized comorbidity associated with adverse COVID-19 outcomes is the presence of cardiovascular disease. In this work, we explored the inclusion of a deep learning-based CVD severity score to improve the prediction accuracy of patient outcome.

Some key results from the experiments showed a significant increase in performance, which was seen in the random forest classifier after including the CVD severity score as a feature for prediction. Concurrently, a significant performance improvement was observed in the logistic regression when using all features and the CVD severity score with the latter model being the best performing of all configurations. The results confirmed that the inclusion of the CVD severity score benefits the models by increasing their ability to predict the patients' outcome. Additionally, while the sample size used was relatively small for machine learning models, the presented results were strengthened as the experiments were validated on a combination of the datasets. Thus, showing the relevance of including the CVD severity score for COVID-19 patient outcome prediction regardless of the patient population. This concurs with the clinical studies, presented in the introduction, that show cardiovascular diseases as relevant comorbidities for COVID-19 patients and suggest to consider cardiovascular diseases when assigning treatment to a COVID-19 patient.

Moreover, while several of the tested features and model configurations resulted in significant performance increases, these increases are not overwhelmingly high. As discussed in the exploratory analysis results, this could be due to shared information across the multiple variables employed for the outcome prediction as shown by the factor loadings. Moreover, while the manual CAC score was also tested, the CVD severity score outperformed the CAC score results across all experiments as expected from the non-significant differences in the distributions of the CAC scores. This could be due to mainly two reasons, one being, the CVD severity score might include additional relevant information of the presence of the comorbidity extracted from the CT-scans, rather than just the calcification of the coronary artery. Furthermore, as CAC is a categorical variable, contrary to the continuous values of the CVD severity score, the machine learning methods could be hindered in their prediction performance.

We hope that this work can shed more light on assigning a better prognosis for COVID-19 patients, as well as for improved clinical resource management. The main takeaway is to add evidence into a series of clinical and computational investigations, as discussed in the introduction, that indicate a significance on the impact of cardiovascular diseases as a risk factor for COVID-19 patients.

References

1. Bartlett, M.S.: Properties of sufficiency and statistical tests. Proc. Roy. Soc. London Ser. A Math. Phys. Sci. **160**(901), 268–282 (1937)
2. Chao, H., et al.: Integrative analysis for COVID-19 patient outcome prediction. Med. Image Anal. **67**, 101844 (2020)
3. Chao, H., et al.: Deep learning predicts cardiovascular disease risks from lung cancer screening low dose computed tomography. arXiv:2008.06997 [cs, eess] (2020)
4. Fang, X., et al.: Association of AI quantified COVID-19 chest CT and patient outcome. Int. J. Comput. Assist. Radiol. Surg. **16**(3), 435–445 (2021). https://doi.org/10.1007/s11548-020-02299-5
5. Fang, X., Yan, P.: Multi-organ segmentation over partially labeled datasets with multi-scale feature abstraction. IEEE Trans. Med. Imaging **39**(11), 3619–3629 (2020). https://doi.org/10.1109/TMI.2020.3001036
6. Figliozzi, S., et al.: Predictors of adverse prognosis in COVID-19: a systematic review and meta-analysis. Eur. J. Clin. Invest. **50**(10) (2020)
7. Guan, et al.: Clinical characteristics of coronavirus disease 2019 in China. N. Engl. J. Med. **382**(18), 1708–1720 (2020). https://doi.org/10.1056/NEJMoa2002032
8. Jain, V., Yuan, J.-M.: Predictive symptoms and comorbidities for severe COVID-19 and intensive care unit admission: a systematic review and meta-analysis. Int. J. Public Health **65**(5), 533–546 (2020). https://doi.org/10.1007/s00038-020-01390-7
9. Kaiser, H.F., Rice, J.: Little jiffy, mark iv. Educ. Psychol. Measur. **34**(1), 111–117 (1974). https://doi.org/10.1177/001316447403400115
10. Liu, J., et al.: A comparative overview of COVID-19, MERS and SARS: review article. Int. J. Surg. **81**, 1–8 (2020). https://doi.org/10.1016/j.ijsu.2020.07.032
11. Moore, D.H.: Classification and regression trees, by Leo Breiman, Jerome H. Friedman, Richard A. Olshen, and Charles J. Stone. Brooks/Cole Publishing, Monterey, 1984,358 Pages, $27.95. Cytometry **8**(5), 534–535 (1987)
12. Nishiga, M., Wang, D.W., Han, Y., Lewis, D.B., Wu, J.C.: COVID-19 and cardiovascular disease: from basic mechanisms to clinical perspectives. Nat. Rev. Cardiol. **17**(9), 543–558 (2020). https://doi.org/10.1038/s41569-020-0413-9
13. North, B.J., Sinclair, D.A.: The intersection between aging and cardiovascular disease. Circ. Res. **110**(8), 1097–1108 (2012)
14. Shi, F., et al.: Review of artificial intelligence techniques in imaging data acquisition, segmentation, and diagnosis for COVID-19. IEEE Rev. Biomed. Eng. **14**, 4–15 (2021). https://doi.org/10.1109/RBME.2020.2987975
15. Siddiqi, H.K., Mehra, M.R.: COVID-19 illness in native and immunosuppressed states: a clinical-therapeutic staging proposal. J. Heart Lung Transplant. **39**(5), 405–407 (2020). https://doi.org/10.1016/j.healun.2020.03.012
16. Spearman, C.: "general intelligence," objectively determined and measured. Am. J. Psychol. **15**(2), 201 (1904)
17. Ssentongo, P., Ssentongo, A.E., Heilbrunn, E.S., Ba, D.M., Chinchilli, V.M.: Association of cardiovascular disease and 10 other pre-existing comorbidities with COVID-19 mortality: a systematic review and meta-analysis. PLOS ONE **15**(8), e0238215 (2020). https://doi.org/10.1371/journal.pone.0238215
18. Tang, Z., et al.: Severity assessment of coronavirus disease 2019 (COVID-19) using quantitative features from chest CT images. arXiv:2003.11988 [cs, eess] (2020)

19. Yang, J., et al.: Prevalence of comorbidities and its effects in patients infected with SARS-CoV-2: a systematic review and meta-analysis. Int. J. Infect. Dis. **94**, 91–95 (2020). https://doi.org/10.1016/j.ijid.2020.03.017
20. Zhou, F., et al.: Clinical course and risk factors for mortality of adult inpatients with COVID-19 in Wuhan, China: a retrospective cohort study. Lancet **395**(10229), 1054–1062 (2020). https://doi.org/10.1016/S0140-6736(20)30566-3

Self-supervision Based Dual-Transformation Learning for Stain Normalization, Classification and Segmentation

Shiv Gehlot[(⊠)] and Anubha Gupta

SBILab, Department of ECE, Indraprastha Institute of Information Technology-Delhi (IIIT-D), Delhi, India
{shivg,anubha}@iiitd.ac.in

Abstract. Stain color variation s across images are common in the medical imaging domain. However, such variations among the training and test datasets may lead to unsatisfactory performance on the latter in any desired task. This paper proposes a novel coupled-network composed of two U-Net type architectures that utilize self-supervised learning. The first subnetwork (N1) learns an identity transformation, while the second (N2) learns a transformation to perform stain normalization. We also introduce classification heads in the subnetworks, trained along with the stain normalization task. To the best of our knowledge, the proposed coupling framework, where the information from the encoders of both the subnetworks is utilized by the decoders of both subnetworks as well as trained in a coupled fashion, is introduced in this domain for the first time. Interestingly, the coupling of N1 (for identity transformation) and N2 (for stain normalization) helps N2 learn the stain normalization task while being cognizant of the features essential to reconstruct images. Similarly, N1 learns to extract relevant features for reconstruction invariant to stain color variations due to its coupling with N2. Thus, the two subnetworks help each other, leading to improved performance on the subsequent task of classification. Further, it is shown that the proposed architecture can also be used for segmentation, making it applicable for all three applications: stain normalization, classification, and segmentation. Experiments are carried out on four datasets to show the efficacy of the proposed architecture.

Keywords: Self-supervised learning · Stain-normalization · Classification

1 Introduction

Staining is a step in microscopic slide preparation that involves staining chemicals to highlight the regions of interest in the tissue/blood sample or bone mar-

Shiv Gehlot would like to thank University Grant Commission (UGC), Govt. of India for the UGC-Senior Research Fellowship. We also acknowledge the Infosys Center for Artificial Intelligence, IIIT-Delhi for our research work.

© Springer Nature Switzerland AG 2021
C. Lian et al. (Eds.): MLMI 2021, LNCS 12966, pp. 477–486, 2021.
https://doi.org/10.1007/978-3-030-87589-3_49

row smear. Stain color variations in the images across datasets collected from various hospitals/centers may arise due to varying illumination conditions, stain chemicals, and staining time. Thus, the diagnostic tool trained on one center's dataset may perform non-optimally on another center's dataset. To counter this problem, stain color normalization is employed. Some of the widely used stain normalization approaches are based on histogram equalization [6,8,18], color deconvolution [9,11,15,17], and color transfer methods [10,13,14]. In histogram equalization methods, the probability density function of the reference image and source image colors are matched for each R, G, and B channel. In [10], color transfer between the two images is used to achieve stain normalization, while in [13], mean and standard deviation of the source and reference images are matched. However, these methods do not utilize histological information. Color deconvolution methods are based on negative matrix factorization (NMF) or singular value decomposition (SVD) to find the stain vectors. In [4] SVD, along with the robust aligning of the reference and query images' Cartesian frames is utilized to counter stain color variation. A limitation of all these methods is that the performance depends on the reference image's choice and may change considerably with the change of reference image. Of late, deep learning (DL) methods are gaining significance in this area. Some generative model-based DL methods for stain normalization are discussed in [16,21,22]. In [22], InfoGAN is utilized, while in [21], variational-autoencoder and deep convolutional Gaussian mixture model are used for stain normalization. In [16], a CycleGAN based method is used to preserve the structure information while applying the color transformation. However, this method requires reference center's data that may not always be available. As a remedy, self-supervised learning can be employed that does not require GT [7]. It is based on first training the network on some pretext tasks and later utilizing for the other downstream tasks. In [19], a network is trained to carry out stain normalization by learning to map the color (stain) augmented images to the corresponding original images. This can be considered as self-supervised learning, wherein the pretext task is to learn to transform the color augmented images to the original images. Color transformation of test images to that of training data can be considered as the downstream task. This paper utilizes this approach but uses two coupled subnetworks (N1 and N2) instead of a single network and learns dual transformation (pretext tasks). In addition, classification heads are added to the encoders of both the subnetworks [5,12]. The first subnetwork (N1) learns an identity transformation, while the second (N2) learns to perform stain normalization. N1 helps N2 learn the context-aware stain normalization task, while N2 helps N1 learn the stain-invariant features for the reconstruction task. Thus, both N1 and N2 assist each other, leading to improved stain-invariant performance on the subsequent classification task. We also experimented with segmentation as a downstream task. Since the proposed architecture can be utilized for stain normalization, classification, and segmentation, we name it an all-in-one network (AION). Elaborate experiments are presented on four datasets for each of the three tasks.

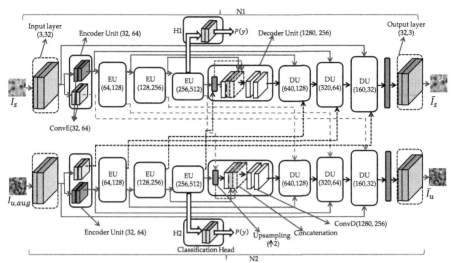

ConvE(A,B)=Conv2D(A,B,F=3,S=2,P=1)+BatchNorm2D+LeakyReLu; **ConvD(A,B)**=Conv2D(A,B,F=3,S=1,P=1)+BatchNorm2D+LeakyReLu
Decoder Unit (DU)=Upsampling by 2 + concatenation + ConvD(A,B); **Input Layer (A,B)**=Conv2D(A,B,F=3, S=2,P=1) +BatchNorm2D+LeakyReLu
Output layer(A,B)=Layer=Conv2D(A,B,F=3, S=1,P=1) +tanh; **Classification Head**=Conv2D(A=512, B=512,F=3,S=2,P=1) + BatchNorm2D + GAP +
Linear+Softmax; A=no. of input filters, **B**=no. of output filters, F=filters of size F × F, S=stride, P=padding;

Fig. 1. AION: An architecture with two coupled-networks and classification heads for learning two transformations; one identity and another for stain normalization. AION can also be used for stain-invariant classification and segmentation.

2   Methods

Consider the source image dataset and the corresponding augmented dataset with color variations. The aim is to perform stain normalization such that the stain color profile of the augmented dataset is as close as possible to the source dataset. If trained successfully, the trained model can map one Center's images to those of another Centre with a matching stain color profile. To accomplish this, we modify the architecture in [19] (here named as AION^{--}) to arrive at AION, presented in Fig. 1.

Normalization by the AION Architecture: The AION architecture can be seen as a self-supervised learning architecture consisting of two parallel U-Net types AION^{--} coupled through cross-connections running from the encoder units of one to the decoder units of the other. The first subnetwork, N1 is trained to learn the identity transformation on the source dataset. In contrast, the second subnetwork, N2 is supposed to map the augmented image dataset's stain color profile to that of the source dataset. Each subnetwork consists of an input layer, four encoder units (EUs), four decoder units (DUs), and one output layer. Input layer and encoder units constitute the encoder, while decoder units and output layer constitute the decoder. The input layer consists of 32 filters, while the successive EUs consist of twice the filters than the preceding layer/unit. Hence, the first EU consists of 64 filters, and the fourth EU consists of 512 filters. The size of each filter is 3 × 3. Each convolutional operation in the encoder is performed with a stride of two and a padding of one. This leads to a reduction of output size by two after each convolutional operation. Each EU consists of two parallel convolutions layers and provides a coupling mechanism for N1 and N2. These parallel layers receive the same input from the preceding EU. However, the output from one goes to the succeeding EU of the same subnetwork and the output of the other goes to the decoder unit of another subnetwork. Similar cross-connections are followed for both the subnetworks (Fig. 1).

Each DU of N1 concatenates the input from the preceding layer/unit, the output from the corresponding EU of N1, and the output from the corresponding EU of N2. DUs also involve upsampling by two by the nearest neighbors approach to match the spatial size for concatenation. The concatenated output is given to the convolutional layers within DUs. The final output is forwarded to the next DU. Similar is the decoder structure of N2. Each DU contains half the number of filters of the preceding DU, with the first DU containing 256 filters and the fourth containing 32 filters. Finally, the output layer provides a three-channel output. Each convolutional operation in the network is followed by batch normalization (BatchNorm) and leaky ReLu except for the output layer, which has tanh activation and no BatchNorm.

Classification and Segmentation: To introduce the classification capability, we add two classification heads (CHs), one in each subnetwork on top of the last EU. Each CH consists of a single convolutional layer with 512 filters followed by BatchNorm, ReLu activation, global averaging pooling, and a classification layer. The CH of N1 predicts the class of images of the source dataset, while CH of N2 predicts the class of color augmented dataset. The AION can also be used for the segmentation. For this case, both N1 and N2 are fed with the source dataset, and the classification heads are not trained.

Training Methodology: To train the AION for stain normalization, an image I_s from the source dataset is given as input to N1. Another image $I_{u,aug}$, obtained through some transformation $\phi(\cdot)$ on image I_u from the source dataset, is given as input to N2, where indices u and s may or may not be equal. N1 aims to reconstruct the input image, while N2 attempts to match I_u. If we denote the output of N1 as $\bar{I}_s$ and of N2 as $\bar{I}_u$, the loss function is given as:

$$\mathcal{L}_1 = MSE(\bar{I}_s, I_s) + MSE(\bar{I}_u, I_u), \tag{1}$$

where $MSE(\cdot)$ is the mean square error. Hence, N1 is learning an identity transformation, and N2 is learning ϕ^{-1}, assuming such inverse exists. Once trained, N2 should transform the stain (color) of the input data to that of the data input to N1. The classification heads are also trained simultaneously to predict the labels of the input images. Hence, the complete loss function is given as:

$$\mathcal{L} = MSE(\hat{I}_s, I_s) + MSE(\bar{I}_u, I_u) + NLL(y_1, \hat{y}_1) + NLL(y_2, \hat{y}_2), \tag{2}$$

where NLL is the negative log-likelihood, y is the true label, and $\hat{y}$ is the predicted label by the classification head. The subscripts 1 and 2 denote the classification head of Network-1 and Network-2, respectively. *It should be noted that*

Table 1. Description of the four dataset. Highlighted cells for Camelyon17 represent the sum of train and val dataset of the respective columns as the architecture trained on one center's is tested on another center's whole data.

Application	Stain Normalization and Classification						Segmentation	
Datasets	Camelyon17 (Patches)					PCam	Databowl	CVCDB
	C0	C1	C2	C3	C4	(Patches)	(Patches)	(Images)
Train	53538	31169	54834	94635	16776	262146	8640	488
Val	6000	4000	4000	15002	4000	32770	1360	62
Test	59538	35169	58834	109637	20776	32770	2503	62

Fig. 2. Sample images from the datasets. (Set-1: left to right) Image from C0, C1, C2, C3, C4, and C5 of Camelyon17. (Set-2) augmented images from C3 obtained with transformation $\phi(\cdot)$. (Set-3) PCam. (Set-4) DSB, (Set-5) CVC.

classification Head-1 is trained on the source image data, while classification Head-2 is trained on the augmented data.

For segmentation, the tanh activations at the output layers are replaced with sigmoid activations. The input of both the networks is the same, and the classification heads are not trained. Also, the original (not augmented) data is used for training. The respective segmented mask is predicted from both the networks, and both the networks are trained simultaneously using the binary cross-entropy loss.

3   Results

In this work, we use four publicly available datasets. Camelyon17 challenge dataset [3] is used for stain normalization, PatchCamelyon (PCam) [20] for classification, and Data Science Bowl (DSB) [1] & CVC-ClinicDB (CVC) [2] for segmentation. Camelyon17 consists of hematoxylin and eosin (H&E) stained whole slide images (WSIs) of lymph node sections collected from five centers labeled as C_0 to C_4. For training and testing, the patches of size 128×128 are extracted from the level-0 of the WSIs. PCam have 96×96 patches having a binary label indicating the presence or absence of metastatic tissue. There is stain-variation among the patches, which makes the classification task challenging. The DSB is a cell segmentation dataset having images of varying sizes. We have extracted 128×128 patches for training and testing. CVC is a polyps segmentation dataset with 612 images of size 384×288 pixels. For training with CVC, we have augmented the training set using random rotations. The distribution of the patches/images for all the datasets is provided in Table 1, and the sample images are shown in Fig. 2.

Stain-Normalization: For Camelyon17, there are variations in the colors of stained images of the five centers (Fig. 2). Intuitively, a classifier trained on one center's data may perform unsatisfactorily on the remaining centers' data. We train a classifier on a particular center's data and test on the remaining centers' data for stain-normalization. The stain normalization quality can then be assessed through improvement in the classification performance over the unnormalized data. We begin our analysis with C3 as the training data. For comparison, we chose Macenko [9] and Reinhard [13] as non-DL methods, and StainGAN [16] and AION^{--} [19] as DL methods. StainGAN [16] is trained using C3 and C0 because it requires two training datasets. Results are also reported by adding a classification head to AION^{--}. We name this architecture as AION^{--}+H. Both AION and AION^{--}+H are trained with the same training methodology as discussed in Sect. 2. To obtain the augmented images, $\phi(\cdot)$ consists of saturation in the range [0, 2.5] and hue in the range [0.5,0.5] with a probability of 0.95. The sample augmented images are shown in Fig. 2. Some results are shown with an external classifier (EC) that consists of 10 convolutional layers, batch normalization and Leaky ReLU as activation, and a classification layer.

All the models are implemented using PyTorch 1.1.0 and trained using RTX2080 GPU. Except for classifiers, all the models are trained using Adam optimizer for 80 epochs, batch size of 32, and initial learning rate of 0.001, which is reduced to one-tenth of the present value if there is no change in the validation performance for seven epochs. For classifiers, an SGD optimizer and a batch size of 64 are used for 150 epochs. The initial learning rate of 0.001 is reduced to one-tenth at 80th, 120th, and 140th epochs. The following results are reported with AION^{--}+H and AION:

1. **AION^{--}+H (H):** Results from the classification head that is added to AION^{--}
2. **AION (H1):** Results with N1's classification head of AION
3. **AION (H2):** Results with N2's classification Head of AION
4. **AION^{--}+H (EC):** Results with an external classifier trained on a center, and tested on normalized images obtained from AION^{--}+H
5. **AION (N2+EC):** Results with an external classifier trained on a center, and tested on normalized images obtained from N2 of AION

Results with C3 as training data are shown in Table 2 in terms of balanced accuracy (BAC), the area under the curve (AUC), and the weighted F1-score (WF1). The mean performance of the methods requiring reference images is inferior to other methods, showing their dependency on the reference image's choice. It can be seen that stain normalization leads to improved classification performance over original images for all the centers. For C0, AION (N2+EC) gives the top performance with a significant margin in terms of BAC and WF1. For C1 and C4, AION^{--}+H (EC) and AION (N2+EC) are having approximately similar performances, and for C2, AION^{--} is giving the best results. Apart from C2, both AION^{--}+H (EC) and AION (N2+EC) are performing better than AION^{--}. This shows the contributions of the added components to the AION^{--}. Another significant observation is the better performance of AION

(H1) than original images. This shows that AION (H1) has also become stain-invariant, even though it is trained with the unnormalized images. This capability may have been introduced due to the coupling in the AION. AION^{--}+H (H) and AION (H2) are trained with augmented images and have seen varying stain colors. Hence, the performance improvement with these architectures is apparent. Table 3 shows the inter-center train-test results in terms of AUC. For all the centers, there is performance enhancement after stain normalization. Also, for most cases, results with AION are better than AION^{--}+H, which are better than AION^{--}.

Table 2. Results with C3 as training data, and other centers as the test data. The shaded cells represent mean results with five various reference images.

Dataset	C0			C1			C2			C4		
Metric/ Method	AUC	BAC	WF1	AUC	BAC	WF1	AUC	BAC	WF1	AUC	BAC	WF1
Original	0.7835	0.5960	0.5422	0.7581	0.7013	0.6948	0.4026	0.3338	0.3187	0.8628	0.7826	0.7824
Macenko [9]	0.5530	0.5025	0.4539	0.6117	0.5410	0.4767	0.5499	0.5428	0.5203	0.6096	0.5907	0.5490
Reinhard [13]	0.5465	0.4898	0.3738	0.5833	0.5087	0.3791	0.4699	0.4759	0.2521	0.4353	0.4867	0.3532
StainGAN [16]	0.7926	0.5714	0.5023	0.8097	0.7217	0.7175	0.6882	0.6072	0.5241	0.5887	0.5097	0.365
AION^{--} [19]	**0.8611**	0.7192	0.7042	0.8257	0.7577	0.7553	**0.9390**	**0.8067**	**0.7535**	0.8115	0.7206	0.7137
AION^{--}+H (H)	0.7975	0.7199	0.7172	0.8327	0.7433	0.7398	0.7457	0.6751	0.6046	0.8739	0.8040	0.8039
AION^{--}+H (EC)	0.8438	0.7538	0.7497	0.8600	**0.7805**	**0.7794**	0.8174	0.7240	0.6567	**0.9289**	**0.8661**	**0.8659**
AION (H1)	0.7767	0.6941	0.6915	0.8040	0.7279	0.7248	0.6419	0.6072	0.6958	0.9286	0.8562	0.8562
AION (H2)	0.8019	0.7186	0.7161	0.8370	0.7395	0.7349	0.7833	0.6828	0.6008	0.8851	0.8129	0.8125
AION (N2+EC)	**0.8661**	**0.7733**	**0.7712**	**0.8605**	0.7781	0.7769	0.8579	0.7534	0.6937	0.9218	0.8627	0.8625

Table 3. AUC with inter-center training-testing

AUC								
Train	C0				C1			
Test	C1	C2	C3	C4	C0	C2	C3	C4
Original	0.8607	0.5076	0.7157	0.5080	0.8065	0.1852	0.6490	0.1052
AION^{--} [19]	**0.8714**	**0.9341**	0.9344	0.7714	**0.7169**	•.8880	0.7305	0.4088
AION^{--}+H (H)	0.8368	0.8915	0.9114	0.8973	0.7093	0.8448	0.8761	0.7907
AION^{--}+H (EC)	0.8183	0.9157	0.9304	0.9137	0.6674	0.8870	0.7787	0.4132
AION (H2)	0.8412	0.8956	0.9141	0.9029	0.7128	0.8465	**0.8774**	**0.8221**
AION (N2+EC)	0.8359	0.9130	**0.9362**	**0.9206**	0.6551	0.8680	0.8088	0.4920
Train	C2				C4			
Test	C0	C1	C3	C4	C0	C1	C2	C3
Original	0.6055	0.5722	0.3577	0.3617	0.4506	0.6532	0.7667	0.4012
AION^{--} [19]	0.6884	**0.8269**	0.7648	0.4211	0.5983	0.5797	0.8019	0.7130
AION^{--}+H (H)	0.6706	0.7722	0.7732	0.7732	0.5708	0.6920	0.7699	0.7466
AION^{--}+H (EC)	0.6772	0.8005	0.8455	0.5670	**0.6125**	0.6269	**0.8264**	**0.7817**
AION (H2)	**0.6995**	0.7800	0.7967	**0.7940**	0.5869	**0.7121**	0.7761	0.7529
AION (N2+EC)	0.6926	0.8034	**0.8508**	0.6623	0.6080	0.6418	0.7821	0.7576

Classification: To highlight AION's classification capability, we use the binary class dataset of PCam with inter-image stain variations. We have also compared the results with AION⁻⁻+H to show the proposed architecture's contribution. Results are summarized in Table 4.I. A methodology similar to stain normalization is used for training with PCam. First, we trained EC on the original PCam training set. This classifier provided an AUC of 0.8448 on the original PCam test set. The AION⁻⁻+H (H), AION (H1), and AION (H2) provide a gain of 4%, 4.61%, and 3.66%, respectively. Even though these classifiers are shallow (6 layers) than EC (10 layers), they provided a significant gain. This gain could be more with deeper heads. Hence, the proposed architecture can produce stain-invariant classifiers as a byproduct. To explore further, we stain-normalized PCam using AION⁻⁻+H and AION trained on C3 of Camelyon17. Training and testing with the resultant dataset provided an AUC of 0.8948 and 0.9105 with AION⁻⁻+H and AION (N2), respectively, which again proves AION's usefulness. We also replaced the custom classifier with ResNet-24 and DenseNet-161. The best AUC is achieved with AION (N2) for both classifiers, which is 0.9356 with ResNet-34, and 0.9275 with DenseNet-161.

Table 4. (I) PCam Classification results, (II) CVC and DSB Segmentation results

(I) PCam Classification Results					
Classifier	Method	AUC	Classifier	Method	AUC
AION⁻⁻+H (H)	AION⁻⁻+H	0.8848	Resnet-34	Original	0.8948
AION (H1)	AION	**0.8909**		AION⁻⁻ [19]	0.9059
AION (H2)	AION	0.8814		AION⁻⁻+H	**0.9181**
				AION (N2)	**0.9356**
EC	Original	0.8448	DensNet-161	Original	0.8691
	AION⁻⁻ [19]	0.9054		AION⁻⁻ [19]	0.9137
	AION⁻⁻+H	0.8948		AION⁻⁻+H	0.9091
	AION (N2)	**0.9105**		AION (N2)	**0.9275**

(II) Segmentation results on CVC and DSB with AION⁻⁻+H and AION					
Dice Similarity Coefficient (DSC)					
Dataset	CVC			DSB		
Initialization/Method	AION⁻⁻+H	AION (N1)	AION (N2)	AION⁻⁻+H	AION (N1)	AION (N2)
C3-Pretrained	0.6897	**0.7453**	0.6896	0.8670	0.8533	**0.8682**
Random	0.8243	0.8198	**0.8328**	0.8712	**0.8723**	0.8705
Intersection over Union (IoU)						
Dataset	CVC			DSB		
Initialization/Method	AION⁻⁻+H	AION (N1)	AION (N2)	AION⁻⁻+H	AION (N1)	AION (N2)
C3-Pretrained	0.7406	**0.7765**	0.7407	0.8527	0.8378	**0.8542**
Random	0.8361	0.8327	**0.8432**	0.8568	**0.8579**	0.8560

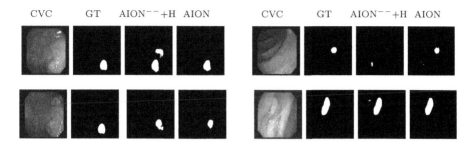

Fig. 3. Sample segmentation masks from AION^{--}+H and AION (H2) from CVC.

Segmentation: In one set of experiments, we considered segmentation as the downstream task and initialized AION^{--}+H and AION with the network's weights trained on C3 (C3-Pretrained). In contrast, in the other experiment, they are initialized randomly. Also, as the classification head is not trained, AION^{--} and AION^{--}+H are the same for segmentation. Both AION^{--}+H and AION gave an inferior performance with C3-pretrained networks as compared to random initialization (Table 4.II). This shows that stain-normalization-specific features are, perhaps, not helpful in the segmentation task on this dataset. Also, the performance difference between the two initializations is large for CVC as compared to DSB. This is due to the similarity in the imaging modality of DSB and C3. Also, AION has better performance than AION^{--}+H for both datasets. The qualitative results are also presented in Fig. 3.

4   Conclusion

In this work, we proposed a novel self-supervision based dual-transformation coupled-network architecture for stain normalization. The architecture is also equipped with classification heads that can achieve stain-invariant classification. The utility is also shown for the downstream task of segmentation. The architecture can perform well on all three applications of stain normalization, classification, and segmentation.

References

1. data science bowl. https://www.kaggle.com/c/data-science-bowl-2018. Accessed 5 Feb 2021
2. Bernal, J., Sánchez, F.J., Fernández-Esparrach, G., Gil, D., Rodríguez, C., Vilariño, F.: WM-DOVA maps for accurate polyp highlighting in colonoscopy: validation vs. saliency maps from physicians. Comput. Med. Imaging Graph. **43**, 99–111 (2015)
3. Bándi, P., et al.: From detection of individual metastases to classification of lymph node status at the patient level: the camelyon17 challenge. IEEE Trans. Med. Imaging **38**(2), 550–560 (2019)

4. Gupta, A., et al.: GCTI-SN: geometry-inspired chemical and tissue invariant stain normalization of microscopic medical images. Med. Image Anal. **65**, 101788 (2020)
5. Haghighi, F., Hosseinzadeh Taher, M.R., Zhou, Z., Gotway, M.B., Liang, J.: Learning semantics-enriched representation via self-discovery, self-classification, and self-restoration. In: Martel, A.L., et al. (eds.) MICCAI 2020. LNCS, vol. 12261, pp. 137–147. Springer, Cham (2020). https://doi.org/10.1007/978-3-030-59710-8_14
6. Jain, A.K.: Fundamentals of Digital Image Processing. Prentice-Hall, Inc. (1989)
7. Jing, L., Tian, Y.: Self-supervised visual feature learning with deep neural networks: a survey (2019)
8. Kothari, S., et al.: Automatic batch-invariant color segmentation of histological cancer images. In: From Nano to Macro, 2011 IEEE International Symposium on Biomedical Imaging, pp. 657–660 (2011)
9. Macenko, M., et al.: A method for normalizing histology slides for quantitative analysis. In: ISBI, pp. 1107–1110 (2009)
10. Magee, D., et al.: Colour normalisation in digital histopathology images. In: Proceedings Optical Tissue Image analysis in Microscopy, Histopathology and Endoscopy (MICCAI Workshop), vol. 100 (2009)
11. McCann, M.T., Majumdar, J., Peng, C., Castro, C.A., Kovačević, J.: Algorithm and benchmark dataset for stain separation in histology images. In: 2014 IEEE International Conference on Image Processing (ICIP), pp. 3953–3957 (2014)
12. Mehta, S., Mercan, E., Bartlett, J., Weaver, D., Elmore, J.G., Shapiro, L.: Y-net: joint segmentation and classification for diagnosis of breast biopsy images. In: Frangi, A.F., Schnabel, J.A., Davatzikos, C., Alberola-López, C., Fichtinger, G. (eds.) MICCAI 2018. LNCS, vol. 11071, pp. 893–901. Springer, Cham (2018). https://doi.org/10.1007/978-3-030-00934-2_99
13. Reinhard, E., Adhikhmin, M., Gooch, B., Shirley, P.: Color transfer between images. IEEE Comput. Graphics Appl. **5**, 34–41 (2001)
14. Ruderman, D.L., Cronin, T.W., Chiao, C.C.: Statistics of cone responses to natural images: implications for visual coding. JOSA A **15**(8), 2036–2045 (1998)
15. Ruifrok, A., Ruifrok, D.: Quantification of histochemical staining by color deconvolution. Anal. Quant. Cytol. Histol. /Int. Acad. Cytol. [and] Am. Soc. Cytol. **23**(4), 291–299 (2001)
16. Shaban, M.T., Baur, C., Navab, N., Albarqouni, S.: StainGAN: stain style transfer for digital histological images. arXiv preprint arXiv:1804.01601 (2018)
17. Abe, T., Murakami, Y., Yamaguchi, M.: Color correction of pathological images based on dye amount quantification. Opt. Rev. **12**(4), 293–300 (2005)
18. Tabesh, A., et al.: Multifeature prostate cancer diagnosis and Gleason grading of histological images. IEEE Trans. Med. Imaging **26**(10), 1366–1378 (2007)
19. Tellez, D., et al.: Quantifying the effects of data augmentation and stain color normalization in convolutional neural networks for computational pathology. Med. Image Anal. **58**, 101544 (2019)
20. Veeling, B.S., Linmans, J., Winkens, J., Cohen, T., Welling, M.: Rotation equivariant CNNs for digital pathology (2018)
21. Zanjani, F.G., Zinger, S., Bejnordi, B.E., van der Laak, J.A.W.M.: Histopathology stain-color normalization using deep generative models. In: 1st Conference on Medical Imaging with Deep Learning (MIDL 2018), pp. 1–11 (2018)
22. Zanjani, F.G., Zinger, S., Bejnordi, B.E., van der Laak, J.A.W.M., de With, P.H.N.: Stain normalization of histopathology images using generative adversarial networks. In: 2018 IEEE 15th International Symposium on Biomedical Imaging (ISBI 2018), pp. 573–577, April 2018.https://doi.org/10.1109/ISBI.2018.8363641

Deep Representation Learning for Image-Based Cell Profiling

Wenzhao Wei[1], Sacha Haidinger[2], John Lock[3], and Erik Meijering[1(✉)]

[1] School of Computer Science and Engineering, University of New South Wales, Sydney, Australia
erik.meijering@unsw.edu.au
[2] Department of Bioengineering, Swiss Federal Institute of Technology Lausanne, Lausanne, Switzerland
[3] School of Medical Sciences, University of New South Wales, Sydney, Australia

Abstract. High-content, microscopic image-based screening data are widely used in cell profiling to characterize cell phenotype diversity and extract explanatory biomarkers differentiating cell phenotypes induced by experimental perturbations or disease states. In recent years, high-throughput manifold embedding techniques such as t-distributed stochastic neighbor embedding (t-SNE), uniform manifold approximation and projection (UMAP), and generative networks have been increasingly applied to interpret cell profiling data. However, the resulting representations may not exploit the full image information, as these techniques are typically applied to quantitative image features defined by human experts. Here we propose a novel framework to analyze cell profiling data, based on two-stage deep representation learning using variational autoencoders (VAEs). We present quantitative and qualitative evaluations of the learned cell representations on two datasets. The results show that our framework can yield better representations than the currently popular methods. Also, our framework provides researchers with a more flexible tool to analyze underlying cell phenotypes and interpret the automatically defined cell features effectively.

Keywords: Variational autoencoders · Manifold embedding · Representation learning · Cell profiling · Fluorescence microscopy

1 Introduction

The advent of high-content fluorescent microscopic image-based screening technology in recent years has greatly advanced cell profiling. High-resolution cell images are a rich source of visual information, which enables systematic evaluation of cellular phenotypes induced by experimental cell perturbations (e.g. drugs, genetic manipulations) or disease states (e.g. cancer, neurodegenerative disorders) [1,2]. To obtain characteristic cell profiles, the most explanatory features describing

---

W. Wei and S. Haidinger—Joint first authors.
J. Lock and E. Meijering—Joint senior authors.

© Springer Nature Switzerland AG 2021
C. Lian et al. (Eds.): MLMI 2021, LNCS 12966, pp. 487–497, 2021.
https://doi.org/10.1007/978-3-030-87589-3_50

cellular phenotypes need to be extracted from the images. One way to do this is by visually examining hundreds of cell states. Examples of human-defined phenotypic features include cell staining patterns, spatial relationships, cell-cycle phases, signaling states, intensity-based features, micro-environment and context features, et cetera [1]. Such features have been mathematically defined and implemented in common bioimage analysis software packages such as CellProfiler [3], ImageJ [4], Ilastik [5] and more, which can be used by biologists to generate exclusive cell morphological profiles.

However, defining representative features for cell profiling is challenging and often case-driven. As the range of possible cell states may not be known a priori, and differences between states can be subtle, human observers may introduce implicit or explicit biases or even miss relevant features. Making the right assumptions is key in representation learning [6–8]. For cell profiling, features should ideally be disentangled [9], capture a hierarchy of spatial properties at increasing scales [10], model continuously varying perturbations as smooth transitions on the cellular phenotype manifold in the feature space [11], and make different cell subpopulations appear as clearly distinguishable clusters [12]. Following cell segmentation and feature extraction, the next task in cell profiling is dimensionality reduction for downstream manifold analysis and visualization [3,13,14]. Popular methods for this task are t-distributed stochastic neighbor embedding (t-SNE) [15] and uniform manifold approximation and projection (UMAP) [16], but they are typically applied to custom-designed image features. Integrated approaches to joint feature design, extraction, and dimensionality reduction for image-based cell profiling are currently lacking.

In this paper, we present an unsupervised representation learning framework that facilitates cell profiling from fluorescence microscopy images (Fig. 1). Building on the concept of variational autoencoders (VAEs) [17–20], it provides an end-to-end and fully automated solution to capturing the relevant high-dimensional image information and presenting cells to the user in a low-dimensional space suitable for interactive visualization and exploration. More specifically, we propose a flexible two-stage VAE (2SVAE) pipeline that allows users to combine automatically extracted features and handcrafted features if needed. Potential risks of over-regularization and posterior collapse of VAEs are mitigated by maximizing the mutual information between the input and the latent space and using a Gaussian mixture model as the prior. In the following, we describe the methods used and present quantitative and qualitative experimental results on two public datasets, showing the superior performance of our pipeline in comparison with several alternative approaches.

2   Methods

2.1   Framework Summary

The proposed deep representation learning framework (Fig. 1) consists of two levels of VAE. At the top level, VAE1 encodes features from single-cell images, transforming the images to 100 latent variables. This allows high-fidelity image reconstruction via the decoder and ensures the latent variables are as descriptive

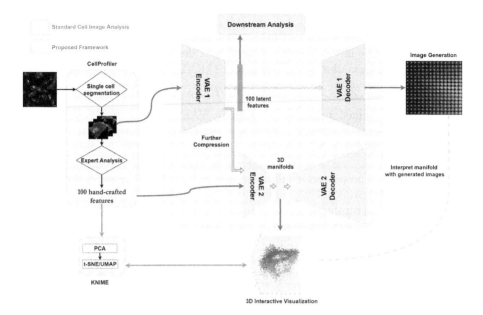

Fig. 1. Graphical overview of our cell profiling pipeline. The beige block highlights existing cell image analysis methods. We use CellProfiler for cell segmentation and obtaining single-cell image patches for training. The software also allows to compute handcrafted cell feature extraction if needed, which we do not use in this paper. We do use KNIME [13] for comparing our results with t-SNE and UMAP. The blue block shows the proposed two-stage VAE framework for deep representation learning. (Color figure online)

of the input as possible and can be used for downstream analysis. The 'meaning' of these variables can be grasped by exploring the output of the decoder for a range of interpolated positions in the latent space. At the bottom level, VAE2 further encodes the information to a 3D latent space, which allows to visualize and explore cell phenotype heterogeneity in a more human-interpretable way [21]. The two VAEs are trained jointly using stochastic gradient descent.

2.2   Implementation Details

We employ a convolutional neural network (CNN) architecture as the backbone of the two VAEs. Each convolutional layer utilizes a 3×3 pixel kernel, a stride of 2, and the spatial size of the feature map is halved after each layer while the depth (number of channels) is doubled, starting from 32. In our setup, the size of the single-cell input RGB-images is currently fixed to 64×64 pixels. In total 5 convolutional layers are applied to ensure the receptive field in the deepest layer matches the input size. Batch normalization is used between each layer.

Two key improvements are introduced to mitigate the common risks of over-regularization and posterior collapse of vanilla VAEs [8,19]. First, we use mutual information (MI) as a proxy for determining the degree of information retained between the input images (X) and the latent variables (Z) [22]. Reliable estimation of the lower bound of MI is possible using a deep neural network (T) known as the mutual information neural estimator (MINE) [23]:

$$I(X, Z) = E_{q_\phi(x,z)}[T(x, z)] - E_{q_\phi(x)q_\phi(z)}\left[e^{T(x,z)-1}\right] \tag{1}$$

Second, the presumed prior distribution in VAEs plays a significant role in balancing between the qualities of image reconstruction and latent representation, and the typically used isotropic Gaussian distribution is unimodal and does not allow mapping to more complex representations [24], resulting in over-regularization of the posterior and a poor manifold [25]. Therefore, we instead use a multimodal prior [24–26], namely a Gaussian mixture model (GMM).

To ensure structurally meaningful features are captured during encoding, we pretrain the encoders and decoders of both VAEs using the structural similarity measure (SSIM) and mean absolute error (MAE) [27] to penalize the reconstruction error without sampling from the prior distribution:

$$\mathcal{L}_{\text{pretrain}} = \underbrace{(1 - \text{SSIM}(x, \widehat{x}_1) + |x - \widehat{x}_1|)}_{\text{Reconstruction Loss VAE1}} \atop \underbrace{- \gamma * (1 - \text{SSIM}(x, \widehat{x}_2) + |x - \widehat{x}_2|)}_{\text{Reconstruction Loss VAE2}} \tag{2}$$

At the end of the pretraining phase, the cell images are projected to the 3D latent space, and the optimal number c of distributions for the GMM is found in a range of 1 to 10 by minimizing the Akaike information criterion (AIC) [28] and Bayesian information criterion (BIC) [29] scores. Next, starting with the pretrained weights of the encoders and decoders and incorporating MINE and the GMM prior, the framework is trained using an objective function that is a weighted ensemble of the variational lower bound (ELBO) from both VAEs:

$$\mathcal{L} = \underbrace{\text{Recon}(x, \tilde{x}_2) - \beta_2 \text{KL}\left[q_\phi(z_2, c \mid x) \| p(z_2 \mid c)\right] + \alpha I_{q_\phi}(x, z_2)}_{\text{ELBO VAE2}} \atop + \gamma \underbrace{\left[\text{Recon}(x, \tilde{x}_1) - \beta_1 \text{KL}\left[q_\phi(z_1 \mid x) \| p(z_1)\right] + \alpha I_{q_\phi}(x, z_1)\right]}_{\text{ELBO VAE1}} \tag{3}$$

where α determines the influence of the MI terms, β_1 and β_2 weight the Kullback-Leibler (KL) terms, and γ balances the two ELBOs. In the experiments, these parameters were fixed to $\alpha = 1$, $\beta_1 = \beta_2 = 5$, and $\gamma = 0.8$.

3   Experiments

3.1   Benchmark Methods

We evaluated several variants of the VAE approach to representation learning in comparison with t-SNE and UMAP (Table 1). Specifically, we considered three variants of both the two-stage design (VAE1+VAE2 as in Fig. 1) and the one-stage design (a merged VAE mapping directly to a 3D manifold), which differ in whether they use MI regularization and/or the GMM prior.

Table 1. Method variants compared in the experiments.

Method	Human-defined features	MI regularization	GMM prior	Two-stage framework
t-SNE [15]	✓	–	–	–
UMAP [16]	✓	–	–	–
VAE [20]	–	–	–	–
InfoMaxVAE [22]	–	✓	–	–
VaDE [26]	–	–	✓	–
2SVAE	–	–	–	✓
2SInfoVAE	–	✓	–	✓
2SInfoVaDE	–	✓	✓	✓

3.2   Benchmark Datasets

Two quite different datasets were used to evaluate the robustness and generalizability of the methods. The first is dSprites [30], a programmatically generated shape dataset commonly used for benchmarking representation and manifold learning methods. It contains 73,728 images of size 64×64 pixels with shapes generated from various independent latent factors (base shape, scale, orientation, position). To keep the dataset size manageable, we used a smaller range for the position variables, resulting in 18,000 images evenly distributed among three discrete shape classes (square, ellipse, heart). The second dataset is from the Broad Bioimage Benchmark Collection (BBBC031v1) [31] and contains synthetic cell images for testing methods on learning cell latent features with defined phenotypes. The 7,387 single-cell images are generated from six underlying biological processes associated with various shapes and colors (Fig. 2). This is a challenging dataset, as many cells are hard to distinguish due to the large overlap of cell features across phenotypic processes.

3.3   Evaluation Metrics

A total of six evaluation metrics from three categories were used in the experiments. The first category consists of unsupervised metrics based on the so-called co-ranking matrix to assess the degree of neighbourhood context preservation from the (higher-dimensional) input data to the (lower-dimensional) latent representations. We used three such metrics: trustworthiness, continuity, and local continuity meta-criterion (LCMC) [32,33]. Each of these metrics were summarized by area-under-the-curve (AUC) scores from 1% to 20% neighbourhood sizes. The second category consists of two disentanglement metrics: interpretability [34] and modularity [35]. These quantify the degree of independence and relevance of latent representations. As a third category we used a classifier metric: accuracy. It assesses the clusterability of the latent representations. To compute this metric, by way of example, we used a simple two-layer perceptron, and to obtain clearer discrete phenotypic states in the BBBC dataset we grouped clusters into pairs (1&2, 3&4, 5&6) and removed the overlapping initial states. Similarly, this metric was also used to classify discrete shapes in the dSprite dataset.

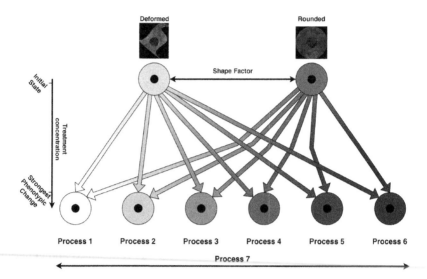

Fig. 2. Overview of the different cell phenotypes in the BBBC dataset. Starting from two possible initial states (top), six possible final states (bottom) corresponding to different cell phenotypes are generated by different biological processes. Linear interpolation is used to modulate the strength of the underlying perturbation process, producing phenotypes that vary continuously along a manifold for each process. A seventh process is sampled uniformly across all other six processes

3.4   Evaluation Results

We performed both quantitative (using the six evaluation metrics) and qualitative (visual inspection) experiments on both datasets. From the quantitative results (Table 2) we observe that our two-stage framework using MI and GMM (2SInfoVAE and 2SInfoVaDE variants) performs favorably. For qualitative analysis we selected 2SInfoVaDE and compared the results visually with t-SNE and UMAP. Inspecting the visualizations of the dSprite dataset (Fig. 3) we observe that the latter two both generate a large number of discrete clusters corresponding to continuous latent attributes (positions and rotations) and the discrete shapes (indicated by different colors) are not distinguished. In contrast, the 2SInfoVaDE projection shows both circular symmetrical patterns and better separated colored clusters. From the BBBC dataset (Fig. 4) we see that all three methods can produce reasonably interpretable latent projections. For both datasets, our 2SInfoVaDE method seems to yield a more separable embedding, as also reflected by its higher quantitative scores (Table 2).

Table 2. Results of the quantitative evaluation. For all metrics, higher values are better. Bold indicates best performance for the given metric and dataset.

Data	Model	Trust.	Cont.	LCMC	Inter.	Modul.	Acc.
dSprite	t-SNE	9.95	9.96	0.30	0.12	89.90	33.94
	UMAP	9.97	9.97	1.20	0.14	94.60	33.70
	VAE	17.31	17.35	7.15	0.24	87.14	48.72
	InfoMaxVAE	16.22	16.42	5.66	0.14	93.13	47.79
	VaDE	17.84	17.78	9.03	0.28	89.40	65.92
	2SVAE	17.80	17.60	9.20	0.25	89.71	**68.14**
	2SInfoVAE	**18.04**	**18.07**	**9.73**	0.29	88.44	65.63
	2SInfoVaDE	17.83	17.83	9.27	**0.40**	**96.20**	67.85
BBBC	t-SNE	15.15	17.33	6.88	0.37	75.10	92.58
	UMAP	14.95	9.97	6.10	0.37	65.50	87.70
	VAE	17.23	17.35	6.73	0.28	66.35	76.88
	InfoMaxVAE	17.37	17.48	7.00	0.31	70.72	78.13
	VaDE	17.67	17.68	8.04	**0.42**	**80.97**	87.37
	2SVAE	17.44	17.56	8.12	0.32	73.80	86.52
	2SInfoVAE	17.85	17.83	8.35	0.40	78.64	92.14
	2SInfoVaDE	**17.86**	**17.96**	**8.70**	0.39	80.02	**94.67**

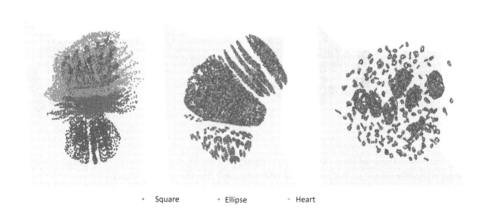

· Square   · Ellipse   · Heart

Fig. 3. Qualitative results for the dSprite dataset. The examples show 3D projections produced by 2SInfoVaDE (left), t-SNE (middle), and UMAP (right).

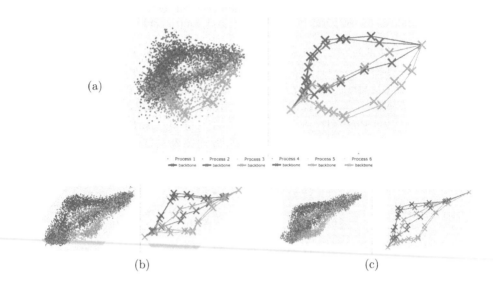

Fig. 4. Qualitative results for the BBBC dataset. The examples show 3D projections produced by (a) 2SInfoVaDE, (b) t-SNE, and (c) UMAP. The shown backbones, computed using nearest-neighbor fitting of fixed sample points from the ground-truth data to the generated results, summarize the phenotypic trajectories and reflect the continuous variation in the strength of the six underlying processes.

Fig. 5. Visual inspection of VAE-generated cell images for various interpolated positions in the latent space for (a) the dSprite and (b) the BBBC dataset.

4   Conclusions

Our two-stage VAE-based deep representation learning framework offers several major advantages for image-based cell profiling over the popular t-SNE and UMAP methods. First, it automatically finds relevant image features, relieving the user of the task to define them manually while still allowing the incorporation of handcrafted features if needed. Second, as our preliminary results indicate, our framework appears more robust and generalizable across applications. And third, it provides ways to interpret the automatically found optimal features by varying the values of the latent variables and inspecting the resulting generated cell images (Fig. 5). The two stages of the framework and our pretraining strategy allow the user to trade off the cell image reconstruction quality and the latent manifold quality. Altogether, we conclude that our framework provides researchers with a more flexible tool to analyze underlying cell phenotypes and interpret the automatically defined cell features effectively. Planned future work includes extending the experiments to a wider range of datasets and using the framework for cell profiling in our drug screening studies.

References

1. Caicedo, J.C., et al.: Data-analysis strategies for image-based cell profiling. Nat. Methods **14**(9), 849–863 (2017). https://doi.org/10.1038/nmeth.4397
2. Ljosa, V., et al.: Comparison of methods for image-based profiling of cellular morphological responses to small-molecule treatment. J. Biomol. Screen. **18**(10), 1321–1329 (2013). https://doi.org/10.1177/1087057113503553
3. McQuin, C., et al.: Cell Profiler 3.0: next-generation image processing for biology. PLoS Biol. **16**(7), e2005970 (2018). https://doi.org/10.1371/journal.pbio.2005970
4. Schindelin, J., Rueden, C.T., Hiner, M.C., Eliceiri, K.W.: The ImageJ ecosystem: an open platform for biomedical image analysis. Mol. Reprod. Dev. **82**(7–8), 518–529 (2015). https://doi.org/10.1002/mrd.22489
5. Berg, S., et al.: Ilastik: interactive machine learning for (bio)image analysis. Nat. Methods **16**(12), 1226–1232 (2019). https://doi.org/10.1038/s41592-019-0582-9
6. Bengio, Y., Courville, A., Vincent, P.: Representation learning: a review and new perspectives. IEEE Trans. Pattern Anal. Mach. Intell. **35**(8), 1798–1828 (2013). https://doi.org/10.1109/TPAMI.2013.50
7. Locatello, F., et al.: Challenging common assumptions in the unsupervised learning of disentangled representations. Proc. Mach. Learn. Res. **97**, 4114–4124 (2019). http://proceedings.mlr.press/v97/locatello19a.html
8. Tschannen, M., Bachem, O., Lucic, M.: Recent advances in autoencoder-based representation learning. In: NeurIPS Workshop on Bayesian Deep Learning (2018). https://arxiv.org/abs/1812.05069
9. Kimmel, J.C.: Disentangling latent representations of single cell RNA-seq experiments. bioRxiv 972166 (2020). https://doi.org/10.1101/2020.03.04.972166
10. Godinez, W.J., et al.: A multi-scale convolutional neural network for phenotyping high-content cellular images. Bioinformatics **33**(13), 2010–2019 (2017). https://doi.org/10.1093/bioinformatics/btx069
11. Way, G.P., Greene, C.S.: Extracting a biologically relevant latent space from cancer transcriptomes with variational autoencoders. In: Proceedings of the Pacific Symposium on Biocomputing (PSB 2018) (2018). https://doi.org/10.1142/9789813235533_0008
12. Neumann, B., et al.: Phenotypic profiling of the human genome by time-lapse microscopy reveals cell division genes. Nature **464**(7289), 721–727 (2010). https://doi.org/10.1038/nature08869
13. Fillbrunn, A., et al.: KNIME for reproducible cross-domain analysis of life science data. J. Biotechnol. **261**, 149–156 (2017). https://doi.org/10.1016/j.jbiotec.2017.07.028

14. Pawlowski, N., Caicedo, J.C., Singh, S., Carpenter, A.E., Storkey, A.: Automating morphological profiling with generic deep convolutional networks. bioRxiv 085118 (2016). https://doi.org/10.1101/085118
15. Maaten, L., van der, Hinton, G.: Visualizing data using t-SNE. J. Mach. Learn. Res. **9**, 2579–2605 (2008). https://www.jmlr.org/papers/v9/vandermaaten08a.html
16. McInnes, L., Healy, J., Melville, J.: UMAP: uniform manifold approximation and projection for dimension reduction. arXiv:1802.03426 (2018). https://arxiv.org/abs/1802.03426
17. Lafarge, M.W., et al.: Capturing single-cell phenotypic variation via unsupervised representation learning. Proc. Mach. Learn. Res. **102**, 315–325 (2019). https://proceedings.mlr.press/v102/lafarge19a.html
18. Park, E.: Manifold learning with variational auto-encoder for medical image analysis. Technical Report, University of North Carolina at Chapel Hill (2015). https://www.cs.unc.edu/~eunbyung/papers/manifold_variational.pdf
19. Higgins, I., et al.: β-VAE: learning basic visual concepts with a constrained variational framework. In: Proceedings of the International Conference on Learning Representations (ICLR 2017) (2017). https://openreview.net/forum?id=Sy2fzU9gl
20. Kingma, D.P., Welling, M.: Auto-encoding variational Bayes. arXiv:1312.6114 (2014). https://arxiv.org/abs/1312.6114
21. Lock, J.G., et al.: Visual analytics of single cell microscopy data using a collaborative immersive environment. In: Proceedings of the 16th ACM SIGGRAPH International Conference on Virtual-Reality Continuum and its Applications in Industry (VRCAI 2018) (2018). https://doi.org/10.1145/3284398.3284412
22. Rezaabad, A.L., Vishwanath, S.: Learning representations by maximizing mutual information in variational autoencoders. arXiv:1912.13361 (2020). https://arxiv.org/abs/1912.13361
23. Belghazi, M.I., et al.: MINE: mutual information neural estimation. arXiv: 1801.04062 (2018). https://arxiv.org/abs/1801.04062
24. Dilokthanakul, N., et al.: Deep unsupervised clustering with Gaussian mixture variational autoencoders. arXiv:1611.02648 (2017). https://arxiv.org/abs/1611.02648
25. Tomczak, J.M., Welling, M.: VAE with a VampPrior. Proc. Mach. Learn. Res. **84**, 1214–1223 (2018). https://arxiv.org/abs/1705.07120
26. Jiang, Z., et al.: Variational deep embedding: an unsupervised and generative approach to clustering. arXiv:1611.05148 (2017). https://arxiv.org/abs/1611.05148
27. Zhao, H., et al.: Loss functions for neural networks for image processing. arXiv:1511.08861 (2018). https://arxiv.org/abs/1511.08861
28. Akaike, H.: A new look at the statistical model identification. IEEE Trans. Autom. Control **19**(6), 716–723 (1974). https://doi.org/10.1109/TAC.1974.1100705
29. Drton, M., Plummer, M.: A Bayesian information criterion for singular models. J. Roy. Stat. Soc. Ser. B (Stat. Methodol.) **79**(2), 323–380 (2017). https://doi.org/10.1111/rssb.12187
30. Matthey, L., Higgins, I., Hassabis, D., Lerchner, A.: dSprites - disentanglement testing sprites dataset. GitHub (2017). https://github.com/deepmind/dsprites-dataset/
31. Ljosa, V., Sokolnicki, K.L., Carpenter, A.E.: Annotated high-throughput microscopy image sets for validation. Nat. Methods **9**(7), 637–637 (2012). https://doi.org/10.1038/nmeth.2083
32. Lueks, W., et al.: How to evaluate dimensionality reduction? Improving the co-ranking matrix. arXiv:1110.3917 (2011). https://arxiv.org/abs/1110.3917

33. Pandit, R., et al.: A principled comparative analysis of dimensionality reduction techniques on protein structure decoy data. In: Proceedings of the International Conference on Bioinformatics and Computational Biology (ICBCB 2016) (2016). https://cs.gmu.edu/~ashehu/
34. Adel, T., Ghahramani, Z., Weller, A.: Discovering interpretable representations for both deep generative and discriminative models. Proc. Mach. Learn. Res. **80**, 50–59 (2018). https://proceedings.mlr.press/v80/adel18a.html
35. Ridgeway, K., Mozer, M.C.: Learning deep disentangled embeddings with the F-statistic loss. arXiv:1802.05312 (2018). https://arxiv.org/abs/1802.05312

Detecting Extremely Small Lesions in Mouse Brain MRI with Point Annotations via Multi-task Learning

Xiaoyang Han[1,2], Yuting Zhai[1,2], Ziqi Yu[1,2], Tingying Peng[3],
and Xiao-Yong Zhang[1,2(✉)]

[1] Institute of Science and Technology for Brain-Inspired Intelligence, Fudan University,
Shanghai 200433, People's Republic of China
xiaoyong_zhang@fudan.edu.cn
[2] Key Laboratory of Computational Neuroscience and Brain-Inspired Intelligence, Fudan
University, Ministry of Education, Shanghai 200433, People's Republic of China
[3] Helmholtz AI, Helmholtz Zentrum, München, Germany
tingying.peng@tum.de

Abstract. Detection of small lesions in magnetic resonance imaging (MRI) images is one of the most challenging tasks. Compared with detection in natural images, small lesion detection in MRI images faces two major problems: First, small lesions only occupy a small fraction of voxels within an image, yielding insufficient features and information for them to be distinguished from the surrounding tissues. Second, an accurate outline of these small lesions manually is time-consuming and inefficient even for medical experts in pathology. Hence, existing methods cannot accurately detect lesions with such a limited amount of information. To solve these problems, we propose a novel multi-task convolutional neural network (CNN), which simultaneously performs regression of lesion number and detection of lesion location. Both lesion number and location can be obtained through point annotations, which is much easier and efficient than a full segmentation of lesion manually. We use an encoder-decoder structure that outputs a distance map of each pixel to the nearest lesion centers. Additionally, a regression branch is added after the encoder to learn the counting of lesion numbers, thus providing an extra regularization. Note that these two tasks share the same encoder weights. We demonstrate that our model enables the counting and locating of extremely small lesions within 3–5 voxels (300×300 voxels per image) with a recall of 72.66% on a large mouse brain MRI image dataset (more than 1000 images), and outperforms other methods.

Keywords: MRI · Mouse brain · Lesion detection · Point annotations · Multi-task learning

1 Introduction

For the detection of lesions in MRI images, deep convolutional neural networks (CNN) have achieved excellent performances [1–3]. Despite these achievements, it is still difficult to detect small lesions in MRI images. Compared with detection in natural images,

© Springer Nature Switzerland AG 2021
C. Lian et al. (Eds.): MLMI 2021, LNCS 12966, pp. 498–506, 2021.
https://doi.org/10.1007/978-3-030-87589-3_51

small lesions detection in MRI images faces two major challenges. (1) The lesions can have an extremely small size, which only occupies a small fraction of voxels. And they have a similar appearance to other tissues. (2) The manual annotation of lesions is time-consuming and expensive because it requires pathologists who have medical expertise. Additionally, manual labeling of lesions varies substantially between different annotators due to individual medical experiences. Even if two human experts count the same number of lesions, the locations they annotate can be different.

Some existing methods use ensemble learning [4, 5] for small lesion detection by cropping different sizes of ROI, yet they are slow due to the enormous amount of computation. By contrast, fully convolutional methods [6, 7] are proposed to train an end-to-end network to detect small lesions, which are much faster in terms of computation time. In this paper, we propose a novel multi-task learning convolutional neural network, which simultaneously performs regression of the lesion number and detection of lesion location. Both lesion numbers and locations can be obtained through point annotation, i.e., pointing to a dot on each lesion. Compared to a full segmentation of lesions, the point annotation is more efficient and consistent [8, 9]. We use an encoder-decoder structure and construct our network in a fully convolutional architecture that outputs the location of each lesion. A regression branch is added after the encoder to learn the count of lesions. The addition of a counting regression branch is inspired by the previous work that the count can help the network understand the feature of the lesions [10–12]. In our model, both tasks share the same encoder weights and can be mutually benefited from each other. We evaluate our model on a point annotation dataset for PM2.5 particles, particulate matter with aerodynamic diameter $\leq 2.5\ \mu m$, in mouse brain MRI images. Moreover, we release our code and trained models as an open-source tool so that a broad research community can freely use it.

2 Methods: Multi-task Learning

Our multi-task convolutional neural network with the encoder-decoder structure is shown in Fig. 1. Our network architecture is similar to U-net but a count regression branch is added after the last convolutional layer in the encoder. The multi-task loss function consists of a location loss that compares the ground-truth distance map with the generated one, and a regression loss that quantifies the error in counting lesion numbers.

2.1 Distance Map Branch

For point annotations, Euclidean distance maps and geodesic distance maps are calculated using the method proposed in [9]. After the last layer in the decoder of the network, a convolutional layer with 1×1 kernels and a sigmoid activation function is added to learn the distant map. We use mean square error loss between output map and distance map as the loss function.

$$L_{dm} = \frac{1}{N} \sum_{n=1}^{N} \left(M_n - \hat{M}_n \right)^2 \tag{1}$$

2.2 Count Regression Branch

Our count regression network consists of a block of consecutively stacked convolutional layers with 3×3 kernels, followed by a global max-pooling and fully connected layers to output a scalar. For regression of lesion number, we use mean square error loss as the optimization function.

$$L_{cnt} = \frac{1}{N} \sum_{n=1}^{N} \left(y_n - \hat{y}_n \right)^2 \qquad (2)$$

The final loss function for multi-task learning is defined by

$$L_{multi} = L_{cnt} + \lambda L_{dm} \qquad (3)$$

with $\lambda \in [0, 1]$.

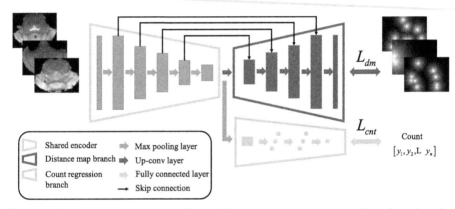

Fig. 1. Multi-task learning structure. Our multi-task learning structure consists of two learning targets: a distance map branch (blue) that learns the distance of each pixel to the nearest lesion and a count regression branch (green) to count the number of lesions. Both branches share the same encoder (orange). The distance map branch uses a rectified learning unit activation after each convolutional layer whilst the count regression branch uses a global max-pooling layer followed by fully connected layers. (Color figure online)

3 Experiments

3.1 Data Description

Particulate matters with aerodynamic diameter $\leq 2.5\ \mu m$ (PM2.5 particle) were delivered into the mouse brain by tail intravenous injection. In vivo imaging was performed using a small animal MRI. T2-weighted imaging (T2WI) was used to acquire brain anatomical images. Susceptibility weighted imaging (SWI) was used to acquire brain images containing the information of PM2.5 particles as shown in Fig. 2. PM2.5 particles in SWI images are shown as hypointensitive lesions. The main parameters for T2WI and SWI are as follows: field of view 18 mm × 18 mm, matrix size 300 × 300, slice thickness 0.5 mm, 30 slices covering the whole mouse brain. The voxel size is 0.06 × 0.06 × 0.5 mm³.

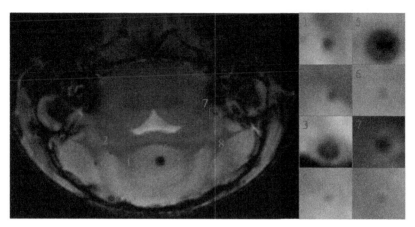

Fig. 2. Representative PM2.5 particles in the mouse brain. In the left image, PM2.5 particles are circled in red in mouse brain MRI and numbered from left to right. In the right columns, each PM2.5 particle is cropped as a size of 11 × 11 voxels patch and zoomed in. (Color figure online)

Our dataset consists of 1312 mouse brain SWI images, including 1202 images with point annotations as a training dataset and only 110 images with full segmentation annotations as testing dataset. For point annotations, labels only need to annotate a voxel for each lesion, whilst full masks on all voxels are available for segmentation annotations. T2WI scans are used to obtain a binary mask to remove other irrelevant tissues.

In total, 21622 lesions are labeled on 1202 images in the point annotation dataset and 1735 lesions in the test dataset. As shown in Table 1, 80.52% of lesions are less than 5 voxels in the testing dataset. Note, the voxel size is extremely small, only 0.06 × 0.06 × 0.5 mm³ per voxel and therefore it is challenging to detect these lesions with such small size.

Table 1. Statistics of lesion size in the testing dataset.

Lesion type	Lesion size[a]	Lesion count
Extreme small lesion	Less than 5 voxels	1397 (80.52%)
Small lesion	5–10 voxels	253 (14.58%)
Large lesion	More than 10 voxels	85 (4.90%)

[a]The voxel size is 0.06 × 0.06 × 0.5 mm³

3.2 Preprocessing

Since the existing methods for mouse brain extraction are not effective, we additionally trained a U-net model with T2WI images to obtain binary masks for brain extraction. The binary mask is dilated, smoothed by a Gaussian kernel and multiplied with the original image to remove other irrelevant tissues such as the skull. The mouse brain areas near the annotations are cropped to obtain a fixed size of 96×128 ROI images. ROI images are normalized to [0, 1] by min-max normalization.

3.3 Experimental Setup

The weights of the CNN are initialized by the normal initializer proposed by He et al. [13]. During the training, each mini-batch has 128 samples. We use Adam [14] optimization with a learning rate of $1e - 4$ as the optimizer. Each model is trained for 5000 epochs. We implement our methods in Python and Keras with TensorFlow as backend and run the experiments on a Nvidia GeForce GTX 1080 GPU.

3.4 Results

Baseline Methods. We implemented three existing approaches as baseline methods, using the same preprocessing: (a) a regression network [11] without the up-sampling path. (b) GP-Unet [10] combines a fully convolutional architecture with global pooling to solve the regression problem. (c) DM-FCN [9] applies a fully convolutional architecture to regress the distance maps. All three methods can count the number of lesions, yet only (a) and (c) are able to output the location of each lesion. All methods are trained on the point annotation dataset and evaluated on the same test dataset.

Count Regression Results. Table 2 reports the count regression results of our method in comparison to the baseline results. Our multi-task learning method (d) performs better than all three baseline methods in the regression task by a large margin, as demonstrated by a statistically lower mean square error (MSE), mean absolute error (MAE) and mean absolute percentage error (MAPE).

Distance Map Localization Results. We set a series of thresholds on the output of the distance map branch to obtain the same number of connected regions as the lesion number output from the count regression branch. We regard the obtained connected regions on the distance map as the detection results of the distance map branch. Since our models are trained on point annotation dataset and evaluated on full segmentation dataset, we define the detection results that have IOU (Intersection over Union) more than 0.3 with any lesion region of ground truth as the true positive. For regression network (a), saliency maps [15] are computed with respect to the input image using keras-vis [16] as the localization results. In Table 3 we report recall, precision and average false positive per image (FPavg). Trained on the point annotation dataset, our method (d) reaches a higher recall and precision than baseline methods, achieving a recall of 72.66% and a precision of 45.81%. The representative localization results are shown in Fig. 3.

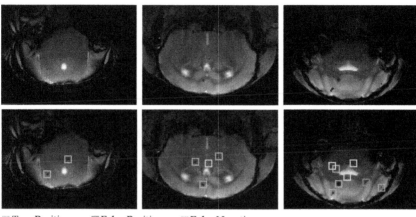

☐ True Positive   ☐ False Positive   ☐ False Negative

Fig. 3. Localization results. Top row: ground truth; bottom row: lesion localization prediction. In the lesion localization prediction (bottom): red mask for localization results, green rectangle for true positive, blue rectangle for false positive and yellow rectangle for false negative. (Color figure online)

Table 2. Accuracy in counting the lesion number. MSE stands for mean squared error, MAE stands for mean absolute error and MAPE stands for mean absolute percentage error.

Method	MSE	MAE	MAPE
Regression network (a)	7.32	1.89	45.06
GP-Unet (b)	15.73	3.18	52.80
DM-FCN (c)	8.23	2.04	95.92
Our network (d)	**5.64**	**1.58**	**37.07**

Table 3. Accuracy in lesion localization. FPavg is the number of false positives per image.

Method	Recall	Precision	FPavg
Regression network (a)	51.80	15.98	4.38
GP-Unet (b)	\	\	\
DM-FCN (c)	67.37	32.16	3.80
Our network (d)	**72.66**	**45.81**	2.67

Ablation Studies. Euclidean distance maps and Geodesic distance maps are calculated for point annotation dataset to provide extra information for localization as shown in Fig. 4. Results in Table 4 show that using Euclidean distance maps perform better than geodesic distance maps. This indicates that Euclidean distance is the most discriminative information in distance maps.

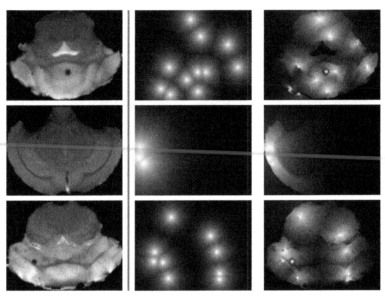

Fig. 4. Euclidean and Geodesic distance maps. From left to right: preprocessed input mouse brain images, Euclidean distance maps, Geodesic distance maps.

Table 4. Accuracy in counting the lesion number and lesion localization with different distance maps.

Distance map	MSE	MAE	Sensitivity	Precision	FPavg
Geodesic	6.04	1.80	50.31	38.07	2.82
Euclidean	**5.64**	**1.58**	**72.66**	**45.81**	**2.67**

Fine-Tune and Training Strategies. Fine-tune methods are used when training multi-task networks. Since the network consists of two branches for counting regression task and distance map localization task, we separately train the counting first model (e) and localization first model (f) as comparison with simultaneously trained model(d) shown in Table 5. The results show that the simultaneously trained models perform better than fine-tune models, indicating that the lesion count can help the network understand the features of the lesions.

Table 5. Accuracy of fine-tune models compared to simultaneously trained models.

Training strategy	MSE	MAE	Recall	Precision	FPavg
Counting first (e)	6.21	1.78	51.22	40.35	4.88
Localization first (f)	6.78	1.92	66.81	42.59	3.30
Combination (d)	**5.64**	**1.58**	**72.66**	**45.81**	**2.67**

4 Conclusion

We developed a multi-task learning model for extremely small lesion detection tasks, i.e., to detect PM2.5 particles, in MRI images. The proposed end-to-end network can simultaneously count the number of extremely small lesions and detect the location of each lesion within the mouse brain. For this challenging task, our model achieves a recall of 73%, which outperforms other state-of-the-art methods. Although the precision achieved by our method is about 13% higher than other state-of-the-art methods, it is still relatively low (46%), indicating that extremely small lesion segmentation in MRI images remains a very difficult task. Note that the PM2.5 lesions in this work are mostly under 5 voxels, which is much smaller than conventional disease lesions, thus making our task even more challenging. Therefore, we will also release our point annotation datasets as well as our tool to attract a broader research community working in this challenging yet biologically important problem.

Acknowledgements. This study was supported in part by Shanghai Municipal Science and Technology Major Project (No. 2018SHZDZX01), ZJLab, and Shanghai Center for Brain Science and Brain-Inspired Technology, the National Natural Science Foundation of China (81873893), Science and Technology Commission of Shanghai Municipality (20ZR1407800), the 111 Project (B18015), and the Major Research plan of the National Natural Science Foundation of China (KRF201923).

References

1. Yan, K., Bagheri, M., Summers, R.M.: 3D context enhanced region-based convolutional neural network for end-to-end lesion detection. In: Frangi, A.F., Schnabel, J.A., Davatzikos, C., Alberola-López, C., Fichtinger, G. (eds.) MICCAI 2018. LNCS, vol. 11070, pp. 511–519. Springer, Cham (2018). https://doi.org/10.1007/978-3-030-00928-1_58
2. Tao, Q., Ge, Z., Cai, J., Yin, J., See, S.: Improving deep lesion detection using 3D contextual and spatial attention. In: Shen, D., et al. (eds.) MICCAI 2019. LNCS, vol. 11769, pp. 185–193. Springer, Cham (2019). https://doi.org/10.1007/978-3-030-32226-7_21
3. Atlason, H.E., Love, A., Sigurdsson, S., Gudnason, V., Ellingsen, L.M.: SegAE: unsupervised white matter lesion segmentation from brain MRIs using a CNN autoencoder. NeuroImage Clin. **24**, 102085 (2019). https://doi.org/10.1016/j.nicl.2019.102085
4. Kamnitsas, K., et al.: Efficient multi-scale 3D CNN with fully connected CRF for accurate brain lesion segmentation. Med. Image Anal. **36**, 61–78 (2017). https://doi.org/10.1016/j.media.2016.10.004

5. Savelli, B., Bria, A., Molinara, M., Marrocco, C., Tortorella, F.: A multi-context CNN ensemble for small lesion detection. Artif. Intell. Med. **103**, 101749 (2020). https://doi.org/10.1016/j.artmed.2019.101749

6. Ronneberger, O., Fischer, P., Brox, T.: U-net: convolutional networks for biomedical image segmentation. In: Navab, N., Hornegger, J., Wells, W.M., Frangi, A.F. (eds.) MICCAI 2015. LNCS, vol. 9351, pp. 234–241. Springer, Cham (2015). https://doi.org/10.1007/978-3-319-24574-4_28

7. Oktay, O., et al.: Attention U-Net: Learning Where to Look for the Pancreas (2018). https://arxiv.org/abs/1804.03999

8. Yoo, I., Yoo, D., Paeng, K.: PseudoEdgeNet: nuclei segmentation only with point annotations. In: Shen, D., et al. (eds.) MICCAI 2019. LNCS, vol. 11764, pp. 731–739. Springer, Cham (2019). https://doi.org/10.1007/978-3-030-32239-7_81

9. van Wijnen, K.M.H., et al.: Automated lesion detection by regressing intensity-based distance with a neural network. In: Shen, D., et al. (eds.) MICCAI 2019. LNCS, vol. 11767, pp. 234–242. Springer, Cham (2019). https://doi.org/10.1007/978-3-030-32251-9_26

10. Dubost, F., et al.: GP-Unet: lesion detection from weak labels with a 3D regression network. In: Descoteaux, M., Maier-Hein, L., Franz, A., Jannin, P., Collins, D.L., Duchesne, S. (eds.) MICCAI 2017. LNCS, vol. 10435, pp. 214–221. Springer, Cham (2017). https://doi.org/10.1007/978-3-319-66179-7_25

11. Dubost, F., et al.: 3D regression neural network for the quantification of enlarged perivascular spaces in brain MRI. Med. Image Anal. **51**, 89–100 (2019). https://doi.org/10.1016/j.media.2018.10.008

12. Dubost, F., et al.: Enlarged perivascular spaces in brain MRI: automated quantification in four regions. Neuroimage **185**, 534–544 (2019). https://doi.org/10.1016/j.neuroimage.2018.10.026

13. He, K., Zhang, X., Ren, S., Sun, J.: Delving deep into rectifiers: surpassing human-level performance on ImageNet classification. In: 2015 IEEE International Conference on Computer Vision (ICCV), Santiago, Chile. pp. 1026–1034. IEEE (2015). https://doi.org/10.1109/ICCV.2015.123

14. Kingma, D.P., Ba, J.: Adam: A Method for Stochastic Optimization (2017). https://arxiv.org/abs/1412.6980

15. Simonyan, K., Vedaldi, A., Zisserman, A.: Deep Inside Convolutional Networks: Visualising Image Classification Models and Saliency Maps (2014). https://arxiv.org/abs/1312.6034

16. Kotikalapudi, R.: Contributors: keras-vis. https://github.com/raghakot/keras-vis

Morphology-Guided Prostate MRI Segmentation with Multi-slice Association

Jianping Li[1], Zhiming Cui[2], Shuai Wang[3], Jie Wei[4], Jun Feng[1(✉)], Shu Liao[5], and Dinggang Shen[5,6(✉)]

[1] School of Information Science and Technology, Northwest University, Xi'an, China
fengjun@nwu.edu.cn
[2] Department of Computer Science, The University of Hong Kong, Hong Kong, China
[3] School of Electrical and Information Engineering, Shandong University, Weihai, China
[4] School of Computer Science and Engineering, Northwestern Polytechnical University, Xi'an, China
[5] Shanghai United Imaging Intelligence Co., Ltd., Shanghai, China
[6] School of Biomedical Engineering, ShanghaiTech University, Shanghai, China
dgshen@shanghaitech.edu.cn

Abstract. Prostate segmentation from magnetic resonance (MR) images plays an important role in prostate cancer diagnosis and treatment. Previous works typically overlooked large variations of prostate shapes, especially on the boundary area. Furthermore, the small glandular areas at ending slices also make the task very challenging. To overcome these problems, this paper presents a two-stage framework that explicitly utilizes prostate morphological representations (e.g., point, boundary) to accurately localize the prostate region with a coarse volumetric segmentation. Based on the 3D coarse outputs of the first stage, a 2D segmentation network with multi-slice association is further introduced to produce more reliable and accurate segmentation, due to large slice thickness in prostate MR images. Besides, several novel loss functions are further designed to enhance the consistency of prostate boundaries. Extensive experiments on large prostate MRI dataset show superior performance of our proposed method compared to several state-of-the-art methods.

Keywords: Prostate MRI segmentation · Morphological representation · Multi-slice association

1 Introduction

Prostate cancer is one of the most common types of cancer, and magnetic resonance imaging (MRI) is widely used as a non-invasive method to assist physicians in the diagnosis and treatment of prostate cancer [1]. Among various sequences of MRI (T2W, DWI, ADC, etc.), T2-weighted (T2W) MRI has been widely used

© Springer Nature Switzerland AG 2021
C. Lian et al. (Eds.): MLMI 2021, LNCS 12966, pp. 507–516, 2021.
https://doi.org/10.1007/978-3-030-87589-3_52

in prostate segmentation, because it has the most abundant anatomical structures [2] for clinical prostate volume measurement [3, 4]. However, for clinicians, it is extremely time-consuming and laborious to manually annotate prostate masks from MRI images. Therefore, it is a practical demand to develop an automatic and accurate prostate segmentation method for T2W MR images.

Although considerable efforts have been taken to improve prostate segmentation performance, it is still a challenging task to accurately extract prostate from MR images, due to the following reasons. (1) Since the prostate itself is small and tends to press against surrounding organs, such as bladder and rectum, the prostate boundaries in MR images are usually blurred [5], with low tissue contrast as shown in Fig. 1(a), thus making segmentation difficult. (2) The shapes and tissue textures across different individuals are with large variations as shown in Fig. 1(a) and Fig. 1(b). (3) Large slice thickness of the prostate MR images usually leads to small prostate areas in the ending slices, where the prostate boundary contours are often irregular and hard to segment as shown in Fig. 1(c) and Fig. 1(d). Therefore, it is essential to develop a prostate segmentation algorithm that can widely adapt to various conditions.

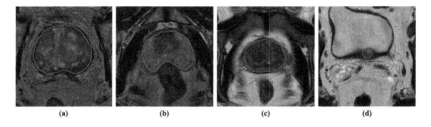

Fig. 1. Sample images from the dataset used in this paper. (a–b) show variation of T2W across different individuals, and (c–d) show different slices from the same individual.

With successful applications of deep learning in image processing and analysis, learning-based methods using convolution neural networks for prostate segmentation have been proposed increasingly in recent years. Milletari et al. [6] proposed an encoder-decoder architecture network (V-Net) that can be applied to 3D images, whose basic structure is very similar to U-Net [7]. Yu et al. [8] proposed to add the residual module to the 3D segmentation network for prostate segmentation. Zhu et al. [9] proposed a boundary-weighted domain adaptive neural network to segment the prostate from MRI. Liu et al. [10] performed prostate segmentation using meta-learning and compactness constraint. The average performance of the 3D-based segmentation network is lower than that of the 2D-based segmentation network due to large slice thickness [11]. However, 2D segmentation cannot maintain generalization for the slices with different glandular regions. Hence, it is necessary to combine their advantages by explicitly utilizing the prostate typological representations in 3D space and simultaneously considering the multi-slice association in 2D space.

In this paper, we propose a novel two-stage morphology-guided prostate MRI segmentation network with the multi-slice association. In the first stage, to tackle the large variations of the prostate shapes, we initially employ a multi-task 3D segmentation network to simultaneously predict the prostate volume and corresponding morphological features, including the prostate endpoints of three orthogonal planes and the boundary. We consider that the three tasks are intrinsically related from a geometric perspective. In the second stage, a 2D segmentation network with a slice-attention mechanism is designed to enhance the associated information between slices, which benefits the segmentation in return, especially on the slice with poor prediction at the first stage. Moreover, several novel loss functions are introduced, including a 3D compact loss to improve the smoothness of the predicted prostate shape and a 2D-weighted Dice loss to alleviate the problem of missing segmentation at ending slices. We have extensively evaluated our method on the prostate MRI dataset collected from real-world clinics. The experimental results show that the proposed approach achieves superior performance and outperforms several state-of-the-art performances by a large margin.

2   Our Method

The proposed method is a two-stage framework as shown in Fig. 2. Morphology-guided multi-task learning has been widely used in medical image segmentation tasks [12,13]. Considering that more morphological information can be obtained in the 3D space, we first use a 3D coarse segmentation network to locate and extract the boundary of the prostate, and then refine segmentation for each slice through the 2D network. In the first stage, the segmentation network uses a V-Net-like densely supervised Encoder-Decoder network in a multi-task manner, which consists of three branches to predict key points, boundary, and volume mask, respectively. We aim to constrain the overall shape of the prostate through three morphological information to avoid incorrect segmentation outside the prostate. Then, we combine the probability map in the first stage and the original input image into a 2-channel 2D image according to each slice, and use it as the input of the second stage network. In the second stage, the fine segmentation network performs 2D segmentation for the output from the previous stage, which applies a slice-attention mechanism to enhance the associated information among slices.

2.1   Morphology-Guided Coarse Segmentation

Existing methods of prostate segmentation tend to obtain over-segmentation of the prostate, due to their lack of monitoring of the prostate position and boundary. In order to get a more accurate shape of the prostate, we use a modified V-Net with multi-task consistency to generate uniform prediction at different morphological levels of three branches, including point, boundary, and volume branches.

Modified V-Net. The modified V-Net consists of a top-down contracting path and a bottom-up expansive path, which enables to capture both rich contextual

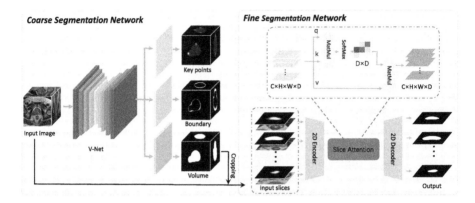

Fig. 2. The architecture of our two-stage prostate MRI segmentation network. In the first stage, the key point prediction branch and boundary prediction branch are added to the modified V-Net, and they use different loss functions to locate and constrain the prostate. The second stage is a 2D fine segmentation network, which uses the slice-attention mechanism to guide the segmentation of ending slices by weighting the deep features.

information and accurate localization information. The contracting path follows the repeated application of down-sampling blocks, while the expansive path consists of repeated up-sampling blocks. The down-sampling block is a $3 \times 3 \times 3$ convolution with stride 2 and doubles the output channels, and the up-sampling block is a $3 \times 3 \times 3$ transposed convolution with stride 2 and reduces the output channels by half. Each convolution layer in the network is followed by a batch normalization layer and a rectified linear unit layer.

Key Point Prediction Branch. To detect the localization of the prostate, we add a key point prediction branch to predict six outermost positions of the prostate in the 3D space. In addition, it can help to identify small area in the slice, and guide the segmentation of the two ends of the prostate. Specifically, we use Gaussian filter on the output to smooth the probability map and reduce the prediction variance. The loss function of key point prediction branch is the mean squared error (MSE) loss, and denote it as $\mathcal{L}_{\text{MSE}-\text{KeyPoint}}$.

Boundary Prediction Branch. The artifacts generated by squeezing between the prostate and surrounding organs, such as bladder and rectum, usually result in inaccurate boundary contour of segmentation result. Such distorted overall shape of the prostate is difficult for pathologists to accept. To make the network pay more attention to the fuzzy boundary region, we introduce a boundary prediction branch to generate more accurate boundary of the prostate. It uses a single convolution layer to predict the boundary, so that the features of network learning can be more strongly expressed in the boundary region. The ground truth of boundary prediction is acquired from morphological operation. The loss function of boundary prediction branch is also MSE loss, and denote it as $\mathcal{L}_{\text{MSE}-\text{Boundary}}$.

Volume Mask Prediction Branch. The prediction of volume mask is the most important segmentation task with the help of the first two branches. For prostate MRI sequences, the thickness among slices is usually large. Thus, the segmentation results of slice, containing prostate with small areas, cannot be improved by using 3D segmentation model. Based on the traditional Dice loss $\mathcal{L}_{\text{Dice}} = 1 - (2|\hat{y} \cap y| + 1)/(|\hat{y}| + |y| + 1)$, we design a novel 2D-weighted Dice loss to consider the contribution of slice with different glandular areas, defined as:

$$\mathcal{L}_{\text{Weighted-Dice}} = \sum_{i=0}^{m} \alpha_i \mathcal{L}_{\text{Dice}}^i \tag{1}$$

where α_i and $\mathcal{L}_{\text{Dice}}^i$ represent the weight and Dice loss of the i-th slice, respectively. For slices with large glandular areas, the segmentation results are usually good, so the weights for the slices with big glandular areas should be decreased, and the weights for the slices with small glandular areas should be increased. The weight of each slice is negatively correlated with its glandular area. When the ground truth of the i-th slice has prostate glands, α_i is expressed as Eq. (2); otherwise it is $1/D$. D, E, S_i, respectively, represent the number of slices in a sequence, the number of slices with glands in the sequence, and the area of the glands on the slice.

$$\alpha_i = \frac{E}{D} \times \frac{\exp(-S_i)}{\sum_{i=1}^{E} \exp(-S_i)} \tag{2}$$

The observation that the prostate region is usually compact in shape can also serve as a prior knowledge to guide network to eliminate differences in the shapes of the prostates in different individuals. We define a compact constraint loss, inspired from [10], as follows:

$$\mathcal{L}_{\text{Compact}} = \frac{A^3}{36\pi V^2} = \frac{\sum_{i \in \Omega} \sqrt{(\nabla p_{u_i})^2 + (\nabla p_{v_i})^2 + (\nabla p_{w_i})^2 + \epsilon}}{36\pi \left(\sum_{i \in \Omega} |p_i| + \epsilon\right)} \tag{3}$$

where A and V are the area and volume of the shape of prostate, respectively. Compared with the traditional pixel-based segmentation loss function, the compact loss provides a better global constraint for a complete and compact shape such as the prostate.

Loss Function. The total loss of coarse segmentation in the first stage is composed of four items, as shown below:

$$\mathcal{L}_{\text{Total}} = \mathcal{L}_{\text{Weighted-Dice}} + \lambda \mathcal{L}_{\text{Compact}} + \mathcal{L}_{\text{MSE-KeyPoint}} + \mathcal{L}_{\text{MSE-Boundary}} \tag{4}$$

where λ is the weighting trade-off, and set as 0.005 in all our experiments.

2.2 Fine Segmentation with Multi-slice Association

The coarse segmentation stage gets a good prediction for the overall profile of the prostate. However, the results of the slices with small part of prostate at both ends of the prostate, are still crude. Thus, we aim to make more fine segmentation on the basis of the first stage. Specifically, in the second stage, we use a 2D network for the fine segmentation of each slice, whose input is the combination of original image and probability map from the coarse segmentation. The architecture of 2D fine segmentation network is illustrated in Fig. 2, which contains three parts: encoder, decoder, and slice-attention module. The slice-attention module aims to enhance multi-slice association to improve the segmentation performance of ending slices.

Encoder and Decoder. The encoder part of the fine segmentation network has four blocks. A block consists of two 3×3 convolution layers, each being followed by a rectified liner unit and down-sampling layer with the stride of 2. The number of channels for feature maps is doubled after each block. For a sequence of D slices, the encoder will give D feature maps, which represent the high-level features of these slices. The decoder part works symmetrically with the encoder, in which the up-sampling is composed of 3×3 deconvolution.

Slice-Attention. Self-attention refers to an attention mechanism, which has been widely used in natural language processing [14], object detection and segmentation [15,16], etc. To obtain the global relationship among slices, we refer to the non-local self-attention module to extract weighted features for better segmentation. The key component of slice-attention is calculated as follows:

$$A = \text{Softmax}\left(q(F)^T k(F)\right) \in R^{D \times D} \tag{5}$$

$$F' = v(F)A \in R^{C \times H \times W \times D} \tag{6}$$

where $F \in R^{C \times H \times W \times D}$ is the input of the structure and F' is the output with the same size. $q(F)$ is a packed Query matrix. $k(F)$ and $v(F)$ stand for Key-Value pairs. We introduce slice-attention to enhance the continuity information between slices. The structure of the slice-attention is shown in Fig. 2. Different from traditional self-attention mechanism, the attention map of our slice-attention means the weight dependence between different slices. We calculate the dot product of Queries and Keys on the slice dimension, to obtain the attention map $A \in R^{D \times D}$, where D represents the number of slices. Then we apply the attention map to weight the Values to obtain the weighted feature maps for each slice, and use these feature maps as the input of decoder.

3   Implementation and Experiments

3.1   Data and Implementation Details

The proposed method of automatic prostate MRI segmentation is trained on 388 cases and tested on 97 cases. Each case is acquired from real-world clinics, and the ground truth of each case is produced by manual annotation of a board certified radiologist. Due to differences in voxel spacing and MRI intensity of the data acquired from different clinics, we preprocessed all the data, including bias field correction, spacing adjustment, and resolution adjustment. MR scans often display intensity non-uniformities due to variations in the magnetic field. We first use the ANTs tool for bias field correction and then adjusted the voxel spacing and resolution by SimpleITK. After processing, the voxel spacing of every MRI sequence is isotropic as $0.3\,mm \times 0.3\,mm \times 3\,mm$, and the image size is $384 \times 384 \times 32$. All of the training and testing images are also normalized by using a mean and standard deviation of training images' intensities.

3.2   Experimental Results

Table 1. The effect of different modules in our proposed method. The evaluation metrics used include Dice coefficient (DSC), Precision, Hausdorff distance (HD), and average surface distance (ASD).

Method	Metric			
	DSC (%)	Precision (%)	HD (mm)	ASD (mm)
V-Net (baseline)	85.66	85.19	0.356	1.318
V-Net+Mt	90.39	90.76	0.178	**0.726**
V-Net+Mt+FS	**92.65**	**93.56**	**0.154**	0.847

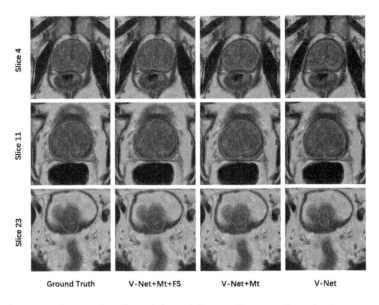

Fig. 3. Segmentation results. From left to right are the ground truth, the segmentation results of our proposed method, the segmentation results of V-Net with multi-task learning, and the segmentation results of V-Net.

Ablation Study. To evaluate the effectiveness of the added modules in our proposed method, we performed a set of ablation experiments. Results are shown in Table 1. We adopt the V-Net as the baseline. Multi-task prediction is abbreviated as Mt. Fine segmentation is abbreviated as FS. The morphological information is important since the model performance is worse when the multi-task prediction branches behind V-Net are disabled. Furthermore, the result of fine segmentation is better than multi-task learning in multiple indicators. The results in Table 1 reveal that the slice-attention mechanism can leverage multi-slice association more effectively.

Qualitative Comparison. Figure 3 shows the segmentation results of three slices from the same case under different ablation conditions. It can be observed that the segmentation performance is improved with the proposed components. The three slices in Fig. 3 are taken from the same individual's prostate MRI sequence. For slices with complete prostate shapes and clear boundaries, V-Net can achieve better results. When the prostate and surrounding organs are squeezed, the multi-task prediction module can locate prostate position more accurately. For the slice at the end, the size of the prostate is already small and irregular. Observation shows that the result of our fine segmentation using the slice-attention module is the closest to the ground truth.

Table 2. Quantitative comparison between the proposed method and other methods.

Method	Metric			
	DSC (%)	Precision (%)	HD (mm)	ASD (mm)
U-Net [7]	86.33	87.10	0.340	1.426
Att U-Net [17]	88.57	90.25	0.241	**0.830**
U-Net++ [18]	89.64	89.04	0.208	0.833
3D U-ResNet [8]	87.12	87.72	0.313	1.292
DenseVoxelNet [19]	88.71	88.55	0.310	1.255
ResNet3DMed [20]	90.18	91.01	0.201	0.913
Our Method	**92.65**	**93.56**	**0.154**	0.847

Comparison with Other Methods. We compared our method with common segmentation networks and other 3D volumetric MRI segmentation methods, as shown in Table 2. Our method surpasses these approaches, especially on Dice coefficient. We believe that most of the above segmentation methods are based on baseline, and some modules such as residual connection or dense connection are adopted for feature encoding or decoding structure. This is not a good solution to the challenge of the segmentation problem itself, such as the blurred boundaries of the prostate and the small-area slices. Overall, our method further improves the original V-Net model, which demonstrates the effectiveness of our multi-task prediction and slice-attention module. From Table 2, we can see our method achieves better results than other methods under all same settings.

4   Conclusion

We present a novel two-stage morphology-guided segmentation framework with multi-slice association for prostate MRI segmentation. Our method adopts 3D full convolutional architecture as the backbone network, and then performs multi-task prediction in different morphology levels. In the fine segmentation stage, the 2D segmentation network is adopted to enhance the association among slices for acquiring more robust segmentation results. In addition to verifying these two points through ablation study, we also compare with other methods that directly perform 3D segmentation, and obtain better results in multiple indicators on our dataset. Future investigations include assessing our method on more 3D volumetric data and further exploring multi-task prediction on morphology and slice-attention mechanism.

Acknowledgment. This work was supported by the National Natural Science Foundation of China under Grants 62073260.

References

1. Ahmed, H.U., et al.: Diagnostic accuracy of multi-parametric MRI and TRUS biopsy in prostate cancer (PROMIS): a paired validating confirmatory study. Lancet **389**(10071), 815–822 (2017)
2. Barentsz, J.O., et al.: ESUR prostate MR guidelines 2012. Eur. Radiol. **22**(4), 746–757 (2012)
3. Hoeks, C.M., et al.: Transition zone prostate cancer: detection and localization with 3-T multiparametric MR imaging. Radiology **266**(1), 207–217 (2013)
4. Toth, R., et al.: Accurate prostate volume estimation using multifeature active shape models on T2-weighted MRI. Acad. Radiol. **18**(6), 745–754 (2011)
5. Zhu, Q., Du, B., Turkbey, B., Choyke, P.L., Yan, P.: Deeply-supervised CNN for prostate segmentation. In: International Joint Conference on Neural Networks (IJCNN), pp. 178–184. IEEE (2017)
6. Milletari, F., Navab, N., Ahmadi, S.A.: V-Net: fully convolutional neural networks for volumetric medical image segmentation. In: Fourth International Conference on 3D Vision (3DV), pp. 565–571. IEEE (2016)
7. Ronneberger, O., Fischer, P., Brox, T.: U-net: convolutional networks for biomedical image segmentation. In: Navab, N., Hornegger, J., Wells, W.M., Frangi, A.F. (eds.) MICCAI 2015. LNCS, vol. 9351, pp. 234–241. Springer, Cham (2015). https://doi.org/10.1007/978-3-319-24574-4_28
8. Yu, L., Yang, X., Chen, H., Qin, J., Heng, P.A.: Volumetric convnets with mixed residual connections for automated prostate segmentation from 3D MR images. In: Proceedings of the AAAI Conference on Artificial Intelligence, vol. 31, no. (1) (2017)
9. Zhu, Q., Du, B., Yan, P.: Boundary-weighted domain adaptive neural network for prostate MR image segmentation. IEEE Trans. Med. Imaging **39**(3), 753–763 (2019)
10. Liu, Q., Dou, Q., Heng, P.-A.: Shape-aware meta-learning for generalizing prostate MRI segmentation to unseen domains. In: Martel, A.L., et al. (eds.) MICCAI 2020. LNCS, vol. 12262, pp. 475–485. Springer, Cham (2020). https://doi.org/10.1007/978-3-030-59713-9_46

11. Wang, B., et al.: Deeply supervised 3D fully convolutional networks with group dilated convolution for automatic MRI prostate segmentation. Med. Phys. **46**(4), 1707–1718 (2019)
12. He, K., Cao, X., Shi, Y., Nie, D., Gao, Y., Shen, D.: Pelvic organ segmentation using distinctive curve guided fully convolutional networks. IEEE Trans. Med. Imaging **38**(2), 585–595 (2018)
13. Luo, X., Chen, J., Song, T., Chen, Y., Wang, G., Zhang, S.: Semi-supervised medical image segmentation through dual-task consistency. arXiv preprint arXiv:2009.04448 (2020)
14. Vaswani, A., et al.: Attention is all you need. arXiv preprint arXiv:1706.03762 (2017)
15. Hu, H., Gu, J., Zhang, Z., Dai, J., Wei, Y.: Relation networks for object detection. In: Proceedings of the IEEE Conference on Computer Vision and Pattern Recognition, pp. 3588–3597 (2018)
16. Wang, X., Girshick, R., Gupta, A., He, K.: Non-local neural networks. In: Proceedings of the IEEE Conference on Computer Vision and Pattern Recognition, pp. 7794–7803 (2018)
17. Oktay, O., et al.: Attention U-Net: learning where to look for the pancreas. arXiv preprint arXiv:1804.03999 (2018)
18. Zhou, Z., Siddiquee, M.M.R., Tajbakhsh, N., Liang, J.: Unet++: redesigning skip connections to exploit multiscale features in image segmentation. IEEE Trans. Med. Imaging **39**(6), 1856–1867 (2019)
19. Yu, L., et al.: Automatic 3D cardiovascular MR segmentation with densely-connected volumetric ConvNets. In: Descoteaux, M., Maier-Hein, L., Franz, A., Jannin, P., Collins, D.L., Duchesne, S. (eds.) MICCAI 2017. LNCS, vol. 10434, pp. 287–295. Springer, Cham (2017). https://doi.org/10.1007/978-3-319-66185-8_33
20. Chen, S., Ma, K., Zheng, Y.: Med3D: transfer learning for 3D medical image analysis. arXiv preprint arXiv:1904.00625 (2019)

Unsupervised Cross-modality Cardiac Image Segmentation via Disentangled Representation Learning and Consistency Regularization

Runze Wang and Guoyan Zheng[✉]

Institute of Medical Robotics, School of Biomedical Engineering, Shanghai Jiao Tong University, No. 800, Dongchuan Road, Shanghai 200240, China
guoyan.zheng@sjtu.edu.cn

Abstract. Deep neural networks based approaches for medical image segmentation rely heavily on the availability of large amount of annotated data, which sometimes is difficult to obtain due to time, logistic effort and the requirement of expertise knowledge. Unpaired image translation enables a cross-modality segmentation network to be trained in an annotation-poor target domain by leveraging an annotation-rich source domain but most existing methods separate the image translation stage from the image segmentation stage and are not trained end-to-end. In this paper, we propose an end-to-end unsupervised cross-modality cardiac image segmentation method, taking advantage of diverse image translation via disentangled representation learning and consistency regularization in one network. Different from learning one-to-one mapping, our method characterizes the complex relationship between domains as many-to-many mapping. A novel diverse inter-domain semantic consistency loss is then proposed to regularize the cross-modality segmentation process. We additionally introduce an intra-domain semantic consistency loss to encourage the segmentation consistency between the original input and the image after cross-cycle reconstruction. We conduct comprehensive experiments on two publicly available datasets to evaluate the effectiveness of the proposed method. The experimental results demonstrate the efficacy of the present approach.

Keywords: Cardiac image segmentation · Diverse image translation · End-to-end · Unsupervised cross-modality segmentation

1 Introduction

Cardiac image segmentation is a prerequisite for many clinical applications including disease diagnosis, surgical planning and computer assisted interventions. In clinics, multi-modality cardiac images such as magnetic resonance imaging (MRI), computed tomography (CT) and ultrasound are widely used as different imaging modalities provide complementary information [1]. Hence, it is

© Springer Nature Switzerland AG 2021
C. Lian et al. (Eds.): MLMI 2021, LNCS 12966, pp. 517–526, 2021.
https://doi.org/10.1007/978-3-030-87589-3_53

important to develop methods for automatic segmentation of multi-modality cardiac images. Deep learning methods, such as convolutional neural networks (CNNs), have demonstrated superior performance in supervised cardiac image segmentation [1–3]. However, supervised learning usually relies on a large amount of training data with pixel-wise annotations [4], which can be labor-intensive, time-consuming and suffers from intra- and inter-observer variability [5], especially when considering annotation of heart structures of all modalities.

One way to mitigate this issue is to leverage unsupervised cross-modality synthesis [6]. This technique typically learns the structural information from an annotation-rich imaging modality (denoted as source domain), and then transfers the knowledge to the segmentation of images from an annotation-poor imaging modality (denoted as target domain), under an assumption that different modalities share the same anatomical structure information. Existing cross-modality segmentation methods are based on either feature alignment [7,8] or Image Translation (IT) [9–12]. Typically, IT-based methods consist of two stages: a cross-modality image synthesis stage and an image segmentation stage. Thus, they are not trained end-to-end such that (1) any error generated in the cross-modality image synthesis stage will be passed to the image segmentation stage [10]; and (2) the cross-modality image synthesis stage cannot benefit from the high-level semantic information obtained from the image segmentation stage [13,14].

In this paper, we propose an end-to-end unsupervised method for cross-modality cardiac image segmentation, taking advantage of diverse image translation via disentangled representation learning and consistency regularization in one network. Instead of learning one-to-one mapping as in CycleGAN [15], our method characterizes the complex relationship between domains as many-to-many mapping [16], where each modality is disentangled into domain-invariant geometry structure features and domain-specific appearance features.

Our contributions can be summarized as follows: (1) we propose an end-to-end unsupervised cross-modality image segmentation framework integrating diverse image translation with semantic image segmentation into one network such that the two parts can benefit from each other, i.e., better image translation will improve cross-modality image segmentation and conversely, better image segmentation will regularize cross-modality image translation; (2) we introduce two consistency losses, i.e., the diverse inter-domain semantic consistency loss and the intra-domain semantic consistency loss, to regularize the cross-modality segmentation process; (3) we demonstrate on two public datasets that the proposed end-to-end network, which takes both anatomical shape information and diverse appearance information into account, produce better segmentation than state-of-the-art methods.

2   Method

Let $x \in X$ and $y \in Y$ be images from two domains, and $m_x \in M_X$ and $m_y \in M_Y$ be corresponding labels to x and y. Note that x and y are not necessarily paired, and we have no access to M_Y in the training phase. Our goal is to design a network to segment unlabeled images in the target domain Y by leveraging X and M_X in the source domain. Our end-to-end unsupervised cross-modality segmentation framework is composed of two modules as shown in Fig. 1: a diverse image translation (DIT) module and a domain-specific segmentation (DSS) module, as detailed below.

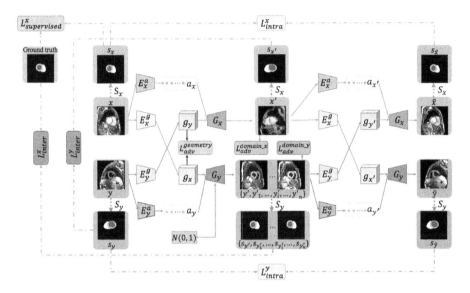

Fig. 1. Overview of our proposed end-to-end unsupervised cross-modality cardiac image segmentation framework. Green and blue colored blocks show parts related with source domain and target domain, respectively, while solid lines and dotted lines link components of DIT and DSS modules, respectively. (Color figure online)

2.1 Diverse Image Translation

Following the diverse image-to-image translation framework [17], the DIT module consists of two main components, a variational autoencoder (VAE) for reconstruction and a generative adversarial network (GAN) for adversarial learning. The VAE components are trained for intra-domain reconstruction, where the reconstruction loss is minimized to encourage the encoders and generators to invert one another. The GAN components are trained for two purposes. The first one is for disentangled representation learning, i.e., the latent space is decomposed into geometry and appearance subspace. The second one is to promote diverse cross-domain synthesized images as realistic as possible. The DIT consists of two geometry encoders $\{E_x^g, E_y^g\}$, two appearance encoders $\{E_x^a, E_y^a\}$, two generators $\{G_x, G_y\}$, a geometry discriminator D_g, and two domain discriminators $\{D_x, D_y\}$. Taking domain x as an example, we can obtain its disentangled geometry feature $g_x = E_x^g(x)$, and appearance code $a_x = E_x^a(x)$, which is set as a 8-bit vector. The appearance encoder is trained to embed images into a latent space that matches the Gaussian distribution $N(0,1)$ by minimizing Kullback-Leibler (KL) divergence between the two. Then the translated image x' is generated by combining the geometry feature from y and the appearance code from x, i.e., $x' = G_x(g_y, a_x)$. The above process can be used further to obtain the cross-cycle reconstruction image $\hat{x} = G_x(E_y^g(y'), E_x^a(x'))$, which is expected to have the same geometry and appearance as x. Note that to achieve diverse image translation in the target domain, the geometry features g_x are combined with multiple appearances codes sampled from the Gaussian distribution $N(0,1)$. The n diversely generated images can be represented as $\{y_i' = G_y(g_x, z_i)\}$, where $z_i \sim N(0,1), i \in \{1, 2, \ldots, n\}$. In addition, a diversity-seeking regularization method [16] is incorporated to encourage the generator G_y to be diversity-sensitive. The overall objective loss function of the DIT module is defined as the weighted sum of three components:

$$L_{DIT} = \lambda_1 L_{recon} + \lambda_2 L_{adv} + \lambda_3 L_{latent} \tag{1}$$

where λ_1, λ_2 and λ_3 are parameters controlling the relatively weights of different losses, which will be detailed below.

Intra-domain Reconstruction Loss L_{recon}. It consists of self-reconstruction loss L_{self_recon}, and cross-cycle reconstruction loss of both source and target domains L_{cc_recon}, i.e., $L_{recon} = L_{self\_recon}^x + L_{self\_recon}^y + L_{cc\_recon}^x + L_{cc\_recon}^y$:

$$L_{self\_recon}^x = \mathbb{E}_x[\|G_x(g_x, a_x) - x\|_1], \quad L_{cc\_recon}^x = \mathbb{E}_x[\|\hat{x} - x\|_1] \tag{2}$$

Adversarial Loss L_{adv}. It consists of domain adversarial loss L_{adv}^{domain} and geometry adversarial loss $L_{adv}^{geometry}$, i.e., $L_{adv} = L_{adv}^{domain\_x} + L_{adv}^{domain\_y} + L_{adv}^{geometry}$

$$L_{adv}^{domain\_x} = \mathbb{E}_x[log D_x(x)] + \mathbb{E}_{x'}[log(1 - D_x(x'))] \tag{3}$$

$$
\begin{aligned}
L_{adv}^{geometry} &= \mathbb{E}_x[\frac{1}{2} \log D^g(g_x) + \frac{1}{2} \log(1 - D^g(g_x))] \\
&+ \mathbb{E}_y[\frac{1}{2} \log D^g(g_y) + \frac{1}{2} \log(1 - D^g(g_y))]
\end{aligned} \tag{4}
$$

Latent Space Loss L_{latent}. It contains latent appearance space reconstruction loss L_{recon} and diversity seeking loss $L_{diversity}$, $L_{latent} = L_{latent}^x + L_{latent}^y + L_{diversity}$,

$$L_{latent}^x = \|E_x^a(G_x(g_x, z)) - z\|_1, \ where \ z \sim N(0,1) \tag{5}$$

$$L_{diversity} = -\frac{\|G_y(g_x, \tilde{z}) - G_y(g_x, \tilde{z}^*)\|_1}{\|\tilde{z} - \tilde{z}^*\|_1}, \ where \ (\tilde{z}, \tilde{z}^*) \sim N(0,1) \tag{6}$$

2.2   Domain-Specific Segmentation

The DSS module contains two segmentation networks, S_x and S_y, for the source domain and the target domain, respectively. The main idea of DSS is to use supervised source domain segmentation to guide unsupervised target domain segmentation, via the diverse image translation. Inspired by [18], we proposed a new consistency loss $L_{consistency}$ to regularize the cross-modality segmentation process in addition to the supervised learning in the source domain. The overall loss function for the DSS module is:

$$L_{DSS} = L_{supervised} + \mu L_{consistency} \tag{7}$$

where μ is a parameter controlling the relative weights between supervised learning and consistency regularization.

Before we present details on $L_{supervised}$ and $L_{consistency}$, we first introduce a generalized segmentation loss function L_{seg}, combining cross entropy loss with Dice loss:

$$L_{seg}(\alpha, \beta, S) = -\sum_{k=1}^{K} \left[\beta_j^k \log(S(\alpha)_j^k) + \frac{2\sum_{j=1}^{N} \beta_j^k \hat{S}(\alpha)_j^k}{\sum_{j=1}^{N} \beta_j^k \beta_j^k + \hat{S}(\alpha)_j^k \hat{S}(\alpha)_j^k} \right] \tag{8}$$

where α denotes the input image, β refers to the target segmentation, S represents the segmentation network. $S(\alpha)_j^k$ and $\hat{S}(\alpha)_j^k$ respectively denote the probability prediction and one-hot output of voxel j for class k.

Supervised Segmentation Loss. It is defined on the source domain as:

$$L_{supervised} = \mathbb{E}_x[L_{seg}(x, m_x, S_x)] \tag{9}$$

Consistency Loss. It consists of a novel diverse inter-domain semantic consistency loss L_{inter} and an intra-domain semantic consistency loss L_{intra}. Specifically, the original image x and the diversely translated image $\{y', y_1', \ldots, y_i', \ldots, y_n'\}$ should have the same semantic structures although they belong to different domains. The diverse image generation can be regarded as online data augmentation, aiming to improve the accuracy and robustness of cross-domain segmentation. L_{inter} is used to constrain the segmentation results between x and $\{y', y_1', \ldots, y_i', \ldots, y_n'\}$ to be consistent. Similarly, L_{intra} is used to enforce the segmentation consistency between the original input x and the cross-cycle reconstructed image $\hat{x}$. The consistency loss is represented as $L_{consistency} = L_{inter} + L_{intra}$:

$$L_{inter} = \mathbb{E}_x[L_{seg}(y', m_x, S_y) + \sum_{i=1}^{n} L_{seg}(y_i', m_x, S_y)] + \mathbb{E}_y[L_{seg}(y, S_x(x'), S_y)] \tag{10}$$

$$L_{intra} = \mathbb{E}_x[L_{seg}(\hat{x}, S_x(x), S_x)] + \mathbb{E}_y[L_{seg}(\hat{y}, S_y(y), S_y)] \tag{11}$$

2.3 Implementation Details

The proposed model is implemented in PyTorch and trained with a Tesla V100 graphics card. The geometry encoders consist of three convolutional layers and four residual blocks followed by batch normalization, while appearance encoders contain five convolutional layers followed by a fully connected layer. The generators are composed of two convolutional layers followed by three transposed convolutional layers The discriminators are composed of convolutional layers for binary classification. The segmentation networks are a standard PSPNet [19]. For training, we use the Adam optimizer with a batch size of 2, and exponential decay rates $(\beta_1, \beta_2) = (0.5, 0.999)$. The initial learning rate is set to 0.0001 and 0.001 for DIT and DSS, respectively. We empirically set $\lambda_1 = \lambda_3 = 10$, $\lambda_2 = 1$, and $\mu = 1$. Considering a trade-off between performance and resource, we empirically choose the number $n = 3$ for the on-line diverse image translation.

3 Experiments and Results

Dataset Description. Two publicly available cardiac image datasets are used in our study: multi-sequence cardiac MR (MS-CMR) dataset from the MS-CMRSeg challenge [20] and CT-MR dataset from the MM-WHS challenge [3]. The MS-CMRSeg challenge dataset contains 45 paired BSSFP CMR and LGE CMR with ground truth segmentation of left ventricular cavity (LV), left ventricular myocardium (MYO), and right ventricle blood cavity (RV). We shuffled all slices and made the training data unpaired. The MM-WHS challenge dataset contains 20 unpaired CT and MR images with the ground truth segmentation. For each 3D image, following the practice in [6], we sampled 16 slices from the long-axis view around the center of LV and cropped them around the heart. Both datasets were divided into a training set and a test set in a ratio of 4 to 1 depending on subjects, i.e., data belonging to the same subject will be only used either as training data or as test data. Based on these two dataset, we compare our method with the state-of-the-art methods introduced in [8,15,17,18].

Table 1. Comparison of segmentation results of different methods.

Methods	CT-MR dataset		MS-CMR dataset		
	LV	MYO	LV	MYO	RV
No adaptation	11.5 ± 15.2	8.5 ± 12.9	33.7 ± 37.8	25.2 ± 34	24.1 ± 32.3
CycleGAN [15]	60.5 ± 10.9	48.1 ± 13.7	52.2 ± 33.9	38.0 ± 26.8	33.6 ± 33.4
CyCADA [18]	62.7 ± 9.4	51.6 ± 4.3	75.2 ± 18.5	51.0 ± 19.4	72.9 ± 19.1
DRIT++ [17]	77.0 ± 9.3	57.2 ± 11.3	63.2 ± 32.9	40.9 ± 23.2	40.0 ± 36.9
SIFA [8]	83.3 ± 7.6	72.0 ± 8.7	81.7 ± 9.1	71.7 ± 13.9	73.3 ± 13.0
Ours	$\mathbf{89.1 \pm 5.3}$	$\mathbf{74.7 \pm 6.0}$	$\mathbf{85.8 \pm 9.5}$	$\mathbf{72.3 \pm 11.3}$	$\mathbf{78.6 \pm 14.5}$

Results on the MM-WHS Challenge Dataset. For this dataset, MR imaging is treated as the target domain while CT imaging is regarded as the source domain. We aim to segment LV and MYO from MR images using the annotation from the CT images. The results are shown in the middle-left of Table 1. Specifically, training a PSPNet with the annotation from the source domain and then applying the trained model to target domain, lead to poor segmentation results. By using CycleGAN to synthesize source domain images from target domain images and then applying trained segmentation network, average Dice coefficients for the LV and the MYO are improved from 11.5% and 8.5% to 60.5% and 48.1%, respectively, indicating the effectiveness of cross-domain image generation. By enforcing both the cycle-consistency and semantics losses, CyCADA [18] achieved a slightly better results than the CycleGAN. Replacing CycleGAN by diverse image generation with disentangled representation learning as introduced in [17], the segmentation results get further improved to 77.0% and 57.2%, respectively. SIFA [8] integrates both image and feature alignment for medical image segmentation and achieved the second-best results. By integrating online diverse image augmentation and consistency regularization, our method achieved the best results. Specifically, an average Dice coefficient of 89.1% and 74.7% was obtained for LV and MYO, respectively.

Results on the MS-CMRSeg Challenge Dataset. Here LGE CMR imaging is used as the target domain while the BSSFP CMR imaging is used as the source domain. We aim to segment three heart structures, i.e., LV, MYO and RV, from the LGE CMR images using the annotations from the BSSFP CMR images. The results are presented in the right of Table 1. From this table, one can see that the worse results are obtained when no image translation is used. With image translation using CycleGAN, average Dice coefficients for the three heart structures are improved to 52.2%, 38.0% and 33.6%, respectively. Replacing CycleGAN by the method introduced in [17], the segmentation results get improved to 63.2%, 40.9% and 40.0%, respectively. CyCADA [18] achieved significantly better results and the average Dice coefficients for the three heart structures are 75.2%, 51.0% and 72.9%, respectively. Again, the second-best results was achieved by SIFA [8] and our method achieved the best results with an average Dice coefficient of 85.8%, 72.3% and 78.6%, respectively.

Table 2. Results of ablation study.

Methods	Diversity	L_{inter}	L_{intra}	DICE	
				LV	MYO
CycleGAN[15]	–	–	–	60.5	48.1
Ours w/o L_{inter}+ L_{intra}	✓	–	–	77.0	57.2
Ours w/o L_{intra}	✓	✓	–	85.0	71.9
Ours	✓	✓	✓	**89.1**	**74.7**

Ablation Study. To investigate the effectiveness of different components introduced in our method, we conducted an ablation study on the MM-WHS challenge dataset. The quantitative and qualitative comparison results are presented in Table 2 and Fig. 2, respectively. It is not surprising to find that the worst results are obtained when CycleGAN is used for image generation. By replacing Cycle-GAN with diverse image generation based on disentangled representation learning but not using semantic consistency regularization, an average Dice coefficient of 77.0% and 57.2% is obtained for LV and MYO segmentation, respectively. By only incorporating the diverse inter-domain semantic consistency loss into our network, we achieved on average an improvement of 8.0% and 14.7% Dice coefficients for LV and MYO segmentation, respectively, indicating the effectiveness of inter-domain semantic consistency regularization for cross-modality semantic segmentation. Further adding the intra-domain semantic consistency loss, we achieved on average an improvement of 4.1% and 2.8% Dice coefficients for LV and MYO segmentation, respectively, which indicates that both diverse inter-domain semantic consistency loss and intra-domain semantic consistency loss are boosting the performance of cross-modality image segmentation. Figure 2 shows segmentation examples obtained from the ablation study.

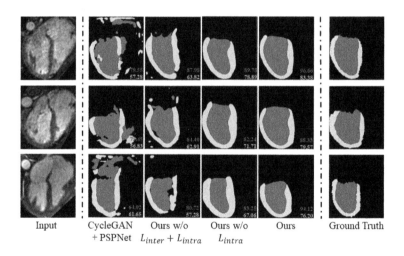

| Input | CycleGAN + PSPNet | Ours w/o $L_{inter} + L_{intra}$ | Ours w/o L_{intra} | Ours | Ground Truth |

Fig. 2. Qualitative comparison of segmentation results obtained from ablation study. Each row is one example.

4   Conclusion

In this paper, we proposed an end-to-end unsupervised cross-modality cardiac image segmentation method, taking advantage of diverse image translation via disentangled representation learning and consistency regularization in one network. Experimental results demonstrated that the present method achieved better results than state-of-the-art methods.

Acknowledgments. This study was partially supported by Shanghai Municipal Science and Technology Commission via Project 20511105205 and 20DZ2220400, and by the Natural Science Foundation of China via project U20A20199.

References

1. Chen, C., et al.: Deep learning for cardiac image segmentation: a review. Front. Cardiovasc. Med. **7**, 25 (2020)
2. Shi, Z., et al.: Bayesian VoxDRN: a probabilistic deep voxelwise dilated residual network for whole heart segmentation from 3D MR images. In: Frangi, A.F., Schnabel, J.A., Davatzikos, C., Alberola-López, C., Fichtinger, G. (eds.) MICCAI 2018. LNCS, vol. 11073, pp. 569–577. Springer, Cham (2018). https://doi.org/10.1007/978-3-030-00937-3_65
3. Zhuang, X., et al.: Evaluation of algorithms for multi-modality whole heart segmentation: an open-access grand challenge. Med. Image Anal. **58**, 101537 (2019)
4. Cheplygina, V., de Bruijne, M., Pluim, J.P.: Not-so-supervised: a survey of semi-supervised, multi-instance, and transfer learning in medical image analysis. Med. Image Anal. **54**, 280–296 (2019)
5. Petitjean, C., Dacher, J.N.: A review of segmentation methods in short axis cardiac MR images. Med. Image Anal. **15**(2), 169–184 (2011)
6. Wu, F., Zhuang, X.: CF distance: a new domain discrepancy metric and application to explicit domain adaptation for cross-modality cardiac image segmentation. IEEE Trans. Med. Imaging **39**(12), 4274–4285 (2020)
7. Dou, Q., et al.: PnP-AdaNet: plug-and-play adversarial domain adaptation network at unpaired cross-modality cardiac segmentation. IEEE Access **7**, 99065–99076 (2019)
8. Chen, C., Dou, Q., Chen, H., Qin, J., Heng, P.A.: Unsupervised bidirectional cross-modality adaptation via deeply synergistic image and feature alignment for medical image segmentation. IEEE Trans. Med. Imaging **39**(7), 2494–2505 (2020)
9. Huo, Y., et al.: SynSeg-Net: synthetic segmentation without target modality ground truth. IEEE Trans. Med. Imaging **38**(4), 1016–1025 (2018)
10. Zhang, Z., Yang, L., Zheng, Y.: Translating and segmenting multimodal medical volumes with cycle-and shape-consistency generative adversarial network. In: Proceedings of the IEEE Conference on Computer Vision and Pattern Recognition, pp. 9242–9251 (2018)
11. Yang, J., Dvornek, N.C., Zhang, F., Chapiro, J., Lin, M.D., Duncan, J.S.: Unsupervised domain adaptation via disentangled representations: application to cross-modality liver segmentation. In: Shen, D., et al. (eds.) MICCAI 2019. LNCS, vol. 11765, pp. 255–263. Springer, Cham (2019). https://doi.org/10.1007/978-3-030-32245-8_29
12. Chen, C., et al.: Unsupervised multi-modal style transfer for cardiac MR segmentation. In: Pop, M., et al. (eds.) STACOM 2019. LNCS, vol. 12009, pp. 209–219. Springer, Cham (2020). https://doi.org/10.1007/978-3-030-39074-7_22
13. Chen, X., et al.: Anatomy-regularized representation learning for cross-modality medical image segmentation. IEEE Trans. Med. Imaging **40**(1), 274–285 (2020)
14. Chen, X., et al.: Diverse data augmentation for learning image segmentation with cross-modality annotations. Med. Image Anal. **71**, 102060 (2021)
15. Zhu, J.Y., Park, T., Isola, P., Efros, A.A.: Unpaired image-to-image translation using cycle-consistent adversarial networks. In: Proceedings of the IEEE International Conference on Computer Vision, pp. 2223–2232 (2017)

16. Mao, Q., Lee, H.Y., Tseng, H.Y., Ma, S., Yang, M.H.: Mode seeking generative adversarial networks for diverse image synthesis. In: Proceedings of the IEEE/CVF Conference on Computer Vision and Pattern Recognition, pp. 1429–1437 (2019)
17. Lee, H.Y., et al.: DRIT++: diverse image-to-image translation via disentangled representations. Int. J. Comput. Vis. **128**(10), 2402–2417 (2020)
18. Hoffman, J., et al.: Cycada: cycle-consistent adversarial domain adaptation. In: International Conference on Machine Learning, PMLR, pp. 1989–1998 (2018)
19. Zhao, H., Shi, J., Qi, X., Wang, X., Jia, J.: Pyramid scene parsing network. In: Proceedings of the IEEE Conference on Computer Vision and Pattern Recognition, pp. 2881–2890 (2017)
20. Zhuang, X.: Multivariate mixture model for myocardial segmentation combining multi-source images. IEEE Trans. Pattern Anal. Mach. Intell. **41**(12), 2933–2946 (2018)

Landmark-Guided Rigid Registration for Temporomandibular Joint MRI-CBCT Images with Large Field-of-View Difference

Jupeng Li[1(✉)], Yinghui Wang[2], Shuai Wang[1], Kai Zhang[1], and Gang Li[2]

[1] School of Electronic and Information Engineering,
Beijing Jiaotong University, Beijing 100044, China
lijupeng@bjtu.edu.cn
[2] Department of Oral and Maxillofacial Radiology, Peking University School
and Hospital of Stomatology, Beijing 100081, China

Abstract. Fused MRI-CBCT images provide desirable complementary information of the articular disc and condyle surface for optimum diagnosis, has been shown to be accurate and reliable in Temporomandibular Disorders (TMD) assessment. But field-of-view difference between multi-modality images brings challenges to conventional registration algorithms. In this paper, we proposed a landmark-guided learning method for Temporomandibular Joint (TMJ) MRI-CBCT images registration. First, end-to-end landmark localization network was used to detect correspondence landmark pairs in the different modality images to generate the landmark guidance information. Then taking image patches centered landmarks as input, an unsupervised learning network regresses the rigid transformation matrix using mutual information as a measure of similarity between image patches. Finally combined landmarks coordinates with the rigid transformation matrix, the whole image registration can be realized. Experiment results demonstrate that our approach achieves better overall performance on registration of images from different patients and modalities with 100x speed-up in execution time.

Keywords: Multi-modality image registration · Temporomandibular joint · MRI-CBCT images · Large field-of-view difference · Landmark-guided

1 Introduction

TMJ is a synovial joint that contains an articular disc (shown in Fig. 1a), which allows for hinge and sliding movements [1]. TMD is an umbrella term covering pain and dysfunction of the muscles of mastication (the muscles that move the jaw) and the TMJ. TMD is common in adults; as many as one third of adults report having one or more symptoms, which include jaw or neck pain, headache, and clicking sound or grating within the joint. Although TMD is not life-threatening, it can be detrimental to quality of life; this is because the symptoms can become chronic and difficult to manage. In addition to the observer's expertise, clear image information is a substantial factor that leads to correct

© Springer Nature Switzerland AG 2021
C. Lian et al. (Eds.): MLMI 2021, LNCS 12966, pp. 527–536, 2021.
https://doi.org/10.1007/978-3-030-87589-3_54

diagnosis of this intractable disease [2]. The articular disc is best depicted on magnetic resonance imaging (MRI) and osseous surfaces are best seen in cone beam CT (CBCT). The fused MRI-CBCT image (see Fig. 1b) provides desirable complementary informa-tion of the articular disc and condyle surfaces for optimum diagnosis. The registration process to generate fused images has been shown to be accurate and reliable in TMD assessment [3].

Unlike registration of single-modality images, multi-modality images registration between MRI and CBCT is challenging due to significant differences in voxel size, pixel intensity, anatomical structure identification, image orientation and field-of-view (see Fig. 1c–e). Only few articles discussing MRI and CT (CBCT) image registration for TMJ visualization or assessment were published within the last 7 years [4]. Lin was the first to explore the 3D rendering of mandible from MRI and CT registered images with 12 fiducial markers attached to the facial skin-surface [5]. In a brief clinical report, Dai chose one sagittal slice of TMJ MRI and CT images from a previous study, as an example, to illustrate a hybrid image of TMJ via Photoshop® software [6]. Al-Saleh published the first study that employed MRI and CBCT registered images to assess diagnostic reliability of TMJ pathology [3]. They evaluated the quality of two techniques of image registration, extrinsic (fiducial marker-based) versus intrinsic (voxel value mutual information based) in 20 TMJ images. In a latest report, Ma, one author of this article, imported the DICOM format data of CT/CBCT and MRI into Amira® to realize automatic/semi-automatic registration of multi-modality images by adjusting the registration parameters [7].

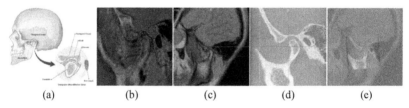

(a)   (b)   (c)   (d)   (e)

Fig. 1. Anatomical structure and images registration of TMJ. (a) Anatomical structure, (b) fused MRI-CBCT image, (c) MRI image, (d) CBCT image, and (e) registered image. From these images, we can see huge field-of-view difference between different modality images.

Related Works: Deep learning methods have been shown strong advantages in medical image registration [8]. In order to evaluate the posture and position of the implant during surgery, Miao proposed a hierarchical regression model to achieve six transformation parameters for real-time 2D/3D registration, in which ground truth data is synthesized by transforming aligned data [9]. Chee proposed a self-supervised affine image registration network (AIRNet) for 3D medical images that is designed to directly estimate the trans-formation parameters between two input images, in which the synthetic dataset was used for the training of the model [10]. The difficult nature of the acquisition of reliable ground truth has motivated research groups to explore unsupervised approaches for image regis-tration transformation estimation [8]. Kori proposed an unsupervised image registration framework for multi-modality MRI image affine registration. Pre-trained VGG-19 was used for feature extraction followed by a key point detector. These key points were fed to the Multi-Layered Perceptron (MLP) based regression module so as to estimate the affine transformation parameters trained by generated set of random data points [11]. In order to register arbitrarily oriented reconstructed images of fetuses scanned in-utero at a wide gestational age range to a standard atlas space, Salehi proposed regression CNNs that learn to predict the angle-axis representation of 3D rotations and translations using image features. They compared mean square error and geodesic loss to train regression CNNs for 3D pose estimation in slice-to-volume registration and volume-to-volume registration [12]. Combined with unsupervised network, coarse-to-fine multi-scale iterative framework and image deformation, Shu proposed an unsupervised network for microscopic image rigid registration. The network optimizes its parameters directly by minimizing the mean square error loss between registered image and reference image without ground truth [13]. As far as we know, no research work involves the problem of different field-of-views in multi-modality medical image registration, which bring difficulties to the current learning-based registration methods.

Contribution: The main contributions of this work are summarized as follows: (1) *Landmark-guided mechanism* was introduced to effectively register MRI-CBCT images of TMJ with large different field-of-views, without any prior assumption on the image pairs. (2) Compared with affine matrix learning methods for rigid images registration, our image spatial transform regression network predicts *real rigid transformation* for multi-modality images.

2 Method

2.1 Overall Framework

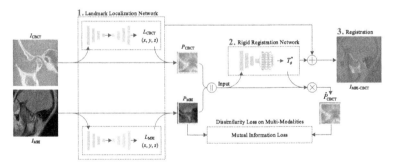

Fig. 2. The workflow of our landmark-guided rigid registration framework applied for TMJ MRI-CBCT images with different field-of-views. We highlight that all registrations are done in 3D throughout this paper. For clarity and simplicity, we depict the 2D formulation of our method in this paper.

As shown in Fig. 2, our overall framework for MRI-CBCT images registration includes three stages. Firstly, landmark numerical coordinate regression network takes I_{MRI} and

I_{CBCT} as input and estimates landmarks' coordinate $L_{MRI}(x, y, z)$ and $L_{CBCT}(x, y, z)$ respectively. Then the spatial transform network regress the rigid transformation matrix T_θ between two image patches P_{MRI} and P_{CBCT} centered landmarks. Finally, combined the rigid transformation matrix T_θ^* with the landmark-guided information, the rigid registration of MRI-CBCT images is achieved.

2.2 Landmark Localization Network

Inspired by landmark localization in human pose estimation [14], we proposed an end-to-end landmark localization network for 3D medical images (L^2Net) by converting heat-map regression into coordinate regression task. L^2Net consists of feature extraction network (U-Net) and coordinate regression layer (Fig. 3). For more technical details you can read our previous work in reference [15].

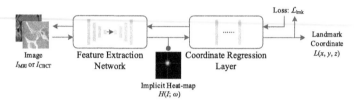

Fig. 3. Architecture of L^2Net for 3D medical images: a feature extraction network that extracts modality independent feature as implicit heat-map $H(I; \omega)$; and a coordinate regression layer that map the feature $H(I; \omega)$ to landmark coordinate $L(x, y, z)$.

Given a MRI/CBCT image I with size of $v = m \times n \times k$, U-Net learning the image feature to output the same size implicit normalized heat-map $H(I; \omega)$. By taking the probabilistic interpretation of $H(I; \omega)$, we can represent the landmark coordinates, $L(x, y, z)$ as center of mass (centroid) function defined as

$$L = (x, y, z) \sum_{(x,y,z) \in v} (x, y, z) * H(I; \omega) \Big/ \sum_v H(I; \omega) \qquad (1)$$

Loss Function: Since the coordinate regression layer outputs numerical coordinates, it is possible to directly calculate Euclidean distance between the predicted coordinate $L_{inf}(x, y, z)$ and ground truth $L_{gt}(x, y, z)$. We take advantage of this fact to formulate the main term of landmark localization loss function (Eq. 2).

$$\mathcal{L}_{euc}(L_{gt}, L_{inf}) = \left\| L_{inf} - L_{gt} \right\|_2 \qquad (2)$$

The shape of the implicit heat-map also affects the regression accuracy of landmark coordinate [16]. More specifically, to force the implicit heat-map to resemble a spherical Gaussian distribution, we can minimize the divergence between the heat-map $H(I; \omega)$ and an appropriate target normal distribution $N(L_{inf}, \sigma_t^2)$. Equation 3 defines distribution regularization, where $D(\cdot \| \cdot)$ is the Jensen-Shannon divergence.

$$\mathcal{L}_{reg}(H(I; \omega), L_{inf}, \sigma_t) = D(H(I; \omega) \| N(L_{inf}, \sigma_t^2)) \qquad (3)$$

Equation 4 shows how regularization is incorporated into the Euclidean distance function. A regularization coefficient, λ, is used to set strength of the regularizer, $\mathcal{L}_{reg}$.

$$\mathcal{L}_{lmk} = \mathcal{L}_{euc}(L_{gt}, L_{inf}) + \lambda \cdot \mathcal{L}_{reg}(H(I; \omega), L_{inf}, \sigma_t) \qquad (4)$$

2.3 Rigid Registration Network

The architecture of spatial transform regression network used for rigid registration is shown in Fig. 4. The input is the concatenated patch pairs (P_{CBCT} ‖ P_{MRI}) that centered landmarks and the output is the transform matrix T_θ which indicates the spatial relationship between two image patches. Each convolution layer is zero-padded and is followed by ReLU activations. After max pooling for two times, two fully connected (FC) layers with ReLU activation function are used to gather information from the entire images to give the *rigid transform matrix*: $M = [\theta x, \theta y, \theta z, \Delta x, \Delta y, \Delta z,]$.

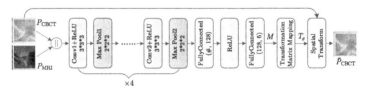

Fig. 4. Architecture of multi-modality image rigid transformation matrix regression network.

Transformation Matrix Mapping: This layer converts the transform M into exact rigid transformation matrix, instead of affine matrix used in reference [10, 11, 13]. Therefore, the shearing transformation caused by affine transformation can be eliminated, so as to improve the accuracy of rigid registration. To make the entire registration network training process back-propagating, this parameters mapping process must meet the derivability requirements. For rotation we can get rotation matrix R_x as following,

$$R_x = \cos(\theta_x) * \begin{bmatrix} 0&0&0&0 \\ 0&1&0&0 \\ 0&0&1&0 \\ 0&0&0&0 \end{bmatrix} + \sin(\theta_x) * \begin{bmatrix} 0&0&0&0 \\ 0&0&-1&0 \\ 0&1&0&0 \\ 0&0&0&0 \end{bmatrix} + \begin{bmatrix} 1&0&0&0 \\ 0&0&0&0 \\ 0&0&0&0 \\ 0&0&0&1 \end{bmatrix}$$

$$= \begin{bmatrix} 1 & 0 & 0 & 0 \\ 0 & \cos(\theta_x) & -\sin(\theta_x) & 0 \\ 0 & \sin(\theta_x) & \cos(\theta_x) & 0 \\ 0 & 0 & 0 & 1 \end{bmatrix}$$

The rotations matrices R_y and R_z (similar definitions with R_x), specified in this way, determine an amount of rotation about each of the individual axes of the coordinate system. And for translation matrix,

$$D = \Delta x * \begin{bmatrix} 0 & 0 & 0 & 1 \\ 0 & 0 & 0 & 0 \\ 0 & 0 & 0 & 0 \\ 0 & 0 & 0 & 0 \end{bmatrix} + \Delta y * \begin{bmatrix} 0 & 0 & 0 & 0 \\ 0 & 0 & 0 & 1 \\ 0 & 0 & 0 & 0 \\ 0 & 0 & 0 & 0 \end{bmatrix} + \Delta z * \begin{bmatrix} 0 & 0 & 0 & 0 \\ 0 & 0 & 0 & 0 \\ 0 & 0 & 0 & 1 \\ 0 & 0 & 0 & 0 \end{bmatrix}$$

$$+ \begin{bmatrix} 1 & 0 & 0 & 0 \\ 0 & 1 & 0 & 0 \\ 0 & 0 & 1 & 0 \\ 0 & 0 & 0 & 1 \end{bmatrix} = \begin{bmatrix} 1 & 0 & 0 & \Delta x \\ 0 & 1 & 0 & \Delta y \\ 0 & 0 & 1 & \Delta z \\ 0 & 0 & 0 & 1 \end{bmatrix}$$

Combined these rotation and translation matrices together in certain order, we can get the rigid transformation matrix T_θ,

$$T_\theta = R_x * R_y * R_z + D \tag{5}$$

Once the transformation matrix T_θ is obtained, a spatial transformation layer [14] is used to warp the moving image using the deformation field T_θ. Each voxel in the warped image $\tilde{P}_{\text{CBCT}}$ is calculated by bi-linear interpolation in the corresponding location, as given by the displacement vector, in the subject image P_{CBCT}:

$$\tilde{P}_{\text{CBCT}}(p) = \sum_{q \in N(p+\omega)} P_{\text{CBCT}}(q)(1 - |p + \omega - q|_2^2) \tag{6}$$

Where p and q are coordinates on the image, and ω is the displacement of p, $N(p + \omega)$ is the set of 8-pixel cubic neighbors of $p + \omega$.

Loss Function: Mutual information (MI) has become a common loss function for (especially multi-modality) image registration [17]. Formally, the mutual information between our image patches P_{MRI} and $\tilde{P}_{\text{CBCT}}$ is defined as the following,

$$\text{MI}(P_{\text{MRI}}, \tilde{P}_{\text{CBCT}}) = \sum_x \sum_y p_{\text{MRI,CBCT}}(x, y) \log \frac{p_{\text{MRI,CBCT}}(x, y)}{p_{\text{MRI}}(x)p_{\text{CBCT}}(y)} \tag{7}$$

In order to realize the training of end-to-end image registration network, here we use *Parzen* windowing [18] to calculate differentiable MI for loss function,

$$\mathcal{L}_{\text{sim}}(P_{\text{MRI}}, \tilde{P}_{\text{CBCT}}) = -\text{MI}(P_{\text{MRI}}, \tilde{P}_{\text{CBCT}}) \tag{8}$$

The optimal rigid transformation matrix T_θ^* is finally obtained through the network training. Finally, the coordinate offset between landmarks $L_{\text{MRI}}(x, y, z)$ and $L_{\text{CBCT}}(x, y, z)$ is mapped to the transformation matrix T_θ^*, accordingly the spatial transformation matrix between the I_{MRI} and I_{CBCT} is obtained.

3 Experiments

3.1 Data

The TMJ dataset consists of 204 CBCT and paired MRI images from 102 patients in Peking University School and Hospital of Stomatology. CBCT images are of size 481 × 481 × 481, where each voxel is of size $0.125 \times 0.125 \times 0.125$ mm^3. MRI images are $256 \times 256 \times (7\text{--}11)$ with voxel size of $0.546875 \times 0.546875 \times 3.3$ mm^3. We group the intensity of each image according to the modality and perform histogram matching.

3.2 Training Details

Network Training. Localization networks were trained using mini-batches of 1-sample each. The implicit heat-map is the same size as the resampled input image with $\sigma = 5$. In our experiments, we picked $\lambda = 1$ using cross validation. The models were optimized with RMSProp using an initial learning rate of 2.5×10^{-4}. Each model was trained for 120 epochs, with the learning rate reduced by a factor of 10 at epochs 60 and 90 (an epoch is one complete pass over the training set). In order to increase the ability of registration network to capture displacement of the input image patches, along the x, y, and z directions the landmark coordinates are randomly added by an offset $[-60 \sim + 60]$ as the center position of image patches. We set the learning rate to 1.0e-2, and use exponential decay to adjust the learning rate parameters (ExponentialLR) method, where the basis coefficient *gamma* is set to 0.95. We implemented our method using Pytorch, and used a workstation equipped with single Maxwell-architecture NVIDIA Titan X GPU.

Landmark Localization Results. The mean radial error (MRE, in mm) is the commonly used evaluation index for medical image landmark detection task [15]. Compared with heat-map based method, the MRE result of our networks is lower obviously and modality independent (Fig. 5).

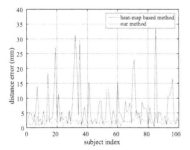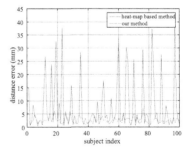

Fig. 5. Distance error of the landmark localization. MRI (left): 6.6803 ± 7.0876 mm (heat-map based method), 2.0244 ± 1.0635 mm (our method); CBCT (right): 7.6375 ± 10.0229 mm (heat-map based method), 2.6371 ± 1.2982 mm (our method).

Rigid Registration Results. In order to evaluate the performance of our proposed unsupervised learning rigid registration network, here we choose to compare the methods including SimpleElastix[1], ANTs[2], two software packages based on traditional iterative optimization, and an affine transformation matrix regression based on convolutional neural networks such as reference [10].

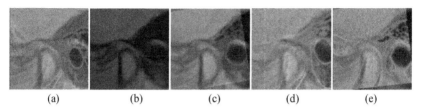

| (a) | (b) | (c) | (d) | (e) |

Fig. 6. Comparison of image patch registration results of various methods. (a) Inital image, (b) ANTs-Affine, (c) Simple Elastix, (d) Learning Affine, (e) Our method.

Figure 6 shows the registration results of multiple methods on the same image patches. In these figures, there are two layers: the bottom layer is an MRI image and the top layer is CBCT image with a color overlay. The first column shows the whole TMJs at the center of the image patch. The two initial images have obvious position and angle misalignment in Fig. 6a. ANTs software package cannot effectively complete the registration task (Fig. 6b). The SimpleElastix obtained good alignment (Fig. 6c) after iterative optimization. The shearing transformation in the affine registration makes the spatial position relationship tilted (Fig. 6d). The mutual information and structural similarity (SSIM) [18] between the registered patches is the most commonly used index to measure the alignment. Table 1 gives quantitative evaluation results of the registration methods.

Table 1. Comparison MI and SSIM of the registration results of different methods.

Methods	Mutual information		Structural similarity		Time
	Mean	Std	Mean	Std	Mean
Intial image	0.36	0.037	0.012	0.0087	–
Manual method	–	–	–	–	>2.0 h
Ants-rigid [2]	0.43	0.202	0.560	0.3390	1.465 s
Elastix-rigid [1]	0.59	0.061	0.128	0.0451	13.765 s
Affine learning [10]	0.48	0.051	0.044	0.0223	0.016 s
Our method	0.57	0.076	0.068	0.0209	0.016 s

---

[1] Medical Image Registration Library – SimpleElastix: https://simpleelastix.github.io/.
[2] Advanced Normalization Tools – ANTs: https://stnava.github.io/ANTs/.

Combined landmark localization with unsupervised image registration stages together, the registration result for the whole MRI-CBCT images is shown in Fig. 7. Conversely, SimpleElastix and Ants software packages, neither of them can achieve registration processing result successfully.

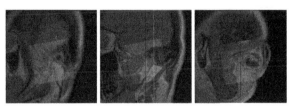

Fig. 7. Rigid registration result for the whole TMJ MRI-CBCT images. The center square area is the result of MRI and CBCT superposition.

4 Conclusion

We proposed an rigid registration network guided by landmarks for the common clinical application of multi-modality medical image registration problems. End-to-end landmark localization network effectively solves the influence of field-of-view difference between different modality images, and rigid transformation regression improves the registration accuracy and speed. We conclude that our method can effectively solve similar image registration applications.

Acknowledgments. This work was supported by the Fundamental Research Funds for the Central Universities (2021JBM003) and the National Natural Science Foundation of China with Project (81671034). Computations used the Department of Radiology and Biomedical Research Imaging Center (BRIC), University of North Carolina at Chapel Hill facility.

References

1. Asim, K.B., Santhosh, G., Aparna, S., et al.: Imaging of the temporomandibular joint: an update. World J. Radiol. **6**(8), 567–582 (2014). https://doi.org/10.4329/wjr.v6.i8.567
2. Al-Saleh M.A, Punithakumar K., Lagravere M., et al.: Three-dimensional assessment of temporomandibular joint using MRI-CBCT image registration, PLoS One **12**(1), e0169555 (2017). https://doi.org/10.1371/journal.pone.0169555
3. Al-Saleh M.A., Jaremko J.L., Alsufyani N., et al.: Assessing the reliability of MRI-CBCT image registration to visualize temporomandibular joints. Dentomaxillofac. Radiol. **44**(6), 20140244 (2015). https://doi.org/10.1259/dmfr.2014024
4. Al-Saleh, M.A., Punithakumar, K., Jaremko, J.L., et al.: Accuracy of magnetic resonance imaging-cone beam computed tomography rigid registration of the head: an in-vitro study. Oral Surg. Oral Med. Oral Pathol. Oral Radiol. Endod. **121**(3), 316–321 (2016). https://doi.org/10.1016/j.oooo.2015.10.029

5. Lin, Y., Liu, Y., Wang, D., et al.: Three-dimensional reconstruction of temporomandibular joint with CT and MRI medical image fusion technology. Hua Xi Kou Qiang Yi Xue Za Zhi **26**(2), 140–143 (2008)

6. Dai, J., Dong, Y., Shen, S.: Merging the computed tomography and magnetic resonance imaging images for the visualization of temporomandibular joint disk. J. Craniofac. Surg. **23**(6), e647–e648 (2012). https://doi.org/10.1097/SCS.0b013e3182710517

7. Ma, R., Li, G., Sun, Y., et al.: Application of fused image in detecting abnormalities of temporomandibular joint. Dentomaxillofac. Radiol. **48**(3), 20180129 (2019). https://doi.org/10.1259/dmfr.20180129

8. Haskins, G., Kruger, U., Yan, P.: Deep learning in medical image registration: a survey. Mach. Vis. Appl. **31**(1–2), 1–18 (2020). https://doi.org/10.1007/s00138-020-01060-x

9. Miao, S., Wang, Z., Liao, R.: A CNN regression approach for real-time 2D/3D registration. IEEE Trans. Med. Imaging **35**(5), 1352–1363 (2016). https://doi.org/10.1109/TMI.2016.252 1800

10. Chee, E., Wu, J.: AIRNet: self-supervised affine registration for 3D medical images using neural networks. arXiv:1810.02583 (2018)

11. Kori, A., Krishnamurthi, G.: Zero shot learning for multi-modal real time image registration. arXiv:1908.06213 (2019)

12. Salehi, S.S.M., Khan, S., Erdogmus, D.: Real-time deep pose estimation with geodesic loss for image-to-template rigid registration. IEEE Trans. Med. Imaging **38**(2), 470–481 (2019). https://doi.org/10.1109/TMI.2018.2866442

13. Shu, C., Chen, X., Xie, Q., et al.: An unsupervised network for fast microscopic image registration, In: Tomaszewski, J.E., Gurcan, M.N. (eds.) Medical Imaging 2018: Digital Pathology, vol. 10581, 105811D. International Society for Optics and Photonics (2018). https://doi.org/10.1117/12.2293264

14. Nibali, A., He, Z., Morgan, S., et al.: Numerical coordinate regression with convolutional neural networks. arXiv:1801.07372 (2018)

15. Li, J., Wang, Y., Mao, J., Li, G., Ma, R.: End-to-end coordinate regression model with attention-guided mechanism for landmark localization in 3D medical images. In: Liu, M., Yan, P., Lian, C., Cao, X. (eds.) MLMI 2020. LNCS, vol. 12436, pp. 624–633. Springer, Cham (2020). https://doi.org/10.1007/978-3-030-59861-7_63

16. Payer, C., Štern, D., Bischof, H., et al.: Integrating spatial configuration into heatmap regression based CNNs for landmark localization. Med. Image Anal. **54**, 207–219 (2019). https://doi.org/10.1016/j.media.2019.03.007

17. Huang, Y., Song, T., Xu, J., et al.: KLDivNet: an unsupervised neural network for multi-modality image registration. arXiv:1908.08767 (2019)

18. Pluim, J.P.W., Maintz, J.B.A., Viergever, M.A.: Mutual-information-based registration of medical images: a survey. IEEE Trans. Med. Imaging **22**(8), 986–1004 (2003). https://doi.org/10.1109/TMI.2003.815867

Spine-Rib Segmentation and Labeling via Hierarchical Matching and Rib-Guided Registration

Caiwen Jiang[1], Zhiming Cui[1], Dongming Wei[1], Yuhang Sun[1], Jiameng Liu[1],
Jie Wei[1], Qun Chen[1,2], Dijia Wu[1,2], and Dinggang Shen[1,2(✉)]

[1] School of Biomedical Engineering, ShanghaiTech University, Shanghai, China
dgshen@shanghaitech.edu.cn
[2] Shanghai United Imaging Intelligence Co., Ltd., Shanghai, China

Abstract. Accurate segmentation and labeling of spine-rib are of great importance for clinical spine and rib diagnosis and treatment. In clinical applications, the spine-rib segmentation and labeling are often challenging, as the shape and appearance of vertebrae are complicated. Previous segmentation and labeling methods usually face considerable difficulties when coping with spine CT images with abnormal curvature spines and implanted metal. In this paper, we propose a multi-stage spine-rib segmentation and labeling method that can be applied to various spine-rib CT images. Our proposed method consists of three steps. First, a 3D U-Net is used to obtain a initial segmentation mask of the spine and rib. Then, the subject information, including gender, age, and the shape of the spine and rib, is used for hierarchically selecting the templates with similar physiological structures from the pre-constructed template library. Finally, the segmentation mask and label from the templates are transferred to the subject via rib-guided registration to achieve correction of the initial results. We evaluated the proposed method on a clinical dataset, and obtained significantly better and robust performance than the state-of-the-art method.

Keywords: Spine-rib segmentation and labeling · Hierarchical matching · Rib-guided registration · Templates

1   Introduction

Spine-rib segmentation and labeling are important for image-guided diagnosis, pre-operative planning, and post-operative evaluation [1,2]. In conventional clinical diagnosis, the doctor needs to determine the type of vertebrae and ribs based on experience, and then segment them slice by slice. Thus, manual segmentation and labeling of vertebrae and ribs in CT images is laborious and subjective [3]. Therefore, it is necessary to propose an automatic spine-rib segmentation and labeling algorithm to improve efficiency and reliability.

© Springer Nature Switzerland AG 2021
C. Lian et al. (Eds.): MLMI 2021, LNCS 12966, pp. 537–545, 2021.
https://doi.org/10.1007/978-3-030-87589-3_55

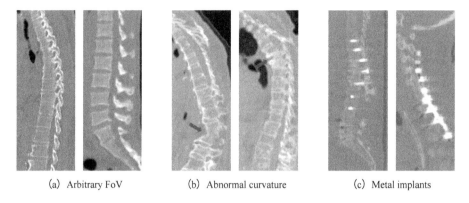

(a) Arbitrary FoV   (b) Abnormal curvature   (c) Metal implants

Fig. 1. Sagittal plane of six cases. (a) The FoV of images varies largely, and the appearance of adjacent vertebrae is very similar. (b) Abnormal curvature of the spine. (c) The surgical metal implants cause peculiar image artifacts.

Automatic spine-rib segmentation and labeling have been applied in various clinical applications, such as detection of vertebra and rib fractures [4], assessment of spinal deformities [5], and computer-assisted surgical interventions [6]. But there are still many challenges in the clinical stage to design an automated spine-rib segmentation and labeling algorithm. As shown in Fig. 1(a), the field of view (FoV) of spine CT images varies largely, and the appearance of adjacent vertebrae is too similar to distinguish. Various pathological circumstances, including scoliosis, vertebra fractures, and lumbarization, increase the difficulty of vertebrae identification, as shown in Fig. 1(b). Moreover, as shown in Fig. 1(c), the presence of surgical metal implants usually causes severe blurring of the vertebral boundary.

Recently, many approaches have been proposed to solve the problems mentioned above, which can be divided into three categories. The first combines machine learning and statistical models [7–9], which are robust to be applied to various spine-rib CT images. However, this category of the method is hard to be used in the application stage, due to the requirement of hand-crafted image features. The second category is based on multi-stage neural networks [10–14], which can effectively extract global context information of vertebrae and ribs to solve the arbitrary FoV problem. But this category of methods can not robustly handle pathological or abnormal spinal images. The third category of methods is based on template matching [15–17]. However, the diversity of templates used in previous methods are limited and cannot cover the complex situation.

In this paper, to address the above-mentioned limitations, we present an accurate and stable spine-rib segmentation and labeling method via hierarchical matching and rib-guided registration. In the first stage, we construct a representative template library by collecting numerous spine-rib CT images. Then given a testing CT image, we first use a 3D U-Net to obtain the semantic segmentation of vertebrae and ribs. In addition, we use information such as gender, age, spine, and rib shape to perform hierarchical matching, and select templates that have similar structures to the input object. Finally, we obtain the corrected spine and

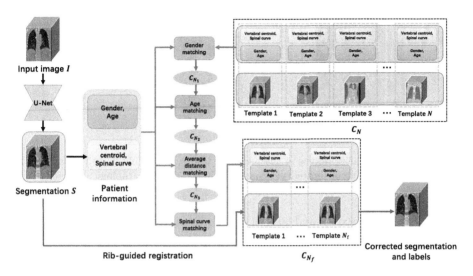

Fig. 2. The overall architecture of the proposed method, where C_N, C_{N_1}, C_{N_2}, C_{N_3} and C_{N_f} denote the template set, and N, N_1, N_2, N_3 and N_f denote the numbers of templates where $(N_f \leqslant N_3 \leqslant N_2 \leqslant N_1 \leqslant N)$.

rib segmentation masks and labels by transferring the annotation from selected templates to the subject via rib-guided registration. We extensively evaluate our method on a clinical dataset with various pathological or abnormal spinal CT images, and results are significantly better than the state-of-the-art methods.

2   Method

As shown in Fig. 2, our proposed method consists of three stages. First, we use a 3D U-Net to obtain initial semantic segmentation S of the vertebrae and ribs, from which spinal structure information such as centroids of labels (vertebrae) and spinal curve can be directly obtained. Then, the spine structure information and physiological information such as gender and age are used to select a template set C_{N_f}, have in which the vertebra and rib from template library C_{N_f} has the most similar structures with S by hierarchical matching. Finally, through rib-guided registration, we register vertebra and rib segmentation masks in C_{N_f} to S for obtaining the corrected segmentation results and labels.

2.1   Template Library Construction

We construct the template library C_N based on a representative clinical dataset, where three examples are shown in Fig. 3. The templates in C_N are selected by three steps. First, the templates are selected to cover all genders, ages, and physiques. Then, both normal and abnormal subjects (with abnormal curvature spine and metal implants) are included in the library to maintain the diversity.

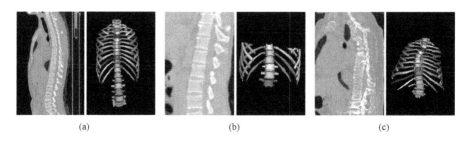

(a)	(b)	(c)

Fig. 3. Three typical examples of the template library C_N, including (a) image of the whole view, (b) image of limited view, and (c) image of the pathological spine.

Finally, the segmentation and label of vertebrae and ribs are manually annotated by doctors.

2.2 Hierarchical Matching

In this section, we first use gender and age to perform initial selection to obtain C_{N_1}. Then, the average distance between the vertebrae is used to select templates that have the similar physique to the input subject. Finally, we select C_f from C_N according to the similarity of spinal curves.

Gender and Age Screening. First, we use gender screening to get C_{N_1} from C_N, where the selected templates have the same gender as input image I. Second, the templates C_{N_2} within three years gap are screened from C_{N_1}. Note that it is unnecessary to filter age precisely, as the spine structure should be consistent within several years.

Average Distance Screening. As individuals of the same gender and age still have different physique, we can select C_{N_3} with the similar physique as I from C_{N_2} by average distance matching. First, we approximately calculate the centroids of labels by averaging the coordinates corresponding to same vertebrae. Then, the length of the entire spinal segment can be calculated by using the coordinates of the starting centroid and the ending centroid in the spine. Finally, the length of the entire spinal segment is divided by the number of vertebrae to obtain the average distance d_p.

The spine sequence in S and templates C_{N_2} may be different, therefore, we need to select the templates containing the same corresponding spine segment as S, and then calculate the template average distance d_t at the same spine segment. Through extensive experiments, we find that, although the labels of vertebrae in S are unreliable, the spine segment in S can be approximately determined by these labels. By comparing d_p and d_t, we can perform further selection to obtain C_{N_3}.

Spinal Curve Screening. Further, we select C_f with similar spinal curvatures as I from C_3 by spinal curve matching. First, the spinal curve can be obtained by performing cubic spline interpolations based on the extracted vertebral centroid positions. The Chamfer distance [13] is calculated from the rigidly aligned spinal curves of I to spinal curves of templates in C_3, and N_f templates with the shortest distance are selected to form C_{N_f}.

2.3 Rib-Guided Registration

After hierarchical matching, the spine physiological structures of templates in C_{N_f} are similar to I. Then, we perform registration between S and templates in C_{N_f} for obtaining corrected segmentation masks and labels. However, considering the shape of adjacent vertebrae are similar and difficult to distinguish, direct registration between S and templates in C_f may lead to misalignment. To tackle this issue, we find that the ribs connected to the vertebrae are easily recognized due to the clear difference in rib length. Therefore, the rib is used to guide the registration with two stages. First, the ribs segmentation masks in S and C_{N_f} are aligned. Then, the registration of vertebrae and ribs is jointly performed between S and C_{N_f}.

Note that the semantic segmentation of I is obtained through a pretrained 3D U-Net. Therefore, we can obtain the segmentation and labeling (S_{rib}) of each rib from S according to the rib length. For the template, we can perform the same operation to obtain segmentation masks and labels of ribs T_{rib}. The segmentation masks of vertebrae and ribs in the templates are denoted as T. We register T_{rib} to S_{rib} for obtaining the deformation field ϕ_{rib}. The registration loss is defined over the rib masks as follows:

$$\mathcal{L}_{rib} = \mathcal{L}_{sim}(T_{rib}(\phi_{rib}), S_{rib}) + \lambda \mathcal{L}_{smooth}(\phi_{rib}). \tag{1}$$

where $\mathcal{L}_{sim}(\cdot, \cdot)$ measures the mean square error (MSE) between $S_{rib}(\phi_{rib})$ and T_{rib}. $\mathcal{L}_{smooth}(\cdot)$ is a regularization term to constrain the deformation field to be smooth, and λ is the balance weight for the regularization.

Through registration between S_{rib} and T_{rib}, the corresponding vertebrae in S and T have been roughly aligned. Then, performing registration between S and T can effectively avoid the interference caused by the similarity of adjacent vertebrae. The deformation field obtained is denoted as ϕ_{rib_vert}. Then the whole rib-guided registration process can be formulated by the following equation:

$$\mathcal{L}_{rib\_guided} = \mathcal{L}_{sim}(T(\phi_{rib} \circ \phi_{rib\_vert}), S) + \lambda \mathcal{L}_{smooth}(\phi_{rib} \circ \phi_{rib\_vert}). \tag{2}$$

There are N_f templates in C_{N_f}, and the rib-guided registration is performed N_f times. Using MSE as the evaluation criterion, the template with the minimum MSE is selected as the final reference template. The corrected label can obtain directly from the reference template, and the corrected segmentation masks by $T(\phi_{rib} \circ \phi_{rib\_vert})$.

3   Experiments

3.1   Dataset and Evaluation Metrics

We collect a clinical dataset that includes CT scans of 1500 patients, of which 802 are male and 698 are female, with the ages distribution ranging from 8 to 99. The segmentation masks and labels of the vertebrae and ribs are manually annotated by doctors. There are 800 scans used for training the segmentation network, 600 scans used for template library construction, and 100 scans used for testing.

To quantitatively evaluate the performance of our proposed method, the segmentation performance is evaluated using both Dice and Hausdorff distance (HD). The labeling performance is evaluated by I_{acc} and S_{acc}, where I_{acc} is the percentage of vertebrae that are assigned with the correct label, and S_{acc} is the percentage of whole scans with correct labels for all vertebrae labels.

3.2   Implementation Details

In training U-Net for segmentation, the Adam optimizer is used with an initial learning rate $\lambda = 0.01$ and batch size $= 2$. The intensity is first normalized to $[0, 1]$. And all the images are resampled to be $256 \times 256 \times 256$ and have $2 \times 2 \times 2 mm^3$ voxel size. All experiments are conducted on two NVIDIA Tesla V100 GPUs using the PyTorch platform.

During the hierarchical matching phase, we use 3 years as the gap for age matching, and match the average distance based on condition $|d_p \pm 2mm| \leqslant d_t$. Then, the Chamfer distance is calculated after aligning different spinal curves by coherent point drift (CPD) rigid registration [18], and the 3 templates with the highest overlap rate are selected to form C_{N_f}. In the stage of rib-guided registration, registration is achieved by affine transformation in ANTs [19] library.

3.3   Ablation Studies

In this section, we conduct extensive experiments to validate the effectiveness of key steps in our proposed framework. We compared our method with the segmentation masks of vertebra and rib using the same 3D U-Net network. The quantitative results are shown in Table 1. It can be found that the performance of segmentation and labeling is significantly improved by our proposed method, i.e., 9.36% improvements of Dice score and also 15.25 improvement of I_{acc}. This proves the effectiveness of rib-guided registration.

3.4   Evaluation and Comparison

Labeling results of our proposed method on five challenging cases are shown in Fig. 4, including various FoVs, metal artifacts, and abnormal curvature. Our proposed method achieves accurate labeling results even on the images with various FoVs and metal artifacts, (see Figs. 4(a) and 4(b)).

Table 1. Quantitative results of ablation studies.

Methods	Segmentation		Labeling	
	Dice (%)	HD (mm)	I_{acc} (%)	S_{acc} (%)
U-Net	81.12 ± 2.23	13.27 ± 4.21	82.11	79.00
Ours	$\mathbf{90.51 \pm 1.54}$	$\mathbf{5.79 \pm 1.27}$	$\mathbf{97.36}$	$\mathbf{95.00}$

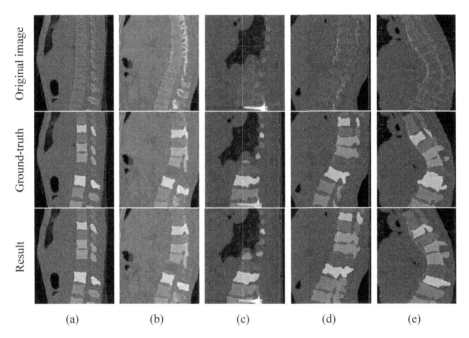

(a)   (b)   (c)   (d)   (e)

Fig. 4. Labeling results of our proposed method on five challenging cases, including (a)(b) limited FoV, (c) metal artifacts, and (d)(e) abnormal curvature. The top row shows are original images, the middle row shows ground-truth, and the bottom row shows our labeling results.

Although our method achieves reliable performance, there are still some limitations of our framework. For example, our proposed method obtains over-segmentation results over the boundaries of rib in Fig. 4(e). The main reason is that the template library does not contain a template very similar to this subject. To address this issue, we plan to collect more representative subjects to enrich the template library in the future.

For quantitative comparison, we compare our proposed method with three related methods proposed by Chen et al. [20], Lessmann et al. [21], and Payer et al. [22], respectively. The results of spine-rib segmentation and labeling from our proposed method and the three comparison methods on our dataset are shown in Table 2. From the results, our proposed method achieves the best performance in both segmentation and labeling tasks.

Table 2. Comparison on segmentation and labeling results

Methods	Segmentation		Labeling	
	Dice (%)	HD (mm)	I_{acc} (%)	S_{acc} (%)
Chen et al. [20]	82.56 ± 2.23	12.27 ± 4.21	85.23	83.00
Lessmann et al. [21]	83.42 ± 1.87	9.87 ± 2.56	88.45	86.00
Payer et al. [22]	88.34 ± 1.67	7.34 ± 1.72	92.83	90.00
Ours	$\mathbf{90.51 \pm 1.54}$	$\mathbf{5.79 \pm 1.27}$	**97.36**	**95.00**

4   Conclusion

In this paper, we have presented a novel method for spine-rib segmentation and labeling. Our method first utilizes a deep learning model to obtain the initial segmentation and label results. Then, a representative spine-rib template library is conducted to match the testing subject, for considering possible mistakes in spine CT images with complicated appearance. Through extensive experiments, our method shows significant performance improvements in spine-rib segmentation and labeling.

References

1. Ben Ayed, I., Punithakumar, K., Minhas, R., Joshi, R., Garvin, G.J.: Vertebral body segmentation in MRI via convex relaxation and distribution matching. In: Ayache, N., Delingette, H., Golland, P., Mori, K. (eds.) MICCAI 2012. LNCS, vol. 7510, pp. 520–527. Springer, Heidelberg (2012). https://doi.org/10.1007/978-3-642-33415-3_64
2. Lecron, F., Boisvert, J., Mahmoudi, S., Labelle, H., Benjelloun, M.: Fast 3D spine reconstruction of postoperative patients using a multilevel statistical model. In: Ayache, N., Delingette, H., Golland, P., Mori, K. (eds.) MICCAI 2012. LNCS, vol. 7511, pp. 446–453. Springer, Heidelberg (2012). https://doi.org/10.1007/978-3-642-33418-4_55
3. Burns, J.E.: Imaging of the spine: a medical and physical perspective. In: Li, S., Yao, J. (eds.) Spinal Imaging and Image Analysis. LNCVB, vol. 18, pp. 3–29. Springer, Cham (2015). https://doi.org/10.1007/978-3-319-12508-4_1
4. Yao, J., Burns, J.E., Munoz, H., Summers, R.M.: Detection of vertebral body fractures based on cortical shell unwrapping. In: Ayache, N., Delingette, H., Golland, P., Mori, K. (eds.) MICCAI 2012. LNCS, vol. 7512, pp. 509–516. Springer, Heidelberg (2012). https://doi.org/10.1007/978-3-642-33454-2_63
5. Forsberg, D., Lundström, C., Andersson, M., Vavruch, L., Tropp, H., Knutsson, H.: Fully automatic measurements of axial vertebral rotation for assessment of spinal deformity in idiopathic scoliosis. Phys. Med. Biol. **58**(6), 1775 (2013)
6. Knez, D., Likar, B., Pernuš, F., Vrtovec, T.: Computer-assisted screw size and insertion trajectory planning for pedicle screw placement surgery. IEEE Trans. Med. Imaging **35**(6), 1420–1430 (2016)
7. Glocker, B., Feulner, J., Criminisi, A., Haynor, D.R., Konukoglu, E.: Automatic localization and identification of vertebrae in arbitrary field-of-view CT scans. In:

Ayache, N., Delingette, H., Golland, P., Mori, K. (eds.) MICCAI 2012. LNCS, vol. 7512, pp. 590–598. Springer, Heidelberg (2012). https://doi.org/10.1007/978-3-642-33454-2_73

8. Glocker, B., Zikic, D., Konukoglu, E., Haynor, D.R., Criminisi, A.: Vertebrae localization in pathological spine CT via dense classification from sparse annotations. In: Mori, K., Sakuma, I., Sato, Y., Barillot, C., Navab, N. (eds.) MICCAI 2013. LNCS, vol. 8150, pp. 262–270. Springer, Heidelberg (2013). https://doi.org/10.1007/978-3-642-40763-5_33

9. Zhan, Y., Maneesh, D., Harder, M., Zhou, X.S.: Robust MR spine detection using hierarchical learning and local articulated model. In: Ayache, N., Delingette, H., Golland, P., Mori, K. (eds.) MICCAI 2012. LNCS, vol. 7510, pp. 141–148. Springer, Heidelberg (2012). https://doi.org/10.1007/978-3-642-33415-3_18

10. Chen, H., et al.: Automatic localization and identification of vertebrae in Spine CT via a joint learning model with deep neural networks. In: Navab, N., Hornegger, J., Wells, W.M., Frangi, A.F. (eds.) MICCAI 2015. LNCS, vol. 9349, pp. 515–522. Springer, Cham (2015). https://doi.org/10.1007/978-3-319-24553-9_63

11. Yang, D., et al.: Deep image-to-image recurrent network with shape basis for automatic vertebra labeling in large-scale 3D CT volumes, 30 July 2019. US Patent 10,366,491

12. Yang, D., et al.: Automatic vertebra labeling in large-scale 3D CT using deep image-to-image network with message passing and sparsity regularization. In: Niethammer, M., et al. (eds.) IPMI 2017. LNCS, vol. 10265, pp. 633–644. Springer, Cham (2017). https://doi.org/10.1007/978-3-319-59050-9_50

13. Cui, Z., et al.: TSegNet: an efficient and accurate tooth segmentation network on 3D dental model. Med. Image Anal. **69**, 101949 (2021)

14. Alomari, R.S., Ghosh, S., Koh, J., Chaudhary, V.: Vertebral column localization, labeling, and segmentation. In: Li, S., Yao, J. (eds.) Spinal Imaging and Image Analysis. LNCVB, vol. 18, pp. 193–229. Springer, Cham (2015). https://doi.org/10.1007/978-3-319-12508-4_7

15. Ullmann, E., Paquette, J.F.P., Thong, W.E., Cohen-Adad, J.: Automatic labeling of vertebral levels using a robust template-based approach. Int. J. Biomed. Imaging **2014**, 719520 (2014)

16. Larhmam, M.A., Benjelloun, M., Mahmoudi, S.: Vertebra identification using template matching modelmp and K-means clustering. Int. J. Comput. Assist. Radiol. Surg. **9**(2), 177–187 (2013)

17. Wu, D.: A learning based deformable template matching method for automatic rib centerline extraction and labeling in CT images. In: 2012 IEEE Conference on Computer Vision and Pattern Recognition, pp. 980–987. IEEE (2012)

18. Myronenko, A., Song, X.: Point set registration: coherent point drift. IEEE Trans. Pattern Anal. Mach. Intell. **32**(12), 2262–2275 (2010)

19. Avants, B.B., Tustison, N., Song, G.: Advanced normalization tools (ants). Insight J. **2**(365), 1–35 (2009)

20. Sekuboyina, A., et al.: Verse: a vertebrae labelling and segmentation benchmark. arXiv preprint arXiv:2001.09193 (2020)

21. Lessmann, N., Van Ginneken, B., De Jong, P.A., Išgum, I.: Iterative fully convolutional neural networks for automatic vertebra segmentation and identification. Med. Image Anal. **53**, 142–155 (2019)

22. Payer, C., Stern, D., Bischof, H., Urschler, M.: Coarse to fine vertebrae localization and segmentation with spatial configuration-net and u-net. In: VISIGRAPP (5: VISAPP), pp. 124–133 (2020)

Multi-scale Segmentation Network for Rib Fracture Classification from CT Images

Jiameng Liu[1], Zhiming Cui[2], Yuhang Sun[3], Caiwen Jiang[1], Zirong Chen[4], Hao Yang[4], Yuyao Zhang[5], Dijia Wu[4], and Dinggang Shen[1,4(✉)]

[1] School of Biomedical Engineering, ShanghaiTech University, Shanghai, China
dgshen@shanghaitech.edu.cn
[2] Department of Computer Science, The University of Hong Kong, Hong Kong, China
[3] School of Biomedical Engineering, Southern Medical University, Guangzhou, China
[4] Shanghai United Imaging Intelligence Co., Ltd., Shanghai, China
[5] School of Information Science and Technology, ShanghaiTech University, Shanghai, China

Abstract. As the most common thoracic trauma, rib fracture classification is essential for clinical evaluation and treatment planning. However, it is challenging for manual identification and classification, due to the tiny size and blurriness of rib fracture in CT images. For automatic classification of rib fractures, conventional methods using hand-crafted features are low in robustness and generalizability. Though previous deep learning-based method shows improved the performance, they empirically normalized all fractures using one size, which ended up in alteration of fracture patterns. Moreover, these methods mainly employed macroscale features with little attention to details, which degrades the classification accuracy, as rib fracture type is essentially determined by tiny fracture details. To address all these issues, we propose a novel framework to classify rib fractures, where we first introduce a multi-scale network to integrate multiple sizes of fractures to minimize size alteration, and further formulate fracture classification problem as a segmentation problem to enforce network attention to tiny fracture details, so as to increase the classification accuracy. Our method has been evaluated on a large dataset (with 53045 cases) with four types of fractures, including acute displaced fracture, acute non-displaced fracture, acute buckle fracture, and chronic fracture. The results are compared with state-of-the-art methods, which suggest that our proposed method achieves the best performance. The capability of our multi-scale segmentation strategy is also verified by experimental results, especially in handling huge size variation of rib fractures during fracture classification.

Keywords: Rib fracture classification · Multi-scale network · Segmentation for classification

© Springer Nature Switzerland AG 2021
C. Lian et al. (Eds.): MLMI 2021, LNCS 12966, pp. 546–554, 2021.
https://doi.org/10.1007/978-3-030-87589-3_56

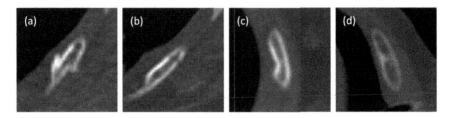

Fig. 1. Four types of rib fractures, including (a) dislocation fracture, (b) slight fracture, (c) cortical twist, (d) callus.

1   Introduction

Rib fracture is the most common injury of thoracic trauma, and its amount and type are vital indicators for fracture grading [2]. Computed tomography (CT) images are most widely used in clinic with manual screening, including extensively identifying and classifying all fractures and determining the subsequent treatment plan. Given the fact that the rib fractures are usually tiny and blurry in CT images, the manual process is extremely time-consuming, labor-intensive, and sometimes results in overlook of small fractures. Consequently, there is an urge need for automatic and precise rib fracture identification and classification for clinical purpose.

Conventional methods mainly rely on handcrafted rib fracture features including fracture location, propagation pattern, amount, and severity [3], and are low in robustness and generalizability. Recently, with the rapid development of deep learning theory, many deep learning-based methods achieved success in disease diagnosis and lesion detection [11,14]. However, only a few studies focus on identification and classification of rib fractures using deep learning. Zhou et al. [19] developed the first rib fracture detection and classification method using convolutional neural network (CNN) which improved the classification performance compared to conventional methods. However, they normalized all fractures using the same scale, which inevitably altered fracture pattern given large variation in fracture sizes. Moreover, their class labels are limited, in which they did not distinguish dislocation fractures, slight fractures, or calluses, whereas these types of fractures have significant differences in clinical practice. Another issue is that fracture types are essentially related to local tiny details, yet previous methods are mainly based on macroscale features, without specific attention to tiny fracture details, which leads to redundant and less discriminative features, and consequently low classification accuracy (Fig.1).

To address the aforementioned issues, we propose a novel multi-scale *segmentation for classification* framework for rib fractures classification. In our task, we aim to classify four types of rib fractures as defined in [15], including dislocation fracture (acute displaced fracture), slight fracture (acute non-displaced fracture), cortical twist (acute buckle fracture), and callus (chronic fracture) as shown in Fig. 1. First, inspired by [8], we formulate our network into multiple scales, where each rib fracture is first normalized to several standard sizes, so that our method will be tolerant to fracture size variations with minimal size alteration. Next, inspired by the dense classification network [9], the scaled fractures are fed into a *segmentation for classification* network, where we impose voxelwise supervision for the classification based on V-Net [12]. Thus we can enforce the network attention to the details of fractures, which are essential information for fracture classification. The final classification result of one rib fracture is obtained by combining the predicted classification probability maps in different scales. Furthermore, we introduce a consistency constraint of the latent vectors between different scaled segmentation networks, which improves both classification accuracy and training efficiency by utilizing high-level semantic information.

We conduct extensive experiments on 53045 rib fractures, and the results show that our method achieves the best performance with 80.41% classification accuracy and 80.31% mean F1-score, respectively, which demonstrates the robustness and accuracy of our proposed method.

2   Methods

The overall framework is shown in Fig. 2. The training pipeline of our method consists of two stages: (1) a modified cascade Faster R-CNN [1,13] for rib fracture detection and generating fracture patches from thoracic CT images, and (2) a two-scale *segmentation for classification* network for generating multi-scale classification probability maps, followed by a fully connected (FC) layer [7] for joint classification.

2.1   Rib Fracture Detection

In the rib fracture detection, we employed a modified cascade Faster R-CNN [1], which is known for its success in object detection in natural images. We dropped the ROI pooling layer from the origin Faster R-CNN structures [18] as all rib fractures share a similar appearance. Similar to [13], we used the anchor boxes strategy as in the Faster R-CNN to train the detection network. An anchor was assigned to a positive label if the Intersection of Union (IoU) is bigger than 0.7 with any ground-truth box manually-annotated by doctors. Whereas, an anchor box was assigned to negative if the IoU is smaller than 0.3 and combined with any ground truth. Those anchor boxes that are neither negative nor positive will not be used for training. In the objective function, we adopt the weighted multi-task loss function as in [13], with focal loss for classification as in [10] to address the class unbalancing.

2.2   Rib Fracture Classification

The framework of the proposed multi-scale *segmentation for classification* network is illustrated in Fig. 2(b). The inputs for the network are the cropped rib fracture patches from chest CT images. Each cropped rib fracture patch I_c

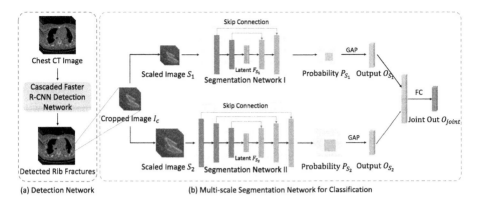

(a) Detection Network   (b) Multi-scale Segmentation Network for Classification

Fig. 2. The framework of our proposed multi-scale segmentation network for rib fracture classification. (a) The pipeline of cascaded Faster R-CNN rib fracture detection network. The inputs are chest CT images and the outputs are the detected bounding boxes of rib fractures. (b) For our proposed multi-scale segmentation network for rib fracture classification, the inputs are the cropped rib fracture patches obtained by the detected bounding box. GAP denotes the global average pooling operator, and FC denotes the fully connected layers.

was first scaled to two selected standard scales S_1 and S_2 (i.e., 24*24*24 and 48*48*48, in our experiments), which are determined according to the physical size distribution in the training data. We employed the segmentation network for fracture classification in each scale, and the prediction results of two segmentation networks will be concatenated to obtain the classification results. In addition, a similarity constraint was employed to keep the consistency between the two encoded latent vectors obtained from the two scaled segmentation networks, so as to improve both classification accuracy and training efficiency by utilizing high-level semantic information.

Segmentation for Classification Network. We utilized a modified V-Net [12] architecture as the *segmentation for classification* networks in each scale. First, the segmentation network was applied to obtain a multi-label probability map P_{S_i} and the corresponding encoded latent vector F_{S_i} for each scaled input S_i. As rib fracture patches are cropped around the centroid in the detection process, the center of the fracture patch is more relevant for classification. Therefore, we only enforce supervision in the middle area of P_{S_i}, denoted as P'_{S_i}, to reduce redundancy. Then, a global average pooling (GAP) [6] operator was employed on P'_{S_i} to squeeze the probability map and obtain the predicted probability O_{S_i} for each fracture type.

We adopt the segmentation loss $\mathcal{L}_{seg}(\cdot)$ on the predicted probability map P'_{S_i} in the segmentation network to impose relationship between voxel and fracture types, so as to enforce attention to the details in fracture. An extra classification loss $\mathcal{L}_{cls}(\cdot)$ was implemented on the squeezed probability vector O_{S_i} to guide the prediction of probability values for each fracture type. The objective function $\mathcal{L}_{scale_i}$ can be formulated as follows:

$$\mathcal{L}_{scale_i} = \mathcal{L}_{seg}(P'_{S_i}, P^{GT}_{S_i}) + \mathcal{L}_{cls}(O_{S_i}, Label_{I_c}), \tag{1}$$

where the $Label_{I_c}$ denotes the ground truth of rib fracture type, and $P^{GT}_{S_i}$ is segmentation ground-truth vector of the same size as the input image S_i with same fracture types as the ground-truth. Different from [12], we use the focal loss in $\mathcal{L}_{seg}(\cdot)$ to solve data unbalancing between rib fractures. Moreover, the focal loss is also used in the classification loss $\mathcal{L}_{cls}(\cdot)$ to address the data unbalancing.

Multi-scale Training Strategy. As shown in Fig. 2(b), first, we scale each cropped patch to two selected standard scales S_1 and S_2. Should the original scale of fracture be close to S_1, its classification probability map would be more reliable, and vice versa. Therefore, we concatenated the classification results of each scale and then utilized the FC layer to automatically learn the contributions of classification results of different scales with respect to the final prediction.

In addition, the latent vector in different scales represent the same fracture, so that they should share same high-level semantic information. To this end, we design a consistent constraint using the mean square error (MSE) loss $\mathcal{L}_{smi}(\cdot)$ between the encoded latent space vector F_{S_1} and F_{S_2} of different scales. And a focal loss $\mathcal{L}_{joint}(\cdot)$ is employed over O_{joint} and $Label_{I_c}$ for the joint classification.

Overall, the training loss $\mathcal{L}_{total}$ for classification is defined as:

$$\mathcal{L}_{total} = \sum_{i=1}^{2}(\mathcal{L}_{seg}(P'_{S_i}, P^{GT}_{S_i}) + \mathcal{L}_{cls}(O_{S_i}, Label_{I_c}))$$
$$+ \mathcal{L}_{joint}(O_{joint}, Label_{I_c}) + \mathcal{L}_{smi}(F_{S_1}, F_{S_2}). \tag{2}$$

The whole objective function for supervising our proposed method is consists of three parts: 1) $\mathcal{L}_{seg}(\cdot)$ and $\mathcal{L}_{cls}(\cdot)$ for training the *segmentation for classification* network at each scale; 2) $\mathcal{L}_{joint}$ for the fusion of classification results of two segmentation networks; 3) $\mathcal{L}_{smi}(F_{S_1}, F_{S_2})$ for maintaining the latent vector consistency between two scale rib fracture patches.

3   Experiments

3.1   Dataset and Evaluation Metrics

In our experiment, we collected 3382 CT thoracic scans from 10 hospitals with a total of 15082 CT thoracic rib fractures to train our rib fracture detection network, of which 2855 CT thoracic scans have at least one rib fractures annotated. The cascaded Faster R-CNN network takes the thoracic CT image as input and outputs a set of detected rib fracture bounding boxes. For the rib fracture classification tasks, we collected 11010 thoracic CT images from hospital, and located 53045 rib fracture patches from trained detection network. For each patch, the rib fracture and its type were verified and classified by experienced radiologists to four categories, including dislocation fracture, slight fracture, cortical twist and callus.

For classification, the dataset was randomly split into a training set of 46804 fractures patches for training and 6241 for testing of rib fracture classification. For different fracture types, the distributions of training and testing sets in our experiment are shown in Table 1. For evaluation, we adopted recall, precision, and F1-score to evaluate the performance for each rib fracture type, and the average value on three metrics was employed to measure the overall performance of our proposed model.

Table 1. Distributions of training and testing sets in four fracture types.

	Dislocation	Slight	Cortical Twist	Callus
Training	10496	14275	8661	13372
Testing	1313	1780	1336	1812

3.2 Implementation Details

We employed V-Net as the *segmentation for classification* network for each scale. All fracture patches were normalized to two selected standard sizes 24*24*24 denoted as S_1 and 48*48*48 denoted as S_2. We used two-downsampling and two-upsampling V-Net for S_1, and three-downsampling and three-upsampling V-Net for S_2. The model was implemented using Pytorch 1.5.1, using four GeForce GTX 1080 TI GPU, and the inference time for single fracture classification is around 2.06 s. We adopted Adam optimizer with momentum set to 0.99 and weight decay to 0.001, and learning rate of 0.0001 (which was reduced by 90% for every 50 epoch). We set windows of CT to [0, 1000] and normalized the HU of all CTs to [−1, 1] before training.

3.3 Results and Discussion

Ablation Study of Different Strategies We conducted a series of experiments to test the effectiveness of the key components of our proposed method. First, we built the baseline networks for rib fracture classification, denoted as $ResNet_{24}$ and $ResNet_{48}$, respectively. $ResNet_{24}$ directly regresses the rib fracture types from input image size of 24*24*24 using two-downsampling 3D residual blocks, and $ResNet_{48}$ takes fracture size of 48*48*48 using three-downsampling 3D sampling blocks. All alternative networks are established by adding different components to the baseline networks.

Our *segmentation for classification* network is denoted as $SegNet_{24}$ and $SegNet_{48}$, respectively. $SegNet_{24}$ is modified by adding two-upsampling layers to the $ResNet_{24}$ and a skip connection is applied here to maintain the low-level information. $SegNet_{48}$ is established by adding three-upsampling layers to $ResNet_{48}$ and a skip connection between downsampling layers and upsampling

Table 2. Comparison of classification results by single-scale segmentation network, classification networks, and our proposed multi-scale *segmentation for classification* network.

Metrics	Methods	Dislocation	Slight	Cortical Twist	Callus	Average
Recall	$ResNet_{24}$	0.7825	0.7164	0.6382	0.8444	0.7453
	$ResNet_{48}$	0.7405	0.6900	0.6785	**0.8988**	0.7519
	$SegNet_{24}$	0.8221	0.7530	0.6469	0.8496	0.7679
	$SegNet_{48}$	0.8294	0.7065	0.6911	0.8093	0.7793
	$FullNet$	**0.8340**	**0.7640**	**0.7201**	0.8841	**0.8005**
Precision	$ResNet_{24}$	0.8080	0.6731	0.6693	**0.8550**	0.7514
	$ResNet_{48}$	0.8511	0.7041	0.6601	0.8112	0.7566
	$SegNet_{24}$	0.8245	0.7064	0.7278	0.8525	0.7778
	$SegNet_{48}$	0.8278	**0.7480**	0.7283	0.8214	0.7813
	$FullNet$	**0.8684**	0.7375	**0.7764**	0.8445	**0.8067**
F1-score	$ResNet_{24}$	0.7950	0.6941	0.6534	0.8497	0.7480
	$ResNet_{48}$	0.7920	0.6970	0.6692	0.8528	0.7527
	$SegNet_{24}$	0.8233	0.7289	0.6850	0.8510	0.7721
	$SegNet_{48}$	0.8286	0.7267	0.7092	0.8545	0.7797
	$FullNet$	**0.8508**	**0.7506**	**0.7472**	**0.8638**	**0.8031**

layers is also adopted to reserve more information. The experimental results of different settings is shown in Table 2.

As shown in Table 2, compared with the baseline networks, our proposed network can significantly improve the classification performance. The average recall, precision and F1-score in *segmentation for classification* networks $SegNet_{24}$ and $SegNet_{48}$ are higher than directly regressing fracture types network $ResNet_{24}$ and $ResNet_{48}$. Especially on average F1-score, $SegNet_{24}$ increases from 74.80% to 77.21% compared with $ResNet_{24}$. These results suggest that the segmentation network can effectively improve the classification robustness and accuracy.

To validate the effectiveness of multi-scale strategy, we concatenate the two classification results of segmentation networks $SegNet_{24}$ and $SegNet_{48}$, denoted as $FullNet$. As shown in Table. 2, the average F1-score of $FullNet$ is increased from 77.97% to 80.31% compared with single-scale segmentation network $SegNet_{48}$, and the F1-score of all fracture types in $FullNet$ are higher than any single scale networks. These results illustrate that the multi-scale method can effectively improve the classification accuracy. And we can also see that the $FullNet$ can significantly improve the precision score by remarkable 4.81% on cortical twist, compared with $SegNet_{48}$. All these results demonstrate that our proposed method can effectively improve the classification robustness, especially on handling huge size variations of rib fractures by introducing multi-scale inputs and *segmentation for classification* network.

Table 3. Comparison with state-of-the-art methods.

Methods	Recall	Precision	F1-score
NoduleNet	0.7617	0.7572	0.7592
ResNet50+AC	0.7850	0.7775	0.7801
Se-ResNext-50	0.7947	0.7861	0.7898
Ours	**0.8005**	**0.8067**	**0.8031**

Comparison with State-of-the-Art Methods. Compared to public dataset like MICCAI 2020 RibFrac challenge, we collected a large in-house dataset from real-world clinics, including 11010 thoracic CT images from 10 hospital, and identified 53045 rib fracture patches which is more representive than the public dataset. To have a fair comparison with state-of-the-art (SOTA) method, we implement the methods they used in rib fracture classification, including NoduleNet [16], ResNet50+Auto-context (ResNet50+AC) [4,17], and Se-ResNexT-50 [5]. We train and test all the networks on our collected dataset which is more representative with a larger sample size, and use the same experimental setting as our proposed method. The quantitative results are shown in Table 3. Our proposed method achieves the best performance in average Recall, Precision and F1-score, indicating the effectiveness of our network with multi-scale inputs and *segmentation for classification* network.

4   Conclusion

In this study, we proposed a multi-scale segmentation network for rib fracture classification from CT images. We used multi-scale strategy to deal with drastic size variations in fracture size, and designed a *segmentation for classification* network to enhance the network learning in fracture details, which are essential information for classification. We adopted extensive experiments to test our model, and the results suggested that our multi-scale network can effectively address the size variation compared to single-scale networks, and also our *segmentation for classification* network indeed improved robustness and classification performance.

References

1. Cai, Z., Vasconcelos, N.: Cascade R-CNN: delving into high quality object detection. In: Proceedings of the IEEE Conference on Computer Vision and Pattern Recognition, pp. 6154–6162 (2018)
2. De Waele, J.J., Calle, P.A., Blondeel, L., Vermassen, F.E.: Blunt cardiac injury in patients with isolated sternal fractures: the importance of fracture grading. Eur. J. Trauma **28**(3), 178–182 (2002)
3. Harden, A.L., Kang, Y.S., Agnew, A.M.: Rib fractures: validation of an interdisciplinary classification system. Forensic Anthropol. **2**(3), 158–167 (2019)

4. He, K., Zhang, X., Ren, S., Sun, J.: Deep residual learning for image recognition. In: Proceedings of the IEEE Conference on Computer Vision and Pattern Recognition, pp. 770–778 (2016)
5. Hu, J., Shen, L., Sun, G.: Squeeze-and-excitation networks. In: Proceedings of the IEEE Conference on Computer Vision and Pattern Recognition, pp. 7132–7141 (2018)
6. Kalchbrenner, N., Grefenstette, E., Blunsom, P.: A convolutional neural network for modelling sentences. arXiv preprint arXiv:1404.2188 (2014)
7. Krizhevsky, A., Sutskever, I., Hinton, G.E.: ImageNet classification with deep convolutional neural networks. Adv. Neural Inf. Process. Syst. **25**, 1097–1105 (2012)
8. Li, S., Zhu, X., Bao, J.: Hierarchical multi-scale convolutional neural networks for hyperspectral image classification. Sensors **19**(7), 1714 (2019)
9. Lifchitz, Y., Avrithis, Y., Picard, S., Bursuc, A.: Dense classification and implanting for few-shot learning. In: Proceedings of the IEEE/CVF Conference on Computer Vision and Pattern Recognition, pp. 9258–9267 (2019)
10. Lin, T.Y., Goyal, P., Girshick, R., He, K., Dollár, P.: Focal loss for dense object detection. In: Proceedings of the IEEE International Conference on Computer Vision, pp. 2980–2988 (2017)
11. Litjens, G., Kooi, T., Bejnordi, B.E., Setio, A.A.A., Ciompi, F., Ghafoorian, M., Van Der Laak, J.A., Van Ginneken, B., Sánchez, C.I.: A survey on deep learning in medical image analysis. Med. Image Anal. **42**, 60–88 (2017)
12. Milletari, F., Navab, N., Ahmadi, S.A.: V-Net: fully convolutional neural networks for volumetric medical image segmentation. In: 2016 Fourth International Conference on 3D vision (3DV), pp. 565–571. IEEE (2016)
13. Ren, S., He, K., Girshick, R., Sun, J.: Faster R-CNN: towards real-time object detection with region proposal networks. arXiv preprint arXiv:1506.01497 (2015)
14. Shen, D., Wu, G., Suk, H.I.: Deep learning in medical image analysis. Annual Rev. Biomed. Eng. **19**, 221–248 (2017)
15. Talbot, B.S., Gange, C.P., Jr., Chaturvedi, A., Klionsky, N., Hobbs, S.K., Chaturvedi, A.: Traumatic rib injury: patterns, imaging pitfalls, complications, and treatment. Radiographics **37**(2), 628–651 (2017)
16. Tang, H., Zhang, C., Xie, X.: NoduleNet: decoupled false positive reduction for pulmonary nodule detection and segmentation. In: Shen, D., et al. (eds.) MICCAI 2019. LNCS, vol. 11769, pp. 266–274. Springer, Cham (2019). https://doi.org/10.1007/978-3-030-32226-7_30
17. Tu, Z., Bai, X.: Auto-context and its application to high-level vision tasks and 3D brain image segmentation. IEEE Trans. Pattern Anal. Mach. Intell. **32**(10), 1744–1757 (2009)
18. Xu, X., Zhou, F., Liu, B., Fu, D., Bai, X.: Efficient multiple organ localization in CT image using 3D region proposal network. IEEE Trans. Med. Imaging **38**(8), 1885–1898 (2019)
19. Zhou, Q.Q., et al.: Automatic detection and classification of rib fractures based on patients' CT images and clinical information via convolutional neural network. Eur. Radiol. **31**(6), 3815–3825 (2020)

Knowledge-Guided Multiview Deep Curriculum Learning for Elbow Fracture Classification

Jun Luo[1], Gene Kitamura[2], Dooman Arefan[2], Emine Doganay[2],
Ashok Panigrahy[2,3], and Shandong Wu[1,2,4(✉)]

[1] Intelligent Systems Program, School of Computing and Information,
University of Pittsburgh, Pittsburgh, PA, USA
jul117@pitt.edu
[2] Department of Radiology, School of Medicine, University of Pittsburgh,
Pittsburgh, PA, USA
[3] University of Pittsburgh Medical Center Children's Hospital of Pittsburgh,
Pittsburgh, PA, USA
[4] Department of Biomedical Informatics and Department of Bioengineering,
University of Pittsburgh, Pittsburgh, PA, USA
wus3@upmc.edu

Abstract. Elbow fracture diagnosis often requires patients to take both frontal and lateral views of elbow X-ray radiographs. In this paper, we propose a multiview deep learning method for an elbow fracture subtype classification task. Our strategy leverages transfer learning by first training two single-view models, one for frontal view and the other for lateral view, and then transferring the weights to the corresponding layers in the proposed multiview network architecture. Meanwhile, quantitative medical knowledge was integrated into the training process through a curriculum learning framework, which enables the model to first learn from "easier" samples and then transition to "harder" samples to reach better performance. In addition, our multiview network can work both in a dual-view setting and with a single view as input. We evaluate our method through extensive experiments on a classification task of elbow fracture with a dataset of 1,964 images. Results show that our method outperforms two related methods on bone fracture study in multiple settings, and our technique is able to boost the performance of the compared methods. The code is available at https://github.com/ljaiverson/multiview-curriculum.

Keywords: Multiview learning · Deep learning · Curriculum learning · Elbow fracture · Clinical knowledge

1 Introduction

Human's cognitive ability relies deeply on integrating information from different views of the objects. This is particularly the case for elbow fracture diagnosis

© Springer Nature Switzerland AG 2021
C. Lian et al. (Eds.): MLMI 2021, LNCS 12966, pp. 555–564, 2021.
https://doi.org/10.1007/978-3-030-87589-3_57

where patients are often required to take both the frontal view (i.e. Anterior-Posterior view) and lateral view of elbow X-ray radiographs for diagnosis. This is because some fracture subtypes might be more visible from a certain perspective: the frontal view projects the distal humerus, the proximal ulna and the radius [7, 21,22], while the lateral view shows the coronoid process and the olecranon process [9,18,22]. In practice, it is also common that some patients only have a single view radiograph acquired, or have a missing view for various reasons.

In recent years, the advance of deep learning has been facilitating the automation of bone fracture diagnosis [3,10,12] through multiple views of X-ray images, which shows faster speed and decent accuracy compared to human experts [13,14,17]. However, few methods leverage multiview information, which provide more visual information from different perspectives for elbow fracture diagnosis.

In this work, we propose a novel multiview deep learning network architecture for elbow fracture subtype classification that takes frontal view and lateral view elbow radiographs as input. While the proposed model is a dual-view (frontal and lateral) architecture, it is flexible as it does not strictly require a dual-view input during inference. Furthermore, our training strategy for the multiview model takes advantage of transfer learning by first training two single-view models, one for frontal view and the other for lateral view, and then transferring the trained weights to the corresponding layers in the proposed multiview network architecture. In addition, we investigate the utilities of integrating medical knowledge of different views into the training via a curriculum learning scheme, which enables the model to first learn from "easier" samples and then transition to "harder" samples to reach better performance.

To evaluate our method, we conduct experiments on a classification task of three classes of elbow fractures that shown in Fig. 1. We compare our method to multiple settings including the single-view models, different combinations of the transfer learning strategy and the knowledge-guided curriculum learning. Our method is also compared to a previous method [11]. Results show that our proposed method outperforms the compared methods, and our method functions seamlessly on a multiview and a single-view settings.

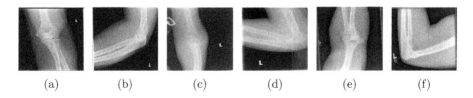

| (a) | (b) | (c) | (d) | (e) | (f) |

Fig. 1. Example images from the three categories from our dataset for classification task: (a) and (b) show the frontal and lateral non-fracture category respectively; (c) and (d) show the frontal and lateral ulnar fracture category respectively; (e) and (f) show the frontal and lateral radial fracture category respectively.

2   Related Work

Multiview learning [23] takes advantage of data with multiple views of the same objects. Co-training [2,16,20] style algorithms were a group of traditional multiview learning algorithms originally focusing on semi-supervised learning, where multiple views of data were iteratively added to the labeled set and learned by the classifier. Another group of multiview learning algorithms explore Multiple Kernel Learning (MKL), which was originally proposed to restrict the search space of kernels [4,6]. Recent work on multiview learning based modeling shows promising effects for medical fields such as bone fracture and breast cancer detection [8,13,17].

Curriculum learning is also an area of active research. It was first introduced by Bengio et al. in [1] to enable the machine learning to mimic human learning by training a machine learning model first with "easier" samples and then transition to "harder" samples. Some existing work focus on integrating domain knowledge into the training process through curriculum learning. For example, [11,15] integrate domain knowledge by using the classification difficulty level of different classes.

3   Methods

3.1   Multiview Model Architecture

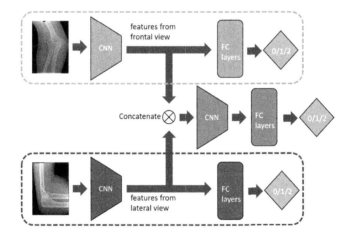

Fig. 2. The proposed multiview model architecture. The green and blue dotted line box represent the frontal and lateral view modules, respectively. Yellow diamonds are the predicted labels, 0, 1, 2 corresponding to non-fracture, ulnar fracture, radial fracture respectively (Color figure online)

To incorporate information from both frontal and lateral view for the elbow X-ray images while maintaining the flexibility of being able to output predictions with one view as input, we propose a novel multiview model architecture shown in Fig. 2. In this architecture, during training, pairs of frontal and lateral view images are fed into their corresponding modules for feature extraction by the convolutional neural networks (CNNs). After the feature extraction, the model splits into three branches as shown in Fig. 2. The top and bottom branches take the corresponding single-view features to the fully connected (FC) layers for classification, while the middle branch takes the concatenated features from both views as input to further extract features and then conducts classification.

Consider a data sample triplet $\mathcal{D}_i = \{x_i^{(F)}, x_i^{(L)}, y_i\}$ where $\mathcal{D}_i$ represents the i-th data sample, $x_i^{(F)}$, and $x_i^{(L)}$ are its images from the frontal and lateral view, and $y_i \in \{0, 1, 2\}$ is its ground truth label with 0, 1, 2 corresponding to non-fracture, ulnar fracture, radial fracture respectively. We denote the three predicted labels from the three branches of our multiview model as $\mathcal{F}(x_i^{(F)})$, $\mathcal{L}(x_i^{(L)})$, and $\mathcal{M}(x_i^{(F)}, x_i^{(L)})$, where $\mathcal{F}$, $\mathcal{L}$, $\mathcal{M}$ represent the *frontal view module*, the *lateral view module*, and the *"merge module"* that contains the two CNN blocks from the frontal and lateral module, the CNN as well as the FC layers in the middle branch.

During training, we minimize the objective function over the i-th data sample computed by Eq. (1) where θ, $\theta_{\mathcal{F}}$, $\theta_{\mathcal{L}}$, and $\theta_{\mathcal{M}}$ represent the parameters in the entire model, the frontal view module, the lateral view module, and the merge module. As shown in Eq. (1) (with C being the number of classes), for each module, the loss is computed with cross entropy loss over the corresponding predicted label and ground truth y_i in a one-hot representation.

$$
\begin{aligned}
J_\theta(x_i^{(F)}, x_i^{(L)}, y_i) &= J_{\theta_{\mathcal{F}}}(x_i^{(F)}, y_i) + J_{\theta_{\mathcal{L}}}(x_i^{(L)}, y_i) + J_{\theta_{\mathcal{M}}}(x_i^{(F)}, x_i^{(L)}, y_i) \\
&= -\sum_{c=1}^{C} \left(y_{i,c} \left(\log(\mathcal{F}(x_i^{(F)})_c) + \log(\mathcal{L}(x_i^{(L)})_c) + \log(\mathcal{M}(x_i^{(F)}, x_i^{(L)})_c) \right) \right)
\end{aligned}
\tag{1}
$$

During test phase, if a frontal view image and a lateral view image are both presented, the default final predicted label is the one predicted from the merge module, i.e. $\mathcal{M}(x_i^{(F)}, x_i^{(L)})$. Alternatively, if there is only one view, the model will still output a predicted label from the module of the corresponding view credited to the designed architecture of our model.

3.2   Transfer Learning from Pretrained Single-View Models

In most medical applications with deep learning, researchers use the ImageNet [5] pretrained model as a way of transfer learning. However, a great number of deep learning models do not have publicly available pretrained weights, especially for self-designed models. Here, we investigate a homogeneous way of transfer learning as shown in Fig. 3: we first train two single-view models (using the same training set as the one for the multiview model) that have identical structure as the frontal view and lateral view module in the multiview architecture. Then, we transfer the trained weights of the CNNs and FC layers from the single view models to the counterparts of the multiview model (refer to the links in Fig. 3). For the middle branch (the gray CNN and LC layers blocks in Fig. 2) in the merge module, we randomly initialize their weights. We make all weights trainable in the multiview model.

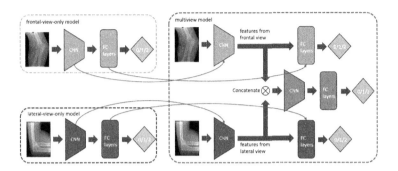

Fig. 3. Transfer learning from pretrained single-view models.

3.3 Knowledge-Guided Curriculum Learning

For the model training, we propose a knowledge-guided curriculum learning to enhance learning effects. The idea of curriculum learning is to enable the training process to follow an "easy-to-hard" order, where the easier samples will be fed into the model for training earlier than the harder samples. To do so, we implemented a multiview-based curriculum learning by adapting the method from [15]. We quantify and integrate medical knowledge by scoring the classification difficulty levels of each category of elbow fracture with board-certified radiologist's expertise. Table 1 shows the quantitative scores reflecting the classification difficulty based on experience of expert radiologists. Note that we use the "Both views" scores to train the multiview model, and use "Frontal/Lateral view only" for homogeneous transfer learning.

Table 1. Quantitative classification difficulty levels for each category of elbow fracture (1-hardest; 100-easiest), which enables the integration of medical knowledge into curriculum learning.

	Non-fracture (normal)	Ulnar fracture	Radial fracture
Frontal view only	30	30	30
Lateral view only	35	60	45
Both views	45	65	55

These scores are used to initialize the sampling probability for each training data point according to Eq. (2) with $e = 1$, where $p_i^{(1)}$ is the initial sampling probability for data point $\mathcal{D}_i$, s_i is its score, s_k is the score of the data point $\mathcal{D}_k$, and N is the number of data points in the dataset. Using the sampling probabilities, at the beginning of every epoch, we permute the training set by sampling all the data points without replacement.

$$p_i^{(e)} = \begin{cases} \frac{s_i}{\sum_{k=1}^{N} s_k} & e = 1, \\ p_i^{(e-1)} \cdot \sqrt[E']{\frac{1/N}{p_i^{(0)}}} & 2 \leq e \leq E', \\ 1/N & E' < e \leq E \end{cases} \tag{2}$$

This enables the easier samples to have a higher chance of being presented before the harder samples. This chance will be exponentially reduced by updating the sampling probabilities for each data point according to Eq. (2). In this equation, e is the current epoch, E' is the last epoch that we update the sampling probabilities. For the rest of the training ($E' < e \leq E$) the sampling probabilities will be fixed to $1/N$.

4 Experiments and Results

4.1 Experiment Settings

Dataset and Implementation Details. This study includes a private dataset of 982 subjects of elbow fractures in an Institutional Review Board-approved retrospective study. The subjects are categorized into three classes: 500 non-fracture (normal) cases, 98 ulnar fracture cases, and 384 radial fracture cases. Each subject includes one frontal and one lateral elbow X-ray image, which makes it a total of 1,964 elbow X-ray images. To increase the robustness of our results, we conduct 8-fold cross validation. For each split of the entire dataset, one fold was used as the hold-out test set. Within the remaining seven folds, we randomly select one fold as the validation set for hyperparameter tuning. The remaining folds are used as the training set. All separations of the dataset are in a stratified manner, which maintains the ratio over different classes. The reported results are averages over the 8 disjoint held-out test sets.

VGG16 [19] is used as the backbone for the two single-view models, and the frontal and lateral modules in the multiview model. We customize the middle branch two $3 \times 3 \times 512$ convolutional layers with max pooling layers, followed by VGG16's classifier for the FC layers. The hyperparameters are selected based on the best validation AUCs. We use the following hyperparameters for the proposed model: batch size 64, learning rate 10^{-4} for the Adam optimizer, and after 16 epochs every sample is treated as having an equal difficulty score. All models were trained on an NVIDIA Tesla V100 GPU. The code is available at https://github.com/ljaiverson/multiview-curriculum.

Metrics. The metrics for the 3-class classification task include accuracy and area under receiver operating characteristic curve (AUC). We also compute a

Table 2. Model performance with both views. The bold numbers correspond to the highest value for each metric (TL: proposed transfer learning from single view models; CL: proposed knowledge-guided curriculum learning).

Model	Accuracy	AUC	Balanced accuracy	Binary task accuracy	Binary task AUC
Single-view-frontal	0.683	0.807	0.570	0.732	0.813
Single-view-lateral	0.856	0.954	0.807	0.895	0.959
Multiview	0.854	0.958	0.796	0.884	0.964
Multiview + TL	**0.891**	0.966	0.847	**0.916**	0.973
Multiview + [11]	0.818	0.939	0.746	0.864	0.952
Multiview + [11] + TL	0.870	0.961	0.811	0.898	0.973
Multiview + CL	0.889	0.970	0.847	0.908	**0.978**
Multiview + CL + TL	0.889	**0.974**	**0.864**	0.910	0.976

balanced accuracy by averaging the ratios between the number of true positives and the total number of samples with respect to each class, which reduces the effect induced by data imbalance. In addition, we evaluate the models' overall ability to distinguish fracture against non-fracture images. This is done by binarizing the ground truth and predicted labels by assigning 0 to them if they originally are 0, and assigning 1 otherwise. We compute the binary task accuracy and the AUC as two additional measures.

4.2 Results

As shown in Table 2, we compare our proposed multiview model with curriculum learning method (CL) and transfer learning (TL) with the following six types of models: 1) two single-view models (frontal/lateral view only), referred as Single-view-frontal/lateral; 2) multiview model with regular training, referred as Multiview; 3) multiview model with only transfer learning strategy, referred as Multiview + TL; 4) multiview model with a previous curriculum training method [11], referred as Multiview + [11]; 5) multiview model with [11] and our proposed transfer learning strategy, referred as Multiview + [11] + TL; and 6) multiview model with only our curriculum learning method, referred as Multiview + CL. We use the output from the middle branch, as the predicted label.

Attributed to the multiple branches of our model and the customized loss function, our model has the flexibility of generating the prediction with a single view as input. In Table 3, we show the results of the performance from the frontal view module and lateral view module separately. Different from [11], our curriculum updates the difficulty score of every sample after every epoch, which benefits the multiview model. Table 2 shows that with both views presented in the test phase, our method achieves the highest AUC and balanced accuracy with a margin of up to 0.118 compared to the state-of-the-art performance. In settings with missing views, however, our strategy does not always perform the

Table 3. Model performance with a single view as input

Model	Input view	Accuracy	AUC	Balanced accuracy	Binary task accuracy	Binary task AUC
Single-view	Frontal	0.720	0.828	0.593	0.761	0.844
Single-view + CL [15]	Frontal	0.683	0.807	0.570	0.732	0.813
Multiview	Frontal	0.658	0.749	0.514	0.702	0.766
Multiview + TL	Frontal	0.738	0.827	0.617	0.774	0.829
Multiview + [11]	Frontal	0.566	0.675	0.396	0.575	0.648
Multiview + [11] + TL	Frontal	0.737	0.815	0.605	0.773	0.831
Multiview + CL	Frontal	0.723	0.814	0.602	0.761	0.823
Multiview + CL + TL	Frontal	**0.756**	**0.829**	**0.636**	**0.786**	**0.846**
Single-view	Lateral	0.856	0.954	0.807	**0.895**	0.959
Single-view + CL [15]	Lateral	0.840	0.946	0.809	0.872	0.948
Multiview	Lateral	0.844	0.951	0.800	0.870	0.956
Multiview + TL	Lateral	0.848	0.954	0.804	0.876	0.961
Multiview + [11]	Lateral	0.837	0.945	0.779	0.870	0.949
Multiview + [11] + TL	Lateral	**0.857**	**0.960**	**0.819**	0.885	**0.969**
Multiview + CL	Lateral	0.838	0.956	0.807	0.867	0.956
Multiview + CL + TL	Lateral	0.840	0.955	0.794	0.874	0.960

best. Table 3 shows that with frontal view as the only input view, our method outperforms all the compared methods per each metric, but with the lateral view as the only input view, our method achieves slightly lower performance than the best results.

5   Conclusion

In this work, we propose a novel multiview deep learning method for elbow fracture subtype classification from frontal and lateral view X-ray images. We leverage transfer learning by first pretraining two single-view models. Meanwhile, medical knowledge was quantified and incorporated in the training process through curriculum learning. The results show that our multiview model outperforms the compared methods, and we achieved improved results over the previously published curriculum training strategies. As future work, we plan to further integrate other domain knowledge with respect to different views and explore curriculum learning in the output space.

Acknowledgements. This work was supported in part by National Institutes of Health grants (1R01CA193603 and 1R01CA218405), the Stimulation Pilot Research Program of the Pittsburgh Center for AI Innovation in Medical Imaging and the associated Pitt Momentum Funds of a Scaling grant from the University of Pittsburgh (2020), and an Amazon Machine Learning Research Award.

References

1. Bengio, Y., Louradour, J., Collobert, R., Weston, J.: Curriculum learning. In: Proceedings of the 26th Annual International Conference on Machine Learning, pp. 41–48 (2009)
2. Blum, A., Mitchell, T.: Combining labeled and unlabeled data with co-training. In: Proceedings of the Eleventh Annual Conference on Computational Learning Theory, pp. 92–100 (1998)
3. Cheng, C.T., et al.: A scalable physician-level deep learning algorithm detects universal trauma on pelvic radiographs. Nat. Commun. **12**(1), 1–10 (2021)
4. Cortes, C., Mohri, M., Rostamizadeh, A.: Learning non-linear combinations of kernels. In: Bengio, Y., Schuurmans, D., Lafferty, J., Williams, C., Culotta, A. (eds.) Advances in Neural Information Processing Systems, vol. 22. Curran Associates, Inc. (2009)
5. Deng, J., Dong, W., Socher, R., Li, L.J., Li, K., Fei-Fei, L.: Imagenet: a large-scale hierarchical image database. In: 2009 IEEE Conference on Computer Vision and Pattern Recognition, pp. 248–255. IEEE (2009)
6. Duffy, N., Helmbold, D.P.: Leveraging for regression. In: COLT, pp. 208–219 (2000)
7. El-Khoury, G.Y., Daniel, W.W., Kathol, M.H.: Acute and chronic avulsive injuries. Radiol. Clin. North Am. **35**(3), 747–766 (1997)
8. Geras, K.J., et al.: High-resolution breast cancer screening with multi-view deep convolutional neural networks. arXiv preprint arXiv:1703.07047 (2017)
9. Goldfarb, C.A., Patterson, J.M.M., Sutter, M., Krauss, M., Steffen, J.A., Galatz, L.: Elbow radiographic anatomy: measurement techniques and normative data. J. Shoulder Elbow Surg. **21**(9), 1236–1246 (2012)
10. Guan, B., Zhang, G., Yao, J., Wang, X., Wang, M.: Arm fracture detection in x-rays based on improved deep convolutional neural network. Comput. Electr. Eng. **81**, 106530 (2020)
11. Jiménez-Sánchez, A., et al.: Medical-based deep curriculum learning for improved fracture classification. In: Shen, D., et al. (eds.) MICCAI 2019. LNCS, vol. 11769, pp. 694–702. Springer, Cham (2019). https://doi.org/10.1007/978-3-030-32226-7_77
12. Kalmet, P.H., et al.: Deep learning in fracture detection: a narrative review. Acta Orthopaedica **91**(2), 215–220 (2020)
13. Kitamura, G., Chung, C.Y., Moore, B.E.: Ankle fracture detection utilizing a convolutional neural network ensemble implemented with a small sample, de novo training, and multiview incorporation. J. Digit. Imaging **32**(4), 672–677 (2019)
14. Krogue, J.D., et al.: Automatic hip fracture identification and functional subclassification with deep learning. Radiol. Artif. Intell. **2**(2), e190023 (2020)
15. Luo, J., Kitamura, G., Doganay, E., Arefan, D., Wu, S.: Medical knowledge-guided deep curriculum learning for elbow fracture diagnosis from x-ray images. In: Medical Imaging 2021: Computer-Aided Diagnosis, vol. 11597, p. 1159712. International Society for Optics and Photonics (2021)
16. Nigam, K., Ghani, R.: Analyzing the effectiveness and applicability of co-training. In: Proceedings of the Ninth International Conference Information Knowledge Management, pp. 86–93 (2000)
17. Rayan, J.C., Reddy, N., Kan, J.H., Zhang, W., Annapragada, A.: Binomial classification of pediatric elbow fractures using a deep learning multiview approach emulating radiologist decision making. Radiol. Artif. Intell. **1**(1), e180015 (2019)

18. Sandman, E., Canet, F., Petit, Y., Laflamme, G.Y., Athwal, G.S., Rouleau, D.M.: Effect of elbow position on radiographic measurements of radio-capitellar alignment. World J. Orthop. **7**(2), 117 (2016)
19. Simonyan, K., Zisserman, A.: Very deep convolutional networks for large-scale image recognition. arXiv preprint arXiv:1409.1556 (2014)
20. Sindhwani, V., Niyogi, P., Belkin, M.: A co-regularization approach to semi-supervised learning with multiple views. In: Proceedings of ICML Workshop on Learning with Multiple Views, vol. 2005, pp. 74–79. Citeseer (2005)
21. Stevens, M.A., El-Khoury, G.Y., Kathol, M.H., Brandser, E.A., Chow, S.: Imaging features of avulsion injuries. Radiographics **19**(3), 655–672 (1999)
22. Whitley, A.S., Jefferson, G., Holmes, K., Sloane, C., Anderson, C., Hoadley, G.: Clark's Positioning in Radiography 13E. CRC Press Boca Raton (2015)
23. Xu, C., Tao, D., Xu, C.: A survey on multi-view learning. arXiv preprint arXiv:1304.5634 (2013)

Contrastive Learning of Single-Cell Phenotypic Representations for Treatment Classification

Alexis Perakis[1]([✉]), Ali Gorji[2], Samriddhi Jain[2], Krishna Chaitanya[3], Simone Rizza[2], and Ender Konukoglu[3]

[1] KOF Swiss Economic Institute, ETH Zurich, Zurich, Switzerland
perakis@kof.ethz.ch
[2] ETH Zurich, Zurich, Switzerland
[3] Computer Vision Lab, ETH Zurich, Zurich, Switzerland

Abstract. Learning robust representations to discriminate cell phenotypes based on microscopy images is important for drug discovery. Drug development efforts typically analyse thousands of cell images to screen for potential treatments. Early works focus on creating hand-engineered features from these images or learn such features with deep neural networks in a fully or weakly-supervised framework. Both require prior knowledge or labelled datasets. Therefore, subsequent works propose unsupervised approaches based on generative models to learn these representations. Recently, representations learned with self-supervised contrastive loss-based methods have yielded state-of-the-art results on various imaging tasks compared to earlier unsupervised approaches. In this work, we leverage a contrastive learning framework to learn appropriate representations from single-cell fluorescent microscopy images for the task of Mechanism-of-Action classification. The proposed work is evaluated on the annotated BBBC021 dataset, and we obtain state-of-the-art results in NSC, NCSB and drop metrics for an unsupervised approach. We observe an improvement of 10% in NCSB accuracy and 11% in NSC-NSCB drop over the previously best unsupervised method. Moreover, the performance of our unsupervised approach ties with the best supervised approach. Additionally, we observe that our framework performs well even without post-processing, unlike earlier methods. With this, we conclude that one can learn robust cell representations with contrastive learning. We make the code available on GitHub (https://github.com/SamriddhiJain/SimCLR-for-cell-profiling).

Keywords: Fluorescent microscopy · Phenotypes · Profiling · Cell images · Cell representations · Unsupervised learning · Contrastive learning

A. Perakis, A. Gorji and S. Jain—These authors contributed equally to the paper.

Electronic supplementary material The online version of this chapter (https://doi.org/10.1007/978-3-030-87589-3_58) contains supplementary material, which is available to authorized users.

© Springer Nature Switzerland AG 2021
C. Lian et al. (Eds.): MLMI 2021, LNCS 12966, pp. 565–575, 2021.
https://doi.org/10.1007/978-3-030-87589-3_58

1 Introduction

An effective approach in the field of drug discovery is to relate treatments under development to existing ones. Comparing an unknown to a known treatment enables finding desired similarities and avoiding unwanted effects. Cells of interest are exposed to chemical compounds and then imaged using various microscopy techniques. A cell's response mechanism upon treatment is called Mechanism-of-Action (MOA). Such response mechanisms modify a cell's phenotype (morphology) to various degrees. To compare different treatments we can classify them into different MOAs. To perform this classification, we require cell representations that accurately capture the cell morphology. Morphological cell profiling uses image recognition techniques to construct these cell representations. Typically, cell profiling-based MOA classification is done on thousands of images in a transductive learning setting in the literature, where the whole data are included in the training and evaluation process.

In the literature, many works have focused on automating the tedious task of learning meaningful single-cell representations and utilize them downstream for MOA assignment. Initial works like [24,27] rely on creating expert engineered features. Later, deep neural networks are used to learn these features. Early works making use of neural networks rely on fully supervised approaches [11,21] to learn these features, where [11] implements multi-scale convolutional neural networks to extract cell morphology directly from images. Alternatively, [3] proposes a weakly supervised learning method. Other works as in [12,28] use the metadata information to devise pseudo-labels for the network supervision. All the fully or weakly supervised methods above suffer from the requirement of labeled sets, and annotating such images by experts leads to high costs and time-consuming efforts. Even in [12,28], the labels acquired from metadata can be prone to imprecise labeling due to the nature of the data and treatment analysis techniques. Some other works use transfer learning techniques [1,26]. With these approaches, it may not be feasible to acquire appropriate labeled datasets for pre-training.

These reasons encourage the usage of unsupervised learning approaches that provide the following advantages: large unlabeled datasets can be directly used in training, and no pre-training labelled datasets are required. In approaches as above [5,13,19,22], the aim is to learn robust single-cell representations. In [19] the authors perform clustering on the learned cell representations and use the cluster assignments as labels, where the implementation is similar to [5]. Alternatively, some works use generative models like variational autoencoders [22] or generative adversarial networks [13] to learn such representations.

For representation learning, many recent works propose self-supervised learning methods using unlabeled data. Recently, self-supervised approaches based on a contrastive loss [14] have yielded state of the art performance for imaging tasks such as classification, object detection, segmentation [7,15,17,33] on benchmark natural image datasets as well as for medical imaging tasks on MR [6], CT [34,35], X-ray [2,29,32,36], dermatology [2], electron microscopy [18] image datasets and ECG signals [20]. These contrastive loss based approaches outperform traditional self-supervision based pretext tasks (e.g., rotation [10], inpainting [25]) and generative models [8,9].

In cell profiling, the datasets contain thousands or millions of images. Unsupervised methods are promising and provide more viable solutions for such applications than supervised methods. Hence, in the proposed work, we leverage the popular contrastive learning framework [7] for learning robust cell representations using only unlabeled data. In contrastive loss-based learning [14], representations are learned by contrasting positive examples to negative examples. As in [7], we train the network to pull the representations of positive examples to be close in the latent space and push the negative examples representations to be far away from positive examples.

Our contributions are:

- We are the first to learn single-cell representations using a contrastive learning framework in an unsupervised setting.
- For a cell profiling dataset, we evaluate and find the most important components and hyper-parameters used in the contrastive framework such as: (i) encoder network size, (ii) data augmentation strategies, (iii) projection head size, (iv) batch size value and (v) temperature parameter.
- We achieve state-of-the-art NSCB accuracy and NSC-NSCB drop for an unsupervised method. Our unsupervised results in NSC, NSCB, and drop metrics match the state-of-the-art for a supervised method.
- The learned single-cell representations perform well even without any post-processing, unlike earlier works.

2 Methods

We divide this section into two parts: (a) representation learning and (b) MOA classification. In the first part, we learn representations in a contrastive learning framework as proposed in [7]. In the second part, we use these representations for the downstream task of Mechanism-of-Action classification of the treatment profiles.

(a) **Representation Learning:** In the contrastive framework, we learn a global representation for each input image as illustrated in Fig. 1. We follow [7] and sample a mini-batch of images of size N from the whole dataset X. Then, for each sampled image x we apply two random transformations t_i and t_j (sampled from a set of transformations T). The transformed images are denoted by $\tilde{x}_i = t_i(x)$ and $\tilde{x}_j = t_j(x)$ to obtain $2N$ images in the batch, as shown in Fig. 1. For the encoder network, we use a ResNet [16] denoted by f to get the representation h where $h_i = f(x_i)$, which is followed by a projection head g. The output latent representation is given by z where $z_i = g(f(x_i))$. The two transformed images arising from a given image x are denoted as the positive pair. The remaining $2(N-1)$ images in the batch act as the negative pairs and form the negative images set Ω^-. The contrastive loss applied on the positive pairs of output latent representations is defined as follows:

$$l(\tilde{x}_i, \tilde{x}_j) = -\log \frac{e^{\text{sim}(z_i,z_j)/\tau}}{e^{\text{sim}(z_i,z_j)/\tau} + \sum_{x_n \in \Omega^-} e^{\text{sim}(z_i,g(f(x_n)))/\tau}} \tag{1}$$

where z_i, z_j are the output latent representations of the positive pair (x_i, x_j) and x_n are the corresponding negative images from the set Ω^-. τ denotes the temperature parameter. The similarity between two representations is computed using cosine similarity, which is defined as $\text{sim}(z_i, z_j) = z_i^T z_j / \|z_i\| \|z_j\|$.
The net contrastive loss across all positive pairs in the batch is given below:

$$L_{net} = \frac{1}{2N} \sum_{k=1}^{N} [l(t_i(x_k), t_j(x_k)) + l(t_j(x_k), t_i(x_k))] \tag{2}$$

By optimizing the loss L_{net}, we enable the network to learn representations of a positive pair for a given image to be similar under different transformations such as crop, rotation, color jitter, etc. Also, they should be dissimilar to representations of the remaining images in the batch that constitute the negative set. With this optimization, we aim to learn robust and meaningful global representations $h = f(.)$ that can be used for downstream tasks.

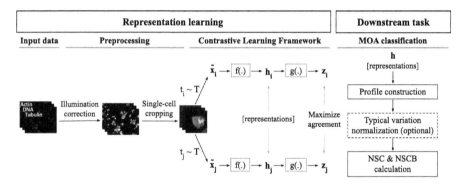

Fig. 1. Our experimental pipeline can be split into two parts: representation learning and downstream task. In representation learning, we feed single-cell images into the contrastive learning framework [7], where single-cell representations are extracted. Subsequently, these representations are aggregated into treatment profiles and post-processed (optional step) in the downstream task. We use the resulting profiles for MOA classification and calculate NSC and NSCB accuracies.

(b) MOA Classification: Our proposed method is evaluated on a downstream MOA classification task. The final latent representations z_i and projection head $g(.)$ are only used for the representation learning stage and are discarded for the classification step. The Mechanism-of-Action classification is performed on the learned cell representations h_i that are output from the encoder network.

We use two widely applied procedures for evaluating how MOA are assigned to treatments: not-same-compound (NSC) matching from [24] and not-same-compound-and-batch (NSCB) matching from [1]. In a first step, single-cell representations are aggregated into *treatment profiles* as described in [24]. This aggregation results in a treatment profile vector for each treatment in the dataset.

In the second step, one MOA is assigned to each treatment profile. Both NSC and NSCB scores are 1st nearest neighbour MOA classification accuracies where the distance measure used is cosine distance. Treatments are assigned an MOA label one at a time. In NSC matching, for a given treatment, we search the representation space for the nearest neighbour not containing the same compound. Recall that a treatment is a compound-concentration pair. In NSCB matching, the search further excludes all treatments from the same experimental batch.

3   Datasets and Network Details

Dataset: Our work uses the annotated part of the BBBC021 dataset [4] available from the Broad Bioimage Benchmark Collection [23]. BBBC021 consists of multi-channel images captured from human MCF-7 breast cancer cells exposed to chemical compounds for 24 h. Cells are imaged by fluorescent microscopy. The 3 grey-scale channels represent DNA, B-tubulin and F-actin. The proposed method is evaluated on the subset of BBBC021 that has previously been labelled for MOAs by [24]. There are 12 distinct MOA present in this subset for a total of 103 treatments and 38 compounds. Each treatment corresponds to a compound-concentration combination. The control cells are treated with DMSO (dimethyl sulfoxide). We solely use the ground truth annotated part of BBBC021 for our study, in line with the evaluation strategy applied in all the works in the literature [1, 3, 11–13, 19, 21, 22, 24, 26, 27, 31].

Data Pre-processing: We apply the illumination correction algorithm [27] on the original images as performed in [3]. Then, we crop all the single-cell instances from these images to fixed image dimensions of 96×96, based on the cell locations introduced by [24].

Post-processing: We use typical variation normalization (TVN) [3, 19, 22, 31] to reduce batch effects and improve profiles. It comprises of a *whitening* step and a *correlation alignment* (CORAL) [30] step. We also evaluate the effects of using only whitening [19, 22, 31] as well as no post-processing.

Network Details: We choose ResNet [16] for the encoder network $f(.)$ and MLP layers for the projection head network $g(.)$.

Training Details: We train the network with the following parameter settings. We choose Adam as the optimizer with a learning rate of $3e^{-4}$ and a weight decay of $1e^{-5}$. We run the initial hyper-parameter evaluation for 150 epochs on a subset of 15% of the annotated images. We run the final experiments with the optimal set of hyper-parameters for up to 600 epochs on the complete set of annotated data. All our experiments are performed on a Nvidia Titan Xp GPU.

4   Experiments

Experimental Setup: The part of BBBC021 annotated for MOAs contains 2'526 original images which translates into 454'793 single-cells. First, we tune our hyperparameters on a subset of the annotated BBBC021 dataset containing 15% of the images as performed by [21]. This significantly reduces our training time and compute resources. In a second step, we train our framework on the whole annotated BBBC021 dataset and report our final results. Note that we perform training and evaluation in a transductive learning setting on the whole annotated data in line with the evaluation strategy used in the earlier works.

Evaluation: We use NSC, NSCB scores and the NSC-NSCB drop to measure the performance.

Ablation Study of Hyper-parameters: For the ablation study of hyper-parameters, we only use 15% of the annotated images. We report all results for these experiments with TVN post-processing. (Refer to the Supplementary for results using only whitening or without post-processing.) We evaluate the following hyper-parameters to analyze their effect on downstream performance.

(i) **Encoder Network Size** $f(.)$: For the encoder, the following ResNet sizes are evaluated: ResNet18, ResNet50, ResNet101. The remaining hyper-parameters are investigated for a ResNet50 encoder as it yielded the best results.

(ii) **Data Augmentation strategy**: We explore different data augmentations (T) such as crop, flip, rotations by 90°, color jitter, grey-distortion, and Gaussian blur. In the default setting, we have all augmentations. We experiment by removing one augmentation at a time from this main set of augmentations to analyze the importance of each.

(iii) **Projection head** $g(.)$: We explore three types of projection heads: the identity, a linear projection and a non-linear projection (two layer MLP).

(iv) **Batch Size**: We evaluate the following batch sizes used for each training iteration: 64, 128, 256.

(v) **Temperature coefficient**: As done in [7], we evaluate the effect of temperature for the following values: $\tau = 0.05, 0.1, 0.5, 1$ for different combinations of other hyper-parameters.

Final Experiments and Comparison: We choose the hyper-parameters for the final set of experiments from the ablation results. They are a ResNet50 as the encoder network, a two-layer MLP as the projection head, an augmentation strategy without the grey distortion, a batch size of 256 and a temperature parameter of $\tau = 0.5$. Here, the evaluation is performed in a transductive learning setting on the whole annotated dataset as described earlier.

TVN Post-processing : We also evaluate if there is any difference in performance with and without TVN post-processing applied to the learned representations.

5 Results

Ablation Study of Hyper-parameters
(i) Encoder size: In Table 1, we present the results for different ResNet architectures evaluated as the encoder network. We observe that ResNet18 slightly outperforms both ResNet50 and ResNet101. However, when we compare ResNet50 to ResNet18, we observe smaller values for NSC-NSCB drop with and without post-processing for ResNet50 (Refer to Table 4 in Supplementary). Hence, we choose ResNet50 for the remaining hyper-parameter evaluation.
(ii) Data augmentation strategies: In Table 1, we present our results for the different augmentation strategies adopted. We drop one augmentation at a time to analyze its importance. We observe that dropping the grey color distortion significantly improves NSC and NSCB performance. This could be due to grey-scale averaging over the three channels. It may not be meaningful for such datasets as they contain relatively independent pixel intensities in each channel, with different channels capturing different parts of a cell. Typical RGB channels contain dependent information. The most important augmentation was found to be cropping followed by color jitter. This is also observed in earlier works [7] on natural images.

Other hyper-parameter evaluations are presented in the Supplementary Material in Tables 5, 6 and 7. We summarize our findings to be the following: (iii) a two-layer MLP (non-linear projection) and linear projection yield similar NSC and NSCB scores. They both yield higher scores than the identity. (iv) A batch size of 256 yields higher NSC and NSCB scores on average over batch sizes of 64 and 128. (v) The temperature coefficients of 0.1 and 0.5 yield higher scores compared to a large value of 1 and very small value of 0.05.

Table 1. Model performance is shown for different encoder network sizes ((a) f size) and for one augmentation removed at a time ((b) Augmentation removal) under TVN post-processing conditions. Jitter, Grey, and Blur refer to color-jitter, Grey color distortion, Gaussian Blur, respectively.

Metric	(a) f size			(b) Augmentation removal						
	ResNet18	ResNet50	ResNet101	None	Crop	Flip	Rotation	Jitter	Grey	Blur
NSC	95%	93%	92%	94%	83%	93%	94%	91%	96%	92%
NSCB	91%	91%	90%	91%	70%	88%	87%	88%	94%	88%

Final Experiments and Comparison: In Table 2, we present our results from the final set of experiments on the whole annotated dataset in a transductive setting as evaluated in earlier works. We observe that the proposed unsupervised contrastive learning framework yields better results than earlier unsupervised works with an improvement over the best method [19] of 10% and 11% in NSCB scores and drop respectively. We also observe that the proposed framework yields similar scores to supervised counterparts [1] where large number of annotations or suitable pre-training datasets are required to achieve such high performance.

Table 2. MOA classification accuracy. * indicates results without post-processing.

Supervised				Unsupervised			
Method	NSC	NSCB	Drop	Method	NSC	NSCB	Drop
Ljosa et al. [24]	94%	77%	17%	Janssens et al. [19]	**97%**	**85%**	12%
Singh et al. [27]	90%	85%	5%	Lafarge et al. [22]	93%	82%	**11%**
Ando et al. [1]	**96%**	**95%**	1%	Lafarge et al. [22] *	92%	72%	20%
Pawlowski et al. [26]	91%	NA	NA	Our work *	**97%**	94%	3%
Caicedo et al. [3]	95%	89%	6%	Our work + Whitening	96%	**95%**	1%
				Our work + TVN	**97%**	92%	5%

TVN Post-processing: We observe that our method performs well even without applying post-processing such as TVN or whitening. Earlier works are sensitive to the post-processing step where a 10% decrease in NSCB is observed for [22]. Our work has only a 1% decrease in NSCB when removing the whitening step, as shown in Table 2.

t-SNE Plot: Figure 2 shows a t-SNE visualization of the treatment profiles obtained with our final experiment. Here, the treatments are classified into the 12 MOAs available. In Fig. 2, we can observe that only 4 out of 103 and 5 out of 92 treatments are classified incorrectly during NSC and NSCB MOA assignment respectively. NSC and NSCB mis-classifications are marked with black squares and black diamonds respectively.

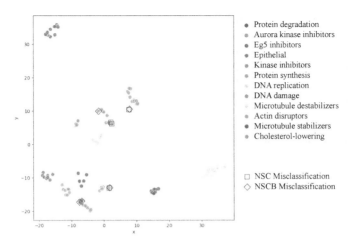

Fig. 2. t-SNE plot for treatment profiles subject to whitening post-processing.

6   Conclusion

In this work, we conclude that using an unsupervised approach with contrastive learning on single-cell images leads to excellent representations in morphological cell profiling. With this framework, we demonstrate state of the art results in an unsupervised learning setting for the downstream task of MOA classification. In the unsupervised setting, our NSCB score of 95% for MOA matching is the highest classification accuracy reported and similarly the drop of 1% between our NSC and NSCB scores is the best reported. Furthermore, our unsupervised results are identical to the state of the art transfer learning approach [1] and higher than the supervised approach [3], both relying on labels. We perform an ablation study of hyper-parameters and conclude that encoder size as well as data augmentations are the most crucial hyper-parameters for obtaining maximum improvements on cell profiling datasets. Finally, we also show that the performance of the resulting representations does not deteriorate even when removing post-processing techniques.

References

1. Ando, D.M., McLean, C.Y., Berndl, M.: Improving phenotypic measurements in high-content imaging screens. BioRxiv, p. 161422 (2017)
2. Azizi, S., et al.: Big self-supervised models advance medical image classification. arXiv preprint arXiv:2101.05224 (2021)
3. Caicedo, J.C., McQuin, C., Goodman, A., Singh, S., Carpenter, A.E.: Weakly supervised learning of single-cell feature embeddings. In: Proceedings of the IEEE Conference on Computer Vision and Pattern Recognition, pp. 9309–9318 (2018)
4. Caie, P.D., et al.: High-content phenotypic profiling of drug response signatures across distinct cancer cells. Mol. Cancer Ther. **9**(6), 1913–1926 (2010)

5. Caron, M., Bojanowski, P., Joulin, A., Douze, M.: Deep clustering for unsupervised learning of visual features. In: Proceedings of the European Conference on Computer Vision (ECCV), pp. 132–149 (2018)
6. Chaitanya, K., Erdil, E., Karani, N., Konukoglu, E.: Contrastive learning of global and local features for medical image segmentation with limited annotations. In: Advances in Neural Information Processing Systems, vol. 33 (2020)
7. Chen, T., Kornblith, S., Norouzi, M., Hinton, G.: A simple framework for contrastive learning of visual representations. In: International Conference on Machine Learning, pp. 1597–1607. PMLR (2020)
8. Donahue, J., Krähenbühl, P., Darrell, T.: Adversarial feature learning. ICLR (2017)
9. Donahue, J., Simonyan, K.: Large scale adversarial representation learning. In: Advances in Neural Information Processing Systems, vol. 32 (2019)
10. Gidaris, S., Singh, P., Komodakis, N.: Unsupervised representation learning by predicting image rotations. arXiv preprint arXiv:1803.07728 (2018)
11. Godinez, W.J., Hossain, I., Lazic, S.E., Davies, J.W., Zhang, X.: A multi-scale convolutional neural network for phenotyping high-content cellular images. Bioinformatics **33**(13), 2010–2019 (2017)
12. Godinez, W.J., Hossain, I., Zhang, X.: Unsupervised phenotypic analysis of cellular images with multi-scale convolutional neural networks. BioRxiv, p. 361410 (2018)
13. Goldsborough, P., Pawlowski, N., Caicedo, J.C., Singh, S., Carpenter, A.E.: CytoGAN: generative modeling of cell images. BioRxiv, p. 227645 (2017)
14. Hadsell, R., Chopra, S., LeCun, Y.: Dimensionality reduction by learning an invariant mapping. In: 2006 IEEE Computer Society Conference on Computer Vision and Pattern Recognition (CVPR 2006), vol. 2, pp. 1735–1742. IEEE (2006)
15. He, K., Fan, H., Wu, Y., Xie, S., Girshick, R.: Momentum contrast for unsupervised visual representation learning. In: Proceedings of the IEEE/CVF Conference on Computer Vision and Pattern Recognition, pp. 9729–9738 (2020)
16. He, K., Zhang, X., Ren, S., Sun, J.: Deep residual learning for image recognition. In: Proceedings of the IEEE Conference on Computer Vision and Pattern Recognition, pp. 770–778 (2016)
17. Hjelm, R.D., et al.: Learning deep representations by mutual information estimation and maximization. ICLR (2019)
18. Huang, G.B., Yang, H.F., Takemura, S.y., Rivlin, P., Plaza, S.M.: Latent feature representation via unsupervised learning for pattern discovery in massive electron microscopy image volumes. arXiv preprint arXiv:2012.12175 (2020)
19. Janssens, R., Zhang, X., Kauffmann, A., de Weck, A., Durand, E.Y.: Fully unsupervised deep mode of action learning for phenotyping high-content cellular images. BioRxiv, p. 215459 (2020)
20. Kiyasseh, D., Zhu, T., Clifton, D.A.: CLOCS: contrastive learning of cardiac signals. arXiv preprint arXiv:2005.13249 (2020)
21. Kraus, O.Z., Ba, J.L., Frey, B.J.: Classifying and segmenting microscopy images with deep multiple instance learning. Bioinformatics **32**(12), 52–59 (2016). https://doi.org/10.1093/bioinformatics/btw252
22. Lafarge, M.W., Caicedo, J.C., Carpenter, A.E., Pluim, J.P.W., Singh, S., Veta, M.: Capturing single-cell phenotypic variation via unsupervised representation learning. In: International Conference on Medical Imaging with Deep Learning, pp. 315–325. PMLR (2019)
23. Ljosa, V., Sokolnicki, K., Carpenter, A.E.: Annotated high-throughput microscopy image sets for validation. Nat. Meth. **9**, 637 (2012). https://doi.org/10.1038/nmeth.2083

24. Ljosa, V., et al.: Comparison of methods for image-based profiling of cellular morphological responses to small-molecule treatment. J. Biomol. Screen. **18**(10), 1321–1329 (2013)
25. Pathak, D., Krahenbuhl, P., Donahue, J., Darrell, T., Efros, A.A.: Context encoders: feature learning by inpainting. In: Proceedings of the IEEE Conference on Computer Vision and Pattern Recognition, pp. 2536–2544 (2016)
26. Pawlowski, N., Caicedo, J.C., Singh, S., Carpenter, A.E., Storkey, A.: Automating morphological profiling with generic deep convolutional networks. BioRxiv, p. 085118 (2016)
27. Singh, S., Bray, M.A., Jones, T., Carpenter, A.: Pipeline for illumination correction of images for high-throughput microscopy. J. Microsc. **256**(3), 231–236 (2014)
28. Spiegel, S., Hossain, I., Ball, C., Zhang, X.: Metadata-guided visual representation learning for biomedical images. BioRxiv, p. 725754 (2019)
29. Sriram, A., et al.: COVID-19 prognosis via self-supervised representation learning and multi-image prediction. arXiv preprint arXiv:2101.04909 (2021)
30. Sun, B., Feng, J., Saenko, K.: Correlation Alignment for Unsupervised Domain Adaptation. In: Csurka, G. (ed.) Domain Adaptation in Computer Vision Applications. ACVPR, pp. 153–171. Springer, Cham (2017). https://doi.org/10.1007/978-3-319-58347-1_8
31. Tabak, G., Fan, M., Yang, S., Hoyer, S., Davis, G.: Correcting nuisance variation using Wasserstein distance. PeerJ **8**, e8594 (2020)
32. Vu, Y.N.T., Wang, R., Balachandar, N., Liu, C., Ng, A.Y., Rajpurkar, P.: Contrastive learning leveraging patient metadata improves representations for chest x-ray interpretation. arXiv preprint arXiv:2102.10663 (2021)
33. Wu, Z., Xiong, Y., Yu, S.X., Lin, D.: Unsupervised feature learning via nonparametric instance discrimination. In: Proceedings of the IEEE Conference on Computer Vision and Pattern Recognition, pp. 3733–3742 (2018)
34. Xie, Y., Zhang, J., Liao, Z., Xia, Y., Shen, C.: PGL: prior-guided local self-supervised learning for 3D medical image segmentation. arXiv preprint arXiv:2011.12640 (2020)
35. Yan, K., et al.: Self-supervised learning of pixel-wise anatomical embeddings in radiological images. arXiv preprint arXiv:2012.02383 (2020)
36. Zhang, Y., Jiang, H., Miura, Y., Manning, C.D., Langlotz, C.P.: Contrastive learning of medical visual representations from paired images and text. arXiv preprint arXiv:2010.00747 (2020)

CorLab-Net: Anatomical Dependency-Aware Point-Cloud Learning for Automatic Labeling of Coronary Arteries

Xiao Zhang[1], Zhiming Cui[2], Jun Feng[1(✉)], Yanli Song[3], Dijia Wu[3], and Dinggang Shen[3,4(✉)]

[1] School of Information Science and Technology, Northwest University, Xi'an, China
fengjun@nwu.edu.cn
[2] Department of Computer Science, The University of Hong Kong, Hong Kong, China
[3] Shanghai United Imaging Intelligence Co., Ltd., Shanghai, China
[4] School of Biomedical Engineering, ShanghaiTech University, Shanghai, China
dgshen@shanghaitech.edu.cn

Abstract. Automatic coronary artery labeling is essential yet challenging step in coronary artery disease diagnosis for clinician. Previous methods typically overlooked rich relationships with heart chamber and also morphological features of coronary artery. In this paper, we propose a novel point-cloud learning method (called CorLab-Net), which comprehensively captures both inter-organ and intra-artery spatial dependencies as explicit guidance to assist the labeling of these challenging coronary vessels. Specifically, given a 3D point cloud extracted from the segmented coronary artery, our CorLab-Net improves artery labeling from three aspects: First, it encodes the inter-organ anatomical dependency between vessels and heart chambers (in terms of spatial distance field) to effectively locate the blood vessels. Second, it extracts the intra-artery anatomical dependency between vessel points and key joint points (in terms of morphological distance field) to precisely identify different vessel branches at the junctions. Third, it enhances the intra-artery local dependency between neighboring points (by using graph convolutional modules) to correct labeling outliers and improve consistency, especially at the vascular endings. We evaluated our method on a real-clinical dataset. Extensive experiments show that CorLab-Net significantly outperformed the state-of-the-art methods in labeling coronary arteries with large appearance-variance.

Keywords: Coronary artery labeling · Anatomical distance field · Morphological distance field · 3D point cloud

1 Introduction

The diagnosis of coronary heart diseases needs a 3D model of the coronary artery with precisely labeled vessels, which provides necessary information for many

© Springer Nature Switzerland AG 2021
C. Lian et al. (Eds.): MLMI 2021, LNCS 12966, pp. 576–585, 2021.
https://doi.org/10.1007/978-3-030-87589-3_59

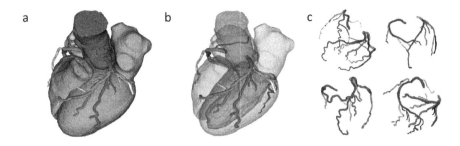

Fig. 1. Sample heart consisting of aorta, atrium, ventricles and coronary arteries. (a) A chamber along with major coronary arteries; (b) A 3D heart shows the blood vessels intertwining and attaching around the organ; (c) The coronary arteries structure varies among individuals, as well as the thin and messy side branches.

subsequent procedures, such as locating stenosis/occlusion, generating medical reports, calculating statistics for important anatomical branches and visualizing regions of interest [1,2]. As shown in Fig. 1, coronary arteries are intertwined with or attached to multiple organs, i.e., the heart chambers, aorta, and heart surface. Precisely labeling them is practically challenging, mainly due to 1) coronary vessels are typically tortuous dendritic tubular structures with diverse lengths; 2) coronary vessels have dramatically changed shape appearance across individuals and spatial positions.

Several automatic methods have been proposed for coronary artery (or similar 3D structure) labeling, e.g., based on registration techniques, CNNs, and point-cloud deep learning. Traditional methods [3,4] adopt registration techniques to label coronary arteries via main branch matching, which cannot sensitively identify detailed vessels. CNN-based methods can perform labeling more efficiently, while general convolutional kernels defined in the grid space are hard to capture the contextual information along with the tortuous tree-like structures of the coronary vessels [5]. Inspired by the pioneering works in 3D computer vision and graphics [6–8], more recent works attempted to apply point-cloud networks or graph convolutional networks (GCNs) for tree-like vascular structures labeling, achieving promising resulting in specific applications [9–12]. However, since coronary vessels have relatively more complicated structures, deep networks working solely with point-coordinates information may fail to capture discriminative shape details for fine-grained vessel labeling.

In this paper, to identify detailed coronary vessels with varying lengths and shape appearance, we propose to leverage comprehensive spatial/anatomical dependencies as explicit guidance in developing task-specific point-cloud deep networks. That is, a CorLab-Net method is designed to integrate both inter-organ and intra-artery contextual information with point-coordinate information for accurate point-wise labeling. Our CorLab-Net has three key technical contributions: 1) It defines a spatial distance field between vessels and heart chambers, serving as the inter-organ spatial guidance to assist the localization of the blood vessels. 2) It predicts a morphological distance field between vessel points and key joint points (i.e., landmarks), providing the intra-artery anatomical guidance to help identify different vessel branches at the junctions. 3) It integrates GCN modules to capture the high-order correlations between neighboring points, rendering the intra-artery local guidance to enhance labeling consistency and correct prediction outliers, especially at the vascular endings.

2   Method

Our CorLab-Net method mainly includes three components: anatomical distance field module, morphological distance field module, and a point-cloud network integrating GCN modules, such as the schematic diagram shown in Fig. 2. The two distance field modules form an effective enriched-feature space by capturing the inter-organ and intra-artery spatial dependencies to provide detailed contextual information. Based on the outputs of these two modules, the point-cloud network uses PointNet++ [7] as the backbone and integrates the GCN modules in its decoding part to conduct locally-consistent point-wise labeling.

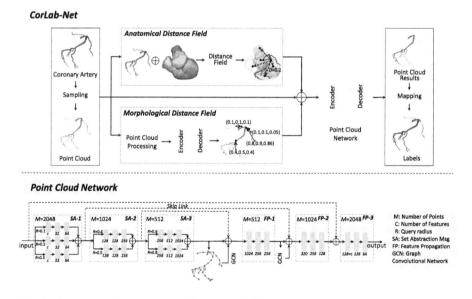

Fig. 2. Overview of the proposed framework. (a) Entire processing pipeline of our network; (b) The point cloud network learns the principle of consistency between neighboring points based on PointNet++ and GCN module.

2.1   Anatomical Distance Field

As can be seen from the examples in Fig. 1, coronary vessels start from the aorta, wrap around and attach to the surfaces of the left and right atria and ventricle. In practice, clinicians annotate different artery vessels according to the spatial associations between the artery and specific surrounding organs. It implies that the inter-organ anatomical dependencies convey critical contextual information to help locate different vessels. Unfortunately, such anatomical dependencies were typically ignored by existing automatic methods [10,13], as they work solely on point coordinates for blood vessel identification.

Inspired by the clinical experience, we assume each branch of a coronary artery is roughly distributed at a fixed position of the heart, according to which an anatomical distance field is defined to model the inter-organ dependencies between different coronary branches and surrounding heart parts. Specifically, we denote the points of coronary arteries P as $P = [p_1, p_2, \cdots p_N]$ with $N = 2048$ points, $p_i \in R^3$, and surface points of heart chambers C as $C = [c_1, c_2, \cdots c_M]$ with $M \in [8000, 14000]$ points, $c_j \in R^3$. The details to calculate such a distance field are as follows:

$$S_i = \min_{\substack{i=1,\cdots,N \\ j=1,\cdots,M}} \sqrt{(p_{i,1} - c_{j,1})^2 + (p_{i,2} - c_{j,2})^2 + (p_{i,3} - c_{j,3})^2}, \tag{1}$$

$$Dist_i = \begin{cases} \frac{S_i}{Th}, & if \ \frac{S_i}{Th} < 1, \\ 1, & else, \end{cases} \tag{2}$$

where Th is an empirical number, and $Th = 50$ in this study.

In practice, we calculate the minimum distance from each point of the coronary artery branch to the surfaces of the aorta, left and right atria and left and right ventricle, respectively, and then merge these distance constraints with point coordinate information to form the whole distance field, as shown in Fig. 2. Finally, these discriminative information and other effective vectors are fed into point cloud network to enrich the whole input feature space.

2.2 Morphological Distance Field

Traditional point-cloud networks treat all input points equally [6,7], ignoring the important contributions of key points in defining the shape of an input point cloud. For example, in coronary labeling, the identification of a joint point can provide strong clues regarding the semantic labels of all other points, from the proximal one to the distal one, on the corresponding artery branch.

To integrate such intra-artery contextual information into our CorLab-Net method, a morphological distance field is defined accordingly. Specifically, we assume that different points have different contributions to the network; since joint points contain more semantic information, the network should pay more attention to them. Therefore, as shown in Fig. 2, we first apply a landmark-detection point-cloud network to the original coronary data to regress the joint points of each branch. Then, the spatial displacements from each point on the branch to the respective joint points are further calculated, forming the morphological distance field.

In the implementation process, we directly regress the distance vectors from each point to the closest joint points to replace the above process. Note that these two processes are equivalent. Specifically, the input to this module is coronary artery points P. The output Q are the vector differences between each point and the closest joint point of sub-branches, where $Q = [q_1, q_2, \cdots q_K]$ with $K = 2048$ points, $q_r \in R^3$. To train the landmark regression network, we define our regression loss based on the chamfer distance [14,15] to constrain the predicted vectors Q to be close to the real vector differences T. The regression loss L_{reg} for network training is the chamfer distance between the predicted vector differences Q and ground-truth vectors T:

$$L_{reg}(Q, T) = \frac{1}{|Q|} \sum_{q_k \in Q} \min_{t_w \in T} \|q_k - t_w\|_2^2 + \frac{1}{|T|} \sum_{t_w \in T} \min_{q_k \in Q} \|t_w - q_k\|_2^2 , \quad (3)$$

2.3   Point-Cloud Deep Network

Since there is no order information in the point cloud data, the network captures context information between neighboring points by using various aggregating neighboring points in the encoder stage, such as Multi-Scale Grouping (MSG) and Multi-Resolution Grouping (MRG) [7]. The decoder stage is often overlooked. It generates high-level semantic vectors, which has a strong mapping relationship with the label.

In labeling task of coronary arteries, the neighboring points generally belong to the same category, especially the points at vascular endings. If the network can match the label relationship of neighboring points with its corresponding high-level features, the labeling consistency relationship will be established through spatial distribution between each point and neighboring points. The consistency can better guide the network to make decisions.

We build a graph structure to capture the consistent relationship between neighboring points, vertices of graph are 3D coordinates of point cloud, and edges are determined by Euclidean distance between the coordinates. If two points are close together in space, they are connected, and the threshold d is set to measure whether the two vertices are connected.

In this way, the high-level features of all points in the neighborhood are fused by the graph structure. The features of each points have a strong mapping relationship with the label, and the fused features have enriched feature representation.

3   Experiments and Results

3.1   Dataset and Evaluation Metrics

In our experiments, we collected a private database of 100 subjects, and all of them are annotated by three experts. Our data included not only voxel label for each coronary artery vessel, but also labels for atrium and ventricles. These subjects and corresponding annotations compose the experimental dataset.

The evaluation is performed on coronary artery segments by the predicted label and ground truth. The precision rate for each branch is calculated by

$P = \frac{TP}{TP+FN}$. The Recall is Recall $= \frac{TP}{TP+FP}$. The F1 score is F1 $= 2 *$ $\frac{Precision \times Recall}{Precision + Recall}$. Since the numbers of coronary branches in different subjects are imbalance, we also adopt the mean metrics of all branch segments.

3.2   Implementation Details

Point Cloud Processing. Due to the differences of subjects, our image sequence is not fixed, and the size is $512 \times 512 \times S_e$, where $S_e \in [155, 353]$. Therefore, before the image is sampled into point cloud, it needs to be normalized. There are mainly three steps in normalization: (1) the input image is cropped according to the position of the segmented coronary artery in the image, and only the part containing coronary vessels is retained. (2) Since the cropped sizes of each subject are inconsistent, the cropped image coordinates are divided by the size for normalization. (3) The 3D image is sampled as point cloud, and the point cloud data has been normalized to the range $[0, 1]$.

Parameters Setting. The numbers of vertices H in the two GCNs are set to 512 and 1024, respectively, and GCN parameters are the same as [16]. The threshold d for edge connection is an empirical number, the two GCNs are set to 0.1 and 0.05 in this study, respectively. Five-fold cross-validation is used in the experiment, and we train the network for 200 epochs using Adam optimizer with an initial learning rate of 0.001. The model is implemented using PyTorch with an NVIDIA Tesla V100S GPU, and each mini-batch contains 4 point cloud samples. In addition, chamfer distance and cross-entropy are, respectively, used as loss functions in regression and classification network.

3.3   Comparison with Existing Methods

To verify the effectiveness of our framework, several point cloud algorithms were tested on dataset, including PointNet [6], PointNet++ [7] and HN-Net [10]. They all directly take the 3D point cloud as input to predict the labels of coronary arteries. The statistic results and visual presentations are presented in Table. 1 and Fig. 3, respectively.

Quantitative Results. The overall coronary labeling results are summarised in Table.1, where the proposed framework shows significant improvement to other methods. Concretely, compared with the backbone PointNet++, our method increases by 11%, 10.6% and 11.3% in meanPrecision, meanRecall and F1 score, respectively, which demonstrates the effectiveness of the anatomical dependency-aware point-cloud learning framework. At the same time, it can be seen that the metrics of PointNet and HN-Net are far below our method, which further indicates the advantages of all proposed module in our framework.

Qualitative Results. The visual presentations are shown in Fig. 3 for three typical subjects. It can be observed that labeling results produced by our method

Table 1. Comparisons of PointNet (PN), PointNet++ (PN++), HN-Net (HN) and CorLab-Net on our dataset. Recall, Precision(P) and F1 score are used as the evaluation metrics (%).

Method	Metric	LM	LAD	D	LCX	OM	RAMUS	S	RCA	R-PDA	R-PLB	AM	Avg
PN	P	63.0	75.6	68.9	77.5	76.6	59.8	61.2	89.2	56.6	96.8	80.9	73.3
	Recall	72.0	71.8	63.3	74.3	77.4	95.5	63.4	99.8	64.6	68.6	89.9	73.7
	F1	58.3	73.7	65.9	73.8	77.0	62.5	55.1	94.2	58.5	73.9	85.1	70.9
PN++	P	78.5	91.9	83.8	94.1	82.9	63.3	70.2	98.3	83.1	92.7	93.6	84.8
	Recall	78.4	87.6	76.7	95.0	76.0	65.8	71.5	97.4	85.3	93.2	95.5	83.9
	F1	70.4	89.7	80.6	95.6	81.9	64.4	64.1	97.9	87.0	94.9	94.6	83.7
HN	P	86.5	91.5	85.2	96.5	80.5	71.3	79.5	97.6	88.0	97.4	90.8	87.7
	Recall	81.5	89.8	88.6	97.2	88.5	73.0	78.5	96.9	89.1	92.8	98.1	88.5
	F1	77.0	90.6	76.0	96.9	84.3	76.5	79.0	97.2	88.6	95.0	94.3	86.9
Ours	P	93.0	99.2	90.7	98.9	98.7	83.7	94.3	99.9	95.2	99.7	100	**95.8**
	Recall	99.8	97.1	95.6	99.8	87.1	74.3	88.7	100	99.3	98.6	99.0	**94.5**
	F1	96.3	98.1	93.1	99.4	92.6	78.1	91.4	99.8	97.2	99.2	99.5	**95.0**

match better with the ground truth, especially for typical parts, such as vascular bifurcation and vascular endings (as highlighted by red arrows in Fig. 3).

The first column is the ground truth, followed by the comparison result of our method and others. Notably, the other three methods not only fail to produce satisfactory results in these difficult areas, but also have a large number of label errors in the main branches. Our method can avoid this problem because it adds enriched feature vectors and point consistency constraint. The qualitative results shown in Fig. 3 are consistent with the quantitative comparison, which further demonstrates the effectiveness and efficiency of our framework for automatic labeling of coronary arteries.

3.4   Ablation Study

We conduct extensive experiments to validate the effectiveness of our model components. First, we build a baseline network, denoted as bNet, which is our network backbone (PointNet++) for automatic labeling of coronary arteries. All the alternative networks are derived by augmenting the baseline network with different network components. We describe the details and results in the following.

Anatomical Distance Field. The anatomical distance field represent relative location of vessels and chambers, which constrains the vessel execution at the region level. To validate its benefits, we augment the baseline network bNet with the anatomical distance field (bNet-A) and compare prediction results of both networks in Table 2. By appending the module, the meanPrecision, meaRrecall and meanF1 increase by 5.5%, 4.2%, 5.5%, respectively.

Morphological Distance Field. The morphological distance field focuses on the important contribution of key points to the network, which represents the

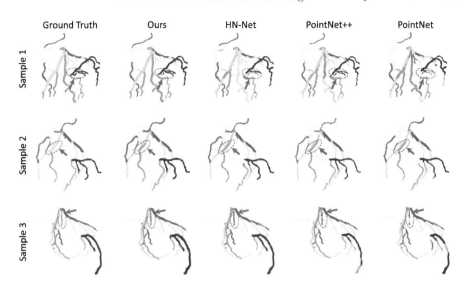

Fig. 3. Comparison of labeling results for three typical subjects by four different methods. Different colors represent different categories of coronary vessels, and there are 11 classes in total. The red circle with arrow are artificially designed for visual presentation, highlighting the typical parts. (Color figure online)

contextual relationships of points at the part level. To validate its benefit, we add the module to labeling vessel (bNet-A-M), and the metrics go up over 3.6%.

Point Consistency based on GCN. To further minimize the errors of network at vascular endings, we add the graph network module to capture the category consistency of the neighboring points, which belongs to the constraint relationship of the point level. We have verified all modules on the dataset and the metrics have been improved with varying degrees.

Table 2. Part of the ablation study results for our method. The anatomical distance field, morphological distance field, as well as point consistency based on GCN are all essential parts of our CorLab-Net.

Method	meanPrecision (%)	meanRecall (%)	meanF1 (%)
bNet	84.8	83.9	83.7
bNet-A	90.3	88.1	89.2
bNet-A-M	93.9	92.4	93.5
Ours	95.8	94.5	95.0

4   Conclusion

In this study, we propose an anatomical dependency-aware point-cloud learning framework (CorLab-Net) for coronary artery labeling. Specifically, we develop three modules, including anatomical distance field, morphological distance field, and point consistency block, to accurately label the coronary arteries. Compared with the existing point cloud networks, our method achieves superior performance and outperforms the state-of-the-art performance by a large margin, which suggests the potential applicability of our framework in real-world clinical scenarios. Furthermore, our work addressed an essential task in clinical practice. Automatic and accurate labeling of the coronary artery has a great benefit on the diagnosis of coronary heart diseases, which provides necessary information for many subsequent procedures, such as generating medical reports, calculating statistics for important anatomical branches, and visualizing regions of interest.

Acknowledgment. This work was supported by the National Natural Science Foundation of China under Grants 62073260.

References

1. Zhang, D., et al.: Direct quantification of coronary artery stenosis through hierarchical attentive multi-view learning. IEEE Trans. Med. Imaging **39**(12), 4322–4334 (2020)
2. Chen, S.Y., Carroll, J.D., Messenger, J.C.: Quantitative analysis of reconstructed 3-D coronary arterial tree and intracoronary devices. IEEE Trans. Med. Imaging **21**(7), 724–740 (2002)
3. Yang, G., et al.: Automatic coronary artery tree labeling in coronary computed tomographic angiography datasets. In: Computing in Cardiology, vol. 2011, pp. 109–112. IEEE (2011)
4. Cao, Q., et al.: Automatic identification of coronary tree anatomy in coronary computed tomography angiography. Int. J. Cardiovasc. Imaging **33**(11), 1809–1819 (2017)
5. Wu, D., et al.: Automated anatomical labeling of coronary arteries via bidirectional tree LSTMs. Int. J. Comput. Assist. Radiol. Surg. **14**(2), 271–280 (2019)
6. Qi, C.R., Su, H., Mo, K., Guibas, L.J.: PointNet: deep learning on point sets for 3D classification and segmentation. In: Proceedings of the IEEE Conference on Computer Vision and Pattern Recognition, pp. 652–660 (2017)
7. Qi, C.R., Yi, L., Su, H., Guibas, L.J.: PointNet++ deep hierarchical feature learning on point sets in a metric space. In: Proceedings of the 31st International Conference on Neural Information Processing Systems, pp. 5105–5114 (2017)
8. Guo, Y., Wang, H., Hu, Q., et al.: Deep learning for 3D point clouds: a survey. IEEE Trans. Pattern Anal. Mach. Intell. **PP**(99), 1 (2020)
9. Yang, X., Xia, D., Kin, T., Igarashi, T.: Surface-based 3D deep learning framework for segmentation of intracranial aneurysms from TOF-MRA images. arXiv preprint arXiv:2006.16161 (2020)
10. Yao, L., et al.: Graph convolutional network based point cloud for head and neck vessel labeling. In: Liu, M., Yan, P., Lian, C., Cao, X. (eds.) MLMI 2020. LNCS, vol. 12436, pp. 474–483. Springer, Cham (2020). https://doi.org/10.1007/978-3-030-59861-7_48

11. Wang, S., Dai, W., Xu, M., Li, C., Zou, J., Xiong, H.: Structure-aware graph construction for point cloud segmentation with graph convolutional networks. In: 2020 IEEE International Conference on Multimedia and Expo (ICME), pp. 1–6. IEEE (2020)
12. Lin, Z.H., Huang, S.Y., Wang, Y.C.F.: Convolution in the cloud: learning deformable kernels in 3D graph convolution networks for point cloud analysis. In: Proceedings of the IEEE/CVF Conference on Computer Vision and Pattern Recognition, pp. 1800–1809 (2020)
13. Yang, H., Zhen, X., Chi, Y., Zhang, L., Hua, X.S.: CPR-GCN: conditional partial-residual graph convolutional network in automated anatomical labeling of coronary arteries. In: Proceedings of the IEEE/CVF Conference on Computer Vision and Pattern Recognition, pp. 3803–3811 (2020)
14. Fan, H., Su, H., Guibas, L.J.: A point set generation network for 3D object reconstruction from a single image. In: Proceedings of the IEEE Conference on Computer Vision and Pattern Recognition, pp. 605–613 (2017)
15. Cui, Z., et al.: TsegNet: an efficient and accurate tooth segmentation network on 3D dental model. Med. Image Anal. **69**, 101949 (2021)
16. Veličković, P., Cucurull, G., Casanova, A., Romero, A., Lio, P., Bengio, Y.: Graph attention networks. arXiv preprint arXiv:1710.10903 (2017)

A Hybrid Deep Registration of MR Scans to Interventional Ultrasound for Neurosurgical Guidance

Ramy A. Zeineldin[1,2,3]([✉]), Mohamed E. Karar[2], Franziska Mathis-Ullrich[3], and Oliver Burgert[1]

[1] Research Group Computer Assisted Medicine (CaMed), Reutlingen University, Reutlingen, Germany
Ramy.Zeineldin@Reutlingen-University.DE
[2] Faculty of Electronic Engineering (FEE), Menoufia University, Menofia Governorate, Egypt
[3] Health Robotics and Automation (HERA), Karlsruhe Institute of Technology, Karlsruhe, Germany

Abstract. Despite the recent advances in image-guided neurosurgery, reliable and accurate estimation of the brain shift still remains one of the key challenges. In this paper, we propose an automated multimodal deformable registration method using hybrid learning-based and classical approaches to improve neurosurgical procedures. Initially, the moving and fixed images are aligned using classical affine transformation (MINC toolkit), and then the result is provided to the convolutional neural network, which predicts the deformation field using backpropagation. Subsequently, the moving image is transformed using the resultant deformation into a moved image. Our model was evaluated on two publicly available datasets: the retrospective evaluation of cerebral tumors (RESECT) and brain images of tumors for evaluation (BITE). The mean target registration errors have been reduced from 5.35 ± 4.29 to 0.99 ± 0.22 mm in the RESECT and from 4.18 ± 1.91 to 1.68 ± 0.65 mm in the BITE. Experimental results showed that our method improved the state-of-the-art in terms of both accuracy and runtime speed (170 ms on average). Hence, the proposed method provides a fast runtime for 3D MRI to intraoperative US pair in a GPU-based implementation, which shows a promise for its applicability in assisting the neurosurgical procedures compensating for brain shift.

Keywords: Brain shift · Computer-aided diagnosis · Deformable · MRI-US registration · Deep learning

1 Introduction

Image-guided neurosurgery (IGN) has proven to be a valuable tool for assisting neurosurgeons in the planning, interventional, and post-operative clinical phases [1, 2]. Yet, achieving accurate lesion localization and differentiation from the surrounding anatomical structures remains a challenging task in neurosurgery. This challenge is related to

© Springer Nature Switzerland AG 2021
C. Lian et al. (Eds.): MLMI 2021, LNCS 12966, pp. 586–595, 2021.
https://doi.org/10.1007/978-3-030-87589-3_60

the difficulty of visually defining these pathologic structures from healthy tissue in addition to the brain movements, known as "brain shift", due to neurosurgical manipulation, gravity, and anesthesia [3].

Hence, intra-operative magnetic resonance images (iMRI) and intra-operative ultrasound images (iUS) have been used for compensation of brain shift during surgery [4]. The iMRI scanner however limits the physician's access to the operative field and special surgical tools are required, which may be associated with high costs. iUS is portable, inexpensive, requires little preparation, and provides fast data acquisition. Though iUS can visualize interior soft tissue and structures, it has the difficulty of imaging through bones, and its high dependency on the inter-operator interpretation may result in image inconsistency. Consequently, the fusion of pre-interventional MRI images with the iUS data acquired intra-operatively is proposed to compensate for the brain shift to enable guided surgery.

Over the past years, many approaches have been applied for medical image registration that can be classified into non-learning- and learning-based approaches [5, 6]. Basically, classical or non-learning methods are formulated as an iterative pair-wise optimization problem that requires proper feature extraction, choosing a similarity measurement, defining the used transformation model, and finally an optimization mechanism to investigate the search space. Over time, extensive literature has developed using diverse combinations of the aforementioned elements [7–10]. Still, the traditional iterative process is computationally expensive, requiring long processing times ranging from tens of minutes to hours even with an efficient implementation on a regular central processing unit (CPU) or modern graphical processing unit (GPU).

To overcome the limitations of classical methods, learning-based approaches have been proposed in recent years. Learning methods formulate the classical optimization problem into a problem of loss function estimation. Rather than optimizing for every input pair of images individually, deep learning tends to find a function that takes many pairs of images and directly computes the transformation field. Some neural networks were proposed for the registration of a pre-operative MRI to the iUS volumes for brain shift correction [11, 12]. To better cope with inaccurate ground truth data and to eliminate the time required for dataset annotation, unsupervised learning was introduced [13].

In this work, we propose a real-time automated deformable MRI-iUS registration method using a mixture of deep learning and traditional registration tool. By combing both methods, our method intends to provide considerably improved robustness and computational performance for assisting neurosurgeons intra-operatively. The main contributions of this paper are as follows.

– We introduce our hybrid learning-based and traditional approach (see Fig. 1) for MRI-iUS deformation field estimation.
– We validate the performance of our model on data from 36 patients from two publicly available multi-site datasets: BITE and RESECT and compare it to the state-of-the-art non-learning- and learning-based registration algorithms.
– To the best of our knowledge, this is the first real-time non-linear pre-operative MRI to iUS registration method using hybrid learning-based and classical approaches towards brain shift compensation.

2   Material and Methods

2.1   Dataset

In this study, we have used two public accessible multi-site datasets, namely BITE [14] and RESECT [15]. These datasets contain pre-operative MRI and 3D iUS images from 14 patients and 22 patients, respectively. Expert-labeled anatomical landmarks, provided for each MRI-iUS pair, are utilized for ground truth evaluations. For the BITE dataset, we use the landmarks chosen by the first two experts while the third expert's annotation in the dataset was excluded for consistency since they provided data for the first six patients only.

2.2   Proposed Workflow

Traditional Image Registration. Medical image registration is the process of aligning two or more sets of imaging data acquired using mono- or multi-modalities into a common coordinate system. Let I_F and I_M denote the fixed and the moving images, respectively, and Let ϕ be the deformation field that relates the two images. Then, our goal is to find the minimum cost function C as:

$$C = D(I_F, I_M.\phi) + R(\phi) \tag{1}$$

where $(I_M.\phi)$ is the moving image I_M warped by the deformation field ϕ, the dissimilarity metric is denoted by D, and $R(\phi)$ represents the regularization parameter. In this work, MRI and iUS scans are utilized as the moving and fixed images, respectively, since our goal is to reflect the brain shift in the pre-operative MRI data.

Learning-Based Registration. Figure 1 presents an outline of the proposed non-rigid registration method. Our model consists of two steps: first, I_F and I_M are fed into our convolutional neural network (CNN) that then predicts ϕ. Second, I_M is transformed into a warped image $(I_M.\phi)$ using a spatial re-sampler. The developed CNN architecture utilized in experiments is based on U-Net [16] and our previous enhancement [17]. Using backpropagation, which is a feedback loop that estimates the network weighting parameters, the network can automatically learn the optimal features and the deformation field.

Our CNN contains two main parts: a feature extractor (or encoder) as well as a deformation field estimator (or decoder). 3D convolutions are applied in both encoder and decoder parts instead of the 2D convolutions used in the original U-Net architecture. Table 1 lists the detailed implementation of each layer in our CNN. The encoder consists of two consecutive 3D convolutional layers, each followed by a rectified linear unit (ReLU) and 3D spatial max pooling. A stride of 2 is employed to reduce the spatial dimension in each layer by half, similar to the traditional pyramid registration scheme. In the decoding path, each step consists of a 3D up-sampling, a concatenation with the corresponding features from the encoder, 3D up-convolutions, a batch normalization layer, followed by a rectified linear unit (ReLU). Finally, a 1 x 1 x 1 convolution layer is applied to map the resultant feature vector map into ϕ.

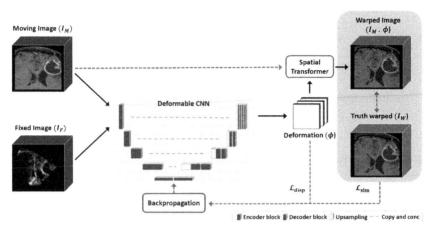

Fig. 1. An overview of the proposed workflow for 3D MRI to iUS image deformable registration. Dashed *red* arrows show the processes applied in the training stage only (Color figure online).

Table 1. Our deformable CNN architecture details.

#	Operation	Output	#	Operation	Output
0	Input	128 × 128 × 128 × 1	8	UpSampling3D	32 × 32 × 32 × 256
	Input	128 × 128 × 128 × 1		Concatenate	32 × 32 × 32 × 320
	Concatenate	128 × 128 × 128 × 2		Conv3D/BatchNorm/ReLU	32 × 32 × 32 × 128
1	Conv3D/BatchNorm/ReLU	64 × 64 × 64 × 32		Conv3D/BatchNorm/ReLU	32 × 32 × 32 × 128
2	Conv3D/BatchNorm/ReLU	32 × 32 × 32 × 64	9	UpSampling3D	64 × 64 × 64 × 128
3	Conv3D/BatchNorm/ReLU	16 × 16 × 16 × 128		Concatenate	64 × 64 × 64 × 160
4	Conv3D/BatchNorm/ReLU	8 × 8 × 8 × 256		Conv3D/BatchNorm/ReLU	64 × 64 × 64 × 64
5	Conv3D/BatchNorm/ReLU	8 × 8 × 8 × 512		Conv3D/BatchNorm/ReLU	64 × 64 × 64 × 64
	Conv3D/BatchNorm/ReLU	8 × 8 × 8 × 512	10	UpSampling3D	128 × 128 × 128 × 32
	Dropout	8 × 8 × 8 × 512		Concatenate	128 × 128 × 128 × 34
7	UpSampling3D	16 × 16 × 16 × 512		Conv3D/BatchNorm/ReLU	128 × 128 × 128 × 16
	Concatenate	16 × 16 × 16 × 640		Conv3D/BatchNorm/ReLU	128 × 128 × 128 × 16

(continued)

Table 1. (continued)

#	Operation	Output	#	Operation	Output
	Conv3D/BatchNorm/ReLU	$16 \times 16 \times$ 16×256	11	Conv3D/BatchNorm/ReLU	$128 \times 128 \times$ 128×16
	Conv3D/BatchNorm/ReLU	$16 \times 16 \times$ 16×256	12	Conv3D	$128 \times 128 \times$ 128×3

Loss Function. Owing to the applied two-step approach, the overall loss function $\mathcal{L}$ has two components, as Shown in Eq. 2. $\mathcal{L}$ computes the image similarity between the warped image $(\phi . I_M)$ and the ground truth warped image I_W, whereas $\mathcal{L}_{disp}$ corresponds to the deformation field gradient error.

$$\mathcal{L} = \mathcal{L}_{sim} + \mathcal{L}_{disp} \qquad (2)$$

where $\mathcal{L}_{sim}$ employs the similarity metric of the local normalized correlation coefficient (*NCC*), which are calculated as follows:

$$\mathcal{L}_{sim} = \text{NCC}(\mathbf{I_W}, \phi . \mathbf{I_W}) = \frac{1}{N} \sum_{\mathbf{p} \in X} \frac{\sum_i (\mathbf{I_W(p)} - \overline{\mathbf{I_W(p)}}) \sum_i (\phi . \mathbf{I_M(p)} - \overline{\phi . \mathbf{I_M(p)}})}{\sqrt{\sum_i (\mathbf{I_W(p)} - \overline{\mathbf{I_W(p)}})^2} \sqrt{\sum_i (\phi . \mathbf{I_M(p)} - \overline{\phi . \mathbf{I_M(p)}})^2}} \qquad (3)$$

where $(\phi . I_M(p))$ and $I_W(p)$ are the voxel intensities of a corresponding patch p in the warped image and the ground truth, respectively, whereas $(\overline{\phi . I_M(p)})$ and $\overline{I_W(p)}$ are the mean pixel intensities for both images. $\mathcal{L}_{disp}$ measures spatial gradients differences in the predicted displacement d as follows:

$$\mathcal{L}_{disp} = \sum_{\mathbf{p} \in X} \| \nabla d(p) \| \qquad (4)$$

3 Experimental Results

3.1 Experimental Setup

Due to the large differences between the two databases in terms of study characteristics and the followed MRI and iUS protocols, a pre-processing step is crucial. First, for each patient, the iUS images were resampled to the voxel resolution of the MRI of $1 \times 1 \times 1$ mm^3. Then, the MRI images were cropped to the field of view (FOV) of the iUS. After that, all images were resized into an image resolution of $128 \times 128 \times 128$ pixels to be applicable by the proposed deep learning model. An affine alignment on MRI and iUS volumes was achieved using the MINC toolkit (https://github.com/BIC-MNI/minc-tools). Finally, we obtained the ground truth warped MRI by applying thin-plate spline transformation to the input MRI and the expert labeled MRI-iUS tags using the MINC toolkit.

As the number of cases is rather limited, we use intensive data augmentation to help prevent the model from overfitting and improve the registration results. This involves random 3D flipping, 3D rotations [0–30 degrees], random gamma intensity transformation

[0.8–1.2], and elastic deformation. For the experiments, our model was implemented in Python using the TensorFlow library. The experiments were run on an Intel Xeon Gold 6248 (27.5M Cache, 2.50 GHz) CPU with 8 GB RAM and a single NVIDIA Tesla V100 GPU 32GB. For training our network, we divide the cases into two sets 78% for the training set and 22% for the validation set, the ADAM optimizer with an initial learning rate of 0.0001, and a batch size of 4 was used. To compare with other studies, we use the mean target registration error (mTRE), which represents the average distance between the corresponding landmarks in each MRI-iUS pair after registration. The evaluation of our experiments was performed using the same approach reported in [18].

3.2   Registration Results

Figure 2 shows seven examples of aligning MRI to iUS for different patients using our proposed registration method. Each column corresponds to an individual patient. From the visual results, it can be seen that overlaid MRI-iUS pairs are significantly improved after applying our method. Table 2 and Table 3 depict the pre-and post-registration results for all trained cases for the BITE and the RESECT datasets, individually. In both tables, the minimum achievable affine is the minimum mTRE we can achieve using an affine transformation for the registration. In the last column, an average of the results over the listed cases for each dataset as well as their standard deviation (std dev) is reported. For the BITE database, our model reduced the initial mTRE (provided in the dataset) from 4. 18 ± 1.91 mm to 1.68 ± 0.65 mm. Similarly, an mTRE of 0.99 ± 0.22 mm was achieved on the RESECT database starting with an initial mTRE value of 5.35 ± 4.29. This result highlights that our method delivers better results than initial alignment and similar results to the minimum achievable truth affine registration on average. In a few cases, the proposed approach performs similar or slightly worse compared to the ground-truth

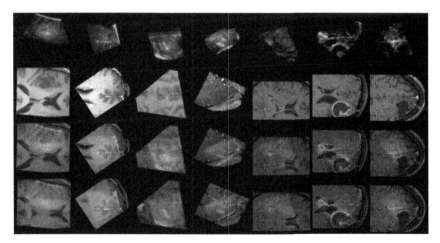

Fig. 2. Examples of MRI to iUS registration. From the top row, iUS images (*green*), preoperative T2-FLAIR MRI (*grey*), initial overlay of iUS on MRI, and final deformable registration, respectively. Columns cases of BITE #5, #6, #14 and RESECT #5, #9, #15, #17, and #23, respectively (Color figure online).

affine results which indicate that the optimal transformation has been achieved using the truth rigid registration. Overall, this analysis confirms that the number of available training cases affects the accuracy and robustness of the CNN. This was indicated by the superior performance of our method on the RESECT dataset (22 cases) over the BITE dataset (14 cases).

Table 2. Details of the MRI-iUS registration for each case in the BITE dataset. Underlined bold represent cases used during the validation stage.

Case	1	2	3	4	5	6	**7**	**Std dev**
# Landmarks	27	25	29	22	21	27	19	-
Before registration	5.88	6.06	8.91	3.87	2.57	2.24	3.02	1.91
Ours (affine + deformable)	1.29	1.08	1.38	1.03	1.45	1.29	2.18	0.65
Minimum achievable affine	1.84	1.42	2.11	1.45	2.04	1.50	1.99	0.45
Case	8	9	**10**	**11**	12	13	14	**Average**
# Landmarks	23	21	25	25	21	23	23	24
Before registration	3.75	5.08	2.99	1.51	3.68	5.13	3.78	4.18
Ours (affine + deformable)	1.55	1.68	1.88	1.72	3.65	1.79	1.52	1.68
Minimum achievable affine	2.19	2.43	1.44	1.32	2.67	2.46	2.30	1.94

3.3   Comparison with the State-of-the-Art Methods

The initial and final landmarks errors for the proposed method and approaches found in the literature for MRI-iUS registration are displayed in Fig. 3(a). For the BITE database, our method are compared with LC^2 [7], SSC [8], SeSaMI [9], miLBP [19], Laplacian Commutators [20], cDRAMMS [10], and ARENA [21]. The results obtained indicate that our method outperforms other evaluated competing techniques, providing mTRE of 1.68 ± 0.65 mm with about 0.40 mm margin smaller than cDRAMMS.

Furthermore, our method was applied to the RESECT database as well (Fig. 3(b)). Here, we compare our results with conventional methods: LC^2 [22], SSC [18], NiftyReg [23], cDRAMMS, ARENA [21] in addition to learning methods: FAX [11], CNN + STN [12]. As illustrated in Fig. 3 (b), our model ranks first with mTRE of 0.99 ± 0.22 mm followed by the learning-based method FAX with mTRE of 1.21 ± 0.55 mm. Though team FAX reported comparable results, our method predicts a 3D pair of MRI-iUS images with a runtime of 170 ms compared to 1.77 s (team FAX), and thus 10 times faster. Experimental findings found clear support for the potential of using learning-based registration methods in neurosurgery.

Table 3. Details of the MRI-iUS registration for each case in the RESECT dataset. Underlined bold represent cases used during the validation stage.

Case	**1**	2	3	**4**	5	6	7	**8**
# Landmarks	15	15	15	15	15	15	15	15
Before registration	1.81	5.70	9.56	2.45	12.03	3.25	1.86	2.65
Ours (affine + deformable)	1.09	1.04	0.93	1.04	0.95	0.88	1.05	1.39
Minimum achievable affine	1.13	1.11	0.94	1.04	1.00	0.91	1.26	1.21
Case	12	13	**14**	15	16	17	18	**Std dev**
# Landmarks	15	15	15	15	15	15	15	-
Before registration	19.71	4.56	3.02	3.23	3.39	6.37	3.57	4.29
Ours (affine + deformable)	0.91	0.99	1.04	1.26	0.87	0.91	0.76	0.22
Minimum achievable affine	0.96	1.01	1.05	1.34	0.87	1.02	0.81	0.16
Case	19	21	23	24	**25**	26	27	**Average**
# Landmarks	15	15	15	15	15	15	15	15
Before registration	3.29	4.56	7.02	1.09	10.06	2.82	5.77	5.35
Ours (affine + deformable)	0.74	0.92	0.71	0.69	1.66	0.94	1.09	0.99
Minimum achievable affine	0.8	0.98	0.71	0.71	0.94	0.97	1.15	1.00

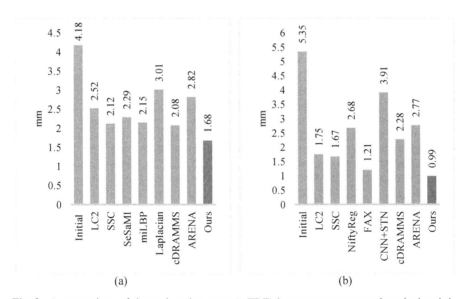

(a)   (b)

Fig. 3. A comparison of the registration error (mTRE) between our proposed method and the state-of-the-art methods on the BITE dataset (a) and the RESECT dataset (b).

4 Conclusion

This study presented an automated fast and robust non-linear approach for pre-operative MRI to iUS registration to assist intraoperative neurosurgical procedures. In our experiments, the performance of our proposed method has been evaluated using 36 patients from two multi-location databases. Outstandingly, our model outperforms the state-of-the-art in terms of both performance and computational efficiency. Furthermore, the qualitative results indicate that the registered MRI-iUS pairs have a significant improvement over the initial alignment. Therefore, the results of our proposed registration method are promising and can be applied for clinical use during future work.

Acknowledgments. The first author is supported by the German Academic Exchange Service (DAAD) [scholarship number 91705803].

References

1. Miner, R.C.: Image-guided neurosurgery. J. Med. Imaging Radiat. Sci. **48**, 328–335 (2017)
2. Coburger, J., Wirtz, C.R.: Fluorescence guided surgery by 5-ALA and intraoperative MRI in high grade glioma: a systematic review. J. Neurooncol. **141**(3), 533–546 (2018). https://doi.org/10.1007/s11060-018-03052-4
3. De Momi, E., et al.: A method for the assessment of time-varying brain shift during navigated epilepsy surgery. Int. J. Comput. Assist. Radiol. Surg. **11**(3), 473–481 (2016). https://doi.org/10.1007/s11548-015-1259-1
4. Delorenzo, C., Papademetris, X., Staib, L.H., Vives, K.P., Spencer, D.D., Duncan, J.S.: Image-guided intraoperative cortical deformation recovery using game theory: application to neocortical epilepsy surgery. IEEE Trans. Med. Imaging **29**, 322–338 (2010)
5. Liu, J., et al.: Image registration in medical robotics and intelligent systems: fundamentals and applications. Adv. Intell. Syst. **1**, (2019)
6. Haskins, G., Kruger, U., Yan, P.: Deep learning in medical image registration: a survey. Mach. Vis. Appl. **31**(1–2), 1–18 (2020). https://doi.org/10.1007/s00138-020-01060-x
7. Wein, W., Ladikos, A., Fuerst, B., Shah, A., Sharma, K., Navab, N.: Global registration of ultrasound to mri using the LC2 metric for enabling neurosurgical guidance. In: Mori, K., Sakuma, I., Sato, Y., Barillot, C., Navab, N. (eds.) MICCAI 2013. LNCS, vol. 8149, pp. 34–41. Springer, Heidelberg (2013). https://doi.org/10.1007/978-3-642-40811-3_5
8. Heinrich, M.P., Jenkinson, M., Papież, B.W., Brady, S.M., Schnabel, J.A.: Towards real-time multimodal fusion for image-guided interventions using self-similarities. In: Mori, K., Sakuma, I., Sato, Y., Barillot, C., Navab, N. (eds.) MICCAI 2013. LNCS, vol. 8149, pp. 187–194. Springer, Heidelberg (2013). https://doi.org/10.1007/978-3-642-40811-3_24
9. Rivaz, H., Karimaghaloo, Z., Collins, D.L.: Self-similarity weighted mutual information: a new nonrigid image registration metric. Med. Image Anal. **18**, 343–358 (2014)
10. Machado, I., et al.: Deformable MRI-Ultrasound registration using correlation-based attribute matching for brain shift correction: accuracy and generality in multi-site data. Neuroimage **202**, 116094 (2019)
11. Zhong, X., et al.: Resolve intraoperative brain shift as imitation game. In: Stoyanov, D., et al. (eds.) POCUS/BIVPCS/CuRIOUS/CPM -2018. LNCS, vol. 11042, pp. 129–137. Springer, Cham (2018). https://doi.org/10.1007/978-3-030-01045-4_15

A Hybrid Deep Registration of MR Scans to Interventional Ultrasound 595

12. Sun, L., Zhang, S.: Deformable MRI-ultrasound registration using 3D convolutional neural network. In: Stoyanov, D., et al. (eds.) POCUS/BIVPCS/CuRIOUS/CPM -2018. LNCS, vol. 11042, pp. 152–158. Springer, Cham (2018). https://doi.org/10.1007/978-3-030-01045-4_18

13. Balakrishnan, G., Zhao, A., Sabuncu, M.R., Guttag, J., Dalca, A.V.: VoxelMorph: A learning framework for deformable medical image registration. IEEE Trans. Med. Imaging (2019)

14. Mercier, L., Del Maestro, R.F., Petrecca, K., Araujo, D., Haegelen, C., Collins, D.L.: Online database of clinical MR and ultrasound images of brain tumors. Med. Phys. **39**, 3253–3261 (2012)

15. Xiao, Y., Fortin, M., Unsgard, G., Rivaz, H., Reinertsen, I.: REtroSpective evaluation of cerebral tumors (RESECT): a clinical database of pre-operative MRI and intra-operative ultrasound in low-grade glioma surgeries. Med. Phys. **44**, 3875–3882 (2017)

16. Ronneberger, O., Fischer, P., Brox, T.: U-Net: Convolutional networks for biomedical image segmentation. In: Navab, N., Hornegger, J., Wells, W.M., Frangi, A.F. (eds.) MICCAI 2015. LNCS, vol. 9351, pp. 234–241. Springer, Cham (2015). https://doi.org/10.1007/978-3-319-24574-4_28

17. Zeineldin, R.A., Karar, M.E., Coburger, J., Wirtz, C.R., Burgert, O.: DeepSeg: deep neural network framework for automatic brain tumor segmentation using magnetic resonance FLAIR images. Int. J. Comput. Assist. Radiol. Surg. **15**(6), 909–920 (2020). https://doi.org/10.1007/s11548-020-02186-z

18. Heinrich, M.P.: Intra-operative ultrasound to MRI fusion with a public multimodal discrete registration tool. In: Stoyanov, D., et al. (eds.) POCUS/BIVPCS/CuRIOUS/CPM -2018. LNCS, vol. 11042, pp. 159–164. Springer, Cham (2018). https://doi.org/10.1007/978-3-030-01045-4_19

19. Jiang, D., Shi, Y., Yao, D., Wang, M., Song, Z.: miLBP: a robust and fast modality-independent 3D LBP for multimodal deformable registration. Int. J. Comput. Assist. Radiol. Surg. **11**(6), 997–1005 (2016). https://doi.org/10.1007/s11548-016-1407-2

20. Zimmer, V.A., González Ballester, M.Á., Piella, G.: Multimodal image registration using Laplacian commutators. Inf. Fus. **49**, 130–145 (2019)

21. Masoumi, N., Xiao, Y., Rivaz, H.: ARENA: Inter-modality affine registration using evolutionary strategy. Int. J. Comput. Assist. Radiol. Surg. **14**(3), 441–450 (2018). https://doi.org/10.1007/s11548-018-1897-1

22. Wein, W.: Brain-shift correction with image-based registration and landmark accuracy evaluation. In: Stoyanov, D., et al. (eds.) POCUS/BIVPCS/CuRIOUS/CPM -2018. LNCS, vol. 11042, pp. 146–151. Springer, Cham (2018). https://doi.org/10.1007/978-3-030-01045-4_17

23. Drobny, D., Ranzini, M., Ourselin, S., Vercauteren, T., Modat, M.: Landmark-based evaluation of a block-matching registration framework on the RESECT pre- and intra-operative brain image data set. In: Zhou, L., et al. (eds.) LABELS/HAL-MICCAI/CuRIOUS -2019. LNCS, vol. 11851, pp. 136–144. Springer, Cham (2019). https://doi.org/10.1007/978-3-030-33642-4_15

Segmentation of Peripancreatic Arteries in Multispectral Computed Tomography Imaging

Alina Dima[1,2(✉)], Johannes C. Paetzold[3,4], Friederike Jungmann[1],
Tristan Lemke[1], Philipp Raffler[1], Georgios Kaissis[1,2], Daniel Rueckert[2],
and Rickmer Braren[1]

[1] Institute of Diagnostic and Interventional Radiology,
Technical University of Munich, Munich, Germany
[2] AI in Medicine and Healthcare, Technical University of Munich, Munich, Germany
alina.dima@tum.de
[3] Image-Based Biomedical Modeling, Technical University of Munich,
Munich, Germany
[4] ITERM Institute Helmholtz Zentrum München, Neuherberg, Germany

Abstract. Pancreatic ductal adenocarcinoma is an aggressive form of cancer with a poor prognosis, where the operability and hence chance of survival is strongly affected by the tumor infiltration of the arteries. In an effort to enable an automated analysis of the relationship between the local arteries and the tumor, we propose a method for segmenting the peripancreatic arteries in multispectral CT images in the arterial phase. A clinical dataset was collected, and we designed a fast semi-manual annotation procedure, which requires around 20 min of annotation time per case. Next, we trained a U-Net based model to perform binary segmentation of the peripancreatic arteries, where we obtained a near perfect segmentation with a Dice score of 95.05% in our best performing model. Furthermore, we designed a clinical evaluation procedure for our models; performed by two radiologists, yielding a complete segmentation of 85.31% of the clinically relevant arteries, thereby confirming the clinical relevance of our method.

Keywords: Arterial segmentation · Vessels · Annotation · PDAC

1 Introduction

Clinical Motivation. Pancreatic ductal adenocarcinoma (PDAC) is the most prevalent malignant pancreatic tumour, accounting for over 90% of pancreatic malignancies [23]. It is an aggressive cancer with patients often developing metastases. Its overall 5-year survival rate is less than 9% [23]. In current clinical practice, a total tumor resection (surgically removing the tumor tissue) offers the best

Electronic supplementary material The online version of this chapter (https://doi.org/10.1007/978-3-030-87589-3_61) contains supplementary material, which is available to authorized users.

prospects for a patient diagnosed with PDAC [9]. Therefore it is attempted to be applied to as many patients as possible. However, peripancreatic vascular anatomy is complex and tumors often tend to infiltrate both arteries and veins. Although central veins can typically be reconstructed, tumor infiltration of certain arteries or arterial segments is not amendable to surgical resection. Our short-term goal is to enable an automated identification of the relevant arteries, whereas our long-term goal aims at an automated definition of tumor-to-artery contact zones and the prediction of tumor resectability.

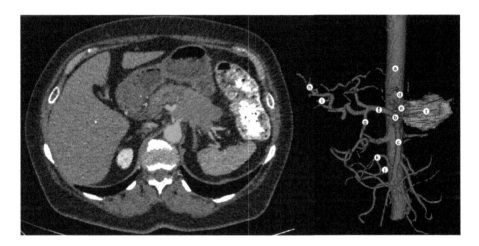

Fig. 1. Example case where PDAC infiltrates the arterial tree, visualized on an arterial CT. The left image represents the axial view, while the right image shows a 3D rendering of the arterial tree and tumor. Legend: (a) aorta, (b) celiac trunk, (c) superior mesenteric artery, (d) left gastric artery, (e) splenic artery, (f) common hepatic artery, (g) gastroduodenal artery, (h) left hepatic artery, (i) right hepatic artery, (j, k) pancreaticoduodenal arteries, (t) tumor

Vascular infiltration occurs when a tumor invades the vessel wall. Tumor resectability in case of vessel wall contact or invasion depends on the function of the specific artery and the possibility to reconstruct the vessel. Specifically, vascular infiltration was traditionally deemed inoperable, but recent medical advancements in surgical procedures such as graft reconstruction and preoperative treatment such as the administration of neo-adjuvant chemotherapy [2] have allowed complete tumor resection in patients with limited arterial involvement – referred to as *borderline-resectable* – patients [5]. Consequently, there is a need to develop new ways to predict tumor resectability in borderline-resectable PDAC patients.

Previous work looked into predicting tumor resectability for borderline-resectable PDAC patients with venous infiltration [7], and discovered that portal venous confluence contour irregularity can predict tumor cell infiltration of the vessel wall. We want to take this work a step further and predict tumor resectability for PDAC patients with the more challenging arterial infiltration, motivating our goal to achieve full segmentation of the peripancreatic arteries. This segmentation is also an initial step to quantify tumor-to-artery contact zones.

Vessel segmentation is an extensively studied task in medical image analysis. Pixelwise delineation of vascular segments has clinical relevance on many scales. Among others, the segmentation of retinal vessels for the study of cardiovascular and ophthalmologic diseases [13], the segmentation of three dimensional volumetric angiography images for stroke diagnosis [14,21]; the reconstruction of the portal and hepatic veins for preoperative liver surgery simulation [8]; the segmentation of the celiac trunk for gastrectomy planing [20]; and even the segmentation of optoacoustic skin images to monitor diabetes progression [4].

The advent of convolutional neural networks (CNNs) has led to an unparalleled precision and speed in vessel segmentation, especially compared to the older Hessian-based approaches such as [3,11], which, while generic and robust, lacked the ability to segment individual datasets at human level accuracy. In recent years the U-Net [18] has established itself as the *de facto* architecture in medical segmentation problems, with existing evidence that a well-optimized U-Net offers a baseline which is difficult to beat even by custom designed architectures [6,12,16,22]. Another area of interest for the problem of vessel segmentation has been the research of loss functions better suited for the graph structure inherent in vessels. One contribution in this area is clDice [19], which incorporates efficiently computed quasi-skeletons of vessels.

Our Contribution. In this work we introduce the task of peripancreatic artery segmentation to provide the basis for the quantification of tumor-to-artery contact zones. This will enable future patient stratification with regard to specific treatment options and may provide a new biomarker for diagnosis and treatment planning. To this end, we collected a private dataset consisting of 143 multispectral CT images in the arterial phase and developed a procedure to quickly annotate this dataset. Furthermore, we trained a highly accurate segmentation model for the task of peripancreatic arterial segmentation. Finally, we provided a comprehensive clinical evaluation to mitigate the noisy ground truth, illustrating the clinical relevance of our method.

2   Method and Experiments

2.1   Dataset

The clinical and imaging data were collected retrospectively according to the principles set forward in the Declaration of Helsinki and Good Clinical Practice. Written conformed consent was waived and the study approved by the institutional ethics review board of the Technical University of Munich, Faculty of Medicine (Protocol Number 180/17S, 2017). In total, a dataset consisting of pre-treatment spectral CT images from 143 patients imaged between September 2016 and July 2020 was retrospectively collected. All patients underwent contrast enhanced CT in the arterial phase (bolus tracking via trigger-ROI in the aortic arch after injection of 70 ml contrast agent) using a Philips IQON Spectral CT scanner (Philips Healthcare, Best, The Netherlands). Iodine maps – an image type in which the pixel values represent iodine quantities [17] – were reconstructed using the Philips IntelliSpace Portal software (version 11.0) to observe better contrast between arteries and the surrounding tissue. The resolution of all images ranges between 0.6 and 1 mm for the x and y dimensions, and 0.9 in the z dimension.

For cross-validation, the imaging data were partitioned into four even folds yielding a size of 35 or 36 samples per subset. For each cross-validation, two subsets served as train set, one as validation set and the remaining one as test set, rotated such that each partition was used for testing and validation exactly once and twice for training. Age at diagnosis and sex were obtained for all patients using the hospital's information system. The mean age of the patients analyzed in this study was 69.1 ± 9.2 years and the cohort consisted in equal parts of male and female patients (71 male, 72 female).

2.2 Labeling Strategy

The annotation and curation of new data are a quintessential part of many research projects due to the scarcity of public data in the field. Particularly, establishing a good annotation strategy is essential. Despite the existence of a number of free public annotation tools, a steep learning curve leads their under-utilisation, researchers in some cases opting for annotation procedures involving a high degree of manual labour. We experimented with data annotation tools and found a relatively fast annotation process using 3D Slicer [15] without the need to manually delineate vessels up to pixel accuracy. Two medical experts were involved in the annotation process, each annotating one half of the dataset. The average time spent per annotated image was 20 min.

The annotation process relies on creating three separate segments for different tissues: arteries, bones and everything else including: intestines, veins, parenchymal organs and foreign material like catheters. The three segmentations are generated through seed growth, by manually placing seeds for each segmentation class until the parapancreatic arteries are sufficiently separated from the rest. A detailed description of the manual annotation process can be found in the Appendix.

We focus our annotation effort on the arteries around the pancreas which are most relevant for the task of identifying contact between PDAC and the arteries: where contact is most likely to occur while the patient still being operable. These vessels are: the aorta (A), the celiac trunk (CT), the left gastric artery (LGA), the common hepatic artery (CHA), the right and left hepatic arteries (RHA, LHA), the gastroduodenal artery (GDA), the splenic artery (SA), the superior mesenteric artery (SMA), as well as the two most proximal pancreaticoduodenal arteries (PDA1 and PDA2). If present, we also segment a variant of the right hepatic artery called truncus hepatomesentericus dexter (THD), where the right hepatic lobe is supplied via the SMA, which has surgical implications. Figure 1 illustrates an example image annotated with the relevant arteries.

The other arteries visible on the scan were included in the ground truth if they were connected to the core segmented area following the aforementioned labelling process; otherwise they were not manually added. As an example, the renal arteries were clipped before connecting to the kidneys; however the proximity of this clipping point with respect to the kidneys substantially varied. This variation is motivated by the fact that these arteries are not of clinical relevance.

The annotation process results in a relatively noisy dataset where some vessels appear in some of the labels but not in others. We argue that our dataset is sufficient to satisfy the clinical requirements of the task, while also spending minimal annotation effort and ultimately saving clinicians' time.

2.3   Training the Neural Network

Inspired by the work of [6] we train a 3D U-Net [18] for the binary segmentation of the peripancreatic arteries.

Network Architecture: Our network consists of four U-Net layers, each consisting of a covolutional block; followed by a max pooling operation in the encoder part and an upsampling block concatenated with the output from the corresponding encoder in the decoder part. Each convolutional block consists of two series of one $3 \times 3 \times 3$ 3D convolution, batch normalization and a ReLU nonlinearity. The number of output channels for each layer follows: 16, 32, 48 and 64. The last U-Net layer is followed by a $1 \times 1 \times 1$ 3D convolution and a logistic sigmoid non-linearity to obtain the final predictions.

Implementation Details: The input was rescaled to the interval $[-1, 1]$ and cropped to a volume of $256 \times 256 \times 128$ pixels, centered around the ostium of the superior mesenteric artery. Starting from randomly initialized weights, we trained our models using the Cross-Entropy loss with the Adam optimizer [10] for 200 epochs. The learning rate was kept constant for the first 150 epochs and then dropped by a factor of 10 in the following 50 epochs, with the initial learning rate adjusted according to the loss function. The model selection was done based on the Dice score on the validation set. For augmentation we used expert-controlled random 3D rotations up to $10°$; 3D translations up to 20 pixels in-plane and 5 pixels out-of-plane; scaling up to 20%; gamma changes up to 10%, as well as Gaussian noise injection with $\sigma \in [0.05, 0.1]$ to match the variation seen in the data. We implemented our model in Tensorflow [1], and performed all of our experiments on an Nvidia Quadro RTX 8000. The code is available at https://github.com/alinafdima/peripancreatic-segmentation.

2.4   Results

Quantitative Evaluation.   We trained a U-Net using as input the iodine maps, the arterial images or both. Additionally, we compared training using the soft Dice and the clDice losses. The model selection was based on the Dice score performance on the validation set, and we performed evaluation using the Dice score, precision and recall. We obtained near perfect Dice scores across all types of data and training losses. In terms of the training loss the Cross-Entropy performed better, not suffering from the stability issues associated with the other soft Dice-based losses. Input-wise we find that the iodine maps result in better performance than the arterial images, indicating more signal to noise ratio due to improved contrast than their arterial counterparts for vessel segmentation tasks. Providing both arterial and iodine maps as input to the network yielded slightly better results than using only the iodine maps, while also reducing the standard deviation.

Table 1. Quantitative Results: We trained a U-Net using different settings. For all of the settings we achieved a very high segmentation performance.

Network	Image Input	Dice Score	Precision	Recall
Model-iodine	Iodine map	94.89 ± 2.36	**96.52 ± 2.67**	93.46 ± 3.96
Model-arterial	Arterial	93.70 ± 4.78	95.67 ± 2.61	92.26 ± 7.22
Model-combined	Iodine + arterial	**95.05 ± 1.88**	96.2 ± 2.63	94.01 ± 2.91
Model-Dice	Iodine map	92.57 ± 4.44	92.2 ± 7.41	93.43 ± 4.48
Model-clDice	Iodine map	93.85 ± 2.56	93.68 ± 4.14	**94.26 ± 4.1**

Qualitative Evaluation. We present a few example predictions from our model trained on the iodine maps in Fig. 2. Overall the segmentations are highly accurate, delineating almost perfectly the arterial walls in the segmented portions. Many of the mistakes are either over- or under-segmentations of distal segments which are not clinically relevant. Occasionally we see false negatives

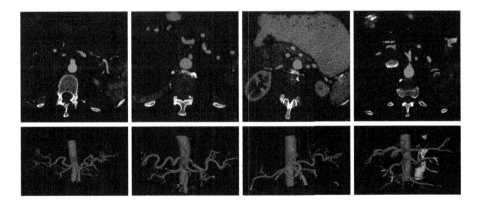

Fig. 2. Qualitative results: In the first row our segmentations are presented in the axial view; in the second row as a 3D rendering. Magenta represents true positives, blue false negatives and yellow false positives. We observe a mostly consistent and accurate segmentation across images with varying contrast.

Table 2. Clinical evaluation: We evaluated the segmentations of the individual arteries according to the segmentation quality, for both the ground truth and our main model. The scores range from 0 to 2, where a 2 indicates a perfect segmentation of the clinically relevant region of the vessel and a 0 indicating a failure case.

Vessel	Score 2		Score 1		Score 0		Total cases
	Pred	GT	Pred	GT	Pred	GT	
A	97.2%	100.0%	2.8%	0.0%	0.0%	0.0%	143
–CT	97.9%	100.0%	2.1%	0.0%	0.0%	0.0%	143
—LGA	81.82%	94.41%	15.38%	5.59%	2.8%	0.0%	143
—CHA	90.91%	100.0%	8.39%	0.0%	0.7%	0.0%	143
——LHA	86.71%	90.91%	4.9%	6.99%	8.39%	2.1%	143
——RHA	86.71%	93.01%	6.29%	4.9%	6.99%	2.1%	143
——THD	92.0%	100.0%	8.0%	0.0%	0.0%	0.0%	25
——GDA	64.34%	82.52%	25.17%	11.89%	10.49%	5.59%	143
—SA	79.72%	93.71%	19.58%	6.29%	0.7%	0.0%	143
–SMA	96.5%	98.6%	3.5%	1.4%	0.0%	0.0%	143
—PDA1	74.83%	82.52%	15.38%	9.09%	9.79%	8.39%	143
—PDA2	78.32%	81.12%	9.79%	8.39%	11.89%	10.49%	143

comprising of entire vessel segments like in the third example, or large false positives as can be seen in the fourth example, where a vein was segmented.

Inter-rater Variability. In a simple cross rater evaluation experiment we quantitatively reviewed the ground truth segmentations of each annotator, illustrated in Table 2. Each of the clinically relevant arteries received a score on a scale of 0 to 2, where a score of 2 represents a complete segmentation of the clinically relevant portion, 1 a partial segmentation, and 0 represents a failure case. 92.55% of the clinically relevant arteries were perfectly segmented in the ground truth, with most mistakes at the smaller arteries. The two annotators also had different annotation and rating styles, one obtaining an inter-rater score of 88.89%, while the other a score of 96.03%. We view this as an upper-bound for the predictions produced by our network.

Clinical Evaluation. We asked our raters to also clinically evaluate the predictions of our iodine model trained with the Cross-Entropy loss using the same criteria as in the inter-rater experiment, summarized on Table 2. We observe an overall good accuracy of complete segmentations across all types of arteries and arterial sizes, obtaining complete segmentations of the clinically relevant section in 85.31% of all arteries. Excluding the gastroduodenal artery which is not always of clinical relevance, this score increases to 87.14%. As could be expected, the larger arteries such as the aorta, celiac trunk and the superior mesenteric artery were the most accurately segmented, being under-segmented in only 2 − 4% of cases. The smallest arteries were the least accurately segmented in the predictions, but also in the ground truth, thus being objected to noisier annotations. The gastroduodenal artery was the most likely to be under-segmented, and together with the two most proximal pancreaticoduodenal arteries were the most likely to be failure cases. The truncus hepatomesentericus dexter was successfully segmented in almost all cases where it occurs.

3   Discussion and Conclusion

In this work we introduce the task of peripancreatic artery segmentation in multispectral contrast CT images, motivated by the high relevance of these vessels for surgery planning in PDAC. A private clinical dataset was collected and efficiently annotated; using these annotations we trained a CNN for segmentation of peripancreatic arteries. We obtained near perfect segmentation accuracy with a Dice score of over 95.05% in our best performing model. Many of the errors of the network were erroneous artery detections not present in the ground truth but which are anatomically correct, or arterial under-segmentations for arteries that were not consistently annotated in the ground truth. Under the context of a noisy data labelling process, we consider that the U-Net segmentation performance reaches near human segmentation accuracy. From our clinical evaluation we also derived 74% complete segmentations for all but one arterial branch relevant for the clinical task, speaking for the clinical usefulness of our method. We obtained complete segmentations in 85.31% of total arteries, with many of the lower-performing arteries also having noisy annotations as quantified by inter-rater evaluation.

We experimented with two input types: conventional arterial CTs and iodine maps and discovered that the models trained on the iodine maps performed on average 1% better than those trained on the arterial images. This is particularly surprising because the annotations were performed on the arterial images, which could have introduced a bias to favor the arterial image in the dataset. The higher performance on the iodine maps can be attributed to the higher contrast present in the image type, and it suggests that the iodine maps could be employed for better segmentation of the peripancreatic arteries, but possibly also other vessels.

In the future we plan to extend this work to predict arterial contact and infiltration in PDAC, by quantifying the contact between the peripancreatic arteries and the tumor, potentially combining the current arterial phase images with their venous phase counterparts. We also plan to increase the diversity in our data by incorporating arterial phase CT images from multiple scanners. Furthermore, given the noise inherent to manual labelling, another future direction could be incorporating uncertainty into our models.

References

1. Abadi, M., et al.: Tensorflow: a system for large-scale machine learning. In: 12th {USENIX} Symposium on Operating Systems Design and Implementation ({OSDI} 16), pp. 265–283 (2016)

2. Blazer, M., Wu, C., et al.: Neoadjuvant modified (m) folfirinox for locally advanced unresectable (lapc) and borderline resectable (brpc) adenocarcinoma of the pancreas. Ann. Surg. Oncol. **22**(4), 1153–1159 (2015)
3. Frangi, A.F., Niessen, W.J., Vincken, K.L., Viergever, M.A.: Multiscale vessel enhancement filtering. In: Wells, W.M., Colchester, A., Delp, S. (eds.) MICCAI 1998. LNCS, vol. 1496, pp. 130–137. Springer, Heidelberg (1998). https://doi.org/10.1007/BFb0056195
4. Gerl, S., et al.: A distance-based loss for smooth and continuous skin layer segmentation in optoacoustic images. In: Martel, A.L., et al. (eds.) MICCAI 2020. LNCS, vol. 12266, pp. 309–319. Springer, Cham (2020). https://doi.org/10.1007/978-3-030-59725-2_30
5. Isaji, S., Mizuno, S., et al.: International consensus on definition and criteria of borderline resectable pancreatic ductal adenocarcinoma 2017. Pancreatology **18**(1), 2–11 (2018)
6. Isensee, F., Kickingereder, P., Wick, W., Bendszus, M., Maier-Hein, K.H.: No new-net. In: Crimi, A., Bakas, S., Kuijf, H., Keyvan, F., Reyes, M., van Walsum, T. (eds.) BrainLes 2018. LNCS, vol. 11384, pp. 234–244. Springer, Cham (2019). https://doi.org/10.1007/978-3-030-11726-9_21
7. Kaissis, G.A., Lohöfer, F.K., et al.: Borderline-resectable pancreatic adenocarcinoma: contour irregularity of the venous confluence in pre-operative computed tomography predicts histopathological infiltration. PloS one **14**(1), e0208717 (2019)
8. Keshwani, D., Kitamura, Y., Ihara, S., Iizuka, S., Simo-Serra, E.: TopNet: topology preserving metric learning for vessel tree reconstruction and labelling. In: Martel, A.L., et al. (eds.) MICCAI 2020. LNCS, vol. 12266, pp. 14–23. Springer, Cham (2020). https://doi.org/10.1007/978-3-030-59725-2_2
9. Kimura, K., Amano, R., et al.: Clinical and pathological features of five-year survivors after pancreatectomy for pancreatic adenocarcinoma. World J. Surg. Oncol. **12**(1), 1–8 (2014)
10. Kingma, D.P., Ba, J.: Adam: a method for stochastic optimization. In: 3rd International Conference on Learning Representations, ICLR 2015 (2015)
11. Law, M.W.K., Chung, A.C.S.: Three dimensional curvilinear structure detection using optimally oriented flux. In: Forsyth, D., Torr, P., Zisserman, A. (eds.) ECCV 2008. LNCS, vol. 5305, pp. 368–382. Springer, Heidelberg (2008). https://doi.org/10.1007/978-3-540-88693-8_27
12. Navarro, F., et al.: Shape-aware complementary-task learning for multi-organ segmentation. In: Suk, H.-I, Liu, M., Yan, P., Lian, C. (eds.) MLMI 2019. LNCS, vol. 11861, pp. 620–627. Springer, Cham (2019). https://doi.org/10.1007/978-3-030-32692-0_71
13. Niemeijer, M., Staal, J., et al.: Comparative study of retinal vessel segmentation methods on a new publicly available database. In: Medical Imaging 2004: Image Processing, vol. 5370, pp. 648–656. International Society for Optics and Photonics (2004)
14. Paetzold, J.C., Schoppe, O., et al.: Transfer learning from synthetic data reduces need for labels to segment brain vasculature and neural pathways in 3d. In: International Conference on Medical Imaging with Deep Learning-Extended Abstract Track (2019)
15. Pieper, S., Halle, M., Kikinis, R.: 3D Slicer. In: 2004 2nd IEEE International Symposium on Biomedical Imaging: From Nano to Macro (IEEE Cat No. 04EX821), pp. 632–635. IEEE (2004)

16. Qasim, A.B., et al.: Red-gan: attacking class imbalance via conditioned generation. yet another medical imaging perspective. In: Medical Imaging with Deep Learning, pp. 655–668. PMLR (2020)
17. Rassouli, N., Etesami, M., Dhanantwari, A., Rajiah, P.: Detector-based spectral ct with a novel dual-layer technology: principles and applications. Insights Into Imaging **8**(6), 589–598 (2017)
18. Ronneberger, O., Fischer, P., Brox, T.: U-Net: convolutional networks for biomedical image segmentation. In: Navab, N., Hornegger, J., Wells, W.M., Frangi, A.F. (eds.) MICCAI 2015. LNCS, vol. 9351, pp. 234–241. Springer, Cham (2015). https://doi.org/10.1007/978-3-319-24574-4_28
19. Shit, S., Paetzold, J.C., et al.: clDice - a novel topology-preserving loss function for tubular structure segmentation. In: CVPR (2021)
20. Tang, X., Huang, B., et al.: Celiac trunk segmentation incorporating with additional contour constraint. Applied Intelligence, pp. 1–12 (2021)
21. Tetteh, G., Efremov, V., et al.: Deepvesselnet: vessel segmentation, centerline prediction, and bifurcation detection in 3-D angiographic volumes. Front. Neurosci. **14**, 1285 (2020)
22. Todorov, M.I., Paetzold, J.C., et al.: Machine learning analysis of whole mouse brain vasculature. Nat. Methods **17**(4), 442–449 (2020)
23. Wild, C.P., Weiderpass, E., Stewart, B.W.: World Cancer Report 2020. International Agency for Research on Cancer, Lyon, France (2020)

Skullengine: A Multi-stage CNN Framework for Collaborative CBCT Image Segmentation and Landmark Detection

Qin Liu[1], Han Deng[2], Chunfeng Lian[1], Xiaoyang Chen[1], Deqiang Xiao[1],
Lei Ma[1], Xu Chen[1], Tianshu Kuang[2], Jaime Gateno[2], Pew-Thian Yap[1(✉)],
and James J. Xia[2(✉)]

[1] Department of Radiology and Biomedical Research Imaging Center (BRIC),
University of North Carolina at Chapel Hill, Chapel Hill, NC, USA
qinliu19@email.unc.edu, ptyap@med.unc.edu
[2] Department of Oral and Maxillofacial Surgery, Houston Methodist Hospital,
Houston, TX, USA
jxia@houstonmethodist.org

Abstract. Accurate bone segmentation and landmark detection are
two essential preparation tasks in computer-aided surgical planning for
patients with craniomaxillofacial (CMF) deformities. Surgeons typically
have to complete the two tasks manually, spending ∼12 h for each set of
CBCT or ∼5 h for CT. To tackle these problems, we propose a multi-
stage coarse-to-fine CNN-based framework, called SkullEngine, for high-
resolution segmentation and large-scale landmark detection through a
collaborative, integrated, and scalable JSD model and three segmentation and landmark detection refinement models. We evaluated our framework on a clinical dataset consisting of 170 CBCT/CT images for the task
of segmenting 2 bones (midface and mandible) and detecting 175 clinically common landmarks on bones, teeth, and soft tissues. Experimental
results show that SkullEngine significantly improves segmentation quality, especially in regions where the bone is thin. In addition, SkullEngine
also efficiently and accurately detect all of the 175 landmarks. Both tasks
were completed simultaneously within 3 min regardless of CBCT or CT
with high segmentation quality. Currently, SkullEngine has been integrated into a clinical workflow to further evaluate its clinical efficiency.

Keywords: Cone-Beam Computed Tomography (CBCT) Image ·
Segmentation · Landmark Detection

1 Introduction

Accurate bone segmentation and landmark detection are two fundamental tasks
in preparing cone-beam computed tomography (CBCT) or computed tomography (CT)[1] scans for use in computer-aided surgical simulation to treat patients

---

[1] For brevity, in the rest of the paper CBCT refers to both CBCT and CT. CBCT is
more frequently used clinically.

© Springer Nature Switzerland AG 2021
C. Lian et al. (Eds.): MLMI 2021, LNCS 12966, pp. 606–614, 2021.
https://doi.org/10.1007/978-3-030-87589-3_62

with craniomaxillofacial (CMF) deformities. In current clinical practice, it takes at least a day and a half for a surgeon or a trained operator to manually perform both tasks to obtain the CBCT segmentation masks and landmark coordinates, which are time-consuming, labor-intensive, and error-prone. Therefore, there is an urgent need to develop a reliable automatic segmentation and landmark detection method for clinical use.

Automatic CMF bone segmentation and landmark detection are practically challenging due to the complex CMF anatomy, significant variations in appearance (especially in patients with severe deformities), and large image sizes (up to $768 \times 768 \times 576$). Most existing methods, including conventional methods [3,13,15] and CNN-based methods [6,10,14,17], formulate the segmentation and the landmark detection as two independent tasks without considering their inherent relationship (e.g., landmarks usually lie on the boundaries of segmented bone regions).

In recent years, CNN-based joint segmentation and landmark detection (JSD) approaches [7,16] have been proposed to combine the two tasks via multi-task learning. In [7], the authors proposed a multi-task dynamic transformer network (DTNet) for concurrently segmenting mandible and detecting 64 landmarks. In [16], the authors proposed a context-guided multi-task fully convolutional network for jointly segmenting two bony structures (i.e., midface and mandible) and 15 boney landmarks. However, these approaches have three major drawbacks that hinder them from being integrated into clinical practice: 1) they cannot detect large-scale landmarks (e.g., over 100 landmarks) due to the limited graphics processing unit (GPU) memory, 2) they cannot specifically refine segmentation in regions that are important for surgical planning (e.g., regions with thin bones), 3) they are non-scalable because the segmentation and landmark detection tasks are highly coupled in a single network.

To tackle these issues, we propose a coarse-to-fine CNN-based framework, the SkullEngine, for high-resolution bone segmentation and large-scale landmark detection through a collaborative, integrated, and scalable JSD model and three refinement models. The goal of SkullEngine is to segment 2 bones (i.e., midface and mandible) and 175 landmarks (i.e., 66 for bones, 68 for teeth, and 41 for soft tissues) from a CBCT image. SkullEngine achieves this goal in two sequential stages: coarse and refinement. In the coarse stage, a scalable JSD model, a combination of 3D U-Net-based segmentation and landmark detection models, takes a down-sampled image as input for coarse segmentation and global landmark detection (i.e., all landmarks except the tooth landmarks). The tooth landmarks are not detected in this stage because they are close together in a small region and therefore need to detected in a higher resolution. In the refinement stage, based on the coarse segmentation mask and global landmarks achieved with the previous stage, region-of-interest volumes are cropped from the original CBCT image for further segmentation refinement and tooth landmark detection.

A major technical contribution of the proposed SkullEngine is that our new JSD model is scalable and modular compared with previous JSDs. In addition, a major clinical contribution of SkullEngine is its clinical effectiveness. The use of

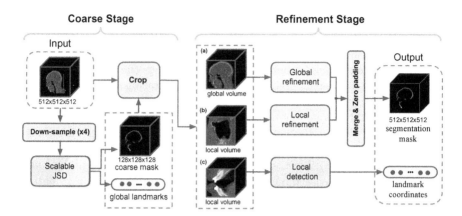

Fig. 1. The framework of SkullEngine. The volumes (a) and (b) are cropped for segmentation refinement, while the volume (c) is cropped for tooth landmark detection. The size of each volume on the picture is just for illustration and may be various for different cases.

SkullEngine can significantly reduce CBCT data preparation time from ∼12 h, or CT from ∼5 h, to 3 min for both clinically challenging tasks of high-resolution CBCT segmentation and large-scale landmark detection. We demonstrate in this study the accuracy of SkullEngine using a clinical dataset containing 92 CBCT scans and 78 CT scans of patients with CMF deformities. We have already integrated SkullEngine into a clinical workflow to continue to evaluate its efficiency in daily clinical practice.

2   Methods

SkullEngine is a coarse-to-fine CNN-based framework that consists of two stages: first coarse then refinement (Fig. 1). In Sect. 2.1, we describe the coarse stage, in which a scalable JSD model is developed for coarse segmentation and global landmark detection (i.e., bony and facial landmarks). In Sect. 2.2, we describe the refinement stage, in which two segmentation models and one landmark detection model are developed for bone segmentation refinement and local tooth landmark detection, respectively.

2.1   Scalable JSD Model for Coarse Segmentation and Global Landmark Detection

Existing JSD models [7,16] are non-scalable due to their ad-hoc network design for predefined segmentation and landmark detection tasks using different loss functions. With the breakthrough of U-Net [1,12], as well as its variants [4,9], it is now possible to use U-Net as a building block to develop a scalable JSD model for unifying the segmentation and landmark detection tasks by taking

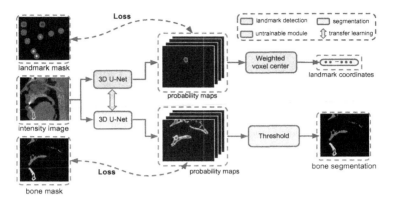

Fig. 2. The training and inference framework for the scalable JSD model. The segmentation model and landmark detection model can be trained as a unified voxel-classification task. In the picture, we only show one segmentation model and one detection model. The transfer learning arrow between the two models means one model can be initialized by the weight from the other model instead of training from scratch.

both of them as a voxel-classification task. In addition, if the segmentation and landmark detection models have the same network structure, we can use transfer learning techniques [2] to train the models more efficiently. Inspired by these ideas, we propose a simple and scalable JSD model (Fig. 2). For segmentation, the ground truth is a voxel-level multi-class mask (Fig. 2 bottom-left). For landmark detection, the ground truth is also a voxel-level landmark mask (Fig. 2 top-left), which are generated from landmark-level coordinates. We generate the landmark mask by assigning voxels to the label of a given landmark (e.g., the i-th label represents the i-th landmark) if these voxels belonging to the neighborhood of that landmark. We define the neighborhood of each landmark as a sphere with a predefined radius, which is a hyper-parameter we empirically set to 3 voxels.

Construction of the Scalable JSD Model. Our JSD model consists of a segmentation model and two landmark detection models. The segmentation model is used for coarse bone segmentation of the midface and the mandible, whereas the two landmark detection models are used for detecting 66 bony and 41 facial global landmarks, respectively. However, the tooth landmark detection model is not included in the scalable JSD because the resolution of the down-sampled volume is too coarse to achieve any meaningful results.

Training and Inference. In the training phase, the input images are down-sampled to a fixed resolution (e.g., 2.0 mm^3 in our experiments) for training both the segmentation and landmark detection models. The segmentation model is first trained from scratch. The two landmark detection models are then initialized and fine-tuned using the parameters from the segmentation model. All are trained using the Focal loss function [8]. In the inference phase, the input image

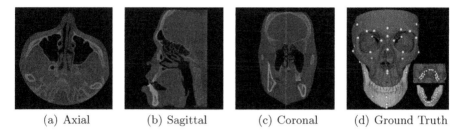

(a) Axial	(b) Sagittal	(c) Coronal	(d) Ground Truth

Fig. 3. An example of CBCT and the ground-truth landmarks. Red label represents the midface, and green represents the mandible. (a) Axial, (b) sagittal, and (c) coronal views, and (d) 3D view of the bony and tooth landmarks marked by spheres.

is down-sampled to the same resolution as in the training phase. We then run the three models independently for inference.

2.2 Bone Segmentation Refinement and Local Landmark Detection

The refinement stage aims to refine the bone segmentation results based on the coarse segmentation results and to detect the tooth landmarks that are undetectable in the coarse stage. As shown in Fig. 1 (right), the original image is cropped based on the coarse mask and global landmarks. Volume (a) in Fig. 1 is a global volume from cropped from the original image that contains the whole skull. The global refinement model uses a patch-based training and inference method for high-resolution segmentation in volume (a) (e.g., by cropping patches with a resolution of 0.4 mm³). Volume (b) in Fig. 1 shows a volume with thin facial bone. In fact, we need to crop two such volumes (for both the left and right facial bones) from the CBCT image using the paired right and left bony landmarks in the thin bone region as centers. Volume (c) in Fig. 1 is the tooth volume that is cropped also based on the bony landmarks. We train a tooth landmark detection model only in this region with a relative higher resolution (e.g., 0.8 mm³). During inference, we first crop the tooth volume that has the same size and resolution as the training patch based on the already detected bony landmarks, and then feed the cropped volume for tooth landmark detection. Finally, the segmentation mask obtained by the global and local refinement models are merged and zero-padded to the original size.

3 Experiments and Results

3.1 Materials

We evaluated the effectiveness of the proposed framework using 92 CBCT and 78 CT scans, which were randomly selected from our digital archive of 170 patients who had already undergone surgery in treating their CMF deformities (Table 1).

Table 1. Summary of the CBCT/CT dataset.

	CBCT/CT Dataset
Number of scans	170 (CBCT: 92, CT: 78)
Dataset splitting	Training 70%, validation 10%, testing 20%
Median spacing (mm^3)	0.39 × 0.39 × 1.0
Median size	512 × 512 × 418
Manual segmentations	Midface, mandible (both including teeth)
Number of landmarks per scan	175 (bone: 66, teeth: 68, face: 41)

The study was approved by Institutional Review Board (Pro00013802). All personal information were de-identified prior to the study. Following clinical standard of care protocol, experienced CMF surgeons used currently available tools to generate segmentation masks and annotate landmark for all 170 scans as ground truth [11]. The segmentation labels included two bony masks: midface and mandible. The 175 landmarks included 66 on both midface and mandible, 68 on both upper and lower teeth, and 41 on facial soft tissues. The average time of creating ground truth was 12 h for each set of CBCT and 5 h for each set of CT. Figure 3 shows a random example of CBCT scan and its ground truth.

We randomly divided the dataset into three groups: 119 scans (70%) for training, 17 (10%) for validation, and 34 (20%) for testing. We applied a stratified sampling strategy to ensure each group included a balanced portion of CBCT and CT scans. The segmentation and landmark detection tasks were completed using our proposed coarse-to-fine SkullEngine framework. Finally, the archived results were compared with two state-of-the-art competing methods: 3D U-Net [1] and its upgraded variant based on PointRend [5]. We implemented the two methods for both segmentation and landmark detection tasks.

During the evaluation, the computational speed was calculated, starting from the input of CBCT image into SkullEngine until the output of the segmentation masks and landmark coordinates was completed. For segmentation task, we used Dice similarity coefficient (DSC), sensitivity (SEN), and positive prediction value (PPV) to evaluate the segmentation accuracy using means and standard deviations (SDs). For landmark detection task, we used root mean squared error (RMSE) and true positive rate (TPR) as metrics to evaluate landmark detection accuracy.

3.2 Results

For each testing dataset, both segmentation and landmark detection tasks were completed simultaneously within 3 min regardless of CBCT or CT. In contrast, a trained operator spent ∼12 h for a set of CBCT and ∼5 h for CT when the current clinical standard method was used. It indicates SkullEngine has a high degree of efficiency.

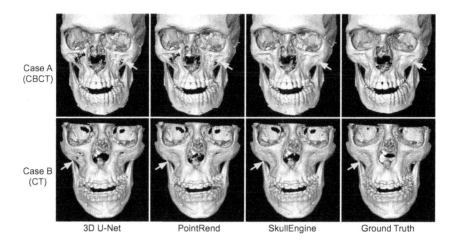

Fig. 4. Comparison of segmentation results for two randomly selected cases. The thin bone areas are pointed by the yellow arrows. (Color figure online)

Table 2 shows the comparison results of segmentation accuracy among the three methods. The results clearly show that our SkullEngine is superior. Figure 4 shows 2 randomly selected examples. Note that in the thin bone areas are often misidentified as holes due to the low bony density. Since SkullEngine has a refinement model for segmentation, it is capable of solving the "hole" problem in the thin bone regions, unlike the other two competing methods. Table 3 shows the comparison results of landmark detection accuracy between the 3 methods. The results clearly show that our SkullEngine is more accurate. Figure 5 (left) shows the detection results of SkullEngine on bones, teeth, and facial soft-tissue landmarks, respectively. Figure 5 (right) shows the error distribution of all 175 landmarks.

Table 2. Comparison results of bone segmentation (mean±SD).

Method	Holes on thin bone	Midface			Mandible		
		DSC (%)	SEN (%)	PPV (%)	DSC (%)	SEN (%)	PPV (%)
3D U-Net [1]	Yes	87.9 ± 7.4	83.2 ± 9.5	89.2 ± 7.9	89.4 ± 4.1	91.7 ± 7.1	93.7 ± 4.9
PointRend [5]	Yes	88.3 ± 7.2	84.7 ± 8.9	88.9 ± 8.5	92.4 ± 3.4	91.2 ± 7.4	94.3 ± 5.1
SkullEngine	No	**88.5 ± 6.9**	**85.3 ± 9.3**	**91.8 ± 7.3**	**93.5 ± 3.4**	**92.2 ± 6.1**	**95.1 ± 4.5**

3.3   Implementation Details

We conducted our experiments on a standard workstation equipped with Intel dual-Xeon E5 CPUs, and a single NVidia Titan XP GPU with 12 GB memory. The SkullEngine was implemented and trained using Python 3.7 and Pytorch 1.7.

Table 3. Comparison results of landmark detection (mean ± SD).

Method	Bony landmarks		Tooth landmarks		Facial landmarks	
	RMSE	TPR (%)	RMSE	TPR (%)	RMSE	TPR (%)
3D U-Net [1]	3.17 ± 1.79	96.7 ± 3.1	2.58 ± 3.03	97.3 ± 4.1	3.46 ± 3.31	96.4 ± 4.2
PointRend [5]	3.23 ± 2.10	95.0 ± 4.7	2.36 ± 2.96	97.4 ± 3.9	3.28 ± 3.15	97.3 ± 3.7
SkullEngine	**3.03 ± 1.96**	**98.5 ± 2.5**	**2.10 ± 2.89**	**98.6 ± 3.5**	**3.34 ± 3.20**	**97.5 ± 3.9**

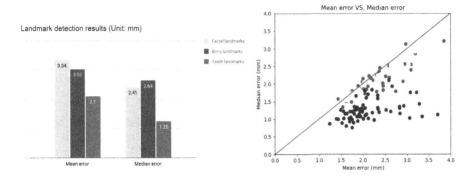

Fig. 5. Results of landmark detection of all 175 landmarks using SkullEngine, including 66 bony, 68 teeth, and 41 facial landmarks.

4 Conclusion

In this work, we have proposed a clinical practical framework for high-quality segmentation and large-scale landmark detection from skull CBCT/CT scans. Our multi-stage framework, the SkullEngine, collaboratively integrates segmentation and landmark detection models to maximize the overall performance. The experimental results showed its superior accuracy when compared to the state-of-the-art methods. The results also showed a significant reduction in labor and time spent on CBCT data preparation from 12 h to less than 3 min with high-quality segmentation results, highlighting the practical value of SkullEngine. Currently, we have integrated SkullEngine into our clinical workflow for further evaluating its clinical efficiency.

Acknowledgement. This work was supported in part by United States National Institutes of Health (NIH) grants R01 DE022676, R01 DE027251, and R01 DE021863.

References

1. Çiçek, Ö., Abdulkadir, A., Lienkamp, S.S., Brox, T., Ronneberger, O.: 3D U-Net: learning dense volumetric segmentation from sparse annotation. In: Ourselin, S., Joskowicz, L., Sabuncu, M.R., Unal, G., Wells, W. (eds.) MICCAI 2016. LNCS, vol. 9901, pp. 424–432. Springer, Cham (2016). https://doi.org/10.1007/978-3-319-46723-8_49

2. Dodge, J., Ilharco, G., Schwartz, R., Farhadi, A., Hajishirzi, H., Smith, N.: Fine-tuning pretrained language models: weight initializations, data orders, and early stopping. arXiv preprint arXiv:2002.06305 (2020)
3. Gupta, A., Kharbanda, O.P., Sardana, V., Balachandran, R., Sardana, H.K.: A knowledge-based algorithm for automatic detection of cephalometric landmarks on CBCT images. Int. J. Comput. Assist. Radiol. Surg. 10(11), 1737–1752 (2015). https://doi.org/10.1007/s11548-015-1173-6
4. Isensee, F., Petersen, J., Kohl, S.A.A., Jäger, P.F., Maier-Hein, K.H.: nnU-Net: breaking the spell on successful medical image segmentation. arXiv preprint arXiv:1904.08128 (2019)
5. Kirillov, A., Wu, Y., He, K., Girshick, R.: PointRend: image segmentation as rendering. arXiv preprint arXiv:2012.08193v2 (2020)
6. Lang, Y., et al.: Automatic localization of landmarks in craniomaxillofacial CBCT images using a local attention-based graph convolution network. In: Martel, A.L., et al. (eds.) MICCAI 2020. LNCS, vol. 12264, pp. 817–826. Springer, Cham (2020). https://doi.org/10.1007/978-3-030-59719-1_79
7. Lian, C., et al.: Multi-task dynamic transformer network for concurrent bone segmentation and large-scale landmark localization with dental CBCT. In: Martel, A.L., et al. (eds.) MICCAI 2020. LNCS, vol. 12264, pp. 807–816. Springer, Cham (2020). https://doi.org/10.1007/978-3-030-59719-1_78
8. Lin, T.Y., Goyal, P., Girshick, R., He, K., Dollar, P.: Focal loss for dense object detection. In: IEEE International Conference on Computer Vision (ICCV), pp. 2999–3007 (2017)
9. Milletari, F., Navab, N., Ahmadi, S.A.: V-Net: fully convolutional neural networks for volumetric medical image segmentation. In: 2016 Fourth International Conference on 3D Vision (3DV), pp. 565–571. IEEE (2016)
10. Minnema, J., et al.: Segmentation of dental cone-beam CT scans affected by metal artifacts using a mixed-scale dense convolutional neural network. Med. Phys. 46(11), 5027–5035 (2019)
11. Xia, J.J., et al.: Algorithm for planning a double-jaw orthognathic surgery using computer-aided surgical simulation (CASS) protocol: 2. 3D cephalometry. Int. J. Oral. Maxillofac. Surg. 44(12), 1441–50 (2015)
12. Ronneberger, O., Fischer, P., Brox, T.: U-Net: convolutional networks for biomedical image segmentation. In: Navab, N., Hornegger, J., Wells, W.M., Frangi, A.F. (eds.) MICCAI 2015. LNCS, vol. 9351, pp. 234–241. Springer, Cham (2015). https://doi.org/10.1007/978-3-319-24574-4_28
13. Shahidi, S., et al.: The accuracy of a designed software for automated localization of craniofacial landmarks on CBCT images. BMC Med. Imaging 14(1), 1–8 (2014)
14. Torosdagli, N., Liberton, D.K., Verma, P., Sincan, M., Lee, J.S., Bagci, U.: Deep geodesic learning for segmentation and anatomical landmarking. IEEE Trans. Med. Imaging 38(4), 919–931 (2018)
15. Wang, L., et al.: Automated segmentation of dental CBCT image with prior-guided sequential random forests. Med. Phys. 43(1), 336–346 (2016)
16. Zhang, J., et al.: Context-guided fully convolutional networks for joint craniomaxillofacial bone segmentation and landmark digitization. Med. Image Anal. 60, 101621 (2020)
17. Zheng, Y., Liu, D., Georgescu, B., Nguyen, H., Comaniciu, D.: 3D deep learning for efficient and robust landmark detection in volumetric data. In: Navab, N., Hornegger, J., Wells, W.M., Frangi, A.F. (eds.) MICCAI 2015. LNCS, vol. 9349, pp. 565–572. Springer, Cham (2015). https://doi.org/10.1007/978-3-319-24553-9_69

Skull Segmentation from CBCT Images via Voxel-Based Rendering

Qin Liu[1], Chunfeng Lian[1], Deqiang Xiao[1], Lei Ma[1], Han Deng[2], Xu Chen[1], Dinggang Shen[1], Pew-Thian Yap[1(✉)], and James J. Xia[2(✉)]

[1] Department of Radiology and Biomedical Research Imaging Center (BRIC), University of North Carolina at Chapel Hill, Chapel Hill, NC, USA
qinliu19@email.unc.edu, ptyap@med.unc.edu
[2] Department of Oral and Maxillofacial Surgery, Houston Methodist Hospital, Houston, TX, USA
jxia@houstonmethodist.org

Abstract. Skull segmentation from three-dimensional (3D) cone-beam computed tomography (CBCT) images is critical for the diagnosis and treatment planning of the patients with craniomaxillofacial (CMF) deformities. Convolutional neural network (CNN)-based methods are currently dominating volumetric image segmentation, but these methods suffer from the limited GPU memory and the large image size (*e.g.*, 512 × 512 × 448). Typical ad-hoc strategies, such as down-sampling or patch cropping, will degrade segmentation accuracy due to insufficient capturing of local fine details or global contextual information. Other methods such as Global-Local Networks (GLNet) are focusing on the improvement of neural networks, aiming to combine the local details and the global contextual information in a GPU memory-efficient manner. However, all these methods are operating on regular grids, which are computationally inefficient for volumetric image segmentation. In this work, we propose a novel VoxelRend-based network (VR-U-Net) by combining a memory-efficient variant of 3D U-Net with a voxel-based rendering (VoxelRend) module that refines local details via voxel-based predictions on non-regular grids. Establishing on relatively coarse feature maps, the VoxelRend module achieves significant improvement of segmentation accuracy with a fraction of GPU memory consumption. We evaluate our proposed VR-U-Net in the skull segmentation task on a high-resolution CBCT dataset collected from local hospitals. Experimental results show that the proposed VR-U-Net yields high-quality segmentation results in a memory-efficient manner, highlighting the practical value of our method.

Keywords: VoxelRend · CBCT image · High-resolution segmentation

1 Introduction

Segmentation of high-resolution cone-beam computed tomography (CBCT) images plays an essential role in generating three-dimensional (3D) models in the diagnosis and treatment planning for patients with craniomaxillofacial (CMF)

© Springer Nature Switzerland AG 2021
C. Lian et al. (Eds.): MLMI 2021, LNCS 12966, pp. 615–623, 2021.
https://doi.org/10.1007/978-3-030-87589-3_63

deformities [13,14]. Convolutional neural network (CNN)-based methods are currently dominating the 3D medical image segmentation. Among these CNNs, U-Net [12] is a breakthrough and has evolved to a commonly used benchmark in various segmentation challenges [1,7,11,16]. However, CNN-based volumetric segmentation often requires massive memory of graphics processing units (GPUs) to process large image size dataset (e.g., millions of voxels). On the other hand, typical ad-hoc strategies, e.g., down-sampling [15] or patch cropping [4], can reduce the GPU memory usage but they also degrade segmentation quality due to insufficient capture of local fine details or global contextual information. Therefore, it is challenging to achieve a well-balanced trade-off between the segmentation accuracy and the GPU memory usage for CBCT images.

To tackle with this challenge, some related methods have been proposed in recent years. Data-swapping method [6] was proposed to swap the feature maps from GPU to central processing unit (CPU) memory. After the forward propagation process was finished, the feature maps were swapped back to GPU memory from CPU before executing the layer in the backward propagation process. However, this method requires additional overhead due to frequent swap computation. The fast segmentation methods like the Global-Local Network [2] is composed of a global branch and a local branch, taking the entire down-sampled image and its cropped local patches as respective inputs. With less than 2 GB GPU memory used, GLNet can yield decent segmentation results for ultra-high resolution images. The 3D Regions of Interest (RoI)-aware U-Net (3D RU-Net) [5] consists of a global image encoder and a GPU memory-efficient local decoder, performing a global-to-local multi-task learning procedure for joint RoI localization and in-region segmentation where the two tasks share one backbone encoder network. However, all the previous proposed neural networks operate on regular grids, which is convenient but not computationally efficient for image segmentation, in which most of the computation will unnecessarily focus on the low-frequency regions, e.g., the background, instead of the high-frequency regions, e.g., sparse boundaries between objects.

More recently, a method of treating image segmentation as a rendering problem was proposed in [9] for high-quality segmentation of 2D images. From this perspective, a fine-resolution segmentation can be rendered from a coarse-resolution segmentation by performing point-wise predictions on certain locations with high prediction uncertainties. In this work, we propose to extend this general perspective in the domain of the CBCT segmentation. Specifically, we propose a novel neural network, VoxelRend-based U-Net (VR-U-Net), which integrates a GPU memory-efficient variant of 3D U-Net with a voxel-based rendering (VoxelRend) module for high-resolution volumetric segmentation. The VoxelRend module is a flexible and slight-weight neural network derived from PointRend [9] to perform efficient voxel-wise predictions at challenging voxel locations. The proposed VR-U-Net was evaluated in the segmentation of multi-class bony structures on a high-resolution CMF CBCT dataset. It is challenging to segment the CBCT images due to their high-resolution (up to 100 million voxels) and the wild spread artifacts caused by noise, beam hardening, and inhomogeneity. With only one single Titan XP GPU and less than 4 GB memory used, our VR-U-Net yields higher segmentation quality and achieves much more competitive accuracy memory usage trade-offs compared to state-of-the-arts. The experimental results have revealed that our proposed method has the huge potential for extension towards other volumetric segmentation tasks due to its inherent generalizability. Therefore, we make the code publicly available to encourage more researches on this topic.

2   3D VoxelRend-Based U-Net

In this section, we start by analyzing computation memory usage of the 3D U-Net [3] for binary segmentation with experimental statistics. Then, we present the network structure of the proposed VR-U-Net, which consists of a GPU memory-efficient variant of 3D U-Net serving as the backbone and a VoxelRend module established upon the backbone network. A brief description of the training and inference phase of the VR-U-Net is also described at the end of this section.

2.1   Memory Analysis of 3D U-Net

The memory issue of training and inference for 3D U-Net [3] is exacerbated by the encoder-decoder framework of 3D U-Net. The encoder part of 3D U-Net compresses the input image into low-resolution feature maps, while the decoder part decompresses these low-resolution feature maps until their original resolution is reached. In the training phase, all feature maps need to be retained for the backward computation. In the inference phase, the feature maps in the encoder part need to be retrained in the forward computation due to the skip connections between the encoder and decoder. Table 1 summarizes the experimental statistics on memory consumption of each layer of the 3D U-Net for binary segmentation. L-Stage 1 denotes the top layer of the encoder, and L-Stage 1 denotes the top layer of the decoder. From this table, we observe that the memory consumption of R-Stage 1 takes up more than one third of the overall memory consumption (marked in bold), and the memory consumption of R-Stage 1 together with that of the R-Stage 2 take up more than half of the overall memory consumption. Therefore, we build a GPU memory-efficient variant of 3D U-Net by discarding both the R-Stage 1 layer and R-Stage 2 layers in the 3D U-Net so that the R-Stage 3 layer is directly followed by the output layer, as shown in Fig. 1.

2.2   Network Architecture

Figure 1 shows the architecture of our proposed VR-U-Net, which consists of two modules. The first module is a GPU memory-efficient variant of 3D U-Net [3] serving as the backbone network, and the other one is a VoxelRend module established on the multi-layer feature maps of the backbone network. The backbone network is modified from 3D U-Net by discarding the top two layers of the decoder of 3D U-Net due to their heavy memory consumption.

Table 1. Summary of memory usage in each layer of 3D U-Net.

Layer	Input size	Output size	#Channels	#Conv.	Memory (%)
L-Stage 1	$128 \times 128 \times 64$	$64 \times 64 \times 32$	16	1	18.7
L-Stage 2	$64 \times 64 \times 32$	$32 \times 32 \times 16$	32	2	9.36
L-Stage 3	$32 \times 32 \times 16$	$16 \times 16 \times 8$	64	3	3.51
L-Stage 4	$16 \times 16 \times 8$	$8 \times 8 \times 4$	128	3	0.88
L-Stage 5	$8 \times 8 \times 4$	$16 \times 16 \times 8$	256	3	0.22
R-Stage 4	$16 \times 16 \times 8$	$32 \times 32 \times 16$	256	3	1.76
R-Stage 3	$32 \times 32 \times 16$	$64 \times 64 \times 32$	128	3	7.02
R-Stage 2	$64 \times 64 \times 32$	$128 \times 128 \times 64$	64	2	18.7
R-Stage 1	$128 \times 128 \times 64$	$128 \times 128 \times 64$	32	1	**37.5**
Output	$128 \times 128 \times 64$	$128 \times 128 \times 64$	2	1	2.34

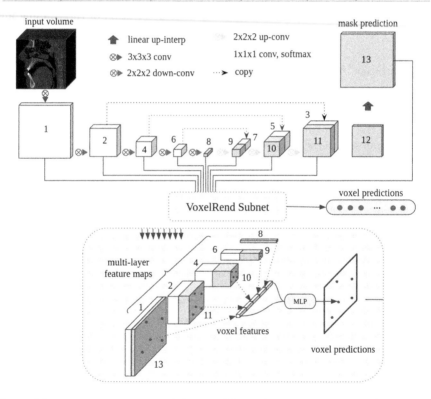

Fig. 1. Schematic representation of 3D VR-U-Net. The network consists of a backbone network and a VoxelRend module. The backbone network is a GPU memory-efficient variant of 3D U-Net that yields a mask prediction at a coarser resolution than the input volume. The VoxelRend module refines the coarse segmentation by performing voxel-wise predictions at adaptively selected locations (red dots) using a small MLP. The MLP uses interpolated features computed at these voxels (dashed arrows) from multi-layer feature maps generated by the backbone. (Color figure online)

Therefore, the output layer is directly following the third layer of the decoder part to produce the mask prediction at a relatively coarse resolution. The feature maps marked as 12 in the Fig. 1 represent the probability maps of the mask predictions. These probability maps will up-sampled to the original resolution using trilinear interpolation, which is memory-efficient due to no convolutional operations. The loss function of the backbone module is computed based on the up-sampled segmentation results. In order to improve the segmentation quality, the VoxelRend selects a set of voxels (red dots) and makes prediction for each voxel independently with a small multi-layer perceptron (MLP). This MLP uses interpolated features computed at these voxels (dashed arrows) from the multi-layer feature maps provided by the backbone network.

Similar to PointRend [9], the VoxelRend module consists of three components: 1) A *voxel selection sub-module* selects a few of real-value voxels to make predictions on. The voxels with higher prediction uncertainty (*e.g.*, those with probabilities more closer to 0.5 for binary segmentation) are more likely to be selected. 2) A *voxel-wise feature extraction sub-module* computes feature representation for each selected voxel. The features for a real-value voxel is computed by trilinear interpolation of the feature maps. 3) A *voxel prediction sub-module* is trained to predict the class of all the selected voxels in one pass based on their feature representations.

2.3   Training and Inference

In the training phase, the backbone module and the VoxelRend module are learned jointly. The backbone module takes in the training volumes, and performs coarse mask prediction, which will be up-sampled to the original resolution for loss computation. The VoxelRend model performs voxel-wise predictions at locations where the predictions from the backbone module are ambiguous. For example, a voxel is more likely to be selected if its prediction probability is more closer to 0.5 for binary segmentation. Therefore, the loss function of the proposed VR-U-Net consists of the mask prediction loss L_m, and the voxel predictions loss L_v. The overall loss function L is the linear combination of L_m and L_v, which is formulated in Eq. 1.

$$L = \alpha L_m + (1 - \alpha) L_v, \alpha \in [0, 1] \tag{1}$$

Both L_m and L_v are implemented with Focal Loss [10] in this work. In the inference phase, the mask prediction is conducted first to obtain the coarse segmentation. Then, voxel predictions is performed on the N most uncertain voxels in the mask prediction. The number of N is a hyper-parameter. In our implementation, we select 5% of total voxels in a given volume.

3   Experiments and Results

3.1   Dataset

The effectiveness of the proposed VR-U-Net was evaluated using 92 sets of CBCT and 78 sets of CT images of normal subjects and patients with CMF deformities

(Table 2). The study was approved by Institutional Review Board, and all personal information was de-identified. Images were divided into 2 groups: 136 for training and 34 for test images using stratified sampling. All images in the dataset were acquired for orthographic surgical planning, and thus suffer from the clinical variability encountered in clinical settings. An example of the typical volume from the dataset is shown in Fig. 2.

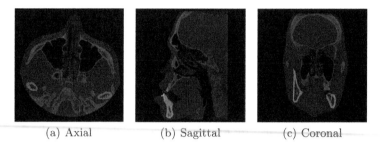

(a) Axial   (b) Sagittal   (c) Coronal

Fig. 2. An example CBCT image in our dataset with corresponding manual segmentation. Left: Axial view. Middle: Sagittal view. Right: Coronal view. The label in red is the midface, and the label in green is the mandible. (Color figure online)

Table 2. Summary of the dataset. All images were acquired once per subject.

	Dataset
Number of scans	170
Subject type	Normal (26), patient (144)
Median spacing (mm^3)	0.39 × 0.39 × 1.0
Median resolution	512 × 512 × 418
Manual segmentations	Midface, mandible

3.2 Accuracy and Memory Usage Comparison

Table 3 shows the comparison results between the proposed VR-U-Net and the 3D U-Net [3] in terms of segmentation accuracy and memory usage in the inference phase. To investigate the effectiveness of the proposed VoxelRend module, we removed it from the VR-U-Net to obtain a degraded version of VR-U-Net for reference. The experimental results of this degraded VR-U-Net were also included in the Table 3. Our network enabled the largest input patch size of 224 × 224 × 224, which was impractical for 3D U-Net [3]. To thoroughly compare the performance of our proposed method with the state-of-the-art 3D U-Net, we set different groups of experiment settings in terms of patch size and patch spacing.

Table 3. The comparison of accuracy and memory usage. The VoxelRend module is removed in the degraded VR-U-Net.

Model	Patch size	Spacing (mm^3)	Avg. Dice (%)		Memory (MB)
			Midface	Mandible	
3D U-Net	128^3	0.4^3	87.6 ± 7.1	89.4 ± 4.1	2607
	128^3	0.8^3	87.9 ± 6.4	89.3 ± 5.3	2607
	224^3	0.4^3	–	–	Out of memory
	224^3	0.8^3	–	–	Out of memory
Degraded VR-U-Net	128^3	0.4^3	84.8 ± 5.2	88.2 ± 5.9	1165
	128^3	0.8^3	84.4 ± 6.8	87.9 ± 5.4	1165
	224^3	0.4^3	85.0 ± 5.8	88.9 ± 5.7	3599
	224^3	0.8^3	84.7 ± 6.5	88.7 ± 6.4	3599
VR-U-Net	128^3	0.4^3	$\mathbf{88.3 \pm 6.6}$	92.2 ± 4.1	1165
	128^3	0.8^3	86.8 ± 4.0	90.8 ± 6.7	1165
	224^3	0.4^3	87.9 ± 4.8	$\mathbf{92.4 \pm 3.4}$	3599
	224^3	0.8^3	87.3 ± 6.4	91.3 ± 3.7	3599

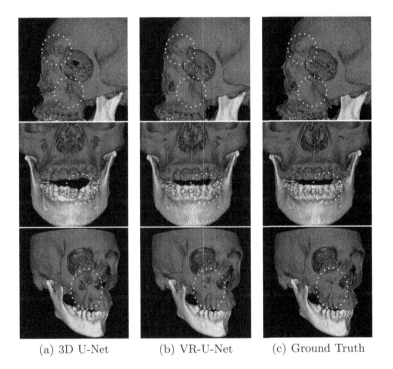

(a) 3D U-Net   (b) VR-U-Net   (c) Ground Truth

Fig. 3. Illustration of the performance of VoxelRend module. Each row represents a test case. (a) Segmentation results of 3D U-Net. (b) Segmentation results of VR-U-Net. (c) Manual segmentation.

The Dice Similarity Coefficient (DSC) is used in our experiments as a measurement for the segmentation accuracy. The experimental results showed that by adding the VoxelRend module, the segmentation accuracy of VR-U-Net was improved. Figure 3 showed the improvements on three test cases. The experimental results also show that VR-U-Net achieved higher segmentation accuracy than the 3D U-Net using only half of the memory consumption compared with that of 3D U-Net.

Our experiments and models were implemented using Python 3.7 and Pytorch 1.4, with a standard workstation equipped with two Intel Xeon E5-2620 v4 CPUs working at 2.10 GHz, and a single NVidia Titan XP GPU with 12 GB memory. For each experiment, the model was trained for 2000 epochs using the Adam optimizer [8] with a momentum of 0.99 and an initial learning rate of 0.0001. We set the α in Eq. 1 to 0.5 by default. During each training iteration, the patches in the mini-batch were randomly cropped from the image domain with Z-score normalization. All inference was done patch-based as well that patches were cropped from the original image with no overlap. The average inference time for each experiment was less than 3 min.

4   Conclusion

In this work, we proposed a novel neural network for skull segmentation from high-resolution CBCT images. Our proposed network integrates a GPU memory-efficient variant of 3D U-Net with a VoxelRend module that performs voxel-based predictions on adaptively selected locations. We evaluated the proposed network on a challenging high-resolution CBCT image dataset. The experimental results show that our method yields high-quality segmentation results in a memory-efficient manner, highlighting the practical value of our method.

Acknowledgement. This work was supported in part by United States National Institutes of Health (NIH) grants R01 DE022676, R01 DE027251, and R01 DE021863.

References

1. Brügger, R., Baumgartner, C.F., Konukoglu, E.: A partially reversible U-Net for memory-efficient volumetric image segmentation. In: Shen, D., et al. (eds.) MICCAI 2019. LNCS, vol. 11766, pp. 429–437. Springer, Cham (2019). https://doi.org/10.1007/978-3-030-32248-9_48
2. Chen, W., Jiang, Z., Wang, Z., Cui, K., Qian, X.: Collaborative global-local networks for memory-efficient segmentation of ultra-high resolution images. arXiv preprint arXiv:1905.06368v2 (2019)
3. Çiçek, Ö., Abdulkadir, A., Lienkamp, S.S., Brox, T., Ronneberger, O.: 3D U-Net: learning dense volumetric segmentation from sparse annotation. In: Ourselin, S., Joskowicz, L., Sabuncu, M.R., Unal, G., Wells, W. (eds.) MICCAI 2016. LNCS, vol. 9901, pp. 424–432. Springer, Cham (2016). https://doi.org/10.1007/978-3-319-46723-8_49

4. Cui, Z., Yang, J., Qiao, Y.: Brain MRI segmentation with patch-based CNN approach. In: 2016 35th Chinese Control Conference (CCC), pp. 7026–7031. IEEE (2016)
5. Huang, Y.J., et al.: 3D RoI-aware U-Net for accurate and efficient colorectal tumor segmentation. arXiv preprint arXiv:1806.10342v5 (2019)
6. Imai, H., Matzek, S., Le, T.D., Negishi, Y., Kawachiya, K.: High resolution medical image segmentation using data-swapping method. In: Shen, D., et al. (eds.) MICCAI 2019. LNCS, vol. 11766, pp. 238–246. Springer, Cham (2019). https://doi.org/10.1007/978-3-030-32248-9_27
7. Isensee, F., Petersen, J., Kohl, S.A.A., Jäger, P.F., Maier-Hein, K.H.: nnU-Net: breaking the spell on successful medical image segmentation. arXiv preprint arXiv:1904.08128 (2019)
8. Kingma, D.P., Ba, J.L.: Adam: a method for stochastic optimization. arXiv preprint arXiv:1412.6980 (2014)
9. Kirillov, A., Wu, Y., He, K., Girshick, R.: PointRend: image segmentation as rendering. arXiv preprint arXiv:2012.08193v2 (2020)
10. Lin, T.Y., Goyal, P., Girshick, R., He, K., Dollár, P.: Focal loss for dense object detection. In: IEEE International Conference on Computer Vision (ICCV), pp. 2999–3007 (2017)
11. Milletari, F., Navab, N., Ahmadi, S.A.: V-Net: fully convolutional neural networks for volumetric medical image segmentation. In: 2016 Fourth International Conference on 3D Vision (3DV), pp. 565–571. IEEE (2016)
12. Ronneberger, O., Fischer, P., Brox, T.: U-Net: convolutional networks for biomedical image segmentation. In: Navab, N., Hornegger, J., Wells, W.M., Frangi, A.F. (eds.) MICCAI 2015. LNCS, vol. 9351, pp. 234–241. Springer, Cham (2015). https://doi.org/10.1007/978-3-319-24574-4_28
13. Wang, L., et al.: Automated bone segmentation from dental CBCT images using patch-based sparse representation and convex optimization. Med. Phys. 41, 043503 (2014)
14. Xia, J., et al.: Algorithm for planning a double-jaw orthognathic surgery using a computer-aided surgical simulation (CASS) protocol. Part 1: planning sequence. Int. J. Oral Maxillofac. Surg. 44(12), 1431–1440 (2015)
15. Yu, C., Wang, J., Peng, C., Gao, C., Yu, G., Sang, N.: BiSeNet: bilateral segmentation network for real-time semantic segmentation. In: Proceedings of the European conference on computer vision (ECCV), pp. 325–341 (2018)
16. Zhou, Z., Rahman Siddiquee, M.M., Tajbakhsh, N., Liang, J.: UNet++: a nested U-Net architecture for medical image segmentation. In: Stoyanov, D., et al. (eds.) DLMIA/ML-CDS -2018. LNCS, vol. 11045, pp. 3–11. Springer, Cham (2018). https://doi.org/10.1007/978-3-030-00889-5_1

Alzheimer's Disease Diagnosis via Deep Factorization Machine Models

Raphael Ronge[1], Kwangsik Nho[2], Christian Wachinger[1]🆔,
and Sebastian Pölsterl[1(✉)]🆔

[1] Artificial Intelligence in Medical Imaging (AI-Med), Department of Child
and Adolescent Psychiatry, Ludwig-Maximilians-Universität, Munich, Germany
sebastian.poelsterl@med.uni-muenchen.de
[2] Department of Radiology and Imaging Sciences, and The Indiana Alzheimer's
Disease Research Center, Indiana University School of Medicine,
Indianapolis, IN, USA

Abstract. The current state-of-the-art deep neural networks (Dnns)
for Alzheimer's Disease diagnosis use different biomarker combinations
to classify patients, but do not allow extracting knowledge about the
interactions of biomarkers. However, to improve our understanding of
the disease, it is paramount to extract such knowledge from the learned
model. In this paper, we propose a Deep Factorization Machine model
that combines the ability of DNNs to learn complex relationships and
the ease of interpretability of a linear model. The proposed model has
three parts: (i) an embedding layer to deal with sparse categorical data,
(ii) a Factorization Machine to efficiently learn pairwise interactions, and
(iii) a DNN to implicitly model higher order interactions. In our experiments on data from the Alzheimer's Disease Neuroimaging Initiative, we
demonstrate that our proposed model classifies cognitive normal, mild
cognitive impaired, and demented patients more accurately than competing models. In addition, we show that valuable knowledge about the
interactions among biomarkers can be obtained.

Keywords: Alzheimer's disease · Biomarkers · Interactions ·
Factorization machines

1 Introduction

Alzheimer's Disease (AD) patients account for 60–80% of all dementia cases [3].
Worldwide, 50 million patients have dementia and their number is estimated
to triple by 2050 [17]. AD is a neurodegenerative disease whose progression is
highly heterogeneous and not yet fully understood [22]. Mild cognitive impairment (MCI) is a pre-dementia stage which results in cognitive decline, but not
to an extent that it impairs patients' daily live [18]. Subjects with MCI are at
an increased risk of developing dementia due to AD, which would make them
completely dependent upon caregivers [18]. This transition is complex and not
yet fully understood. Therefore, research in the last decade focused on identifying biomarkers to infer which stage of the disease a patient is in [11]. Important

© Springer Nature Switzerland AG 2021
C. Lian et al. (Eds.): MLMI 2021, LNCS 12966, pp. 624–633, 2021.
https://doi.org/10.1007/978-3-030-87589-3_64

biomarkers include demographics, brain atrophy measured by magnetic resonance images (MRI), and predispositions due to genetic alterations in the form of single nucleotide polymorphisms (SNPs) (see [22] for a detailed overview).

For accurate patient stratification it is important to also consider the interrelationships between biomarkers and model their interactions. Deep learning techniques excel at implicitly learning complex interactions, but extracting this knowledge is challenging due to their black-box nature [1]. At the other end of the spectrum are linear models that are highly interpretable, but only account for interactions when those are explicitly specified. Hence, approaches that can model complex interactions while preserving interpretability are required to further improve our understanding of the interaction between biomarkers.

In this work, we propose a model that is able to utilize both low- and high-order feature interactions. Our model comprises two parts: (i) a Factorization Machine (FM) that explicitly learns pairwise feature interactions without the need of feature engineering, and (ii) a deep neural network (DNN) that can learn arbitrary low- and high-order feature interactions implicitly. Consequently, our model preserves the best of both worlds: the interpretability of linear models – via the FM – and the discriminatory power of DNNs. In our experiments, we demonstrate that our proposed model outperforms competing methods for classifying healthy controls, patients with MCI, and patients with AD.

2   Related Work

Several existing works study the fusion of multi-modal data for AD diagnosis and the interaction between features. Zhang et al. [33] use MRI, FDG-PET, and biomarkers derived from cerebrospinal fluid (CSF). They use a multiple-kernel SVM that uses one kernel per modality and combines modalities by a weighted sum of modality-specific kernels. This way, interactions can only be addressed implicitly by absorbing them into the sum of kernels and interpretability is lost.

Tong et al. [28] introduce a non-linear graph fusion approach for multimodal AD diagnosis. Their approach can assign an overall importance value to each modality in a manner that scales independently of the number of features. However, this does not allow for modelling interactions between single features. Khatri et al. [12] use an Extreme Learning Machine (ELM) – a single layer feed-forward neural network (NN). They use regional volume and thickness measurements, CSF biomarkers ApoE allele information, and the Mini-Mental State Examination (MMSE) cognitive score. Because of the use of MMSE, their model is not solely based on biological measurements, but includes diagnostic information, which usually is among the variables of interest. Moreover, ELMs are a type of neural network, which makes their interpretation difficult [16,29]. In [30], a "Multimodal deep learning [model] for early detection of Alzheimer's Disease stage" is proposed. They account for the lack of interpretability of DNNs, by running the model multiple times with one feature masked at a time. The sharper the drop in performance, the higher they rank the importance of the masked feature. While this provides a measure of importance on a per-feature

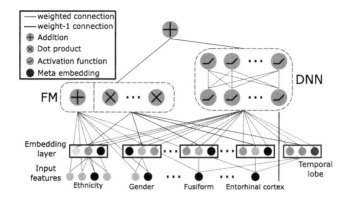

Fig. 1. Overview of the proposed model.

level, it ignores how the model utilizes feature interactions. Ning et al. [15] use a two hidden layer NN to perform AD diagnosis based on MRI-derived features and genetics. They attempt to determine the importance of features and feature interactions via back-propagation based on the partial derivatives method [5]. The authors point out that it remains to be tested how well the computed importance measure reflects the actual prediction computation by their model.

Finally, we want to emphasize that all the existing approaches only study feature interactions by trying to disentangle what the model learned in a post-hoc manner, but not how changes to the model architecture can make the model itself more interpretable.

3   Methods

The proposed model comprises three major parts for improved AD diagnosis (see Fig. 1). The first part is an embedding layer to deal with sparse data [6]. The second part is based on the Factorization Machine [20], which models pairwise feature interactions as an inner product of latent vectors from the embedding layer. The third part is a Deep Neural Network (DNN) that has the potential to implicitly learn complex feature interactions. The combined model is closely related to the DeepFM [6] for click-through-rate prediction. We train our model to differentiate between three groups: AD, MCI and CN patients.

3.1   Embedding Layer

The first layer is an embedding layer, similar to the one in [6]. The layer serves two purposes: First, DNNs are unable to train on sparse data, and second, one feature can span multiple columns if it is e.g. a one-hot encoded categorical feature. Via the embedding, each feature is represented as one dense vector, which leads to a more comprehensible representation. The embedding layer condenses each d-dimensional feature $\mathbf{x}_i$ into a vector $\mathbf{e}_i$ with fixed length m: $\mathbf{e}_i = \mathbf{A}_i \mathbf{x}_i$, where $\mathbf{A}_i \in \mathbb{R}^{m \times d}$ is the learned embedding matrix for feature $i \in \{1, \ldots, n\}$. The embedding vector $\mathbf{e}_i$ represents the entire feature and eliminates the problems that arise when training neural networks on sparse data – in particular, when categorical features with many categories are present, because the embedding layer reduces the dimensionality compared to a one-hot encoding.

Another advantage of the embedding layer is that it can be used to combine sets of features describing whole brain areas. For instance, one can combine the volume measurements of all regions belonging to the temporal lobe into one embedding vector. Therefore, the embedding layer can be used as a mean to incorporate domain knowledge about the structural or functional relationship between features. As AD is a highly heterogeneous disease and brain regions are strongly interrelated, this can improve predictive performance as well as interpretability of interaction effects, which we will discuss next.

3.2 Factorization Machine

The Factorization Machine (FM; [20]) consists of three parts: a bias, a linear predictor, and a pairwise-interaction term (for simplicity, we omit the bias in Fig. 1). For an n-dimensional feature vector $\mathbf{x}$, the FM for class c is defined as:

$$\hat{y}_{\text{FM}}^c(\mathbf{x}) = \underbrace{w_0}_{\text{bias}} + \underbrace{\sum_{i=1}^n w_i x_i}_{\text{linear predictor}} + \underbrace{\sum_{i=1}^n \sum_{j=i+1}^n \langle \mathbf{e}_i, \mathbf{e}_j \rangle x_i x_j}_{\text{interaction term}} \tag{1}$$

The key idea of the FM is to not learn interaction weights explicitly, which would scale quadratically in the number of features, but implicitly through the dot product $\langle \mathbf{e}_i, \mathbf{e}_j \rangle$. Hence, weights are shared across interaction terms and one has to learn n embedding matrices $\mathbf{A}_i$ instead of n^2 weights that would be required for explicit interaction modelling. To preserve the linear complexity of the linear model, while still accounting for pairwise interactions, we reformulate the pairwise-interaction computation as in [20], resulting in a $\mathcal{O}(kn)$ runtime.

3.3 Deep Factorization Machine

So far, our model only accounts for linear and pairwise interactions. We account for high-order interactions implicitly by employing a DNN alongside the FM from above [6] (see Fig. 1). The DNN receives the concatenated embedding vectors $\mathbf{e}_i$, and thus is equipped to learn from high-dimensional sparse data. The DNN contains two hidden layers with ReLU activation function $\sigma(x) = \max(0, x)$:

$$\hat{y}_{\text{DNN}}^c(\mathbf{x}) = \sigma(\mathbf{W}^{(1)} \cdot \sigma(\mathbf{W}^{(0)} \cdot \text{CONCAT}(\mathbf{e}_1, \ldots, \mathbf{e}_n) + \mathbf{b}^{(0)}) + \mathbf{b}^{(1)}), \tag{2}$$

where $\mathbf{W}^{(k)}$ and $\mathbf{b}^{(k)}$ are the weight matrix and bias of the k-th layer, respectively. Finally, the overall prediction for class c of our DeepFM model is:

$$\hat{y}^c(\mathbf{x}) = \text{Softmax}\left(\hat{y}_{\text{FM}}^c(\mathbf{x}) + \hat{y}_{\text{DNN}}^c(\mathbf{x})\right), \tag{3}$$

where $\hat{y}_{\text{FM}}^c$ is the factorization machine defined in Eq. (1). During training, we optimize the weights of the FM part $(w_0, \ldots, w_n)$, the embedding matrices $\mathbf{A}_1, \ldots, \mathbf{A}_n$, which are shared among the FM and deep part of our model, and the parameters of the DNN $(\mathbf{W}^{(0)}, \mathbf{W}^{(1)}, \mathbf{b}^{(0)}, \mathbf{b}^{(1)})$.

Table 1. Overview of the data used in our experiments (MMSE is not used as feature).

Feature	AD-patients			MCI-patients			NC-patients		
	min	mean	max	min	mean	max	min	mean	max
MMSE	10.0	21.9	30.0	10.0	27.5	30.0	20.0	29	30.0
Age	55.0	74.4	90.9	54.4	73.1	91.4	55.0	73.9	90.1
Education (Years)	4.0	15.45	20.0	6.0	16.03	20.0	6.0	16.43	20.0
Gender	58.3 % Male			61.8% Male			52.3 % Male		

4 Experiments

We evaluated the proposed model on data provided by the Alzheimer's Disease Neuroimaging Initiative [10]. Table 1 summarizes the data. Our dataset contains a total of 1492 patients with 6844 visits and three class labels: AD (1536 visits), MCI (3131 visits), cognitive normal (CN; 2177 visits). In addition to demographic data, we collected for each patient MRIs and processed them with FreeSurfer [4] to obtain 20 volume and 34 thickness measurements. Moreover, we collected Amyloid-β (Aβ), Tau, and phosphorylated Tau (pTau) concentration in CSF. Finally, we collected 41 genetic markers, previously shown to be associated with AD and atrophy [8,13], as described in [31]. Except for CSF measurements, which are only available for 1863 visits, each modality is available for all patients. Aβ, Tau, and pTau are important biomarkers and in order to not heavily reduce the available data, we keep them and handle missing values as zero, to get as good as a prediction as possible. In total, we used 109 features.

To avoid data leakage due to confounding effects of age and sex [32], we split the data into 5 non-overlapping folds using only baseline visits such that diagnosis, age and sex are balanced across folds [9]. We used one fold as test set and combined the remaining folds such that 80% of it comprise the training set and 20% the validation set. We extended the training set, but not validation or test, by including each patient's longitudinal data.

We optimized the models' hyperparameters (see Table 2) for each of the five folds separately via Bayesian black-box optimization on the validation set [7]. We compare the proposed DeepFM to a standalone DNN, the FM [20], and a linear logistic regression model that explicitly accounts for all pairwise interactions. Each model is evaluated by the balanced accuracy on the respective test set.

Table 2. Hyperparameter Search Space. $\mathcal{U}$ uniform-/$\mathcal{LU}$ log-uniform-distribution.

Hyperparameter	DeepFM	FM	DNN	Linear
Neurons Layer 1	$\mathcal{U}(1,400)$	–	$\mathcal{U}(1,400)$	–
Neurons Layer 2	$\mathcal{U}(1,400)$	–	$\mathcal{U}(1,400)$	–
Neurons Layer 3	–	–	$\mathcal{U}(0,400)$	–
Length Embedding Vector	$\mathcal{U}(1,20)$	$\mathcal{U}(1,20)$	$\mathcal{U}(1,20)$	–
Learning Rate	$\mathcal{LU}(1e^{-04},0.9)$	$\mathcal{LU}(1e^{-04},0.9)$	$\mathcal{LU}(1e^{-04},0.9)$	$\mathcal{LU}(1e^{-04},0.9)$
L1-Regu. Weight	$\mathcal{LU}(1e^{-04},0.9)$	$\mathcal{LU}(1e^{-04},0.9)$	$\mathcal{LU}(1e^{-04},0.9)$	$\mathcal{LU}(1e^{-04},9)$
L2-Regu. Weight	$\mathcal{LU}(1e^{-04},0.9)$	$\mathcal{LU}(1e^{-04},0.9)$	$\mathcal{LU}(1e^{-04},0.9)$	$\mathcal{LU}(1e^{-04},9)$
Dropout	$\mathcal{LU}(0.1,0.9)$	–	$\mathcal{LU}(0.1,0.9)$	–

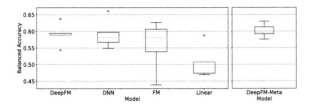

Fig. 2. Balanced accuracy comparison

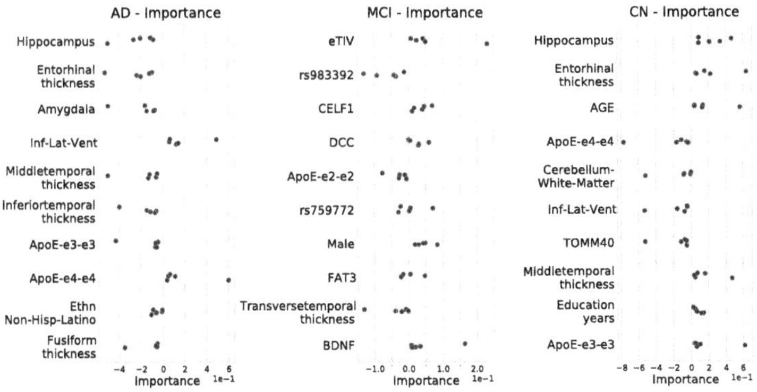

Fig. 3. DeepFM - 10 most important linear features

5   Results and Discussion

Performance. With a median balanced accuracy of 0.589, DeepFM has the highest performance of all models (see Fig. 2). It is slightly better than DNN ($acc_{med} = 0.582$) and FM ($acc_{med} = 0.581$). DeepFM improves over the DNN thanks to the added FM model. Solely the linear model is unable to achieve similar performance. This shows the effectiveness of the FM approach and its capability to learn interactions among many features. In addition, we explored meta-embeddings, which combine volume measurements of larger brain regions into a single embedding vector. We only combine brain volume features as those are the largest feature group and combining them by brain region is medically reasonable. Our results demonstrate that this leads to a slightly higher balanced accuracy of 0.596, but most importantly lowers the variance across folds (Fig. 2, right). By combining larger brain regions, the pairwise interaction space shrinks and the model is less prone to overfitting.

Feature Importance. We get a direct measure of feature importance by looking at the weights of the linear part of the DeepFM in (1). Figure 3 displays the weights of the ten most important features for each fold (sorted by their mean weight over all folds). Because weights are class-specific, we obtain three rankings. As the top ten features for the AD (CN) model account for 42% (39.8%) of the total feature importance, we only analyze those in more detail.

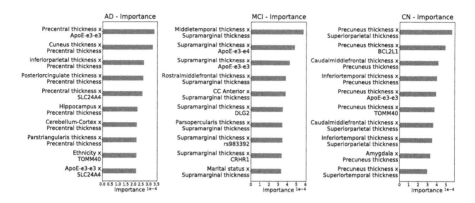

Fig. 4. DeepFM - 10 most important feature interactions

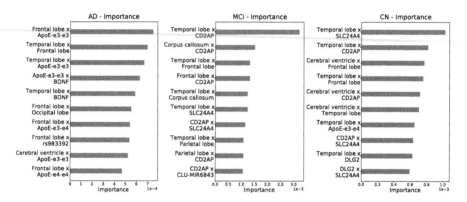

Fig. 5. DeepFM-Meta - 10 most important feature interactions

The first and third most important features for AD are the volumes of Hippocampus and Amygdala, both lie in the temporal lobe, a region typically affected by AD [23,27]. The thickness measures in the top features for AD lie in the parietal and temporal lobe. Both regions were previously shown to exhibit atrophy in AD patients [2,24]. The volume of the lateral ventricle is dependent on the atrophy of the brain regions surrounding it. Two variants of the genetic marker ApoE, e3–e3 and e4–e4, are important for prediction, which is reassuring because it is an important marker for AD [19,21]. Being not Hispanic, or not Latino has a slightly positive influence on the AD risk, which is supported by previous studies [25,26].

It is striking that the learned impact of all the top features for AD and CN prediction is in line with medical findings. Consequently, the weights' signs for the prediction of CN have the opposite sign compared to those for AD. For MCI, the picture is less clear. As MCI patients are part of a complex transition, this group is more heterogeneous. While for AD and CN a few very important features are sufficient for prediction, for MCI, the top ten features only make up for 23.7% of the total feature importance and learned weights are incoherent.

Feature Interactions. The second information that is easily accessible is the importance of feature interactions (see Fig. 4). This gives DeepFM an advantage over DNNs, for which extracting information about learned interactions is challenging. The information of a feature in an interaction is interconnected with their own and their partners embedding matrix. We extract the importance information by running the model with the test data and computing the relative importance of each single interaction and the output of the rest of the model for every patient. We can then plot the importance value for each interaction as the mean over all patients. The 109 features per patient accumulate to 5886 pairwise interactions. Thus, a single interaction has a relatively small contribution.

AD affects multiple brain regions at every stage of disease progression. The difficulty of pinpointing AD to a specific region becomes apparent for the pairwise interactions, where multiple brain regions interact. Numerous interactions consist of regions in or near the Hippocampus. Some genetic markers also appear. The interactions between them and different brain regions are an interesting finding that needs to be further explored in the future. However, in this setup the brain is scattered in many volumes and small differences or measurement errors become much more pronounced. This makes the interpretation difficult.

At this point, the advantage of the embedding layer comes into play. While interactions between single small brain regions are hard to interpret, the embedding layer can be used to embed larger regions into a so-called *meta embedding*. The interactions between these larger regions give a better picture of important regions (see Fig. 5). Small deviations have less influence and overfitting is reduced. Using meta embedding, DeepFM is especially utilizing interactions between ApoE and temporal or frontal lobe volumes. Comparing the most important feature interactions to distinguish AD from the rest, with the ones used for CN patients, it becomes even clearer that DeepFM learns medically explainable features. The CN-model focuses especially on the temporal lobe and cerebral ventricle. Both regions are known to be affected early on in the disease [14,27].

6   Conclusion

We proposed a Deep Factorization Machine model that combines the strength of deep neural networks to implicitly learn feature interactions and the ease of interpretability of a linear model. Our experiments on Alzheimer's Disease diagnosis demonstrated that the proposed architecture is able accurately classify patients than competing methods and can reveal valuable insights about the interaction between biomarkers.

Acknowledgements. This research was supported by the Bavarian State Ministry of Science and the Arts and coordinated by the Bavarian Research Institute for Digital Transformation, and the Federal Ministry of Education and Research in the call for Computational Life Sciences (DeepMentia, 031L0200A).

References

1. Barredo Arrieta, A., Díaz-Rodríguez, N., Del Ser, J., Bennetot, A., Tabik, S., et al.: Explainable Artificial Intelligence (XAI): concepts, taxonomies, opportunities and challenges toward responsible AI. Inf. c **58**, 82–115 (2020)
2. Dickerson, B.C., Bakkour, A., Salat, D.H., Feczko, E., Pacheco, J., et al.: The cortical signature of Alzheimer's disease: regionally specific cortical thinning relates to symptom severity in very mild to mild AD dementia and is detectable in asymptomatic amyloid-positive individuals. Cereb. Cortex **19**(3), 497–510 (2008)
3. Fan, L., Mao, C., Hu, X., Zhang, S., Yang, Z., et al.: New insights into the pathogenesis of Alzheimer's disease. Front. Neurol. **10**, 1312 (2020)
4. Fischl, B.: FreeSurfer. NeuroImage **62**(2), 774–781 (2012)
5. Gevrey, M., Dimopoulos, I., Lek, S.: Review and comparison of methods to study the contribution of variables in artificial neural network models. Ecol. Model. **160**(3), 249–264 (2003)
6. Guo, H., Tang, R., Ye, Y., Li, Z., He, X.: DeepFM: a factorization-machine based neural network for CTR prediction. In: Proceedings of the 26th International Joint Conference on Artificial Intelligence, pp. 1725–1731 (2017)
7. Head, T., Kumar, M., Nahrstaedt, H., Louppe, G., Shcherbatyi, I.: scikit-optimize/scikit-optimize v0.8.1, September 2020. https://doi.org/10.5281/zenodo.4014775
8. Hibar, D.P., Stein, J.L., Renteria, M.E., Arias-Vasquez, A., Desrivières, S., et al.: Common genetic variants influence human subcortical brain structures. Nature **520**(7546), 224–229 (2015)
9. Ho, D.E., Imai, K., King, G., Stuart, E.A.: Matching as nonparametric preprocessing for reducing model dependence in parametric causal inference. Polit. Anal. **15**(3), 199–236 (2007)
10. Jack, C.R., Bernstein, M.A., Fox, N.C., Thompson, P., Alexander, G., et al.: The Alzheimer's disease neuroimaging initiative (ADNI): MRI methods. J. Magn. Reson. Imaging **27**(4), 685–691 (2008)
11. Jack, C.R., et al.: Tracking pathophysiological processes in Alzheimer's disease: an updated hypothetical model of dynamic biomarkers. Lancet Neurol. **12**(2), 207–216 (2013)
12. Khatri, U., Kwon, G.R.: An efficient combination among sMRI, CSF, cognitive score, and APOE ϵ4 biomarkers for classification of AD and MCI using extreme learning machine. Comput. Intell. Neurosci. **2020**, 1–18 (2020)
13. Lambert, J.C., Ibrahim-Verbaas, C.A., Harold, D., Naj, A.C., Sims, R., et al.: Meta-analysis of 74,046 individuals identifies 11 new susceptibility loci for Alzheimer's disease. Nat. Genet. **45**(12), 1452–1458 (2013)
14. Nestor, S.M., Rupsingh, R., Borrie, M., Smith, M., Accomazzi, V., et al.: Ventricular enlargement as a possible measure of Alzheimer's disease progression validated using the Alzheimer's disease neuroimaging initiative database. Brain **131**(9), 2443–2454 (2008)
15. Ning, K., et al.: Classifying Alzheimer's disease with brain imaging and genetic data using a neural network framework. Neurobiol. Aging **68**, 151–158 (2018)
16. Olden, J.D., Joy, M.K., Death, R.G.: An accurate comparison of methods for quantifying variable importance in artificial neural networks using simulated data. Ecol. Model. **178**(3–4), 389–397 (2004)
17. Patterson, C., et al.: World Alzheimer report 2018. Technical Report, Alzheimer's Disease International (2018)

18. Petersen, R.C.: Mild cognitive impairment. N. Engl. J. Med. **364**(23), 2227–2234 (2011)
19. Reiman, E.M., Arboleda-Velasquez, J.F., Quiroz, Y.T., Huentelman, M.J., Beach, T.G., et al.: Exceptionally low likelihood of Alzheimer's dementia in APOE2 homozygotes from a 5,000-person neuropathological study. Nat. Commun. **11**, 1–11 (2020)
20. Rendle, S., Schmidt-Thieme, L.: Pairwise interaction tensor factorization for personalized tag recommendation. In: Proceedings of the Third ACM International Conference on Web Search and Data Mining, pp. 81–90 (2010)
21. Saykin, A.J., Shen, L., Foroud, T.M., Potkin, S.G., et al.: Alzheimer's disease neuroimaging initiative biomarkers as quantitative phenotypes: genetics core aims, progress, and plans. Alzheimers Dement. **6**(3), 265–273 (2010)
22. Scheltens, P., et al.: Alzheimer's disease. Lancet **388**(10043), 505–517 (2016)
23. Scott, S.A., DeKosky, S.T., Scheff, S.W.: Volumetric atrophy of the amygdala in Alzheimer's disease: quantitative serial reconstruction. Neurology **41**(3), 351–351 (1991)
24. Singh, V., Chertkow, H., Lerch, J.P., Evans, A.C., Dorr, A.E., Kabani, N.J.: Spatial patterns of cortical thinning in mild cognitive impairment and Alzheimer's disease. Brain **129**(11), 2885–2893 (2006)
25. Tang, M.X., et al.: Incidence of AD in African-Americans, Caribbean Hispanics, and Caucasians in Northern Manhattan. Neurology **56**(1), 49–56 (2001)
26. Tang, M.X., et al.: The APOE-ϵ4 Allele and the risk of Alzheimer disease among African Americans, Whites, and Hispanics. JAMA **279**(10), 751–755 (1998)
27. Teipel, S.J., Pruessner, J.C., Faltraco, F., Born, C., Rocha-Unold, M., et al.: Comprehensive dissection of the medial temporal lobe in AD: measurement of hippocampus, amygdala, entorhinal, perirhinal and parahippocampal cortices using MRI. J. Neurol. **253**(6), 794–800 (2006)
28. Tong, T., Gray, K., Gao, Q., Chen, L., Rueckert, D.: Nonlinear graph fusion for multi-modal classification of Alzheimer's disease. In: Zhou, L., Wang, L., Wang, Q., Shi, Y. (eds.) MLMI 2015. LNCS, vol. 9352, pp. 77–84. Springer, Cham (2015). https://doi.org/10.1007/978-3-319-24888-2_10
29. Tsang, M., Cheng, D., Liu, Y.: Detecting statistical interactions from neural network weights. In: International Conference on Learning Representations (2018)
30. Venugopalan, J., Tong, L., Hassanzadeh, H.R., Wang, M.D.: Multimodal deep learning models for early detection of Alzheimer's disease stage. Sci. Rep. **11**(1), 1–3 (2021)
31. Wachinger, C., Nho, K., Saykin, A.J., Reuter, M., Rieckmann, A.: A longitudinal imaging genetics study of neuroanatomical asymmetry in Alzheimer's disease. Biol. Psychiatry **84**(7), 522–530 (2018)
32. Wen, J., Thibeau-Sutre, E., Diaz-Melo, M., Samper-González, J., et al.: Convolutional neural networks for classification of Alzheimer's disease: overview and reproducible evaluation. Med. Image Anal. **63**, 101694 (2020)
33. Zhang, D., Wang, Y., Zhou, L., Yuan, H., Shen, D.: Multimodal classification of Alzheimer's disease and mild cognitive impairment. NeuroImage **55**(3), 856–867 (2011)

3D Temporomandibular Joint CBCT Image Segmentation via Multi-directional Resampling Ensemble Learning Network

Kai Zhang[1], Jupeng Li[1(✉)], Ruohan Ma[2], and Gang Li[2]

[1] School of Electronic and Information Engineering,
Beijing Jiaotong University, Beijing 100044, China
lijupeng@bjtu.edu.cn
[2] Department of Oral and Maxillofacial Radiology, Peking University School
and Hospital of Stomatology, Beijing 100081, China

Abstract. Accurate segmentation of temporomandibular joint (TMJ) from dental cone beam CT (CBCT) images is basis of for early diagnosis of TMJ-related diseases such as temporomandibular disorders (TMD). Fully convolutional networks (FCN) have achieved the state-of-the-art performance in medical image segmentation field. Both enough contextual information as well as rich spatial semantic information is required to obtain accurate segmentation, however, due to the limited GPU memories, high-resolution 3D volume cannot be directly input to these models. In this paper, we propose *Multi-directional Resampling Ensemble Learning Network* for 3D TMJ-CBCT image segmentation. This model extracts four semantic features from multi-directional resampled volumes, and then integrates features via ensemble learning network to achieve accurate segmentation. We implement extensive evaluations of the proposed method on a clinical images dataset, including images acquired from 89 patients. Our method achieves the Mean DSC value of 0.9814 ± 0.0054, the Mean Hausdorff Distance of 1.5711 ± 1.0252 mm, and the Mean Average Surface Distance of 0.0555 ± 0.0198 mm.

Keywords: Medical image segmentation · GPU memory limitation · Multi-directional resampling · Ensemble learning network · TMJ-CBCT image

1 Introduction

TMJ is the most complex structures in human body [1]. Millions of people suffer from TMD which is a common TMJ disease with high prevalence and closely related to multi-discipline [2]. Cone beam CT (CBCT) is popular with radiologists due to its high resolution, short data acquisition time and low-dose radiation [3]. The segmentation of TMJ-CBCT images helps doctors better diagnose, consequently provides quantitative measurements and aids surgical treatments. However, manual segmentation is not only time-consuming but also subject to inter- and intra-operator variability.

Deep learning techniques have been widely employed in 3D volumetric segmentation [4–6]. Plenty of state-of-the-art methods have been proposed and can be divided into

© Springer Nature Switzerland AG 2021
C. Lian et al. (Eds.): MLMI 2021, LNCS 12966, pp. 634–643, 2021.
https://doi.org/10.1007/978-3-030-87589-3_65

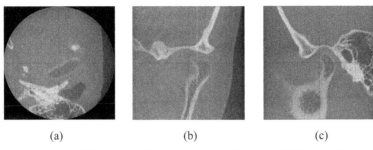

(a)   (b)   (c)

Fig. 1. An example of TMJ segmentation in transverse (a), coronal (b), sagittal (c) views. The mandibular condyle around the temporal bone is shown in red region (Color figure online).

two categories. The first category is patch-based method which inputs blocks around pixels into the traditional convolutional neural networks (CNNs) to realize the voxel-wise classification, e.g. DeepMedic [7]. But repeated calculations and limited receptive fields lead to the restriction of these methods. Another category is fully convolutional method, e.g. FCN [8], U-Net [4], which typically includes an encoding path to extract features and a decoding path to recover spatial resolution. However, due to the extremely computational calculation of 3D convolutions [9] and the limited GPU memories, the source image is often processed in two ways: 1) cropped into overlapping 3D patches [5], which cannot make better use of global knowledge because of the limited receptive fields; 2) resized to a volume with lower resolution [10, 15], which will cause the loss of spatial detail information, especially when processing high-resolution images.

The transverse, sagittal and coronal planes of TMJ-CBCT image are shown in Fig. 1. Lower radiation dose and lower density of the mandibular condyle result in weak contrast in TMJ-CBCT images. Considering the complex bone structure around the condyle, a large enough receptive field is necessary, and the patch-based training with limited receptive fields is obviously not suitable for TMJ-CBCT segmentation [18]. Since both large enough receptive field and rich spatial information are required to obtain accurate segmentation result [11], increasing the amount of input information as much as possible while ensuring the enough global contextual information input is the key point of this research.

Due to huge amount of 3D volume data, we cannot input the entire high-resolution volume to any network without pre-processing, such as down-sampling. Fortunately, we found that the 3D images contain different details in the three fault views. In addition, there are some researches using multi-view feature extraction to obtain image information from different views [12, 13]. Different from traditional ensemble strategies (average [6, 14], expectation maximization method [12], majority voting algorithm, etc.), we propose a novel network based on 3D U-Net [5], named Multi-directional Resampling Ensemble Network. Firstly, four weak learners extract transverse, sagittal, coronal and global information respectively. Then, the ensemble network fuses four predictions and an initial gray contextual data to generate final segmentation result. Benefits from our network architecture, our network pays more attention on the TMJ area using limited

GPU memory, which is similar to the spatial attention mechanism [15, 16]. We performed comparison experiments to show that our method achieves the state-of-the-art performance on the clinical image dataset.

2 Methods

2.1 Overview

The overall framework for TMJ-CBCT image segmentation is shown in Fig. 2. The initial 3D TM-CBCT image is abstracted into I with size $W \times H \times D$. For down-sampled volume I_n, we define $f_n(\cdot)$ as the first four 3D U-Nets (Net-1 ~ Net ~ 4), where $n \in \{t, s, c, d\}$ and $f_e(\cdot)$ as the ensemble network (Net-5). So, the network prediction R can be expressed as:

$$R(I) = f_e(\mathrm{U}(f_t(I_t)) \oplus \mathrm{U}(f_s(I_s)) \oplus \mathrm{U}(f_c(I_c)) \oplus \mathrm{U}(f_d(I_d)) \oplus I) \tag{1}$$

Where $\oplus$ denotes channel connection; $\mathrm{U}(\cdot)$ represents up-sampling, which aim to ensure the same size of output feature maps for Net-5.

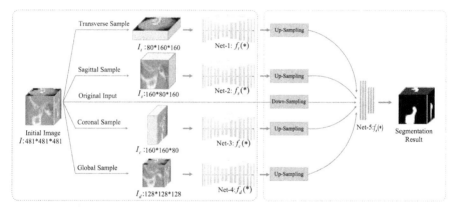

Fig. 2. The workflow of our multi-directional resampling ensemble learning network applied for TMJ-CBCT image segmentation. This model contains four independent 3D U-Nets to extract semantic knowledge in multi-views. An ensemble network is implemented to fuse the segmentation results of the previous stage to achieve the final result.

2.2 Multi-directional Resampling Feature Extraction Network

As illustrated in Fig. 2, in order to extract more detailed information in the transverse (t), sagittal (s) and coronal (c) views, the initial image are down-sampled along three different tomographic views. Since the results of the above three resamplings are deformed in internal spatial location, a branch for extracting undeformed global features is indispensable. In the following ablation experiments, the experimental results also show that adding the global branch is beneficial to enhance the eventual segmentation performance.

3D U-Nets with the same structural design are employed for multi-directional feature extraction. The encoder includes 6 3D convolution layers with kernel size $3 \times 3 \times 3$ followed by the Rectified Linear Unit (ReLU), which are mainly employed to gain the contextual information, and 3 max-pooling layers which are implemented to reduce the spatial resolution of the feature by half, doubling the receptive field of the network simultaneously. The decoder is composed of 6 3D $3 \times 3 \times 3$ convolutions interleaved with 3 up-sampling layers which double the size of the spatial dimension by the trilinear interpolation. At the end of the decoder, a $1 \times 1 \times 1$ convolution is performed to decrease the number of output channels to 1, followed by a sigmoid function which maps the gray value of the output to 0 to 1. The skip connections between the encoder and decoder layers integrate multi-scale features from the low-level to the high-level and accelerate the network convergence [4].

2.3 Ensemble Learning Network

The structure of the ensemble learning network is simplified on the basis of 3D U-Net, such as reducing the feature map channels to deal with larger input, and increasing middle convolutional layers to enhance the ability to extract high-level features. The overall ensemble process is shown in Fig. 3. Specifically, the training images of the ensemble stage are firstly fed to the multi-directions feature extracting networks, then the predictions are up-sampled to $160 \times 160 \times 160$ for facilitating combination, finally the up-sampled images and the down-sampled original images are combined together as the inputs of our ensemble network. Simply integrating the predictions is not enough, we import an original branch as an additional channel, which is designed to provide original gray information for the feature extraction. That is, the original branch can provide abundant spatial location information for other branches, meanwhile the multi-directional sampling branches can also supply a rough segmentation reference for the original branch, so that the network makes use of the multi-directional predictions, focusing on the TMJ regions to yield better results. The integrated heatmap (h) is a neat and elegant picture, removing the interference of irrelevant regions. The edges of result and ground truth almost coincide in Fig. 3(e).

2.4 Loss Function

We employed the Binary Cross Entropy (BCE) to compute the loss function. The cross-entropy loss function L is:

$$L(g_i, r_i) = -\frac{1}{N} \sum_{i=1}^{N} (g_i \log r_i + (1 - g_i) \log(1 - r_i)) \tag{2}$$

Where r_i denotes the prediction, g_i denoted the Manual segmentation result. N is the batch size.

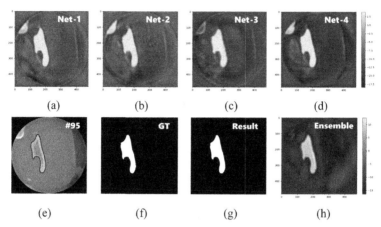

Fig. 3. Visualization of the ensemble process on heatmaps. (a), (b), (c) and (d) represent the predictions of the multi-directional resampling network respectively. (h) is the heatmap of the ensemble network. We set the threshold to 0 to get the final segmentation result (g). (f) denote the ground truth. (e) is a slice of TMJ-CBCT image, here we use the red line to represent contour of the segmentation result, and the blue line to represent contour of the ground truth (Color figure online).

3 Experiments and Results

3.1 Dataset and Evaluation Metrics

The dataset used here was collected from Peking University School of Stomatology, including 89 TMJ-CBCT clinical images captured by 3D Accuitomo 170 (J. Morita MFG. Corp., Kyoto, Japan) for left TMJ of 89 patients. 72 images were used as training data (36 images for stage 1, another 36 images for the fusion network) and 17 were used for testing. The volume resolution is $0.125 \times 0.125 \times 0.125$ mm^3 with size of $481 \times 481 \times 481$. To quantitatively measure the segmentation accuracy, three measures are employed, Dice Similarity Coefficient (DSC), Average Surface Distance (ASD), and Hausdorff Distance (HD) respectively.

3.2 Training Details

Preprocessing. Due to the inconsistent distribution of the images generated by the CBCT instrument under complicated circumstances, the uneven intensity distribution in the training data is not conducive to the convergence of the model. To mitigate the risk of over-fitting, we performed histogram matching on the original data.

Network Training. Five 3D U-Net networks were trained using Stochastic Gradient Descent (SGD) with a momentum of 0.9. In order to accelerate the convergence speed and the stability of the next training stage, we set a piece-wise decay learning rate which is decreased from 0.1 to 0.001. In addition, we trained the network of stage 1 using 150 epochs, and the stage 2 using 300 epochs. Our network was implemented on the popular open source framework PyTorch on a Nvidia 2080TI GPU with 11GB memories.

3.3 Ablation Study

As shown in Table 1, to demonstrate the efficacy of each branch in the first stage, we arranged an ablation experiment, which remove one branch in turn to verify its effectiveness. Overall, any branch contributes to the final segmentation performance, which emphasizes that integrating the 3D U-Nets effectively gains the robust and accurate predictions. Moreover, the results prove that the 5th branch named original branch is able to improve segmentation performance, as directly seen from the last two rows of Table 2. Not just for fusion, the purpose of adding the original input is to utilize the predictions of the first stage to further guide the next fine segmentation. Likewise, it can be observed that the global branch is indispensable from the 4th row and the bottom row, which is able to provide adequate aggregation of spatial information for the fusion network to improve the overall segmentation performance.

Table 1. Ablation study to demonstrate the efficacy of five branches. All the results have the same hyperparameter settings, and no postprocessing is performed.

Net-1	Net-2	Net-3	Net-4	Gray	DSC	ASD (mm)	HD (mm)
	√	√	√	√	0.9790 ± 0.0058	0.08 ± 0.06	4.11 ± 6.01
√		√	√	√	0.9798 ± 0.0063	0.08 ± 0.09	2.37 ± 2.90
√	√		√	√	0.9792 ± 0.0049	0.07 ± 0.04	3.78 ± 6.19
√	√	√		√	0.9790 ± 0.0056	0.07 ± 0.03	4.80 ± 5.64
√	√	√	√		0.9801 ± 0.0056	0.08 ± 0.09	4.95 ± 6.60
√	√	√	√	√	**0.9804** ± 0.0049	**0.06** ± 0.02	**2.28** ± 2.34

3.4 Results Analysis

We implemented two popular methods to compare with our proposed method, the 3D U-Net [17] and nnU-Net [6]. It is worth noting that nnU-Net is an out-of-the-box tool, which has achieved the best results in many segmentation tasks. Some comparison results are shown in Fig. 4. It is observed that our method can produce smoother boundaries than 3D U-Net and fewer misclassifications (#95) compared with nnU-Net. The reason for the misclassifying of nnU-Net might still be its limited receptive field, even though its network structure integrates the Cascade U-Net which is proposed to deal with high-resolution images. For #99, our network can be able to resist the interference of the condyle bone remodeling which does not usually have to be marked in clinical practice.

In the segmentation of the tip area of bone structure, our method typically performs best among these three methods. As presented in Fig. 5, the bitty tip area and the ambiguous border pose a huge challenge, 3D U-Net cannot segment well due to the loss of too much image detail information. Compared with 3D U-Net, the segmentation performance of nnU-Net is worse, the reason may be that the limited receptive field limits the segmentation ability of the model.

To improve the segmentation performance, we applied 5-fold cross-validation ensemble strategy to integrate 5 models to predict the final segmentation results. The

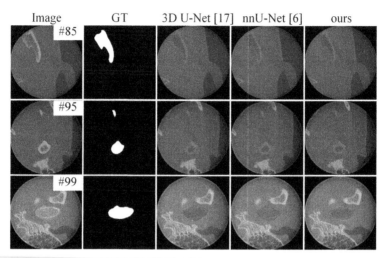

Image   GT   3D U-Net [17]   nnU-Net [6]   ours

Fig. 4. Qualitative performance comparison between U-Net [17], nnU-Net [6] and our method.

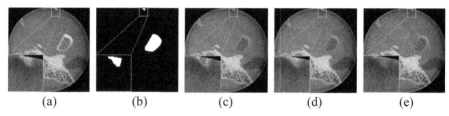

(a)   (b)   (c)   (d)   (e)

Fig. 5. Segmentation comparisons in the tip area of bone structure between 3D U-Net, nnU-Net and our method. (a) Cross-sectional tomographic image of original CBCT image. (b) Ground truth of corresponding slice. (c) Segmentation result of 3D U-Net. (d) Segmentation result of nnU-Net. (e) Segmentation result of our method.

quantitative comparison results are shown in Table 2. To highlight the effectiveness of ensemble learning, we set up MRNet, which employs the average strategy to simply integrate the first stage segmentation results. The experimental result shows that our proposed method realizes the best performance in the mean of ASD and HD. In addition, the proposed network achieves a mean DSC of 0.9814 ± 0.0054, which is slightly lower than 0.0028 compared with nnU-Net. In short, the above comparisons demonstrate our method is very competitive with the state-of-the-art methods.

Figure 6 shows the testing occupied memory and prediction time of the three methods respectively, which are all computed without 5-fold cross-validation ensemble. It is noticed that 3D U-Net has the shortest testing time because of the simple structure, but its segmentation performance is still room for improvement as shown in Table 2; nnU-Net has the lowest GPU memory consumption, but it spends nearly 10 min per-CBCT. Although the occupied memory (6521 MB) of our method is much more than nnU-Net (2113 MB), the testing time of our method is only 74.2 s. In clinical applications, efficiency is an important factor in considering whether one method can be applied.

Obviously, our method retains excellent segmentation performance and higher efficiency, so it is easier to implement clinical applications.

Table 2. Quantitative comparison between the proposed method and other methods. The mean and the standard deviation of the DSC, ASD and HD values are reported in pairs.

Methods	DSC	ASD (mm)	HD (mm)
3D U-Net [17]	0.9741 ± 0.0058	0.0949 ± 0.0684	2.9505 ± 3.9406
nnU-Net [6]	**0.9842** ± 0.0065	0.0572 ± 0.0592	1.8263 ± 2.7062
MRNet	0.9793 ± 0.0052	0.0743 ± 0.0495	2.5133 ± 2.7123
ours	0.9814 ± 0.0054	**0.0555** ± 0.0198	**1.5711** ± 1.0252

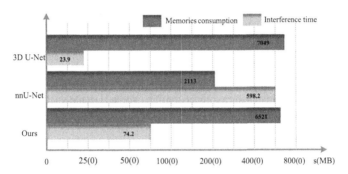

Fig. 6. Comparison of memories consumption and interference time among different methods.

4 Conclusions

To solve information loss caused by the limited GPU memories, we proposed the multi-directional resampling ensemble learning Network for 3D TMJ-CBCT volume segmentation. This model extracts different semantic features from multi-directional resample volumes, then integrates features via ensemble network to achieve accurate segmentation results. Experiments on the clinical TMJ-CBCT images demonstrate that our method can obtain accurate and robust segmentation with faster processing speed, which owns high superiority to apply to computer-aided diagnosis.

Acknowledgments. This work was supported by the Fundamental Research Funds for the Central Universities No. 2021JBM003 and the National Natural Science Foundation of China with Project No. 81671034.

References

1. Belikova, K., Rogov, O., Rybakov, A., et al.: Deep Negative Volume Segmentation. arXiv preprint arXiv:2006.12430 (2020)
2. Geiger, D., Bae, W.C., Statum, S., Du, J., Chung, C.B.: Quantitative 3D ultrashort time-to-echo (UTE) MRI and micro-CT (µCT) evaluation of the temporomandibular joint (TMJ) condylar morphology. Skeletal Radiol. **43**(1), 19–25 (2013). https://doi.org/10.1007/s00256-013-1738-9
3. Bayram, M., et al.: Volumetric analysis of the mandibular condyle using cone beam computed tomography. Eur. J. Radiol. **81**(8), 1812–1816 (2012). https://doi.org/10.1016/j.ejrad.2011.04.070
4. Ronneberger, O., Fischer, P., Brox, T.: U-Net: convolutional networks for biomedical image segmentation. In: Navab, N., Hornegger, J., Wells, W.M., Frangi, A.F. (eds.) MICCAI 2015. LNCS, vol. 9351, pp. 234–241. Springer, Cham (2015). https://doi.org/10.1007/978-3-319-24574-4_28
5. Çiçek, Ö., Abdulkadir, A., Lienkamp, S.S., Brox, T., Ronneberger, O.: 3D U-Net: learning dense volumetric segmentation from sparse annotation. In: Ourselin, S., Joskowicz, L., Sabuncu, M.R., Unal, G., Wells, W. (eds.) MICCAI 2016. LNCS, vol. 9901, pp. 424–432. Springer, Cham (2016). https://doi.org/10.1007/978-3-319-46723-8_49
6. Isensee, F., Kickingereder, P., Wick, W., Bendszus, M., Maier-Hein, K.H.: No New-Net. In: Crimi, A., Bakas, S., Kuijf, H., Keyvan, F., Reyes, M., van Walsum, T. (eds.) BrainLes 2018. LNCS, vol. 11384, pp. 234–244. Springer, Cham (2019). https://doi.org/10.1007/978-3-030-11726-9_21
7. Patel, T.R., et al.: Multi-resolution CNN for brain vessel segmentation from cerebrovascular images of intracranial aneurysm: a comparison of U-Net and DeepMedic. In: Medical Imaging 2020: Computer-Aided Diagnosis. 124142W, Texas (2020). https://doi.org/10.1117/12.2549761
8. Long, J., Shelhamer, E., Darrell, T.: Fully convolutional networks for semantic segmentation. In: Proceedings of the IEEE Conference on Computer Vision and Pattern Recognition, pp. 3431–3440 (2015)
9. Chen, C., Liu, X., Ding, M., Zheng, J., Li, J.: 3D dilated multi-fiber network for real-time brain tumor segmentation in MRI. In: Shen, D., et al. (eds.) MICCAI 2019. LNCS, vol. 11766, pp. 184–192. Springer, Cham (2019). https://doi.org/10.1007/978-3-030-32248-9_21
10. Roth, H.R., et al.: Deep learning and its application to medical image segmentation. Med. Imaging Technol. **36**(2), 63–71 (2018). https://doi.org/10.11409/mit.36.63
11. Wang, B., Qiu, S., He, H.: Dual encoding U-Net for retinal vessel segmentation. In: Shen, D., et al. (eds.) MICCAI 2019. LNCS, vol. 11764, pp. 84–92. Springer, Cham (2019). https://doi.org/10.1007/978-3-030-32239-7_10
12. Wang, G., Li, W., Ourselin, S., Vercauteren, T.: Automatic brain tumor segmentation using cascaded anisotropic convolutional neural networks. In: Crimi, A., Bakas, S., Kuijf, H., Menze, B., Reyes, M. (eds.) BrainLes 2017. LNCS, vol. 10670, pp. 178–190. Springer, Cham (2018). https://doi.org/10.1007/978-3-319-75238-9_16
13. Wang, Y., Zhou, Y., Shen, W., Park, S., Fishman, E.K., Yuille, A.L.: Abdominal multi-organ segmentation with organ-attention networks and statistical fusion. Med. Image Anal. **55**, 88–102 (2019). https://doi.org/10.1016/j.media.2019.04.005
14. Kamnitsas, K., et al.: Ensembles of multiple models and architectures for robust brain tumour segmentation. In: Crimi, A., Bakas, S., Kuijf, H., Menze, B., Reyes, M. (eds.) BrainLes 2017. LNCS, vol. 10670, pp. 450–462. Springer, Cham (2018). https://doi.org/10.1007/978-3-319-75238-9_38

15. Sinha, A., Dolz, J.: Multi-scale self-guided attention for medical image segmentation. IEEE J. Biomed. Health Inf. **25**(1), 121–130 (2019). https://ieeexplore.ieee.org/document/9066969

16. Li, X., Zhong, Z., Wu, J., et al.: Expectation-maximization attention networks for semantic segmentation. In: Proceedings of the IEEE/CVF International Conference on Computer Vision, pp. 9167–9176 (2019)

17. Zhang, K., Li, J., Ma, R., Li, G.: An end-to-end segmentation network for the temporomandibular joints CBCT image based on 3D U-Net. In: Proceedings of the 2020 13th International Congress on Image and Signal Processing, pp. 664–668 (2020)

18. Fang, C., Li, G., Pan, C., Li, Y., Yu, Y.: Globally guided progressive fusion network for 3D pancreas segmentation. In: Shen, D., et al. (eds.) MICCAI 2019. LNCS, vol. 11765, pp. 210–218. Springer, Cham (2019). https://doi.org/10.1007/978-3-030-32245-8_24

Vox2Surf: Implicit Surface Reconstruction from Volumetric Data

Yoonmi Hong, Sahar Ahmad, Ye Wu, Siyuan Liu, and Pew-Thian Yap[✉]

Department of Radiology and Biomedical Research Imaging Center (BRIC),
University of North Carolina, Chapel Hill, NC, USA
ptyap@med.unc.edu

Abstract. Surface reconstruction from volumetric T1-weighted and T2-weighted images is a time-consuming multi-step process that often involves careful parameter fine-tuning, hindering a more wide-spread utilization of surface-based analysis particularly in large-scale studies. In this work, we propose a fast surface reconstruction method that is based on directly learning a continuous-valued signed distance function (SDF) as implicit surface representation. This continuous representation implicitly encodes the boundary of the surface as the zero isosurface. Given the predicted SDF, the target 3D surface is reconstructed by applying the marching cubes algorithm. Our implicit reconstruction method concurrently predicts the surfaces of the brain parenchyma, the white matter and pial surfaces, the subcortical structures, and the ventricles. Evaluation based on data from the Human Connectome Project indicates that surface reconstruction of a total of 22 cortical and subcortical structures can be completed in less than 20 min.

Keywords: Cortical surface reconstruction · Implicit representation · Deep learning

1 Introduction

Reconstruction of human cortical and subcortical surfaces from magnetic resonance (MR) images enables the quantification of morphometric properties of the cerebral cortex and the subcortical structures. Anatomical surface models are pivotal for investigating neurodegenerative, neurological, and psychiatric disorders [13,23]. The reconstruction process typically involves (i) skull stripping; (ii) brain tissue and structure segmentation; (iii) surface reconstruction; and (iv) surface topology correction [14]. A major limitation of this typical multi-step process is that error in each step can propagate to downstream processing steps. For example, erroneous segmentation or defective topology can result in inaccurate anatomical surfaces. To overcome this problem, we aim to reconstruct cortical and subcortical surfaces directly from volumetric data, bypassing the segmentation step and allowing the whole process to be optimized end-to-end.

This work was supported in part by United States National Institutes of Health (NIH) grants EB008374 and EB006733.

C. Lian et al. (Eds.): MLMI 2021, LNCS 12966, pp. 644–653, 2021.
https://doi.org/10.1007/978-3-030-87589-3_66

Existing 3D shape learning methods can be classified into three categories: (i) point-based; (ii) mesh-based; and (iii) voxel-based methods. Point-based methods utilize neural networks for 3D shape learning on a point cloud [4,21]. One such method PointNet [4], which extracts global shape feature using max-pool operation, is widely used as an encoder for point generation networks. However, point clouds inherently lack topological information and are not suitable for generating watertight surfaces.

Mesh-based methods use either predefined template meshes [1,16] or parameterization techniques [18,22] to generate 3D shape meshes. The drawback of these methods is that the predefined template meshes can only model shapes with fixed mesh topology, and the parameterization technique is sensitive to the input mesh quality and cutting strategies. Deep networks are employed to improve the performance of parameterization technique [26], but they require multiple planes to describe complex topologies and the generated mesh is not closed. A solution is to utilize sphere parameterization to generate closed meshes [3], but the outcome is limited to genus-zero surfaces.

Voxel-based methods can use a dense occupancy grid (occupied/unoccupied) for learning 3D shapes. These methods are limited by the computational and memory requirements for handling high resolution (512^3 or above) data, resulting in poor preservation of fine-grained shape details [6,24]. Instead of binary occupancy values [19], continuous signed distance function can be employed [20], corresponding to soft labeling in segmentation tasks [15].

In this paper, we introduce a method that uses global and local features from structural MR images to reconstruct surfaces. The input of the network is the 3D coordinates and intensity of a query point and the intensities of its neighborhood. A learned implicit function can be evaluated at any continuous point with arbitrary resolution. Our method works directly on volumetric MR images to predict the surfaces of the brain parenchyma, the white matter and pial surfaces, the subcortical structures, and the ventricles. The key advantages of our method, called Vox2Surf, are as follows:

1. Direct volume-to-surface mapping using implicit surface representation in terms of the signed distance function (SDF).
2. Explicit surface-driven optimization using SDFs, allowing information beyond the original image resolution to be used for detail preservation.
3. Mitigate the need for time-consuming subvoxel annotation.
4. Harnessing information from multiple structures to be used simultaneously in a multi-task learning framework for mutual guidance in surface prediction.
5. Explicit use of global contextual information and local detail information to improve surface prediction.
6. Surface prediction for cortical and subcortical structures.

Evaluation based on T1-weighted (T1w) and T2-weighted (T2w) images from the Human Connectome Project (HCP) indicates surfaces of 22 structures can be reconstructed in less than 20 min using an Nvidia Titan Xp GPU graphics card. Vox2Surf therefore greatly improves the feasibility of surface-based analysis in large-scale studies.

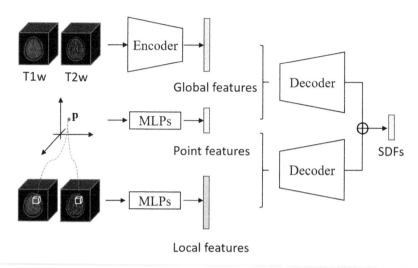

Fig. 1. Overview of the proposed method. MLP: Multi-layer perceptron.

2   Methods

We illustrate our surface reconstruction method in Fig. 1. We employ global and local features extracted from input T1w and T2w MR images to generate surfaces. The global features are extracted from an encoder and combined with point features, and then fed to a global feature decoder. To reconstruct fine-grained details of the surface, a module for extraction of local features is employed [25]. We extract local features from T1w and T2w image intensities, and then feed them into the local feature decoder. The final output is computed as the sum of the outputs of two decoders.

2.1   Problem Formulation

For an arbitrary point $\mathbf{p} \in \mathbb{R}^3$, SDF is a continuous function that assigns a real value indicating the distance from point $\mathbf{p}$ to the surface, and a sign determining whether $\mathbf{p}$ is inside or outside the surface. The SDF gives a positive value if $\mathbf{p}$ is outside the surface, and a negative value if $\mathbf{p}$ is inside the surface. Our aim is to learn a vector-valued SDF f such that

$$f(\mathbf{p}) = \mathbf{s_p}, \tag{1}$$

where $\mathbf{s_p}$ is a vector of signed distance values at point $\mathbf{p}$ associated with multiple target surfaces. We aim to reconstruct the surfaces of the brain parenchyma, left/right (L/R) cerebral cortex, subcortical structures, and ventricles with the learned function f. Instead of learning the SDF for each target surface separately, we jointly predict multiple surfaces in a single network, similar to [7].

2.2   Implicit Surface Reconstruction

Inspired by [25], we combine global and local features extracted from T1w and T2w images for reconstruction of surfaces with greater precision. Let $X_i \in \mathbb{R}^{N \times N \times N}, i = \{1, 2\}, N \in \mathbb{N}$, be T1w and T2w images, respectively. We first obtain latent vector $\mathbf{z} \in \mathbb{R}^n$ from an encoder consisting of several 3D convolutional layers, instance normalization, and Leaky ReLU activation functions. This encoded latent vector represents global features of the images. We additionally extract local features from corresponding intensities of the 3D point $\mathbf{p}$ and its neighboring points $\mathcal{N}_{\mathbf{p}}$. Here, we consider 2-voxel neighborhood information.

The encoded global and local features are the inputs to the point-wise decoders, which share a common architecture realized via multi-layer perceptrons (MLPs). The decoder architecture is based on the implicit decoder in IM-NET [5], consisting of a sequence of seven fully connected layers with skip connections. Each fully connected layer is followed by Leaky ReLU activations except for the last layer. The nonlinearity function of the last layer is 'tanh'. The element-wise summation of the decoder outputs gives the SDF for each target surface. We then employ a topology correction algorithm [2] to fix mispredictions via spherical level-set evolution on the implicit surface representation and to ensure the spherical topology of each target surface independently before applying the marching cubes algorithm [17] to render the surface meshes.

2.3   Loss Function and Implementation Details

We employ the standard mean squared error (MSE) as the loss function. For a given mini-batch $\mathcal{B}$, the objective function is defined as

$$\mathcal{L} := \sum_{j \in \mathcal{B}} \| f(\mathbf{p}_j) - \mathbf{s}_{\mathbf{p}_j} \|_2^2. \tag{2}$$

This choice is appropriate since SDFs are typically spatially smooth.

Sampling strategy is critical to the shape reconstruction problem. Simple and uniform sampling in a bounding box has been found to yield the best results [19]. However, we found that dense sampling near boundary is more effective, as adopted in [5,7,20,25]. We extract more samples near surface boundaries and the total number of extracted samples are 1M for each subject.

Our network is implemented in TensorFlow 1.13.1, and trained using the Adam optimizer with a learning rate of 5×10^{-5} without weight decay, and a batch size of 65536.

3   Experimental Results

3.1   Dataset and Preprocessing

We evaluated our proposed method using minimally preprocessed Human Connectome Project (HCP) dataset [10–12]. We resampled the T1w and T2w images to size $364 \times 435 \times 364$ with resolution $0.5\,\mathrm{mm}^3$. We padded the images to 512^3 for training. The ground-truth (GT) SDFs were clipped to $[-5, 5]$ and rescaled to $[-1, 1]$ for training. The number of subjects used for training, validation, and testing were 15, 1, and 6, respectively.

We used cortical surfaces available as part of the HCP minimally-preprocessed dataset. We utilized the available whole-brain masks and volumetric anatomical parcellations to generate surfaces of the whole-brain and subcortical structures. For each subcortical structure, i.e., the (L/R) putamen, caudate, pallidum, hippocampus, amygdala, thalamus, accumbens, lateral ventricles and third ventricle, we obtained a binary image that was used to extract the isosurface. The irregular triangulated surfaces were then smoothed using a mesh fairing algorithm based on Laplacian smoothing [8]. Finally, all the surfaces were employed to compute SDFs with the Connectome Workbench[1].

3.2   Quantitative Evaluation

We used the modified Hausdorff distance (MHD) [9] for evaluating the predicted surfaces. The MHD is used to measure the similarity of two point sets and is defined as

$$d_{\mathrm{MHD}}(X, Y) = \frac{1}{2} \left[\frac{1}{|X|} \sum_{x \in X} \min_{y \in Y} d(x, y) + \frac{1}{|Y|} \sum_{y \in Y} \min_{x \in X} d(y, x) \right], \qquad (3)$$

where X is the point set of the predicted surface and Y is the point set of the GT surface. The MHD for the six testing subjects are summarized in Table 1. We also computed the mean curvature and surface area of the cortical surfaces, and summarized the mean absolute error (MAE) results in Table 2 and Table 3, respectively. The quantitative results demonstrate that fine-grid predictions give lower errors.

3.3   Qualitative Comparison

We compared the reconstructed surfaces generated from different image grids. Figure 2 shows the subcortical structures and the ventricles generated with 128^3, 256^3, and 512^3 voxel resolution as well as the GT surfaces. It is evident that the surfaces, especially the ventricles, reconstructed at the highest resolution preserve the most details. We present the reconstructed pial, white matter, and whole-brain surfaces in Fig. 3. The proposed method reconstructed the surfaces with detailed gyral and sulcal patterns at the highest voxel resolution. More cortical convolutions are preserved with increasing voxel resolution.

---

[1] Available at https://www.humanconnectome.org/software/connectome-workbench.

Table 1. Summary statistics (mean ± standard deviation) of modified Hausdorff distance (mm) for all surfaces evaluated at different resolutions.

Surfaces	128^3	256^3	512^3
L/R Accumbens	2.448 ± 0.661	1.821 ± 0.639	**1.609 ± 0.748**
L/R Amygdala	2.454 ± 0.451	1.653 ± 0.372	**1.331 ± 0.305**
L/R Caudate	3.406 ± 0.334	2.157 ± 0.280	**1.572 ± 0.447**
L/R Hippocampus	2.376 ± 0.500	1.470 ± 0.367	**1.071 ± 0.238**
L/R Pallidum	2.464 ± 0.245	1.855 ± 0.352	**1.611 ± 0.371**
L/R Putamen	2.307 ± 0.265	1.583 ± 0.335	**1.306 ± 0.345**
L/R Thalamus	2.286 ± 0.208	1.565 ± 0.247	**1.329 ± 0.283**
L/R Lateral Ventricle	8.811 ± 3.879	1.745 ± 0.273	**0.951 ± 0.276**
L/R White Matter	2.974 ± 0.663	1.312 ± 0.266	**0.644 ± 0.082**
L/R Pial	2.989 ± 0.332	1.469 ± 0.340	**0.819 ± 0.135**
Third Ventricle	N/A	3.849 ± 1.084	**1.977 ± 0.765**
Whole-Brain	2.696 ± 0.671	1.713 ± 0.676	**1.387 ± 0.627**

Table 2. MAE values (mean ± standard deviation) of mean curvatures for cortical surfaces at different resolutions.

Surfaces	128^3	256^3	512^3
Left white matter	0.211 ± 0.146	**0.022 ± 0.014**	0.035 ± 0.034
Right white matter	0.077 ± 0.092	**0.018 ± 0.011**	0.048 ± 0.048
Left pial	0.391 ± 0.537	0.105 ± 0.113	**0.083 ± 0.056**
Right pial	2.286 ± 5.376	0.821 ± 1.748	**0.209 ± 0.175**

Table 3. MAE values (mean ± standard deviation) of surface areas (cm^2) and percentage errors for cortical surfaces at different resolutions.

Surfaces	128^3	256^3	512^3
Left white matter	532.4 ± 94.4 (152%)	263.9 ± 101.0 (44%)	**53.0 ± 36.9** (6%)
Right white matter	564.7 ± 96.6 (174%)	290.6 ± 106.7 (50%)	**57.0 ± 47.6** (7%)
Left pial	643.1 ± 107.1 (142%)	391.9 ± 117.9 (58%)	**116.1 ± 132.4** (14%)
Right pial	641.7 ± 84.8 (140%)	403.3 ± 89.2 (60%)	**121.2 ± 117.2** (14%)

Representative results for white matter surfaces are shown in Fig. 4, indicating that mean curvature of the reconstructed surfaces are close to the ground-truth surfaces.

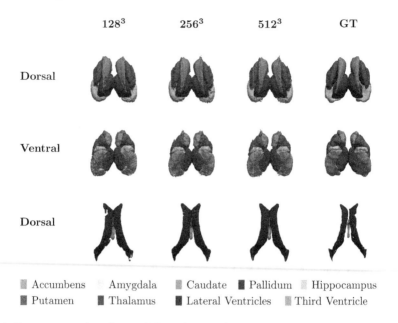

Fig. 2. Reconstructed surfaces of the subcortical structures and ventricles at different resolutions.

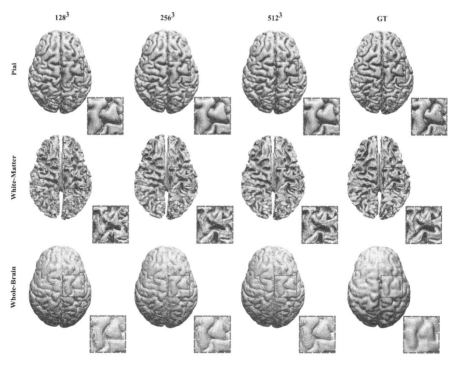

Fig. 3. Reconstructed pial, white matter, and whole-brain surfaces at different resolutions.

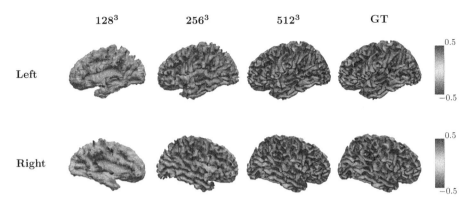

Fig. 4. White matter surfaces color-coded with mean curvatures at different resolutions.

4   Conclusion

In this paper, we proposed a multi-task learning based surface reconstruction method based on the prediction of continuous vector-valued SDFs that implicitly represent the surfaces of different cortical and subcortical structures. We integrated both global and local features from T1w and T2w images. We generated the surfaces of the (L/R) cerebral cortex, subcortical structures, and ventricles from the predicted SDFs with the marching cubes algorithm. The results show that our reconstructed surfaces are accurate with respect to ground-truth surfaces.

References

1. Bagautdinov, T., Wu, C., Saragih, J., Fua, P., Sheikh, Y.: Modeling facial geometry using compositional VAEs. In: IEEE Conference on Computer Vision and Pattern Recognition (CVPR) (2018). https://doi.org/10.1109/cvpr.2018.00408
2. Bazin, P.L., Pham, D.L.: Topology correction of segmented medical images using a fast marching algorithm. Comput. Methods Program. Biomed. **88**(2), 182–190 (2007). https://doi.org/10.1016/j.cmpb.2007.08.006
3. Ben-Hamu, H., Maron, H., Kezurer, I., Avineri, G., Lipman, Y.: Multi-chart generative surface modeling. ACM Trans. Graph. **37**(6), 1–15 (2019). https://doi.org/10.1145/3272127.3275052
4. Charles, R.Q., Su, H., Kaichun, M., Guibas, L.J.: PointNet: deep learning on point sets for 3D classification and segmentation. In: IEEE Conference on Computer Vision and Pattern Recognition (CVPR) (2017). https://doi.org/10.1109/cvpr.2017.16
5. Chen, Z., Zhang, H.: Learning implicit fields for generative shape modeling. In: IEEE Conference on Computer Vision and Pattern Recognition (CVPR), pp. 5939–5948 (2019)

6. Choy, C.B., Xu, D., Gwak, J.Y., Chen, K., Savarese, S.: 3D-R2N2: a unified approach for single and multi-view 3D object reconstruction. In: Leibe, B., Matas, J., Sebe, N., Welling, M. (eds.) ECCV 2016. LNCS, vol. 9912, pp. 628–644. Springer, Cham (2016). https://doi.org/10.1007/978-3-319-46484-8_38

7. Cruz, R.S., Lebrat, L., Bourgeat, P., Fookes, C., Fripp, J., Salvado, O.: DeepCSR: a 3D deep learning approach for cortical surface reconstruction. arXiv preprint arXiv:2010.11423 (2020)

8. Desbrun, M., Meyer, M., Schröder, P., Barr, A.H.: Implicit fairing of irregular meshes using diffusion and curvature flow (1999)

9. Dubuisson, M.P., Jain, A.K.: A modified hausdorff distance for object matching. In: Proceedings of the International Conference on Pattern Recognition (ICPR), vol. 1, pp. 566–568 (1994)

10. Fischl, B.: FreeSurfer. NeuroImage **62**(2), 774–781 (2012). https://doi.org/10.1016/j.neuroimage.2012.01.021

11. Glasser, M.F., et al.: The minimal preprocessing pipelines for the Human Connectome Project. NeuroImage **80**, 105–124 (2013). https://doi.org/10.1016/j.neuroimage.2013.04.127

12. Jenkinson, M., Beckmann, C.F., Behrens, T.E., Woolrich, M.W., Smith, S.M.: FSL. NeuroImage **62**(2), 782–790 (2012). https://doi.org/10.1016/j.neuroimage.2011.09.015

13. Kuperberg, G.R., et al.: Regionally localized thinning of the cerebral cortex in schizophrenia. Arch. Gen. Psychiatry **60**(9), 878–888 (2003). https://doi.org/10.1001/archpsyc.60.9.878

14. Li, G., Nie, J., Wu, G., Wang, Y., Shen, D., Initiative, A.D.N., et al.: Consistent reconstruction of cortical surfaces from longitudinal brain MR images. Neuroimage **59**(4), 3805–3820 (2012)

15. Li, H., Wei, D., Cao, S., Ma, K., Wang, L., Zheng, Y.: Superpixel-guided label softening for medical image segmentation. In: Martel, A.L., et al. (eds.) MICCAI 2020. LNCS, vol. 12264, pp. 227–237. Springer, Cham (2020). https://doi.org/10.1007/978-3-030-59719-1_23

16. Litany, O., Bronstein, A., Bronstein, M., Makadia, A.: Deformable shape completion with graph convolutional autoencoders. In: IEEE Conference on Computer Vision and Pattern Recognition (CVPR) (2018). https://doi.org/10.1109/cvpr.2018.00202

17. Lorensen, W.E., Cline, H.E.: Marching cubes: a high resolution 3D surface construction algorithm. ACM SIGGRAPH Comput. Graph. **21**(4), 163–169 (1987)

18. Maron, H.: Convolutional neural networks on surfaces via seamless toric covers. ACM Trans. Graph. **36**(4), 1–10 (2017). https://doi.org/10.1145/3072959.3073616

19. Mescheder, L., Oechsle, M., Niemeyer, M., Nowozin, S., Geiger, A.: Occupancy networks: learning 3D reconstruction in function space. In: IEEE Conference on Computer Vision and Pattern Recognition (CVPR), pp. 4460–4470 (2019)

20. Park, J.J., Florence, P., Straub, J., Newcombe, R., Lovegrove, S.: DeepSDF: learning continuous signed distance functions for shape representation. In: IEEE Conference on Computer Vision and Pattern Recognition (CVPR), pp. 165–174 (2019)

21. Qi, C.R., Yi, L., Su, H., Guibas, L.J.: PointNet++: deep hierarchical feature learning on point sets in a metric space. In: Guyon, I., Luxburg, U.V., Bengio, S., Wallach, H., Fergus, R., Vishwanathan, S., Garnett, R. (eds.) Advances in Neural Information Processing Systems, vol. 30, pp. 5099–5108 (2017)

22. Sinha, A., Bai, J., Ramani, K.: Deep learning 3D shape surfaces using geometry images. In: Leibe, B., Matas, J., Sebe, N., Welling, M. (eds.) ECCV 2016. LNCS, vol. 9910, pp. 223–240. Springer, Cham (2016). https://doi.org/10.1007/978-3-319-46466-4_14

23. Tang, X., et al.: Regional subcortical shape analysis in premanifest huntington's disease. Hum. Brain Mapp. **40**(5), 1419–1433 (2019). https://doi.org/10.1002/hbm.24456

24. Wu, Z., Song, S., Khosla, A., Yu, F., Zhang, L., Tang, X., Xiao, J.: 3D ShapeNets: a deep representation for volumetric shapes. In: IEEE Conference on Computer Vision and Pattern Recognition (CVPR) (2015). https://doi.org/10.1109/cvpr.2015.7298801

25. Xu, Q., Wang, W., Ceylan, D., Mech, R., Neumann, U.: DISN: deep implicit surface network for high-quality single-view 3D reconstruction. In: Advances in Neural Information Processing Systems, pp. 492–502 (2019)

26. Yang, Y., Feng, C., Shen, Y., Tian, D.: FoldingNet: point cloud auto-encoder via deep grid deformation. In: IEEE Conference on Computer Vision and Pattern Recognition (CVPR) (2018). https://doi.org/10.1109/cvpr.2018.00029

Clinically Correct Report Generation from Chest X-Rays Using Templates

Pablo Pino[1,3]($\boxtimes$)(ID), Denis Parra[1,3](ID), Cecilia Besa[2,4], and Claudio Lagos[2,4](ID)

[1] Department of Computer Science, Pontificia Universidad Católica de Chile,
Santiago, Chile
pdpino@uc.cl, dparra@ing.puc.cl
[2] School of Medicine, Pontificia Universidad Católica de Chile, Santiago, Chile
{cbesa,crlagos}@uc.cl
[3] Millennium Institute Foundational Research on Data, ANID, Santiago, Chile
[4] Millennium Nucleus in Cardiovascular Magnetic Resonance, ANID, Santiago, Chile

Abstract. We address the task of automatically generating a medical report from chest X-rays. Many authors have proposed deep learning models to solve this task, but they focus mainly on improving NLP metrics, such as BLEU and CIDEr, which are not suitable to measure clinical correctness in clinical reports. In this work, we propose CNN-TRG, a Template-based Report Generation model that detects a set of abnormalities and verbalizes them via fixed sentences, which is much simpler than other state-of-the-art NLG methods and achieves better results in medical correctness metrics.

We benchmark our model in the IU X-ray and MIMIC-CXR datasets against naive baselines as well as deep learning-based models, by employing the Chexpert labeler and MIRQI as clinical correctness evaluations, and NLP metrics as secondary evaluation. We also provide further evidence indicating that traditional NLP metrics are not suitable for this task by presenting their lack of robustness in multiple cases. We show that slightly altering a template-based model can increase NLP metrics considerably while maintaining high clinical performance. Our work contributes by a simple but effective approach for chest X-ray report generation, as well as by supporting a model evaluation focused primarily on clinical correctness metrics and secondarily on NLP metrics.

Keywords: Image report generation · Deep learning · Templates

1 Introduction

Writing a report from medical image studies is an important daily activity for radiologists, yet it is a time-consuming and error-prone task, even for experienced radiologists. AI could alleviate this workload on physicians by providing

---

Electronic supplementary material The online version of this chapter (https://doi.org/10.1007/978-3-030-87589-3_67) contains supplementary material, which is available to authorized users.

© Springer Nature Switzerland AG 2021
C. Lian et al. (Eds.): MLMI 2021, LNCS 12966, pp. 654–663, 2021.
https://doi.org/10.1007/978-3-030-87589-3_67

Findings: Heart size is mildly enlarged. There are diffusely increased interstitial opacities bilaterally. No focal consolidation, pneumothorax, or pleural effusion. No acute bony abnormality.
Impression: Findings concerning for interstitial edema or infection.

Fig. 1. Imaging study example from the IU X-ray dataset

computer-aided diagnosis (CAD) systems that can analyze an imaging study and generate a written report, which could be used as a starting point by a radiologist to iterate until producing a final report. For chest X-rays, typically, the radiologists examine one or more images from a patient, indicate if there are abnormalities, describe their visual characteristics, and provide a diagnostic or conclusion. Figure 1 provides an example from the IU X-ray dataset [4].

Many deep learning models are proposed in the literature to generate written reports from one or more images [1–3,9,10,13,14,17,20,27] employing encoder-decoder architectures or using an image encoder followed by a retrieval or paraphrasing approach. However, it is hard to tell how ready these approaches are for regular clinical use since they are traditionally evaluated by Natural Language Processing (NLP) metrics, such as BLEU [21] or CIDEr-D [28], and these may not be suitable to measure correctness in the medical domain [2,16,17,22,27]. For instance, research on the BLEU metric supports its use for evaluating Machine Translation (MT), but not for other tasks [18,24]. To overcome this problem, some authors have used metrics to evaluate the clinical correctness of the generated reports, such as Chexpert labeler [8] and MIRQI [31], although these have not been tested with expert clinicians nor widely used yet. Moreover, there is a lack of studies on explainability of these systems. This is a highly relevant aspect, since the decisions made from the system predictions will have a direct impact on patients in a clinical setting [25].

In this work, we address the task of report generation from chest X-rays and we make two main contributions. First, we propose CNN-TRG, a deep learning model that detects the presence or absence of abnormalities and then generates the report by relying on a set of pre-defined templates, achieving better performance than state-of-the-art methods in terms of clinical correctness. We also design our model to be simpler and more transparent than the typical encoder-decoder approaches, allowing more control in terms of interpretation. Second, we provide evidence that some traditional NLP metrics are not suitable for evaluating this task by showing they are not robust to textual changes in the reports. Thus, we show that our model can improve its performance in these metrics without affecting clinical correctness.

2 Background and Related Work

Data: Report Structure and Content. Literature [19] shows that the two main datasets used in this task are IU X-ray [4] and MIMIC-CXR [12]. Both contain chest X-rays and their reports written by radiologists, which have two sections of interest: *findings* and *impression*. In *findings*, the radiologist indicates the presence or absence of abnormalities and describes visual characteristics of the positive findings, such as location and severity, among others. In *impression*, the radiologist summarizes the observations into a diagnostic or conclusion. See the example in the Fig. 1.

Table 1. Example of a ground-truth *findings* sections and three generated reports, with BLEU (B), ROUGE-L (R-L) and chexpert metrics calculated. Correct and incorrect sentences are **bold** and *italics*, respectively.

Report	NLP		Chexpert		
	B	R-L	F-1	P	R
Ground-truth: Heart size is mildly enlarged. Small right pneumothorax is seen.	–	–	–	–	–
Heart size is normal. No pneumothorax is seen.	0.493	0.715	0	0	0
The cardiac silhouette is enlarged. *No pneumothorax.*	0.146	0.464	0.5	0.5	0.5
Mild cardiomegaly. Pneumothorax on right lung.	0.075	0.289	1	1	1

Authors addressing the task of report generation [19] choose one or both of these sections to be generated automatically. We argue that the main information required to write the *findings* section can be observed directly from an image. On the contrary, writing the *impression* may require additional information, such as analyzing multiple views together (frontal and lateral), checking patient symptoms, comparing with prior imaging exams or using medical knowledge that cannot necessarily be inferred from the images alone. Hence, in this article we start by proposing a method to generate the *findings* section of the reports.

Metrics and Clinical Correctness. Most works evaluate the report generation performance using NLP metrics, such as BLEU [21], CIDEr-D, [28] and ROUGE-L [15], which measure n-gram matching between the ground truth and a generated text. These metrics are very popular in machine translation and other NLP tasks; however, they may not be suitable to measure correctness in clinical reports [2,16,17,22,27] or in other tasks [18,24]. To overcome this, some authors have used other metrics to measure the medical accuracy of generated reports. In six previous works [2,3,13,16,17,20], the authors employed the Chexpert labeler [8], a rule-based tool that detects a set of 13 abnormalities from the generated and ground truth reports, and then evaluated these findings using classification metrics. Similarly, Zhang et al. [31] proposed MIRQI, which labels 20 abnormalities and captures visual characteristics described (location, size, etc.). Some authors [1,7,9,30] have used other methods for correctness evaluation, but they do not provide an implementation. To the best of our knowledge, none of these metrics have been validated with expert clinicians, but they aim at clinical accuracy, unlike NLP metrics.

Consider the examples from Table 1, showing a ground truth example, three generated reports, and the performance they achieve using some of these metrics. A generated sample can be clinically incorrect and achieve high NLP scores, or be correct and achieve low NLP scores. Thus, we emphasize the importance of clinical correctness metrics in this work using the Chexpert labeler and MIRQI as primary metrics above traditional NLP metrics.

Models. The most common approach in the literature derives from the general domain image captioning task with encoder-decoder architectures. Most works use common Convolutional Neural Networks (CNNs) as encoder (e.g. Densenet [6]), and as decoder: a single LSTM to generate word by word [2]; two LSTMs arranged hierarchically to generate sentences and words [9,10,16,31]; or a Transformer-based network [3,17,29]. Some authors [1,13,14,20,27] have employed retrieval or hybrid retrieval-generation approaches for text generation. Compared to other methods, our proposed model CNN-TRG is much simpler, as it uses fewer templates and a more straightforward retrieval process; we test it more thoroughly in the two main datasets available and primarily using medical correctness metrics; and is able to achieve much higher clinical performance.

3   Template-Based Report Generation: CNN-TRG

We propose CNN-TRG, a template-based model that detects abnormalities in the image using a CNN as a classifier and relies on fixed sentences as templates for the text generation. To detect abnormalities, we implement a CNN that receives a chest X-ray and performs multi-label classification of the presence of 13 abnormalities (the chexpert set of labels except for *"No Finding"*). We use Densenet-121 [6], which has shown good results in report generation [19] as well as other medical-related tasks [23]. We trained it using a binary cross-entropy loss for 40 epochs, and applied early stopping by optimizing the PR-AUC classification metric. We initialized the network with the pre-trained weights from ImageNet [5], then trained on the Chexpert dataset [8] for the same classification task, and lastly fine-tuned in the target dataset (IU X-ray or MIMIC-CXR). We used the PyTorch framework[1] in our implementation[2].

For text generation, we manually curated a set of two sentences per abnormality indicating presence and absence, totaling 26 sentences. We built the templates by examining the reports and picking existing sentences or creating new ones. To generate the full report, the image is fed to the CNN to compute the binary classification, then the corresponding absence or presence template is chosen for each abnormality, and the sentences are concatenated into the final report. Figure 2 shows the process, and the full details follow in the supplementary material.

We tested the model using two template sets: *single* and *grouped*. Both provide the same meaning clinically (in terms of the presence of the 13 abnormalities), but are written differently.

---

[1] https://pytorch.org/.
[2] https://pdpino.github.io/clinically-correct.

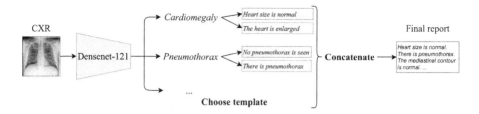

Fig. 2. CNN-TRG model for report-generation

Single. Concise sentences that indicate the presence or absence directly, for example: *"No pleural effusion"* and *"Pleural effusion is seen"*. The presence templates do not provide detailed visual characteristics (location, size, etc.), since the classification model does not predict this information.

Grouped. To resemble more the reports from each dataset, we grouped multiple abnormalities into common sentences from the training set. For example, in IU X-ray, if all the lung-related abnormalities are classified as absent, the template chosen is *"The lungs are clear"*, instead of using their individual absence templates. If at least one of the abnormalities does not match the group, the model falls back to the *single* set of individual sentences.

4   Experiments

4.1   Datasets

We perform the experiments with two publicly available datasets: IU X-ray[3] [4] and MIMIC-CXR[4] [11,12]. Both contain frontal and lateral chest X-rays, IU X-ray has 7,470 images and 3,955 reports, whilst MIMIC-CXR has 377,110 images and 227,827 reports. We used the official train-validation-test split for MIMIC-CXR, and we split the IU X-ray dataset in 80%-10%-10% proportions. We used the *findings* section of the reports and kept only frontal X-rays, leaving a total of 3,311 images in IU X-ray and 243,326 in MIMIC-CXR. To train the CNN in the classification task, we used the chexpert labels provided in MIMIC-CXR, and computed them for IU X-ray applying the Chexpert labeler [8] to the reports.

4.2   Metrics

We used three NLP metrics: BLEU (denoted as B, calculated as the average of BLEU 1-4), ROUGE-L (R-L), and CIDEr-D (C-D), implemented in a publicly available python library[5]. CIDEr-D ranges from 0 (worst) to 10 (best), while the others from 0 (worst) to 1 (best). As clinical correctness metrics, we used

---

[3] https://openi.nlm.nih.gov/faq.
[4] https://physionet.org/content/mimic-cxr-jpg/2.0.0/.
[5] https://github.com/salaniz/pycocoevalcap.

Chexpert labeler[6] [8] and MIRQI[7] [31], which were detailed in Sect. 2. In both cases, we provide F1-score (F-1), precision (P), and recall (R). The chexpert values are the macro average across the 14 labels.

4.3 Baselines

Naive Models. We implement three simple baselines that are not clinically useful, but provide a reference value for the metrics. *Constant*: returns the same report for all the images, manually curated using common sentences from the dataset describing a healthy subject. *Random*: returns a random report from the training set. *1-nn (nearest-neighbor)*: returns the report from the nearest image in the training set, using CNN extracted features from the images as feature space. We used the same CNN as the CNN-TRG model.

Encoder-Decoder Model. We use the CNN from CNN-TRG as encoder and a LSTM with a visual attention mechanism as decoder. The model is trained to generate the full report word by word from the input images. We froze the CNN weights during the report-generation training to avoid over-fitting, and applied early stopping by optimizing the chexpert F-1 score in the validation set. For the LSTM, we used a hidden size of 512 and word embeddings of size 100 initialized with the pre-trained RadGlove [32].

Literature Models. We compare our approach with the results from eleven models [1–3,10,13,14,16,17,20,27,31]. We re-implemented the CoAtt model [10], and for the rest we show the results from their papers.

5 Results

Table 2 shows a benchmark of our model against all baselines in both datasets, using the test split. We discuss the results next.

Template Sets. As expected, the clinical performance is the same for both *single* and *grouped* sets, since their clinical meaning is unchanged, but the *grouped* set achieves higher NLP performance, particularly in the IU X-ray dataset. Thus, we show that we can improve NLP metrics only by using more common sentences while preserving the clinical correctness in terms of the seen abnormalities.

CNN-TRG Clinical Correctness. Our template-based models outperform all other models in terms of clinical correctness, both in chexpert and MIRQI F-1 scores. Specifically, our model achieves much better performance than (1) the naive methods, showing it surpasses a first lower standard; and (2) the deep learning models, proving our approach to be more effective while simpler. We also present the results in chexpert F-1 score for each disease in the supplementary material, showing that our model surpasses all other models in every abnormality.

---

[6] https://github.com/stanfordmlgroup/chexpert-labeler.
[7] https://github.com/xiaosongwang/MIRQI.

Table 2. Results in IU X-ray and MIMIC-CXR. Chexpert metrics are macro-averaged across labels. [f+i] indicates they generated both *findings* and *impression* sections concatenated, while the rest generated *findings* only; [*] indicates we re-implemented the code; [Ab] indicates they used a subset of the data only with reports that have one or more abnormal findings; super script letters R, T and L indicate Retrieval, Transformer and LSTM-based approaches.

	Model	NLP			Chexpert			MIRQI		
		B	R-L	C-D	F-1	P	R	F-1	P	R
IU X-ray	Constant	0.297	0.366	0.307	0.038	0.026	0.071	0.469	0.462	0.481
	Random	0.202	0.284	0.145	0.066	0.065	0.068	0.374	0.378	0.384
	1-nn	0.220	0.301	0.245	0.145	0.150	0.144	0.497	0.508	0.500
	CNN-LSTM-att[L]	0.202	0.319	0.208	0.140	0.159	0.148	0.484	0.492	0.487
	CoAtt*[10][L]	0.231	0.316	0.221	0.144	0.162	0.147	0.491	0.503	0.491
	Zhang et al.[31][L,f+i]	0.271	0.367	0.304	–	–	–	0.478	0.490	0.483
	CLARA [1][R]	0.302	–	**0.359**	–	–	–	–	–	–
	KERP [14][R]	0.299	0.339	0.280	–	–	–	–	–	–
	RTEX [13][R]	–	0.202	–	–	0.193	0.222	–	–	–
	S-M et al.[27][R,f+i]	**0.515**	**0.580**	–	–	–	–	–	–	–
	CNN-TRG single	0.167	0.282	0.030	**0.239**	**0.225**	**0.357**	**0.529**	0.534	**0.540**
	CNN-TRG grouped	0.273	0.352	0.249	**0.239**	**0.225**	**0.357**	**0.529**	0.535	0.540
MIMIC-CXR	Constant	0.137	0.201	0.059	0.021	0.012	0.071	0.163	0.158	0.176
	Random	0.073	0.142	0.078	0.163	0.186	0.151	0.359	0.372	0.362
	1-nn	0.119	0.193	0.151	0.320	0.325	0.319	0.635	0.645	0.641
	CNN-LSTM-att [L]	0.103	0.244	0.479	0.308	0.378	0.297	0.644	0.652	**0.648**
	CoAtt*[10][L]	0.120	0.252	0.401	0.201	0.356	0.198	0.544	0.551	0.545
	Boag et al. [2][L]	0.184	–	0.850	0.186	0.304	–	–	–	–
	Liu et al. [16][L]	0.192	0.306	**1.046**	–	0.309	0.134	–	–	–
	Chen et al. [3][T]	0.205	0.277	–	0.276	0.333	0.273	–	–	–
	Lovelace et al. [17][T]	**0.257**	**0.318**	0.316	0.228	0.333	0.217	–	–	–
	CVSE [20][R,Ab]	–	0.153	–	0.253	0.317	0.224	–	–	–
	RTEX [13][R]	–	0.205	–	–	0.229	0.284	–	–	–
	CNN-TRG single	0.080	0.151	0.026	**0.428**	**0.381**	**0.531**	**0.668**	**0.749**	0.640
	CNN-TRG grouped	0.094	0.185	0.238	**0.428**	**0.381**	**0.531**	0.666	0.746	0.637

NLP vs Clinical Correctness. Naive models achieve higher NLP performance than CNN-TRG and comparable to some literature models, even though they are not clinically useful by design. On the other hand, naive models achieve very low performance on chexpert and MIRQI, whereas the *CNN-LSTM-att*, literature and CNN-TRG models show higher values. This suggests that these clinical correctness metrics are better to differentiate automated systems than NLP metrics.

Model Transparency. An advantage of the CNN-TRG model over an end-to-end deep learning approach is the increased transparency. For example, the CNN-LSTM-att baseline performs abnormality detection and text generation inside a black box, while our template-based uses a fully transparent text generation process. Furthermore, by design, our method allows providing a local explanation for each disease independently. Consider Fig. 3 showing an input image, the ground truth and generated report, and a Grad-CAM [26] heatmap indicating feature importance for *Cardiomegaly*, the only abnormality found.

CXR	Generated by CNN-TRG	Ground Truth
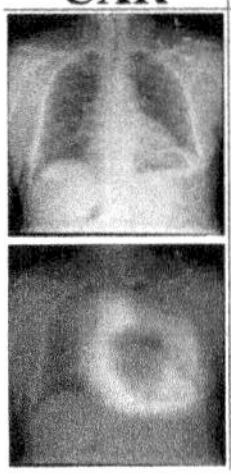	The heart is enlarged. The mediastinal contour is normal. No focal consolidation. The lungs are free of focal airspace disease. No atelectasis. No pleural effusion. No fibrosis. No pneumonia. No pneumothorax is seen. No pulmonary edema. No pulmonary nodules or mass lesions identified. No fracture is seen.	The heart is mildly enlarged. Left hemidiaphragm is elevated. There is no acute infiltrate or pleural effusion. The mediastinum is unremarkable.

Fig. 3. Example of a report generated with the CNN-TRG using the *single* set of templates, and a Grad-CAM heatmap indicating the activations for the *Cardiomegaly* classification. The colors indicate correct sentences. Best viewed in color. (Color figure online)

6   Limitations

The main limitation of our work is that we mostly report the results presented in the original articles. The comparison with other methods could then be improved, since most articles do not provide clinical correctness metrics, and the evaluation protocols may vary. Only MIMIC-CXR has an official train-test split, so the IU X-ray dataset could be more affected by this problem. Additionally, both our templates and the chexpert metric are limited by the set of 13 abnormalities, disregarding their visual characteristics and other chest pathologies. Lastly, our templates are specific to chest X-ray datasets. Hence, in order to use our method with other image modalities or body parts, we would have to manually curate a set of templates covering relevant abnormalities.

7   Conclusions and Future Work

We address the task of automatically generating a text report from chest X-rays and establish a new state-of-the-art in terms of clinical correctness. We present report examples and naive models which challenge the reliability of some traditional NLP metrics to measure model performance, suggesting that text similarity measures might not be suitable in this task. We believe this field should shift to favor clinical correctness instead of traditional NLP metrics to evaluate the systems more appropriately.

As future work, we will replicate implementations from some papers to evaluate and compare their performance under the same experimental conditions. Additionally, we will improve the template-based model by detecting more abnormalities and their visual characteristics, such as location, severity, and more, to provide a more detailed description. We will leverage the templates available at the Radiological Society of North America website[8]. Lastly, we will further study the clinical correctness evaluation problem by studying the existing metrics, proposing new ones, and validating them with expert radiologists.

---

[8] https://radreport.org/.

Acknowledgments. This work was partially funded by ANID, Millennium Science Initiative Program, Code ICN17_002 and by ANID, Fondecyt grant 1191791.

References

1. Biswal, S., Xiao, C., Glass, L.M., Westover, B., Sun, J.: Clara: clinical report auto-completion. In: The Web Conference (2020). https://doi.org/10.1145/3366423.3380137
2. Boag, W., Hsu, T.M.H., Mcdermott, M., Berner, G., Alesentzer, E., Szolovits, P.: Baselines for chest X-ray report generation. In: ML4H at NeurIPS (2020)
3. Chen, Z., Song, Y., Chang, T.H., Wan, X.: Generating radiology reports via memory-driven transformer. In: EMNLP (2020). https://doi.org/10.18653/v1/2020.emnlp-main.112
4. Demner-Fushman, D., et al.: Preparing a collection of radiology examinations for distribution and retrieval. JAMIA (2015). https://doi.org/10.1093/jamia/ocv080
5. Deng, J., Dong, W., Socher, R., Li, L., Li, K., Fei-Fei, L.: ImageNet: a large-scale hierarchical image database. In: CVPR (2009). https://doi.org/10.1109/CVPR.2009.5206848
6. Huang, G., Liu, Z., Van Der Maaten, L., Weinberger, K.Q.: Densely connected convolutional networks. In: CVPR (2017). https://doi.org/10.1109/CVPR.2017.243
7. Huang, X., Yan, F., Xu, W., Li, M.: Multi-attention and incorporating background information model for chest x-ray image report generation. IEEE Access (2019). https://doi.org/10.1109/ACCESS.2019.2947134
8. Irvin, J., et al.: CheXpert: a large chest radiograph dataset with uncertainty labels and expert comparison. In: AAAI Conference on Artificial Intelligence (2019). https://doi.org/10.1609/aaai.v33i01.3301590
9. Jing, B., Wang, Z., Xing, E.: Show, describe and conclude: on exploiting the structure information of chest x-ray reports. In: ACL (2019). https://doi.org/10.18653/v1/P19-1657
10. Jing, B., Xie, P., Xing, E.: On the automatic generation of medical imaging reports. In: ACL (2018). https://doi.org/10.18653/v1/P18-1240
11. Johnson, A., et al.: MIMIC-CXR-JPG-chest radiographs with structured labels (version 2.0.0). PhysioNet (2019). https://doi.org/10.13026/8360-t248
12. Johnson, A.E.W., et al.: MIMIC-CXR, a de-identified publicly available database of chest radiographs with free-text reports. Sci. Data (2019). https://doi.org/10.1038/s41597-019-0322-0
13. Kougia, V., Pavlopoulos, J., Papapetrou, P., Gordon, M.: RTEX: a novel framework for ranking, tagging, and explanatory diagnostic captioning of radiography exams. JAMIA (2021). https://doi.org/10.1093/jamia/ocab046
14. Li, C.Y., Liang, X., Hu, Z., Xing, E.P.: Knowledge-driven encode, retrieve, paraphrase for medical image report generation. In: AAAI Conference on Artificial Intelligence (2019). https://doi.org/10.1609/aaai.v33i01.33016666
15. Lin, C.Y.: ROUGE: a package for automatic evaluation of summaries. In: Text Summarization Branches Out (2004)
16. Liu, G., et al.: Clinically accurate chest x-ray report generation. In: ML4H (2019)
17. Lovelace, J., Mortazavi, B.: Learning to generate clinically coherent chest X-ray reports. In: EMNLP (2020). https://doi.org/10.18653/v1/2020.findings-emnlp.110

18. Mathur, N., Baldwin, T., Cohn, T.: Tangled up in BLEU: Reevaluating the evaluation of automatic machine translation evaluation metrics. In: ACL (2020). https://doi.org/10.18653/v1/2020.acl-main.448
19. Messina, P., et al.: A survey on deep learning and explainability for automatic image-based medical report generation (2020)
20. Ni, J., Hsu, C.N., Gentili, A., McAuley, J.: Learning visual-semantic embeddings for reporting abnormal findings on chest X-rays. In: EMNLP (2020). https://doi.org/10.18653/v1/2020.findings-emnlp.176
21. Papineni, K., Roukos, S., Ward, T., Zhu, W.J.: BLEU: a method for automatic evaluation of machine translation. In: ACL (2002). https://doi.org/10.3115/1073083.1073135
22. Pino, P., Parra, D., Messina, P., Besa, C., Uribe, S.: Inspecting state of the art performance and NLP metrics in image-based medical report generation. arXiv preprint arXiv:2011.09257 (2020). In LXAI at NeurIPS 2020
23. Rajpurkar, P., et al.: CheXNet: radiologist-level pneumonia detection on chest x-rays with deep learning (2017)
24. Reiter, E.: A structured review of the validity of BLEU. Comput. Linguist. (2018). https://doi.org/10.1162/coli_a_00322
25. Reyes, M., et al.: On the interpretability of artificial intelligence in radiology: Challenges and opportunities. Radiol. Artif. Intell. (2020). https://doi.org/10.1148/ryai.2020190043
26. Selvaraju, R.R., Cogswell, M., Das, A., Vedantam, R., Parikh, D., Batra, D.: Grad-CAM: Visual explanations from deep networks via gradient-based localization. In: ICCV, pp. 618–626 (2017). https://doi.org/10.1109/ICCV.2017.74
27. Syeda-Mahmood, T., et al.: Chest X-ray report generation through fine-grained label learning. In: Martel, A.L., et al. (eds.) MICCAI 2020. LNCS, vol. 12262, pp. 561–571. Springer, Cham (2020). https://doi.org/10.1007/978-3-030-59713-9_54
28. Vedantam, R., Lawrence Zitnick, C., Parikh, D.: CIDEr: consensus-based image description evaluation. In: CVPR (2015). https://doi.org/10.1109/CVPR.2015.7299087
29. Xiong, Y., Du, B., Yan, P.: Reinforced transformer for medical image captioning. In: MLMI (2019). https://doi.org/10.1007/978-3-030-32692-0_77
30. Xue, Y., et al.: Multimodal recurrent model with attention for automated radiology report generation. In: Frangi, A.F., Schnabel, J.A., Davatzikos, C., Alberola-López, C., Fichtinger, G. (eds.) MICCAI 2018. LNCS, vol. 11070, pp. 457–466. Springer, Cham (2018). https://doi.org/10.1007/978-3-030-00928-1_52
31. Zhang, Y., Wang, X., Xu, Z., Yu, Q., Yuille, A., Xu, D.: When radiology report generation meets knowledge graph. In: AAAI Conference on Artificial Intelligence (2020). https://doi.org/10.1609/aaai.v34i07.6989
32. Zhang, Y., Ding, D.Y., Qian, T., Manning, C.D., Langlotz, C.P.: Learning to summarize radiology findings. In: LOUHI at NeurIPS (2018). https://doi.org/10.18653/v1/W18-5623

Extracting Sequential Features from Dynamic Connectivity Network with rs-fMRI Data for AD Classification

Kai Lin[1], Biao Jie[1(✉)], Peng Dong[1], Xintao Ding[1], Weixin Bian[1], and Mingxia Liu[2]

[1] School of Computer and Information, Anhui Normal University,
Wuhu 241003, Anhui, China
jbiao@ahnu.edu.cn

[2] Department of Radiology and BRIC, University of North Carolina at Chapel Hill,
Chapel Hill, NC 27599, USA

Abstract. Dynamic functional connectivity (dFC) networks based on resting-state functional magnetic resonance imaging (rs-fMRI) can help us understand the function of brain better, and have been applied to brain disease identification, such as Alzheimer's disease (AD) and its early stages (*i.e.*, mild cognitive impairment, MCI). Deep learning (*e.g.*, convolutional neural network, CNN) methods have been recently applied to dynamic FC network analysis, and achieve good performance compared to traditional machine learning methods. Existing studies usually ignore sequence information of temporal features from dynamic FC networks. To this end, in this paper, we propose a recurrent neural network-based learning framework to extract sequential features from dynamic FC networks with rs-fMRI data for brain disease classification. Experimental results on 174 subjects with baseline resting-state functional MRI (rs-fMRI) data from ADNI demonstrate the effectiveness of our proposed method in binary and multi-category classification tasks.

1 Introduction

Alzheimer's disease (AD) is a common neurodegenerative disease in the elderly and the most common cause of dementia, characterized by progressive cognitive impairment [1]. Its prodromal stage, called mild cognitive impairment (MCI), has received widespread attention because of its high possibility of developing into AD [2–4]. The accurate progression prediction of AD-related disorders is of great significance for early treatment and delaying the deterioration of the disease. Resting-state functional magnetic resonance imaging (rs-fMRI) provides a non-invasive solution to objectively measure the activity of human brain neurons [5]. Functional connectivity (FC) networks based on rs-fMRI can characterize neural interactions between brain regions and have been applied to various AD-related brain disease diagnosis [6–10].

The original version of this chapter was revised: The Acknowledgment section has been updated. The correction to this chapter is available at
https://doi.org/10.1007/978-3-030-87589-3_72

© Springer Nature Switzerland AG 2021, corrected publication 2022
C. Lian et al. (Eds.): MLMI 2021, LNCS 12966, pp. 664–673, 2021.
https://doi.org/10.1007/978-3-030-87589-3_68

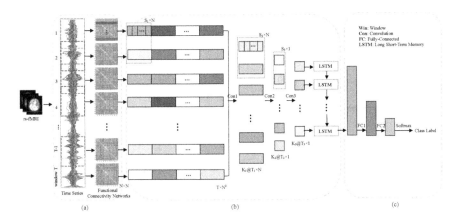

Fig. 1. Illustration of the proposed RNN-based learning framework for sequential feature extraction and classification with rs-fMRI data, including three parts: (a) dynamic functional connection network construction, (b) temporal features and sequential features extraction, and (c) classification.

Existing research on FC networks is generally based on temporal correlation between distributed brain regions, implicitly assuming that functional connectivity of the human brain is static during the entire fMRI recording period [11]. Thus, these studies ignore the dynamic characteristics of brain networks. Recent studies have shown that functional connectivity exhibits significant dynamic changes [12,13]. Therefore, many studies focus on analyzing dynamic FC networks, and have a deeper understanding of the basic characteristics of the brain network [14] and dysfunction [15]. Recently, deep learning methods have been applied to the analysis and classification of dynamic FC networks. For example, constitutional neural network-based methods [16,17] have been applied to brain disease diagnosis, and achieve good performance compared to traditional machine learning methods. However, sequence information of temporal features from dynamic FC networks is usually ignored in these studies.

To this end, in this paper, we propose a recurrent neural network (RNN) based learning framework to extract sequential features from dynamic FC networks with rs-fMRI data for AD-related brain disease classification. Figure 1 illustrates the proposed RNN framework for sequential feature extraction and classification with rs-fMRI data, including three parts: (a) dynamic functional connection network construction, (b) temporal features and sequential features extraction, and (c) classification. Specifically, we first divide rs-fMRI time series into multiple overlapping segments using a fixed-size sliding window. For each time series segment, an FC network was constructed by calculating the Pearson's correlation coefficient of blood oxygen level dependent (BOLD) signals from paired brain regions. Then, each FC network is expanded into a row vector and then spliced into a matrix. Then, we construct three convolutional layers to extract high-level features from low-level functional connectivity, and we use the long short-term memory (LSTM) layer to capture long-term temporal dynamics. Finally, two fully connected layers and a softmax layer are used to classify brain diseases. We evaluate the proposed method on 174 subjects with 563 rs-fMRI

Table 1. Characteristics of the studied subjects (Mean ± Standard Deviation). MMSE: Mini-Mental State Examination.

Group	AD	lMCI	eMCI	NC
Male/Female	16/15	27/18	20/30	20/28
Age	74.7 ± 7.4	72.3 ± 8.1	72.4 ± 7.1	76.0 ± 6.8
MMSE	21.8 ± 3.3	27.1 ± 2.1	28.1 ± 1.6	28.8 ± 1.4

scans from the Alzheimer's Disease Neuroimaging Initiative (ADNI) database.[1] The experimental results demonstrate that our proposed method helps improve diagnostic performance.

2   Method

2.1   Subjects and Data Preprocessing

The rs-fMRI data obtained from the ADNI database were studied. We use rs-fMRI data from 174 subjects, including 31 AD, 45 late MCI (lMCI), 50 early MCI (eMCI), and 48 normal controls (NCs). It is worth noting that one subject may have one or more scans at intervals of 6 months to 1 year. There include 99, 145, 165, 154 scans for AD, lMCI, eMCI and NC subject groups, respectively. The specifications of the data acquired for each scan are as follows: the in-plane image resolution is 2.29–3.31 mm, the slice thickness is 3.31 mm, TE (echo time) = 30 ms, TR (repetition time) = 2.2–3.1 s, and there are 140 volumes (time points) for each subject. The demographic and clinical information of these subjects is summarized in Table 1.

The rs-fMRI data are preprocessed using a standard pipeline in the FSL FEAT software package: (1) removal of the first 3 rs-fMRI volumes, (2) slice timing correction, (3) head motion correction, (4) bandpass filtering and (5) regression of white matter, cerebrospinal fluid, and motion parameters. Using the AAL atlas [18] and a deformable registration method [19], the brain space of fMRI scans for each subject is partitioned into 116 regions-of-interest (ROIs). For each ROI, the average rs-fMRI time series is calculated from BOLD signals in all voxels within specific ROIs, which is used as the input for our method.

2.2   Proposed RNN-Based Learning Framework

As shown in Fig. 1, our framework consists of three parts: (a) dynamic functional connection network construction, (b) temporal features and sequential features extraction, and (c) classification. More details can be found below.

Dynamic FC Network Construction. As shown in Fig. 1 (a), with the mean time series of ROIs, we constructed a dynamic FC network based on continuous and overlapping time windows. For each subject, we first segment all rs-fMRI time series into T continuous and overlapping time windows with the constant length L. Then, an FC network (corresponding to an adjacency matrix)

---

[1] http://adni.loni.usc.edu.

$\mathbf{M}^t \in \mathbb{R}^{N \times N}(t = 1, \cdots, T)$ is constructed by calculating Pearson's correlation coefficient between BOLD signals of paired ROIs at the t-th time window as:

$$\mathbf{M}^t(i, j) = \frac{covar(x_i^t, x_j^t)}{\sigma_{x_i^t} \sigma_{x_j^t}} \qquad (1)$$

where $covar$ represents the covariance between two vectors, and $\sigma_{x_i^t}$ represents the standard deviation of the vector x_i^t. Here, x_i^t and x_j^t represent the BOLD signal segments of the i-th and j-th ROI in the t-th time window, respectively.

According to the definition in Eq. (1), $\mathbf{M}^t(i, j)$ is used to measure the FC between a pair of brain regions. Therefore, given T time windows/segments, we can generate a set of FC networks $\mathcal{M} = \{\mathbf{M}^1, \mathbf{M}^2, \cdots, \mathbf{M}^T\} \in \mathbb{R}^{T \times N \times N}$ to deliver rich dynamic characteristics of brain FC networks.

Temporal Features and Sequential Features Extraction. Based on the dynamic FC network, we employ three convolutional layers to further extract higher-level brain network features, as shown in Fig. 1 (b). Specifically, we first expand the FC network of each time window into a vector of $1 \times N^2$, so the dynamic FC network is transformed into a matrix $\mathcal{G} \in \mathbb{R}^{T \times N^2}$. For each subject, the matrix $\mathcal{G}$ of T time windows is used as the input of the proposed network. Then, we set up three convolutional layers, and set the size of the three-layer convolution kernel to $S_1 \times N$, $S_2 \times N$ and $S_3 \times 1$, and set the stride of each layer along the time dimension and the space dimension to $(1, N)$, $(1, 1)$ and $(2, 1)$, respectively. Each convolutional layer is followed by batch normalization, rectified linear unit (ReLU) activation, and 0.25 dropout.

For temporal and sequential feature extraction, the convolution along the spatial dimension is the feature mapping for each ROI, which can be understood as calculating a single output value for each ROI by calculating the weighted combination of the functional connectivity between each ROI and all ROIs. This is expected to reflect *spatial changes* centered at each specific ROI. The convolution along the time dimension corresponds to different feature mappings of the same ROI, which can be understood as calculating a single output value for each specific ROI by calculating the weighted combination of the functional connectivity between the same ROI and all ROIs in adjacent time windows. This is expected to reflect *temporal changes* of each specific ROI. It should be emphasized that after each layer of convolution operation, higher-level network dynamic characteristics will be obtained. Therefore, given K_1, K_2 and K_3 channels for these three convolutions, we will get a $T_1 \times N \times K_1$ tensor, a $T_2 \times 1 \times K_2$ tensor and a $T_3 \times 1 \times K_3$ tensor in turn.

To capture the interaction between consecutive adjacent time segments, we use recurrent networks to simulate the temporal dynamic pattern of brain activity. Specifically, we first convert the high-level brain network features learned in the previous layer into an ordered sequence, which will then be handed over to the recurrent network for processing. We use the long short-term memory (LSTM) network to capture the time series patterns and dig deeper into the different contributions between the time series. The LSTM architecture used in this study is shown in Fig. 1 (b), including an LSTM layer. In this way, the error signal can easily propagate to the bottom layer of the LSTM, thereby reducing the disappearance of the gradient. The LSTM layer (containing 64 neurons) is a representation of the overall functional characteristics and is used to learn time dynamics along the time step.

Classification. As shown in Fig. 1 (c), with the output of LSTM as input, we employ two fully connected layers and a softmax layer for prediction. The first fully connected layer contains 32 neurons. The second fully connected layer contains 16 neurons. There are 2 and 4 neurons in the last fully connected layer for binary and four-category classification, respectively.

Implementation. For the proposed network shown in Fig. 1, we empirically set the parameters as follows: $N = 116, L = 70, T = 34, T_1 = 33, T_2 = 32, T_3 = 13,$ $S_1 = 2, S_2 = 2, S_3 = 8, K_1 = 8, K_2 = 16$ and $K_3 = 32$. The Adam optimizer with recommended parameters is used for training, and the number of epochs and batch size are empirically set as 200 and 16, respectively.

3   Experiment

3.1   Experimental Setting

In this study, we employed a 5-fold *subject-level* cross-validation (CV) strategy to ensure that there is no overlap between training and test data. Both binary and multi-class classification experiments are performed, including 1) eMCI vs. NC classification, 2) AD vs. NC classification, and 3) AD vs. lMCI vs. eMCI vs. NC classification. Specifically, for each classification task, all subjects are partitioned into 5 subsets (the size of each subset is roughly the same). Each subset is sequentially selected as the test set, while the remaining four subsets are combined to construct the training set. We further select 20% of the training subjects as validation data to determine the optimal parameters of the model. It is worth noting that, to enhance the model's generalization ability, each scan of each subject is treated as an independent sample, but all scans of the same subject have the same class label. We evaluate the performance by calculating the overall accuracy of all categories and the accuracy of each category.

We compare our method with the following four methods. 1) **Baseline**: In this method, a stationary FC network is first constructed for each subject by computing the Pearson correlation coefficient between the time series of any pair of ROIs. Then, the connectivity strengths of stationary FC networks are used as features. A t-test method with a threshold (*i.e.*, p-value < 0.05) is used for feature selection, followed by a linear SVM with default parameters for classification. 2) **SVM**: In this method, a stationary FC network is first constructed for each subject. Then, local clustering coefficients of the stationary FC network are extracted as features, where t-test and a linear SVM with default parameters are also used for feature selection and classification, respectively. 3) **DFCN-mean** [20]: In this method, a dynamic FC network is first constructed for each subject. Then, the

Table 2. Performance of five methods in two binary classification tasks, *i.e.*, eMCI vs. NC and AD vs. NC classifications. ACC = Accuracy.

Method	eMCI vs. NC (%)			AD vs. NC (%)		
	ACC	ACC_{NC}	ACC_{eMCI}	ACC	ACC_{NC}	ACC_{AD}
Baseline	57.1	48.1	65.6	73.3	77.8	66.7
SVM	63.6	50.0	75.0	75.0	80.0	66.7
DFCN-mean	67.7	47.3	84.7	76.4	**100.0**	33.3
CNN	76.2	77.3	75.2	87.8	92.0	80.0
Proposed	**84.5**	**84.0**	**84.8**	**92.8**	96.7	**86.7**

Table 3. Performance of five methods in the multi-class classification task, *i.e.*, AD vs. lMCI vs. eMCI vs. NC classification. ACC = Accuracy.

Method	AD vs. lMCI vs. eMCI vs. NC (%)				
	ACC	ACC_{NC}	ACC_{eMCI}	ACC_{lMCI}	ACC_{AD}
Baseline	30.6	20.0	38.9	30.0	33.3
SVM	35.0	22.0	69.5	21.0	6.7
DFCN-mean	44.0	36.0	**87.6**	22.0	0.0
CNN	52.8	44.7	47.6	**65.0**	40.0
Proposed	**61.7**	**57.3**	57.1	56.0	**46.7**

temporal and spatial mean features of dFC network are extracted. The manifold regularized multi-task feature learning (M^2TFL) and multi-kernel SVM are used for feature selection and classification [21], respectively. 4) **CNN**: As a variant of our method, this method has a similar network architecture, but is implemented without considering the temporal dynamics along with time steps. That is, we replace the LSTM layer in our method with an average pooling layer.

3.2 Classification Performance

The quantitative results achieved by different methods in two binary and one multi-class classification tasks are reported in Table 2 and Table 3, respectively. As can be seen from Table 2 and Table 3, our proposed method outperforms the competing methods most cases. For instance, our proposed method yields the accuracy of 84.5% and 92.8% for eMCI vs. NC classification and AD vs. NC classification, respectively, while the best accuracies obtained by the competing methods are 76.2% and 87.8%, respectively. For the challenging AD vs. lMCI vs. eMCI vs. NC classification task, our proposed method achieves the overall best accuracy of 61.7%, while the second-best overall accuracy of four competing methods is 52.8%. These results suggest the effectiveness of our proposed method in rs-fMRI based brain disease classification.

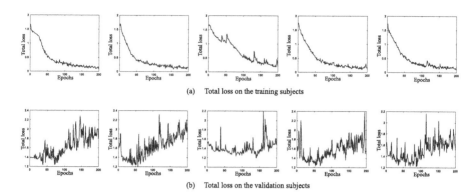

(a)   Total loss on the training subjects

(b)   Total loss on the validation subjects

Fig. 2. Total loss of the proposed method with 200 epochs in each fold cross-validation (from left to right) for AD vs. lMCI vs. eMCI vs. NC classification task. Here, (a) total loss on training data, and (b) total loss on validation data.

From Table 2 and Table 3, one can also have more interesting observations. *First*, CNN-based methods (*i.e.*, CNN and our proposed method) generally outperform traditional learning methods (*i.e.*, Baseline, SVM and DFCN-mean). This suggests that CNN can capture the underlying properties of brain networks, and thus can be applied to various tasks of brain network analysis. *Second*, compared with the CNN method, the proposed method can achieve higher performance, which proves the advantage of exploring the temporal information from functional connectivity networks. *Finally*, compared with stationary FC network-based methods (*i.e.*, Baseline and SVM), dynamic FC network-based methods (*i.e.*, DFCN-mean, CNN and our method) can achieve better accuracies, indicating that the dynamics of FC networks can provide useful information for better understanding the pathology of brain diseases.

On the other hand, Fig. 2 presents the total loss on the training subjects and validation subjects in each fold of cross-validation for AD vs. lMCI vs. eMCI vs. NC classification. From Fig. 2, our method can converge fast within 80 epochs.

3.3   Visual Illustration of Discriminative Functional Connectivity

We further conduct an experiment to identify the discriminative brain regions that contribute the most to the specific classification task, and identify the important connectivity between discriminative brain regions.

Specifically, in the brain network high-level feature extraction layer, the output of the first convolutional layer denotes the feature vector of each subject in the T_1 time segments. Since there are $K_1 = 8$ channels in the first convolutional layer, we can construct $K_1 = 8$ feature vectors for each subject, with each feature vector corresponding to a specific channel. For simplicity, we average the feature vectors of all time periods for each channel. Then, using the standard t-test, we measure the group difference of eMCI vs. NC and AD vs. NC, respectively. It is worth noting that, since the obtained feature vectors are different in each fold

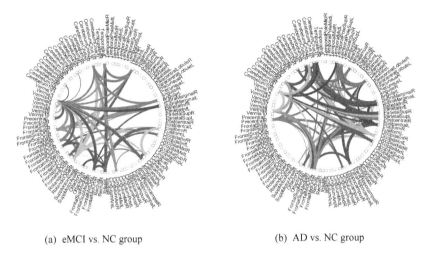

(a) eMCI vs. NC group   (b) AD vs. NC group

Fig. 3. Discriminative features for (a) eMCI vs. NC and (b) AD vs. NC classification. Each arc shows the selected feature between two ROIs, where colors are randomly allocated for a better visualization, and the thickness of each arc indicates its discriminative power that is inversely proportional to the corresponding p-value in t-test.

of cross-validation, for each channel, we use the standard t-test for each fold cross-validation, integrating all brain regions with p-values less than 0.05 in the 5-fold cross-validation, and selecting brain regions with 3 or more occurrences as the discriminative brain regions. We further perform the standard t-test on functional connectivity of discriminative brain regions to obtain discriminative features. Figure 3 reports the most discriminative features on the 5th and 7th channels for (a) eMCI vs. NC group and (b) AD vs. NC group, respectively.

As shown in Fig. 3, for eMCI vs. NC classification, the discriminative brain regions we selected include the left fusiform gyrus, the left lobule VI of cerebellar hemisphere, and lobule VII vermis, which are consistent with previous studies [22,23]. For AD vs. NC classification, there are two discriminative brain regions selected by our method, including the left crus I of cerebellar hemisphere and the right lobule IV, V of cerebellar hemisphere. According to previous studies [24], these two brain regions may be biologically associated with AD. These results further validate that our method is potentially helpful in discovering fMRI biomarkers for MCI and AD identification.

4   Conclusion

In this paper, we propose an RNN-based learning framework for AD-related brain disease classification using rs-fMRI time series data. Specifically, a dynamic functional connectivity network is constructed by calculating the correlation between paired brain regions. Then we use three convolutional layers to extract temporal features from dynamic FC networks. After that, we further employ the LSTM to capture the sequential information of temporal features along with multiple time segments, and employ two fully connected layers for classification. Experimental results on 174 subjects with rs-fMRI data from the ADNI dataset demonstrate the effectiveness of our proposed method.

Acknowledgment. K. Lin, B. Jie, P. Dong, X. Ding and W. Bian were supported in part by NSFC (Nos. 61976006, 61573023, 61902003), Anhui-NSFC (Nos. 1708085MF145, 1808085MF171), AHNU-FOYHE (No. gxyqZD2017010) and CERNET Innovation Project (NGII20190621).

References

1. Fan, L., et al.: New insights into the pathogenesis of Alzheimer's disease. Front. Neurol. **10**, 1312 (2020)
2. Reiman, E.M., Langbaum, J.B., Tariot, P.N.: Alzheimer's prevention initiative: a proposal to evaluate presymptomatic treatments as quickly as possible. Biomark. Med. **4**(1), 3–14 (2010). PMID: 20383319
3. Liu, M., Zhang, J., Adeli, E., Shen, D.: Landmark-based deep multi-instance learning for brain disease diagnosis. Med. Image Anal. **43**, 157–168 (2018)
4. Zhang, L., Wang, M., Liu, M., Zhang, D.: A survey on deep learning for neuroimaging-based brain disorder analysis. Front. Neurosci. **14** (2020)
5. Lee, M., Smyser, C., Shimony, J.: Resting-state fMRI: a review of methods and clinical applications. Am. J. Neuroradiol. **34**(10), 1866–1872 (2013)
6. Jie, B., Zhang, D., Gao, W., Wang, Q., Wee, C.Y., Shen, D.: Integration of network topological and connectivity properties for neuroimaging classification. IEEE Trans. Biomed. Eng. **61**(2), 576–589 (2014)
7. Shen, H., Wang, L., Liu, Y., Hu, D.: Discriminative analysis of resting-state functional connectivity patterns of schizophrenia using low dimensional embedding of fMRI. NeuroImage **49**(4), 3110–3121 (2010)
8. Jie, B., Liu, M., Zhang, D., Shen, D.: Sub-network kernels for measuring similarity of brain connectivity networks in disease diagnosis. IEEE Trans. Image Process. **27**(5), 2340–2353 (2018)
9. Wang, M., Lian, C., Yao, D., Zhang, D., Liu, M., Shen, D.: Spatial-temporal dependency modeling and network hub detection for functional MRI analysis via convolutional-recurrent network. IEEE Trans. Biomed. Eng. **67**(8), 2241–2252 (2020)
10. Wang, M., Zhang, D., Huang, J., Yap, P.T., Shen, D., Liu, M.: Identifying autism spectrum disorder with multi-site fMRI via low-rank domain adaptation. IEEE Trans. Med. Imaging **39**(3), 644–655 (2019)
11. Sporns, O.: The human connectome: a complex network. Ann. N. Y. Acad. Sci. **1224**(1), 109–125 (2011)
12. Hutchison, R.M., et al.: Dynamic functional connectivity: promise, issues, and interpretations. NeuroImage **80**, 360–378 (2013)
13. Zhang, J., et al.: Neural, electrophysiological and anatomical basis of brain-network variability and its characteristic changes in mental disorders. Brain **139**(8), 2307–2321 (2016)
14. Kudela, M., Harezlak, J., Lindquist, M.A.: Assessing uncertainty in dynamic functional connectivity. NeuroImage **149**, 165–177 (2017)

15. Damaraju, E., et al.: Dynamic functional connectivity analysis reveals transient states of dysconnectivity in schizophrenia. NeuroImage: Clin. **5**, 298–308 (2014)
16. Jie, B., Liu, M., Lian, C., Shi, F., Shen, D.: Designing weighted correlation kernels in convolutional neural networks for functional connectivity based brain disease diagnosis. Med. Image Anal. **63**, 1–14 (2020)
17. Kawahara, J., et al.: BrainNetCNN: Convolutional neural networks for brain networks. Towards predicting neurodevelopment. NeuroImage **146**, 1038–1049 (2016)
18. Tzourio-Mazoyer, N., et al.: Automated anatomical labeling of activations in SPM using a macroscopic anatomical parcellation of the MNI MRI single-subject brain. NeuroImage **15**(1), 273–289 (2002)
19. Vercauteren, T., Pennec, X., Perchant, A., Ayache, N.: Diffeomorphic demons: efficient non-parametric image registration. NeuroImage **45**(1), S61–S72 (2009)
20. Jie, B., Liu, M., Shen, D.: Integration of temporal and spatial properties of dynamic connectivity networks for automatic diagnosis of brain disease. Med. Image Anal. **47**, 81–94 (2018)
21. Zhang, D., Shen, D.: Multi-modal multi-task learning for joint prediction of multiple regression and classification variables in Alzheimer's disease. NeuroImage **59**(2), 895–907 (2012)
22. Bokde, A.L.W., et al.: Functional connectivity of the fusiform gyrus during a face-matching task in subjects with mild cognitive impairment. Brain **129**(5), 1113–1124 (2006)
23. Thomann, P.A., Schläfer, C., Seidl, U., Santos, V.D., Essig, M., Schröder, J.: The cerebellum in mild cognitive impairment and Alzheimer's disease - a structural MRI study. J. Psychiatr. Res. **42**(14), 1198–1202 (2008)
24. Suk, H.I., Wee, C.Y., Lee, S.W., Shen, D.: Supervised discriminative group sparse representation for mild cognitive impairment diagnosis. Neuroinformatics **13**, 277–295 (2015)

Integration of Handcrafted and Embedded Features from Functional Connectivity Network with rs-fMRI for Brain Disease Classification

Peng Dong[1], Biao Jie[1(⊠)], Lin Kai[1], Xintao Ding[1], Weixin Bian[1], and Mingxia Liu[2]

[1] School of Computer and Information, Anhui Normal University, Wuhu 241003, Anhui, China
jbiao@ahnu.edu.cn

[2] Department of Radiology and BRIC, University of North Carolina at Chapel Hill, Chapel Hill, NC 27599, USA

Abstract. Functional connectivity networks (FCNs) based on the resting-state functional magnetic imaging (rs-fMRI) can help to enhance our knowledge and understanding of brain function, and have been applied to diagnosis of brain diseases, such as Alzheimer's disease (AD) and its prodromal stage, *i.e.*, mild cognitive impairment (MCI). Traditional methods usually extract meaningful measures (*e.g.*, local clustering coefficients) from FCNs as (handcrafted) features for training the model. Recently, deep neural networks (DNNs) have been used to learn (embedded) features from FCNs for classification. However, few work explores to integrate both kinds of features (*i.e.*, handcrafted features from traditional methods and embedded features from DNN methods), although these features may convey complementary information for further improving the classification performance. Accordingly, in this paper, we propose a novel learning framework to integrate the handcrafted features from traditional method and embedded features from DNN method for classification of brain disease with rs-fMRI data. Experimental results on 174 subjects with baseline rs-fMRI data from the ADNI demonstrate the superiority of the proposed methods against several existing methods.

1 Introduction

The human brain is a complex structure, which contains multi-level and multi-functional substructure. Functional magnetic resonance imaging (fMRI) Neuroimaging technology provides a important tool to explore the mechanism and cognitive processing of the brain [1], and have been widely applied to analysis of brain disease, including Alzheimer's disease (AD) and its early stages, *i.e.*, mild cognitive impairment (MCI) [2]. Recently, function connectivity networks (FCN)

The original version of this chapter was revised: The Acknowledgment section has been updated. The correction to this chapter is available at
https://doi.org/10.1007/978-3-030-87589-3_72

C. Lian et al. (Eds.): MLMI 2021, LNCS 12966, pp. 674–681, 2021.
https://doi.org/10.1007/978-3-030-87589-3_69

based on resting-state fMRI (rs-fMRI) data, which characterizes the interaction between distributed brain regions, and thus help to better understand the pathology of brain diseases, have been widely applied to analysis of brain diseases [3]. Recently, studies have successfully applied FCNs to the task of automated brain disease classification [4].

In these studies, each FCN is usually considered as a graph with each node denoting a brain region and edge corresponding to relation between nodes. Therefore, graph object provides a important way for exploring the association between brain functional deficits and the underlying structure [5]. In FCN classification task, traditional methods usually extract some meaningful graph measures (e.g.local clustering coefficients [6], regional homogeneity [3] and connectivity strengths [7]) from the FCNs as (handcrafted) features for training a learning model, and feature selection methods (e.g.t-test [6], LASSO [8]) were performed to select more discriminative features for improving the performance of the learning model. These studies have helped us to better understand pathology of the brain disease and improve the performance of brain disease diagnosis.

As a powerful learning method, deep learning technology, e.g., convolutional neural network (CNN), is able to automatically learn (embedded) features from raw sensory data, has been successfully applied to many fields of medical image analysis [9], including computer-aided breast lesion detection [10], lung nodules [11] and histopathological diagnosis [12]. Recently, studies have applied deep neural network (DNN) methods to learning high-level features from FCNs for brain connectivity analysis. For example, (Jie et al. , 2020) [13] defined a unified CNN-based learning framework to extract hierarchical connectivities from FCNs for disease diagnosis based on rs-fMRI data. Compared with handcrafted feature based methods, deep leaning based methods can achieve a good disease classification performance. However, few work explores to integrate both kinds of features (i.e., handcrafted features from traditional methods and embedded features from DNN methods), although these features may convey complementary information for further improving the classification performance. Through this method the generalization error rate is low, the accuracy is high, and it can be applied to most classifiers without adjusting parameters.

Accordingly, in this paper, we propose a novel learning framework to integrate the handcrafted features from traditional method and embedded features from CNN method for classification of brain disease with rs-fMRI data. Figure 1 illustrates the proposed learning framework. Specifically, we first build the FCN of each subject using Pearson correlation coefficient (PCC) as connectivity measure. Then, based on constructed FCN, we extract the clustering coefficient as handcrafted features by using traditional graph object based method, and learn embedded features by using CNN-based method. Finally, we integrate two kinds of features for brain disease classification. Experimental results on 174 subjects with baseline rs-fMRI data from the ADNI demonstrate the superiority of the proposed methods against several existing methods.

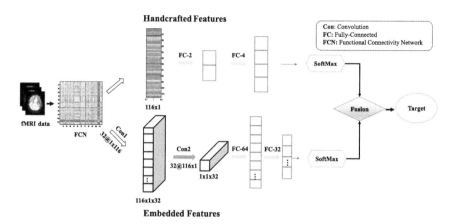

Fig. 1. Illustration of the proposed learning framework integrating handcrafted and embedded features of FCNs based on rs-fMRI data for brain disease classification.

Table 1. Characteristics of the studied subjects from ADNI (Mean±Standard Deviation). MMSE: Mini-Mental State Examination.

Group	HC	eMCI	lMCI	AD
Male/Female	20/28	20/30	27/18	16/15
Age	76.0 ± 6.8	72.4 ± 7.1	72.3 ± 8.1	74.7 ± 7.4
MMSE	28.8 ± 1.4	28.1 ± 1.6	27.1 ± 2.1	21.8 ± 3.3

2 Method

2.1 Subjects

In this study, we use 174 subjects with rs-fMRI data from the ADNI database[1], including 48 healthy controls (HCs), 50 early MCI (eMCI), 45 late MCI (lMCI) and 31 AD. Data acquisition was performed as follows: the image resolution is 2.29–3.31 mm for inplane, and slice thickness is 3.31 mm, TE = 30 ms and TR = 2.2–3.1 s. The clinical and demographic information of subjects is given in Table 1.

2.2 Image Preprocessing and Network Construction

Following [14], we use the standard pipeline to preprocess the rs-fMRI data, including (1) discarding the first 10 rs-fMRI volumes, (2) slice timing correction, and (3) head motion correction. The brain space of fMRI scans is partitioned into 116 regions-of-interest (ROIs) using the Automated Anatomical Labeling (AAL) template [15] with a deformable registration method [16]. The band-pass filtering is performed within a frequency interval of [0.025 Hz, 0.100 Hz]. The

---

[1] http://adni.loni.usc.edu.

BOLD signals from the gray matter tissue are extracted, and the mean time series of each ROI is calculated to construct the FC network.

Based on the mean time series of ROIs, we construct the FCN using PCC as measure of the functional connectivity between the ROIs. For each subject, a FCN is built with the nodes denoting brain ROIs and the strength of connections corresponding to the correlation coefficents.

2.3   Feature Learning

In order to improve the classification performance of brain disease with rs-fMRI data, we proposed to integrate the handcrafted features from traditional method and embedded features from CNN method, as illustrated in Fig. 1. In traditional method, the local weighted clustering coefficients are extracted from FCNs as handcrafted features, and followed by an efficient shallow networks, where a rectified linear unit (ReLU) is used as the activation function of each layer, and include two fully connected (FC) layers with 2 and 4 unites, respectively. In CNN method, there are two convolutional layers for feature extraction, and followed by two FC layers with 64 and 32 units, respectively. In the proposed framework, motived by a recent work [17], we put two separated FC layers followed by a softmax layer on each learning network.

Handcrafted Feature Extraction. Considering our aim is to classify patients from HC, the local weighted clustering coefficients [18], which characterize the topological information of specific nodes and have been applied to brain disease classification [4], are extracted from constructed FCNs. Specifically, given the FCN (matrix) $F^n \in R^{N \times N}$ of the nth subject, where N is the number of ROIs, the local weighted clustering coefficient of the ith ROI is defined as follows.

$$c_i = \frac{2}{d_i(d_i - 1)} \sum_{j,q=1}^{N} (F_{ij}^n F_{jq}^n F_{iq}^n)^{\frac{1}{3}}, \tag{1}$$

where $d_i = \sum_{j=1}^{N} F_{ij}$ is the weighted degree of node i. These extracted local weighted clustering coefficents are considered as handcrafted features and applied to subsequent classification.

Embedded Feature Extraction. In this paper, we extract embedded features from the constructed FCNs by using CNN-based method. Specifically, we adopt the kernels with the size of $1 \times N$, and set the stride size along width and height dimension to $(1, 1)$. Therefore, the convolution along the width dimension is a feature mapping for each ROI, and characterize the connectivities of specific ROI. Then, we use the kernel with the size of $N \times 1$ and the stride of $(1, 1)$. Therefore, the convolution in this layer is a feature mapping for the whole FCN. The features obtained in this layer are used for subsequence classification.

2.4   Classification

Deep neural networks usually yield class probabilities using "softmax" function on the output of the neurons, $i.e.$,

$$p_i(y|x) = \frac{e^{\mathbf{z}_i}}{\sum_j e^{\mathbf{z}_j}} \tag{2}$$

where $\mathbf{z}_i$ is the output of the ith neuron. Finally, we integrate both kinds of features for classifying patients from HCs by computing the average value of the class probabilities produced by the shallow network and CNN method.

3   Experiment and Results

3.1   Experimental Setting

To investigate the performance of the proposed method, we perform two groups of experiments, including a binary classification task, $i.e.$, eMCI vs. HC classifications, and a multi-class classification task, $i.e.$, AD vs. lMCI vs. eMCI vs. HC classification, by using a 5-fold cross-validation. Specifically, for each classification task, the set of subjects is equivalently partitioned into five subsets. One subset is used as the testing set. The remaining four subsets are combined as training set. In addition, in each cross validation we select 15% training subjects as the validation data for tuning the parameters.

3.2   Methods for Comparison

We first compare the proposed method with the support vector machine (SVM) based methods denoting as baseline and SVM. In baseline method, we directly use the connectivity strengths of FCNs as features. In the SVM method, we extract the local weighted clustering coefficients from FCNs as features. For both methods, we perform the t-test method for feature selection and a linear SVM with default parameters for classification. Here, a one-to-all is used for multi-class classification task. In additional, we compare the method only using handcrafted features with the shallow network (denoted as handcraft), and method only using embeded features from the CNN (denoted as CNN).

3.3   Classification Performance

We evaluate the performance of all methods by computing the overall accuracy of all categories, and the accuracy of each category. Tables 2–3 presents the results of all methods in classification tasks. As we can see from Tables 2–3, our proposed method can achieve better classification performance compared with all competing methods. For example, the proposed method, respectively, achieves the ACC values of 78.4% and 57.1% for eMCI vs. HC classification and AD vs. lMCI vs. eMCI vs. HC classification, while the best overall accuracy

Table 2. Performance of all methods for eMCI vs. HC classification. ACC: Accuracy.

Method	eMCI vs. HC (%)		
	ACC	ACC_{HC}	ACC_{eMCI}
Baseline	57.1	48.1	65.6
SVM	63.6	50.0	75.0
Handcraft	65.7	50.7	77.6
CNN	72.5	68.7	75.7
Proposed	**78.4**	**79.1**	**83.9**

Table 3. Performance of all methods for AD vs. lMCI vs. eMCI vs. HC classification. ACC: Accuracy.

Method	AD vs. lMCI vs. eMCI vs.HC (%)				
	ACC	ACC_{HC}	ACC_{eMCI}	ACC_{lMCI}	ACC_{AD}
Baseline	30.6	20.0	38.9	30.0	33.3
SVM	35.0	22.0	69.5	21.0	6.7
Handcraft	47.9	22.6	65.6	73.3	77.8
CNN	48.5	64.7	59.5	25.0	26.7
Proposed	**57.1**	**72.0**	**75.7**	**34.1**	**40.0**

of competing methods is 72.5% and 48.5%, suggesting the effectiveness of our proposed method. In additional, the proposed method consistently outperform the methods only using one kind of features (*i.e.*, handcraft and CNN), indicating that the both kinds features can convey the complementary information and should be integrated to further improve the classification performance.

Furthermore, to investigate the complementarity of two kinds of features, for each classification task, we extract and choice the maximum values of softmax function in each network (*i.e.*, the shallow network and CNN network) for testing subjects in the choosed the first fold cross-validation. Figure 2 gives the obtained results. As we can see from Fig. 2, the predicted class probabilities predicted by the shallow network approach and CNN approach are complementary, indicating that two kinds of features can convey complementary information for jointly improving the classification performance.

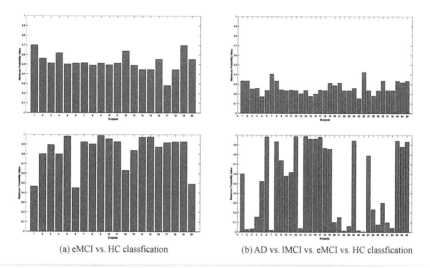

(a) eMCI vs. HC classfication   (b) AD vs. lMCI vs. eMCI vs. HC classfication

Fig. 2. Predicted maximum class probability value of testing subjects in the first fold cross-validation for the shallow network (up) and CNN (down)

4 Conclusion

In this paper, we propose a novel learning framework that integrates two kinds of features (*i.e.*, the handcrafted features from traditional method and embedded features from CNN method) for classification of brain disease with rs-fMRI data. Experimental results on 174 subjects with baseline rs-fMRI data from the ADNI show that the proposed method can achieve better classification performance that several existing methods.

Acknowledgment. P. Dong, B. Jie, K. Lin, X. Ding and W. Bian were supported in part by NSFC (Nos. 61976006, 61573023, 61902003), Anhui-NSFC (Nos. 1708085M F145, 1808085MF171), AHNU-FOYHE (No. gxyqZD2017010) and CERNET Innovation Project (NGII20190621).

References

1. Caria, A., Sitaram, R., Birbaumer, N.: Real-time f*mri*: a tool for local brain regulation. Neuroscientist **18**(5), 487–501 (2012)
2. Supekar, K., Menon, V., Rubin, D., Musen, M., Greicius, M., Sporns, O.: Network analysis of intrinsic functional brain connectivity in Alzheimer's disease. PLOS Comput. Biol. **4**(6), e1000100 (2008)
3. Wang, X., Yun, J., Tang, T., Wang, H., Lu, Z.: Altered regional homogeneity patterns in adults with attention-deficit hyperactivity disorder. Eur. J. Radiol. **82**(9), 1552–1557 (2013)
4. Wee, C.Y., et al.: Enriched white matter connectivity networks for accurate identification of MCI patients. Neuroimage **54**(3), 1812–1822 (2011)

5. van den Heuvel, M.P., Hulshoff Pol, H.E.: Exploring the brain network: a review on resting-state FMRI functional connectivity. Eur. Neuropsychopharmacol. **20**(8), 519–534 (2010)
6. Wee, C.Y., et al.: Identification of mci individuals using structural and functional connectivity networks. Neuroimage **59**(3), 2045–2056 (2012)
7. Liu, F., et al.: Multivariate classification of social anxiety disorder using whole brain functional connectivity. Brain Struct. Funct. **220**(1), 101–115 (2013). https://doi.org/10.1007/s00429-013-0641-4
8. Jie, B., Zhang, D., Gao, W., Wang, Q.: Integration of network topological and connectivity properties for neuroimaging classification. IEEE Trans. Biomed. Eng. **61**(2), 576–589 (2014)
9. June-Goo, L.: Deep learning in medical imaging: general overview. Korean J. Radiol. **18**(4), 570–584 (2017)
10. Wang, D., Khosla, A., Gargeya, R., Irshad, H., Beck, A.H.: Deep learning for identifying metastatic breast cancer (2016)
11. Walsh, S.L.F., Calandriello, L., Silva, M., Sverzellati, N.: Deep learning for classifying fibrotic lung disease on high-resolution computed tomography: a case-cohort study. Lancet Respir. Med. **6**(11), 837–845 (2018)
12. Litjens, G., et al.: Deep learning as a tool for increased accuracy and efficiency of histopathological diagnosis. Sci. Rep. **6**, 26286 (2016)
13. Jie, B., Liu, M., Lian, C., Shi, F., Shen, D.: Designing weighted correlation kernels in convolutional neural networks for functional connectivity based brain disease diagnosis. Med. Image Anal. **63**, 1–14 (2020)
14. Jie, B., Liu, M., Shen, D.: Integration of temporal and spatial properties of dynamic connectivity networks for automatic diagnosis of brain disease. Med. Image Anal. **47**, 81–94 (2018)
15. Tzourio-Mazoyer, N., et al.: Automated anatomical labeling of activations in SPM using a macroscopic anatomical parcellation of the MNI MRI single-subject brain. Neuroimage **15**(1), 273–289 (2002)
16. Vercauteren, T., Pennec, X., Perchant, A., Ayache, N.: Diffeomorphic demons: efficient non-parametric image registration. Neuroimage **45**(1), S61–S72 (2009)
17. Chen, Z., Le, Z., Cao, Z., Jing, G.: Distilling the knowledge from handcrafted features for human activity recognition. IEEE Trans. Industr. Inf. **14**(10), 4334–4342 (2018)
18. Rubinov, M., Sporns, O.: Complex network measures of brain connectivity: uses and interpretations. Neuroimage **52**(3), 1059–1069 (2010)

Detection of Lymph Nodes in T2 MRI Using Neural Network Ensembles

Tejas Sudharshan Mathai[1]([✉]), Sungwon Lee[1], Daniel C. Elton[1],
Thomas C. Shen[1], Yifan Peng[2], Zhiyong Lu[2], and Ronald M. Summers[1]

[1] Imaging Biomarkers and Computer-Aided Diagnosis Laboratory, Radiology and
Imaging Sciences, Clinical Center, National Institutes of Health, Bethesda, MD, USA
tejas.mathai@nih.gov
[2] National Center for Biotechnology Information, National Library of Medicine,
National Institutes of Health, Bethesda, MD, USA

Abstract. Reliable localization of abnormal lymph nodes in T2 Magnetic Resonance Imaging (MRI) scans is needed for staging and treatment of lymphoproliferative diseases. Radiologists need to accurately characterize the size and shape of the lymph nodes and may require an additional contrast sequence such as diffusion weighted imaging (DWI) for staging confirmation. The varied appearance of lymph nodes in T2 MRI makes staging for metastasis challenging. Moreover, radiologists often times miss smaller lymph nodes that could be malignant over the course of a busy clinical day. To address these imaging and workflow issues, in this pilot work we aim to localize potentially suspicious lymph nodes for staging. We use state-of-the-art detection neural networks to localize lymph nodes in T2 MRI scans acquired through a variety of scanners and exam protocols, and employ bounding box fusion techniques to reduce false positives (FP) and boost detection accuracy. We construct an ensemble of the best detection models to identify potential lymph node candidates for staging, obtaining a 71.75% precision and 91.96% sensitivity at 4 FP per image. To the best of our knowledge, our results improve upon the current state-of-the-art techniques for lymph node detection in T2 MRI scans.

Keywords: MRI · T2 · Lymph node · Detection · Deep learning

1 Introduction

Lymph nodes (LN) are a part of the lymphatic system, and they contain immune cells to help the body fight infection by filtering foreign substances in the lymphatic fluid that flows through them. Localization of lymph nodes in the abdomen is crucial as it allows enlarged and metastatic lymph nodes to be distinguished from non-metastatic lymph nodes [1–3]. It is especially paramount to identify enlarged LN if they are found at sites that do not correspond to the first site of lymphatic spread as this signals distant metastasis. Nodal location

© Springer Nature Switzerland AG 2021
C. Lian et al. (Eds.): MLMI 2021, LNCS 12966, pp. 682–691, 2021.
https://doi.org/10.1007/978-3-030-87589-3_70

and size are important aspects of the AJCC tumor, lymph node, metastasis (TNM) system [2], which provides guidelines for the diagnosis, management and treatment strategy of cancer and lymphoproliferative disorders [1,2].

Multi-parametric MRI images are becoming widely used for cancer and lymph node staging. Among the various MRI sequences, T2 and Diffusion Weighted Imaging (DWI) are commonly preferred to image the LN [1,3–5]. Unfortunately, accurate staging is difficult due to their irregular shapes, heterogeneous anatomical location, diverse yet similar appearance to other tissue structures (e.g. fat), and size. As shown in Fig. 1(a), lymph nodes in T2 MRI appear iso-intense (or slightly hyper-intense) relative to the surrounding fat [1] making it difficult to distinguish them, and contrast agents are often administered to reveal intensity differences between metastatic and normal LN tissue [1,3]. Further confounding the analysis of LN is the multitude of MRI scanners and exam protocols being used at different institutions to visualize abnormalities.

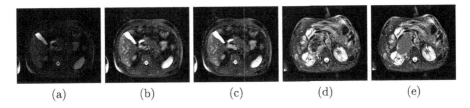

Fig. 1. (a) A T2 MRI image and (b) its normalized and histogram equalized counterpart are shown. There are 3 annotated LN regions as shown in (c), which is the result of our proposed ensemble detector; green boxes are the ground truth, yellow are the true positives, and red are the false positives. (d) A large lymph node annotated with both short and long axis diameters. (e) Failed segmentation result shown in red after generating pseudo-mask using GrabCut from the annotations. Note the incomplete segmentation of LN boundary (bottom left) that provides wrong training signals to a detection + segmentation network. (Color figure online)

Nodal size is the widely used criteria to determine malignancy [1,6]. Radiologists in clinical practice first localize the lymph nodes by scrolling back and forth through the T2 MR slices, and measure an identified node by using two orthogonal lines: the long and short axis diameters (LAD and SAD) [6]. Lymph nodes with a mean SAD of 3–10 mm are considered normal when imaged using conventional MRI, while those with a SAD of ≥10 mm are deemed suspicious for metastasis [1]. At our institution however, the accepted guideline is for radiologists to complete nodal sizing with either both diameters or the LAD only. This guideline can vary across institutions, thereby adding another obfuscating element to the analysis of LN. Furthermore, as radiologists size lymph nodes often during the course of a busy clinical day, despite their diligence, small LN might be missed (≥3 mm) [3]. Thus, given the morbidity surrounding lymphoproliferative diseases and the imaging and workflow challenges, there is a need for automated LN detection in T2 MRI images for subsequent sizing and staging.

Prior research has focused on the identification of lymph nodes in CT scans [9–15] with limited work done on MRI scans to localize lymph nodes [3,7,8]. In [3], a Mask-RCNN network was used to identify and segment lymph node regions in a multi-parametric pelvic MRI image (2 DWI + 1 T2 image). They achieved a sensitivity of 62.6% on their external testing dataset obtained from different hospitals and different scanners with 8.2 FP being identified per volume. In [7], initial lymph node candidate regions were detected using a GentleBoost classifier trained on image features followed by a convolutional neural network (CNN) to reduce false positives. Multiple views were incorporated into the CNN training to provide 3D structure, and a sensitivity of 85% was achieved at the rate of 5–10 FP per image. Finally, a Faster-RCNN was used in [8] to identify metastatic pelvic LN with an AUC of 91.2%.

As T2 MRI is arduous enough to identify and stage lymph nodes with a combination of T2 and DWI scans being preferred clinically, radiologist complete their annotations on one of the two (or more) sequences. This requires the registration of one sequence to another (e.g. DWI to T2) for reliable annotation usage in algorithmic development. However, as documented in [3], insufficient registration can result in errors that can lower the accuracy of LN detection. Therefore, in this pilot work, we first opt to use challenging T2 MRI sequences acquired using different scanners and exam protocols, and detect lymph nodes with state-of-the-art detection networks [16–23]. The intention is that the reasonable T2-based detection performance can be supplemented by the addition of accurately registered DWI scans. We improve upon the detection performance by using bounding box fusion techniques [24] that significantly reduce the number of FP detected by these networks. We detail results that either improve upon or are on par with the performance of previously published lymph node detection approaches, despite a lower quantity of training data. Finally, to replicate clinical use, we propose an ensemble of the best detection models and show its applicability towards lymph node detection.

Contribution. 1) For the LN detection task, we use state-of-the-art models to localize lymph nodes in T2 MRI and employ bounding box fusion techniques to reduce false positives. 2) To mimic clinical usage, we propose a ensemble of the best detection models to identify potential lymph node candidates for staging.

2   Methods

State-of-the-Art Object Detectors. We first quantified the performance of state-of-the-art object detectors on the LN detection task in T2 MRI: 1) Faster R-CNN [16], 2) DetectoRS [17], 3) YOLOv3 [18], 4) SSD [19], 5) RetinaNet [20], 6) FCOS [21], 7) FoveaBox [22], and 8) VFNet [23]. We further subdivided them into two categories: two-stage and one-stage detectors. Two-stage detectors (e.g. Faster R-CNN, DetectoRS) are region-based; the first stage generates region proposals corresponding to objects of interest, and the second stage classifies these region proposals and regresses the object bounding box coordinates.

These types of detectors are slow and computationally expensive. DetectorRS uses a Recursive Feature Pyramid (RFP) to add extra gradient feedback at the Feature Pyramid Network (FPN) [27] bottom-up layers, and Switchable Atrous Convolutions (SAC) to achieve peak object detection accuracy.

On the other hand, one-stage detectors [18–23] are faster as they skip the region proposal stage, densely sample all possible locations for objects, and directly predict the bounding box coordinates and class probabilities for different categories in a single pass. One-stage detectors can further be subdivided into anchor-based and anchor-free methods; anchor-based methods include YOLOv3 [18], SSD [19], and RetinaNet [20], while anchor-free methods include FCOS [21], FoveaBox [22], and VFNet [23]. YOLOv3 and SSD were attempts at fast and efficient object detection in images, but they were not suitable for detecting small objects. RetinaNet [20], on the other hand, overcame the common class imbalance problem plaguing detection tasks by using a focal loss function, along with the FPN to represent objects at different scales. However, these methods (including Faster R-CNN) require optimization of the sizes, aspect ratios and number of anchor boxes to achieve optimal object detection performance, involving additional computation and hyperparameter optimization [21–23].

FCOS [21] navigates away from anchor-based configurations by incorporating a FPN-based multi-level prediction inside the Fully Convolutional Network (FCN) [28], and a centerness score computed from the classification score to reduce the FP that are far away from the target object center. VFNet combines FCOS (without centerness branch) with an Adaptive Training Sample Selection (ATSS), sets the IoU between the ground truth and the prediction as the classification score, integrates it into a novel IoU-aware Varifocal loss, develops a star-shaped bounding box representation, and refines the box predictions. FoveaBox consists of a backbone to compute features from the input and a fovea head network that estimates the object occurrence possibility through per pixel classification on the backbone's output, and predicts the box at each position in the image that may be potentially covered by an object. We chose the anchor-free detectors in our experiment as they achieve superior results over anchor-based and even two-stage detectors.

Weighted Boxes Fusion. Object detectors typically predict an object's location with bounding box coordinates and provide confidence scores for them. Often times, ensembling over multiple selected epochs of a single detector or over different object detection models adds generalization [24] to a detection task, and yields more accurate results in contrast to a single model. Many strategies [24,31,32] have been proposed to combine the predicted boxes to yield accurate results, and we utilize weighted boxes fusion [24] as a post-processing step.

Ensemble of Best Detection Models. In order to add generalizability to the predictions of the detection networks, a threshold of 60% on the precision was set. Models that met the cut-off were ensembled together, leading to a higher detection performance in comparison to a single model.

Data Description. The lymph node dataset consisted of abdominal MRI studies that were acquired between January 2015 and September 2019, and they were downloaded from the National Institutes of Health (NIH) Picture Archiving and Communication System (PACS). Initially, 584 T2-weighted MRI scans and associated radiology reports from different patients ($n = 584$) were downloaded. As mentioned in Sect. 1, radiologists completed nodal sizing with both the SAD and LAD, or with just the LAD. The lymph node size measured and annotated by the radiologist on a slice in a scan is linked to the radiology report using "bookmark" hyperlinks. NLP was used to find only lymph node related bookmarks [33]. The lymph node extent and size measurements were extracted from these bookmarks and considered the gold standard annotation.

Next, an experienced radiologist checked the collected data and filtered those patients with scans that had incorrect annotations (e.g. kidney masses). Patients with scans that also contained only LAD measurements were removed; the intent was to standardize the data and analysis with the AJCC TNM guidelines [2]. This process resulted in a total of 376 T2 MRI scans ($n = 376$ patients) with 520 distinct lymph nodes that had been annotated with both the LAD and SAD measurements. The voxels in the scans were normalized to be within the 1%–99% of their intensity range [25]. As seen in Fig. 1, it increases the contrast between the bright and dark structures in T2 MRI scans, and mitigates the effect of outlier intensities from the image sensors present in various MRI scanners.

Following normalization, contrast enhancement of important structures, such as lymph nodes, was achieved through histogram equalization [26] without the excessive enhancement of image noise. The final dataset was then randomly divided into training (60%, 225 scans), validation (20%, 76 scans), and test (20%, 75 scans) splits at the patient-level. The resulting scans had dimensions in the range from $(256–640) \times (192–640) \times (18–60)$ voxels. In contrast to prior detection and segmentation approaches [3,29], we do not rely on pseudo masks (e.g. generated through GrabCut); as seen in Fig. 1, their synthesis from radiologist annotations can often be incorrect and affect the training of the detection and segmentation networks, such as Mask RCNN [30].

Implementation Details. Radiologists focus their LN sizing efforts over a few slices of specific MRI sequences (e.g. T2), and generally corroborate their finding with another sequence (e.g. DWI). Prior work suggests that either 1- or 3- slice(s) are sufficient for the sizing task, with in-plane slice providing the most information [7]. With this in mind, we generated 3-slice T2 MRI images with the center slice containing the radiologist annotated LN, and used these 3-slices as the input for the detection networks. We used the framework proposed in [34] to implement the various one-stage and two-stage detectors used in this work. ResNet50 was used as the backbone for Faster R-CNN, DetectoRS, RetinaNet, FCOS, FoveaBox, and VFNet, while YOLOv3 used DarkNet53 and SSD used the VGG16 backbone. All the models were trained starting with the pre-trained MS COCO weights. Data augmentation included random flipping, random crops, random shifts and rotations in the range of [0, 32] pixels, and [0, 10] degrees respectively. A grid search was run on the batch size and learning rate for each model resulting in a batch size of 2 samples. The learning rate for YOLOv3, DetectoRS, SSD, FCOS, FoveaBox, and VFNet was 1e-3, while it was 25e-4 for

Faster R-CNN and RetinaNet. Each model was trained for 24 epochs, and the 5 epochs with the lowest validation loss were ensembled together and used for LN detection. In contrast, the models in the final detection ensemble only had 1 epoch with the lowest validation loss. These results are shown in Table 1. All experiments were run on a workstation running Ubuntu 16.04LTS and containing a NVIDIA Tesla V100 GPU.

Baseline Comparisons. We compare our lymph node detection results against those obtained by [3] on multi-parametric pelvic MRI, [7] on T1-weighted pelvic MRI, [9,10] on mediastinal LN in chest CT, and [11] on abdominal and mediastinal LN in CT.

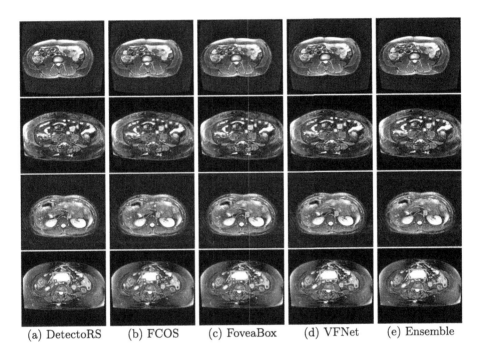

(a) DetectoRS   (b) FCOS   (c) FoveaBox   (d) VFNet   (e) Ensemble

Fig. 2. Columns (a)–(d) show the lymph node detection results of the best detectors (DetectoRS, FCOS, FoveaBox, VFNet) incorporated into our ensemble on four different MRI images. Column (e) displays the result of our ensemble detector following Weighted Boxes Fusion. Green boxes are the ground truth, yellow are the true positives, and red are the false positives. Note that some detectors (e.g. DetectoRS) miss detecting lymph nodes (rows 1 and 2, col 1), but the ensemble detector benefits from the prediction of the remaining networks to yield the final detections.

3   Results and Discussion

Results. A clinically acceptable result [3] for lymph node detection is a precision of $\geq 60\%$, sensitivity of 85%, and 4–6 FP per image. Missing potentially

Table 1. Detection performance of various state-of-the-art detectors and our proposed ensemble method. "S" stands for Sensitivity @[0.5, 1, 2, 4, 6, 8, 16] FP

Method	mAP	S@0.5	S@1	S@2	S@4	S@6	S@8	S@16
Faster R-CNN [16]	58.72	61.90	70.23	80.95	83.33	83.33	83.33	83.33
DetectoRS [17]	61.83	66.67	73.81	77.38	80.95	80.95	80.95	80.95
YOLOv3 [18]	56.42	65.47	66.67	77.38	77.38	77.38	77.38	77.38
SSD [19]	40.21	35.71	57.14	70.23	79.76	79.76	79.76	82.14
RetinaNet [20]	57.36	53.57	65.47	77.38	83.33	86.90	88.09	89.28
FCOS [21]	60.09	61.90	77.38	83.33	88.09	89.28	89.28	89.28
FoveaBox [22]	61.67	61.90	76.19	79.76	84.52	88.09	89.28	89.28
VFNet [23]	63.91	67.85	75	80.95	83.33	83.33	83.33	83.33
Ensemble (VFNet + FoveaBox + FCOS + DetectoRS)	**71.75**	**73.81**	**79.76**	**85.71**	**91.66**	**91.66**	**91.66**	**91.66**
Ensemble (SAD < 10 mm)	61.77	64.51	77.41	87.09	87.09	87.09	87.09	87.09
Ensemble (SAD ≥ 10 mm)	74.30	77.35	81.13	84.91	94.34	94.34	94.34	94.34

metastatic lymph nodes would be problematic, so we strive for a reasonable trade-off between precision and recall. Detection results are presented in Fig. 2 and Table 1 in terms of mean average precision (mAP) and sensitivity on the test dataset. YOLOv3 and SSD perform the worst in terms of LN detection with a precision of ≤60% and sensitivity of ≤80% indicating that they had significant issues with detecting LN ≤10 mm. RetinaNet and Faster R-CNN had higher sensitivities (≥80%) at 4 FP per image, but their mAP values were still low and below our precision cut-off for ensembling. The detectors surpassing the ≥60% precision threshold included FCOS, FoveaBox, VFNet, and DetectoRS, with the maximum mAP of 63.91% achieved by VFNet. The one-stage anchor-free detectors significantly outperformed the anchor-based and two-stage detectors. Although promising, it was difficult to establish a clear winner amongst the anchor-free detectors. To this end, we ensembled the models passing the cut-off to obtain the highest LN detection performance with a mAP of 71.75% and 91.66% sensitivity at 4 FP per image. Against [3], our mAP is 71.75% vs 64.5%, and recall is 91.66% vs 62.6% at 8 FP. Against [7], our sensitivity is 91.66% vs 80% at 8 FP per image. Compared with [9–11], we obtain sensitivities of 91.66% at 4 FP vs 66% at 4 FP, 88% at 6 FP, and 90% at 6 FP respectively.

We further analyzed the behavior of the ensemble detector on lymph nodes when they were stratified by size. The ensemble model posts a moderate performance of 61.77% mAP and sensitivity of 87.09% at 4 FP per image when tested on LN with a SAD of ≤10 mm. One reason for the lower mAP is the ensemble detects many FP (as seen in Fig. 2, 2nd row) due to the small size of the LN, thereby lowering the mAP. Our results are similar to those of [3] at ~65% sensitivity, but we still achieve a significantly higher recall *without the use of DWI sequences*. On the other hand, we attain a mAP of 74.30% and sensitivity of 94.34% at 4 FP per image on lymph nodes with SAD ≥10 mm. These results are again consistent with past literature [3,12,13], yet we considerably outperform their results with a ≥10% increase in sensitivity. Our ensemble detector executes in 285 ms/22 s per image/volume vs. 862 ms/67 s from DetectoRS, 218 ms/17 s from FCOS, 134 ms/10 s from FoveaBox, and 276 ms/22 s from VFNet respectively.

Discussion. As described above, the two-stage Faster R-CNN and one-stage anchor-based detectors (YOLOv3, SSD, RetinaNet) resulted in significantly inferior detection performance and did not match our clinical implementation standards. Potential reasons for their performance include anchor ratio, size, and box optimization, and detection difficulties when encountering LN candidates with small size (≤ 5 mm). In contrast, two-stage DetectoRS and the anchor-free detectors (FCOS, FoveaBox, VFNet) met our clinical use goals, and yielded significantly improved detection results as shown in Table 1. Weighted Boxes Fusion was the only post-processing step that was undertaken as it was necessary to fuse multiple detections from different models, but it is a small price to pay for the substantial increase in mAP and sensitivity. Of note, our results were achieved by taking off-the-shelf object detectors, and retraining them on a difficult T2 MRI dataset. The results were unencumbered by optimization of anchor parameters and free of region proposal generation. These make the networks simpler with fewer network weights to be learned, thereby reducing the chances of overfitting to the small training dataset. In fact, the only major parameter tuning that was done was with respect to the batch size and optimizer learning rate. We believe that this shows the power of the anchor-free detection networks without the need for a complicated training pipeline design [35].

We do not use DWI scans and this is a limitation of our work as it is routine practice for radiologists to confirm a finding on T2 MR with DWI. The ensemble performs satisfactorily on lymph nodes of size (≤ 10 mm), but smaller lymph nodes (≤ 3 mm) can be potentially metastatic [36], and a multi-parametric MRI input to the network might boost the detection performance of smaller LN. Future work is oriented towards the utilization of different MRI sequences (e.g. DWI and T1) to localize LN better. A structured report on the lymph node characteristics can also be created detailing the location, exam protocol, malignancy status, etc., thereby reducing this tedious analysis workload of the radiologists.

4   Conclusion

In this paper, we first quantify the performance of state-of-the-art object detectors on localizing lymph nodes in challenging T2 MRI scans. Weighted Boxes Fusion was used to then fuse the bounding box predictions from multiple epochs of a model to boost the detection performance. Consequently, a max mAP of 63.91% for VFNet and max sensitivities of 89.28% were achieved by FCOS and FoveaBox respectively. Next, we ensembled the best performing detectors together, which were VFNet, FCOS, FoveaBox and DetectoRS, to yield the highest mAP of 71.65% and sensitivity of 91.66% at 4 FP per image. We also stratified the results based on lymph node size, and found the ensemble performance to be significantly better than the current state-of-the-art in lymph node detection.

Acknowledgements. This work was supported by the Intramural Research Programs of the NIH National Library of Medicine and NIH Clinical Center. We also thank Jaclyn Burge for the helpful comments and suggestions.

References

1. Taupitz, M.: Imaging of lymph nodes — MRI and CT. In: Hamm, B., Forstner, R. (eds.) MRI and CT of the Female Pelvis. MR, pp. 321–329. Springer, Heidelberg (2007). https://doi.org/10.1007/978-3-540-68212-7_15
2. Amin, M.B., et al.: The eighth edition AJCC cancer staging manual: continuing to build a bridge from a population-based to a more "Personalized" approach to cancer staging, CA Cancer J. Clin. **67**(2), 93–99 (2017)
3. Zhao, X., et al.: Deep learning based fully automated detection and segmentation of lymph nodes on multiparametric MRI for rectal cancer: a multicentre study. eBioMedicine **56** (2020)
4. Caglic, I., et al.: Diffusion-weighted imaging (DWI) in lymph node staging for prostate cancer. Transl. Androl. Urol. **7**(5), 814–823 (2018)
5. Heijnen, L.A., et al.: Diffusion-weighted MR imaging in primary rectal cancer staging demonstrates but does not Characterise lymph nodes. Eur. Radiol. **23**, 3354–3360 (2013)
6. Ganeshalingam, S., et al.: Nodal staging. Cancer Imaging **9**(1), 104–111 (2009)
7. Debats, O.A., et al.: Lymph node detection in MR lymphography: false positive reduction using multi-view convolutional neural networks. PeerJ **7**, e8052 (2019)
8. Lu, Y., et al.: Identification of metastatic lymph nodes in MR imaging with faster region-based convolutional neural networks. Cancer Res. **78**(17), 5135–5143 (2018)
9. Liu, J., et al.: Mediastinal lymph node detection and station mapping on chest CT using spatial priors and random forest. Med. Phys. **43**, 4362–4374 (2016)
10. Seff, A., Lu, L., Barbu, A., Roth, H., Shin, H.-C., Summers, R.M.: Leveraging mid-level semantic boundary cues for automated lymph node detection. In: Navab, N., Hornegger, J., Wells, W.M., Frangi, A.F. (eds.) MICCAI 2015. LNCS, vol. 9350, pp. 53–61. Springer, Cham (2015). https://doi.org/10.1007/978-3-319-24571-3_7
11. Roth, H.R., et al.: A new 2.5D representation for lymph node detection using random sets of deep convolutional neural network observations. In: Golland, P., Hata, N., Barillot, C., Hornegger, J., Howe, R. (eds.) MICCAI 2014. LNCS, vol. 8673, pp. 520–527. Springer, Cham (2014). https://doi.org/10.1007/978-3-319-10404-1_65
12. Barbu, A., et al.: Automatic detection and segmentation of lymph nodes from CT data. IEEE Trans. Med. Imaging **31**, 240–250 (2012)
13. Feulner, J., et al.: Lymph node detection and segmentation in chest CT data using discriminative learning and a spatial prior. Med. Image Anal. **17** (2012)
14. Feuerstein, M., et al.: Automatic mediastinal lymph node detection in chest CT. Med. Imaging (2009)
15. Kitasaka, T., et al.: Automated extraction of lymph nodes from 3-D abdominal CT images using 3-D minimum directional difference filter. In: Ayache, N., Ourselin, S., Maeder, A. (eds.) MICCAI 2007. LNCS, vol. 4792, pp. 336–343. Springer, Heidelberg (2007). https://doi.org/10.1007/978-3-540-75759-7_41
16. Ren, S., et al.: Faster R-CNN: towards real-time object detection with region proposal networks. IEEE PAMI **39**(6), 1137–1149 (2017)
17. Qiao, S., et al.: DetectoRS: detecting objects with recursive feature pyramid and switchable Atrous convolution, arXiv (2020)
18. Redmon, J., et al.: YOLOv3: an incremental improvement, arXiv (2018)
19. Liu, W., et al.: SSD: single shot MultiBox detector. In: Leibe, B., Matas, J., Sebe, N., Welling, M. (eds.) ECCV 2016. LNCS, vol. 9905, pp. 21–37. Springer, Cham (2016). https://doi.org/10.1007/978-3-319-46448-0_2

20. Lin, T.Y., et al.: Focal loss for dense object detection. In: ICCV, pp. 2999–3007 (2017)
21. Tian, Z., et al.: FCOS: fully convolutional one-stage object detection. In: ICCV, pp. 9627–9636 (2019)
22. Kong, T., et al.: FoveaBox: beyond anchor-based object detector, arXiv (2019)
23. Zhang, H., et al.: VarifocalNet: an IoU-aware dense object detector. In: CVPR, pp. 8514–8523 (2021)
24. Solovyev, R., et al.: Weighted boxes fusion: ensembling boxes from different object detection models. Img. Vis. Comp. **107** (2021)
25. Kociolek, M., et al.: Does image normalization and intensity resolution impact texture classification? Comput. Med. Imaging Graph. **81** (2020)
26. Chen, C.-M., et al.: Automatic contrast enhancement of brain MR images using hierarchical correlation histogram analysis. J. Med. Biol. Eng. **35**(6), 724–734 (2015). https://doi.org/10.1007/s40846-015-0096-6
27. Lin, T.Y., et al.: Feature pyramid networks for object detection. In: CVPR, pp. 2117–2125 (2017)
28. Long, J., et al.: Fully convolutional networks for semantic segmentation. In: CVPR, pp. 3431–3440 (2015)
29. Zlocha, M., Dou, Q., Glocker, B.: Improving RetinaNet for CT lesion detection with dense masks from weak RECIST labels. In: Shen, D., et al. (eds.) MICCAI 2019. LNCS, vol. 11769, pp. 402–410. Springer, Cham (2019). https://doi.org/10.1007/978-3-030-32226-7_45
30. Yan, K., et al.: MULAN: multitask universal lesion analysis network for joint lesion detection, tagging, and segmentation. In: Shen, D., et al. (eds.) MICCAI 2019. LNCS, vol. 11769, pp. 194–202. Springer, Cham (2019). https://doi.org/10.1007/978-3-030-32226-7_22
31. Gidaris, S., et al.: Object detection via a multi-region and semantic segmentation-aware CNN model. In: ICCV, pp. 1134–1142 (2015)
32. Jaeger, P., et al.: Retina U-net: embarrassingly simple exploitation of segmentation supervision for medical object detection. In: ML4H at NeurIPS, pp. 1–12 (2019)
33. Peng, Y., et al.: Automatic recognition of abdominal lymph nodes from clinical text. Clin. Nat. Lang. Proc. (2020)
34. Chen, K., et al.: MMDetection: open MMLab detection toolbox and benchmark, arXiv (2019)
35. Yan, K., et al.: Learning from multiple datasets with heterogeneous and partial labels for universal lesion detection in CT. IEEE Trans. Med. Imag. (2020)
36. Langman, G., et al.: Size and distribution of lymph nodes in rectal cancer resection specimens. Dis. Col. Rec. **58**(4), 406–414 (2015)

Seeking an Optimal Approach for Computer-Aided Pulmonary Embolism Detection

Nahid Ul Islam[1], Shiv Gehlot[1], Zongwei Zhou[1], Michael B. Gotway[2], and Jianming Liang[1]([✉])

[1] Arizona State University, Tempe, AZ 85281, USA
{nuislam,sgehlot,zongweiz,jianming.liang}@asu.edu
[2] Mayo Clinic, Scottsdale, AZ 85259, USA
Gotway.Michael@mayo.edu

Abstract. Pulmonary embolism (PE) represents a thrombus ("blood clot"), usually originating from a lower extremity vein, that travels to the blood vessels in the lung, causing vascular obstruction and in some patients, death. This disorder is commonly diagnosed using CT pulmonary angiography (CTPA). Deep learning holds great promise for the computer-aided CTPA diagnosis (CAD) of PE. However, numerous competing methods for a given task in the deep learning literature exist, causing great confusion regarding the development of a CAD PE system. To address this confusion, we present a comprehensive analysis of competing deep learning methods applicable to PE diagnosis using CTPA at the both image and exam levels. At the image level, we compare convolutional neural networks (CNNs) with vision transformers, and contrast self-supervised learning (SSL) with supervised learning, followed by an evaluation of transfer learning compared with training from scratch. At the exam level, we focus on comparing conventional classification (CC) with multiple instance learning (MIL). Our extensive experiments consistently show: (1) transfer learning consistently boosts performance despite differences between natural images and CT scans, (2) transfer learning with SSL surpasses its supervised counterparts; (3) CNNs outperform vision transformers, which otherwise show satisfactory performance; and (4) CC is, surprisingly, superior to MIL. Compared with the state of the art, our optimal approach provides an AUC gain of 0.2% and 1.05% for image-level and exam-level, respectively.

Keywords: Pulmonary embolism · CNNs · Vision transformers · Transfer learning · Self-supervised learning · Multiple instance learning

1 Introduction

Pulmonary embolism (PE) represents a thrombus (occasionally colloquially, and incorrectly, referred to as a "blood clot"), usually originating from a lower extremity or pelvic vein, that travels to the blood vessels in the lung, causing

© Springer Nature Switzerland AG 2021
C. Lian et al. (Eds.): MLMI 2021, LNCS 12966, pp. 692–702, 2021.
https://doi.org/10.1007/978-3-030-87589-3_71

vascular obstruction. PE causes more deaths than lung cancer, breast cancer, and colon cancer combined [1]. The current test of choice for PE diagnosis is CT pulmonary angiogram (CTPA) [2], but studies have shown 14% under-diagnosis and 10% over-diagnosis with CTPA [3]. Computer-aided diagnosis (CAD) has shown great potential for improving the imaging diagnosis of PE [4–13]. However, recent research in deep learning across academia and industry produced numerous architectures, various model initialization, and distinct learning paradigms, resulting in many competing approaches to CAD implementation in medical imaging producing great confusion in the CAD community. To address this confusion and develop an optimal approach, we wish to answer a critical question: *What deep learning architectures, model initialization, and learning paradigms should be used for CAD applications in medical imaging?* To answer the question, we have conducted extensive experiments with various deep learning methods applicable for PE diagnosis at both image and exam levels using a publicly available PE dataset [14].

Convolutional neural networks (CNNs) have been the default architectural choice for classification and segmentation in medical imaging [15,16]. Nevertheless, transformers have proven to be powerful in Natural Language Processing (NLP) [17,18], and have been quickly adopted for image analysis [19–21], leading to vision transformer (ViT) [19,22]. Therefore, to assess architecture performance, we compared ViT with 10 CNNs variants for classifying PE. Regardless of the architecture, training deep models generally requires massive carefully labeled training datasets [23]. However, it is often prohibitive to create such large annotated datasets in medical imaging; therefore, fine-tuning models from ImageNet has become the *de facto* standard [24,25]. As a result, we benchmarked various models pre-trained on ImageNet against training from scratch.

Supervised learning is currently the dominant approach for classification and segmentation in medical imaging, which offers expert-level and sometimes even super-expert-level performance. Self-supervised learning (SSL) has recently garnered attention for its capacity to learn generalizable representations without requiring expert annotation [26,27]. The idea is to pre-train models on pretext tasks and then fine-tune the pre-trained models to the target tasks. We evaluated 14 different SSL methods for PE diagnosis. In contrast to conventional classification (CC), which predicts a label for each instance, multiple instance learning (MIL) makes a single prediction for a bag of instances; that is, multiple instances belonging to the same "bag" are assigned a single label [28]. MIL is label efficient because only a single label is required for each exam (an exam is therefore a "bag" of instances). Therefore, it is important to ascertain the effectiveness of MIL for PE diagnosis at the exam level.

In summary, our work offers three contributions: (1) a comprehensive analysis of competitive deep learning methods for PE diagnosis; (2) extensive experiments that compare architectures, model initialization, and learning paradigms; and (3) an optimal approach for detecting PE, achieving an AUC gain of 0.2% and 1.05% at the image and exam levels, respectively, compared with the state-of-the-art performance.

2   Materials

The Radiological Society of North America (RSNA) Pulmonary Embolism Detection Challenge (RSPED) aims to advance computer-aided diagnosis for pulmonary embolism detection [29]. The dataset consists of 7,279 CTPA exams, with a varying number of images in each exam, using an image size of 512×512 pixels. The test set is created by randomly sampling 1000 exams, and the remaining 6279 exams form the training set. Correspondingly, there are 1,542,144 and 248,480 images in the training and test sets, respectively. This dataset is annotated at both image and exam level; that is, each image has been annotated as either PE presence or PE absence. Each exam has been annotated for an additional nine labels (see Table 2).

Similar to the first place solution for this challenge, lung localization and windowing have been used as pre-processing steps [30]. Lung localization removes the irrelevant tissues and keeps the region of interest in the images, whereas windowing highlights the pixel intensities within the range of [100, 700]. Also, the images are resized to 576×576 pixels. Figure 1 illustrates these pre-processing steps in detail. We considered three adjacent images from an exam as the 3-channel input of the model.

(a)   (b)   (c)   (d)   (e)   (f)

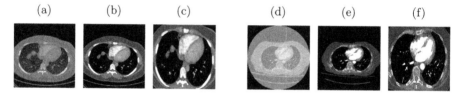

Fig. 1. The pre-processing steps for image-level classification. (a, d) original CT images, (b, e) after windowing, and (c, f) after lung localization. For windowing, pixels above 450 HU and below -250 HU were clipped to 450 HU and -250 HU, respectively.

3   Methods

3.1   Image-Level Classification

Image-level classification refers to determining the presence or absence of PE for each image. In this section, we describe the configurations of supervised and self-supervised transfer learning in our work.

Supervised Learning: The idea is to pre-train models on ImageNet with ground truth and then fine-tune the pre-trained models for PE diagnosis, in which we examine ten different CNN architectures (Fig. 2). Inspired by SeRes-Next50 and SeResNet50, we introduced squeeze and excitation (SE) block to the Xception architecture (SeXception). These CNN architectures were pre-trained on ImageNet[1]. We also explored the usefulness of vision transformer (ViT), where the images are reshaped into a sequence of patches. We experimented with ViT-B_32 and ViT-B_16 utilizing 32×32 and 16×16 patches, respectively [31]. Again, ViT architectures were pre-trained on ImageNet21k. Upscaling the image for a given patch size will effectively increase the number of patches, thereby enlarging the size of the training dataset; models are also trained on different sized images. Similarly, the number of patches increases with a decrease in the patch size.

Self-supervised Learning (SSL): In self-supervised transfer learning, the model is pre-trained on ImageNet without ground truth and then fine-tuned for PE diagnosis. Self-supervised learning has gained recent attention [32–34]. With the assistance of strong augmentation and comparing different contrastive losses, a model can learn meaningful information [35] without annotations. These architectures are first trained for a pretext task; for example, reconstructing the original image from its distorted version. Then the models are fine-tuned for a different task, in our case, PE detection. We pre-train models through 14 different SSL approaches, all of which used ResNet50 as the backbone.

3.2 Exam-Level Classification

Apart from the image-level classification, the RSPED dataset also provides exam-level labels, in which only one label is assigned for each exam. For this task, we used the features extracted from the models trained for image-level PE classification, and explored two learning paradigms as follows:

Conventional Classification (CC): We stacked all the extracted features together resulting in an $N \times M$ feature for each exam, where N and M denote the number of images per exam and the dimension of the image feature, respectively. However, as N varies from exam to exam, the feature was reshaped to $K \times M$. Following [30], we set the K equal to 192 in the experiment. The features were then fed into a bidirectional Gated Recurrent Unit (GRU) followed by pooling and fully connected layers to predict exam-level labels.

Multiple Instance Learning (MIL): MIL is annotation efficient as it does not require annotation for each instance [36]. An essential requirement for MIL is permutation invariant MIL pooling [28]. Both max operators and attention-based operator are used as MIL pooling [28], and we experimented with a combination of these approaches. The MIL approach is innate for handling varying images (N) in the exams and does not require any reshaping operation as does Conventional Classification (CC). For MIL, we exploited the same architecture as in CC by replacing pooling with MIL pooling [28].

---

[1] We pre-trained SeXception on ImageNet and the others are taken from PyTorch.

4 Results and Discussion

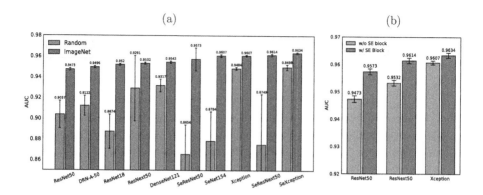

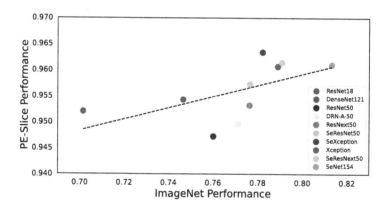

Fig. 2. (a) For all 10 architectures, transfer learning outperformed random ini in image-level PE classification, in spite of the pronounced difference between ImageNet and RSPED. Mean AUC and standard deviation over ten runs are reported for each architecture. Compared with the previous state of the art (SeResNext50), the SeXception architecture achieved a significant improvement ($p = 1.68$E-4). (b) We observed a performance gain with the help of SE block. Note that, all the architectures under comparison were pre-trained from ImageNet.

Fig. 3. There was a positive correlation between the results on ImageNet and RSPED ($R = 0.5914$), suggesting that the transfer learning performance could be inferred by ImageNet pre-training performance.

1. Transfer Learning Significantly Improves the Performance of Image-Level Classification Despite the Modality Difference Between the Source and Target Datasets. Figure 2 shows a significant performance gain for every pre-trained model compared with random initialization. There is also a positive correlation of 0.5914 between ImageNet performance and PE classification performance across different architectures (Fig. 3), indicating that useful weights learned from ImageNet can be successfully transferred to the PE classification task, despite the modality difference between the two datasets. With the help of GradCam++ [37], we also visualized the attention map of SeXception, the best performing architecture. As shown in Fig. 4, the attention map can successfully highlight the potential PE location in the image.

2. Squeeze and Excitation (SE) Block Enhances CNN Performance. Despite fewer parameters compared with many other architectures, SeXception provides an optimal average AUC of 0.9634. SE block enables an architecture to extract informative features by fusing spatial and channel-wise information [38]. Thus, the SE block has led to performance improvements from ResNet50 to SeResNet50, ResNext50 to SeResNext50 and from Xception to SeXception (Fig. 2b).

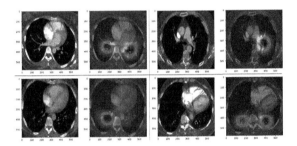

Fig. 4. The SeXception attention map highlighted the potential PE location in the image using GradCam++.

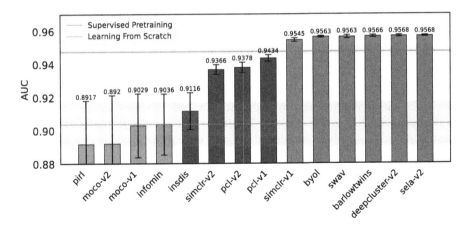

Fig. 5. Self-supervised pre-training extracted more transferable features compared with supervised pre-training. The blue and red dashed lines represent supervised pre-training and learning from scratch with standard deviation (shaded), respectively. 6 out of 14 SSL methods (in green) outperformed the supervised pre-training. All the reported methods had ResNet50 as the backbone. (Color figure online)

3. Transfer Learning with a Self-supervised Paradigm Produces Better Results than its Supervised Counterparts. As summarized in Fig. 5, SeLa-v2 [32] and DeepCluster-v2 [33] achieved the best AUC of 0.9568, followed by Barlow Twins [34]. Six out of fourteen SSL models performed better than supervised pre-trained ResNet50 (Fig. 5). Further experiments can be conducted with other backbones (Fig. 2) to explore if SSL models outperform other supervised counterparts as well, a subject of future work.

Table 1. ViT performs inferiorly compared with CNN for image-level PE classification. For both architectures (ViT-B_32 and ViT-B_16), random initialization provides the worst performance. Both increasing the image size and reducing the patch size can enlarge the training set and therefore lead to an improved performance. Finally, similar to CNNs, initializing ViTs on ImageNet21k provided significant performance gain, indicating the usefulness of transfer learning.

PE AUC with vision transformer (ViT)				
Model	Image size	Patch size	Initialization	Val AUC
SeXception	576	NA	ImageNet	**0.9634**
ViT-B_32	512	32	Random	0.8212
ViT-B_32	224	32	ImageNet21k	0.8456
ViT-B_32	512	32	ImageNet21k	0.8847
ViT-B_16	512	16	Random	0.8385
ViT-B_16	224	16	ImageNet21k	0.8826
ViT-B_16	512	16	ImageNet21k	0.9065
ViT-B_16	576	16	ImageNet21k	*0.9179*

4. CNNs Have Better Performance than ViTs. As shown in Table 1, random initialization provides a significantly lower performance than ImageNet pre-training. The best AUC of 0.9179 is obtained by ViT-B_16 with image size 576×576 and ImageNet21k initialization. However, this performance is inferior to the optimal CNN architecture (SeXception) by a significant margin of approximately 4%. We attribute this result to the absence of convolutional filters in ViTs.

5. Conventional Classification (CC) Marginally Outperforms MIL. The results of CC for exam-level predictions are summarized in Table 2. Although SeXception performed optimally for image-level classification (see Fig. 2), the same is not true for exam-level classification. There is no architecture that performs optimally across all labels, but overall, Xception shows the best AUC across nine labels. The results of MIL for exam-level predictions are summarized in Table 3. Xception achieved the best AUC with a combination of attention and max pooling. Similar to CC approach, no single MIL method performs optimally for all labels. However, Xception shows the best mean AUC of 0.8859 with Attention and Max Pooling across all labels. *Furthermore, the AUC for MIL is*

Table 2. The features extracted by the models trained for image-level classification were helpful for exam-level classification. However, no model performed consistently best for all labels. We report the mean AUC over 10 runs and bold the optimal results for each label. The Xception architecture achieved a significant improvement ($p = $ 5.34E-12) against the previous state of the art[†].

Labels	SeResNext50[†]	Xception	SeXception	DenseNet121	ResNet18	ResNet50
NegExam PE	0.9137	0.9242	**0.9261**	0.9168	0.9141	0.9061
Indetermine	0.8802	0.9168	0.8857	0.9233	0.9014	**0.9278**
Left PE	0.9030	0.9119	0.9100	**0.9120**	0.9000	0.8965
Right PE	0.9368	0.9419	**0.9455**	0.9380	0.9303	0.9254
Central PE	0.9543	0.9500	0.9487	**0.9549**	0.9445	0.9274
RV LV Ratio $\geq$ 1	0.8902	**0.8924**	0.8901	0.8804	0.8682	0.8471
RV LV Ratio < 1	0.8630	0.8722	**0.8771**	0.8708	0.8688	0.8719
Chronic PE	0.7254	**0.7763**	0.7361	0.7460	0.6995	0.6810
Acute& Chronic PE	**0.8598**	0.8352	0.8473	0.8492	0.8287	0.8398
Mean AUC	0.8807	**0.8912**	0.8852	0.8879	0.8728	0.8692

marginally lower than CC (0.8859 vs. 0.8912) but the later requires additional prepossessing steps (Sect. 3.2). More importantly, MIL provides a more flexible approach and can easily handle varying number of images per exam. Based on result #3, the performance of exam-level classification may be improved by incorporating the features from SSL methods.

6. The Optimal Approach: The existing first place solution [30] utilizes SeResNext50 for image-level and CC for exam-level classification. Compared with their solution, our optimal approach achieved an AUC gain of 0.2% and 1.05% for image-level and exam-level PE classification, respectively. Based on our rigorous analysis, the optimal architectures for the tasks of image-level and exam-level classification are SeXception and Xception.

Table 3. The performance varies with pooling strategies for MIL. Attention and Max Pooling (AMP) combines the output of Max Pooling (MP) and Attention Pooling (AP). MIL utilized the feature extracted by the model trained for image-level PE classification. For all three architectures, the best mean AUC is obtained by AMP, highlighting the importance of combining AP and MP.

Architecture	SeResNeXT50			Xception			SeXception		
Labels/Pooling	AMP	MP	AP	AMP	MP	AP	AMP	MP	AP
NegExam PE	0.9138	0.9137	0.9188	**0.9202**	**0.9202**	0.9172	0.9183	0.9137	0.9201
Indetermine	**0.9144**	0.9064	0.8986	0.8793	0.8580	0.8933	0.8616	0.8564	0.8499
Left PE	**0.9122**	0.9059	0.9086	0.9106	0.9100	0.9032	0.9042	0.9004	0.9024
Right PE	0.9340	0.9345	0.9373	0.9397	0.9366	0.9397	0.9403	0.9383	**0.9412**
Central PE	**0.9561**	0.9537	0.9529	0.9487	0.9465	0.9507	0.9472	0.9424	0.9453
RV LV Ratio $\geq$ 1	0.8813	0.8774	0.8822	**0.8920**	0.8871	0.8819	0.8827	0.8779	0.8813
RV LV Ratio < 1	0.8597	0.8606	0.862	**0.8644**	0.8619	0.8567	0.8676	0.8642	**0.8644**
Chronic PE	0.7304	0.7256	0.7233	**0.7788**	0.7664	0.7699	0.7334	0.7168	0.7342
Acute& Chronic PE	**0.8453**	0.8470	0.8228	0.8392	0.8396	0.8350	0.8405	0.8341	0.8367
Mean AUC	0.8830	0.8805	0.8785	**0.8859**	0.8807	0.8831	0.8773	0.8716	0.8751

5   Conclusion

We analyzed different deep learning architectures, model initialization, and learning paradigms for image-level and exam-level PE classification on CTPA scans. We benchmarked CNNs, ViTs, transfer learning, supervised learning, SSL, CC, and MIL, and concluded that transfer learning and CNNs are superior to random initialization and ViTs. Furthermore, SeXception is the optimal architecture for image-level classification, whereas Xception performs best for exam-level classification. A detailed study of SSL methods will be undertaken in future work.

Acknowledgments. This research has been supported partially by ASU and Mayo Clinic through a Seed Grant and an Innovation Grant, and partially by the NIH under Award Number R01HL128785. The content is solely the responsibility of the authors and does not necessarily represent the official views of the NIH. This work has utilized the GPUs provided partially by the ASU Research Computing and partially by the Extreme Science and Engineering Discovery Environment (XSEDE) funded by the National Science Foundation (NSF) under grant number ACI-1548562. We thank Ruibin Feng for helping us some experiments. The content of this paper is covered by patents pending.

References

1. U.S. Department of Health and Human Services Food and Drug Administration: The Surgeon General's Call to Action to Prevent Deep Vein Thrombosis and Pulmonary Embolism (2008)
2. Stein, P.D., et al.: Multidetector computed tomography for acute pulmonary embolism. N. Engl. J. Med. **354**(22), 2317–2327 (2006)
3. Lucassen, W.A.M., et al.: Concerns in using multi-detector computed tomography for diagnosing pulmonary embolism in daily practice. A cross-sectional analysis using expert opinion as reference standard. Thromb. Res. **131**(2), 145–149 (2013)
4. Masutani, Y., MacMahon, H., Doi, K.: Computerized detection of pulmonary embolism in spiral CT angiography based on volumetric image analysis. IEEE TMI **21**(12), 1517–1523 (2002)
5. Liang, J., Bi, J.: Computer aided detection of pulmonary embolism with toboggan-ing and mutiple instance classification in CT pulmonary angiography. In: Karsse-meijer, N., Lelieveldt, B. (eds.) IPMI 2007. LNCS, vol. 4584, pp. 630–641. Springer, Heidelberg (2007). https://doi.org/10.1007/978-3-540-73273-0_52
6. Zhou, C., et al.: Computer-aided detection of pulmonary embolism in computed tomographic pulmonary angiography (CTPA): performance evaluation with independent data sets. Med. Phys. **36**(8), 3385–3396 (2009)
7. Tajbakhsh, N., Gotway, M.B., Liang, J.: Computer-aided pulmonary embolism detection using a novel vessel-aligned multi-planar image representation and convolutional neural networks. In: Navab, N., Hornegger, J., Wells, W.M., Frangi, A.F. (eds.) MICCAI 2015. LNCS, vol. 9350, pp. 62–69. Springer, Cham (2015). https://doi.org/10.1007/978-3-319-24571-3_8
8. Rajan, D., et al.: PI-PE: a pipeline for pulmonary embolism detection using sparsely annotated 3D CT images. In: Proceedings of the Machine Learning for Health NeurIPS Workshop, pp. 220–232. PMLR, 13 December 2020

9. Huang, S.-C., et al.: PENet-a scalable deep-learning model for automated diagnosis of pulmonary embolism using volumetric CT imaging (2020)

10. Zhou, Z., et al.: Models genesis: generic autodidactic models for 3D medical image analysis. In: Shen, D., et al. (eds.) MICCAI 2019. LNCS, vol. 11767, pp. 384–393. Springer, Cham (2019). https://doi.org/10.1007/978-3-030-32251-9_42

11. Zhou, Z.: Towards annotation-efficient deep learning for computer-aided diagnosis. PhD thesis, Arizona State University (2021)

12. Zhou, Z., Shin, J.Y., Gurudu, S.R., Gotway, M.B., Liang, J.: Active, continual fine tuning of convolutional neural networks for reducing annotation efforts. Med. Image Anal., 101997 (2021)

13. Zhou, Z., Shin, J., Zhang, L., Gurudu, S., Gotway, M., Liang, J.: Fine-tuning convolutional neural networks for biomedical image analysis: actively and incrementally. In: CVPR, pp. 7340–7349 (2017)

14. Colak, E., et al.: The RSNA pulmonary embolism CT dataset. Radiol. Artif. Intell. **3**(2) (2021)

15. Litjens, G., et al.: A survey on deep learning in medical image analysis. Med. Image Anal. **42**, 60–88 (2017)

16. Deng, S., et al.: Deep learning in digital pathology image analysis: a survey. Frontiers Med. **14**(4), 470–487 (2020). https://doi.org/10.1007/s11684-020-0782-9

17. Devlin, J., et al.: BERT: pre-training of deep bidirectional transformers for language understanding. arXiv preprint arXiv:1810.04805 (2018)

18. Brown, T.B., et al.: Language models are few-shot learners. arXiv preprint arXiv:2005.14165 (2020)

19. Dosovitskiy, A., et al.: An image is worth 16×16 words: transformers for image recognition at scale. In: ICLR (2021)

20. Han, K., et al.: Transformer in transformer (2021)

21. Touvron, H., et al.: Training data-efficient image transformers & distillation through attention. arXiv preprint arXiv:2012.12877 (2020)

22. Vaswani, A., et al.: Attention is all you need. In: Advances in Neural Information Processing Systems, vol. 30 (2017)

23. Haghighi, F., Taher, M.R.H., Zhou, Z., Gotway, M.B., Liang, J.: Transferable visual words: exploiting the semantics of anatomical patterns for self-supervised learning (2021)

24. Tajbakhsh, N., et al.: Convolutional neural networks for medical image analysis: full training or fine tuning? IEEE TMI **35**(5), 1299–1312 (2016)

25. Shin, H.-C., et al.: Deep convolutional neural networks for computer-aided detection: CNN architectures, dataset characteristics and transfer learning. IEEE TMI **35**(5), 1285–1298 (2016)

26. Jing, L., Tian, Y.: Self-supervised visual feature learning with deep neural networks: a survey. TPAMI, 1 (2020)

27. Haghighi, F., Hosseinzadeh Taher, M.R., Zhou, Z., Gotway, M.B., Liang, J.: Learning semantics-enriched representation via self-discovery, self-classification, and self-restoration. In: Martel, A.L., et al. (eds.) MICCAI 2020. LNCS, vol. 12261, pp. 137–147. Springer, Cham (2020). https://doi.org/10.1007/978-3-030-59710-8_14

28. Ilse, M., Tomczak, J.M., M.: Welling. Attention-based deep multiple instance learning. arXiv preprint arXiv:1802.04712, 2018

29. RSNA STR Pulmonary Embolism Detection (2020). https://www.kaggle.com/c/rsna-str-pulmonary-embolism-detection/overview. Accessed 21 June 2021

30. RSNA STR Pulmonary Embolism Detection (2020). https://www.kaggle.com/c/rsna-str-pulmonary-embolism-detection/discussion/194145. Accessed 21 June 2021

31. Devlin, J., et al.: BERT: Pre-training of deep bidirectional transformers for language understanding. In: Proceedings of the 2019 Conference of the NAACL: Human Language Technologies, Volume 1 (Long and Short Papers), pp. 4171–4186 (2019)
32. Asano, Y.M., et al.: Self-labelling via simultaneous clustering and representation learning. arXiv preprint arXiv:1911.05371 (2019)
33. Caron, M., Bojanowski, P., Joulin, A., Douze, M.: Deep clustering for unsupervised learning of visual features. In: Ferrari, V., Hebert, M., Sminchisescu, C., Weiss, Y. (eds.) Computer Vision – ECCV 2018. LNCS, vol. 11218, pp. 139–156. Springer, Cham (2018). https://doi.org/10.1007/978-3-030-01264-9_9
34. Zbontar, J., et al.: Barlow twins: self-supervised learning via redundancy reduction. arXiv preprint arXiv:2103.03230 (2021)
35. Hu, D., et al.: How well self-supervised pre-training performs with streaming data? arXiv preprint arXiv:2104.12081 (2021)
36. Carbonneau, M.-A., et al.: Multiple instance learning: a survey of problem characteristics and applications. arXiv preprint arXiv:1612.03365 (2016)
37. Gildenblat, J., contributors: Pytorch library for cam methods. https://github.com/jacobgil/pytorch-grad-cam (2021)
38. Hu, J., Shen, L., Sun, G.: Squeeze-and-excitation networks. In: CVPR, pp. 7132–7141 (2018)

Correction to: Machine Learning in Medical Imaging

Chunfeng Lian, Xiaohuan Cao, Islem Rekik, Xuanang Xu,
and Pingkun Yan

Correction to:
C. Lian et al. (Eds.): *Machine Learning in Medical Imaging*,
LNCS 12966, https://doi.org/10.1007/978-3-030-87589-3

In an older version of papers 68 and 69, the CERNET Innovation Project (NGII20190621) had been omitted from the Acknowledgment section. This has been corrected.

The updated version of these chapters can be found at
https://doi.org/10.1007/978-3-030-87589-3_68
https://doi.org/10.1007/978-3-030-87589-3_69

Correction to: A Gaussian Process Model for Unsupervised Analysis of High Dimensional Shape Data

Wenzheng Tao, Riddhish Bhalodia, and Ross Whitaker

Correction to:
Chapter "A Gaussian Process Model for Unsupervised
Analysis of High Dimensional Shape Data"
in: C. Lian et al. (Eds.): *Machine Learning in*
***Medical Imaging*, LNCS 12966,**
https://doi.org/10.1007/978-3-030-87589-3_37

In an older version of this chapter, the acknowledgement was incomplete. This has been corrected.

The updated version of this chapter can be found at
https://doi.org/10.1007/978-3-030-87589-3_37

Author Index

Ahmad, Sahar   644
Arefan, Dooman   555
Azad, Reza   406

Bagci, Ulas   396
Belagiannis, Vasileios   426
Besa, Cecilia   654
Bhalodia, Riddhish   356
Bian, Weixin   664, 674
Braren, Rickmer   596
Buerger, Christian   376
Bulas, Dorothy   366
Burgert, Oliver   586

Carneiro, Gustavo   426
Chaitanya, Krishna   565
Chao, Hanqing   467
Chen, Qun   537
Chen, Xiaoyang   606
Chen, Xu   606, 615
Chen, Zirong   546
Cohen-Adad, Julien   406
Cordeiro, Filipe R.   426
Cui, Zhiming   507, 537, 546, 576

Demir, Ugur   396
Deng, Han   606, 615
Deng, Ruining   437
Dima, Alina   596
Ding, Xintao   664, 674
Doganay, Emine   555
Dong, Peng   664, 674

Elton, Daniel C.   682

Feng, Jun   507, 576
Fricke, Tobias   376

Gateno, Jaime   606
Gehlot, Shiv   477, 692
Glüer, Claus-Christian   376
Gorji, Ali   565
Gotway, Michael B.   692
Gupta, Anubha   477

Hahn, Juergen   467
Haidinger, Sacha   487
Han, Xiaoyang   498
Haque, Ayaan   457
Hoefer, Steven   366
Homayounieh, Fatemeh   467
Hong, Yoonmi   644
Huo, Yuankai   437

Imran, Abdullah-Al-Zubaer   457
Irmakci, Ismail   396
Islam, Nahid Ul   692

Jago, James   366
Jain, Samriddhi   565
Jambawalikar, Sachin   396
Jha, Aadarsh   437
Jiang, Caiwen   537, 546
Jie, Biao   664, 674
Jin, Qun   386
Jungmann, Friederike   596

Kai, Lin   674
Kaissis, Georgios   596
Kalra, Mannudeep K.   467
Karar, Mohamed E.   586
Keles, Elif   396
Kitamura, Gene   555
Konukoglu, Ender   565
Kuang, Tianshu   606

Lagos, Claudio   654
Lee, Sungwon   682
Lemke, Tristan   596
Li, Gang   527, 634
Li, Jianping   507
Li, Jupeng   527, 634
Li, Yunxiang   386
Lian, Chunfeng   606, 615
Liang, Dong   447
Liang, Jianming   692
Liao, Shu   507
Lin, Kai   664
Linguraru, Marius George   366

Liu, Fengbei   426
Liu, Jiameng   537
Liu, Jiameng   546
Liu, Mingxia   664, 674
Liu, Qin   606, 615
Liu, Quan   437
Liu, Siyuan   644
Liu, Wenjun   386
Lock, John   487
Lorenz, Cristian   376
Lu, Zhiyong   682
Luo, Jun   555

Ma, Lei   606, 615
Ma, Ruohan   634
Machado Reyes, Diego   467
Mahadevan-Jansen, Anita   437
Mathai, Tejas Sudharshan   682
Mathis-Ullrich, Franziska   586
Meijering, Erik   487
Meyer, Carsten   376
Millis, Bryan A.   437

Nho, Kwangsik   624

Paetzold, Johannes C.   596
Panigrahy, Ashok   555
Parra, Denis   654
Peña, Jaime   376
Peng, Tingying   498
Peng, Yifan   682
Perakis, Alexis   565
Pino, Pablo   654
Pohl, Hans G.   366
Pölsterl, Sebastian   624

Raffler, Philipp   596
Reid, Ian   426
Rizza, Simone   565
Ronge, Raphael   624
Roshanitabrizi, Pooneh   366
Rouhier, Lucas   406
Rueckert, Daniel   596

Sagar, Md Motiur Rahman   376
Sanchez-Jacob, Ramon   366
Shen, Dinggang   507, 537, 546, 576, 615
Shen, Thomas C.   682
Song, Yanli   576
Spampinato, Concetto   396
Sprague, Bruce Michael   366
Srihari, Sargur N   110

Summers, Ronald M.   682
Sun, Yuhang   537, 546

Tao, Wenzheng   356
Tian, Yu   426
Topcu, Ahmet   396
Turkbey, Baris   396
Turkbey, Evrim   396
Tyska, Matthew J.   437

Wachinger, Christian   624
Wang, Adam   457
Wang, Dayang   416
Wang, Jun   386
Wang, Runze   517
Wang, Shuai   386, 507, 527
Wang, Yaqi   386
Wang, Yinghui   527
Wei, Dongming   537
Wei, Jie   507, 537
Wei, Wenzhao   487
Whitaker, Ross   356
Wu, Dijia   537, 546, 576
Wu, Shandong   555
Wu, Ye   644
Wu, Zhan   416

Xia, James J.   606, 615
Xiao, Deqiang   606, 615
Xu, Ziyue   396

Yan, Pingkun   467
Yang, Hao   546
Yao, Tianyuan   437
Yap, Pew-Thian   606, 615, 644
Yilmaz, Eren Bora   376
Yu, Hengyong   416
Yu, Ziqi   498

Zeineldin, Ramy A.   586
Zember, Jonathan   366
Zeng, Guodong   386
Zhai, Yuting   498
Zhang, Kai   527, 634
Zhang, Qianni   386
Zhang, Xiao   576
Zhang, Xiao-Yong   498
Zhang, Yuyao   546
Zhao, Mengyang   437
Zheng, Guoyan   517
Zhou, Zongwei   692
Zhu, Qingyong   447

Printed in the United States
by Baker & Taylor Publisher Services